Gravity and Magnetic Encyclopedic Dictionary

Serguei Goussev

Geophysical References Series No. 25

Giovanni Florio and John Peirce, volume editors
Ed Biegert, managing editor

SOCIETY OF EXPLORATION GEOPHYSICISTS

Society of Exploration Geophysicists
10300 Town Park Dr., Ste. SE 1000
Houston, TX 77072

Published 2022
Printed in the United States of America

Library of Congress Control Number: 2022914174

ISBN 978-0-931830-47-1 (Series)
ISBN 978-1-56080-386-7 (Volume)

To my wife, Mila

Contents

About the Author

Serguei Goussev graduated from Lomonosov Moscow State University, Russia, with a joint degree in geology and geophysics in 1974. From 1974 to 1995, he worked in the acquisition, processing, and interpretation of geophysical data acquired in the Okhotsk Sea, South China Sea, northern Siberia, southern Yemen, and Central Asia. Since immigrating to Canada in 1995, he has been working in geophysical exploration projects across the globe. In total, he has gained 40+ years of experience in gravity, magnetic, and seismic exploration for oil-and-gas and mineral resources. Between 1997 and 2020, Serguei authored and co-authored 30 gravity and magnetic exploration studies published in geophysical journals and presented at international conventions and workshops. He designed the Goussev filter for high-resolution detection of low-amplitude gravity and magnetic gradient anomalies. He has been a member of the Society of Exploration Geophysicists (SEG) since 1997.

Preface

This Dictionary is compiled as a reference book of terms and definitions used in the theory and practice of gravity and magnetic methods in the hydrocarbon and mineral exploration, geodesy, engineering, environmental, and scientific research applications. The extended entry format takes into consideration the very different geophysical backgrounds that its wide range of readers will have, including university students, geologists, exploration project managers, gravity and magnetic specialists, and other geoscientists. With a few exceptions, Dictionary entries are taken from classical and contemporary gravity and magnetic textbooks, peer-reviewed geophysical journals (*Geophysics*, *Exploration Geophysics*, *Geophysical Prospecting*, *The Leading Edge*, *Interpretation*, *Journal of Geophysical Research*, *Journal of Applied Geophysics*, and others), Expanded Abstracts of Technical Programs and Workshop Presentations at the Annual and Biennial International Meetings of the Society of Exploration Geophysicists (SEG), and similar abstract publications of other national and regional geophysical societies, including the ASEG, CSEG, EAGE, KEGS, SAGA, and SBGf.

This Dictionary follows the naming, symbols, and other conventions implemented in the *Encyclopedic Dictionary of Applied Geophysics, 4th ed.* (Sheriff, 2002). The **bold** typeface within an entry indicates a vector and *italics* denotes a related entry or synonym. Because gravity and magnetic anomalies are interpreted in terms of geological structures, definitions of common structural elements are included in this Dictionary with a brief description of their anomalous gravity and magnetic signatures.

The main text is appended with a Total Index of all entries and an Application Index composed of the entries assembled in the Acquisition, Attributes, Corrections, Filters, Geology, Instruments, Modeling, and Source Depth Estimation sections. Finally, there is a list of more than 500 references used as sources of information in the selection of entries and their descriptions.

From Bohr magneton at the atomic level to van Allen belts in the magnetosphere around the Earth, from an absolute atomic gravimeter to the gravity field curvature, and from Bayesian stochastic inversion to artificial intelligence (AI), this Dictionary contains more than 3,200 entries and presents a terminology-guided summary of the gravity and magnetic theory, measuring instruments, methods of data acquisition, processing, analysis, and interpretation for geophysical studies of the Earth and other planets.

As with any compilation that is wide in scope, this Dictionary is neither fully complete nor flawless. It would be very much appreciated if readers who find errors and omissions or have proposals for new entries would contact me at gmdictionary2021@gmail.com.

I owe a debt of gratitude to my colleagues at GEDCO in Calgary and Fugro Gravity & Magnetic Services in Houston for their contributions to this work. I also extend my grateful thanks to Prof. Giovanni Florio and Dr. John Peirce for serving as technical editors of this book.

Serguei A. Goussev
Ajax, Ontario, 2022

A

absolute accuracy: an instrument characteristic of *quantum magnetometers* that describes their ability to measure a true value of *Earth magnetic field*. Absolute accuracy is a function of several parameters of the magnetometer measuring system, among which the major parameter is the *gyromagnetic ratio* of protons and electrons in a *magnetic sensor*. Because the gyromagnetic ratio can be determined precisely only for a *proton magnetometer*, an *Overhauser magnetometer*, and a *potassium magnetometer*, the measurement *accuracy* of all other types of quantum magnetometers is estimated in terms of their *relative accuracy*.

absolute atomic gravimeter: a *free-fall absolute gravimeter* that utilizes the wave-like quantum nature of atoms and employs them as the identical test masses in a vacuum free fall to measure *absolute gravity*. A sequence of the laser-light pulses and non-uniform magnetic fields manipulate the streams of cold atoms into different pathways in the atom interferometer loop. Due to the different distances traveled by the atom streams, the recombined atom waves will be out of *phase*, wherein their phase shift is directly proportional to *gravitational acceleration* and, thereby, allows a determination of the absolute gravity value. The cooling temperature of atoms is maintained near absolute zero (–273°C) in order to reduce their velocity to approximately 2 m · s^{-1}. Achievable *accuracy* is up to 1–2 *microgal (μGal)*, and it depends on the integration time of the repeated free-fall cycles: the longer the integration time, the higher the accuracy. Also referred to as *atomic gravimeter*, *atom-interferometer gravimeter*, *absolute quantum gravimeter*, *cold-atom absolute gravimeter*, *cold-atom interferometer gravimeter*, *falling-atom gravimeter*, *quantum gravimeter*. See also *falling-mirror gravimeter* and *free-fall method*.

absolute base station: a *base station* with a known value of *absolute gravity*. Absolute base stations are tied to the global network of the absolute gravity values known as the *International Gravity Standardization Network (IGSN71)*. Ties between the IGSN71 and regional or local absolute base stations are established with the use of absolute gravimeters or high-*precision* relative gravimeters. Also referred to as *absolute gravity base*, *primary gravity base*, and *calibrated base station*. See also *absolute gravimeter*, *relative gravimeter*, and *relative gravity base*.

absolute Bouguer anomaly: *Bouguer anomaly* in the *absolute gravity* values obtained by tying the *relative gravity* measurements in the *gravity survey* area to an *absolute base station*. See also *relative Bouguer anomaly*.

absolute gravimeter: a *gravimeter* for measurements of *absolute gravity*. Previously, a *pendulum gravimeter* was the standard instrument for obtaining absolute gravity measurements. At present, a *free-fall absolute gravimeter* is used for these measurements with *precision* of 1 *microgal (μGal)* and achievable *accuracy* up to 1–2 μGal, which is nearly 100 times higher than the best accuracy of measurements by pendulum gravimeters. In addition to providing a reference value, or *gravity datum*, for the *relative gravity* measurements, the absolute gravimeters are used in monitoring the pre-earthquake stress buildup and associated vertical displacement of rocks, oil-and-gas field development and enhanced recovery, CO_2 sequestration, depletion of the geothermal and groundwater reservoirs, tectonic deformations, postglacial rebound, volcanic mass variations, and other effects resulting in the local and regional changes of the absolute gravity. See also *relative gravimeter*, *International Gravity Standardization Network 1971 (IGSN71)*, and *4D gravity monitoring*.

absolute gravimetry: methods and instruments for stationary measurements of *absolute gravity* using either a *pendulum gravimeter* or a *free-fall absolute gravimeter*. The International Bureau of Weights and Measures (BIPM) has proclaimed the *free-fall method* as a primary method for the absolute gravity measurements: it is based on the use of high-*precision* instruments, such as the atomic clock and laser or atom interferometer, capable of determining a time and distance with *accuracy* up to ±0.1 nanosecond (10^{-10} sec.) and ±0.2 nanometer ($2 \cdot 10^{-10}$ m), respectively, and providing the absolute gravity accuracy nearly 100 times higher than the best accuracy of measurements by pendulum gravimeters. See also *absolute gravity corrections*.

absolute gravity: **1.** a numerical value of the *gravitational acceleration* which quantifies *acceleration* of a free-falling body of *mass* under the action of *gravitational force*. **2.** three-component *vector* $\mathbf{g} = (g_x, g_y, g_z)$ whose *magnitude* and orientation on the Earth's surface are defined by a vector sum of the gravitational acceleration due to attraction of the Earth's mass and *centrifugal acceleration* due to the Earth's rotation. **3.** *scalar* quantity $g = |\mathbf{g}|$ representing the length, or magnitude, of the three-dimensional (3D) *Earth gravity vector* $\mathbf{g}$ which defines the strength and direction of *Earth gravity field* at each observation point. **4.** instantaneous gravitational acceleration g measured by an *absolute gravimeter* as opposed to *relative gravity* measured by a conventional, or regular, *gravimeter* known as a *relative gravimeter*. **5.** numerical value of *gravitational attraction* of a planetary body, usually in the units of *nominal gravity g*, such as ~0.38 g for Mars and ~0.17 g for the Moon, where 1 g = 9.8 m · s^{-2}. In the past, the absolute gravity on the Earth was calculated from measurements by a *pendulum gravimeter* using the relationship between the *period* of a swinging pendulum, pendulum length, and absolute gravity value. At present, it is measured directly by the *free-fall method* using the high-*precision* instruments calibrated to the time and length standards.

Absolute gravity is a reference value, or *absolute gravity datum*, for the relative gravity measurements during the ground, marine (shipboard), submarine, and airborne gravity surveys. It also is used to tie the locally acquired relative gravity data to the regional and national databases and global absolute gravity datum known as the *International Gravity Standardization Network 1971* (*IGSN71*). Sometimes referred to as *full-field gravity*. See also *free-fall absolute gravimeter*.

absolute gravity acceleration: same as *absolute gravity* with an emphasis on *acceleration* as a physical *dimension* of the *gravitational field*, such as the *Earth gravity field*.

absolute gravity base: See *absolute base station*.

absolute gravity corrections: corrections applied to *absolute gravity* measurements recorded by a *free-fall absolute gravimeter*, of which there are nine: *atmospheric pressure correction*, *datum transfer correction*, *diffraction correction*, *ocean loading correction*, *polar motion correction*, *self-attraction correction*, *speed-of-light correction*, *tidal correction*, and *tilt correction*.

absolute gravity datum: *absolute gravity* used as a reference value, or *datum*, relative to which all *relative gravity* measurements are made during the ground, marine (shipboard), and airborne gravity surveys. The *International Gravity Standardization Network (IGSN71)* dataset with the *Honkasalo term* removed has been adopted as the global absolute gravity datum.

absolute gravity gradient: the horizontal or vertical *gradient* of the *Earth gravity field* determined by calculating the difference between two measured *absolute gravity* values divided by a *baseline*, which is a distance between measurement points along the horizontal or vertical direction. See *horizontal gradient* and *vertical gradient*.

absolute gravity gradiometer: an assemblage, or *array*, of two or three synchronized *absolute gravimeter* systems operating in the same reference frame for simultaneous measurements of *absolute gravity* and calculation of the *absolute gravity gradient* and *gravity field curvature*. See also *atomic gravity gradiometer* and *gradiometer arm*.

absolute horizontal gradient: the *horizontal gradient* of the *Earth gravity field* determined by direct measurements of a horizontal rate of change of *gravitational acceleration* on the Earth's surface with the use of an *absolute gravity gradiometer*.

absolute magnetic direction: the *vector* direction of the *Earth magnetic field* calculated by using *magnetic inclination* and *magnetic declination* values determined at a *geomagnetic observatory* by theodolite-assisted high-*accuracy* measurements of the vertical and horizontal components of the Earth's magnetic field. See also *geomagnetic vector* and *magnetic vector*.

absolute magnetometer: a *scalar magnetometer* in the context of measurements of the absolute numerical value of the *Earth magnetic field* regardless of its *vector* direction at observation points.

absolute quantum gravimeter: same as *absolute atomic gravimeter* with an emphasis on its operating principle based on the quantum wave-like nature of atoms and their use as the identical test masses in a vacuum free fall to measure *absolute gravity*.

absolute scalar magnetometer (ASM): a high-*sensitivity* laser-pumped *helium magnetometer* utilizing the *magnetic resonance*, atomic spectrometry, and *Zeeman effect* to measure *scalar* magnitudes of the *Earth magnetic field* with *precision* up to 1–2 pT$\sqrt{}$Hz over 100 Hz *bandwidth* and a final corrected *accuracy* of measurements up to 65 *picotesla (pT)*. With its extended bandwidth and three mutually orthogonal sensing coils, ASM also can determine *vector* components of the Earth's magnetic field by superimposing the orthogonal vector measurements modulated by three different frequencies onto the ambient magnetic field $\boldsymbol{B}_0$, so the total magnetic field $\boldsymbol{B}_{tot}$ sensed by an ASM *magnetic sensor* can be expressed as

$$\left|B_{tot}\right| = \left|B_0\right| + \left|\sum \beta_j \cos(2\pi f_{mj} t)\boldsymbol{e}_j\right|,$$

where β_j is vector modulation amplitude; f_{mj} is its respective modulation frequency; $\boldsymbol{e}_j$ is a vector orientation in the ASM reference frame; $j = x, y, z$; and t is measurement time. Vector components are reconstructed by *real time* processing of the scalar output signal containing the principal *harmonics* of the vector modulation from which the vector amplitudes and their respective orientations are derived with precision up to 1 nT$\sqrt{}$Hz over 0.4 Hz bandwidth. Scalar and vector measurements are synchronized and spatially coherent, i.e., they correspond exactly to the same spatial locations. The ASM is installed on each of the three Earth-orbiting satellites in the *Swarm mission*.

absolute tie: referencing, or tying, the *relative gravity* data acquired over local areas to the regional or global network of *absolute gravity* values. Absolute ties allow consideration of local data in the context of regional gravity observations that could be useful in determining the extent of the regional gravity anomalies and the selection of parameters for *regional-residual separation* of detected anomalies. See also *International Gravity Standardization Network (IGSN71)*.

absolute vertical gradient: a *vertical gradient* of *Earth gravity field* determined by direct measurements of *gravitational acceleration* on the Earth's surface with the use of an *absolute gravity gradiometer*.

ac demagnetization: the *demagnetization* of a rock sample by the alternating, or *polarity*-reversing, magnetic field generated by the alternating electric current in a coil or loop around the sample.

acceleration: **1.** the rate of change of a measured quantity defined by the second-order *derivative* calculated in the *domain* of a particular variation. **2.** the change of a moving body velocity, either in its *magnitude* or direction, or in both variation domains, per unit time. **3.** the first-order derivative of velocity of a moving body in the *time domain*. **4.** the physical *dimension* of a *gravitational field*, such as the *Earth gravity field*, which could be described as the *field* of *gravitational acceleration* of a free-falling body under the action of *gravitational force* generated by the attracting Earth mass. See also *angular acceleration*, *centrifugal acceleration*, *centripetal acceleration*, *differential acceleration*, *inertial acceleration*, *total acceleration*, and *secular acceleration*.

acceleration of free fall: *gravitational acceleration* with an emphasis on the *acceleration* of a free-falling body of *mass* in the direction to the attracting Earth or other planet. Its value g_P at the surface of the planet P can be expressed as

$$g_P = GM_P/R_P^2,$$

where G is the *universal gravitational constant,* M_P is the mass of the attracting planet P, and R_P is the planet radius. See also *Newton law of gravitational attraction.*

acceleration of gravity: *gravitational acceleration* with an emphasis on the *acceleration* of a free-falling body of *mass* under the action of *force of gravity* also known as *gravitational force.*

accelerometer: a device whose mechanical or electric output is directly proportional to *acceleration.* Its typical design is based on the use of a spring-suspended *proof mass,* within which is either a mechanical (metal or quartz) spring or a *virtual spring* with the proof mass levitating in the magnetic or electrostatic field. When the proof mass is subjected to acceleration, an unbalanced force emerges in accordance with the *Newton law of motion,* and this force causes the proof mass to move from its original position, whereby the acceleration can be determined by measuring a displacement of the proof mass. In high-*sensitivity* accelerometers, the displacement is measured with the use of a capacitive readout system which registers the amount of the electric current or voltage produced by changes in the capacitance or magnetic field balance to force the proof mass back to its original position. Accelerometers are used to measure: (a) *gravitational acceleration* in *airborne gravimeter* systems; (b) spatial derivatives, or gradients, of the gravitational acceleration in marine, airborne, and satellite *gravity gradiometer* systems; (c) horizontal and vertical inertial accelerations of a moving *observation platform,* so that their values subsequently can be removed from gravity measurements to obtain the gravitational acceleration values due to the attraction of the Earth mass; and (d) common-mode accelerations of the moving observation platform. See also *electromagnetic accelerometer, electrostatic accelerometer, inertial-grade accelerometer, vector accelerometer, vertical accelerometer, common-mode acceleration,* and *inertial acceleration.*

acceptance criteria: 1. *tolerance* in the context of the *quality control (QC)* criteria which the gravity or magnetic *survey* data must satisfy to be accepted. **2.** parameters for eliminating the spurious *noise* solutions obtained by *automated source depth estimation* methods, such as *Euler deconvolution* and *Werner deconvolution.* Commonly, the retained depth solutions are those for which depth variations are limited to a user-defined percentage (often, 15% or less) of the mean (average) value for all depth solutions at each calculation point. See also *clustering.*

accretionary prism: a wedge-shaped mass of the *thrust*-deformed sediments on a continental lithospheric plate near the boundary between the oceanic and continental plates moving toward each other in the subduction zone. It is composed of the oceanic-floor pelagic sediments, oceanic *crust* basalts, and fill-in turbidites that have been scraped off the descending oceanic plate in the process of subduction. See also *lithosphere* and *underplating.*

accuracy: 1. a difference between the measured or approximated value and the actual, or true, value of a particular quantity. **2.** an instrument characteristic that describes its ability to measure the true value of a specific quantity. Often, it is estimated by a statistically determined average deviation of the measured values from the true values. Accuracy is different from (a) *precision,* which defines a range of deviations, or scattering, of the measured values from the true values, and (b) *sensitivity,* which defines the smallest change in the measured quantity, such as a *field* or its *gradient,* that could be detected by the measuring instrument. Estimates of a projected *survey accuracy* are based on the selected *survey specifications* and performance characteristics of measuring instruments. See also *absolute accuracy, relative accuracy, ground truth,* and *rms accuracy.*

ACLAS: the acronym for Automated Coherent Lineament Analysis and Selection, which is a method for combining the *lineaments* obtained by different *enhanced processing* techniques applied to the same gravity or magnetic dataset. It is based on identifying the coherent parts in each set of lineaments and assessing their mutual proximity quantified by a distance threshold, so that only lineaments within a given distance range, or *tolerance,* contribute to the combined output representing the average location of all accepted lineaments. This method suppresses the artificial anomalies generated by individual *data enhancement* techniques and provides structural information which is incorporated into the process of *integrated interpretation* (Cascone et al., 2017). See also *artificial anomaly, coherence,* and *lineament interpretation.*

aclinic line: a continuous line on the *isoclinal map* connecting points where *magnetic inclination* is zero. The aclinic line is coincident with *magnetic equator* that divides the Earth into the northern and southern geomagnetic hemispheres. See also *agonic line.*

acoustic basement: the deepest relatively continuous seismic reflection horizon below which seismic reflections are poor or absent. Often associated with the intrasedimentary depositional unconformity or thick carbonate sequence in the lower sedimentary section, the acoustic basement could be structurally close to or coincident with the *geological basement,* but it also could have a much shallower occurrence. It is used as the uppermost constraint in the *interpretation* of the *magnetic basement.* Also referred to as *seismic basement.* See also *basement* and *constraints.*

acoustic log: same as *sonic log* with an emphasis on measurements of interval velocities of the acoustic wave that are converted into *apparent density* values.

acquisition: 1. measurements of the *Earth gravity field* or *Earth magnetic field* along the regularly spaced lines during marine (shipboard), submarine, and airborne surveys or at the irregularly spaced observation points, or stations, during a ground or sea floor *gravity survey.* **2.** the process of obtaining and recording gravity and magnetic data. See also *survey* and *station.*

acquisition filtering: application of a *low-pass filter* to the time-domain airborne or marine (shipboard) gravity and *gravity gradient* measurements along the *survey* lines in order to suppress *dynamic noise* resulting, primarily, from uncompensated inertial accelerations of a survey aircraft or ship. See also *time domain*, *time filter*, and *inertial acceleration*.

acquisition footprint: **1.** alignments of low-*amplitude* artificial anomalies in the gridded data that follow the orientation of the *survey* lines, such as *corrugations* representing the residual errors in *leveling* and *microleveling*. **2.** systematic offsets, or displacements, of the map contours or color-coded values on the gravity or magnetic *grid* described as a *herringbone effect* caused by the accumulated *positioning* errors and insufficient compensation for *gravimeter drift* or aeromagnetic *heading error* and *lag error*. **3.** patterns of anomalies in the gridded data with a higher frequency content relative to the surrounding area that are caused by *anisotropic sampling* of data during the ground gravity or magnetic *survey*. **4.** *grid aliasing*. Also referred to as *acquisition imprint*. See also *artificial anomaly*, *gridding*, and *pattern*.

acquisition imprint: See *acquisition footprint*.

acquisition line: a *survey* line, or *traverse*, along which measurements of the Earth's gravity or magnetic field are made.

acquisition noise: a generic term applied to all types of *noise* in the gravity and magnetic data recorded along the *acquisition* lines. The noise level in an airborne gravity or magnetic *survey* is monitored by the *real time* calculations of *fourth difference* or *noise envelope* in order to estimate a limit of detectability of the low-*amplitude* anomalies. In the marine (shipboard) and airborne measurements, it is referred to with the same meaning as *dynamic noise*. See also *acquisition* and *acquisition filtering*.

Adams-Williamson equation: the equation approximating a rate of change in the assumed near-surface rock *density* ρ as a function of the Earth's radius R. It is based on velocities of the seismic compressional and shear waves determined from the deep seismic reflection or refraction data and expressed as

$$\partial\rho/\partial R \;= -\rho[g/V_P^2 - 4V_S^2/3)],$$

where $\partial\rho/\partial R$ is the rate of increase in rock density with decreasing Earth radius; g is the *average gravity* value, which is supposed to be constant through the *crust* and upper *mantle*; V_P is the velocity of the seismic compressional wave, or P-wave; and V_S is the velocity of the shear wave, or S-wave. See also *Brocher formula*, *Gardner equation*, *Nafe-Drake relation*, and *Poincare-Prey reduction*.

adaptive quadtree discretization: a method of *mesh discretization* of the continuous surface of a *Digital Terrain Model (DTM)* into the variable-width square prisms in calculation of *airborne terrain correction*. It adapts to the varying data characteristics, such as *terrain* roughness, distance from *observation point*, and required *accuracy* of the terrain correction. Larger prisms are used, where *terrain effect* is below a user-specified survey *noise* level, and small prisms are used to increase *lateral resolution* of the DTM in the areas of rapid terrain variations and strong terrain effects (Li et al., 2010). See also *prism*.

adaptive tilt angle (ATA): the *tilt angle (TA)* modified for the *vertical gravity* derivative components of a *full tensor gravity gradient (FTGG)* measured by a *full tensor gravity gradiometer* or obtained from measurements by a *partial tensor gravity gradiometer*. It is calculated for mapping of the *gradient* anomalies and depth estimates of the gravity sources by

$$\mathrm{ATA} = \arctan[\alpha T_{zz}/(T_{zx}^2 + T_{zy}^2)^{1/2}],$$

where α is a scaling value which is related to the source type in a way analogous to the *Euler structural index (SI)* as: $\alpha = 3$ for *three-dimensional (3D)* sources, e.g., a *sphere*; $\alpha = 1$ for *two-dimensional (2D)* sources, e.g., the horizontal or vertical *thin sheet* and horizontal line of mass; and T_{zz}, T_{zx}, T_{zy} are derivatives of the vertical gravity T_z along the mutually orthogonal x-, y-, z-directions in *Cartesian coordinates*. For the sphere, horizontal line of mass, and vertical thin sheet source approximations, the horizontal position of sources is defined by 90° values on the ATA *grid*, whereas the horizontal thin sheet location is defined by 0° values (Salem et al., 2013). See also *derivative*, *full tensor gradient (FTG)*, and *adaptive tilt-depth method*.

adaptive tilt-depth method: a *grid*-based semi-automatic method of estimating the depth of gravity sources approximated by four types of *two-dimensional (2D)* and *three-dimensional (3D) ideal source* geometries. For *sphere* and vertical *thin sheet*, the source depth is estimated by the distance between 45° and 90° *contour* lines on the *adaptive tilt angle (ATA)* grid. For the horizontal line of mass, the source depth is the distance between the 0° and 90° contours. For the horizontal thin sheet, the depth is the distance between the 0° and 45° contour lines on the ATA *contour grid* (Salem et al., 2013). See also *source geometry* and *tilt-depth method*.

add-on survey: a shipboard gravity or magnetic *marine survey* added to the *two-dimensional (2D)* or *three-dimensional (3D)* seismic *acquisition* program on the same ship. See also *submarine survey*.

admittance function: See *gravitational admittance* and *atmospheric admittance function*.

advanced source-parameter imaging method: a *profile*-based method of estimating the parameters of *two-dimensional (2D)* magnetic sources approximated by the *contact*, *thin sheet*, and *horizontal cylinder* source geometries: depth, location, *model*-approximating shape with its *Euler structural index (SI)*, and *apparent susceptibility*. Estimates are made with the use of a *horizontal local wavenumber*, *vertical local wavenumber*, *first-order ASA* (ASA_1), and *second-order ASA* (ASA_2). Source depth solutions are calculated in multiple ways and spatially averaged over *source body* locations to obtain stable estimates. The obtained results are gridded for visualization and *interpretation* (Smith and Salem, 2005). See also *gridding*, *source parameter imaging (SPI™) method*, and *improved SPI™ method*.

aerogravity: **1.** gravity data obtained from airborne measurements of the relative changes in *Earth gravity field*, or *relative gravity*, with the use of a fixed-wing aircraft (airplane). **2.** *airborne gravimetry*. **3.** *airborne gravity gradiometry*. See also *aeromagnetics*.

aerogravity survey: airborne measurements of *relative gravity* by an *airborne gravimeter* mounted on the inertial *stabilized platform* and installed onboard a fixed-wing aircraft. Gravity measurements are made at the regularly distributed observation points along the *acquisition* lines with the objective of mapping the anomalies associated with elements of the subsurface geological *structure*. Achievable *accuracy* of measurements is 1.0–2.0 *milligal (mGal)* with *half-wavelength resolution* of 1.5–2.0 km. See also *AGG survey*, *gravity survey*, and *helicopter-suspended gravimeter*.

aeromagnetic compensation: elimination of the static and dynamic magnetic interference effects produced by aircraft components and aircraft maneuvering in a variable *Earth magnetic field*, including: (a) permanent, or static, effects of the aircraft engine and other ferromagnetic parts; (b) temporary, or dynamic, induced effects in the iron and paramagnetic parts that are known as *eddy currents*; and (c) electric currents in the onboard electronic systems. There are two types of aeromagnetic compensation: (1) passive compensation by shielding the *magnetometer* sensors with permanent magnets, electric coils, or high *permeability* iron straps and (2) active, or adaptive, *real-time compensation* based on the 3D magnetic model of the *aircraft signature*. Active real-time compensation is presently a primary method of aeromagnetic compensation. Aircraft model parameters and compensation coefficients are determined from the results of pre-survey calibration tests, where a set of three orthogonally mounted fluxgate magnetometers is used to record the changes in amplitudes of the measured *Earth magnetic field* during the standard *pitch*, *roll*, and *yaw* maneuvering in the orthogonal flight headings. Effectiveness of the aeromagnetic compensation is evaluated by a standard pre-survey test flown to obtain *figure-of-merit (FOM)*. See also *heading* and *fluxgate magnetometer*.

aeromagnetic gradiometer: a multi-sensor *magnetic gradient* system mounted on a fixed-wing aircraft (airplane) or attached to a *drone*. Horizontal gradients are measured by the arrangement, or *array*, of two magnetometers mounted in the wingtip extensions and one *magnetometer* in the tail *stinger*. Vertical gradients are measured by the stinger-mounted array of two vertically separated magnetic sensors. A modified array of two or three sensors is towed by helicopter or *drone* in detailed surveys for mineral *exploration* and special research projects. See also *detailed survey*, *longitudinal gradient*, *transverse gradient*, *magnetic sensor*, *slung sensor*, and *towed bird system*.

aeromagnetic survey: airborne measurements of the *Earth magnetic field*, its *gradient*, and *full tensor gradient (FTG)* components along the regularly spaced *traverse lines* and *tie lines* with the use of a specially equipped fixed-wing aircraft (airplane), *drone*, or helicopter. Flight *altitude* and *survey line spacing* are selected in accordance with *survey* objectives. Detailed fixed-wing surveys are usually flown at a low-altitude *drape flight* defined by flight characteristics of the survey aircraft and safety regulations. See also *detailed survey*, *survey layout*, and *survey specifications*.

aeromagnetic survey equipment: the navigation, measurement, magnetic compensation, and pre-processing equipment onboard the survey aircraft (airplane or helicopter) and on the ground. Typically, this includes: (1) high-*sensitivity* magnetometer, (2) tri-axial *fluxgate magnetometer*, (3) real-time *aeromagnetic compensation* system, (4) radar *altimeter* or *LIDAR* system, (5) *barometric altimeter*, (6) *differential GPS (DGPS)* navigation system, (7) time-synchronized *base-station magnetometer*, (8) *GPS base station*, (9) computerized data recording system, and (10) video camera for recording the ground objects below flight lines. See also *survey* and *survey design*.

aeromagnetics: 1. *airborne magnetometry*. **2.** magnetic data obtained from the airborne measurements of *Earth magnetic field*, its *gradient*, *full tensor gradient (FTG)*, and, sometimes, *vector* components with the use of a fixed-wing aircraft (airplane), helicopter, or *drone*. A magnetometer can be mounted in a tail *stinger* behind the airplane, towed in a streamlined container called a *bird*, or suspended below a low-flying drone. Magnetic gradients are measured either by an *array* of three magnetometers, where two units are mounted at the wingtip extensions and one in the tail stinger, or by an assemblage of two or three magnetometers towed below and behind the aircraft or suspended below the drone. See also *high-resolution aeromagnetic survey*, *microfabricated atomic magnetometer (MFAM)*, and *slung sensor*.

AG gravity: *relative gravity* data obtained from measurements by an *airborne gravimeter (AG)* along the regularly spaced *acquisition* lines in the *gravity survey* area. Airborne *observed gravity* Δg_{obs} can be expressed as

$$\Delta g_{obs} = \Delta g_{raw} + \Delta g_{va} - \Delta g_{tilt} - \Delta g_{drift} \pm \Delta g_{Eo},$$

where Δg_{raw} is measured gravity value, or *raw gravity*; Δg_{va} is *vertical acceleration correction*; Δg_{tilt} is *platform tilt correction*, also known as *off-level correction*; Δg_{drift} is *airborne gravimeter drift correction*; and Δg_{Eo} is *Eötvös correction*. Commonly, *base-tie value* is added to Δg_{obs} to tie the airborne survey data to the local or regional *absolute base station* and present the AG gravity data in *absolute gravity* values. See also *AGG gravity*.

AGC filter: a line- or *grid*-based *automatic gain control (AGC)* operator which balances amplitudes of anomalies of different wavelengths by attenuating the high amplitudes and amplifying the low amplitudes. Amplitude balancing is performed with the use of a *moving window* centered at each *observation point*. Prior to application of the AGC filter, the regional component of the data is removed by *detrending* or other *regional-residual separation* method. See also *balanced ASA (ASAB)*.

Agerlin principle: the *modeling* and *data processing* principle stating that a comparison between the observed gravity or magnetic *field* and the field generated by a constructed *model*, i.e., *model response*, should be made for the observed and model data processed by the same techniques (Jacobsen, 1988). For example, after a *low-pass filter* is applied to the flight-line-oriented *airborne gravity* measurements to suppress *dynamic noise*, then the same *filter* must be applied to the calculated model response in order to accurately represent contributions of the model components and estimate *model misfit.* In accordance with this principle, terrain corrections calculated for the airborne gravity and *gravity gradient* data need to be low-pass filtered prior to their application to avoid compensating for the high-frequency *terrain effect* components which were suppressed by *acquisition filtering.* See also *terrain correction.*

AGG gravity: *relative gravity* data reconstructed from measurements by a *airborne gravity gradiometer (AGG).* The AGG gravity differs from the relative gravity, measured with the use of a conventional (regular) *relative gravimeter* on the ground or an *airborne gravimeter (AG)* at some *altitude* above the ground, by the amount of a *regional gravity* field component in the *gravity survey* area. See also *AG gravity, GOCE gravity,* and *residual gravity.*

AGG survey: 1. airborne measurements of the *gradient* and *gradient tensor* components of *Earth gravity field* or its *differential curvature gradients* by an *airborne gravity gradiometer (AGG)* along the regularly spaced *airborne survey* lines. **2.** *FTG survey* with the use of an airborne gravity gradiometer (AGG).

agonic line: a continuous line on an *isogonic map* connecting the points where *magnetic declination* is zero. There are two agonic lines: one runs around the Earth and across both *Earth magnetic poles,* and another runs across the *magnetic north pole* making a closed contour over Europe, Africa, and Asia. See also *aclinic line.*

airborne Bouguer correction: a *complete Bouguer correction* for *airborne gravity* data. Its application involves two *data processing* steps: (1) compensation for the *Bouguer effect* of the air and rock mass below observation points down to a *reference ellipsoid* and (2) compensation for the *terrain effect* due to the topographic *relief* variations in the *airborne survey* and surrounding areas by an *airborne terrain correction.* See also *marine Bouguer correction.*

airborne gradiometer: an instrument for airborne measurements of the *gradient* and *gradient tensor* components of the *Earth gravity* or *magnetic field.* See also *airborne survey, airborne gravity gradiometer (AGG), gravity gradiometer,* and *magnetic gradiometer.*

airborne gravimeter (AG): an airborne system for measurements of *relative gravity.* Its *gravity sensor* design is based on the use of the inertial-grade accelerometers, such as a *vertical accelerometer,* where *proof mass* is either held in a balanced position within the electrical coil by assemblage of permanent magnets and capacitive pick-off rings, or it levitates in the magnetic or electrostatic field. An airborne gravimeter is mounted on the inertial *stabilized platform* with a minimized *cross-coupling effect.* High-*accuracy* positioning data from the *differential Global Positioning System (DGPS)* are used for calculation of the vertical *inertial acceleration* at each *observation point* in the post-flight data processing for suppression of the residual inertial accelerations and elimination of the residual cross-coupling effects. See also *airborne gravity gradiometer (AGG), electromagnetic accelerometer, electrostatic accelerometer,* and *inertial-grade accelerometer.*

airborne gravimeter calibration: a daily procedure established as part of *quality control (QC)* over performance of the *accelerometer*-based *airborne gravimeter (AG)* system. It includes (a) pre-flight automatic alignment and calibration of the gyroscopes and servomotor-coupled accelerometers of the inertial *stabilized platform* and (b) *static test.* See also *ground gravimeter calibration* and *airborne magnetometer calibration.*

airborne gravimeter drift: the *drift* in *airborne gravity* measurements estimated by (a) linear interpolation of the pre-flight and post-flight still readings at the same location on the airfield and (b) prorating the obtained *drift curve* values in accordance with a time increment of each individual airborne measurement. See also *still reading* and *static test.*

airborne gravimeter drift correction: a correction for *airborne gravimeter drift* determined from the results of the daily *static test* and prorated by a linear approximation of *drift* for all airborne observation points. See also *airborne gravimeter (AG)* and *drift.*

airborne gravimetry: methods and instruments to record and process measurements of the relative changes in *Earth gravity field,* or *relative gravity,* with the use of a fixed-wing aircraft (airplane). Airborne gravity measurements are conducted along the pre-planned flight lines over the *airborne survey* area at a nearly constant flight *elevation* and in a low-turbulence environment to obtain the acceptable *accuracy.* High-*precision* measurements by *Global Positioning System (GPS)* or more precise *differential GPS (DGPS)* are required to determine *3D coordinates* and aircraft velocity that are used to calculate all necessary corrections for the acquired gravity data. Also referred to as *aerogravity.* See also *marine gravimetry, submarine gravimetry, sea-bottom gravimetry,* and *satellite gravimetry.*

airborne gravity: 1. *relative gravity* data obtained from measurements by an *airborne gravimeter (AG)* mounted on the inertial *stabilized platform* and installed onboard a fixed-wing aircraft (airplane) or airship (dirigible). Airborne gravity data are low-pass filtered versions of the *ground gravity* data acquired on the Earth's surface due to both the flight *altitude* of the *airborne survey* and *acquisition filtering* applied to suppress a high-frequency *dynamic noise.* Achievable *accuracy* of airborne gravity measurements is up to 1–2 *milligal (mGal)* with *half-wavelength resolution* of 1.5–2.0 km. Also referred to as *AG gravity.* **2.** gravity anomalies obtained from airborne gravity measurements. At each *observation point,* the *complete Bouguer anomaly* Δg_B value is expressed in units of *milligal (mGal)* as

$$\Delta g_B = g_{obs} - \Delta g_{TG} + 0.3086h - 2\pi G\rho_B(h - h_g) \pm \Delta g_T \pm \Delta g_{Eo},$$

where g_{obs} is *observed gravity*; Δg_{TG} is *theoretical gravity correction*; 0.3086 is *free-air factor*; h is *elevation* of the aircraft flight above sea level or *reference ellipsoid*; h_g is height of the aircraft flight above the ground, i.e., altitude; G is *universal gravitational constant*; ρ_B is *Bouguer density*; Δg_T is *terrain correction*; and Δg_{Eo} is

Eötvös correction (Hinze et al., 2013). See also *gravity anomaly*, *ground gravity*, *marine gravity*, *satellite gravity*, and *AGG gravity*.

airborne gravity accuracy: 1. the root-mean-square (rms) accuracy defined by the square root of the average of the sum of squares of differences between the individual values in the *airborne gravity* dataset and their corresponding *ground truth* values. **2.** the root-mean-square of differences between the crossover values calculated after *leveling* of the airborne gravity data. See also *crossover value*.

airborne gravity gradiometer (AGG): a *gravity gradiometer* system for airborne measurements of the *gradient* and *gradient tensor* components of the *Earth gravity field*. It is mounted on the inertial *stabilized platform* and can be installed onboard a fixed-wing aircraft (airplane) or helicopter. The best reported *accuracy* of airborne gravity gradient measurements is within the range of 3–7 *Eötvös (E)* with *half-wavelength resolution* up to 0.3–1.0 km. Two currently operational AGG types are based on the use of a *rotating-disk gravity gradiometer* to measure either *full tensor gradient (FTG)* or *differential curvature gradients* of the *Earth gravity field*. See also *AGG gravity* and *GOCE gravity gradiometer*.

airborne gravity gradiometry: methods and instruments to record and process measurements of the *gradient* and *full tensor gradient (FTG)* components of the *Earth gravity field* or its *differential curvature gradients* with the use of a fixed-wing aircraft (airplane) or helicopter and *airborne gravity gradiometer (AGG)* system. Airborne gravity gradiometry data are low-pass filtered versions of the *gravity gradient* data acquired on the Earth's surface due to both flight *altitude* of *airborne survey* and *acquisition filtering* applied to suppress a high-frequency *dynamic noise*. See also *low-pass filter*, *rotating-disk gravity gradiometer*, and *GOCE gravity gradiometer*.

airborne gravity resolution: the highest achievable *lateral resolution* of *airborne gravity* measurements defined by the *half-wavelength resolution* of the acquired data along the *airborne survey* line. It is a function of (a) aircraft speed, (b) aircraft *altitude*, (c) length of *low-pass filter* applied to suppress *dynamic noise*, and (d) depth to *target*. For the low-flying aircraft and relatively shallow targets, it is approximated as

$$\lambda_{half} \cong vT/2,$$

where v is aircraft speed in meters per second and T is *time filter* length in seconds. At the speed of 50 m · sec^{-1} (180 km per hour) and *filter length* of 80 seconds, the airborne gravity resolution is 2000 m. For deep targets, the resolution estimate is often approximated by the half-*wavelength* of the point-source anomalies and expressed as $\lambda_{half} \approx 1.54 d_{ps}$, where d_{ps} is the estimated depth of a *point source*.

airborne magnetics: *aeromagnetics* in the context of the magnetic data obtained from measurements of the *Earth magnetic field* and its gradients during an *aeromagnetic survey*.

airborne magnetometer: a *magnetometer* system for measurements of the *Earth magnetic field* during an *airborne survey* with the use of a fixed-wing aircraft (airplane), helicopter, or *drone*. See also *aeromagnetic survey equipment*, *aeromagnetic compensation system*, *towed bird system*, and *microfabricated atomic magnetometer (MFAM)*.

airborne magnetometer calibration: a set of *airborne magnetometer* tests conducted prior to an *aeromagnetic survey* and each time when a configuration of the aircraft body or airborne magnetometer system has been altered. Typically, this includes a *compensation test*, a *heading error test*, and a *lag test*. See also *figure-of-merit (FOM)*, *aeromagnetic compensation*, and *airborne gravimeter calibration*.

airborne magnetometry: methods and instruments to record and process measurements of the *Earth magnetic field*, its *gradient*, and *gradient tensor* components with the use of a fixed-wing aircraft (airplane), helicopter, or *drone*. Airborne magnetic measurements are conducted along the pre-planned flight lines over the *airborne survey* area in a *drape flight* mode, wherein *terrain clearance* is chosen based on the *Digital Elevation Model (DEM)* of the survey area, flight characteristics of the aircraft, and safety regulations. High-accuracy *positioning* measurements by *Global Positioning System (GPS)* or more precise *differential GPS (DGPS)* are required to determine *3D coordinates* and aircraft velocity that are used in processing of the acquired magnetic data. See also *acquisition*, *accuracy*, *precision*, *high-resolution aeromagnetic survey*, *marine magnetometry*, *submarine magnetometry*, and *satellite magnetometry*.

airborne survey: measurements of the *Earth gravity field* or *Earth magnetic field*, its *gradient*, and *gradient tensor* components along the regularly spaced flight lines over an area of exploration or research interest with the use of a fixed-wing aircraft (airplane), airship (dirigible), helicopter, or *drone*. See also *drone survey*, *fixed-wing survey*, *helicopter survey*, *ground survey*, *marine survey*, and *submarine survey*.

airborne terrain correction: a *terrain correction* applied to the *airborne gravity* or airborne *gravity gradient* data to compensate for the *terrain effect*. Its advanced space-domain algorithm involves (a) *adaptive quadtree discretization* of a *Digital Terrain Model (DTM)* into the variable-width square prisms, (b) calculation of the gravity or gravity-gradient effect of each *prism* in the *airborne survey* and surrounding areas using the formulas with a constant *terrain density* for all prisms, (c) summing the gravity or gravity-gradient effects of all individuals prisms within a specified range of distances around each *observation point* to determine a correction value at the observation point location, and (d) low-pass *filtering* of the calculated terrain corrections in accordance with the *Agerlin principle* before their application in order to avoid compensating for terrain effect components suppressed by *acquisition filtering*. Terrain corrections are calculated to the distance from observation points at which the combined gravity or gravity-gradient effects of all individual prisms become less than the estimated *noise* level. Optimal *lateral resolution* of the DTM is determined based on *terrain clearance* of the airborne survey and applied acquisition filtering (Li et al., 2010). See also *observed gravity* and *marine terrain correction*.

airborne vector magnetometry: methods and instruments for airborne measurements of the *vector* components of the *Earth magnetic field*. Onboard magnetic instruments include a *total field magnetometer* and a three-component *fluxgate magnetometer* or *absolute scalar magnetometer (ASM)* operating in both the *scalar* and vector modes for measuring the *baseline* values and three mutually orthogonal *magnetic vector* components. Based on the aircraft *attitude* data (heading, bank, and elevation angles) recorded by the independent system of inertial accelerometers, the measured vector data are transformed from the aircraft-based reference frame of observations into the standard *NED reference frame* and compensated for magnetic interference effects due to the aircraft attitude variations, *aircraft signature*, and *heading error*. Post-flight processed vector data are used for calculation of *magnetic vector attributes* and their residual components (Christensen and Dransfield, 2002; Xie et al., 2020). See also *aeromagnetic compensation*, *aeromagnetic survey equipment*, *inertial accelerometer*, and *residual magnetic vector*.

aircraft signature: the gravity or magnetic field of the *airborne survey* aircraft that interferes with measurements of the airborne *gravity gradient* and magnetic measurement systems. Each aircraft has its own signature, so that gravity gradient and magnetic data acquired along the same flight lines with the use of different aircrafts will have a level shift between the measured values. An aircraft signature is determined by three-dimensional (3D) *modeling* of the aircraft frame and onboard electronic and mechanical components, and its gravity gradient and magnetic effects are eliminated by *self-gradient correction* and *aeromagnetic compensation*, respectively. See also *self-gradient effect*.

Airy antiroot: *isostatic antiroot* in the context of the *Airy hypothesis* of *isostasy*.

Airy compensation: *isostatic compensation* in accordance with the *Airy hypothesis* and *Airy-Heiskanen model* of *isostasy*.

Airy depth: the *isostatic root depth* onshore or *isostatic antiroot depth* offshore derived with reference to the depth of the crust-mantle boundary, i.e., *Moho boundary*, predicted in accordance with the *Airy-Heiskanen model* of *isostasy*.

Airy gravity anomaly: 1. the *Airy gravity effect* as it relates to being a regional component of the anomalous gravity field due to *isostatic compensation* in accordance with the *Airy hypothesis*. **2.** *Airy isostatic anomaly*. See also *anomalous field*.

Airy gravity effect: a regional-scale gravity effect of the *Airy root* or *Airy antiroot* calculated with structural parameters of the *Airy-Heiskanen model* of *isostatic compensation*. This gravity effect comprises the amount of *Airy isostatic correction*. Sometimes, it is called *Airy gravity anomaly*. Also referred to as *root gravity effect* on continents and *antiroot gravity effect* in oceans and seas. See also *isostatic root depth* and *isostatic antiroot depth*.

Airy-Heiskanen equations: the mathematical expressions of a relationship between *topography* or *bathymetry* and thickness of *isostatic root* or *isostatic antiroot*, respectively, in accordance with the *Airy-Heiskanen model* of *isostatic compensation*. In oceans and seas, it is defined as

$$T_{TC} = T_{SL} - h_w[1 - (\rho_{ac} - \rho_w)/\Delta\rho],$$

where T_{TC} is total crustal thickness, i.e., isostatic antiroot depth, under the ocean/sea floor; T_{SL} is *reference crustal thickness* at the shoreline *elevation* (sea level); h_w is water depth; ρ_{ac} is average crustal *density*; ρ_w is water density; and $\Delta\rho$ is *density contrast* between the *crust* and upper *mantle*. On continents, it is defined as

$$T_{TC} = T_{SL} - h_{th}[1 + (\rho_{ac}/\Delta\rho)],$$

where T_{TC} is total crustal thickness, i.e., isostatic root depth, under the topographic surface; T_{SL} is reference crustal thickness at sea level; h_{th} is topographic height, or elevation, above sea level; and ρ_{ac} is average crustal density. See also *isostatic antiroot depth*, *isostatic root depth*, and *Jachens-Roberts method*.

Airy-Heiskanen gravity correction: an *isostatic correction* calculated with the use of structural parameters of the *Airy-Heiskanen model* of *isostatic compensation*, i.e., correction for the regional-scale *Airy gravity effect*. Referred to with the same meaning as *Airy isostatic correction*.

Airy-Heiskanen isostasy: a generalization of the *Airy hypothesis* in the mathematical model of *isostatic compensation* known as the *Airy-Heiskanen model*.

Airy-Heiskanen model: the mathematical model of *isostatic compensation* based on the *Airy hypothesis* of *isostasy*. It assumes that (a) isostatic compensation is achieved on a local basis and (b) the lithostatic load of a unit, or block, area of rock masses above sea level is equal to a change in the thickness of the crustal block relative to its thickness at sea level. The basic formula is

$$\rho h = \Delta\rho(T_{TSL} - T_{SL}),$$

where ρ is average crustal *density*; h is topographic height, or *elevation*, above sea level; $\Delta\rho$ is *density contrast* between the *crust* and upper *mantle*; T_{TSL} is total crustal thickness relative to sea level at every observation point, i.e., thickness without h; and T_{SL} is assumed crustal thickness at the shoreline, i.e., thickness without the lithostatic load, also known as *reference crustal thickness*. See also *Pratt-Hayford model* and *Vening Meinesz model*.

Airy hypothesis: one of three basic theories of *isostasy* which describes a geostatic balance, or equilibrium, between the *crust* and upper *mantle*. It assumes that densities in the crustal blocks and upper mantle are laterally constant, and rock mass excesses of the large topographic highs and mountains are balanced, or compensated, by an increased crustal thickness below them, whereas rock mass deficiencies, created by the large topographic depressions and ocean/sea basins, are compensated by a decreased crustal thickness below them. Long-*wavelength* regional-scale gravity anomalies, generated by lateral variations of the crustal thickness, are removed by application of *isostatic correction* to obtain *isostatic gravity anomaly*, which is also referred to as

Airy isostatic anomaly. See also *mass deficiency*, *mass excess*, *Pratt hypothesis*, and *Vening Meinesz hypothesis.*

Airy isostasy: a geostatic balance, or equilibrium, between the *crust* and the upper *mantle* described by the *Airy hypothesis* of *isostasy.*

Airy isostatic anomaly: the *isostatic anomaly* obtained by application of the *isostatic correction* calculated with the use of *isostatic antiroot depth* or *isostatic root depth* which is determined at each *observation point* in accordance with the *Airy-Heiskanen model* of *isostatic compensation* and *Airy-Heiskanen equations.* Referred to with the same meaning as *Airy gravity anomaly.* See also *Pratt isostatic anomaly* and *flexural isostatic anomaly.*

Airy isostatic correction: the *isostatic correction* calculated with parameters of the *Airy-Heiskanen model* of *isostatic compensation.* Calculations assume that (a) the *Moho boundary* mirrors the smoothed topography and bathymetry scaled by *density contrast* between the *crust* and upper *mantle,* (b) density contrast across the Moho boundary is constant, and (c) crustal thickness at the shoreline is a known constant. Sometimes referred to as *Airy-Heiskanen gravity correction.* See also *Jachens-Roberts method.*

Airy isostatic response: a characteristic of the isostatic deformation of the *lithosphere* under topographic or bathymetric load, where the lithosphere is approximated by the bending elastic plate of a low strength, or flexural rigidity. It is opposite to the *Bouguer isostatic response,* where the flexural rigidity of the isostatic model-approximated lithosphere is assumed to be relatively high without a significant flexure.

Airy modeling: *forward modeling* of the *isostatic compensation effect* and regional depth structure of the *Moho boundary* with the use of structural parameters of the *Airy-Heiskanen model.*

Airy root: *isostatic root* in the context of the *Airy hypothesis* of *isostasy.*

AIRYROOT: the computer algorithm developed by the U.S. Geological Survey for calculation of the *isostatic compensation effect* on continents to the distance 166.7 km around each *observation point* in accordance with the *Airy-Heiskanen model* of *isostasy.* It calculates (a) *isostatic root depth* from the topographic data using the average crustal *density*, *reference crustal thickness*, and *density contrast* between the lowermost *crust* and uppermost *mantle*, and (b) *gravitational attraction*, or gravity effect, of the *isostatic root* at sea level (Simpson et al., 1983). Modified versions are used for calculation of three-dimensional (3D) *isostatic correction* onshore and offshore. See also *Jachens-Roberts method.*

alias filter: the *low-pass filter* applied to a continuous signal before *sampling* in order to restrict its *bandwidth* to requirements of the *Nyquist theorem* and prevent *aliasing.* An alias filter removes frequencies that are higher than the *Nyquist frequency* that otherwise would be aliased as false lower frequencies in the process of sampling. Also referred to as *antialias filter.* See also *spatial frequency* and *wavelength.*

aliasing: **1.** an inherent property of the *sampling* process and recovery of amplitudes and shapes of anomalies from the discrete measurements of a continuous gravity or magnetic *field.* When the sampling is less than two samples per the shortest *wavelength*, the field components with spatial frequencies higher than the *Nyquist frequency* will be aliased (i.e., represented by the artificial mirror frequencies) as false lower frequencies: *spatial frequency* $f_N + \Delta f$ will be aliased as the false frequency $f_N - \Delta f$, where f_N is the Nyquist frequency. To prevent data contamination by aliasing, the spatial frequencies higher than the Nyquist frequency are removed by an *alias filter* before sampling. **2.** self-contaminating property of the *discrete Fourier transform (DFT)* due to its periodic nature: instead of one value $\mathcal{F}(k_0)$ for each *wavenumber* k_0, it produces $\mathcal{F}(k_0)$ plus an infinite set of additional values for other wavenumbers k with the period $2\pi/\Delta x$, where Δx is *sample interval.* The effect of the additional values decreases rapidly with increasing k; it is minimized by reducing the sample interval. See also *grid aliasing*, *spatial aliasing*, *wavenumber aliasing*, and *survey design aliasing.*

alkali-vapor magnetometer: same as *optically pumped magnetometer* with an emphasis on the *sensor* type which utilizes the gaseous, or vaporized, chemical elements from the alkali group, such as cesium, potassium, and rubidium.

allochthonous salt: *salt* moved by a viscous flow from its place of deposition to the shallower depths within the sedimentary section under a load of the overlaying younger sediments. The base of the allochthonous salt is a common target for two-dimensional (2D) *modeling* and three-dimensional (3D) *inversion* of the gravity and *full tensor gravity gradient (FTGG)* data.

altimeter: an instrument for determining aircraft height above the Earth's surface, also known as *altitude*, by measuring a traveltime of the transmitted microwave-range electromagnetic pulses, e.g., a *radar altimeter*, or short pulses of a laser light, e.g., a *laser altimeter*, during the airborne surveys and satellite missions. Operational *precision* of altimeters is a function of the Earth's surface slope: the higher the slope, the lower the precision. Airborne radar altimeter measurements are used for calculation of altitude and *orthometric height* under all atmospheric conditions. Satellite radar and laser altimeter measurements are used for calculation of *ellipsoidal height*, also known as *GPS height*, but the laser altimeter cannot measure through thick clouds. The best reported *accuracy* of commercial Satellite Laser Ranging (SLR) systems is approximately 2–5 cm. See also *barometric altimeter* and *LIDAR.*

altimetric gravity: *free-air gravity* calculated from the *satellite altimetry* data acquired over oceans and seas. Referred to with the same meaning as *satellite-altimeter gravity.*

altimetry-derived gravity: See *satellite-altimeter gravity.*

altitude: **1.** the height of an aircraft or satellite above the Earth's surface. The altitude of an *airborne survey* defines the lateral and vertical *resolution* of anomalies: the lower the altitude, the higher the resolution of anomalies. High-altitude measurements, such as those during reconnaissance surveys and satellite missions, are capable of resolving only those anomalies that have wavelengths about the same and longer than the altitude value. In airborne surveys, the altitude is measured by a *radar altimeter* or *LIDAR*. Altitude values over continents are often denoted by the abbreviation AGL: above ground level. Altitude is different from *elevation* which is a height above *mean sea level (MSL)* or, in the case of *GPS*-based observations, above the *reference ellipsoid*. Referred to with the same meaning as *station* elevation in a ground gravity or magnetic *survey* and *terrain clearance* or *ground clearance* in an airborne survey. **2.** an *upward continuation* level in *multiridge analysis*. See also *reconnaissance survey*.

ambient field: a gravity or magnetic *field* which is external to the field generated by a particular *source body*. For gravity and magnetic surveys, the ambient fields are *Earth gravity field* and *Earth magnetic field*, respectively. For *remanent magnetization* of rocks, the ambient field is the Earth's magnetic field at the geological time when those rocks were magnetized. See also *background magnetic field*.

ambient noise: *background noise* in the context of average *noise* level upon which amplitudes of the gravity or magnetic anomalies are superimposed.

ambiguity: 1. the nonuniqueness of *interpretation* in general and *modeling* in particular, wherein different interpretation scenarios and modeling solutions, such as a variety of source or structural shapes, physical properties (*density* or *magnetization*), and depths, could produce an acceptable representation of the *observed field*. Ambiguity is an inherent property of gravity and magnetic methods regardless of the *accuracy* of the measurements and mathematical complexity of *data processing* algorithms. Interpretation or modeling ambiguity is minimized by the independent geological and geophysical information provided by *constraints*. **2.** an inherent property of the potential fields: if the gravity or magnetic field is known on a surface bounding the distribution of sources or above this surface in a source-free space, then there are infinitely many source distributions in the subsurface that could generate the same observed field. Referred to with the same meaning as *nonuniqueness problem*. See also *potential field*.

ampere (A): a unit of electric current, denoted in *SI units* of measurement, by which rock *magnetization* is measured in units of ampere per meter (Am^{-1}) and *magnetic dipole moment* in units of ampere meter squared (Am^2).

Ampere law: the physical law postulating that a moving charged particle creates a *magnetic field*. The flow of the charged particles (i.e., electric current) I through the length of wire ΔL creates the magnetic field $\Delta \mathbf{B}$ at the point P, which can be expressed as

$$\Delta \mathbf{B}(P) = C_m I(\Delta \mathbf{L} \times \hat{\mathbf{r}})/\mathbf{r}^3,$$

where $\Delta \mathbf{L}$ is a *vector* in the direction of electric current flow; $\hat{\mathbf{r}}$ is a unit vector pointing from the element ΔL to point P; r is the distance between ΔL and P; C_m is constant of proportionality whose value depends on the system of measurements: $C_m = 1$ in the cgs system and $C_m = 10^{-7}$ *henry* · meter^{-1} in the SI system; and $\Delta \mathbf{L} \times \hat{\mathbf{r}}$ is the *vector product* which produces a new vector (magnetic field vector), whose direction is orthogonal to both $\mathbf{L}$ and $\hat{\mathbf{r}}$ vectors. The upgrade of this law with the integration of ΔL around a loop of wire is referred to as *Biot-Savart law*.

amplitude: the maximum positive or negative departure of the *observed field* from its zero value or from a *baseline* value in the *survey* area. Amplitude is a linear function of the physical properties of rocks (*density* or *magnetization*), and it is a nonlinear function of the source depth or volume. The amplitude of an *anomaly* is most sensitive to the source depth and distance from the source, then to the source volume, and then to the source *density contrast* or *magnetization contrast*. For two sources of the same depth and volume, the source with a higher density or magnetization *contrast* will generate the anomaly with the higher amplitude. Amplitude decrease with increasing distance from a *source body* is called *amplitude attenuation*, and it can be approximated by the *inverse power law* as

$$A(r) = C/\mathbf{r}^N,$$

where C is constant, r is distance from the source, and N is *fall-off rate* quantified for the ideal sources by the *Euler structural index (SI)*. See also *anomaly amplitude* and *ideal source*.

amplitude attenuation: See *fall-off*.

amplitude balancing: an *enhanced processing* technique that equalizes the observed amplitudes by increasing the low amplitudes and decreasing the high amplitudes of anomalous gravity or magnetic field with the use of (a) *automatic gain control (AGC)*, (b) orthogonal Hilbert transforms in calculation of *balanced ASA (ASAB)*, and (c) the ratio of derivatives in calculation of *tilt angle (TA)*, *theta angle*, and their modifications. Also referred to as *amplitude equalization*. See also *attribute*, *derivative*, *Hilbert transform*, and *normalized derivatives*.

amplitude equalization: See *amplitude balancing*.

amplitude polarity: a positive or negative sign of an *anomaly* controlled by the sign of a *density contrast* or *magnetization contrast* of its source with the laterally adjacent rock formations. See also *amplitude* and *lateral contrast*.

amplitude-ratio method: See *ASA-ratio method*.

amplitude resolution: a quantitative estimate of the smallest *amplitude* of the gravity or magnetic *anomaly* that could be detected above the level of *acquisition noise* or revealed by *enhanced processing* of the acquired data. Algorithms of *enhanced processing* are capable of detecting the *target*-generated anomalous amplitudes that are less than the *dynamic noise* and *background noise* levels. The smaller and the shallower the target, the higher the amplitude resolution requirements. Referred to with the same meaning as *vertical resolution*. See also *lateral resolution*.

amplitude response: 1. a quantitative estimate of amplitudes in the anomalous gravity or magnetic *field*. **2.** a characteristic of a *filter* or any other *transfer function* that defines the *amplitude* of the output signal as a function of the space-domain or spectral-domain parameters of the input data, such as *wavelength* or *wavenumber*, respectively. **3.** estimated changes in amplitudes of the *residual gravity* anomalies caused by changes in *density contrast*, reservoir depth, and fluid content of the reservoir *model* which is built for *reservoir monitoring*.

See also *anomalous field*, *anomaly*, *space domain*, and *spectral domain*.

amplitude spectrum: 1. the *amplitude*-versus-*wavenumber* relationship that describes amplitude components of the Fourier-transformed gravity or magnetic data as functions of their respective wavenumbers in the *spectral domain*. **2.** the wavenumber-dependent function A(*k*) in the *complex spectrum* of the Fourier-transformed data, expressed as

$$\left|A(k)\right| = (RP^2 + IP^2)^{1/2},$$

where RP is a real part representing the *cosine transform* components of the *Fourier transform* and IP is an imaginary part representing the *sine transform* components. The squared amplitude spectrum $|A(k)|^2$ is called the *energy-density spectrum* or *power-density spectrum*. Also known as *Fourier amplitude spectrum*. See also *Fourier transform*, *phase spectrum*, and *power spectrum*.

AN-EUL method: the line- or *grid*-based automatic method of estimating the depth and *Euler structural index (SI)* of *two-dimensional (2D)* magnetic sources in application of *Euler deconvolution*. This method involves a calculation of *analytic signal amplitude (ASA)* and its first- and second-order vertical derivatives representing the second- and third-order vertical derivatives of *total magnetic intensity (TMI)*, respectively. For sources whose structural index N could be inferred from analysis of the shapes of TMI anomalies, the depth *h* is estimated at locations of the ASA maxima using the following formulas:

$$h_{mc} = \left|ASA_0\right|/\left|ASA_1\right| \text{ for } \textit{magnetic contact } (N = 0),$$

$$h_{td} = 2\left|ASA_0\right|/\left|ASA_1\right| \text{ for } \textit{thin dike } (N = 1),$$

$$h_{vhc} = 3\left|ASA_0\right|/\left|ASA_1\right| \text{ for } \textit{vertical cylinder} \text{ or } \textit{horizondal cylinder } (N = 2),$$

$$h_{spd} = 4\left|ASA_0\right|/\left|ASA_1\right| \text{ for } \textit{sphere} \text{ or } \textit{point dipole } (N = 3),$$

where $|ASA_0|$ is ASA of the original TMI field and $|ASA_1|$ is ASA of the first-order *vertical derivative* of the TMI. For magnetic sources with unknown *source geometry*, the estimates of their depth *h* and structural index N are calculated at locations of the ASA maxima as, respectively,

$$h = \left(\left|ASA_0\right| \cdot \left|ASA_1\right|\right)/\left|ASA_0\right| \cdot \left|ASA_2\right| - \left|ASA_1\right|^2,$$

$$N = \left(2\left|ASA_1\right|^2 - \left|ASA_2\right| \cdot \left|ASA_0\right|\right)/\left(\left|ASA_0\right| \cdot \left|ASA_2\right| - \left|ASA_1\right|^2\right),$$

where $|ASA_2|$ is ASA of the second-order vertical derivative of the TMI, and N could have the non-integer values of *fractional structural index (fSI)*. Second- and third-order vertical derivatives make this method sensitive to *noise*, so that *upward continuation* to the lowest empirically determined level is applied prior to their calculation (Salem and Ravat, 2003). See also *ASA-derived depth*, *ASA-ratio method*, and *local wavenumber methods*.

analytic signal (AS): a *complex variable* whose real and imaginary parts are related via *Hilbert transform* and, thereby, represent the *Hilbert transform pair*. The analytic signal of the *one-dimensional (1D)* function *f*(*x*) can be expressed as

$$AS\{f(x)\} = f(x) + iH\{f(x)\},$$

where *f*(*x*) is a real part of the AS, $i = \sqrt{-1}$ is a complex unit number indicating orthogonality, and H{*f*(*x*)} is the Hilbert transform of *f*(*x*) representing the imaginary part of the AS, which is the 90° *phase*-shifted real part and also referred to as *quadrature* or *quadrature signal*. Derivatives of the gravity or magnetic field in the mutually orthogonal directions satisfy requirements of the Hilbert transform pair and represent the AS components. In the general form, AS of the gravity or magnetic *vector field* **F** at the observation point (*x*,*y*) is expressed as

$$\mathbf{AS}(x, y) = (\partial F/\partial x)\mathbf{i} + (\partial F/\partial y)\mathbf{j} + i(\partial F/\partial x)\mathbf{k},$$

where $\partial F/\partial x$, $\partial F/\partial y$, $\partial F/\partial z$ are derivatives of *anomalous field* F in the mutually orthogonal *x*-, *y*-, *z*-directions in *Cartesian coordinates*, and **i**, **j**, **k** are unit vectors parallel to these directions. For the line-oriented and *profile* data, the *horizontal derivative* of the anomalous field F represents the real part and the *vertical derivative* is the imaginary part of the AS referred to as the *2D analytic signal (2DAS)*. For the gridded data, AS is referred to as the *3D analytic signal (3DAS)*, where two horizontal derivatives $\partial F/\partial x$ and $\partial F/\partial y$ in the orthogonal *x*- and *y*-directions represent the real parts and the vertical derivative $\partial F/\partial z$ is the imaginary part of the 3DAS. Generally, the AS could be derived from any *potential function*, such as a component of the anomalous field *vector*, a component of the *gradient vector*, a component of the *gradient tensor*, an *N*th-order *vertical derivative* of the anomalous field, a *derivative* of the *gradient* component, or a low- or band-pass-filtered result of any of these. Absolute value, or *magnitude*, of the AS is known as *analytic signal amplitude (ASA)*, and its numerous versions for the gridded data are widely used in mapping the structural boundaries and automatic estimates of the *source body* parameters, including the depth, *dip*, and *structural index (SI)*. In practice, the AS and ASA are often referred to with the same meaning. The AS applications are the most efficient for regular-spaced and high-*resolution* data, such as high-resolution airborne magnetic and gravity data, because the sparse and irregularly distributed or low-resolution data make the calculation of derivatives unreliable. Sometimes referred to as a *complex analytic signal*. See also *direct analytic signal (DAS)*, *directional analytic signal*, *enhanced analytic signal (eAS)*, *AS-LW relationship*, and *gridding*.

analytic signal amplitude (ASA): the absolute value, or *magnitude*, of an *analytic signal (AS)*. At the observation point (*x*,*y*), ASA of the gridded gravity or magnetic *field* is defined as

$$\left|AS(x, y)\right| = \left|AS\right| = ASA = [(\partial F/\partial x)^2 + (\partial F/\partial y)^2 + (\partial F/\partial)^2]^{1/2},$$

where $\partial F/\partial x$, $\partial F/\partial y$, $\partial F/\partial z$ are derivatives of *anomalous field* F in the mutually orthogonal *x*-, *y*-, *z*-directions, respectively, in *Cartesian coordinates*. An ASA anomaly is a bell-shaped function that peaks above the *source body* and structural boundaries, such as a lithological *contact* or high-*offset* fault. Above *two-dimensional (2D)* sources, the ASA is also called the *envelope amplitude*, as it bounds, or envelopes, the *phase* shifts caused by all possible directions of the *main geomagnetic field* and source *magnetization* and, thereby, makes the 2D ASA *shape* (but not *amplitude*) independent from *magnetic inclination* and *magnetic declination* in the survey area and *skewness* of magnetic anomalies. The ASA calculation produces a *scalar field* of all-positive anomalies over their respective source bodies, wherein the ASA maxima are positioned directly above the *contact* and *thin dike* 2D source geometries regardless of their *dip* and *magnetization direction* (but not for the three-dimensional [3D] sources), and its *half-width* value is equal to a depth of *magnetic contact*. Due to the derivative nature of the ASA anomalies, the short-*wavelength* components in the observed gravity or magnetic field are enhanced, whereas the long-wavelength components are suppressed in a way of the *high-pass filter* effect. For the same reason, various *noise* effects, such as *gridding* artifacts, unedited *cultural magnetic anomalies,* and line *corrugations*, are enhanced too. For 3D sources, ASA is much less sensitive to the direction of the *main geomagnetic field* than the original *total magnetic intensity (TMI)* amplitudes, and it is often used as a high-pass-filtered alternative to the standard *reduction-to-pole (RTP)* or *reduction-to-equator (RTE)* in the low *magnetic latitude* and *magnetic equator* areas. Because of their numerical equivalence, ASA and calculated *total gradient (TG)* are referred to with the same *data enhancement* meaning. The ASA *grid* is commonly called the *analytic signal (AS)* grid. Decay, or *fall-off*, of the gravity or magnetic ASA with the distance r from its source is proportional, respectively, to $1/r^2$ and $1/r^3$ for 2D *contact*-like sources and $1/r^3$ and $1/r^4$ for *point sources*. Also referred to as *analytic signal magnitude*, *analytic signal modulus*, *instantaneous amplitude*, *local amplitude*, and *zero-order ASA* (ASA_0). See also *enhanced analytic signal amplitude (eASA)*, *first-order ASA* (ASA_1), *first-order enhanced ASA* ($eASA_1$), *second-order ASA* (ASA_2), *second-order enhanced ASA* ($eASA_2$), *balanced ASA (ASAB)*, *directional ASA*, *direct analytic signal amplitude (DASA)*, *FTG analytic signal amplitude*, and *ASTA filter*.

analytic signal Euler deconvolution: *Euler deconvolution* applied to the *analytic signal amplitude (ASA)* of line-oriented or gridded *anomalous field* data. Because a regional background is suppressed by the ASA calculations, the overdetermined system of the Euler deconvolution equations for each position of *moving window* along the *survey* line or *grid* could be solved for the local source coordinates (x_0, y_0, z_0) and its *Euler structural index (SI)*. Once the source coordinates and SI are known, then *susceptibility contrast* and source *dip* could be determined (Keating and Pilkington, 2004). See also *ASA vertical derivative*, *AN-EUL method*, and *derivative Euler deconvolution*.

analytic signal filter: an *enhanced processing* operator for the calculation of three mutually orthogonal first-order derivatives, i.e., *vertical derivative* and two horizontal derivatives, of the *anomalous field* with their subsequent integration into the *analytic signal amplitude (ASA)*. See also *analytic signal* and *horizontal derivative*.

analytic signal inclination: an angle between the horizontal plane and direction of the *analytic signal vector*. Equivalent to *tilt angle (TA)* and *theta angle*.

analytic signal magnitude: See *analytic signal amplitude (ASA)*.

analytic signal method: a method of *enhanced processing* and *interpretation* that is based on calculation of *analytic signal amplitude (ASA)* and its derived versions, such as *tilt ASA*, *ASTA filter*, *directional ASA*, *first-order ASA* (ASA_1), and other ASA versions. This method is applied for: (a) detection and mapping of faults, lithological contacts, structural edges, and isolated source bodies, such as kimberlite intrusions, ore bodies, and *unexploded ordnance (UXO)*; (b) estimation of depths and other source parameters of the ASA anomalies using the analytical methods of *Euler deconvolution*, *AN-EUL method*, *ASA-ratio method*, *local wavenumber methods,* and other methods; and (c) calculation of a *high-pass filter* alternative to the standard *reduction-to-pole (RTP)* and *reduction-to-equator (RTE)* at low magnetic latitudes and *magnetic equator* areas. The ASA maxima are positioned directly above sources approximated by *two-dimensional (2D)* source geometries, such as *thin dike* and *contact*, regardless of their structural *dip* and direction of *magnetization*. The difference in locations of the ASA and *horizontal gradient* maxima over a dipping 2D *source body* is used to determine the direction of its dipping and dip value. For the *three-dimensional (3D)* source bodies, the analytic signal method is less dependent on the direction of the *main geomagnetic field* and less subject to distortions by *remanent magnetization* effects than other magnetic *enhanced processing* methods. See also *edge*, *fault*, *kimberlite intrusion*, *magnetization direction*, *ore body*, and *source geometry*.

analytic signal modulus: See *analytic signal amplitude (ASA)*.

analytic signal phase: **1.** a *complex attribute* defined by the arctangent of the ratio of the imaginary part ImP to the real part ReP of an *analytic signal (AS)* as

$$\mathrm{ASPh} = \arctan[\mathrm{ImP}/\mathrm{ReP}].$$

For the gridded field F, where the *vertical derivative* and two mutually orthogonal horizontal derivatives represent the imaginary (90° out-of-*phase*) part and real (in-phase) parts, respectively, the analytic signal phase is equivalent to *tilt angle (TA)* and expressed as

$$\mathrm{ASPh} = \arctan\{(\partial \mathrm{F}/\partial z)/[(\partial \mathrm{F}/\partial x)^2 + (\partial \mathrm{F}/\partial y)^2]^{1/2}\} = \mathrm{TA}.$$

For the *profile*-based calculations with the absolute value of the *horizontal derivative*, known as *horizontal gradient*, the analytic signal phase formula is written as

$$\mathrm{ASPh} = \arctan(\partial \mathrm{F}/\partial z/|\partial \mathrm{F}/\partial x|).$$

When a sign of the horizontal derivative is utilized, the analytic signal phase is referred to as *local phase* or *instantaneous phase*. **2.** an angle between the horizontal plane and the direction of the *analytic signal vector* which is sometimes referred to as *analytic signal inclination*. See also *anomalous field*, *gridding*, and *normalized derivatives*.

analytic signal vector: *3D gradient vector* for which the length, or *magnitude*, is defined by the *3D analytic signal amplitude (3DASA)*, and its *phase*, or angle relative to the horizontal plane, is equivalent to *theta angle* and *tilt angle (TA)*.

anastomosing: a characteristic *signature* of a branching-and-reconnecting of the magnetic and gravity *gradient* anomalies generated by certain types of normal and strike-slip faults, including those which bound a *pull-apart basin*. See also *strike-slip fault*.

angular acceleration: **1.** *acceleration* due to a change in the direction of *angular velocity* per unit time, i.e., acceleration of a body moving in a uniform circular motion. It is expressed as

$$a_{aa} = \omega^2 r_{ar},$$

where ω is angular velocity and r_{ar} is distance from the axis of rotation. Motion-related *tilt* of the *observation platform* introduces a vertical component of angular acceleration which interferes with measurements by paired accelerometers mounted on the horizontal or tilted rotating disk in a *rotating-disk gravity gradiometer*. **2.** acceleration resulting from a self-rotation of an Earth-orbiting satellite and variations of its orbital velocity. Angular accelerations are non-gravitational *vector* components of *gravity gradient* measurements by a rotating-disk gravity gradiometer and Earth-orbiting satellites, and they are compensated by *angular acceleration correction*, which is also known as *centripetal acceleration correction*. Referred to with the same meaning as *centripetal acceleration*. See also *accelerometer* and *GOCE gravity gradiometer*.

angular acceleration correction: **1.** *centripetal acceleration correction*. **2.** correction applied to the *accelerometer*-based *gravity gradient* measurements by an Earth-orbiting satellite to compensate for the angular accelerations resulting from the satellite self-rotation and variations of its orbital velocity. See also *angular acceleration* and *angular velocity*.

angular frequency: the angular displacement per unit time measured in degrees per second or in radians per second ($\text{rad} \cdot \text{s}^{-1}$). Angular frequency ω is related to a repetition rate, or *frequency*, of the periodic function and its *period* as

$$\omega = 2\pi f = 2\pi/T,$$

where f is frequency in units of *hertz (Hz)* and T is period in seconds. Referred to with the same meaning as *angular velocity*. Also known as *circular frequency*. See also *radian*.

angular velocity: the rate of change of the angular position of an instrument on a rotating surface, such as a *gravimeter* on the Earth's surface or paired accelerometers in a *rotating-disk gravity gradiometer*, measured in units of *radian* per second ($\text{rad} \cdot \text{s}^{-1}$). Angular velocity ω and linear velocity v on the Earth's surface are related as

$$\omega = v/\text{R}\cos\theta,$$

where R is Earth radius and θ is geographic latitude. Referred to with the same meaning as *angular frequency*. See also *acceleration* and *accelerometer*.

anisotropic data: **1.** *ground survey* data acquired with *anisotropic sampling*. **2.** gridded gravity or magnetic data with a strong directional bias characterized by the presence of a dominant *trend* in the *anomalous field* values that could be associated with the regional-scale geological *structure*. See also *gridding*, *directional filtering*, and *anisotropic sampling*.

anisotropic indices: See *FTG anisotropic indices*.

anisotropic sampling: **1.** a rectangular *mesh* of the regularly spaced measurement points, e.g., 2 × 1 km, in a *ground gravity* survey with a closer *station spacing* across the long axis of a geological *structure* of the exploration interest. Also referred to as *anisotropic station spacing*. **2.** a characteristic of the ground gravity or magnetic data acquired along roads and unpaved paths (trails). Depending on distances between *acquisition* paths, the subsequent *gridding* of the anisotropically sampled data could introduce a strong directional bias in the *anomalous field*, distort shapes of anomalies, and produce artificial mid- and long-*wavelength* anomalies amplified by *enhanced processing* of the gridded data. See also *acquisition footprint*, *artificial anomaly*, *sampling*, and *undersampling*.

anisotropic station spacing: See *anisotropic sampling*.

anisotropy: **1.** variation of a physical property as a function of the direction of its measurement. Anisotropy of the gravity and magnetic fields is described qualitatively by identification of the dominant trends of anomalies and quantitatively by spatial variations of the *gradient* and *gradient tensor* values measured or calculated in three mutually orthogonal directions, such as those representing the *full tensor gradient (FTG)* components. Anisotropy of the gradient values is related to the structural geometry of the source bodies and their physical properties, such as *density* and *magnetization* contrasts, and its parameters are analyzed by *enhanced processing* of the gravity and magnetic gradient and FTG data. **2.** deviation from *isotropy*. See also *magnetic anisotropy*, *density contrast*, *magnetization contrast*, *source body*, and *FTG anisotropy*.

annihilator: a non-zero distribution of *density* or *magnetization* within a *source body* that could generate a nearly zero gravity or magnetic *anomaly* undetectable within the *ambient field*. See also *Schiaparelli body* and *self-demagnetization*.

anomalous field: **1.** the final result of the *data reduction* process that represents a *scalar field* of the gravity or magnetic anomalies obtained by application of corrections for all predictable planetary and non-geological effects contributing to the measured field values. Also referred to as the *observed field*. **2.** a *vector field* of gravity or magnetic anomalies generated by subsurface distribution of an *anomalous mass* or *anomalous magnetization*. Obtained anomalies do not describe the actual values of rock densities or magnetizations, but rather a distribution of the *density* or *magnetization* contrasts created by laterally adjacent rock units. Two adjacent rock units with a relatively high density or magnetization could generate the same anomaly as a low-density or low-magnetization rock unit surrounded by the less dense or less magnetized rocks. Interference of anomalies generated by closely spaced sources and superposition of anomalies from sources at sufficiently different depths define the composite nature of an anomalous field. The anomalous gravity field is represented by the *free-air anomaly*, *Bouguer anomaly*, and *isostatic anomaly*. The anomalous magnetic field is represented by the *TMI anomaly* and *RTP anomaly*. **3.** a field of the *gradient* and gradient-derived gravity or magnetic anomalies. **4.** the *residual field* obtained by *regional-residual separation* which suppresses all *total field* anomalies that have wavelengths longer than the estimated *wavelength* of the *anomaly* generated by the exploration *target*. See also *density contrast*, *magnetization contrast*, and *interference of anomalies*.

anomalous gradient tensor: a *gradient tensor* representing the *gradient* components of the *anomalous field* generated by a local *anomalous mass* or *anomalous magnetization*.

anomalous gravity: a *field* of the gravity anomalies generated by rock *mass* distributions associated with elements of the geological structures within the Earth's *crust* and upper *mantle*. It is obtained from *observed gravity* by application of *observed gravity corrections* compensating for all predictable planetary and non-geological effects contributing to the measured absolute and relative gravity values, i.e., anomalous gravity is the final result of a *gravity reduction* process. See also *gravity anomaly*, *absolute gravity*, *relative gravity*, and *raw gravity*.

anomalous gravity vector: a three-component *vector* of a *gravity anomaly* in the gridded *anomalous field*. Its direction and length, or *magnitude*, are defined by the *gradient* of the *anomalous potential* of the gravity anomaly. Derivatives of the anomalous *gravity potential* in the mutually orthogonal *x*-, *y*-, *z*-directions represent three components of the anomalous gravity vector. Referred to with the same meaning as *gravity anomaly vector*. See also *gridding* and *gravity vector*.

anomalous magnetic vector: a three-component *vector* of a *magnetic anomaly* in the gridded *anomalous field*. Its direction and length, or *magnitude*, are defined by the *gradient* of the *anomalous potential* of the *RTP magnetic anomaly*. Derivatives of the anomalous *magnetic potential* in the mutually orthogonal *x*-, *y*-, *z*-directions represent three components of the anomalous magnetic vector. Referred to with the same meaning as *magnetic anomaly vector*. See also *gridding* and *reduction-to-pole (RTP)*.

anomalous magnetics: the *field* of magnetic anomalies generated by rock *magnetization* distributions associated with elements of the geological structures primarily in the upper part of the Earth's *crust*. It is obtained from *observed magnetics* by application of *observed magnetic corrections* compensating for all predictable planetary and non-geological effects contributing to the measured values of the *Earth magnetic field*, i.e., anomalous magnetics is a final result of the *magnetic reduction* process. See also *magnetic anomaly* and *raw magnetics*.

anomalous magnetization: the magnetic *source body* parameter defined by (a) *scalar* difference in *magnetic susceptibility* values, or *susceptibility contrast*, between the source body and host rock formations, and (b) *vector* addition of the *remanent magnetization* vector and *induced magnetization* vector, wherein the latter vector is parallel to the ambient *Earth magnetic field*. See also *ambient field* and *magnetization direction*.

anomalous mass: See *mass differential*.

anomalous potential: **1.** a *scalar* quantity representing a *vertical integral* of the gravity or magnetic *anomaly* generated by an isolated *source body*. Its *gradient* defines the direction and length, or *magnitude*, of the anomalous gravity or magnetic *vector*. Derivatives of the anomalous potential in the mutually orthogonal *x*-, *y*-, *z*-directions quantify three components of the *anomalous gravity* or *magnetic vector* at each *grid* point of the gridded *anomalous field*. **2.** a *disturbing potential* as it relates to being a *gravitational potential* for the anomalous gravitational components that are not accounted by the ellipsoidal shape of the Earth and its rotation. See also *derivative*, *gravitational attraction*, *gridding*, *anomalous gravity vector*, and *anomalous magnetic vector*.

anomaly: **1.** a local- or regional-scale difference between the observed and theoretical, or *model*-predicted, values of *Earth gravity field* or *Earth magnetic field*. **2.** the gravity or magnetic *field* obtained by application of data corrections for all predictable planetary and non-geological effects contributing to the measured gravity and magnetic values, e.g., *free-air anomaly* and *Bouguer anomaly* or *total magnetic anomaly (TMA)* and *RTP anomaly*. **3.** the portion of a gravity or magnetic *survey* that is different in its *amplitude* and/or *spatial frequency* content from the survey in general and might be associated with an *ore body*, mineral deposit, or geological *structure* favorable for the accumulation of hydrocarbons. Lithology or structure-related lateral change in the composition or properties of rock formations is required for the existence of an anomaly, i.e., there is no anomaly in the absence of a lateral *density contrast* or *magnetization contrast*. An obtained *anomalous field* represents the sum, or total, of all gravity or magnetic fields generated by multiple sources in the subsurface, wherein the contribution of each *source body* or structural element depends primarily on its distance from the *observation point*, its volume, and the amount of lateral *density* or *magnetization* contrast. See also *anomaly amplitude*, *anomaly energy*, *anomaly frequency*, *anomaly origin*, *anomaly resolution*, *anomaly shape*, *anomaly slope*, *anomaly wavelength*, *interference of anomalies*, and *Schiaparelli body*.

anomaly amplitude: a parameter of the gravity or magnetic *anomaly* quantified by its maximum deviation from the zero level or by its value above *background noise*. The amplitude of a gravity or magnetic anomaly is a function of five major parameters of its source: (1) *density contrast* or *magnetization contrast* with the laterally adjacent rocks: the larger the contrast, the higher the amplitude; (2) *source geometry* defined by source volume and shape: the larger the volume, the higher the amplitude and the more compact the shape, the narrower the amplitude; (3) source depth: the deeper the source, the smaller the amplitude; (4) *observation point* position relative to the source: the highest amplitudes are immediately above the source; and (5) isolation from other sources: *interference of anomalies* from the closely spaced sources of different shapes and volumes could be constructive or destructive. The magnetic anomaly amplitude also depends on *magnetic inclination* and *magnetic declination* of the planetary magnetic field, such as the *Earth magnetic field*. See also *anomaly frequency*, *anomaly resolution*, *anomaly shape*, and *anomaly wavelength*.

anomaly energy: a parameter of the gravity or magnetic *anomaly* quantified by a planar square of the anomaly area on its *profile* or by a volumetric integral over the anomaly area on a *grid*. Anomaly energy is controlled, primarily, by (a) distance from its source, (b) source depth, (c) source volume, and (d) *density contrast* or *magnetization contrast* created by the source.

anomaly frequency: a parameter of the gravity or magnetic *anomaly* that represents its dominant *spatial frequency*, or *wavelength*, wherein the higher frequencies correspond to the shorter wavelengths. For the similar shapes and volumes of sources, the anomaly frequency is a function of the source depth: the deeper the source, the lower the spatial frequency and the longer the wavelength of the anomaly. See also *anomaly wavelength* and *anomaly amplitude*.

anomaly half-width: See *half-amplitude width*.

anomaly origin: a qualitative characteristic relating the observed gravity or magnetic *anomaly* to a particular *source body* or structural element. Because a lateral change in *density* or *magnetization* of rock formations is required for the existence of an anomaly, its origin is associated primarily with: (a) the geological *structure* or structural discontinuity within the sedimentary section, e.g., *anticline* or *syncline*, high-*offset* or low-offset *fault*, *thrust*, *horst*, *graben*, emplacement of *allochthonous salt*, shale, or salt *diapir*, *igneous intrusion*, etc.; (b) the geological structure or laterally contrasting lithological heterogeneity at the top of the *geological basement*; (c) the regional-scale change in the depth structure and composition of the *Moho boundary*; (d) the local-scale change in the lithology and mineral composition of the shallow rocks associated with an *ore body*, *kimberlite intrusion*, mineralization of faults and fractures, *salt* dissolution, accumulations of *detrital magnetite*, *diagenetic magnetite*, and other *diagenetic magnetic minerals*; and (e) *unexploded ordnance (UXO)* and other isolated non-geological objects. See also *contact* and *contrast*.

anomaly resolution: a quantitative characteristic of acquired data that describes (a) the smallest *anomaly amplitude* that exceeds the estimated *accuracy* of measurements and, therefore, could be detected, or resolved, in the obtained *anomalous field*, and (b) the shortest *anomaly wavelength* that reliably could be identified in the line-oriented or gridded data after application of all relevant data corrections. See also *gravity corrections*, *magnetic corrections*, *gridding*, *resolution*, *lateral resolution*, and *vertical resolution*.

anomaly separation: separation of the gravity or magnetic *anomaly* from (a) superposition of the regional components of an *anomalous field*, (b) interfering effects of source bodies or geological structures other than the source of a particular anomaly, and (c) *noise*. Regional components are assumed to be associated with any geological structure or isolated *source body* which is much larger than a lateral extent of the source of the *target anomaly* or too deep for the present exploration interest. See also *interference of anomalies*, *inversion-based separation*, *separation filtering*, and *regional-residual separation*.

anomaly shape (*grav.*): a parameter of the *gravity anomaly* that depends on three main factors: (1) distribution of *density* within the *source body*; (2) *source geometry*, i.e., volume, shape, depth, *dip*, and distance from observation points; and (3) *azimuth* of the *survey* line relative to the *strike* of a non-spherical *source body*. Different from magnetic anomalies, gravity anomalies are monopolar by nature, so the symmetric density distribution generates a symmetric anomaly with its maximum located directly above the source body. Source shape is important at relatively close source-observation distances, because at sufficiently far distances all shapes of the isolated sources have the same gravity effect of a *point source*. Source volume and source depth define the width and slopes of the gravity anomaly: the larger the volume and the deeper the source, the wider the anomaly and the more gentle its slopes. See also *anomaly slope* and *monopolar anomaly*.

anomaly shape (*mag.*): a parameter of the *magnetic anomaly* that depends on six main factors: (1) distribution of *magnetization* within the *source body*; (2) direction and intensity of the *remanent magnetization* of the source body; (3) *source geometry*, i.e., volume, shape, depth, *dip*, and distance from observation points; (4) directional parameters of the *main geomagnetic field*, i.e., *magnetic inclination* and *magnetic declination*; (5) orientation of *strike* of a non-spherical source relative to the direction of a *geomagnetic vector*; and (6) *azimuth* of the *survey* line relative to the strike of a non-spherical source body and orientation of the main geomagnetic field. Because regular, or conventional, magnetometers measure the field component parallel to the non-vertical *Earth magnetic field* with unknown direction of *magnetization* of its sources, there is always a *phase* contribution to the anomalous *signature* that shifts the anomaly maximum relative to the source location and, sometimes, changes the anomaly sign. At a *magnetic pole*, where inclination is 90°, the magnetic anomaly generated by vertical *thin dike* with a positive *susceptibility contrast* and *induced magnetization* has a positive, or *peak*, signature regardless of its strike. At the *magnetic equator*, where inclination is 0°, the magnetic anomaly generated by the same but east–west striking thin dike has a negative, or *trough*, signature, whereas a similar but north–south striking thin dike does not generate a detectable magnetic anomaly regardless of its dip. Source shape is important at the relatively close source-observation distances, because at sufficiently far distances all shapes of the isolated sources have the same magnetic effect of a *point dipole*. Source volume and source depth define the width and slopes of a magnetic anomaly: the larger the volume and the deeper the source, the wider the anomaly and the more gentle its slopes. Different from gravity anomalies, magnetic anomalies are dipolar by nature, so their shapes are asymmetric and composed of the non-equal positive and negative parts, wherein the positive maximum is shifted from the location of a source center toward the magnetic equator and a negative minimum is shifted toward the magnetic pole. After *reduction-to-pole (RTP)*, the symmetric induced magnetization distribution within the source body generates a symmetric magnetic anomaly whose maximum is located directly above the source body. See also *dipolar anomaly* and *scalar magnetometer*.

anomaly slope: a parameter of the observed *anomaly* that is independent from the *anomaly amplitude* and *density contrast* or *magnetization contrast* created by its source and related to the source depth: the shallower the source, the steeper the slopes; the deeper the source, the more gentle the anomaly slopes. The direct relationship between the *ideal source* depth and the anomaly slope forms the basis of *depth rules* and *graphical source depth estimation* techniques. See also *anomaly shape (grav.)* and *anomaly shape (mag.)*.

anomaly strike: the orientation of a long axis of a gravity or magnetic *anomaly* on a map or *grid*. See also *strike* and *cross-track aliasing*.

anomaly wavelength: a *wavelength* of the observed *anomaly* defined by a doubled horizontal distance between the *peak* and *trough* values of the anomaly on a gravity or magnetic *profile* or *grid*. The anomaly wavelength depends primarily on: (a) source depth: the deeper the source, the longer the wavelength; the shallower the source, the shorter the wavelength; and (b) source thickness and lateral extent, i.e., source volume: the larger the volume, the longer the wavelength. Because these factors cannot be separated, source depth estimates based on wavelengths of anomalies are rough approximations at best. Density or magnetization *contrast* has no effect on the wavelength of an anomaly. See also *anomaly amplitude*, *anomaly frequency*, *anomaly origin*, *anomaly resolution*, and *anomaly separation*.

anomaly width: the short axis of an *anomaly* determined as: (a) horizontal distance between inflection points or *half-amplitude width* points on the *profile* of an anomaly; (b) one-half the estimated *anomaly wavelength* on a map or *grid*; or (c) distance between the *horizontal gradient* maxima. Anomaly width is orthogonal to the long axis of an anomaly, or *anomaly strike*, and it depends, primarily, on the source volume and depth: the larger the volume, the wider the anomaly; the deeper the source, the wider its anomaly. See also *inflection point* and *half-wavelength resolution*.

antenna offset: the horizontal distance between the actual position of a *gravimeter* or *magnetometer* and the position of a *GPS* navigation antenna during the shipboard (marine) and airborne surveys. The antenna offset is a major component of an *instrumentation lag*, which is eliminated by application of a *lag correction*. In marine and airborne gravity surveys, it is also referred to as the *lever arm*. See also *lag test* and *lever arm effect*.

antialias filter: *alias filter* in the context of a *low-pass filter* applied to a continuous signal before its *sampling* in order to prevent *aliasing*.

anticline: an upward curved, or convex, intrasedimentary fold with a core composed of the stratigraphically older and, usually, higher *density* rock formations, except for the salt-cored anticlines. Depending on its size and amplitude (closure), the anticlinal fold creates a small or large positive lateral *density contrast* generating the local- or regional-scale *gravity high* in *residual gravity* or original anomalous gravity field, respectively. A salt-cored anticline can generate a local *gravity low* within the larger area of relatively high *anomalous gravity* values. Deep anticlines have no magnetic *signature* due to a low *magnetization* of the sedimentary rocks. When a *magnetite*-rich layer is present in an anticlinal rock formation, the shallow or mid-depth anticline can be identified by the concentric curvilinear anomalous alignments, continuous or segmented, of a high-*frequency* (short-*wavelength*) content. See also *anomalous field*, *residual field*, *salt*, *subcrop*, and *syncline*.

antiferromagnetism: a subset of *ferromagnetism* that represents *magnetization* of the rock-forming minerals the atomic magnetic moments of which are aligned antiparallel and cancel one another, so their *total magnetization* may be nearly zero. Magnetic properties of antiferromagnetic rocks are similar to those of *paramagnetic* rocks, but they exhibit *hysteresis* and their *susceptibility* decreases with temperature up to a mineral-specific value called the *Neel temperature*, above which they become completely paramagnetic. See also *magnetic dipole moment*, *diamagnetism*, and *paramagnetism*.

antiroot: See *isostatic antiroot*.

antiroot gravity effect: the *Airy gravity effect* in oceans, seas, and large continental depressions, i.e., it is a gravity effect in areas of *isostatic antiroot*.

antiroot isostatic effect: the *isostatic compensation effect* in oceans, seas, and large continental depressions in the context of the *Airy-Heiskanen model* of *isostasy*.

apparent: pertaining to quantities obtained from measurements or calculations of a particular rock property, such as *apparent density* or *apparent susceptibility*, under certain approximating assumptions, such as horizontal occurrence and homogeneity of rock formations, to signify its difference from a true value of that property.

apparent curvature attributes: **1.** *curvature attributes* of the gravitational *equipotential surface* calculated in the same fixed Cartesian coordinate frame, with the vertical *z*-axis orthogonal to the horizontal plane, as used for measurements of the *gravity gradient tensor (GGT)* components over the *gradient survey* area. Apparent curvature attributes are less accurate than *true curvature attributes* calculated from the *rotated gradient tensor* data adapted to a *local reference frame* at each observation or *grid* point. Differences between apparent and true curvature attributes could be significant; therefore, it is advisable to perform the point-by-point coordinate axes rotations and adjustments of the measured gradient tensor components to the local coordinate frames prior to calculation of the curvature attributes (Li, 2018). **2.** *curvature* attributes of the anomalous gravity field that are calculated with the use of a *Hessian matrix*.

apparent density: **1.** *density* of rock formations obtained from the *borehole gravity meter (BHGM)*, *acoustic log*, and *gamma-gamma log* measurements in boreholes or from seismic interval velocities via the *Gardner equation* and other seismic-derived relations, such as the *Adams-Williamson equation*, *Brocher formula*, and *Nafe-Drake curve*. Also referred to as *interval density*. **2.** a density distribution obtained by *inverse modeling* of the *residual gravity* field for mapping the geological boundaries below unconsolidated sediments, glacier debris, and dense vegetation using the following assumptions: (a) field-generating sources are vertical homogeneous blocks with the constant densities, (b) sources are in the uppermost part of the sedimentary section, and (c) the Earth's surface is horizontal. **3.** the distribution of rock densities within a targeted sedimentary formation or at the top of *basement* obtained by (a) *3D inversion* of the residual gravity field or (b) *terracing*. Apparent density is not the actual, or true, rock density but rather its valid approximation. Also known as *formation density*. See also *bulk density*.

apparent susceptibility: **1.** *magnetic susceptibility* of rock formations measured in boreholes (wells). **2.** susceptibility of the *two-dimensional (2D)* source bodies obtained by the source-parameter estimation methods based on *Euler deconvolution* and *Werner deconvolution*. **3.** susceptibility of rocks obtained by *inverse modeling* of the residual *total magnetic intensity (TMI)* or *RTP magnetic field* with the objective of mapping the geological boundaries below unconsolidated sediments, glacier debris, and dense vegetation using the following assumptions: (a) field-generating sources are approximated by vertical prisms with *infinite bottom* and lateral dimension of one *grid cell*, (b) *induced magnetization* is present for homogeneous sources, (c) sources are in the uppermost part of the sedimentary section, and (d) a constant depth to the top of all prisms. Grids obtained from aeromagnetic data are continued downward to the ground level. **4.** susceptibility of igneous rocks, such as *volcanics* and *igneous intrusion*, within the sedimentary section or metamorphic and igneous rocks at the top of the *basement* obtained by *3D inversion* of the residual magnetic field. Apparent susceptibility is not the actual, or true, susceptibility of rocks but rather its valid approximation. Also known as *formation susceptibility*. See also *downward continuation*, *prism*, *residual field*, and *advanced source-parameter imaging method*.

aquifer: a heterogeneous rock formation composed of a permeable and saturating rock material capable of maintaining a flow of groundwater. Shallow alluvial aquifers have the distinct *gravity signature* of a residual *gravity low* and, sometimes, a *magnetic signature* of the clustered low-*amplitude* positive short-*wavelength* anomalies generated by localized accumulations of *detrital magnetite*. See also *anomaly* and *residual gravity*.

aquifer gravity: *time-lapse gravity* data obtained from high-*accuracy* measurements of *absolute gravity* and *relative gravity* for monitoring the groundwater storage changes in the natural (alluvial) aquifers. Ground *microgravity* observations are integrated with the high-*resolution* surface *elevation* data and groundwater level measurements in production wells. In addition to standard *gravity data corrections*, the gravity data are corrected for *water-table effect* and for the free-air and Bouguer effects produced by the ground surface subsidence and rising due to expansion and compaction of both water and aquifer rocks. Except for the extreme water storage changes, the additional *terrain effect* is commonly ignored. Absolute gravity measurements are made at the selected observation points unaffected by groundwater level variations to provide the reference, or anchor, points for more widely expanded relative gravity measurements for delineation of a spatial distribution of groundwater mass changes (Pool, 2008). See also *Bouguer effect*, *free-air effect*, *CO_2 sequestration gravity*, *geothermal gravity*, *volcano gravity*, and *InSAR imaging*.

arbitrary gravity base: *relative gravity base* in the context of a pre-selected gravity value to which all *relative gravity* values measured during the *ground gravity* survey are referenced.

archaeomagnetism: the study of the magnetic properties, such as a direction and intensity of *remanent magnetization*, in archaeological materials and objects for the purposes of paleomagnetic dating, including *polarity* and polarity excursion dating. Also referred to as *remanence dating*. See also *magnetic archaeology*, *paleomagnetism*, and *polarity excursions*.

arc-minute: a unit of angular measurement equal to 1/60 of one degree or 1/21,600 of a full circle of 360°. Arc-minute is used as a unit of *spatial resolution* for global compilations of Earth's gravity and magnetic data in the format of *grid*, such as *DNSC08* and *Earth Magnetic Anomaly Grid-2 (EMAG2)*. Map, or planar, *grid cell size* depends on the geographic coordinates, and it is defined by the spatial arc length along a latitude or longitude that represents an integer number of arc-minutes. For *WGS84 ellipsoid*, the length of 1 arc-minute of latitude varies from 1860 m at the north and south geographic poles to 1840 m at the equator, whereas the length of 1 arc-minute of longitude varies from 1850 m at the equator to 0 at the poles. See also *arc-second*.

arc-second: a unit of angular measurement equal to 1/60 of one *arc-minute* or 1/3600 of one degree, i.e., 1 arcsec = 0.000277778°. Arc-second is used in estimates of *sensor tilt* and *verticality error*.

array: **1.** an assemblage of measuring instruments mounted on the same platform. **2.** two or three synchronized *gravimeter* or *magnetometer* systems separated vertically or horizontally and mounted on the same platform for measurements of gravity and magnetic gradients as well as calculation of *absolute gravity gradient*, *gravity field curvature*, and *magnetic field curvature*. Often referred to as *gradient array*. See also *gradient* and *absolute gravity gradiometer*.

artifact: See *artificial anomaly*.

artificial anomaly: **1.** a false *anomaly* produced by a measurement, processing, or *gridding* error, such as *anisotropic sampling*, *bull's-eye effect*, *herringbone effect*, *Gibbs ringing*, and *misleveling*. **2.** *magnetic anomaly* of a non-geological origin from the ferrous man-made objects and electric currents. Sometimes referred to as *artifact*.

artificial intelligence (AI): a mathematical system of interacting computer algorithms that develops the ability to process, analyze, and correctly interpret multidimensional datasets in a way of *integrated interpretation*, i.e., the system can perform operations and make decisions imitating a human intelligence. The concept of AI is implemented in *machine learning (ML)* algorithms applied for the automatic analysis and *interpretation* of geological and geophysical data. Referred to with the same meaning as *artificial neural network (ANN)*. See also *dimension*.

artificial magnetic anomalies: **1.** same as *cultural magnetic anomalies* with an emphasis on their man-made, or artificial, origin. **2.** magnetic anomalies (*boudinage artifact, bull's-eye anomaly, corrugations*, etc.) produced by measurement, processing, and *gridding* errors. Also referred to as *magnetic artifacts*.

artificial neural network (ANN): a data-driven computational system that imitates the functional properties and associative memory of the biological neural network of a human brain. The system is implemented in *machine learning (ML)* algorithms applied for (a) integration of multiple data parameters to find relationships between attributes of the *anomalous field* and elements of the targeted geological structures at different scales of *lateral resolution* and (b) statistical evaluation and classification of anomalies in the process of their integrated analysis and *interpretation*. An ANN is a self-adapting system consisting of interacting computer algorithms that change their composition based on the external and internal information flowing through the network of algorithms during the "training" phase to find patterns in the data. Once it is determined that a specific anomaly *pattern* is associated with a particular *source body* or geological *structure* at one location, then that pattern is used to identify the same or similar anomalies over the whole *survey* or regional study area. Commonly referred to as *artificial intelligence (AI)*. See also *attribute, pattern recognition, target*, and *convolutional neural network (CNN)*.

artificial sun illumination: same as *sunshading* and *false-sun illumination* with an emphasis on an artificial source of light, or "sun", which creates a *three-dimensional (3D)* effect of illumination of the gravity or magnetic *grid* from different directions, as if the grid represents a topographic surface of hills and valleys.

AS grid: a *grid* of calculated values of the *analytic signal amplitude (ASA)* in a magnetic or gravity *survey* area. See also *gridding* and *analytic signal (AS)*.

AS-LW relationship: a relationship of equivalence between the amplitude of *analytic signal (AS)* and *local wavenumber (LW)* and their higher-order versions. Vertical and horizontal local wavenumbers are equivalent to the normalized horizontal and vertical derivatives of the two-dimensional (2D) *analytic signal amplitude (ASA)*, respectively, and can be expressed as

$$k_z = -(\partial|\mathrm{AS}|/\partial x)/|\mathrm{AS}| = -(\partial \mathrm{ASA}/\partial x)/\mathrm{ASA},$$

$$k_x = (\partial|\mathrm{AS}|/\partial z)/|\mathrm{AS}| = (\partial \mathrm{ASA}/\partial z)/\mathrm{ASA},$$

where k_z is *vertical local wavenumber*, k_x is *horizontal local wavenumber*, $|\mathrm{AS}| = \mathrm{ASA} = [(\partial M/\partial x)^2 + (\partial M/\partial z)^2]^{1/2}$, and M is *magnitude* of the anomalous magnetic or gravity field. Second-order versions of the local wavenumber and analytic signal amplitude are related by

$$k_{xx} = (\partial|\mathrm{AS}_2|/\partial z)/|\mathrm{AS}_2| = (\partial \mathrm{ASA}_2/\partial z)/\mathrm{ASA}_2$$

$$k_{zz} = -(\partial|\mathrm{AS}_2|/\partial x)/|\mathrm{AS}_2| = -(\partial \mathrm{ASA}_2/\partial x)/\mathrm{ASA}_2,$$

where k_{xx} and k_{zz} are second-order horizontal and vertical local wavenumbers, respectively, and $|\mathrm{AS}_2| = [(\partial^2M/\partial x\partial z)^2 + (\partial^2M/\partial z^2)^2]^{1/2}$ is amplitude of the 2D *second-order analytic signal (AS_2)* (Pilkington and Keating, 2006). See also *normalized derivatives*.

ASA-derived depth: the magnetic source depth determined from a *principal profile* of the *analytic signal amplitude (ASA)* anomaly by measuring either (a) the width of the ASA anomaly at one-half its maximum value, i.e., *half-amplitude width*, or (b) the distance between inflection points of the ASA anomaly profile, as these points can be less affected by *interference of anomalies* from adjacent sources. For the vertical or sub-vertical *magnetic contact*, the depth to its top is defined as

$$h_{mc} = x_{1/2}/3.46 = x_{ip}/1.414,$$

where $x_{1/2}$ is the half-amplitude width and x_{ip} is the width of the ASA anomaly between inflection points. For the vertical and sub-vertical *thin dike*, the depth to its top is defined as

$$h_{thd} = x_{1/2}/2 = x_{ip}/1.155.$$

For a *horizontal cylinder*, the depth to its center is defined as

$$h_{hc} = x_{1/2}/1.533 = x_{ip}.$$

Formulas are valid for *two-dimensional* (2D) sources with *infinite strike length* and orthogonal orientation of the ASA anomaly profile to the source body *strike* (MacLeod et al., 1993). See also *ASA-ratio method, inflection point*, and *tilt-depth method*.

ASA-ratio method: **1.** the method of estimating the depth to a structural or lithological boundary approximated by the *contact* source geometry. Depth h to the top of a particular boundary is determined from the ratio of the maximum values of *analytic signal amplitude (ASA)* and amplitude of the first-order or second-order *enhanced analytic signal (eAS)* as

$$h = |\mathrm{ASA}|_{\max}/|\mathrm{eAS}_1|_{\max} \quad \text{or}$$

$$h = \{2|\mathrm{ASA}|_{\max}/|\mathrm{eAS}_2|_{\max}\}^{1/2},$$

where $|\mathrm{ASA}|_{\max}$ is the maximum value of the gravity or magnetic ASA anomaly, $|\mathrm{eAS}_1|_{\max}$ is the maximum amplitude of the *first-order enhanced AS (eAS_1)*, and $|\mathrm{eAS}_2|_{\max}$ is the maximum amplitude of the *second-order enhanced analytic signal (eAS_2)* (Hsu et al., 1996). **2.** a *profile*-based semi-automatic method of estimating the depth to magnetic sources approximated by *thin dike*. It is based on the ratio of *zero-order ASA (ASA_0)* to *first-order ASA (ASA_1)*, i.e., $R_{ASA} = ASA_0/ASA_1$, which is calculated for all observation points along a selected profile or line-oriented data. Local minima of this ratio are observed directly above the thin dike locations, and at these points their values are equal to the top dike depth (Cooper, 2015). Sometimes referred to as *amplitude-ratio method*. See also *contact-depth method*, *tilt-depth method*, and *ASA-derived depth*.

ASA vertical derivative: the *vertical derivative* of the *analytic signal amplitude (ASA)* implemented in an *analytic signal Euler deconvolution* for estimates of the *source body* parameters. It is obtained in the *space domain* by the *finite-difference method* involving the calculation of a difference between the original and upward-continued ASA values under the condition that a continuation distance is much less (1/10 or less) than the *sample interval* of the gridded data (Florio et al., 2006). See also *gridding* and *upward continuation*.

ASCII file: the American Standard Code for Information Interchange (ASCII) computer file, which is used in the tabulation of line-oriented and irregularly spaced gravity and magnetic data. The ASCII data files are arranged in a *format* of rows and columns representing all relevant *survey* information, such as numbered survey lines, *fiducials*, *Julian date*, geographic coordinates (latitude and longitude), *elevation* or *radar altimeter* values, *XY-coordinates*, measured gravity or magnetic field values, magnetic *diurnals*, gravity *base station* value, etc., per each observation point. Additional columns are added after completion of the *data reduction* process to include all calculated corrections to measured values and the obtained corrected values. These ASCII files form the basis for a *database* preparation. See also *gravity corrections* and *magnetic corrections*.

ASCII format: the standard *format* of the data description and arrangement used for loading into computer applications. It is based on the American Standard Code for Information Interchange (ASCII) and includes, at least, four files, wherein three (or more) files represent the general information (e.g., text description of the data and *survey*, data format, map *datum*, and projection parameters), and the fourth file contains the measured data values and relevant *acquisition* and reference information per each *observation point*. See also *ASCII file* and *metadata*.

ASTA filter: an edge-detection *filter* utilizing the algorithm based on calculation of the *analytic signal amplitude (ASA)* of the *tilt angle (TA)* and expressed as

$$\mathrm{ASTA} = [(\partial \mathrm{TA}/\partial x)^2 + (\partial \mathrm{TA}/\partial y)^2 + (\partial \mathrm{TA}/\partial z)^2]^{1/2}.$$

An ASTA filter has its maximum values above the *source body* edges. Its *lateral resolution* and *amplitude* response to different depths of sources are superior to those of the ASA and TA calculated separately (Ansari and Alamdar, 2011). See also *edge-detector attribute* and *tilt ASA*.

astatic balance: an unstable dynamic balance of *proof mass* in certain types of *spring gravimeter*. In the astatic balance, the *gravitational force* on the proof mass is balanced by the force of a mechanical spring, while the third force, such as a separate mass, is idle in the balance but it acts, when the unbalance occurs, in the same direction as the gravitational force to drive the proof mass further out-of-balance and, thereby, increase the gravimeter *sensitivity*. See also *neutral balance*.

astatic gravimeter: a *spring gravimeter* where *proof mass* is sustained in an *astatic balance*, so that a small change in *gravitational force* results in an increased deflection of the proof mass from its *null point* and, therefore, provides a high *sensitivity* for this *gravimeter*.

astatic magnetometer: *magnetometer* for measurements of *remanent magnetization* in rock samples. Its design is based on the use of two magnets of the same *magnetic moment* and opposed *magnetic polarity* that are suspended by torsion fiber. The measured angular deflection of magnets above a variously oriented rock sample is directly proportional to its remanent magnetization. See also *spinner magnetometer*.

astatic sensor: a spring-type *gravity sensor* with an increased *sensitivity* to changes in the measured *relative gravity* values. See also *astatic balance* and *spring gravimeter*.

asthenosphere: the layer of a relative weakness, or yielding zone, composed of the partially melted rocks in the upper *mantle* immediately below the nearly rigid *lithosphere*. The asthenosphere is involved in *isostatic compensation*, plate tectonic movements, and *magma* origin. Depth to its top varies from 80–100 to 150–250 km under the continents and from 40–50 km to 80–100 km under the oceans. Its bottom is estimated at 300–400 km below the Earth's surface. See also *isostasy* and *sea-floor spreading*.

astrobleme: a structural depression formed by the blow, or impact, of a meteorite, asteroid, or other high-velocity cosmic body on the planetary surface. Astroblemes are identified by the circular and oval outlines of their impact craters, whose diameters vary from less than one kilometer to tens and hundreds of kilometers. The depression and surrounding ejected rocks create the lateral *density* and *magnetization* contrasts sufficient for generating the coherent low-amplitude *total field* and *gradient* anomalies detectable by calculation of *residual gravity* or *residual magnetics* and application of the *enhanced processing* and imaging techniques, such as *Goussev filter* and *radial sunshading*. Small buried astroblemes have the *signature* of (a) a local *magnetic low* associated with a destroyed and randomized *bulk magnetization* of rocks within the impact depression and bounded by the low-*amplitude* gradient anomalies or (b) a local *gravity low* produced by the younger, lower *density* sediments filling in the impact depression. Large astroblemes have the signature of a local gravity and magnetic highs within a larger area of the gravity or magnetic low that is generated by *uplift* below the center of depression, wherein this uplift is formed by the impact-triggered dynamic rebound of the older, higher density rocks or high-magnetization, high-density *igneous body* intruded into the shallower depth levels from a deep *magma* chamber or from the relatively shallow planetary *mantle* in response to a shock wave generated by the cosmic body impact. Often referred to as *impact structure*. See also *coherence*, *intrusion*, *lateral contrast*, and *forensic magnetism*.

atmospheric admittance function: a proportionality function in the relationship between changes in measured gravity values and atmospheric pressure for the calculation of an *atmospheric pressure correction*. It is defined in units of *microgal (μGal)* per millibar of atmospheric pressure (μGal/mbar) and varies approximately from 0.30 to 0.43 μGal/mbar with a *default value* of 0.30

μGal/mbar. Also referred to as *barometric admittance factor*. See also *atmospheric pressure effect*.

atmospheric correction: an integrated correction applied to *absolute gravity* measurements and high-*accuracy* measurements of *relative gravity* during *ground gravity* surveys to compensate for the *atmospheric effect*. Atmospheric correction includes *atmospheric mass correction* and *atmospheric pressure correction*.

atmospheric effect: a combined gravity effect of the varying air mass and pressure of the Earth's atmosphere represented by an *atmospheric mass effect* and *atmospheric pressure effect*. In the high-*accuracy* ground gravity surveys and *absolute gravity* measurements, the atmospheric effect is compensated by an *atmospheric mass correction* and *atmospheric pressure correction*.

atmospheric mass correction: a correction applied to *ground gravity* data acquired by a high-*accuracy* survey within a broad range of heights in order to compensate for the *atmospheric mass effect*. It is calculated in units of *milligal (mGal)* for elevations up to 10 km as

$$\Delta g_{\mathrm{AM}} = 0.874 - 9.9 \cdot 10^{-5} h + 3.56 \cdot 10^{-9}\, h^2,$$

where h is height, or *elevation*, of the *observation point* above sea level or a *reference ellipsoid* in meters. Application of this correction increases the gravity anomaly *amplitude* by 0.86 mGal at the height of 100 m and 0.77 mGal at 1000 m. The latest updates of *theoretical gravity* include the gravity effect of the atmospheric mass, so the atmospheric mass correction is either subtracted from the theoretical gravity value at each *station* or added to the *observed gravity* value (Hinze et al., 2013). See also *atmospheric pressure correction*.

atmospheric mass effect: the gravity effect of a varying mass of the Earth's atmosphere that changes the *gravitational attraction* at observation points in a way opposite to attraction of the Earth's mass. It is included in *theoretical gravity* as described by the *International Gravity Formula 1980 (IGF1980)* and its subsequent updates of *IGF1987* and *IGF1999*, which include the assumption that the total atmospheric mass is within a *reference ellipsoid*. This assumption becomes invalid over the continental areas with a broad range of topographic heights due to changes in atmosphere thickness and temporal variations of atmospheric pressure. In high-*accuracy ground gravity* surveys, the atmospheric mass effect is compensated by application of an *atmospheric mass correction*. See also *atmospheric pressure effect*.

atmospheric pressure correction: a correction applied to high-*accuracy* ground measurements of *relative gravity* by relative gravimeters and *absolute gravity* by a *free-fall absolute gravimeter* to compensate for the *atmospheric pressure effect* produced by local atmospheric pressure variations which result in changes of atmospheric air mass and, therefore, changes in *gravitational attraction* at observation points during measurements. The correction value Δg_{AP} is calculated for each cycle of measurements in units of *microgal (μGal)* and expressed as

$$\Delta g_{\mathrm{AP}} = -\mathrm{A}(P_{obs} - P_{nom}),$$

where a negative sign indicates the inverse relationship between the pressure and observed gravity values, A is *atmospheric admittance function* in units of μGal per millibar of atmospheric pressure (μGal/mbar), P_{obs} is observed atmospheric (barometric) pressure, and P_{nom} is nominal, or standard, barometric pressure at the *observation point* (Merriam, 1992). Also referred to as *barometric pressure correction*. See also *atmospheric mass correction*.

atmospheric pressure effect: the gravity effect of a varying pressure of the Earth's atmosphere resulting in changes of gravity values measured on the Earth's surface due to the redistribution of atmospheric mass and its surface loading. Regular gravity changes are approximately ±5 *microgal (μGal)* and up to 10–15 μGal for large atmospheric pressure fronts of 30–50 hectopascal (hPa), where 1 hPa = 1 millibar (mbar). In high-*accuracy* ground gravity surveys, the *atmospheric mass effect* is compensated by an *atmospheric pressure correction*.

atom-interferometer gravimeter: same as *absolute atomic gravimeter* with an emphasis on its operating principle based on the use of an atom interferometer and the quantum wave-like nature of atoms.

atomic gravimeter: See *absolute atomic gravimeter*.

atomic gravity gradiometer: an instrument *array* for measurements of *absolute vertical gradient* or *absolute horizontal gradient* of the *Earth gravity field*. It consists of two *absolute atomic gravimeter* systems mounted vertically or horizontally on the same platform. From a differential *phase* shift in both atom gravimeter interferometers, the difference Δg_{abs} between two simultaneous measurements of *absolute gravity* is determined with respect to the common reference frame, so that platform vibrations and laser pulse fluctuations are rejected as a common-mode *noise*. Vertical or horizontal *gradient* is calculated from the obtained difference Δg_{abs} divided by the *baseline*, which is a separation distance between two absolute gravimeters. For the vertical baseline of 10 m, the atom gravity gradiometer *sensitivity* is estimated at approximately $4E/\sqrt{Hz}$, where $1\mathrm{E} = 10^{-9}\mathrm{s}^{-2}$. Also known as a *quantum gravity gradiometer*. See also *common-mode rejection*, *Eötvös (E)*, *gravity gradiometer*, and *hertz (Hz)*.

atomic magnetometer: an *optically pumped magnetometer* utilizing the *magnetic resonance* properties of atoms from the alkali group of elements, such as cesium, potassium, and rubidium, for measurements of the *Earth magnetic field* and its *vector* components. Also referred to as an *atomic-resonance magnetometer*, *alkali-vapor magnetometer* and a *laser-pumped magnetometer*. See also *microfabricated atomic magnetometer (MFAM)*.

atomic-resonance magnetometer: See *atomic magnetometer*.

attenuation: **1.** a decrease in *amplitude* of the gravity or magnetic *anomaly* with increasing distance from its source. Referred to with the same meaning as *fall-off*. **2.** the suppression of *noise* and regional components of an *anomalous field* by *filtering* and *regional-residual separation*. See also *fall-off rate* and *fall-off factor*.

attenuation factor: See *fall-off factor*.

attenuation rate: same as *fall-off rate* with an emphasis on a rate of decline, or *attenuation*, of the *amplitude* of the gravity or magnetic *anomaly* with increasing distance from its source.

attitude: the time-varying position and orientation of a *survey* aircraft or ship relative to the horizontal plane and direction of the *acquisition line* during an airborne or marine (shipboard) survey. See also *altitude*, *crab*, *pitch*, *roll*, and *yaw*.

attitude compensation platform: same as *stabilized platform* with an emphasis on its use to counterbalance variations of *attitude* of a survey ship or aircraft during a marine (shipboard) or airborne *gravity survey*.

attitude noise: errors in the spatial orientation of each *magnetic sensor* onboard an aircraft or Earth-orbiting satellite during three-component magnetic *vector* measurements. The attitude-related orientation error of 0.001° (~4 arc-seconds) results in a vector component measurement error of 1 *nanotesla (nT)* for the *geomagnetic field* intensity of 50,000 nT (Xie et al., 2020). See also *arc-second* and *vector magnetometer*.

attribute: a measurable quantitative characteristic, or parameter, of an *anomalous field* that describes spatial variation of its particular property (*amplitude*, *texture*, vector orientation, etc.) related to a distribution of *density* or *magnetization* in the subsurface. Attributes are computed at the stage of *enhanced processing* by numerical transformations of the anomalous field and its components or by combinations of its derivatives and gradients. Calculation of different attributes highlights different components of the anomalous field associated with elements of the subsurface structures, and the multi-attribute analysis is a part of *integrated interpretation* of gravity and magnetic data. Attributes include but are not limited to: *analytic signal amplitude (ASA)*, *curvature attributes*, *monogenic signal attributes*, *edge-detector attribute*, *invariants*, *local amplitude*, *local orientation*, *local phase*, *normalized source strength (NSS)*, *theta angle*, *tilt angle (TA)*, *vesselness*, and their modifications. Attributes are often referred to as specific filters, such as *tilt filter* and *vesselness filter*, because the calculation of attributes converts the original anomalous field into a *field* of new anomalies unavailable before their calculation. See also *complex attribute*, *derivative*, *gradient*, and *gradient tensor*.

auroral electrojets: temporal components of the *external magnetic field* of the Earth that are effects of interaction between the *main geomagnetic field* of the rotating Earth and *solar wind*. Interaction generates the charged particles which maintain a persistent flow above and around *geomagnetic poles* in *ionosphere*, wherein the instabilities in a flow of the charged particles produce rapid time-varying fluctuations of the external magnetic field up to hundreds of *nanotesla (nT)* on and above the Earth's surface. Magnetic effects of the auroral electrojets are recorded by a *base-station magnetometer* and incorporated into *diurnal correction*. See also *equatorial electrojets*, *temporal magnetic variations*, and *van Allen belts*.

automated anomaly axis correlation: a *grid*-based technique for correlating the axes of linear and curvilinear anomalies. Axes of the mid- and long-*wavelength* anomalies from relatively deep sources can be correlated by the *Blakely-Simpson method*, whereas axes of short-wavelength anomalies of shallower origin are better correlated by *coherent filtering*. See also *skeletonization* and *steerable filter*.

automated source depth estimation: source depth estimates using *profile*- and *grid*-based algorithms which automatically analyze anomalies of the gravity or magnetic field and its derivatives and compute depths to their corresponding sources approximated by simple source geometries, such as *thin dike*, *contact*, *cylinder*, *prism*, and *sphere*. Methods include but are not limited to: *AN-EUL method*, *enhanced local wavenumber method*, *Euler deconvolution*, *improved SPI™ method*, *multimodel method*, *Naudy method*, *source parameter imaging (SPI™) method*, and *Werner deconvolution*. See also *graphical source depth estimation*.

automatic drift compensation: *gravimeter drift compensation* by a software algorithm in automated *spring gravimeter* systems. It is based on a linear approximation of *drift rate* and involves the calculation of a difference between the repeated measurements at control points in the *ground gravity* survey area and distribution, or prorating, the obtained difference values over the *station*-by-station measurements between the control points. Automatic drift compensation is included in *automatic drift correction*. See also *repeat station*, *tie station*, and *loop-base station*.

automatic drift correction: an integrated instrument correction applied to *relative gravity* measurements during a *ground gravity* survey. It includes four *raw gravity corrections*: (1) *drift correction* for the linear component of *gravimeter drift*, (2) *temperature correction* for *temperature drift*, (3) *tilt correction* for deviation of a sensing axis of a *spring gravimeter* from the vertical alignment, and (4) *Earth tide correction*. It is incorporated into the process of measurements by automated spring gravimeter systems and applied at each *station* before *network adjustment*. See also *residual drift*.

automatic Earth tide correction: a correction applied to *relative gravity* measurements to compensate for the *Earth tide effect* during *ground gravity* surveys. As a part of *automatic drift correction*, it is incorporated into the process of measurements by automated *spring gravimeter* systems and applied at each *station*. Calculations are based on the *Longman algorithm*.

automatic gain control (AGC): an *enhanced processing* algorithm that balances amplitudes of the gravity and magnetic anomalies by attenuating the high amplitudes and amplifying the low amplitudes. Its principal operation parameters are the size of a *moving window* and a balancing, or gain, function, which is often defined by the average *amplitude* computed in the moving window centered at each *observation point*. Selection of the moving window size defines a *wavelength* range of anomalies that will be balanced in the output. The AGC algorithm is implemented in an *AGC filter*. See also *amplitude balancing*.

automatic temperature correction: an instrument correction for *temperature drift* of a *relative gravimeter* during *ground gravity* surveys. As a part of *automatic drift correction*, it is incorporated into the process of measurements by automated *spring gravimeter* systems and applied at each *station*.

automatic tilt correction: a correction for *sensor tilt*, i.e., for deviation of a sensing axis of a *spring gravimeter* from the vertical alignment, during *ground gravity* surveys. Sensor tilt within the range of ±200 arc-seconds is compensated at each *station* as a part of *automatic drift correction* incorporated into the process of measurements by automated *spring gravimeter* systems and applied at each *station*. See also *arc-second*.

autonomous underwater vehicle (AUV): an unmanned mini-submarine which navigates by following software-controlled pre-programmed traverses in shallow waters or under remote control by an acoustic communication system in deep water areas. Its reported operational depth range is more than 1500 m. An AUV is used for high-*resolution* submarine measurements of the *Earth magnetic field* or variations of the *Earth gravity field*, i.e., *relative gravity*, and its gradients in deep-sea mineral exploration. Increased vertical and lateral resolutions of anomalies are obtained by measurements at low heights (~50 m) above the sea floor under the reduced *dynamic noise* environment of a deep-sea and slow-moving (~4 km · h^{-1}) AUV. Also referred to as *sea-floor drone*. See also *traverse*, *lateral resolution*, *vertical resolution*, *submarine survey*, and *unmanned aerial vehicle (UAV)*.

AUV gravimeter: a modified *airborne gravimeter (AG)* system adapted to submarine measurements of *relative gravity* with the use of an *autonomous underwater vehicle (AUV)*. It is mounted on the stabilized mini-platform installed inside the 50 cm diameter pressure capsule. Effects of the AUV *tilt* up to ±30° and ±15° in *pitch* and *roll*, respectively, are minimized by *real-time compensation* making the remaining tilt-related increase in *observed gravity* values less than 1 *microgal (μGal)*. The AUV inertial vertical accelerations are determined with use of a high-*precision* depth *sensor* that measures the ambient water pressure and converts it into depth values. Achievable *amplitude resolution* is 1–2 mGal with an *rms crossover difference* of approximately 0.1 *milligal (mGal)* after *leveling* (Ishihara et al., 2018). See also *inertial acceleration*, *stabilized platform*, and *AUV vertical acceleration correction*.

AUV gravity: *relative gravity* data obtained from measurements by an *AUV gravimeter* along a regular network of *acquisition* lines in the submarine *gravity survey* area. AUV *observed gravity* Δg_{obs} is expressed as

$$\Delta g_{obs} = \Delta g_{raw} \pm \Delta g_{va} - \Delta g_{tilt} - \Delta g_{drift} \pm \Delta g_{Eo},$$

where Δg_{raw} is measured gravity value, or *raw gravity*; Δg_{va} is *AUV vertical acceleration correction*; Δg_{tilt} is *tilt correction*, also known as *off-level correction*; Δg_{drift} is *drift correction*; and Δg_{Eo} is *Eötvös correction*. Tilt correction could be small and omitted, whereas the drift correction, including *Earth tide correction*, is incorporated into the process of *leveling*. See also *AG gravity*, *AGG gravity*, and *observed gravity corrections*.

AUV vertical acceleration correction: a correction applied to *relative gravity* data acquired by an *autonomous underwater vehicle (AUV)* to compensate for its vertical inertial accelerations during a *submarine survey*. Calculation involves the conversion of water pressure *sensor* data into AUV depth changes and subsequent adjustments of the obtained *vertical acceleration* values for (a) spatial separation of the depth sensor and *AUV gravimeter*, (b) time delay between receiving depth sensor data and gravimeter data, and (3) variations of the pressure-to-depth conversion factor due to changes in seawater *density* and the non-rigidity of the AUV under deep water pressure (Ishihara et al., 2018). See also *AUV gravity*, *inertial acceleration*, and *moving-platform corrections*.

AUV magnetometer: a high-precision *magnetometer*, typically an *Overhauser magnetometer*, which is towed by an *autonomous underwater vehicle (AUV)* during a submarine *magnetic survey* for detection of deep-sea iron ore accumulations and shallow-water *unexploded ordnance (UXO)*, pipelines, and power and communication cables prior to offshore or coastal construction. See also *magnetic spherule* and *UAV magnetometer*.

average gravity: *nominal gravity* as it relates to being the globally averaged theoretical value of *gravitational acceleration*, or *theoretical gravity*, on the surface of a *reference ellipsoid* approximating the actual Earth shape.

average vertical gravity gradient: the globally averaged vertical rate of change of the *theoretical gravity field* of the Earth defined as 0.3086 mGal · m^{-1} = 0.3086 μGal · mm^{-1} = 3086 E. With the z-axis positive up, the gradient value is negative and, thereby, indicates a decrease of *Earth gravity field* values with increasing *elevation*, and it is equal to the *vertical gravity gradient* at ±45° latitude on a *reference ellipsoid*. Also referred to as *free-air factor*, *nominal free-air gradient*, and *nominal vertical gravity gradient*. See also *milligal (mGal)*, *microgal (μGal)*, and *Eötvös (E)*.

average vertical magnetic gradient: the globally averaged vertical rate of change of the *International Geomagnetic Reference Field (IGRF)* defined as 0.025 *nanotesla (nT)* per meter (nT · m^{-1}). With the z-axis positive up, the gradient value is negative and, thereby, indicates a decrease of IGRF values with increasing *elevation*. Also known as *nominal vertical magnetic gradient*.

averaging filter: a *grid*-based *filter* utilizing an algorithm that replaces each *grid cell* value by the average value of all grid cells within a rectangular or square *moving window* centered at every grid point. Two versions of averaging filter can be applied: (1) a *smoothing filter* that suppresses *noise* and enhances the long-*wavelength* components of the gravity or magnetic field and (2) a *sharpening filter* that increases the *sharpness* of anomalies and enhances edges of subsurface structural elements obtained by *inversion*. The sharpening filter operation is based on subtracting the smoothed grid from the original grid. Depending on grid cell size, a 3 × 3 or 5 × 5 square filter *kernel* with the respective averaging of 9 or 25 grid cell values generally provides sufficient smoothing and sharpening results. Sometimes referred to as *mean filter*. See also *median filter*, *SNN median filter*, and *differential median filtering*.

axial gradiometer: an instrument for measurements of *gradient* of the magnetic or gravity *field* in the direction of the orientation of the axis along which a pair of the synchronized magnetic or gravity sensors comprises the *gradiometer arm*. See also *GOCE gravity gradiometer*, *SQUID magnetic tensor gradiometer*, *gravity sensor*, and *magnetic sensor*.

axial magnetometer: a *vector magnetometer* that measures a *vector* component of the *Earth magnetic field* in the direction of the orientation of its *sensor*. See also *SQUID magnetometer*, *tri-axial magnetometer*, and *magnetic sensor*.

azimuth: **1.** the horizontal angle between a reference direction, such as the geographic meridian, and a given direction. Azimuth is usually measured in degrees clockwise from the geographic north through 360°. Occasionally, it is referenced to the geographic south or to *magnetic north* and measured counterclockwise. Also referred to as *true bearing* and *heading*. **2.** the *phase angle* of a two-dimensional (2D) *geomagnetic vector* **B** formed by the horizontal north and east components B_x and B_y, wherein the phase angle value is defined by *magnetic declination*. **3.** the direction of the *horizontal gradient vector* on a *grid* described by a horizontal angle φ measured from the geographic north and defined as

$$\varphi = \arctan[(\partial \mathrm{F}/y)/(\partial \mathrm{F}/\partial x)]$$

where $\partial \mathrm{F}/\partial y$ and $\partial \mathrm{F}/\partial x$ are horizontal derivatives of the *anomalous field* F in *Cartesian coordinates*. See also *horizontal derivative*.

azimuth imaging: a method of the enhanced visualization and real-time comparative analysis of linear and curvilinear gravity and magnetic anomalies that is based on the interactive *artificial sun illumination* of the gridded *anomalous field* from different directions, or azimuths, and heights of the artificial source of light. Referred to with the same meaning as *sunshading*. See also *anomaly*, *azimuth*, *gridding*, and *real-time mode*.

B

background contrast: the difference between the *density* or *susceptibility* values in a *two-dimensional (2D)* or *three-dimensional (3D)* model and a reference value known as the *background density* or *background susceptibility*, respectively.

background density: a reference value of rock *density* used in calculation of the gravity field generated by a *two-dimensional (2D)* or *three-dimensional (3D)* model. In *modeling* the *Bouguer anomaly*, the value of *Bouguer density* is obtained from the results of the *Nettleton test*, and it is often used as the background density. The concept of background density is based on the postulate that gravity anomalies are defined by mass differences between their sources and surrounding rocks, i.e., they are defined by *mass differential* rather than the total mass of each individual source. Change in the background density value is equivalent to *DC shift*.

background field: the *regional field* in the context of all *wavelength* components of the *anomalous field* that are not associated with the exploration or research *target*.

background magnetic field: **1.** the average value of the *Earth magnetic field* measured at a *base station* during the magnetically quiet time around a local midnight. The background field is used primarily as a reference level, or *magnetic datum*, for measurements of *diurnals* in the magnetic *survey* area and calculation of *diurnal correction*. **2.** the amount of *baseline shift* added to or subtracted from the measured magnetic field values in *data processing*. Also referred to as *DC level* and *DC value*. See also *ambient field* and *IGRF correction*.

background noise: the average *noise* level, regardless of whether it is *coherent noise* or *random noise*, upon which amplitudes of anomalies are superimposed in the observed (anomalous) gravity or magnetic *field*. Also referred to as *ambient noise*.

background susceptibility: a reference value of rock *susceptibility* used in calculation of the magnetic field generated by a *two-dimensional (2D)* or *three-dimensional (3D)* model. The background susceptibility *default value* is commonly zero.

backstripping: See *gravity stripping*.

Backus effect: the effect of inaccuracies in satellite-derived global models of the *main geomagnetic field* that are caused by a non-unique solution of the *Backus problem*, especially in areas around the *magnetic equator*. It is minimized by measurements of the *vector* components in addition to *scalar* measurements of *magnitude* of the main geomagnetic field. See also *scalar magnetometer* and *vector magnetometer*.

Backus problem: the nonuniqueness in reconstruction of a global *magnetic dipole* field (which is a common approximation of the *main geomagnetic field*) from *scalar* magnetic measurements only. The Backus problem can be solved with a minimized nonuniqueness when (a) the current spatial position of the *magnetic equator* is known with sufficient *accuracy* or (b) the global *model* of the main geomagnetic field is based on the use of both scalar and *vector* magnetic data. See also *temporal magnetic variations* and *equatorial electrojets*.

balanced ASA (ASAB): the *analytic signal amplitude (ASA)* of the gridded gravity or magnetic field, where the high and low ASA amplitudes are balanced by the orthogonal Hilbert transforms. The ASAB formula is expressed as

$$\mathrm{ASAB} = |\mathrm{ASA}|/k + \{[H_x|\mathrm{ASA}|]^2 + [H_y|\mathrm{ASA}|]^2 + |\mathrm{ASA}|^2\}^{1/2},$$

where k is an empirical constant that controls the amount of *amplitude balancing*, and H_x and H_y are Hilbert transforms of the ASA in the orthogonal x- and y-directions, respectively (Cooper, 2009). See also *gridding*, *3D Hilbert transform*, *amplitude balancing*, and *automatic gain control (AGC)*.

balanced profile curvature: a *profile curvature* calculated using *amplitude balancing* techniques in order to balance the low and high amplitudes of its first- and second-order horizontal derivatives and increase the *vertical resolution* of profile curvature anomalies associated with the structural and lithological boundaries in the subsurface. See also *balanced ASA (ASAB)* and *horizontal derivative*.

balancing: See *amplitude balancing*.

ballistic absolute gravimeter: See *free-fall absolute gravimeter*.

band: **1.** the range of wavelengths or wavenumbers over which a given *filter* is designed to operate in the pass mode, such as *band-pass filter*, or in the reject mode, such as *band-reject filter*. **2.** an estimated range of spatial frequencies in the *profile* or *grid* of an anomalous gravity or magnetic field. See also *anomalous field*, *bandwidth*, *spatial frequency*, *wavelength*, and *wavenumber*.

band-limited function: **1.** a function for which the *power spectrum* is effectively zero at wavenumbers that are larger than some critical *wavenumber* k_c, i.e., it has negligibly small energy at wavelengths shorter than some critical *wavelength* λ_c. **2.** a function for which the *Fourier transform* has negligibly small components outside some range of spatial frequencies, or wavenumbers, i.e., outside the frequency-limited *band*. Gravity and magnetic anomalies are effectively band-limited, because contributions of the largest wavenumbers to anomalous amplitudes are insignificant due to the exponential decrease of the *amplitude* components with the increasing wavenumbers. See also *spatial frequency*.

band-pass filter: a *filter* that passes without *attenuation* a range, or *band*, of the spectral-domain wavenumbers or space-domain wavelengths between user-specified *cutoff* values. Data components outside the cutoff limits are strongly attenuated, but they cannot be completely excluded from the data. Because the *anomaly wavelength* depends, primarily, on source depth, often a band-pass filter is used for separation of the shallow-sourced anomalies from the deep-sourced anomalies and vice versa. See also *Butterworth filter, high-pass filter, low-pass filter, matched filter, separation filter, space domain*, and *spectral domain*.

band-reject filter: the opposite of a *band-pass filter*. It suppresses a range, or *band*, of the space-domain wavelengths or spectral-domain wavenumbers between user-specified *cutoff* values. See also *wavelength* and *wavenumber*.

banded-contour grid: a *grid* of the gravity and magnetic *field* where the colored or gray-scaled bands representing a user-specified range of values are used instead of contours of the equal values on a *contour grid*. See also *gridding, pseudocolor*, and *color-coded grid*.

banded flight tracks: groups of two, three, and four flight tracks, or flight lines, with a closer separation than that of the regular flight lines. They are used as a cost-saving measure during an airborne *reconnaissance survey* for estimates of the strikes and other parameters of the magnetic and gravity anomalies in the *exploration* areas selected for a more detailed *regional survey*. See also *strike*.

banded iron formation (BIF): a thin-bedded or finely laminated rock formation containing, at least, 15% of the iron-rich minerals of the sedimentary origin, mostly *magnetite* and hematite, in separate bands or sheets mingled with a chert or fine-grained quartz. Typically, the BIF has a very strong *magnetic signature*, and its *magnetic anomaly* can reach hundreds and thousands of *nanotesla (nT)* units, when the BIF contains massive concentrations of the iron ore. The BIF magnetite and hematite are often remanently magnetized with a highly variable *magnetization direction* and orientation of *magnetization vector* within the same formation (Christensen and Dransfield, 2002). Among other methods, BIF accumulations are mapped by calculation of *total gradient (TG)*, which is equivalent to and referred to with the same meaning as *analytic signal amplitude (ASA)*. Due to the high *density* of magnetite, a massive BIF can generate a detectable *residual gravity* anomaly. See also *ore body, remanent magnetization, detrital magnetite*, and *diagenetic magnetite*.

bandwidth: **1.** an airborne or marine (shipboard) *acquisition* parameter representing a number of discrete measurements of the gravity or magnetic field per regular interval of time in units of *hertz (Hz)*. For example, 5 Hz and 0.18 Hz bandwidths represent 5 measurements per second and 18 measurements per 100 seconds, respectively. Selection of the bandwidth defines the highest *waveform* frequency, or shortest *wavelength*, that can be recovered from the acquired data and, therefore, it specifies the highest achievable *lateral resolution* of the airborne and marine surveys, where the nominal along-the-line *sample interval* in meters is determined from the aircraft or ship speed in meters per second divided by the bandwidth value in Hz. Also referred to as *sampling bandwidth* and *Nyquist bandwidth*. **2.** a range of space-domain wavelengths or spectral-domain *wavenumbers* representing a particular *waveform, anomaly*, or anomalous trend in the measured field. **3.** a range of wavelengths or wavenumbers over which a particular *band-pass filter* is designed to operate in its pass or reject mode. **4.** a difference between the highest frequency (the shortest wavelength) and the lowest frequency (the longest wavelength) components in the measured data that is referred to as the *measurement bandwidth*. See also *sampling theorem, spectral domain*, and *wavenumber*.

Barkhausen effect: discrete changes, or jumps, in *magnetization* of a *ferromagnetic* rock, when it is exposed to a smoothly increasing magnetic field. The Barkhausen effect produces pulses of electric current in a coil surrounding a ferromagnetic rock sample, and these pulses are considered as evidence of the existence of ferromagnetic domains, whose separate changes in size and orientation result in jumps of the rock magnetization induced by the externally applied magnetic field. See also *magnetic domain*.

barometric admittance factor: See *atmospheric admittance function*.

barometric altimeter: an instrument installed onboard an aircraft to measure and record the atmospheric pressure which is converted into the aircraft height above sea level, i.e., *elevation*. In theory, the atmospheric pressure decrease could be converted into the elevation with *accuracy* superior to *GPS*-determined height values. In practice, the weather adversely affects barometric altimeter measurements, as they often respond to high and low pressure changes within a moving atmospheric front instead of changes in elevation. Operational *precision* of a barometric altimeter is 5–40 m depending on the atmospheric pressure at a flight *altitude*. See also *altimeter*.

barometric height: the height, or *elevation*, of a *survey* aircraft flight above sea level determined by conversion of *barometric altimeter* measurements.

barometric pressure correction: See *atmospheric pressure correction*.

barycentric image: a *color-coded grid* displaying three calculated attributes of a gridded gravity or magnetic field by assigning the highest intensities of three different colors to apexes of the equilateral triangle which is used instead of a *color bar*. Value-coded color intensities are interpolated in inverse proportion to the distance from a respective apex, so the weighted sum of the color intensities is constant (typically 1 or 100%) at every point inside the *grid*. It is used for visualization of any three attributes which are normalized to make their sum equal to 1, such as in the case of *FTG anisotropic indices*. Also referred to as *ternary image, triangle plot*, and *triangular diagram*. See also *attribute, glyph, normalization*, and *RGB composite image*.

base-check method: a method of the control over *gravimeter drift* during a *ground gravity* survey, where the *gravimeter* is returned to the *base station* at specified time intervals to determine the *drift* accumulated in the process of measurements. Time intervals between the repeated measurements at the base station, or *reoccupations*, are selected to be shorter than the short periodic components of *Earth tides* (typically, 4–6 hours) in order to include the gravity effect of the tide components into *drift correction*. See also *station* and *cross-check method*.

base constant: the difference between the *absolute gravity* value and *relative gravity* value measured by a *relative gravimeter* at a *base station*. Base constant values obtained during a marine (shipboard) survey at different base stations are used to determine a *gravimeter drift* for the whole *marine gravity* survey. The base constant value is added to the *measured gravity* to obtain the absolute gravity values for each *observation point*. See also *still reading*.

base gravity station: the coordinate point in a *gravity survey* area where *absolute gravity* or *relative gravity* value is known. There are two types of base gravity stations: *absolute base station*, also known as *absolute gravity base*, and *relative gravity base*.

base level: **1.** the *absolute gravity* value or *relative gravity* value at a *base station* that is used as a reference level, or *gravity datum*, relative to which all measurements are made in the ground, marine (shipboard), or airborne *gravity survey* area. **2.** a *background magnetic field* that is utilized as a reference magnetic field level, or *magnetic datum*, relative to which *diurnals* are measured by a *base-station magnetometer*.

base magnetometer: See *base-station magnetometer*.

base map: a map showing *survey* location data, such as locations of *acquisition* lines, railroads, highways, pipelines, power lines, wellheads, bridges, and other objects of the industrial and agricultural infrastructure that could generate *cultural magnetic anomalies* and *microseismic noise*. See also *track plot*.

base station: **1.** **(*grav.*)** a gravity *station* where the *absolute gravity* or *relative gravity* value is known and relative to which the gravity values at other stations over the *gravity survey* area are measured. It is also used to determine *gravimeter drift* during ground, marine (shipboard), and airborne gravity surveys and anchor, or tie, the auxiliary geodetic data. Local gravity base stations are tied to the regional or national network of base stations with known absolute gravity values to facilitate the merging of gravity datasets acquired in previous years. See also *International Standardized Gravity Network 1971 (ISGN71)*. **2.** **(*mag.*)** a magnetic station where the *base station magnetometer* is set up to monitor and record *diurnals* and other *external magnetic field* effects. Its location is selected in a magnetically quiet area away from power lines, pipelines, roads, bridges, and other sources of magnetic *noise*. An average of the quiet-time values measured around a local midnight is referred to as the *background magnetic field* or base station *DC level* which is used for calculation of *diurnal correction*. The base station for airborne surveys is set up within 40–60 km from the most distant point of regular measurements in the *magnetic survey* area. Large aeromagnetic surveys can have several base stations for more accurate recording of diurnals. Surveys with *magnetic gradiometer* systems require the base station only for detection of *magnetic storm*. **3.** a stationary *GPS* receiver, or *GPS base station*, used by *Differential GPS (DGPS)* for high-accuracy *positioning* of a moving *observation platform*, such as a *survey* ship or aircraft. See also *elevation base station*.

base-station artifacts: artificial anomalies in the gravity and magnetic measurements at a *base station* caused by passing vehicles, low-flying aircrafts, *microseismic noise*, and other observation disturbances.

base-station loop: a sequence of the ground gravity or magnetic measurements that begins and ends at the same coordinate point of a primary or secondary *base station*. See also *loop*, *primary gravity base*, and *secondary gravity base*.

base-station magnetometer: a stationary *magnetometer* set up at a magnetic *base station* for monitoring and measurements of *diurnals* and other external magnetic field effects during a *magnetic survey*. A base-station magnetometer is synchronized with the survey magnetometer, and it has the same or similar *precision*. It is placed at a magnetically quiet location at a sufficient distance (several hundred meters or more) from large ferrous objects, sources of electrical current and alternating electrical fields, roadway traffic, power lines, pipelines, etc. Typical precision of measurements is 1 *nanotesla (nT)* with the recording intervals from 1 second to 1 minute. Base-station magnetometer measurements are subtracted from regular survey measurements on a *station*-by-station or point-by-point basis to eliminate the interfering effects caused by variations of the *external magnetic field* of the Earth. In the airborne surveys, it is often referred to as *ground magnetometer*. Also known as *base magnetometer*.

base-tie value: a gravity value added to *relative gravity* obtained from ground, marine (shipboard), or airborne gravity data in order to tie the *gravity survey* data to the local or regional *absolute base station* and present the acquired data in *absolute gravity* values. It is calculated from additional measurements by a *relative gravimeter* along a *traverse* connecting the *survey* area with the absolute base station or by a mobile *absolute gravimeter* at any observation point in the survey area. For the *airborne survey* ties, the additional *free-air correction* analogous to *tripod correction* is required to account for the height difference between the onboard and *ground gravity* measurements. See also *acquisition* and *still reading*.

base ties: repeated measurements by a *relative gravimeter* at the location of a *base station* during a *ground gravity* or *marine gravity* survey in order to determine *gravimeter drift*. See also *gravity tie* and *still reading*.

baseline: **1.** the separation distance between gravity or magnetic sensors in a *gradiometer* system. The gradient value is obtained from the difference between the measured gravity or magnetic field values divided by the baseline value. **2.** a reference level of initial measurements against which the subsequently measured values of the *Earth gravity field* are compared, such as *microgravity* measurements during a *time-lapse gravity survey*. **3.** a magnetic field value that is measured by a *scalar magnetometer* and used as a reference value for *vector* measurements by a three-component *fluxgate magnetometer*. **4.** an area-averaged field value relative to which the *amplitude* of a particular *anomaly* is determined. **5.** an asymptotic line at both ends of an anomaly *profile* that represents the regional field level or zero level for a particular anomaly. **6.** the distance between a *GPS base station* and a moving *GPS* receiver in ground or airborne gravity and magnetic surveys.

baseline shift: a change in values of the measured *Earth magnetic field* after application of *IGRF correction* for the latitude- and longitude-dependent variations of the *main geomagnetic field*. Often, the average of the calculated IGRF values in a *magnetic survey* area is added back to the individual measurements to make them closer to the absolute values of the Earth's magnetic field. Baseline shift does not affect *structural interpretation* of relative changes in the measured field, and it is automatically compensated by the *forward modeling* and *inversion* algorithms.

baseline survey: a high-*accuracy* ground *gravity survey* conducted with the objective of establishing a reference level, or *baseline*, for subsequent *time-lapse gravity*, or *4D gravity*, measurements.

basement: a geological *unconformity* comprising a geological foundation on which the sedimentary rock formations have been deposited. It is composed of lithologically heterogeneous crystalline (igneous and metamorphic) rocks, often of Precambrian age, with a highly variable *density* and *magnetization*. Basement rock densities vary mostly within 2.50–2.85 $g \cdot cm^{-3}$ and often overlap with sedimentary density values, whereas the *susceptibility* values of the basement rocks are 50–150 times higher than susceptibilities of the sedimentary rocks. There are four basement types integrated into *interpretation* of gravity and magnetic data: *acoustic basement*, *geological basement*, *gravity basement*, and *magnetic basement*. Three types of geophysically derived basement (acoustic, gravity, and magnetic) can represent different rock formations and not coincide with each other. Among these three types, the magnetic basement is the best approximation of a true, or geological, basement, especially in the geologically complex areas of extensive salt tectonic deformations, thrusting, and active and passive continental margins. The basement is a primary *target* in interpretation of aeromagnetic and marine (shipboard) magnetic data for oil-and-gas *exploration* and analysis of sedimentary basins. Also referred to as *crystalline basement*. See also *basin analysis*.

basin: a large structural depression in the upper *crust* where thick sequences of sedimentary rocks have been accumulated. Basins are often bounded by normal and strike-slip faults which commonly have distinct gravity and magnetic signatures, especially in the *field* of *gradient* anomalies. Depending on the area extent, sedimentary basins are identified by a regional-scale decrease of the *Bouguer gravity* and *RTP magnetic field* values, except for localized depressions with high-*density* and high-*magnetization* igneous intrusions, extensive *volcanics*, and *lava* flows. Visibility of basin-related gravity and magnetic lows is improved by calculation of the *regional field*, application of the *upward continuation*, and other techniques suppressing the anomalies generated by the near-surface and shallow sources. See also *fault*, *igneous intrusion*, *signature*, *strike-slip fault*, *gravity low*, and *magnetic low*.

basin analysis: the integrated geological-geophysical study that involves *enhanced processing* and *interpretation* of gravity and magnetic data to obtain structural information about a particular *basin*: distribution and types of faults, shear and fracture zones, regional-scale and local sedimentary and *basement* structures (anticlines, synclines, horsts, grabens, uplifts, thrusts, etc.), igneous bodies and depths of their occurrence, basement depth, and total thickness of the accumulated sedimentary formations. Gravity and magnetic data provide complementary structural information for the basin analysis: *high-resolution aeromagnetic survey* data are the most efficient in the detection of faults and related structural discontinuities, igneous intrusions, *volcanics*, and their contacts, whereas the ground, marine, and airborne *gravity survey* data are the most efficient in the detection of the geological structures and identification of their types, e.g., *anticline* or *syncline*, *horst* or *graben*, *uplift* or *trough*, *thrust* or *flower structure*, *igneous intrusion* or *reef*. See also *contact*, *fault*, *igneous body*, *shear zone*, and *structure*.

batch processing: a type of non-interactive *data processing* in which a specific sequence of operations, such as the calculation of source depth parameters along all *survey* lines, is completed during a single continuous cycle.

batholith: a large *igneous body* with an area extent of 80–100 km^2 or greater and an unknown bottom. Due to its large volume and high *density* and *magnetization* contrasts with surrounding sedimentary rocks, intrasedimentary batholiths generate high-*amplitude* positive gravity and magnetic anomalies. Granitic batholith surrounded by the higher density metamorphic and igneous rocks in the upper *basement* can have the *gravity signature* of a low-amplitude negative *residual gravity* anomaly. Upper sedimentary batholiths are prolific sources of thermal energy. See also *contrast*, *dike*, *pluton*, *sill*, *stock*, and *magma*.

bathymetric correction: a *terrain correction* that is applied to marine (shipboard) gravity or *gravity gradient* data.

bathymetric effect: a *terrain effect* in the context of a marine (shipboard) gravity or *gravity gradient* survey.

bathymetry: **1.** measurements and recording of the water depth in oceans, seas, and lakes for mapping their bottom *relief*. **2.** a continuous surface representing the shapes and relative positions of the submarine portions of ocean, sea, and lake bottoms. See also *topography*.

bay: a time-varying component of *external magnetic field* of the Earth described as a transient, or short-duration, *magnetic disturbance* with *period* of approximately one hour and appearance of the bay along coastline in the displayed records of *base-station magnetometer*. Bay onset is usually accompanied by a burst of *micropulsations*. Magnetic effects of bays are compensated by application of *diurnal correction*. See also *diurnals*.

Bayes theorem: the theorem that defines the subsequent, or posterior, probability distribution in terms of the previous, or prior, probability and likelihood probability distributions as

$$P(x|y) = P(x) \cdot L(y|x)/P(y),$$

where $P(x|y)$ is a conditional posterior probability distribution that describes a new level of knowledge of unknown parameters of the variable x obtained by incorporating the information derived from the observed data y; $P(x)$ is a prior probability distribution of the variable x derived from the available geological and geophysical information about parameters of x without contribution of the observed data y, i.e., it is derived from the available *constraints*; $L(y|x)$ is a conditional likelihood probability which is a function of x and a measure of the degree of fit between the observed data and model-predicted data, i.e., *model response*; and $P(y)$ is a normalization constant. See also *conditional probability*, *model misfit*, and *stochastic inversion*.

Bayesian stochastic inversion: same as *stochastic inversion* with an emphasis on its basic assumption that the inverted *residual gravity* values and unknown *voxel model* parameters are random variables whose probability distributions satisfy *Bayes theorem*. The solution of an *inverse problem* is obtained for *density* distributions and their boundaries, i.e., subsurface *structure*, by maximizing the posterior probabilities derived from the iteratively optimized distributions of the prior and likelihood probabilities in a relationship between the voxel model variables.

benchmark: the ground geodetic point of a height above sea level, i.e., *elevation*, used as a control coordinate point for high-*accuracy* gravity measurements. Benchmarks are complementary to *GPS*-based *positioning*, because the *reference ellipsoid* model in the GPS software does not take them into account. Also referred to as *elevation base station*. See also *GPS accuracy*.

BHGM density: *apparent density* determined from measurements by a *borehole gravity meter (BHGM)* or borehole-adapted *vibrating-string gravimeter* in drilled wells (boreholes). The difference between gravity readings at two different depths and the *free-air gradient* value of 0.3086 $mGal \cdot m^{-1}$ are used for calculation of the apparent density ρ in grams per cubic centimeter ($g \cdot cm^{-3}$):

$$\rho = 3.6827 - 128.576(\Delta g - \Delta T)/\Delta h,$$

where Δg is gravity difference in units of *milligal (mGal)* between two consecutive measurements along the borehole, Δh is vertical separation of measurement points in meters, and ΔT is change in the surface *terrain effect* between two measurements. The terrain-related gravity effect is taken into account in relatively shallow (<1 km) measurements and high-relief terrain. The BHGM density calculations are based on the assumption that rock layers are horizontal, infinite in all directions, uniformly thick, and constant in *density*. The BHGM density measurements are not influenced by the well casing or other effects that degrade the quality of the *gamma-gamma log* data. See also *Poisson jump* and *reading*.

bi-directional gridding: a method of *gridding* based on *two-dimensional (2D)* interpolation of line-oriented datasets. Discrete measured values are interpolated initially along the *survey* lines and then in the orthogonal direction across the survey lines to create a continuous surface of a *grid*. See also *minimum-curvature gridding* and *gradient-enhanced minimum-curvature*.

Biot-Savart law: the *Ampere law* upgraded by extending its original equation for electric currents through a segment of wire ΔL into the integration around the entire wire loop.

bird: an airtight and streamlined non-magnetic container, or housing, where a *magnetometer* is mounted. It is suspended from a *survey* aircraft to eliminate the aircraft-related magnetic interfering effects during measurements of the *Earth magnetic field*. A bird is towed at 40–50 m below an aircraft on a cable with electrical connections to power and operate the magnetometer. In a present-day *fixed-wing survey*, the magnetic *total field* and *gradient* measurements are made with the use of an *aeromagnetic compensation* system and magnetometers mounted on the *stinger* and wing tips. In a *drone survey* or *helicopter survey*, the cable-suspended single-*sensor* magnetometer or *multi-sensor gradient array* is referred to as a *slung sensor*. See also *fish*.

Bishop model: a *three-dimensional (3D)* model of the *basement* with an area extent of 315 × 315 km and *grid cell* of 200 m. It is derived from the scaled *Digital Terrain Model (DTM)* near Bishop, California. The basement is overlain with the non-magnetic sediments and divided by vertical *contact*-like boundaries into three homogeneous terranes with four high-*susceptibility* vertical igneous intrusions truncated at the basement surface. Depth to the basement increases from 1 km at the NW corner to 9 km at the SE corner. Used for testing of *enhanced processing* algorithms and new gravity and magnetic methods of *inversion* for the basement depth (Williams et al., 2005). See also *terrane* and *igneous intrusion*.

bitmap: 1. a *raster image* file containing the monochrome- or color-coded values for each *pixel*, or bit, of this digital image. Raster image file types include BMP and JPEG. **2.** a digital image file *format*.

Blakely-Simpson method: the polynomial-fitting method of the automatic locating and tracing the maxima of a gridded gravity or magnetic field and its horizontal and vertical gradients. This method is based on comparing each *grid cell* value with eight surrounding cell values in four directions (column, row, and both diagonals) and interpolating the determined maxima by second-order polynomials across three consecutive *grid* cells. See also *gridding* and *skeletonization*.

block structure: a *structure* characterized by its vertical and sub-vertical boundaries, commonly faults of various types, across which the relatively large changes in the physical properties of rocks, such as *density* and *magnetization*, often emerge and, thereby, generate the detectable *gravity* and *magnetic anomalies*. See also *anomaly* and *fault*.

Bohr magneton: the physical constant adopted as a unit, or fundamental, *magnetic dipole moment* of the spinning (rotating about axis of rotation) and orbiting charged subatomic particles. It is denoted by μ_B, and its value represents a *magnitude* of the magnetic dipole moment emerging from, or caused by, the spinning-and-orbiting motion of a single electron around the nucleus, i.e., a Bohr magneton is dipolar by nature. Constructive and destructive *vector* summation of the magnetic moments of elementary magnetic dipoles define the magnetic properties of rock-forming minerals at the atomic level. See also *magnetic dipole, magnetic dipole strength*, and *magnetic dipole potential.*

borehole gravimeter: a remote-controlled instrument for measuring the *relative gravity* values in drilled wells (boreholes). There are two *relative gravimeter* types used for borehole measurements in exploration applications: the *spring gravimeter*, known as a *borehole gravity meter (BHGM)*, and a borehole-adapted version of the *vibrating-string gravimeter*. The difference between gravity measurements at two different depths is used to calculate *apparent density*, which is referred to as *BHGM density*.

borehole gravimetry: methods and instruments for measuring variations of the *Earth gravity field* and its components in drilled wells (boreholes). The main objectives of these measurements are (a) to obtain estimates of *apparent density* of the well-penetrated rock formations and (b) to detect gas-oil and gas-water contacts behind the well casing. Borehole gravity data require *double Bouguer correction* for the gravity effect of the infinite plate between two measurement points, which is analogous to the *Bouguer plate effect*: one correction is for the depth-increased *gravitational attraction* at the upper point and the other correction is for the upward attraction of the *Bouguer plate* at the lower point of the measurements. See also *borehole magnetometry* and *Poincare-Prey reduction.*

borehole gravity meter (BHGM): a remote-controlled *spring gravimeter* designed to fit into a logging tool and operate under high-temperature and high-pressure conditions. Measurements are made at temperatures up to 125°C and borehole deviations from the vertical up to 14°. Regular *accuracy* is 5–20 *microgal* (μGal) and *wavelength resolution* is 7–10 m. The best achievable accuracy is 1.0–1.5 μGal, and it is sufficient for detecting the majority of the gas-oil and gas-water contacts. The BHGM is capable of detecting the bypassed oil-and-gas accumulations, and it is the only logging tool that can reliably measure the *apparent density* of rock formations through the well casing. The BHGM measurements are not affected by near-borehole conditions, such as the drilling mud, formation damage, fluid invasion, and casing or cement heterogeneity. See also *BHGM density* and *resolution.*

borehole magnetometry: measurements of *magnetic susceptibility* and orthogonal components of the *Earth magnetic field* in drilled wells (boreholes). Both types of measurements are used to determine magnetic properties of the well-penetrated rock formations and constrain their geological correlation between wells. Borehole magnetic measurements with the three-component *fluxgate magnetometer* can determine the location of an *igneous body* or *reef* missed by drilling. See also *constraints.*

Bott method: a *profile*-based method of estimating the thickness of sediments above the *geological basement* with the use of *residual gravity* (residual *Bouguer anomaly* or *isostatic anomaly*) and *slab formula*. The method implements the iterative *forward calculation* process in which the initial approximation of the sedimentary thickness is defined by a variable thickness of the horizontal infinite *Bouguer slab* with the assumed *density contrast*:

$$h = \Delta g_{res} / 2\pi G \Delta\rho,$$

where h is plate thickness in meters, Δg_{res} is residual gravity in milligals, G is *universal gravitational constant* in *cgs units*, and $\Delta\rho$ is plate density contrast in $g \cdot cm^{-3}$. The misfit between the observed residual gravity and values calculated from the initial model is converted into corrections of the initial sediment thickness estimates. Iterative process is repeated until the *model misfit* becomes equal to a *stopping criterion* (Bott, 1960). See also *milligal (mGal).*

Bott-Smith rules: See *Smith depth rules.*

boudinage artifact: a gridding *artifact* caused by the *bull's-eye effect* in the application of a standard *minimum-curvature gridding* algorithm: it produces a sequence of the isolated circular or elongated anomalies, similar in appearance to a "string of beads" or "bicycle chain", instead of a continuous short-*wavelength* anomaly. Boudinage artifacts are indicators of *undersampling* of the short-wavelength high-*amplitude* anomalies, and their effects are minimized by a *gradient-enhanced minimum-curvature* method of *gridding.* See also *bull's-eye anomaly.*

Bouguer admittance: the *gravitational admittance* determined with the use of *Bouguer anomaly* data. See also *Bouguer coherence* and *free-air admittance.*

Bouguer anomaly: the *field* of gravity anomalies obtained by application of corrections for all known, or predictable, planetary and non-geological gravity effects contributing to the *observed gravity* values. Depending on the *gravity survey* type (ground, marine, or airborne) and its local or regional scale, the following corrections are applied: (1) *theoretical gravity correction* for the rotating homogeneous Earth and changes in the measured gravity values with latitude; (2) *free-air correction* for height, or *elevation*, of each observation point above the *survey datum*: sea level or *reference ellipsoid*; (3) *Bouguer correction* for rock masses above the survey datum onshore and seawater *mass deficiency* below the survey datum offshore; (4) *terrain correction* for the topographic and bathymetric *relief* variations around each observation point; (5) *Bullard B correction* for *curvature* of the Earth's surface; (6) *atmospheric correction* for the atmospheric air mass and pressure variations over the high-relief and mountainous areas; (7) *Eötvös correction* for movement of the *observation platform* within a coordinate system fixed on the rotating Earth; (8) *tidal correction* for the time-varying combined *gravitational attraction* of the Sun and Moon; and (9) *indirect effect correction* for the use of different survey datums: sea level or reference ellipsoid. Tidal correction for *ground gravity* data is commonly incorporated into *drift correction* applied to *raw gravity*. There are two types of the Bouguer anomaly: *simple Bouguer anomaly*, which is the Bouguer anomaly without the terrain correction, and *complete Bouguer anomaly*, which is the Bouguer anomaly with the terrain correction applied. An obtained anomalous gravity field represents the sum, or total, of all gravity anomalies generated in a sedimentary section, *basement*, lower *crust*, and upper *mantle*. The Bouguer anomaly is a standard *format* for subsequent *gridding*, *enhanced processing*, analysis, and *interpretation* of gravity data. Also referred to as *Bouguer gravity*, *Bouguer gravity anomaly*, and *Bouguer gravity field*. See also *absolute Bouguer anomaly*, *relative Bouguer anomaly*, *tensor Bouguer anomaly*, *isostatic anomaly*, *interference of anomalies*, and *predicted gravity*.

Bouguer coherence: a *gravitational coherence* determined with the use of *Bouguer anomaly* data. See also *Bouguer admittance* and *free-air coherence*.

Bouguer correction: a correction applied to *observed gravity* values to compensate for the gravity effect, or *gravitational attraction*, of rock masses above the *survey datum* (sea level or *reference ellipsoid*) onshore and seawater *mass deficiency* below the survey datum offshore. Both rock and seawater mass distributions are approximated by the infinite and uniformly dense horizontal *Bouguer plate*. Its vertical gravitational attraction is called the *Bouguer effect*, the numerical value of which comprises the amount of the Bouguer correction Δg_B calculated at each *observation point* in units of *milligal (mGal)*:

$$\Delta g_B = 2\pi G\rho h = 0.04193\rho h \quad \text{or}$$
$$\Delta g_B = 2\pi G\rho h = 4.193 \cdot 10^{-5}\rho h,$$

where G is *universal gravitational constant* in *cgs units* or *SI units*, respectively; ρ is rock *density* of the Bouguer plate in grams per cubic centimeter ($g \cdot cm^{-3}$) or in kilograms per cubic meter ($kg \cdot m^{-3}$), and it is referred to as *Bouguer density*; and h is thickness of the Bouguer plate defined by height, or *elevation*, of each *observation point* above sea level or the reference ellipsoid in meters. For the average crustal density of 2.67 $g \cdot cm^{-3}$ (2670 $kg \cdot m^{-3}$), the Bouguer correction value is defined as

$$\Delta g_B = 0.1119h.$$

For ground surveys and onshore airborne surveys, Bouguer correction values are subtracted from the observed gravity values on a point-by-point basis. For marine (shipboard) surveys, the Bouguer correction is defined as

$$\Delta g_B = 0.04193\rho_r h_w,$$

where ρ_r is *replacement density*, h_w is bathymetric water depth, and the obtained correction values are added to the observed marine gravity values to compensate for the seawater mass deficiency. The same as *free-air correction*, Bouguer correction does not move observation points down to the datum level, i.e., observation points remain at their original positions on the ground or in the air, and it does not produce a density discontinuity across the datum level. Bouguer correction is referred to as *simple Bouguer correction*, which emphasizes a compensation for the Bouguer effect only, or *complete Bouguer correction*, when both the Bouguer effect and *terrain effect* are compensated. Simple Bouguer correction is also referred to as *Bouguer plate correction*, *Bouguer slab correction*, *mass correction*, and *slab correction*. See also *3D Bouguer correction*.

Bouguer correction constant: same as *Bouguer correction factor (BCF)* with an emphasis on a constant value for the length of each *traverse* and often for the entire *gravity survey* area.

Bouguer correction density: See *Bouguer density*.

Bouguer correction factor (BCF): the *Bouguer correction* parameter defined in units of *milligal (mGal)* per meter ($mGal \cdot m^{-1}$) as

$$\Delta g_{BCF} = 0.04193\rho \quad \text{or} \quad \Delta g_{BCF} = 4.193 \cdot 10^{-5}\rho,$$

where ρ is *Bouguer density* of rocks above the *survey datum* in grams per cubic centimeter ($g \cdot cm^{-3}$) or in kilograms per cubic meter ($kg \cdot m^{-3}$), respectively. For the average crustal density of 2.67 $g \cdot cm^{-3}$ (2670 $kg \cdot m^{-3}$), the Bouguer correction factor is defined as

$$\Delta g_{BCF} = 0.1119\,\text{mGal} \cdot \text{m}^{-1}.$$

Its multiplication by a topographic height, or *elevation*, h of each *station* comprises the amount of *simple Bouguer correction* for any given coordinate point on the Earth's surface. The BCF can be calculated for every *traverse* in a *ground gravity* survey by the *Siegert method*. Sometimes referred to as *flat-plate Bouguer factor*, *Bouguer correction constant*, and *mass correction factor*. See also *free-air factor* and *elevation correction factor*.

Bouguer density: a *density* value for the horizontal infinite *Bouguer plate* of a finite thickness that approximates rock masses above the *survey datum* in the calculation of *Bouguer correction*. For the regional-scale gravity surveys on the continents, the average crustal density of 2.67 $g \cdot cm^{-3}$ is often used as a *default value*. For local-scale *ground gravity* surveys, Bouguer density is determined by the *Nettleton method*. For the marine (shipboard) surveys, *replacement density* is used in lieu of Bouguer density. Incorrectly selected Bouguer density results in either overcorrecting or undercorrecting the *observed gravity* values and, therefore, in distortion of shapes of anomalies with a negative effect on *enhanced processing* and three-dimensional (3D) *inversion*. The Bouguer correction error rate is 0.004193 $mGal \cdot m^{-1}$ per each 0.1 $g \cdot cm^{-3}$ in an incorrect Bouguer density value. Sometimes referred to as *Bouguer correction density*, *Bouguer reduction density*, and *Bouguer density factor*. See also *anomaly shape*, *Siegert method*, and *terrain correction density*.

Bouguer density factor: See *Bouguer density*.

Bouguer effect: **1.** the gravity effect, or *gravitational attraction*, of rock masses above the *survey datum* onshore and seawater *mass deficiency* below the survey datum offshore. It is quantified by the gravity effect of the infinite and uniformly dense horizontal *Bouguer plate* and expressed in units of *milligal (mGal)* as

$$\Delta g_{BE} = 2\pi G\rho h,$$

where G is *universal gravitational constant* in *cgs units*, ρ is *Bouguer density* in grams per cubic centimeter ($g \cdot cm^{-3}$), and h is Bouguer plate thickness equal to the observation point height, or *elevation*, above the survey datum in meters. For calculation of the Bouguer effect offshore, the elevation h and density ρ are replaced by the water depth h_w and *replacement density* ρ_r. The numerical value, or *magnitude*, of the Bouguer effect comprises the amount of *Bouguer correction* at each *observation point*. **2.** a qualitative gravity effect described as: (a) an increase in *observed gravity* values with increasing elevation onshore caused by the increasing rock mass above the survey datum and (b) a decrease in observed gravity values with increasing bathymetric depth offshore caused by the increasing seawater mass deficiency below the survey datum. The Bouguer effect is compensated by application of *simple Bouguer correction*. Sometimes referred to as *Bouguer plate effect* and *horizontal slab effect*. See also *free-air effect*, *indirect effect*, and *terrain effect*.

Bouguer gravity: See *Bouguer anomaly*.

Bouguer gravity anomaly: See *Bouguer anomaly*.

Bouguer gravity field: See *Bouguer anomaly*.

Bouguer isostatic response: a characteristic of the isostatic deformation of the *lithosphere* under a topographic or bathymetric load, wherein the lithosphere is approximated by a plate of relatively high strength, or rigidity, without a significant flexure. It is opposite to *Airy isostatic response*, which approximates the lithosphere by a bending, or flexing, elastic plate of low strength (flexural rigidity). See also *isostasy*.

Bouguer plate: a horizontal *plate*, or slab, of infinite lateral extent, constant *density*, and finite thickness equal to the height, or *elevation*, of the observation point above the *survey datum*: sea level or *reference ellipsoid*. The Bouguer plate is the first-order approximation of the topographic *relief* around each observation point that forms the basis for calculation of *simple Bouguer correction*. The gravity effect, or *gravitational attraction*, of the Bouguer plate is called *Bouguer effect*. Also known as *Bouguer slab*.

Bouguer plate correction: same as *simple Bouguer correction* with an emphasis on a correction for the gravity effect of the infinite and uniformly dense horizontal *Bouguer plate* approximating a rock volume between the ground surface and *survey datum*: sea level or *reference ellipsoid*.

Bouguer plate effect: same as *Bouguer effect* as it relates to the gravity effect of the *Bouguer plate* at any location above and below it that is compensated by *simple Bouguer correction* and *Poincare-Prey reduction*, respectively.

Bouguer plate formula: the formula for calculation of a *gravitational attraction*, or gravity effect, of the *Bouguer plate*, referred to as *simple Bouguer plate value* and comprising the amount of *simple Bouguer correction*. Also known as *slab formula*. See also *Bouguer correction*.

Bouguer reduction: See *Bouguer correction*.

Bouguer reduction density: See *Bouguer density*.

Bouguer slab: See *Bouguer plate*.

Bouguer slab correction: *simple Bouguer correction* in the context of compensation for the gravity effect of the infinite, horizontal and homogeneous plate, or slab, approximating a rock volume between the ground *station* and *survey datum*.

Bouguer slab effect: **1.** *Bouguer effect* as it pertains to the gravity effect, or *gravitational attraction*, of the infinite uniformly dense horizontal *plate*, or slab, known as the *Bouguer slab* or *Bouguer plate*. **2.** an approximation of the gravity effect of *two-dimensional (2D)* source bodies of finite thickness in 2D *modeling* to determine the source *density* change required to match the observed and model-generated gravity fields. See also *infinite length*.

Bouguer spherical cap: the uppermost layer of the Earth with a constant *density* and thickness that includes all rocks above the *survey datum* (sea level or *reference ellipsoid*) and extends along the curved Earth's surface to the arc distance, or surface radius, of 166.735 km around each *observation point*. The Bouguer spherical cap has the same thickness as the *Bouguer plate*. The difference between the gravity effects of the infinite horizontal Bouguer plate and the finite spherical Bouguer cap is compensated by application of the *Bullard B correction*, also known as *Earth curvature correction*. See also *Cogbill formula* and *LaFehr formula*.

boundary analysis: the *interpretation* methodology based on the analysis of anomalies in the gridded gravity and magnetic data with the objective of mapping the *source body* edges, high-*offset* faults, and other structural and lithological boundaries in the subsurface. Boundary analysis involves the calculation of horizontal derivatives and their high-*resolution* combinations, such as *edge-detector attribute*, *enhanced horizontal derivative (EHD)*, *horizontal tilt angle (TDX)*, *second-order tilt angle*, *tilt ASA*, *tilt angle of horizontal gradient (TAHG)* and others. See also *ASTA filter*, *derivative*, *gridding*, *normalized derivatives*, and *tilt angle (TA)*.

boxcar filter: a theoretical *filter* that has an *amplitude response* defined by a *boxcar function*. It is used in *synthetic modeling* for testing new *enhanced processing* algorithms.

boxcar function: a rectangular *window function* with the same peak value (typically, +1) within the *window* length and zero values beyond the window. Also referred to as *boxcar window*.

boxcar window: same as *boxcar function* with an emphasis on its limited lateral extent.

Breiner method: a method of measuring *remanent magnetization* and *induced magnetization* of rock samples with a subsequent calculation of the *Koenigsberger ratio*. A relatively equidimensional rock sample is placed at a distance approximately 2–3 times the sample diameter from a *scalar magnetometer* in the plane of the *magnetic meridian*, and two sample orientations are found: (1) where the induced and remanent magnetizations are maximally additive and the magnetometer signal, or *reading*, is the highest (S_{max}), and (2) where both magnetizations are maximally subtractive and the magnetometer reading is the lowest (S_{min}). Induced magnetization M_{ind}, remanent magnetization M_{rem}, and Koenigsberger ratio Q are determined, respectively, by

$$M_{ind} = (S_{max} + S_{min} - 2S_{av})/2,$$

$$M_{rem} = (S_{max} - S_{min})/2,$$

$$Q = (S_{max} - S_{min})/(S_{max} + S_{min} - 2S_{av}),$$

where S_{av} is the average of several magnetometer readings without the rock sample.

Brocher formula: a mathematical expression approximating the graphical *Nafe-Drake relation* between velocity V_P of the seismic compressional wave, or P-wave, and rock *density* ρ in the Earth's *crust* for the velocity range from 1.5 to 8.5 km · s^{-1}:

$$\rho = 1.6612V_P - 0.4721V_P^2 + 0.0671V_P^3 - 0.0043V_P^4 + 0.000106V_P^5,$$

where rock density ρ is in grams per cubic centimeter (g · cm^{-3}) and velocity V_P is in kilometers per second (km · s^{-1}). Also referred to as *Brocher regression equation*. See also *Adams-Williamson equation* and *Gardner equation*.

Brocher regression equation: See *Brocher formula*.

Browne correction: a correction applied to the measured *dynamic gravity* values to compensate for their artificial increase caused by deflection of the sensing axis of the *gravimeter* from the vertical orientation, when a moving *observation platform* is subjected to strong horizontal inertial accelerations. The correction value is defined as

$$\Delta g_{Br} \approx -\Delta g_z \beta^2/2,$$

where β is angle of deflection, or tilt, the value of which is obtained from independent measurements of the platform horizontal accelerations, and Δg_z is a true *relative gravity* value measured with the vertical sensing axis orientation. Also referred to as *horizontal acceleration correction*, *off-level correction*, and *tilt correction*. See also *inertial acceleration* and *Harrison correction*.

Bruns equation: an equation describing the relationship between *geoid undulation N*, *disturbing potential T*, and *normal gravity* g_N:

$$N = T/g_N,$$

where g_N is normal gravity defined by the *International Gravity Formula (IGF)*. It is used primarily in geodetic applications for the study of *geoid* undulations and prediction of *free-air gravity*. See also *EGM96 geoid height*.

bulk density: **1.** a dry rock *density* defined as the weight of a rock sample divided by its volume, including the pore space. Bulk density is expressed as

$$\rho_B = \rho_T - V_P\rho_T/100,$$

where ρ_T is *true density* and V_P is the percentage of total pore, or void, space. Bulk density values are controlled, primarily, by the amount of porosity in a rock volume, and they vary within the range of 1.9–2.8 g · cm^{-3}. Bulk density is a laboratory-determined density of rock samples recovered from wells. In exploration applications, it is approximated by *apparent density*, which is calculated from geophysical measurements in drilled wells, or boreholes, by a *borehole gravity meter (BHGM)*, *gamma-gamma log*, and *sonic log*. **2.** an approximate average density of near-surface rock formations in the geotechnical engineering applications of the gravity data. See also *formation density* and *natural density*.

bulk magnetization: the absolute value, or *magnitude*, of *total magnetization* of a particular rock volume or rock formation. Often it is used with the same meaning as *bulk susceptibility*.

bulk susceptibility: **1.** the estimated total value of magnetic susceptibilities of all constituent magnetic materials in a particular rock volume. **2.** a qualitative measure of a high or low amount of magnetic minerals, primarily *magnetite*, that are present in a rock volume. See also *magnetic susceptibility*, *effective susceptibility*, and *volume susceptibility*.

bull's-eye anomaly: **1.** a localized circular-shaped gravity or magnetic *anomaly* represented by several positive or negative values which are much higher or lower, respectively, than the surrounding *anomalous field* values. For example, *kimberlite pipe* or a deep sunken ship could generate a gravity or magnetic anomaly of this type. One-point bull's-eye anomaly is commonly associated with the erroneously high or low measured value at one *station* in a ground gravity or magnetic *survey*. **2.** an *artifact* in the application of *minimum-curvature gridding* caused by (a) single or several erroneous values in the irregularly spaced or line-oriented data, respectively, and (b) *bull's-eye effect*. See also *gridding*.

bull's-eye effect: a *gridding* error caused by *undersampling* in the direction of *strike* of a short-*wavelength* high-*amplitude* anomaly. Instead of a continuous alignment of the anomalous values, the standard gridding algorithm produces a sequence of isolated circular or elongated anomalies that has the appearance of a "string of beads" or "bicycle chain" and referred to as *a boudinage artifact*. The bull's-eye effect is a result of incorrect interpolation of line-oriented data, when both the *sample interval* and *survey line spacing* are larger than the width of an *anomaly* striking across several *acquisition* lines: in the absence of additional cross-line data samples within the radius of interpolation, a *minimum-curvature gridding* algorithm will produce a circle in two horizontal dimensions. Also known as *pill-box effect*. See also *bull's-eye anomaly*.

Bullard A correction: See *simple Bouguer correction*.

Bullard B correction: **1.** the adjustment to a *simple Bouguer correction* that accounts for the difference between *gravitational attraction*, or gravity effect, of the infinite horizontal *Bouguer plate* and that of the finite curved *Bouguer spherical cap* around each *station*. **2.** a correction for *curvature effect*. Bullard B correction adjusts the simple Bouguer correction value to a value corresponding to the Bouguer spherical cap, which has the surface radius of 166.735 km and the same thickness as the Bouguer plate. Two factors contribute to the correction value and sign: (1) Bouguer spherical cap sections underlying the Bouguer plate that dominate up to an *elevation* of 4150 m and pull downwards to increase the measured gravity by values from nearly zero at sea level up to the maximum of 1.518 mGal at the elevation of 2100 m and then back to zero at 4150 m; and (2) the infinite Bouguer plate outside 166.735 km that dominates at elevations above 4150 m and pulls upwards to decrease the measured gravity by values down to −4.768 mGal at the elevation of 6300 m (LaFehr, 1991). For high-*accuracy* local gravity surveys and regional surveys covering a broad range of elevations, the numerical approximation of the *LaFehr formula* is used, where the correction value is defined in units of *milligal (mGal)* as

$$\Delta g_{BB} \cong 1.46308 \cdot 10^{-3} h - 3.52725 \cdot 10^{-7}\, h^2 + 5.1 \cdot 10^{-14}\, h^3.$$

More accurate expression of the Bullard B correction is provided by the *Cogbill formula*:

$$\Delta g_{BB} \cong 1.464139 \cdot 10^{-3} h - 3.533047 \cdot 10^{-7} h^2 + 1.002709 \cdot 10^{-13} h^3 + 3.002407 \cdot 10^{-18} h^4.$$

In both formulas, h is station elevation above sea level in meters. Correction calculations are based on a rock *density* of 2.67 $g \cdot cm^{-3}$ and Earth radius of 6371 km to the sea level at 45° latitude. Also referred to as *curvature correction* and *Earth curvature correction.*

Bullard B effect: See *curvature effect.*

Bullard C correction: a *terrain correction* that accounts for the gravity effect of topographic *relief* variations around each *station* to the distance of 166.735 km in high-accuracy *ground gravity* surveys.

Bullard corrections: three gravity corrections implemented in the *Bullard method* to compensate for the gravity effects of a rock volume above the *survey datum* and topographic *relief* variations to the distance of 166.735 km around each *station.* They include: (1) *Bullard A correction* for the gravity effect of the infinite *Bouguer plate* in the same manner as a *simple Bouguer correction*; (2) *Bullard B correction* for *curvature* of the Earth's surface that is not compensated by the Bullard A correction; and (3) *Bullard C correction* for the gravity effect of topographic relief variations around each station in the same manner as a *terrain correction.*

Bullard method: a method of compensating for the gravity effects of rock masses above the ground gravity *survey datum* (sea level or *reference ellipsoid*), including the topographic *relief* variations to the distance of 166.735 km around each *station*, by application of three *Bullard corrections* to the *observed gravity* values. See also *gravity data corrections.*

Butterworth filter: a spectral-domain *filter* characterized by nearly the same, or flat, *amplitude response* within its *passband* and three modes of operation: *low-pass filter, high-pass filter,* and *band-pass filter.* Its low-pass and high-pass transfer functions are expressed as

$$F_{LP}(k) = 1/\{1 + (k/k_c)^{2n}\} \quad \text{and}$$
$$F_{HP}(k) = 1/\{1 + (k_c/k)^{2n}\},$$

where k is *wavenumber,* k_c is *cutoff wavenumber,* and n is *filter order.* Band-pass mode is obtained by overlapping the *passband* ranges of the low- and high-pass modes defined by different cutoff wavenumbers, i.e., passband range is between the selected cutoff wavenumbers. It is one of the most widely used filters in *enhanced processing* of gravity and magnetic data. Sometimes referred to as *maximally flat filter* and *trapezoidal filter.* See also *spectral domain* and *transfer function.*

C

cable correction: **1.** a correction for the length of a floating cable that tows a *marine magnetometer* during the marine (shipboard) *magnetic survey.* **2.** correction for the length of cable on which an airborne *magnetic gradiometer* system is suspended. The correction is applied to the recorded magnetic data in order to position the magnetometer or magnetic gradiometer readings at coordinates of the *GPS* antenna on a *survey* ship or aircraft. Often it is incorporated into the *heading correction.* Also referred to as *cable-offset correction* and *offset correction.* See also *reading.*

cable-offset correction: same as *cable correction* with an emphasis on a correction for the actual offset, or separation distance, of a cable-towed *marine magnetometer* or cable-suspended airborne *magnetic gradiometer* from the *GPS* antenna on a *survey* ship or aircraft.

calculated gradient: the absolute value of a *derivative* calculated from line-oriented or gridded data. It closely approximates the *magnitude* of the *gradient vector* of an anomalous gravity or magnetic field in the calculation direction. Achievable *lateral resolution* of *gradient* anomalies calculated from the gridded data is lower than that of *measured gradient* anomalies due to (a) a significantly larger *grid interval* in comparison with the *baseline* of a gravity or magnetic *gradiometer,* (b) averaging of the *gridding* process, and (c) superposition of the regional components in the *total field* data. Sometimes referred to as *derived gradient.* See also *anomalous field* and *full tensor gradient (FTG).*

calculated terrain: the topographic surface in an *airborne survey* area obtained by calculating the difference between the *GPS*-based measurements of *elevation,* i.e., flight height above *reference ellipsoid,* and radar *altimeter* or *LIDAR* measurements of *altitude,* i.e., flight height above the ground. Calculated terrain data represent a *Digital Terrain Model (DTM).*

calibrated base station: same as *absolute base station* with an emphasis on a gravity *base station* with a known *absolute gravity* value.

calibrated gravity: measured *relative gravity* values after *unit calibration,* i.e., after conversion from the gravimeter counter, or dial, units into the standard gravity units of *milligal (mGal)* or *microgal (μGal).*

calibration: See *airborne magnetometer calibration, airborne gravimeter calibration, ground gravimeter calibration,* and *unit calibration.*

calibration factor: a multiplication constant for conversion of *gravimeter* readings from its counter, or dial, units into the standard gravity units of *milligal (mGal)* or *microgal (μGal).* Calibration factor is an individual instrument characteristic which is specific for each gravimeter. For worldwide range gravimeters, a table of calibration factors is provided by the manufacturer. Also referred to as *conversion factor* and *milligal constant.* See also *reading* and *unit calibration.*

calibration flight: a pre-survey instrument test flown over or near the airborne *magnetic survey* area to perform *airborne magnetometer calibration* and determine *figure-of-merit (FOM).*

calibration range: two coordinate points with the known gravity values that are used for *ground gravimeter calibration.* The difference between two gravity values, or calibration gravity interval, is approximately 40–60 *milligal (mGal),* and it represents a range of values comparable to the largest range of *relative gravity* variations in a *ground gravity* survey. The calibration range is often established on a gentle slope of a hill or other low topographic elevation in order to provide the required gravity calibration interval.

capacitive accelerometer: same as *electrostatic accelerometer* with an emphasis on electrostatic capacitance devices sensing the movements of *proof mass* in an *accelerometer.* See also *sensor.*

capacitive levitating accelerometer: same as *electrostatic accelerometer* with an emphasis on *electrostatic levitation* and capacitance devices implemented in its design.

cap rock: the high-*density* (up to 2.6–2.9 $g \cdot cm^{-3}$) anhydrite or gypsum rock on top of a *salt dome* or salt *diapir.* It is formed as the result of chemical alterations caused by leaching of soluble materials from a *salt* body. Cap rock is *diamagnetic,* and its presence enhances a negative *magnetic anomaly* generated by the salt body, but it could also contain accumulations of *detrital magnetite* and *diagenetic magnetite* in quantities sufficient for generating the clustered short-*wavelength* positive magnetic anomalies. Due to its high density, an adequate amount of cap rock can generate a local low-*amplitude* positive anomaly within the much broader negative *gravity anomaly* of a salt body, and this anomaly is highlighted by calculation of *residual gravity.* See also *caprock effect* and *clustering.*

caprock effect: a gravity effect resulting from the superposition of a broad negative *gravity anomaly* generated by a whole *salt* body on a local positive gravity anomaly generated by high-density *cap rock* on top of a *salt dome* or salt *diapir.* See also *salt nil zone.*

carbonate buildup: See *reef* or *carbonate platform.*

carbonate platform: a local- or regional-scale area of deposition of shallow-water high-*density* sedimentary formations composed of more than 70% carbonate rocks, such as limestone and dolomite. An isolated intrasedimentary carbonate platform generates a distinct positive *gravity anomaly* in the *residual gravity* field, but it has no detectable *magnetic signature,* because *magnetization* of the carbonates and sedimentary rocks is too small to produce a sufficient *magnetization contrast.* Sometimes, the accumulations of *detrital magnetite* in the fluvial paleochannels along margins of a shallow carbonate platform can generate a detectable *magnetic anomaly.* See also *paleochannel.*

carbonatite: the alkaline igneous rock composed of, at least, 50% carbonate minerals. Carbonatite bodies can reach several kilometers in diameter and host commercial-grade accumulations of *rare earth elements (REE)*. Due to their generally high *density* and *magnetization* contrasts with surrounding rocks, carbonatite bodies have distinct gravity and magnetic signatures highlighted by calculation of the *residual field*. High-resolution airborne *magnetic gradient* and *gravity gradient* measurements have been proven to be efficient in mapping carbonatite edges, internal structure, concentrations of high-density minerals such as beforsite (niobium ore), and mineral alterations along the carbonatite body margins (Drenth, 2014). See also *contrast*, *full tensor gravity gradiometer*, *partial tensor gravity gradiometer*, and *signature*.

cardinal directions: north–south and east–west directions along which gravity and magnetic grids and maps are commonly oriented. See also *grid* and *gridding*.

Cartesian component: a *vector* component of the gravity or magnetic *field* measured along one of the mutually orthogonal x-, y-, z-axes in *Cartesian coordinates*.

Cartesian coordinates: a *three-dimensional (3D)* rectangular *coordinate system* wherein three mutually orthogonal axes are straight lines. The position of any point in space is determined by the linear distances from a given starting point along the orthogonal axes denoted by x and y for the horizontal axes and z for the vertical down or up axis. Vector **A** is expressed in Cartesian coordinates as

$$\mathbf{A} = A_x\mathbf{i} + A_y\mathbf{j} + A_z\mathbf{k}.$$

Vector length, or *magnitude*, is defined as

$$|\mathbf{A}| = (A_x + A_y + A_z)^{1/2},$$

where A_x, A_y, A_z are components of vector **A** that represent its projections into x-, y-, z-axes, respectively, and **i**, **j**, **k** are unit vectors parallel to x-, y-, z-axes, respectively. When the x-axis is directed to geographic north (N), y-axis to geographic east (E), and z-axis is vertical positive down (D), the Cartesian coordinate system is called a *right-sided Cartesian system* or *NED reference frame*. Cartesian coordinates are adopted as a standard *format* for presentation of gridded gravity and magnetic data and results of their *interpretation*. Sometimes referred to as *rectangular coordinates*. See also *dimension*, *gridding*, and *spherical coordinates*.

cascaded filtering: application of two or more *filtering* algorithms in a sequence, where the output of each preceding *filter* is the input for the next filter. For the linear filters, a sequential order of *cascading* does not affect the final result. See also *linear filter*.

cascading: a sequential application of *enhanced processing* techniques, such as *cascaded filtering* and *stabilized downward continuation*, where the output of each preceding procedure is the input for the next one. Cascading improves the *lateral resolution* and *vertical resolution* of gravity and magnetic anomalies, but it also can accumulate *noise* and artifacts of each individual technique. See also *artifact*.

Cauchy-Riemann equations: equations which relate the vertical and horizontal derivatives of a *homogeneous function*, such as the gravity or magnetic field F, via the *Hilbert transform* as

$$\partial F/\partial z = \partial H(F)/\partial x,\ \partial F/\partial x = -\partial H(F)/\partial z,$$

where $H(F)$ is the Hilbert transform of the field F. See also *derivative*.

causative body: a *source body* in the context of a structural element or geological body of laterally contrasting *density* or *magnetization* that is the source, or cause, of a gravity or magnetic *anomaly*.

central amplitude: the maximum or minimum value of a gravity or magnetic *anomaly* at its center. Central *amplitude* is a function of the depth and lateral *density* or *magnetization* contrast of a *causative body*. Its sign (positive or negative) is defined by the sign of the *density contrast* or *magnetization contrast* of the causative body. The theoretical central amplitude and its *fall-off factor* are defined by analytical expressions, or formulas, of the gravity or magnetic field generated by an *ideal source*. See also *principal profile*.

central wavenumber: a spectral-domain *pass filter* parameter representing the *wavenumber* value at which the *amplitude* of a filtered gravity or magnetic field is approximately one-half of its value within the filter *passband*. Referred to with the same meaning as *cutoff wavenumber*. See also *spectral domain*.

centrifugal acceleration: **1.** the outward-directed *acceleration* of a body of *mass* on the Earth's surface under the action of *centrifugal force* resulting from the rotation of Earth about its axis. Centrifugal acceleration a_{cf} is defined by the product of the squared *angular velocity* ω of the rotating Earth and *gyration radius* r_g at the point of measurement:

$$a_{cf} = \omega^2 r_g = \omega^2 \mathrm{R}\cos\theta,$$

where R is Earth's radius, θ is geographic latitude, and $\omega = 7.29211 \cdot 10^{-5}$ *radian* per second ($\mathrm{rad \cdot s^{-1}}$). The vector sum of rotational centrifugal acceleration and *gravitational acceleration* due to the attraction of Earth's mass quantifies the resultant gravitational acceleration at every point on the Earth's surface. The centrifugal acceleration effect is largest at the geographic equator, where its *vector* is opposite to the gravitational acceleration vector, and it is zero at the poles, where the gyration radius is zero. **2.** acceleration of an *observation platform* moving on the curved Earth surface. It is defined as

$$a_{cf} = v^2/\mathrm{R},$$

where v is platform velocity and R is the Earth's radius at the *observation point*. See also *angular acceleration*, *centripetal acceleration*, and *Eötvös acceleration*.

centrifugal force: the outward-directed force of pushing-out, or repulsion, resulting from the rotation of Earth about its axis and from movement of a body of *mass* on the curved Earth surface. Centrifugal force due to the Earth's rotation is largest at the equator and decreasing to zero at the poles. Maximum centrifugal force is less than 1/300 of the *gravitational force* produced by the Earth's mass. In accordance with the *Newton law of motion*, centrifugal force per unit mass defines *centrifugal acceleration* at every point on the Earth's surface. Gravitational force *magnitude* on the Earth's surface is defined as a *vector* sum of the gravitational force of attraction due to the Earth's mass and centrifugal force of repulsion due to the rotation of Earth about its axis.

centrifugal potential: a *scalar* quantity U_r which represents *potential* of the rotational *centrifugal force* f_r in a mathematical sense of satisfying the relation

$$f_r = \nabla U_r,$$

where ∇ is the vector *gradient* operator *del*. Centrifugal potential is a function of *angular velocity* ω of the Earth's rotation and *gyration radius* r_g, and it is expressed as

$$U_r = \omega^2 r_g^2 / 2,$$

where $r_g = \mathrm{R}\cos\theta$, R is radius of *reference ellipsoid* approximating the Earth's shape, and θ is latitude. Centrifugal potential and *gravitational potential* comprise the *Earth gravity potential* at any point on the Earth's surface. Also referred to as *rotational potential*.

centrifugal term: the direction-independent component v^2/R in the formula of *Eötvös correction*, where v is velocity of the *observation platform* and R is the Earth's radius at the *observation point*. It is added to the *observed gravity* values regardless of the observation platform direction of movement within a reference frame fixed on the rotating Earth. In the marine (shipboard) surveys, the centrifugal term is commonly ignored, because it is small in comparison with other measurement effects. See also *Coriolis term*.

centripetal acceleration: 1. *angular acceleration* **2.** the inward-directed radial *acceleration* caused by a change in the direction or value of *angular velocity* of *proof mass* in the sensing accelerometers mounted on a rotating disk. The vertical component of its *vector* is a source of systematic errors in *gravity gradient* measurements by a *rotating-disk gravity gradiometer*, and it is compensated by application of a *centripetal acceleration correction*. See also *accelerometer*, *gravity sensor*, and *equivalence principle*.

centripetal acceleration correction: a correction applied to *gravity gradient* measurements by a *rotating-disk gravity gradiometer* to compensate for the vertical component of *centripetal acceleration* of *proof mass* in the sensing accelerometers. Correction accounts for the systematic errors in measurements with the use of (a) horizontal rotating disks, where errors are caused by the motion-related *tilt* of the *observation platform*, and (b) permanently tilted rotating disks mounted on a subplatform (carousel) which rotates with a constant *angular velocity* about the vertical axis, wherein a vertical component of the centripetal acceleration is associated with a permanent tilt and change in the angular velocity component due to changes in the radius of rotation of the individual accelerometers relative to the axis of the carousel rotation. See also *accelerometer*, *self-calibration*, and *sensor*.

cesium magnetometer: an *optically pumped magnetometer* for measurements of the absolute numerical value, or *magnitude*, of *Earth magnetic field*. Its *magnetic sensor* design is based on the use of the gaseous isotopes of cesium. The cesium-133 isotope has a *gyromagnetic ratio* nearly 80 times higher than that of a regular *proton-precession magnetometer* and, thereby, provides the extra-high *sensitivity* up to 0.001 *nanotesla (nT)*. A cesium magnetometer is omnidirectional, i.e., it has the same sensitivity regardless of its magnetic sensor orientation, which makes it, along with other high-performance characteristics, a preferred instrument for airborne magnetic and *magnetic gradient* measurements. Achievable *accuracy* is up to 0.01 nT. Sometimes referred to as *cesium-vapor magnetometer*. See also *helium magnetometer*.

cesium-vapor magnetometer: same as *cesium magnetometer* with an emphasis on its *magnetic sensor* utilizing the gaseous, or vaporized, cesium isotopes.

cgs units: units in the centimeter-gram-second (cgs) system of measurements, where distance is measured in centimeters, mass in grams, and time in seconds. In gravity applications, the cgs unit of *gravitational acceleration* is *Gal*, which is equal to *acceleration* of one centimeter per squared second, i.e., $1\ \mathrm{Gal} = 1\ \mathrm{cm \cdot s^{-2}}$, and rock density is quoted in grams per cubic centimeter, $\mathrm{g \cdot cm^{-3}}$. In magnetic applications, *magnetic susceptibility* is often quoted in the dimensionless cgs units of 10^{-6} called *microcgs* (μ*cgs*) units. In the vintage literature, the equivalent of this cgs unit is referred to as the electromagnetic unit, or *emu*. See also *SI units*.

CHAMP mission: the Challenging Minisatellite Payload (CHAMP) satellite mission launched in 2000 and completed in 2010. Its objective was to map the global gravity and magnetic fields at the initial *altitude* of 450 km decreasing to approximately 300 km at the mission's end. Its satellite dimensions (L × H × W) were 4.3 × 0.75 × 1.6 m. A modified version of the scalar *Overhauser magnetometer* and two vector fluxgate magnetometers in a *gradiometer* arrangement were mounted on a 4 m long pole, or boom, to measure the *Earth magnetic field* and its *vector* components. *Earth gravity field* measurements were based on the *three-dimensional (3D)* continuous tracking of the satellite free-fall trajectory by the high-orbiting constellation of *Global Positioning System (GPS)* satellites and ground-based laser ranging stations. Deviations of the satellite orbit from a theoretical orbit calculated for a homogeneous Earth are directly proportional to changes in the long-*wavelength* components of the Earth's gravity field. See also *GOCE mission*, *GRACE mission*, *GRAIL mission*, *Swarm mission*, *scalar magnetometer*, and *vector magnetometer*.

Chandler wobble: a slow-changing, or secular, *precession* of the Earth's rotational axis with a *period* of 435 days that is assumed to be caused by movements of the lithospheric plates and convection in the *mantle*. Its effect on *absolute gravity* measurements is compensated by *polar motion correction*. See also *lithosphere* and *polar motion*.

channel: 1. *data channel*. **2.** *paleochannel*.

channel filter: a space-domain *filter* applied to line-oriented data recorded by a particular *data channel* during marine (shipboard) or airborne gravity and magnetic surveys. Referred to with the same meaning as *line filter*. See also *space domain*.

chemical remanent magnetization: a permanent *magnetization* acquired by rocks in the process of their chemical consolidation and crystallization under exposure to externally applied magnetic field, such as *Earth magnetic field*, at temperatures below *Curie point*. See also *combustion magnetism*, *remanent magnetization*, and *induced magnetization*.

chessboard method: a method of converting level-flown data to drape-flown data and vice versa using the *upward continuation* or *downward continuation* techniques. Gridded gravity or magnetic field is continued to several horizontal surfaces at successively higher or lower elevations spanning a full range of the desired uneven surface and mimicking stacked *three-dimensional (3D)* chessboards. Field values at any point on the uneven surface are obtained by vertical interpolation between the corresponding values at the nearest bounding horizontal surfaces. See also *drape correction*, *grid*, and *gridding*.

chimney: a columnar mass of sedimentary rocks immediately above a hydrocarbon reservoir, where a seepage of hydrocarbons results in diagenetic alterations of the rocks with changes of their *magnetization* and, thereby, generates clusters of *micromagnetic anomalies* and *magnetic bright spot* anomalies. See also *diagenetic magnetite*, *SRM anomaly*, and *hydrocarbon seepage effect*.

chopping filter: a *high-pass filter* or *low-pass filter* in the context of cutting off, or chopping, the *wavelength* components of data that are longer (in a high-pass mode of operation) or shorter (in a low-pass mode) than their *cutoff wavelength*.

circular convolution: a property of a *convolution* process in the Fourier-domain *filtering* of gridded data. Because *Fourier transform* assumes that sampled data are periodic or, equivalently, circular beyond the dataset area, the *filter* algorithm convolves the data at the opposite edges of a *grid* and, thereby, produces data distortions known as *wraparound effect*. Distortions are minimized by application of *detrending* prior to Fourier transform. See also *gridding* and *Fourier domain*.

circular frequency: See *angular frequency*.

clinker anomaly: a diagenetic *sedimentary magnetic anomaly* generated by accumulations of *diagenetic magnetite* formed as a result of burning clinkers, which are coal seams and lignite-bearing beds in rock outcrops. Burned *outcrop* and *subcrop* clinkers are sources of the strong short-*wavelength* magnetic anomalies with amplitudes up to hundreds and thousands of *nanotesla (nT)*. See also *amplitude*, *anomaly*, and *combustion magnetism*.

clip filter: a space-domain line- or *profile*-based algorithm that assigns null values to data points outside user-specified upper and lower bounds of the filter. See also *clipping* and *space domain*.

clipping: the error in a display of gridded gravity and magnetic anomalies whose maximum or minimum values exceed the *dynamic range* limit selected for a particular *color-coded grid*. Exceeding values are cut out, or clipped, and, therefore, amplitudes of anomalies are distorted. Referred to with the same meaning as *truncation error*. See also *anomaly amplitude*, *gridding*, and *trimming*.

clover-leaf test: same as *heading error test* with an emphasis on its clover-leaf-shaped flight path for the *survey* aircraft to follow during the test.

closure: 1. the vertical distance between the highest point of a subsurface *structure*, such as an anticlinal fold or *salt dome*, and its lowest closed structural *contour* on a *contour map* or *grid*. Also referred to as *amplitude* of the structure. **2.** *closure error*. See also *anticline* and *thrust*.

closure error: the measurement error accumulated during gravity observations along a single *loop* in a *ground gravity* survey. Also known as *closure*, *loop closure*, and *misclosure*.

clustering: 1. automatic grouping of source depth solutions obtained by *Euler deconvolution* or *Werner deconvolution* in order to reduce the number of spurious *noise* solutions and highlight the depth solutions generated by real sources. Its main parameters are radius of grouping and minimum number of depth solutions per single group (cluster). A valid depth pick is plotted when a user-defined minimum number of depth solutions is obtained within a pre-selected radius of clustering around each source depth solution. **2.** statistical grouping of variables in such a way that variables in the same group are more similar to each other than to those in other groups, such as the one implemented in *terracing* of anomalies. **3.** assemblages of short-*wavelength* magnetic anomalies generated by localized accumulations of *detrital magnetite* or *diagenetic magnetite* in ancient fluvial systems and other depositional environments.

CO_2 sequestration gravity: *time-lapse gravity* and *gravity gradient* data from high-*accuracy* ground and borehole measurements for monitoring the injection and permanent storage of CO_2 in deep aquifers, such as brine-bearing rock formations, and depleted hydrocarbon reservoirs whose *structure* is known from seismic and well (borehole) data. Time-lapse *microgravity* data are referenced to the *baseline* gravity data, obtained from measurements made prior to injections, and integrated with reservoir temperature and pressure data from injection wells to account for CO_2 *density* variations with increasing pressure and temperatures. Except for reservoirs containing extra-light hydrocarbons, CO_2 injection usually creates a negative *density contrast*. Obtained microgravity and gravity *gradient* data are implemented in a 3D *property inversion* for the spatial distribution of time-lapse density changes caused by CO_2 injection and for detection of reservoir leaks (Alnes et al., 2008; Gasperikova and Hoversten, 2008; Wan et al., 2018). See also *aquifer*, *aquifer gravity*, *geothermal gravity*, *volcano gravity*, and *4D gravity monitoring*.

coastline effect: a space- and time-variable magnetic effect, or magnetic disturbance, generated by secondary electric currents induced into a shallow subsurface by variations of *external magnetic field* of the Earth in coastal areas and related to the electric conductivity contrast between the relatively low conducting land and highly conductive sea water.

Cogbill formula: a high-*accuracy* power-series representation of *Bullard B correction* for the Earth's surface *curvature*. It is based on 2.67 g·cm^{-3} rock *density*, Earth radius of 6371 km to sea level at 45° latitude, and expressed in units of *milligal (mGal)*:

$$\Delta g_{BB} = \mathrm{A}h - \mathrm{B}h^2 + \mathrm{C}h^3 + \mathrm{D}h^4,$$

where h is *station* height, or *elevation*, above sea level in meters, A = 1.464139·10^{-3}, B = 3.533047·10^{-7}, C = 1.002709·10^{-13}, and D = 3.002407·10^{-18}. See also *LaFehr formula*.

coherence: **1.** a systematic in-*phase* relationship between adjacent waveforms of gridded data or between profiles of data acquired along closely spaced *survey* lines. **2.** a measure of similarity among parameters of two or more functions or their parts. **3.** a measure of correlation between two different datasets, such as *Bouguer gravity* and topography or *free-air gravity* and bathymetry. See also *gravitational coherence*, *gridding*, *profile*, and *waveform*.

coherent filtering: a *data enhancement* technique applied to gridded gravity or magnetic data to detect spatial trends of correlation, or *coherence*, between the peaks, troughs, and other coherent *waveform* components. Coherent *filtering* is based on calculation of the maxima of the cross-correlation function within a *moving window* of a user-specified length.

coherent noise: correlatable *noise* components that exhibit the systematic *phase* and *amplitude* relations between waveforms of gridded data and between adjacent *survey* lines. Coherent noise includes *corrugations*, errors in *gridding*, *edge effects*, and certain types of *cultural noise*. Also known as *regular noise*. See also *waveform*.

cokriging: a geostatistical method of interpolation and extrapolation based on (a) using the spatial correlation between primary and secondary variables, such as the linearly related *density* and anomalous gravity field, to improve the prediction of a primary variable at the interpolated or extrapolated locations, where the measured values are missing, and (b) assigning weights to measured values in accordance with their separation distance, degree of correlation, and in a minimum-*variance* sense, so the prediction, or cokriging, variance is minimized. Cokriging is applied to optimize the uncertainty in primary variable estimates and provide a smooth solution for an *underdetermined problem*. See also *kriging* and *smoothing*.

colatitude: **1.** the angular distance θ of an *observation point* on the Earth's surface from the North Pole. It is a complement of geographic latitude and defined as

$$\theta = 90° - L,$$

where L is geographic latitude in degrees. **2.** the angular distance θ from the vertical axis in *spherical coordinates*. See also *geomagnetic poles*.

cold-atom absolute gravimeter: same as *absolute atomic gravimeter* with an emphasis on its design utilizing a stream of atoms, cooled down to temperatures close to absolute zero (−273.15°C), in lieu of a free-falling *test mass*.

cold atom gravimetry: instruments and methodology for obtaining measurements of *absolute gravity* with the use of an *absolute atomic gravimeter*, which operates on the principle of monitoring the free fall of a laser-manipulated streams of atoms cooled down to temperatures close to absolute zero (−273.15°C). Achievable *accuracy* of measurements is up to 1–2 *microgal (μGal)*, and it depends on an integration time of the repeated free-fall cycles: the longer the integration time, the higher the accuracy. See also *absolute gravimeter* and *free-fall absolute gravimeter*.

cold-atom interferometer gravimeter: same as *absolute atomic gravimeter* with an emphasis on its design utilizing the atom interferometer system which monitors the free fall of streams of atoms cooled down to temperatures close to absolute zero (−273.15°C).

color bar: a multicolored display of *color-coded grid* values by different colors and their shades for which the "hot" colors, such as red and pink, are assigned to high values, and the "cold" colors, such as dark blue and reddish-purple (magenta), are assigned to low values.

color-coded grid: a *grid* showing the values of a gravity or magnetic field, its gradients, and its attributes in *pseudocolor*, whereby different colors and their shades represent a user-defined range of gravity or magnetic values in the same manner as a *contour map*. By convention, red and other "hot" colors represent high grid values, whereas blue and other "cold" colors represent low grid values. Also referred to as *color-scaled grid*. See also *attribute*, *banded-contour grid*, *gradient*, and *RGB composite image*.

color-scaled grid: See *color-coded grid*.

combined elevation correction: *elevation correction* in the context of its representation by the sum of *free-air correction* and *simple Bouguer correction*. Two corrections are combined, because both are functions of the same height, or *elevation*, of the observation point above the *survey datum*: sea level or *reference ellipsoid*.

combustion magnetism: a strong *magnetization* acquired by organic-rich and iron-bearing sedimentary formations as a result of the burning of their flammable components under exposure to an external magnetic field, such as the *Earth magnetic field*. Fast oxidation by burning can convert the moderate- and low-susceptible rock-forming minerals, such as *paramagnetic* pyrite, into high-susceptible *diagenetic magnetite* and other *diagenetic magnetic minerals*. Burned coal seams and lignite-bearing sedimentary formations as well as the burning of hydrocarbon seepage fluids within the near-surface parts of faults and fracture zones generate the short-*wavelength* magnetic anomalies with amplitudes up to hundreds and thousands of *nanotesla (nT)*. See also *clinker anomaly*, *magnetic susceptibility*, *sedimentary magnetic anomaly*, and *hydrocarbon seepage effect*.

common mode: a characteristic of a signal that emerges simultaneously with the same *phase* and direction parameters at all sensors in the common measurement frame on a moving *observation platform*. In *gradiometer* systems, such signals are considered as common-mode *dynamic noise* and eliminated from gravity and magnetic *gradient* measurements by *common-mode rejection*. Common-mode signal outputs of *accelerometer*-based *airborne gravimeter (AG)* and *airborne* *gravity gradiometer (AGG)* systems are used to control an onboard *stabilized platform*. See also *sensor*.

common-mode acceleration: the non-gravitational *acceleration* which is sensed simultaneously by accelerometers in the *gravity gradiometer* system on a moving *observation platform*, such as a ship, airplane, or satellite. In a *GOCE gravity gradiometer* system, the common-mode acceleration datasets are obtained by calculating the sum of accelerations measured by pairs of accelerometers in the same gradiometer arms oriented in three mutually orthogonal directions, and these data are used to counteract the atmospheric drag and solar radiation pressure by *real-time* adjustments of the ion engine thrust in order to maintain the Earth orbiting satellite in a free-fall motion. See also *accelerometer*, *gradiometer arm*, *differential-mode acceleration*, *inertial acceleration*, and *sensor*.

common-mode rejection: suppression of a common-mode *dynamic noise* by subtracting from one another the values measured by sensors of the same *gradiometer arm*. Common-mode dynamic noise includes *observation platform* vibrations and its inertial accelerations, laser beam fluctuations, and other motion-related disturbances which affect the paired accelerometers or other gravity sensors. Common-mode rejection is a fundamental property of all *gradiometer* systems that makes possible the high-*resolution* gravity gradient measurements in airborne, marine (shipboard), submarine, and satellite settings. In *magnetic gradiometer* systems of paired and synchronized magnetometers, the common-mode rejection eliminates *diurnals*. See also *accelerometer*, *common mode*, *inertial acceleration*, *sensor*, *gravity sensor*, and *magnetic sensor*.

compensating root: same as *isostatic root* with an emphasis on a crustal rock mass under the continents that maintains *isostatic equilibrium* between the Earth's *crust* and upper *mantle* in accordance with the *Airy-Heiskanen model* and the *Vening Meinesz model* of *isostatic compensation*.

compensation: See *aeromagnetic compensation* and *isostatic compensation*.

compensation level: See *isostatic compensation level*.

compensation test: a pre-survey flight test conducted in order to determine the *figure-of-merit (FOM)* and estimate the performance of *aeromagnetic compensation* for the static and dynamic interfering magnetic effects of a *survey* aircraft. Also known as *FOM test*. See also *pre-survey flight tests*.

complementary filter: a space-domain *wavelength filter* that has a *passband* opposite to the passband of another pass filter with the same *cutoff wavelength*, e.g., a 2000 m *low-pass filter* can be complemented by a 2000 m *high-pass filter* to display the short-*wavelength* (high-*frequency*) components of an *anomalous field* suppressed by the low-pass filter. See also *space domain*.

complete Bouguer anomaly: **1.** a terrain-corrected *Bouguer anomaly*. **2.** Bouguer anomaly which differs from a *free-air anomaly* by the additional compensation for three gravity effects: (1) *Bouguer effect* of the rock masses above a *survey datum* onshore and seawater *mass deficiency* below a survey datum offshore, (2) the local-scale *terrain effect* of the topographic and bathymetric variations around observation points, and (3) the regional-scale *curvature effect* due to *curvature* of the Earth's surface. These gravity effects are compensated by application of *simple Bouguer correction*, *terrain correction*, and *Bullard B correction*, respectively. See also *simple Bouguer anomaly* and *predicted gravity*.

complete Bouguer correction: a combined gravity correction which includes (a) *simple Bouguer correction* for the gravity effect of rock masses above a *survey datum* onshore and seawater *mass deficiency* below a survey datum offshore, (b) *terrain correction* for the local-scale gravity effect of topographic or bathymetric variations around observation points, and (c) *Bullard B correction* for the regional-scale gravity effect of the Earth's surface *curvature*. Also referred to as *complete Bouguer reduction* and *total Bouguer reduction*.

complete Bouguer reduction: See *complete Bouguer correction*.

complex analytic signal: same as *analytic signal* that is a *complex variable* composed of real and imaginary parts, wherein the imaginary part represents the 90° *phase*-shifted real part.

complex attribute: a quantitative characteristic of anomalous gravity and magnetic fields that is obtained from the real and imaginary components of a *complex variable*, such as the *analytic signal (AS)* or *monogenic signal (MS)*. The term "complex" is used in the mathematical sense of a variable composed of two parts: real and imaginary. In reference to the complex attributes, the terms "local" and "instantaneous" are applied interchangeably with the same meaning. See also *imaginary component*.

complex function: a mathematical function with real and imaginary parts, such as those representing the odd (sine) and even (cosine) components in Fourier-transformed data and comprise a *Hilbert transform pair*, wherein the imaginary part is the 90° *phase*-shifted real part. See also *Fourier transform*, *sine transform*, *cosine transform*, and *Riesz transform*.

complex monogenic signal: same as *monogenic signal (MS)* that is a *complex attribute* of an anomalous magnetic or gravity *field*.

complex number: a number composed of real and imaginary parts and expressed as

$$z = x + iy = |A|e^{i\theta} = |A|(\cos\theta + i\sin\theta),$$

where x is the real part; y is the imaginary part; $i = \sqrt{-1}$ is a complex unit number indicating the orthogonality; A is *magnitude* of the complex number, also known as *modulus* and defined as $|A| = (x^2 + y^2)^{1/2}$; and θ is angle with a positive real axis on a *complex plane*, so that $x = |A|\cos\theta$ and $y = |A|\sin\theta$. Angle θ represents the *phase* of the complex number, and it shows the orientation of its *vector* relative to the real axis and defined as $\theta = \arctan(y/x)$. Complex number phase is referred to with the same meaning as *local phase*. The length of a complex number vector is defined by the modulus of the complex number. See also *complex spectrum*, *imaginary component*, and *Euler identity*.

complex plane: a plane formed by the mutually orthogonal real and imaginary axes representing a reference frame for the real and imaginary parts of a *complex variable*. Any point on this plane can be identified by a corresponding *complex number*. See also *Hilbert transform* and *Riesz transform*.

complex spectrum: a mathematical representation of the *Fourier transform* $\mathcal{F}(k)$ as a *complex function* for which the real part ReP is a *cosine transform* and the imaginary part ImP is a *sine transform*. It can be expressed as

$$\mathcal{F}(k) = \mathrm{ReP}[\mathcal{F}(k)] + i\,\mathrm{ImP}[\mathcal{F}(k)] = |\mathrm{A}(k)|e^{i\theta(k)},$$

where $i = \sqrt{-1}$ is a complex unit number indicating the orthogonality, k is the *wavenumber*, $|\mathrm{A}(k)| = (\mathrm{ReP}^2 + \mathrm{ImP}^2)^{1/2}$ is the *amplitude spectrum*, and $\theta(k) = \arctan(\mathrm{ImP/ReP})$ is the *phase spectrum*. Squared amplitude spectrum $|\mathrm{A}(k)|^2$ is called *energy-density spectrum*, *power-density spectrum*, and *power spectrum*. See also *spectrum* and *polar form of complex number*.

complex variable: a quantity or quantitative characteristic which varies as a function of the real and imaginary components. See also *complex attribute*, *complex function*, *complex number*, and *imaginary component*.

component filter: a *grid*-based Fourier-domain operator which transforms a *scalar* gravity or magnetic field into its three *vector* components along the mutually orthogonal x-, y-, z-directions in *Cartesian coordinates*. A transformation algorithm is based on *field-to-potential transformation* and calculation of the corresponding directional derivatives of the gravity or magnetic *potential*. Assumptions regarding the *source body* shape and distribution of its *density* or *magnetization* are not required. In the magnetic case, the ambient *main geomagnetic field* direction should be known in order to define the filter parameters. See also *directional derivative*, *Fourier domain*, *ambient field*, and *scalar-to-vector transformation*.

Comprehensive Magnetic Field (CMF): the spatially and temporally continuous global *model* of *Earth magnetic field* that includes the *main geomagnetic field*, its *secular variation*, and the quiet-time long-*wavelength* components of the *external magnetic field* of the Earth. The CMF model is based on the integration of magnetic data from a worldwide network of magnetic observatories and high-*precision* measurements by *vector magnetometer* and *scalar magnetometer* systems installed on Earth-orbiting satellites. See also *magnetic observatory* and *INTERMAGNET*.

conditional probability: a probability of event E on the condition that event C_i has already occurred. It is denoted by $P(E|C_i)$ and implemented in *Bayes theorem* and *stochastic inversion*.

conditioning: See *data conditioning*.

congruency: See *phase congruency*.

conservative field: a physical *field* of force wherein the amount of work, or energy, required to move a particle from one point to another is independent of the path taken between these points. In theory, the net (without losses) work for moving the particle around a closed path, or loop, in the conservative field is equal to zero, i.e., energy expended on moving the particle against the force of this field is equal to the energy saved in the particle movement under its action. Gravity and magnetic fields are conservative fields for which the intensity, or strength, at any space point is defined by a *gradient* of the gravity or magnetic *potential*. See also *potential field*.

constraints: restrictions imposed on the selection of *interpretation* scenarios and two-dimensional (2D) or three-dimensional (3D) *inversion* in order to minimize their inherent nonuniqueness and ensure a geological plausibility of the obtained results. Constraints represent the geological and geophysical information that is independent from the gravity or magnetic data, and it includes but is not limited to: (a) formations tops, i.e., depths of the sedimentary formations and *basement* from drilled wells; (b) well cross-sections; (c) well logs, such as *density log*, *magnetic susceptibility log*, and *sonic log*; (d) lithological columns; (e) tables of rock densities and susceptibilities of rock samples extracted from wells; (f) seismic time and depth sections; (g) seismic interval velocities; (h) seismic horizon maps; and (i) surface geology maps. See also *joint inversion* and *nonuniqueness problem*.

contact: **1.** the *source geometry* approximation of a structural discontinuity: (a) structural boundary, such as a large-offset *fault* separating rocks with different *susceptibility* and *density* values within a sedimentary section or *strike-slip fault* which forms the boundaries of a *pull-apart basin*; (b) lithological boundary between igneous and metamorphic rocks within the *basement*, between igneous intrusions and sedimentary rocks, and between sedimentary formations with laterally contrasting susceptibilities and densities; and (c) *edge* of an isolated *source body*, such as *igneous intrusion*, *salt dome*, and *salt* or shale *diapir*. **2.** the location of a lateral change in *density* or *magnetization* of rocks associated with the structural and lithological boundaries in the subsurface. Contact is identified by the maximum value of *horizontal gradient*, *total horizontal derivative of tilt*, and other anomalous maxima obtained by application of *edge detection* techniques. See also *infinite contact*, *magnetic contact*, *gravity contact*, and *contrast*.

contact-depth method: a *grid*- or *profile*-based automated method of estimating the depth to magnetic sources approximated by the vertical *magnetic contact*. The method assumes a negligible *remanent magnetization* of sources and their vertical *magnetization*, i.e., the *TMI field* is reduced-to-pole. Contact depth h_c is defined by a reciprocal, or inverse, of *total horizontal derivative* of the tangent of *tilt angle (TA)* of the gridded *RTP magnetic field* at the *contact* location as

$$h_c = 1/[(\partial \tan \mathrm{TA}/\partial x)^2 + (\partial \tan \mathrm{TA}/\partial y)^2]^{1/2}.$$

The location of the magnetic contact on a grid is determined by the zero value of the TA. For profile data, the *horizontal derivative* of a tangent of the TA is used. This method utilizes second-order derivatives and, therefore, is sensitive to *noise* which can be reduced by the empirically determined *upward continuation* of the RTP magnetic field prior to calculations (Cooper, 2014b). See also *gridding*, *reduction-to-pole (RTP)*, and *tilt-depth method*.

continental-oceanic boundary (COB): the structural boundary marked by a series of detachment *fault* splays and associated with a variable-width, block-faulted transitional zone between the continental *crust* and oceanic crust along a passive continental margin, i.e., along a regional-scale area where the stretched continental crust has subsided during rifting which preceded the creation of a mid-ocean ridge and *sea-floor spreading* without subduction of the oceanic plate. Differences in the origin, *structure* (thickness of accumulated sediments, distribution of faults and igneous bodies, *Moho depth*, etc.), lithological composition, physical properties (*density* and *magnetization*), and tectonic evolution produce the differing patterns of the gravity and magnetic *total field* and *gradient* anomalies over the continental and oceanic parts of the Earth's crust. These anomalous patterns can be highlighted by *enhanced processing* of *satellite gravity* and *high-resolution aeromagnetic (HRAM)* data. The COB mapping and international demarcation of continental margins are based on *integrated interpretation* of the observed gravity and magnetic anomalies and their calculated attributes with incorporation of the available seismic data (Mjelde et al., 2007; Goussev et al., 2008; Pawlowski, 2008). See also *attribute*, *pattern*, and *trench*.

continuation: calculation of a gravity or magnetic field at an *elevation* which is different from the elevation of the actual measurements. Recomputing of *anomalous field* at higher elevations is known as *upward continuation*, whereas its recomputing at lower elevations is referred to as *downward continuation*. Also known as *vertical continuation*.

continuation filter: a data processing operator for which the algorithm is based on *upward continuation* or *downward continuation* of anomalous gravity and magnetic fields. Upward continuation is analogous to a *low-pass filter*, which emphasizes the long-*wavelength* anomalies generated by the broader and deeper sources and suppresses high-frequency *noise*, whereas downward continuation is analogous to a *high-pass filter*, which emphasizes the short-wavelength anomalies generated by the smaller and shallower sources and amplifies the high-frequency noise. Calculation of the difference between two consecutive upward or downward continuations forms the basis for a *Jacobsen filter*. Continuation filters also are used in *merging* the data from surveys completed with different datums and at different elevations. See also *datum*, *elevation*, and *differential continuation*.

continuation residuals: 1. residuals of a gravity or magnetic *field* obtained by calculating the difference between two upward or downward continuations of this field. **2.** a *residual field* obtained by subtracting either the downward-continued or upward-continued field from the original *anomalous field*. See also *upward continuation*, *downward continuation*, and *Jacobsen filter*.

continuous wavelet transform (CWT): the *wavelet transform (WT)* of line-oriented or gridded gravity and magnetic data, where at each transformation cycle, or *scale*, the iteratively dilated (stretched or compressed) *mother wavelet* is convolved with the data in the same manner as a space-domain *convolution filter* for which the *bandwidth* is defined by a scale-corresponding *wavelet*. After completion of each cycle, the wavelet is rescaled by stretching or compression, and the process of *convolution* of a modified wavelet with the data is repeated. Wavelet dilation and translation (shifting increment) are arbitrary within a range of user-specified scales. Multiple cycles of data convolution with the iteratively dilated and translated wavelets decompose the data into the successive scales, or ranges, of *wavelength resolution*: the larger the stretching, the longer the resolved *wavelength*, and the higher the compression, the shorter the wavelet-resolved wavelengths. The CWT converts the data into weighted linear combinations of different wavelength contents, wherein the weights are called *wavelet coefficients*. Locations of individual wavelength components of the original data are determined by the line or *grid* intervals, where a particular wavelength component and operating wavelet closely match each other, so that the convolution-derived wavelet coefficients reach their maximum values. Obtained wavelet coefficients are (a) displayed on the CWT space-scale plane for visualization of scales contaminated by *random noise* and *coherent noise* and (b) arranged in the format of a *matrix* for subsequent processing of the transformed data (Hornby et al., 1999; Ridsdill-Smith, 2000). See also *discrete wavelength transform (DWT)*.

contour: a continuous line on the *contour map* or *grid* that (a) connects data points with the same values of the mapped or gridded gravity or magnetic *field*, its derivatives, and its attributes, and (b) separates data points that have values lower than the contour value from points with the higher values. Referred to with the same meaning as *isoline*. See also *attribute*, *derivative*, *gridding*, and *isoanomaly*.

contour grid: a *grid* showing the continuous lines, or contours, which connect points with the same values of the anomalous gravity or magnetic *field*, its derivatives, or any other field *attribute*. Contours of the equal values are called isolines. A contour grid is often combined with a color-coded representation of the gravity and magnetic values for a better visualization of anomalies. See also *anomalous field*, *contour interval*, and *color-coded grid*.

contour interval: the difference between values of two adjacent contour lines, or isolines, on a *contour map* or *grid*. A contour interval on a *color-coded grid* defines a range of values assigned to each color shade. See also *isoline*.

contour map: a digital or printed map showing the continuous lines, or contours, which connect data points with the same values, such as the gridded values of an *anomalous field* or its derivatives and attributes, depth values of a geological *structure* (seismic horizon, top of *basement* or salt formation, etc.), and a global-scale representation of the *main geomagnetic field* parameters on the *isogonic map*, *isodynamic map*, and *isomagnetic map*. See also *attribute*, *derivative*, and *gridding*.

contrast: the difference in *density* or *magnetization* values between laterally adjacent rock units or formations. Lateral density and magnetization contrasts in the subsurface are fundamental reasons for, or causes of, observed gravity and magnetic anomalies: if there is no lateral *density contrast* or *magnetization contrast*, then there will be no *anomaly* in the measured field. Anomalous *amplitude* is directly proportional to a contrast value: the higher the contrast, the higher the amplitude of the anomaly. Change in the sign of a contrast reverses the sign of an anomaly. Anomaly-generating contrasts are created primarily by: (a) structural deformations within the *crust*, such as faulting, folding, thrusting, rifting, subduction, and uplifting; (b) lateral lithological contacts within the sedimentary section created by carbonate buildups, paleochannels, incised valleys, emplacement of the *salt* and shale diapirs, igneous intrusions, and volcanic rocks; and (c) lateral lithological contacts between metamorphic and igneous rocks in the *basement* and rock formations in the lower *crust*. Also referred to as *lateral contrast*, because the vertical density and magnetization contrasts, such as those between the homogeneous horizontal layers, do not generate gravity or magnetic anomalies. See also *diapir*, *igneous intrusion*, *incised valley*, *paleochannel*, and *contact*.

control lines: same as *tie lines* that are used to control the *mis-tie* values at intersections of shipboard (marine) and airborne *survey* lines.

conversion factor: same as *calibration factor* with an emphasis on a conversion of *gravimeter* readings from dial (counter) units into standard units of *milligal (mGal)* or *microgal (μGal)*.

convolution: 1. a mathematical operation that combines the input signal and *transfer function* to obtain the modified, or output, signal. In space-domain *filtering*, each input amplitude *sample* is replaced by a value scaled proportionally to the *impulse response* of a particular *filter*, so the total output represents a linear combination, or weighted sum, of all scaled values in accordance with the length of the convolution operator and number of its sample points. In spectral-domain filtering, the output signal represents the result of (a) multiplying the amplitude values in the *amplitude spectrum* of an input signal and corresponding amplitudes of a convolution operator on a *wavenumber*-by-wavenumber basis and (b) adding the *phase* values in the *phase spectrum* of an input signal and corresponding phase values of a convolution operator on a wavenumber-by-wavenumber basis. **2.** a change in the *amplitude* and *wavelength* content of an *anomalous field* resulting from the application of a *convolution filter*. See also *Fourier convolution theorem*, *spectral domain*, and *space domain*.

convolution filter: 1. a space-domain or spectral-domain *filter* for which the output is defined by the *convolution* of the input signal with the *impulse response* of the filtering operator, i.e., *transfer function*. For example, a *Hilbert transform* could be considered a convolution filter which changes the *phase* value of an input signal by 90° without changing its *amplitude*. **2.** a *convolution grid filter* in the context of a space-domain convolution applied for *smoothing* or *residualization* of gridded gravity and magnetic data. See also *grid*, *space domain*, and *spectral domain*.

convolution grid filter: a space-domain *filter* which replaces a *grid* value in each *grid cell* by a linear combination, or sum, of values from the adjacent cells. A filter *kernel* is commonly a square *sliding window*, the size of which is defined by the number of grid cells within the sliding window, such as 3 × 3 or 5 × 5. The larger the window, the longer the *wavelength* components of the *anomalous field* obtained by application of this filter. A convolution grid filter is applied in two modes of operation: *smoothing* and *residualization*. The former removes the high-frequency field components and *noise* in order to enhance the mid- and long-wavelength components, whereas the latter subtracts the smoothed grid from the original grid to highlight the short- and mid-wavelength components. See also *Gaussian filter*, *smoothing filter*, and *space domain*.

convolution matrix: a *kernel* in the context of a particular *convolution* operator presented in the format of a *matrix*.

convolution theorem: See *Fourier convolution theorem*.

convolutional neural network (CNN): the *artificial neural network (ANN)* wherein the interacting computer algorithms recognize the targeted patterns via *convolution* of different attributes with multiple sets, or layers, of the original and enhanced geophysical and geological data gridded over the same area. Convolution kernels are optimized for different representations of the layers and are used to identify relationships between the patterns at different scales of *lateral resolution* of the input data to integrate their multiscale characteristics into the deep learning process (McMillan et al., 2021). See also *attribute*, *enhanced processing*, *gridding*, *kernel*, *pattern*, *target*, and *machine learning (ML)*.

coordinate registration: placing the origin of gravity or magnetic *survey* area coordinates at a specific position which is defined by its *XY coordinates* and held constant for all grids and maps produced in an *interpretation* project. See also *grid* and *georeferencing*.

coordinate system: a reference frame relative to which observation points are identified on the Earth's surface and on a map or *grid*. A geographic coordinate system defines the *three-dimensional (3D)* geographic coordinates, i.e., latitude, longitude, and *elevation*, on the curved Earth's surface, and it is implemented in the satellite-based *Global Positioning System (GPS)*. A projected, or planar, coordinate system defines the horizontal position of observation points on the flat 2D surface of a map or grid. Projected coordinates, or map coordinates, are commonly defined by *XY coordinates*, where the *X*- and *Y*-coordinate values represent the *easting* and *northing* distances, respectively, in the standard *Universal Transverse Mercator (UTM)* planar coordinate system. See also *false easing* and *false northing*.

core field: *main geomagnetic field* in the context of its origin from electric currents generated by the convection-driven conductive fluids inside the liquid outer core of the Earth.

core-field correction: same as *IGRF correction* with an emphasis on a correction for the dominant component of *Earth magnetic field* generated inside the Earth's core and referred to as the *main geomagnetic field* or *core field*.

Coriolis acceleration: the Earth-rotation-derived, velocity-, direction-, and latitude-dependent *inertial acceleration* resulting from the *Coriolis effect* on gravity measurements on an *observation platform* moving within a coordinate system fixed on the rotating Earth's surface. The Coriolis acceleration $\mathbf{a}_{CA}$ is a *vector* quantity with *magnitude* expressed as

$$|\mathbf{a}_{CA}| = a_{CA} = 2\omega V \sin\theta,$$

where ω is *angular velocity* of the Earth's rotation, V is velocity of the observation platform, and θ is latitude. The vertical component of Coriolis acceleration depends on the direction of the platform movement, and its magnitude is defined as

$$a_{VCA} = 2\omega V \cos\theta \sin\alpha,$$

where α is *heading*, i.e., direction of the platform movement relative to geographic north. The vertical component of Coriolis acceleration is a part of the *Eötvös acceleration*. See also *equivalence principle*.

Coriolis effect: the motion-related inertial effect described by the deflection of an *observation platform* moving on the rotating Earth's surface outward from the axis of the Earth's rotation (such as in the platform movement from the geographic poles to the equator) to the direction which is opposite to the Earth's rotation direction. As a result, the platform path on a coordinate plane fixed on the rotating Earth's surface becomes curvilinear, i.e., its straight-line path on the non-rotating surface is transformed by the Coriolis effect into a curvilinear path on the rotating surface. The deflection amount is proportional to the latitude-dependent *angular velocity* of the Earth's rotation and velocity of the observation platform. The Coriolis effect interferes with *centrifugal force* and, therefore, produces the inertial *Coriolis acceleration*. In accordance with the *equivalence principle*, the vertical component of the Coriolis acceleration contributes to the gravity values measured on a moving observation platform, such as a *survey* ship or aircraft.

Coriolis term: the direction-, velocity-, and latitude-dependent vertical component of *Coriolis acceleration* in an analytical expression of the *Eötvös correction*. The Coriolis term is subtracted from *observed gravity* values obtained from measurements on a westward-moving *observation platform* and added to observed gravity values on an eastward-moving platform. See also *centrifugal term* and *Eötvös acceleration*.

correction density: a *density* value which is the best approximation of rock densities above or below the sea level or *reference ellipsoid* that is used in calculation of *Bouguer correction* and *terrain correction*. For local survey data onshore, the correction density is selected from results of the *Nettleton test*, whereas in regional studies the averaged crustal density 2.67 $g \cdot cm^{-3}$ is commonly used. For shipboard (marine) survey data, the seawater density 1.03 $g \cdot cm^{-3}$ is replaced in the Bouguer correction formula by a value which is equal to the difference between the assumed density of the sea-floor sediments and seawater density, and it is called *replacement density*. Also referred to as *reduction density*.

corrections: See *gravity data corrections* and *magnetic data corrections*.

correlation filter: a *profile*-based spectral-domain filter which extracts the correlative components of datasets for the same area by using the calculated correlation coefficient $r(k)$ between the same *wavenumber* (k) components of Fourier-transformed data. The filter is designed to enhance either the direct correlation ($0 < r(k) \leq 1$) or inverse correlation ($-1 \leq r(k) < 0$) between the datasets. A correlation filter provides a quantitative estimate of the correlation between variations in one dataset, such as the *field* of the gravity or magnetic anomalies, with variations in another dataset, such as the *topography*, *bathymetry*, rock outcrops, heat flow, or hydrocarbon seepage, in order to minimize the ambiguity of *interpretation*. Grid-based correlation analysis of gravity and magnetic fields involves (a) calculation of the *first vertical derivative* of a gravity field and (b) application of *reduction-to-pole (RTP)* to the *total magnetic intensity (TMI)* data to provide the same phases of gravity and magnetic anomalies for the same sources. See also *decorrelation method*, *Fourier transform*, *grid*, *outcrop*, *phase*, and *Poisson theorem*.

corrugations: low-*amplitude* wave-like ridges, or stripes, of artificial anomalies on a gravity or magnetic *grid* that are oriented along the airborne and marine (shipboard) *survey* lines. Corrugations represent (a) residual errors in the *leveling* of line-oriented data, (b) systematic offsets, or level shifts, of measured values caused by the mislocation of survey lines and shifts in *survey datum*, and (c) *gridding* errors. Corrugations also can be related to unsuppressed variations of the *external magnetic field* of the Earth, such as *diurnals*. The linear pattern of the wave-like stripes orthogonal to the orientation of closely spaced *traverse lines* is known as the *washboard effect*. Corrugations are eliminated by the *decorrugation* process which involves *directional filtering* and other *microleveling* procedures. Also known as *acquisition footprint*. See also *artificial anomaly*, *herringbone effect*, and *pattern*.

cosine bell filter: same as *cosine taper filter* that has a bell-shaped cosine curve in its versions of various orders, wherein the order defines a steepness of the filter mode transition: the higher the order, the steeper the transition from a pass mode to a reject mode of the filter operation.

cosine depth correction: a correction applied to the source depth value determined from the *profile* of a gravity or magnetic *anomaly* in order to compensate for the deviation of a *survey* line from the direction of the short axis of an anomaly on a *contour grid*. This correction is necessary, because all depth estimates from profiles of anomalies are based on the assumption that the extracted profile is orthogonal to the *strike* of a source body, i.e., it is extracted along the short axis of the anomaly. Cosine depth correction is defined as

$$\Delta D_{CDC} = D \cdot \cos\beta,$$

where D is the initial depth estimate and β is the angle between the profile direction and direction of the short axis of the anomaly. When the angle φ between the profile direction and direction of the source strike is considered, the cosine depth correction is referred to as a *strike correction* and expressed as

$$\Delta D_{SC} = D \cdot \sin\varphi.$$

Correction is commonly not applied when the angle φ exceeds 60°–70° and angle β is less than 30°–40° due to a negligible difference in the obtained depth estimates.

cosine roll-off filter: same as *cosine taper filter* with an emphasis on a gradual decrease to zero, i.e., roll-off of data values, in the transition zone between the pass and reject modes of the filter operation. See also *taper* and *taper filter.*

cosine taper filter: 1. a spectral-domain *grid*-based edge-smoothing filter which implements a cosine curve to provide a smooth transition of the grid values to zero in the grid *taper.* **2.** a spectral-domain high- or *low-pass filter* with a smooth cosine curve in its *roll-off window.* Differing from the regular spectral-domain filters, the cosine taper filter effectively eliminates *Gibbs ringing.* An upgraded low-pass cosine taper filter is used as a *time filter* in *airborne gravity* measurements. Also known as *cosine bell filter* and *cosine roll-off filter.* See also *edge-smoothing filters, high-pass filter, spectral domain,* and *taper filter.*

cosine transform: 1. a *Fourier transform* of the even, or symmetric, part of a *potential field* from its original waveforms in the *space domain* into a series of cosine waves in the *wavenumber domain.* The mathematical expression of a cosine transform is given by the cosine integral in the Fourier transform formula. **2.** the real part of a *complex spectrum.* See also *sine transform* and *even function.*

Coulomb law: the physical law of electric charges formulated for the conceptually equivalent model of the magnetic poles and *magnetic force* between them. It states that the *magnitude* of the magnetic force F_m between two magnetic poles is directly proportional to the strength of each pole and inversely proportional to the square of the distance between them:

$$F_m = C_m \rho_1 \rho_2 / r^2,$$

where ρ_1 and ρ_2 are magnetic pole strengths; r is distance between two poles; C_m is magnetic constant of proportionality with different values and dimensions in the cgs and SI systems – in *cgs units* $C_m = 1$ and it is dimensionless, whereas in *SI units* $C_m = \mu_0/4\pi = 10^{-7}$ *henry*·meter^{-1}; and μ_0 is *magnetic permeability* of a source-free space. Magnetic force is attractive for poles of the opposite sign and repulsive for poles of the same sign. See also *magnetic pole, magnetic pole strength,* and *Newton law of universal gravitation.*

covariance: a statistical measure of correlation between variations in two datasets that is defined in terms of deviation from their corresponding averages relative to each other. Covariance is calculated by summing the results of multiplying all differences between the individual values and their averages for each dataset and dividing the obtained sum by the total number of measured values:

$$covar(x,y) = \Sigma(x_i - m_y)(y_i - m_x)/n,$$

where x_i and y_i are individual values in each dataset; m_x and m_y are the arithmetic mean, or average, values for each dataset; and n is total number of measured values in both datasets. When the calculated covariance value is positive, both datasets show a similar *pattern* of variations: an increase in the values of one dataset correlates with an increase in the values of the other. When the covariance value is negative, the datasets show the opposite pattern: an increase in the values of one dataset correlates with a decrease in the values of the other, and vice versa. Covariance is implemented in the *principal component analysis (PSA)* of multidimensional datasets. See also *dimension* and *variance.*

covariance matrix: a square symmetric *matrix* where the elements are *covariance* values calculated for two separate data dimensions, for which the number of rows and columns is defined by the dimensionality of a particular dataset. For example, a 3 × 3 covariance matrix for the three-dimensional (x, y, z) dataset could be expressed as

$$\begin{aligned} & cov(x,x)\ cov(x,y)\ cov(x,z), \\ cov(x,y,z) = & \ cov(y,x)\ cov(y,y)\ cov(y,z), \\ & cov(z,x)\ cov(z,y)\ cov(z,z). \end{aligned}$$

Matrix elements in the main diagonal are covariances calculated for the same *dimension,* and they represent the corresponding *variance* values for each dataset. For every *observation point,* the covariance matrix is computed by crosscorrelating the data in each dimension with itself and with the data in all other dimensions, so the N-dimensional dataset will have a covariance matrix with $N \times N$ elements. An obtained covariance matrix is decomposed into pairs of *eigenvalues* and *eigenvectors* for every data dimension in the process of *principal component analysis (PCA).*

crab: the steady *heading* of a ship or aircraft at some angle to the *survey* line direction that is maintained in order to compensate for the cross-wind or cross-sea motion effects. Crab involves the rotational motions, known as *pitch, roll,* and *yaw,* producing the inertial accelerations of an *observation platform.* See also *inertial acceleration* and *stabilized platform.*

creep: the gradually increasing deformation of a mechanical (metal or quartz) spring in a *spring gravimeter* caused primarily by thermal expansion, elastic aging, and excessive movement. See also *gravimeter drift* and *tare.*

cross-check method: a method of *quality control (QC)* over *gravimeter drift* during *ground gravity* measurements along long traverses, wherein the repeated measurements at a primary *base station,* or *reoccupations,* are time-consuming. Secondary base stations are established as part of an observation sequence along traverses in order to decrease the traveltime intervals between successive reoccupations. Sometimes referred to as the *traveling-base-station method*. See also *base-check method*, *looping method*, and *traverse*.

cross-correlation corrections: same as *cross-coupling corrections* for which the calculation is based on the cross-correlation analysis of the inertial accelerations of a moving *observation platform* and gravity measurements. See also *inertial acceleration*.

cross-correlation method: a method of estimating the direction of a *total magnetization* vector in a *source body*. For the *two-dimensional (2D)* and *three-dimensional (3D)* dipping sources, it is based on a cross-correlation between the maximum values of the *total gradient (TG)* and the *vertical gradient (VG)* of the reduced-to-pole (RTP) magnetic data. The RTP calculations are made for several *magnetic inclination* values, and a true direction is defined by the inclination value which maximizes the cross-correlation (Dannemiller and Li, 2006). An upgrade of this method for application at low *magnetic latitude* and equator areas is based on the *equivalent layer* technique (Ribeiro-Filho et al., 2020). The true *magnetization direction* of a source body provides the most accurate *reduction-to-pole (RTP)* and subsequent *inversion* of magnetic *gradient* and *gradient tensor* data for parameters of a discovered *ore body*. See also *dip* and *vector*.

cross-correlation remanence: *remanent magnetization* for which the *magnitude* and direction in the *source body* are obtained with the use of (a) the *total magnetization* direction determined by the *cross-correlation method* and (b) the parameters of the *main geomagnetic field* at the time of a particular *magnetic survey*.

cross-coupling corrections: a sequence of corrections applied to gravity values measured on a moving *observation platform* in order to compensate for *cross-coupling errors* resulting from the interference of the motion-related horizontal and vertical inertial accelerations of a *survey* ship or aircraft with the onboard *gravimeter* or *gravity gradiometer* measurements. In an airborne *gravity survey*, corrections are calculated with the use of high-*accuracy* positioning data provided by *differential GPS (DGPS)*. In marine (shipboard) surveys, the ship velocity and inertial accelerations are monitored by several auxiliary channels and recorded simultaneously with the gravity data for the subsequent cross-correlation analysis and calculation of the cross-coupling corrections. Also referred to as *cross-correlation corrections* and *inertial acceleration corrections*. See also *Eötvös effect*, *inertial acceleration*, and *positioning*.

cross-coupling effect: interaction, or coupling, between the measurement system of a *gravimeter* or *gravity gradiometer* and the horizontal and vertical inertial accelerations of a moving *observation platform*. A spring or other mechanical arrangement of the *gravity sensor* translates the unsuppressed, or residual, *inertial acceleration* components into accelerations interfering with the gravity measurements, i.e., measurements of a *gravitational acceleration*. For example, during the downward motion of an *observation platform* the measured gravity values are lower, and during the upward motion they are higher than the true values. Cross-coupling effects on the gravity measurements are minimized by a *stabilized platform*. Measurement errors, or *cross-coupling errors*, are reduced to the level of *random noise* by the application of *cross-coupling corrections*. Referred to with the same meaning as *wave-motion effect* and *turbulence effect*. See also *acceleration*, *electromagnetic accelerometer*, *electrostatic accelerometer*, *equivalence principle*, and *lever arm effect*.

cross-coupling errors: errors in gravity measurements on a moving *observation platform*, such as a *survey* ship or aircraft, resulting from the *cross-coupling effect*. Cross-coupling errors are essentially eliminated by the application of *cross-coupling corrections*. See also *inertial acceleration* and *equivalence principle*.

cross-gradient angular function: an *attribute* in the *cross-gradient method* that quantifies a degree of the structural (morphological) differences between two geophysical datasets obtained in the same area, such as the gravity and magnetic *gradient* grids. At any *grid node*, the cross-gradient angular function θ is defined by the *scalar product* of the gradient vectors of two datasets that corresponds to the angle between the *gradient vector* directions. For the *three-dimensional (3D)* case, it is expressed as

$$\theta(x,y,z) = \cos^{-1}\{\nabla m_1(x,y,z)\cdot\nabla m_2(x,y,z)/ \\ |\nabla m_1(x,y,z)|\cdot|\nabla m_2(x,y,z)|\},$$

where ∇ is gradient vector operator *del* and m_1 and m_2 are two geophysical datasets. Structural differences between datasets are the greatest when $\theta = 90°$, i.e., when the gradient vectors are orthogonal (Gallardo and Meju, 2004). See also *grid* and *cross-gradient function*.

cross-gradient function: an *attribute* in the *cross-gradient method* that quantifies a degree of the structural (morphological) correlation between two geophysical datasets obtained in the same area, such as the gravity and magnetic *gradient* grids. At any *grid node*, the cross-gradient function $\boldsymbol{t}$ is defined by the *vector product* of the gradient vectors of two datasets, and for the *three-dimensional (3D)* case it is expressed as

$$\boldsymbol{t}(x,y,z) = \nabla m_1(x,y,z)\times\nabla m_2(x,y,z),$$

where ∇ is *gradient vector* operator *del* and m_1 and m_2 are two geophysical datasets. Structural correlation between datasets is the greatest when $\boldsymbol{t}(x, y, z) = 0$, i.e., when the juxtaposed, or side by side, gradient vectors are parallel. For 3D datasets, the x-, y-, and z-components of the cross-gradient function are defined as

$$t_x = (\partial m_1/\partial y)(\partial m_2/\partial z) - (\partial m_1/\partial z)(\partial m_2/\partial y),$$
$$t_y = (\partial m_1/\partial z)(\partial m_2/\partial x) - (\partial m_1/\partial x)(\partial m_2/\partial z),$$
$$t_z = (\partial m_1/\partial x)(\partial m_2/\partial y) - (\partial m_1/\partial y)(\partial m_2/\partial x).$$

Each component quantifies a structural correlation in the plane orthogonal to the component direction, e.g., the t_z component is used to measure a degree of correlation between the horizontal gradients of gridded *vertical gravity* and RTP magnetic data (Gallardo and Meju, 2004). See also *gridding*, *reduction-to-pole (RTP)*, and *cross-gradient angular function*.

cross-gradient joint inversion: *joint inversion* of gravity and magnetic data constrained by the *cross-gradient function* which quantifies a degree of structural correlation between the gridded fields of the gravity or *gravity gradient* data and reduced-to-pole (RTP) magnetic data (Fregoso and Gallardo, 2009). See also *constraints*, *gridding*, and *reduction-to-pole (RTP)*.

cross-gradient method: a *grid*-based *enhanced processing* technique utilizing the concept of a structural, or morphological, correlation between different physical properties, such as *density* and *magnetization*, over the same area. At any *grid node*, variations of properties are characterized in terms of *magnitude* and specific direction of *gradient* vectors in the gravity and magnetic fields, so the commonality in distribution of their changes defines a degree of the correlation, or similarity, between both anomalous fields. Structural correlation between the gradient fields is quantified by the *cross-gradient function*. Structural differences are quantified by the *cross-gradient angular function*. Variations of these functions are visualized on two grids called the magnitude image and angle image, respectively. Both images are complementary to each other in enhancing the structural edges and lithological boundaries in the subsurface (Cevallos et al., 2012). Cross-gradients assist in the *integrated interpretation* of data obtained by different geophysical methods even in the absence of a direct relationship between the physical properties utilized by these methods. See also *edge*, *gradient vector*, and *cross-gradients*.

cross-gradients: *vector* quantities which represent gradients of the anomalous fields associated with the distribution of the different physical properties, such as *density* and *magnetization* of rocks, over the same area. Commonality in changes of the corresponding *gradient vector* magnitudes and orientations is assumed to be related to the same elements of a subsurface geological *structure*. The degree of structural, or morphological, cross-correlation is estimated by the *cross-gradient function* and *cross-gradient angular function* in the process of *integrated interpretation* with the use of the *cross-gradient method*. See also *anomalous field*, *gradient*, and *magnitude*.

cross-line gradient: 1. the *gradient* of a gravity or magnetic field calculated from a difference between *total field* values measured at adjacent *survey* lines. **2.** a *full tensor gradient (FTG)* element representing the rate of change of a particular *vector* component of a gravity or magnetic field along two directions orthogonal to the orientation of this vector component. There are six cross-line gradients in the symmetric 9-element *FTG matrix* of a *gravity gradient* field:

$$\begin{aligned} g_{xz} &= \partial g_x/\partial z = \partial^2 U/\partial x\partial z \\ &= g_{zx} = \partial g_z/\partial x = \partial^2 U/\partial z\partial x, \\ g_{yz} &= \partial g_y/\partial z = \partial^2 U/\partial y\partial z \\ &= g_{zy} = \partial g_z/\partial y = \partial^2 U/\partial z\partial y, \\ g_{xy} &= \partial g_x/\partial y = \partial^2 U/\partial x\partial y \\ &= g_{yx} = \partial g_y/\partial x = \partial^2 U/\partial y\partial x, \end{aligned}$$

where g_x, g_y, g_z are mutually orthogonal components of *gravity vector* **g** in *Cartesian coordinates*, and U is *gravitational potential*. See also *in-line gradient*, *longitudinal gradient*, and *transverse gradient*.

cross product: same as *vector product* that utilizes a diagonal cross " × " to denote the multiplication of two vectors.

crossover depth: the vertical depth interval, or zone, within which *salt* and surrounding sediments have nearly the same *density* and, therefore, there is no lateral *density contrast* to generate a *gravity anomaly*. Commonly referred to as *nil zone* or *salt nil zone*.

crossover difference: the difference between the measured gravity or magnetic values at the intersection, or *crossover point*, of traverse and tie lines in an airborne, marine (shipboard), or submarine survey. Calculation of the *standard deviation (SD)* of all crossover differences provides an estimate of the internal consistency of measurements and their error level to determine *survey accuracy*. Because the crossover differences represent the sum of errors of two intersecting lines, the obtained SD value is divided by $\sqrt{2}$ for a quantitative evaluation of the *accuracy* of measurements along each survey line. Referred to with the same meaning as *mis-tie*. See also *traverse lines*, *tie lines*, and *rms crossover difference*.

crossover error: See *crossover difference*.

crossover leveling: *leveling* in the context of a *mis-tie* minimization at the crossover, or tie, points of line-oriented gravity or magnetic *survey* data prior to *gridding*.

crossover point: same as *tie point* with an emphasis on a coordinate point where *traverse lines* and *tie lines* crossover, or intersect, each other.

crossover value: the value of the gravity or magnetic field measured at the *crossover point*, or *tie point*, which is a coordinate point in the *survey* area where *traverse lines* and *tie lines* crossover, or intersect, each other. Crossover values are used in the process of *leveling* and for estimates of the *accuracy* of airborne, marine (shipboard), and submarine surveys. See also *mis-tie*.

cross-track aliasing: *spatial aliasing* in the context of the *gridding* artifacts resulting from *undersampling* of anomalies striking across several parallel ship tracks or flight lines in a marine (shipboard) or airborne *survey*.

cross-track wander: displacement of a towed *marine magnetometer* or *towed deep ocean gravimeter (TOWDOG)* in the direction orthogonal to the orientation of the marine (shipboard) *survey* lines which are also referred to as ship tracks.

crust: the outermost layer of the Earth bounded at its bottom by the *Mohorovičič discontinuity*, or *Moho boundary*, which corresponds to the top of the *mantle*. There are two basic types of Earth crust: continental and oceanic. Continental crust is highly heterogeneous and predominantly granitic, and its thickness varies from approximately 25 km to more than 70 km under high mountain ranges with an average thickness of 35–40 km. The average *density* of continental crust is 2.75–2.80 g·cm^{-3}. Oceanic crust is predominantly basaltic, and its lateral compositional variations (thickness and density) are relatively small, so the total thickness ranges between 6 km and 9 km with a median of 7 km. The average density of oceanic crust is 2.95–3.00 g·cm^{-3}. The estimate of average density in the uppermost mantle is approximately 3.35–3.45 $g \cdot cm^{-3}$. Crust composes less than 0.1% of the total Earth volume. See also *continental-oceanic boundary (COB)* and *magnetic crust*.

CRUST 1.0: the 1° × 1° global *grid* model of the Earth's *crust*. For each *grid cell*, values of the bounding depth, *density*, and compressional V_P and shear V_S seismic velocities are presented for eight model layers: water, ice, three sedimentary crust layers (upper, middle, lower), three metamorphic-igneous, or crystalline, crust layers (upper, middle, lower), and the upper *mantle* below the *Moho boundary*. Moho depth estimates are based on 1° averages of the worldwide crustal thickness data. The crustal database represents, essentially, the interpretive results obtained from seismic reflection and refraction data. The CRUST 1.0 supersedes the CRUST 2.0 model.

crustal antiroot: See *isostatic antiroot*.

crustal correction: a gravity correction which accounts for the gravity effect of the Earth's *crust*. It is calculated with the use of *satellite gravity* data, global crustal models, and average crustal density 2.67 $g \cdot cm^{-3}$, and it is applied to obtain the global-scale lithospheric mantle anomaly known as the *mantle residual anomaly* and *mantle gravity anomaly* (Herceg et al., 2016). See also *CRUST 1.0*.

crustal flexure hypothesis: See *Vening Meinesz hypothesis*.

crustal magnetic field: essentially the static component of the *Earth magnetic field* quantified by a *vector* sum of *induced magnetization* and *remanent magnetization* of rocks within the Earth's *crust*. Crustal magnetic field values are obtained from *observed magnetics* by application of *IGRF correction* and *diurnal correction*. Its *magnitude* varies from a few *nanotesla (nT)* to 50–100 nT over the low-magnetized sedimentary basins and up to several thousand nanoteslas over the highly magnetized crystalline *basement* outcrops with more than 10,000 nT above the near-surface iron ore deposits and certain ultramafic igneous rock complexes. The crustal magnetic field is often referred to with the same meaning as *total magnetic intensity (TMI)* or *total magnetic anomaly (TMA)*, because the area extent of the regular magnetic surveys is sufficient for a proper *sampling* of the anomalous *wavelength* contributions from the upper crust only. When the Earth's magnetic field is represented in terms of *spherical harmonics*, the crustal components dominate in the spherical harmonic degrees 14 and higher (wavelengths $\leq$ 2800–3000 km), whereas the lower degrees are dominated by the longer wavelength components generated in the Earth's *mantle* and core. See also *Curie law*, *external magnetic field*, and *lithospheric magnetic field*.

crustal root: See *isostatic root*.

cryogenic gravimeter: same as *superconducting gravimeter* with an emphasis on the extremely low, or cryogenic, temperatures required to maintain the superconductivity of a *proof mass*.

cryogenic magnetometer: See *SQUID magnetometer*.

cryogenic optical gravimeter: a *superconducting gravimeter* that utilizes a high-*precision* laser optical system for measurements of the vertical displacement of the levitating superconducting sphere from its null position to determine *relative gravity* values. See also *magnetic levitation*.

crystalline basement: same as *basement* with an emphasis on its heterogeneous composition from crystalline (igneous and metamorphic) rocks with highly variable densities and *magnetization*.

cultural anomalies: See *cultural magnetic anomalies*.

cultural data: a digital information about the industrial, agricultural, commercial, and communication infrastructure objects which represent *cultural magnetic sources* in the *magnetic survey* area. See also *base map*.

cultural editing: elimination of *cultural magnetic anomalies* which interfere with *aeromagnetic survey* measurements and contaminate acquired data. There are two main approaches to the cultural editing of the aeromagnetic data: (1) automated line- or *grid*-based *filtering* using the *wavelet transform* or *pattern recognition* algorithms and magnetic signatures of known *cultural magnetic sources* and (2) manual editing of line-oriented data using the flight video records, pilot logs, *base map*, and other available information about the survey area. Also referred to as *deculturing* and *cultural suppression*. See also *acquisition*, *magnetic signature*, and *video reviewing*.

cultural magnetic anomalies: non-geological magnetic anomalies generated by man-made ferrous objects and electric current conductors, such as bridges, power lines, wellheads and well casings, cathodically protected and unprotected pipelines, railways, highways, water and fuel tanks, etc. Cultural anomalies could be orders of magnitude stronger than the magnetic anomalies of a geological origin, and they contaminate *magnetic survey* data. Cultural anomalies are identified by (a) high-frequency, or short-*wavelength*, high-*amplitude* content with spike-like shapes indicating their localized surface origin, and (b) correlation with ground objects identified in flight video records and mapped man-made objects. They are eliminated in the process of *cultural editing*, also known as *deculturing*. Also referred to as *artificial magnetic anomalies*, *magnetic artifacts*, and *cultural noise*. See also *magnetic anomaly*, *despiking*, and *video reviewing*.

cultural magnetic effects: surface and near-surface magnetic effects produced by man-made ferrous objects and electric current conductors generating, mostly, short-*wavelength* high-*amplitude*, often spike-like, magnetic anomalies with widely variable amplitudes up to hundreds of *nanotesla (nT)* that are referred to as *cultural magnetic anomalies*.

cultural magnetic sources: man-made ferrous objects and electric current conductors, such as bridges, power lines, wellheads and well casings, automatic oil-pumping units, cathodically protected and unprotected pipelines, railways, highways, water and fuel tanks, large metal structures, etc., that generate *cultural magnetic anomalies*.

cultural noise: **1.** *cultural magnetic anomalies* in the context of their contaminating effect on *magnetic survey* data. **2.** man-made *noise* caused by *cultural magnetic effects* interfering with *magnetic survey* measurements.

3. microvibrations caused by ground and air traffic that result in errors in *ground gravity* measurements.

cultural suppression: See *cultural editing.*

Curie depth: the depth at which rocks are heated to the *Curie temperature* and, therefore, lose almost all of their magnetic properties to become essentially non-magnetic, or *paramagnetic*. Curie depth is predicted around the lower *crust*-upper *mantle* boundary, i.e., *Moho boundary.* Because different magnetic minerals have different Curie temperatures, the Curie depth might be approximating both the rock composition boundary and the depth of equal temperatures known as *isotherm.* In the tectonically stable continental and oceanic crust regions, the averaged values of the Curie depth are estimated at 30–40 km below sea level and 6–7 km below the sea bottom, respectively. Also referred to as *Curie isotherm depth.*

Curie depth estimation: a regional-scale approximation of the *Curie depth* obtained in the process of *spectral analysis* of magnetic data. Curie depth is assumed to be the bottom, or basal surface, of a randomly magnetized horizontal layer. Its depth estimates are obtained by splitting the *grid* of an *IGRF*-corrected magnetic field into large overlapping sub-regions and analyzing the long-*wavelength* end of the *radial power spectrum* of each sub-region (Dolmaz et al., 2005; Ross et al., 2006). The reference value of the Curie depth at locations with a known geothermal gradient is obtained from the difference between the *Curie temperature* (°C) and the surface temperature (°C) divided by a geothermal gradient value ($°C \cdot km^{-1}$). See also *power spectrum.*

Curie isotherm depth: same as *Curie depth* that represents a depth of equal temperatures, or *isotherm*, at which rocks containing magnetic minerals become essentially non-magnetic, i.e., *paramagnetic.*

Curie law: the law of *paramagnetic susceptibility* that applies to the *magnetization* of all dominant types of rocks. According to this law, *magnetic susceptibility* and, hence, magnetization of rocks are inversely proportional to their temperature, because thermal effects tend to oppose the alignment of the atomic magnetic moments whose aggregated effects define the magnetic properties of the rock-forming minerals. See also *magnetic moment*, *paramagnetism*, and *Curie-Weiss law.*

Curie point: the *Curie temperature* in the context of a particular temperature value above which rocks become essentially non-magnetic, i.e., *paramagnetic.* See also *Neel point.*

Curie temperature: the temperature above which rocks containing magnetic minerals lose almost all of their remanent and induced magnetic properties and become essentially non-magnetic, i.e., *paramagnetic.* Above this temperature, the thermal energy of atoms is higher than the coupling energy of the atomic magnetic moments, whose aggregated effects define the *magnetization* of rocks at the atomic level. Curie temperature for most magnetic minerals is within the range of 550°–580°C to 670°C, where the lower limit represents the Curie temperature of *magnetite.* In the tectonically stable continental and oceanic *crust* regions, the Curie temperatures are reached at 30–40 km below sea level and 6–7 km below the sea bottom, respectively. In the high heat flow continental regions, it may occur at depths less than 10–15 km. Also known as *Curie point* and *demagnetization temperature.* See also *induced magnetization*, *remanent magnetization*, and *magnetic dipole moment.*

Curie-Weiss law: the physical law stating that the *magnetic susceptibility* k of rocks heated above the *Curie temperature* is inversely proportional to the difference between the present temperature of rocks and their Curie temperature; it can be expressed as

$$k = C/(T_P - T_C),$$

where C is a constant of proportionality, T_P is the present rock temperature, and T_C is the Curie temperature. Above the Curie temperature, the great majority of rocks exhibit essentially non-magnetic, or *paramagnetic*, properties. See also *Curie law* and *paramagnetism.*

curl: a *vector* quantity representing the *three-dimensional (3D)* spatial *derivative* of a *vector field.* Curl is a measure of the circulation within a *field.* In *Cartesian coordinates*, it is expressed as

$$\nabla \times \mathbf{A} = (\partial A_z/\partial y - \partial A_y/\partial z)\mathbf{i} + (\partial A_x/\partial z - \partial A_z/\partial x)\mathbf{j} + (\partial A_y/\partial x - \partial A_x/\partial y)\mathbf{k},$$

where ∇ is *gradient* operator *del*; A_x, A_y, A_z are components of the vector field **A** along the mutually orthogonal x-, y-, z-axes, respectively; **i**, **j**, **k** are unit vectors parallel to these axes; and × denotes *vector product.* A vector field with a nonzero curl is called a *vortex field.* A vector field with the zero curle at each space point is called the *irrotational field.* See also *Helmholtz theorem.*

curvature: 1. a measure of deviation of the *two-dimensional (2D)* curve from a straight line or *three-dimensional (3D)* continuous surface from a planar, or flat, surface. The amount of deviation, or bending, at each curve or surface point is defined by the reciprocal of the radius of a circle or sphere, respectively, which is tangent to the 2D curve or 3D surface at a given point. For the *one-dimensional (1D)* function or plane curve defined by $y = f(x)$, the curvature is expressed as

$$K = 1/R = (\partial^2 y/\partial x^2)/[1 + (\partial y/\partial x)^2]^{3/2},$$

where R is a curvature radius. When the first-order derivative $\partial y/\partial x$ is small, the curvature is approximated by the second-order *derivative.* **2.** a measure of the rate of change of the *gradient* of an anomalous gravity or magnetic field quantified by the second-order horizontal derivatives of the *anomalous field* and third-order derivatives of the *gravitational potential* or *magnetic potential.* Curvature is a characteristic of 3D surface deformation at a particular point: the more deformed the surface, the larger the curvature value. In a *local reference frame*, the curvature of a gravitational *equipotential surface* at any direction is proportional to the diagonal component of the *rotated gradient tensor* in that direction and inversely proportional the gravity field strength, which is defined by the length, or *magnitude*, of the *anomalous gravity vector* at the observation point. Equipotential and field curvatures are strongly influenced by the depth of a *source body* and by the *interference of anomalies*: the deeper the source, the smoother the curvature and the stronger the interference of anomalies (Li, X., 2015; Li, Y., 2018). See also *curvature attributes*, *curvature-depth method*, *curvature gradients*, *curvature tensor*, and *gravity field curvature*.

curvature analysis: analysis of calculated and gridded *curvature attributes* for the identification of: (a) *lineaments* associated with structural discontinuities, such as lithological contacts, faults, fracture zones, and boundaries of the local intrasedimentary and *basement* structures for oil-and-gas *exploration*; (b) shapes of detected ore bodies, distribution of *fault* and fracture systems, *dike swarm,* and shallow igneous bodies associated with the prospective mineralization zones for mineral exploration; and (c) intrasedimentary structural shapes, such as anticlines, synclines, domes, basins, and planar surfaces. See also *anticline*, *basin*, *contact*, *gridding*, *igneous body*, *ore body*, and *syncline*.

curvature attributes: quantitative characteristics of the *curvature* of a three-dimensional (3D) *equipotential surface* described by combinations of the second-order horizontal derivatives of the gravity or magnetic *potential* that are known as *curvature gradients*. For the gravitational equipotential surface, there are (a) *apparent curvature attributes* calculated from *gravity gradient tensor (GGT)* components in the same fixed Cartesian coordinate frame used for measurements over the *gradient survey* area and (b) *true curvature attributes* calculated from components of the gravity gradient tensor, whose fixed Cartesian coordinate frame of the original measurements is rotated at each observation or *grid* point to a *local reference frame* which varies from point to point. Curvature attributes are very sensitive to lateral changes in the *density contrast* and *magnetization contrast* of rocks, and their gridded values are implemented in the analysis of source configurations, identification of structural shapes (anticlines, synclines, domes, basins, and planes), and high-*resolution* mapping of *lineaments*. Resolving power, or *lateral resolution*, of curvature attributes and their capability of structural delineation decrease with the increasing depth of sources, because the equipotential curvatures of deep-sourced anomalies are smoother and the *interference of anomalies* is stronger than those of shallow-sourced anomalies. For gravity data, the commonly calculated curvature attributes are *curvedness*, *differential curvature*, *Gaussian curvature*, *maximum curvature*, *minimum curvature*, *mean curvature*, and *shape index*. For magnetic data, the curvature attributes are computed from the *RTP magnetic field* or *pseudogravity* data with the assumption of a negligible *remanent magnetization* of the anomaly-generating sources (Li, X., 2015; Li, Y., 2018). See also *anomalous field*, *anomalous potential*, *Cartesian coordinates*, *local reference frame*, and *rotated gradient tensor*.

curvature correction: same as *Bullard B correction* that represents a gravity correction for the Earth's surface *curvature* which is not compensated by *simple Bouguer correction*.

curvature-depth method: a *grid*-based semi-automated method of estimating the depth to gravity and magnetic sources approximated by the *ideal source* geometries of *contact*, horizontal *thin sheet*, *line source*, and *point source*. The method utilizes gridded special functions, namely *horizontal gradient amplitude (HGA)*, *analytic signal amplitude (ASA)*, and *local wavenumber (LW)*, that have their maxima directly above these types of sources. Source depth is estimated with the use of a *curvature matrix* composed of the coefficients of a local quadratic surface approximating in the least-squares sense nine points of a 3 × 3 grid *window* centered at the maximum value of a particular special function. Coefficients are computed from the closed-form analytical expressions and compose the curvature matrix at each *grid* node. Matrix-derived *eigenvalues* and *eigenvectors* are used to locate the ridges of the *two-dimensional (2D)* sources and estimate their depths and strikes. Obtained discrete source depth locations follow the contacts, faults, and other elements of a subsurface *structure* (Phillips et al., 2007). See also *least-squares fit*, *ridge*, and *curvature tensor*.

curvature effect: the gravity effect of *curvature* of the Earth's surface on gravity measurements and calculation of a *simple Bouguer correction*. It is quantified by the numerical difference between the *horizontal slab effect* and the *spherical cap effect,* and it is compensated by the application of a *Bullard B correction*. The curvature effect is compensated in high-*accuracy* ground gravity surveys and regional-scale gravity surveys covering a broad range of surface elevations. Also referred to as *Bullard B effect*. See also *elevation*, *LaFehr formula*, and *Cogbill formula*.

curvature eigenvalues: 1. *Hessian eigenvalues* obtained by *eigenvalue decomposition* of a *Hessian matrix* at each *grid* point of an anomalous gravity or magnetic *field*. A curvature eigenvalue is positive when the field-approximating local quadratic surface is curved upward along a corresponding eigenvector, and it is negative in the case of downward curving. **2.** *eigenvalues* obtained by eigenvalue decomposition of an *FTG matrix*. For gravity sources with shapes approximated by geometries of the point or spherical *mass* and infinite *horizontal cylinder*, the eigenvalue-paired *eigenvectors* define the orientation of axes in a *local reference frame* to which the *gravity gradient tensor (GGT)* is rotated at each observation or *grid* point for calculation of the *true curvature attributes* of the gravitational *equipotential surface*. See also *anomalous field*, *infinite length*, and *rotated gradient tensor*.

curvature gradients: the horizontal gradients of the horizontal components of an anomalous gravity or magnetic field that quantify the *curvature* parameters of an *equipotential surface*. Mathematically, they are defined by the second-order horizontal derivatives of the gravitational or magnetic *potential P* and represented by the corresponding *gradient tensor* elements as

$$T_{xx} = \partial^2 P/\partial x^2; \; T_{yy} = \partial^2 P/\partial y^2; \; T_{xy} = \partial^2 P/\partial x \partial y.$$

Curvature gradients are implemented, primarily, in the gravity and *gravity gradient* data analysis and *interpretation*, where they are denoted by g_{xx}, g_{yy}, and g_{xy}. Combinations of curvature gradients are used in the calculation of *curvature attributes*. See also *anomalous field*, *derivative*, *gradient*, and *differential curvature gradients*.

curvature gravity gradient tensor: 1. a *gravity gradient tensor (GGT)* composed of the second-order horizontal derivatives of an anomalous *gravity field* that are measured or calculated at each observation or *grid* point in a *survey* area and presented in the *format* of a symmetric two-dimensional (2D) *Hessian matrix*. **2.** a *rotated gradient tensor* for which the components g_{xx}, g_{yy}, g_{xy} are known as *curvature gradients* and used for the calculation of *true curvature attributes* of the gravitational *equipotential surface* at each observation or grid point. See also *anomalous field*, *curvature*, *derivative*, *gradient*, and *apparent curvature attributes*.

curvature matrix: 1. a two-dimensional (2D) *Hessian matrix*. **2.** a symmetric *matrix* with elements that are coefficients of a local quadratic surface which fits in the least-squares sense nine points of a 3 × 3 grid *window* centered at the maximum values of the gridded special functions that have their maxima directly above the *source body*. Computed coefficients are used for (a) estimates of source body depth by the *curvature-depth method* and (b) calculation of *curvature attributes*. See also *curvature* and *gridding*.

curvature of gravity: 1. *gravity field curvature*. **2.** a characteristic of a *gravitational field* that describes the deviations of the *three-dimensional (3D)* shape of its *equipotential surface* from a planar, or flat, surface. Its parameters are estimated with the use of *curvature gradients* and quantified by *curvature attributes*. See also *curvature*, *curvature eigenvalues*, and *curvature gravity gradient tensor*.

curvature tensor: 1. a *tensor* in the *format* of a symmetric two-dimensional (2D) *Hessian matrix* with four elements that are second-order horizontal derivatives T_{xx}, T_{yy}, $T_{xy} = T_{yx}$ of an *anomalous field* and used for estimates of its local *curvature* parameters. **2.** a *rotated gradient tensor* composed of the second-order derivatives of *gravitational potential* that are used for the calculation of *true curvature attributes*. **3.** a tensor in the format of a *curvature matrix* implemented in the *curvature-depth method* for the calculation of a *source body* depth. See also *anomalous field* and *apparent curvature attributes*.

curvature variometer: a *gravity gradiometer* system for measuring the relative changes, or variations, of *differential curvature* components, whose absolute values are quantified by *differential curvature gradients*. See also *horizontal variometer*, *full tensor gravity gradiometer*, and *partial tensor gravity gradiometer*.

curve matching: a method of *forward modeling* that is based on comparing the *model*-calculated gravity or magnetic field of a targeted geological *structure* with the field of a certain structure known from theory or experience. Obtaining a close fit between the compared fields implies that the assumed parameters of the modeled object are similar to those of the known object. Due to the inherent ambiguity of *modeling*, the validity of such an implication depends on the available *constraints*. See also *inverse modeling*, *inverse problem*, and *target*.

curvedness: a *curvature* attribute of the gravitational or magnetic *equipotential surface* that can be expressed as

$$\begin{aligned} K_{cdn} &= [(K_{max}^2 + K_{min}^2)/2]^{1/2} \\ &= (2g_{xx}^2 + 2g_{yy}^2 + 4g_{xy}^2)^{1/2}/2g_z, \\ K_{cdn} &= (2T_{xx}^2 + 2T_{yy}^2 + 4T_{xy}^2)^{1/2}, \end{aligned}$$

where k_{max} is *maximum curvature*; k_{min} is *minimum curvature*; g_{xx}, g_{yy}, g_{xy} are gravity *curvature gradients*; g_z is *vertical gravity* calculated by the *vertical integral* of the g_{zz} component of the *gravity gradient tensor (GGT)* at each observation or *grid* point; and T_{xx}, T_{yy}, T_{xy} are *magnetic gradient tensor* components obtained from measurements by a *magnetic gradiometer* or calculated from the *RTP magnetic field* or *pseudogravity* data with the assumption of a negligible *remanent magnetization* of the anomaly-generating sources (Li, 2015). In integration with other *curvature attributes*, gridded curvedness values are used for the high-*resolution* mapping of structural discontinuities, such as faults and lithological contacts, especially those with low *density* and *magnetization* contrasts across them. See also *attribute*, *contact*, *fault*, *density contrast*, *magnetization contrast*, and *rotated gradient tensor*.

cutoff: a space-domain or spectral-domain *filter* parameter that defines its operational range in the pass mode. For example: (a) a 3–6 km *band-pass filter* has the low cutoff at 3 km and high cutoff at 6 km, so that all *wavelength* components of the gravity or magnetic field between 3 km and 6 km will be passed through this filter without suppression, whereas the wavelengths outside the 3–6 km range will be maximally suppressed; (b) a 5 km *low-pass filter* has its cutoff at 5 km, so that all wavelengths longer than 5 km will be passed without suppression through this filter; and (c) a 10 km *high-pass filter* has its cutoff at 10 km, so that all wavelengths shorter than 10 km will be passed through this filter without suppression. See also *cutoff wavelength* and *cutoff wavenumber*.

cutoff frequency: See *cutoff wavenumber*.

cutoff wavelength: 1. a space-domain *filter* parameter that defines a specific value of *wavelength* in the transition zone between the pass and rejection modes of a filter operation. Wavelengths shorter than the cutoff wavelength will be passed without suppression by a *high-pass filter* and maximally suppressed by a *low-pass filter*. **2.** a space-domain analogue of the spectral-domain *cutoff wavenumber*. See also *space domain* and *spectral domain*.

cutoff wavenumber: 1. a spectral-domain *filter* parameter that defines a specific value of *wavenumber* in the transition zone between the pass and rejection modes of a filter operation. Wavenumbers lower than the cutoff wavenumber will be passed without suppression by a *low-pass filter* and maximally suppressed by a *high-pass filter*. **2.** a spectral-domain analogue of the space-domain *cutoff wavelength*.

cylinder: See *horizontal cylinder* and *vertical cylinder*.

D

damping: a *gravimeter* adjustment procedure applied to reduce the *sensitivity* of its *sensor* under the highly variable dynamic environments of ocean-bottom, marine (shipboard), and airborne gravity measurements. Gravity sensors are damped against any predictable high-frequency vibrations and inertial accelerations. See also *dynamic gravity* and *inertial acceleration*.

data channel: an electronic device used for the digital recording of acquired data in the *format* of a sequence of samples indexed by time and numerical values during a gravity or magnetic *survey*. Multiple data channels record *acquisition* and *positioning* parameters, measured values of the Earth's gravity or magnetic field, *acquisition noise* level, inertial accelerations, and other survey data used in the creation of a *database*, calculation of data and instrument corrections, and subsequent *data processing* operations. See also *inertial acceleration* and *sampling*.

data conditioning: **1.** a sequence of *data processing* operations applied to line-oriented or irregularly spaced datasets in order to prepare them for subsequent *data reduction* and *gridding*. It includes but is not limited to: (a) *data extension* and, if necessary, *resampling* of line-oriented datasets; (b) suppression of *acquisition noise* in recorded data; (c) conversion of marine and airborne records from the original time-based *format* into the distance-based format of standard data processing; (d) compensation for magnetic interference, *drift*, and other instrument and measurement effects; (e) minimization of mis-ties by *leveling* and *microleveling*; (f) filling gaps in irregularly spaced observation values by interpolation; (g) rearranging irregularly spaced or line-oriented datasets into a tabulated file format, such as *ASCII format*, suitable for the creation of a *database*; and (h) *initial gridding*. **2.** *smoothing*, *padding*, *detrending*, *despiking*, and other procedures applied to gridded data in order to prepare grids for *enhanced processing*. Also referred to as *data editing*, *preconditioning*, *preprocessing*, and *prefilter transformation*. See also *aeromagnetic compensation* and *mis-tie*.

data correction: See *data reduction*.

data editing: same as *data conditioning* with an emphasis on removing spikes, filling gaps, *resampling*, *smoothing*, and other editing procedures prior to *data reduction* and *gridding*.

data enhancement: *enhanced processing* in the context of a selective transformation of data with the objective of increasing both the *lateral resolution* and *vertical resolution* of the anomalies of an exploration or research interest.

data extension: a *data conditioning* procedure applied to line-oriented data to create additional values along the *survey* margins, so that the values of the first and last data points of each survey line will decrease smoothly to zero in order to minimize *edge effects*. Referred to with the same meaning as *tapering*. See also *grid expansion*.

data processing: a sequence of mathematical operations applied to measured values of the Earth's gravity or magnetic field with the objectives of (a) compensation for all planetary and non-geological effects that interfere with the measurements and conceal or distort the gravity and magnetic anomalies generated by geological sources and (b) *enhancement* of anomalies of the exploration or research interest for their subsequent analysis and *interpretation*. Data processing operations involve rearranging the original irregular or line-oriented data into a *database*, applying corrections for known interfering effects, *gridding*, *filtering*, calculating derivatives and other attributes, changing the display *format*, etc. Standard data processing includes *data conditioning* and *data reduction* of irregularly spaced, line-oriented, and gridded datasets, and its results are added to the initial database. Subsequent *enhanced processing* is based primarily on the use of the gridded data, and it is focused on the detection and visualization of the regional and residual gravity or magnetic anomalies generated by geological structures and their constituents. See also *attribute*, *observed gravity components*, *observed magnetic components*, *regional field*, and *residual field*.

data reduction: a sequence of *data processing* operations designed to calculate and remove, or compensate for, all predictable local, regional, and planetary non-geological effects contributing to measured gravity and magnetic values. After data reduction, the gravity and magnetic values represent an *anomalous field* which is a field of the regional- and local-scale gravity or magnetic anomalies generated by the elements of geological structures, including those of the exploration and research interests. Data reduction is a final part of the standard data processing. Also referred to as *data correction*. See also *observed gravity*, *observed magnetics*, *gravity data corrections*, and *magnetic data corrections*.

database: a digital arrangement, or digital *format*, of acquired gravity and magnetic data designed for (a) loading data into computer processing algorithms and (b) data exchange, storage, and archiving. The database is created from digital files presented in one of the standard data formats, such as *ASCII format*, that contains the measured data, calculated corrections, and final corrected data, and it includes all relevant information about the *acquisition* parameters, *Geodetic Reference System*, and mapping projections. The database format is specific for each major software package and adapted for the *data conditioning*, *data processing*, and *data enhancement* applications available in a particular package.

datum: **1.** a reference surface, such as a *reference ellipsoid* or sea level, for the calculation of gravity corrections to compensate for (a) the *elevation* of observation points by *free-air correction*, (b) the Earth's *curvature* by *Bullard B correction*, and (c) above-the-datum rocks and below-the-datum seawater *mass deficiency* by *Bouguer correction*. **2.** a *gravity datum* representing a reference value, such as the *absolute gravity* or *relative gravity* value, for all measurements during a *gravity survey*. **3.** a *magnetic datum* represented by the *background magnetic field* which is used as a reference level for measurements of *diurnals*. **4.** the *IGSN71 datum* to which the worldwide *absolute gravity* measurements are tied. **5.** a *geodetic datum* used for the calculation of the *three-dimensional (3D)* coordinates (latitude, longitude, and elevation) of observation points. Sometimes referred to as *reference plane*. See also *horizontal datum* and *vertical datum*.

datum transfer correction: 1. an optional correction applied to *absolute gravity* measurements by the *free-fall method* with the use of a laser interferometer and free-falling corner-cube reflector in order to transfer the obtained values from the height of each drop to a user-specified height. The amount of the correction is determined by (a) calculating the difference between the actual drop height, which is defined as the *elevation* of an *observation point* plus the dropping height inside the instrument, and a specified transfer height, and (b) multiplying the difference value by the *vertical gravity gradient* at each observation point. **2.** an *indirect effect correction* in the context of a correction applied to *observed gravity* to transfer the obtained gravity values from a *gravity datum* of the original *survey* to a user-specified *datum*. Also known as *transfer height correction*.

datum transformation: a *data processing* operation applied for the re-calculation of acquired gravity or magnetic data from one *datum* to another. See also *data conditioning* and *indirect effect correction*.

DC: the abbreviation of a data constant used to indicate: (a) a reference level for measurements, processing, and analysis of anomalies, such as *DC level*; (b) a shift of the reference level by a constant value called *DC shift*; and (c) a constant background magnetic field value in a *magnetic survey* called the *DC value*. See also *DC power*.

DC component: See *DC power*.

DC level: 1. a reference level relative to which all amplitudes of an anomalous gravity or magnetic field are determined. The DC level effect on the obtained anomalous fields, such as the *Bouguer anomaly* or *TMI anomaly*, is analogous to the contribution of the *regional field*. **2.** a *base level* or *background magnetic field* that is used as a reference level for measurements of *relative gravity* or magnetic *diurnals* in a gravity or magnetic *survey* area. Derivatives, gradients, and band-pass-filtered anomalous fields essentially are not affected by the DC level and its shifts. See also *anomalous field*, *band-pass filter*, *derivative*, *gradient*, and *DC shift*.

DC power: a numerical power value assigned to the zero *wavenumber* to represent the upper limit for the energy (squared amplitudes) of the *spectrum* components in the *power spectrum* of a gravity or magnetic field. Also known as *DC component*.

DC shift: 1. a *base level* shift defined by a constant value added to or subtracted from the calculated gravity or magnetic field of a constructed *model* in order to minimize *model misfit*. **2.** a constant value added to or subtracted from one of two grids in the process of *grid merging* to change the *DC level* of a particular *grid*. Because the anomalous gravity and magnetic fields, such as *Bouguer anomaly* and *TMI anomaly*, are composed of relative anomalies, the DC shift changes their original amplitudes and, thereby, introduces errors in the subsequent structural and property *inversion*, especially for deep targets. A change of the DC level in gridded data by application of a DC shift also affects the calculation of complex attributes, such as *monogenic signal (MS)*, *direct analytic signal (DAS)*, and their analogues. See also *complex attribute* and *target*.

DC value: *background magnetic field* or *DC level* in the context of a constant background value in a *magnetic survey* area for the measurement of *diurnals* and calculation of *diurnal correction*.

dead zone: an instrument characteristic of certain types of magnetometers, such as the *proton magnetometer*, that defines the range of the *sensor* orientations relative to the direction of the *Earth magnetic field*, where the *magnetic sensor* produces no signal and the *magnetometer* is inoperative. See also *omnidirectional sensor*.

decay rate: See *fall-off rate*.

decibel (dB): a logarithmic unit for expressing the ratio of the maximum *amplitude* or power (squared amplitude) to the minimum amplitude or power of a measured gravity or magnetic *field*. It also is used to describe the *dynamic range* of a particular type of *gravimeter* or *magnetometer*. The amplitude ratio of 2 (power ratio of 4) is equivalent to 6 dB, the amplitude ratio of 10 (power ratio of 100) is equivalent to 20 dB, and the amplitude ratio of 0.1 (power ratio of 0.01) is equivalent to −20 dB. See also *signal-to-noise ratio (S/N)*.

decimal time: a parameter in the *real-time* mode of operation that represents minutes of a current time as the decimal fractions of an hour, e.g., 18.50 = 1830 hours = 6:30 PM.

declination: See *magnetic declination*.

declination-inclination magnetometer: the instrument assemblage of a *fluxgate magnetometer* and a non-magnetic theodolite, where the sensing axis of the magnetometer and optical axis of the theodolite are collinear. Used primarily at geomagnetic observatories to monitor *temporal magnetic variations*. Several measurements are taken with different theodolite orientations to determine *magnetic declination*, *magnetic inclination*, and absolute numerical value, or *magnitude*, of the *geomagnetic vector* at an observation point. Sometimes referred to as *magnetic theodolite*. See also *geomagnetic observatory*.

decompensation correction: a correction applied to the *isostatic anomaly* to minimize the gravity effect of deep compensating rock masses that isostatically support crustal *density* variations within shallower rock masses without a topographic expression, i.e., it is a correction for the gravity effect of *isostatic compensation* which is not related to variations of *topography* or *bathymetry* and, hence, not accounted by *isostatic correction*. The basic assumption is that every sufficiently large rock mass excess or deficiency within the *crust* is underlain by an equivalent rock mass of the opposite characteristic, similar to how *mass excess* is underlain by *mass deficiency* and vice versa, to maintain *isostatic equilibrium*. Decompensation correction is commonly calculated by *upward continuation* of the isostatic anomaly *grid* to some empirically estimated height (e.g., 30–40 km) to approximate the long-*wavelength* anomalies generated by deep compensating rock masses in the lower crust and upper *mantle*. The obtained upward-continued field is subtracted from the isostatic anomaly to obtain the *decompensative gravity anomaly*. See also *isostasy*.

decompensative gravity anomaly: a residual *isostatic anomaly* that represents the part of an isostatic anomaly remaining after subtracting the *decompensation correction*. The decompensative anomaly approximates the *field* of the gravity anomalies generated, primarily, by the regional-scale *density* variations in the upper 15–20 km of the Earth's *crust* that are isostatically supported by deep compensating rock masses in the lower crust and upper *mantle* without a topographic or bathymetric expression. See also *gravity anomaly*, *isostatic compensation*, and *residual anomaly*.

deconvolution: a *data processing* algorithm which is conceptually equivalent to *inversion* and is designed to determine *source body* parameters, such as depth, location, *dip*, and *density* or *magnetization* contrast, from an *observed field*. An observed anomaly is considered the result of *convolution* of the original field generated by a source body with a *transfer function* that describes the filtering effect of the *acquisition* geometry, *amplitude attenuation*, *wavelength attenuation*, physical properties, and other observation and source parameters. The fundamental *ambiguity* between the observed field and its unique sources requires specific source-approximating assumptions, such as the *ideal source* geometries, for the deconvolution process. See also *Euler deconvolution* and *Werner deconvolution*.

decorrelation: a method for removing the effects of *diurnals* from marine (shipboard) *magnetic survey* data that is based on the axiom that true magnetic anomalies and diurnal effects are not correlated with each other. A line-by-line interactive analysis of the survey data within a *moving window* identifies the anomalous waveforms which correlate in time and shape with the records of a *geomagnetic observatory* or *base magnetometer* onshore. Identified waveforms in the onshore records are, first, scaled to fit their corresponding amplitudes and shapes in the survey data, and then they are subtracted from the survey data. The method is applied when the marine survey is far away from a magnetic *base station* or nearest geomagnetic observatory. See also *waveform*.

decorrugation: elimination of *corrugations* in gridded gravity and magnetic data by the application of *directional filtering* along the traverse and control lines or by other *microleveling* procedures. Sometimes referred to as *destriping*. See also *gridding*, *traverse lines*, and *control lines*.

deculturing: *cultural editing* in the context of removing all *cultural magnetic effects*, i.e., magnetic effects of man-made ferrous objects and electric currents, from aeromagnetic data.

deep-pass filter: a *data enhancement* technique that compensates for the *attenuation* of amplitudes of gravity and magnetic anomalies caused by the increasing seawater depth within the continental margins. It is based on multiplying each *grid* value of an *anomalous field* by an exponential of the bathymetric depth in an *amplitude attenuation* formula (Santos et al., 2012). Application of this filter can improve the performance of other data enhancement techniques, such as *analytic signal amplitude* and *total horizontal derivative*, in a manner of *cascaded filtering*. See also *downward continuation*.

default value: 1. the value assigned automatically by a computer program in the absence of specific instructions. **2.** the value recommended to be chosen from an applicable range of values, such as the *Peters index* for the *Peters half-slope method* or the source index for the *straight-slope method*, when the source geometry and other specific parameters of a calculation are unknown. The default value of *density* for the calculation of the regional-scale *Bouguer correction* is 2.67 g · cm^{-3} in *cgs units* and 2670 kg · m^{-3} in *SI units*.

deficient density structure: a local- or regional-scale geological *structure* composed of rocks having *density* values lower than densities of the laterally adjacent rock formations, e.g., *salt dome* and salt *diapir*, *syncline*, *graben*. Its *gravity signature* represents a local or regional *gravity low*. See also *excess density structure*.

Definitive Geomagnetic Reference Field (DGRF): a mathematical description of the *main geomagnetic field* adopted by the International Committee of International Association of Geomagnetism and Aeronomy (IAGA) for each past *epoch* beginning from 1945. The DGRF is based on the retrospectively updated *International Geomagnetic Reference Field (IGRF)* data, wherein each update accounts for the calculated differences between the predicted and actually measured *geomagnetic field parameters*. The IGRF is not designated as definitive until it becomes unlikely that the available data will be significantly improved, i.e., DGRF parameters are not revised in subsequent generations of IGRF models. See also *DGRF model* and *IGRF-12*.

deflection of the vertical: 1. the angle between the actual, or true, vertical direction determined by a *plumb line* at a ground *observation point* and the theoretical vertical direction defined by a perpendicular to the *reference ellipsoid*. **2.** the angular difference between two perpendiculars: one to the surface of a *geoid* and the other to the surface of a reference ellipsoid, wherein its calculated values define the geoid slopes around each observation point. The amount of deflection is controlled by the heterogeneity of the Earth's mass distribution: the higher the deviation from a homogeneous mass, the higher the deflection angle. Deflections are typically less than a few arc-seconds with the resulting maximum error of *vertical gravity* (which is measured with the assumption that its direction is perpendicular to the geoid or reference ellipsoid) of approximately 0.04 *milligal (mGal)*. See also *arc-second* and *Browne correction*.

del: the vector *gradient* operator denoted by the mathematical symbol ∇. In *Cartesian coordinates*, it is expressed as

$$\nabla = \mathbf{i}\partial/\partial x + \mathbf{j}\partial/\partial y + \mathbf{k}\partial/\partial z,$$

where **i**, **j**, **k** are unit vectors parallel to the mutually orthogonal *x*-, *y*-, *z*-axes. For a *vector field*, the expression $\nabla\mathbf{A}$ defines *divergence*, whereas $\nabla \times \mathbf{A}$ defines the *curl* of vector field **A**. For a *scalar field*, ∇A defines the gradient of scalar field A. Squared del is the mathematical expression of the *Laplacian* written as

$$\nabla^2 = \nabla \cdot \nabla = \partial^2/\partial x^2 + \partial^2/\partial y^2 + \partial^2/\partial z^2.$$

See also *Laplace equation*.

demagnetization: 1. a decrease to zero or a highly randomized weak *magnetization* of the originally magnetized rocks subjected to external pressure, thermal, and electromagnetic effects. **2.** the natural phenomenon of a gradual decrease in rock magnetization with the increasing depth of their occurrence due to a depth-related heating up to the *Curie temperature* above which rocks become essentially non-magnetic, i.e., *paramagnetic*. **3.** a method of estimating the stable component of *remanent magnetization* in rock samples that is based on their partial demagnetization by externally applied alternating magnetic fields. See also *self-demagnetization* and *shock demagnetization*.

demagnetization correction: a correction applied to rock parameters in a two-dimensional (2D) or three-dimensional (3D) *magnetic model* to account for the internal *self-demagnetization* effects of isolated rock bodies, such as *ore body* or *pluton* (Guo et al., 2001).

demagnetization factor: a parameter of the internal *self-demagnetization* of rocks that is used in *forward modeling* and *inversion* for the calculation of the *effective magnetization* of rock bodies of various shapes. Denoted by N, it is specific for each direction of *magnetization* for a particular rock body *shape* and can be described as a proportionality coefficient in the relationship between *demagnetizing field* $\mathbf{F}_{DF}$ and effective magnetization $\mathbf{M}_E$, expressed as

$$\mathbf{F}_{DF} = \mathrm{N} \cdot \mathbf{M}_E.$$

Demagnetization factors for source bodies of simple geometric forms, such as a line of dipoles, *sphere*, or *horizontal cylinder*, are approximated by those of ellipsoids, wherein each of the three principal axes has its own demagnetization factor value, but their sum is a constant equal to 4π. Inside the ellipsoid, the component of the *magnetizing force* $\mathbf{H}_i$ along any principal axis *i* in *Cartesian coordinates* is expressed as

$$\mathbf{H}_i = (\mathbf{H}_{EAF})_i - \mathrm{N}_i(\mathbf{M}_E)_i;$$

where $i = x, y, z$ and $\mathbf{H}_{EAF}$ is the externally applied magnetic field. For a homogeneous sphere, $\mathrm{N} = 4\pi/3$. Estimates of the effective magnetization of arbitrary-shaped source bodies are based on the use of numerical integration methods, where the original body volume is divided into small volumes of simple geometric forms with known demagnetization factors, so the effective magnetization of the whole *source body* is obtained by summing the calculated effective magnetizations of each volume part.

demagnetization temperature: same as *Curie temperature* with an emphasis on its *demagnetization* effect on rock-forming minerals.

demagnetizing field: the magnetic *vector field* within a magnetized rock body produced by pairs of positive and negative magnetic charges, or magnetic poles, emerging on the opposite boundaries of the rock body under exposure to an external magnetic field, such as the *Earth magnetic field*. The demagnetizing field $\mathbf{F}_{DF}$ is proportional to the *effective magnetization* $\mathbf{M}_E$ of the rock body and expressed as

$$\mathbf{F}_{DF} = \mathrm{N} \cdot \mathbf{M}_E,$$

where N is the *demagnetization factor* in the direction of the *magnetization* for a particular *shape* of rock body. In an ellipsoidal body magnetized along its principal axis, the direction of the demagnetizing field is opposite to the external magnetic field. The total magnetic field within a rock body, or *internal magnetic field*, is defined by the *vector* sum of the external magnetic field and internal demagnetizing field. See also *ambient field*, *demagnetization*, and *self-demagnetization*.

demi-pentes method: a graphical method of estimating the depth to a source of an isolated (non-interfering) and reduced-to-pole *magnetic anomaly*. The source depth is defined by the horizontal distance between the intersections of the *maximum-slope line* with two half-maximum *slope* lines drawn through two points of tangency before and after the maximum slope portion on the anomaly *profile* (Hinze et al., 2013). See also *Peters half-slope method* and *reduction-to-pole (RTP)*.

denoising: a selective filtering applied to remove the local high-frequency *noise* with a minimized *filtering* effect on non-contaminated gravity and magnetic anomalies. Denoising differs from *smoothing* which is designed to suppress the abnormally high-*amplitude* components and high frequencies in the entire *grid*, regardless of whether the data belong to the noise or represent anomalies of a geological origin. The most efficient denoising is performed by *wavelet denoising* applied to line-oriented datasets to prevent the propagation of noise components into *gridding* and *enhanced processing*. See also *spatial frequency*.

density: a rock property defined by rock *mass* per unit volume. It is measured in grams per cubic centimeter ($g \cdot cm^{-3}$) in *cgs units* and in kilograms per cubic meter ($kg \cdot m^{-3}$) in *SI units*, where $1\ g \cdot cm^{-3} = 10^3\ kg \cdot m^{-3}$. Rock density is a function of the mineral composition of rocks, their void space (porosity), and the material infill. There are several sources of information about rock densities, including: (1) measurements of borehole-extracted core samples, drill cuttings, and surface rock samples; (2) *borehole gravity* measurements; (3) *gamma-gamma log*; (4) conversion of *acoustic log* data; (5) conversion of seismic P-wave interval velocities with the use of the *Gardner equation*, *Nafe-Drake curve*, and other density-velocity relations; (6) *Nettleton* *method*; (7) *Siegert method*; (8) *Parasnis method*; and (9) *property inversion* constrained by seismic data, well formation tops, and other geological information. Generally, densities vary within 1.6–2.8 g · cm^{-3} for sedimentary rocks, 2.5–3.3 g · cm^{-3} for metamorphic rocks, and 2.4–3.5 g · cm^{-3} for igneous rocks. Fluid densities are 1.00 g · cm^{-3} for fresh water, 1.027 g · cm^{-3} for sea water, and 0.6–0.9 g · cm^{-3} for hydrocarbons. Ice density is 0.917 g · cm^{-3}. Average densities of the continental and oceanic *crust* are 2.75 g · cm^{-3} and 2.95 g · cm^{-3}, respectively. The average density of the Earth is 5.52 g · cm^{-3}. Unlike *magnetization*, the density of rock-forming minerals is a directionally independent, or *scalar*, quantity. The density of 2.67 g · cm^{-3} is often used as a reference value for *gravity stripping* and as a *default value* for calculation of *gravity corrections* in regional-scale gravity surveys. The highest densities in the upper crust are associated with crystalline (metamorphic and igneous) *basement* rocks, intra-sedimentary igneous intrusions, and metallic ores. Due to a considerable overlap in the density ranges for different rocks, density values are not diagnostic of common rock types. See also *apparent density*, *bulk density*, *true density*, and *natural density*.

density basement: a geological *unconformity* in the subsurface with a large positive *density contrast*. Gravity anomalies generated by lateral density contrasts below this unconformity are often lost in *background noise*. Typically, the rocks overlaying the density basement have densities within the range of 1.6–2.3 g · cm^{-3}, whereas the underlying rocks typically have densities ranging from 2.6 g · cm^{-3} to 3.2 g · cm^{-3}. Depending on the sedimentation conditions in a particular *basin*, the density basement corresponds to either (a) an *acoustic basement*, such as a top of thick carbonates, thick basalt *lava* flow, and metamorphic rock sequence in the lower sedimentary section, or (b) a *geological basement* underlying the entire sedimentary section. Also referred to as *gravity basement* and *high-density basement*. See also *basement* and *magnetic basement*.

density contact: 1. the location of a lateral change in the *density* of rocks that is associated with a structural or lithological boundary in the subsurface. **2.** a *source body* approximation in the gravity *modeling* of structural and lithological boundaries that often assumes an *infinite depth extent* of the lateral *contact* between rocks with different density values. **3.** an interface (usually a *fault*) between two blocks of the contrasting densities. See also *infinite contact* and *thick contact*.

density contrast: 1. the difference in *density* values between two laterally adjacent rock units or formations. Lateral density contrast in the subsurface is a fundamental reason for, or cause of, an observed *gravity anomaly*: there is no gravity anomaly in the absence of a lateral density contrast. The degree of lateral density contrast created by a *source body* is one of the primary controls over the *amplitude* of a gravity anomaly: the higher the contrast, the higher the amplitude of the *anomaly*. Density contrast is positive when the source body density is higher than that of the laterally adjacent rock units or formations, and it is negative for the reverse case. A change in the sign of a density contrast reverses the sign of a gravity anomaly. Lateral density contrast can be as high as 0.8 g · cm^{-3}, such as between a low-density *salt* body and high-density carbonate rocks. **2.** the difference in density values across a lithological or structural boundary between two rock formations of a different age: the younger overlying rocks and the older underlying rocks. See also *unconformity* and *magnetization contrast*.

density-depth function: a mathematical expression or graphical plot of changes in rock *density* as a function of rock depth. See also *Adams-Williamson equation*, *Brocher formula*, *Gardner equation*, and *Nafe-Drake curve*.

density-difference curve: a graphical plot, or graph, of the calculated difference between *density* values determined from measurements by a *borehole gravimeter (BHGM)* or borehole-adapted *vibrating-string gravimeter* and *density log*, also known as *gamma-gamma log*, in a drilled well (borehole).

density inversion: 1. *inversion* of *residual gravity* for the distribution of apparent densities in a constructed two-dimensional (2D) *model* or three-dimensional (3D) *voxel model* with the use of available geological and geophysical *constraints*. **2.** constrained *property inversion* for the distribution of density contrasts as a function of the 2D or 3D subsurface position. The obtained *modeling* results reproduce the residual gravity field within a given *tolerance* and with some limitations on the types of possible density structures. See also *apparent density*, *density contrast*, and *gravity inversion*.

density log: a logging tool for measuring the apparent densities of rock formations in drilled wells (boreholes). It contains (a) a primary gamma-ray source, such as radioactive isotope cesium Cs^{137}, and (b) a detector of the secondary, or backscattered, gamma rays generated within a rock formation by the primary gamma-ray source. The intensity of the backscattered radiation depends on the amount of electrons in the rock material, and it is proportional to the *apparent density*. Due to the short radius of gamma-ray source penetration (typically less than 30 cm), density log measurements are often complemented with measurements by a *borehole gravity meter (BHGM)* or borehole-adapted *vibrating-string gravimeter*. Also referred to as *gamma-gamma log*. See also *density*, *density-difference curve*, and *sonic log*.

density model: a two-dimensional (2D) or three-dimensional (3D) *gravity model* representing the distribution of *apparent density* in the subsurface. There are four basic types of *density* models: (1) a model with constant densities of the sedimentary layers and isolated rock bodies estimated with the use of available geological-geophysical information; (2) a model with variable densities of the sedimentary layers obtained from seismic interval velocities via the *Gardner equation*, *Nafe-Drake curve*, and other density-velocity relations; (3) a variable density model, such as a *voxel model*, that is based on the *BHGM density*, *density log*, and *sonic log* measurements in drilled wells with interpolation and extrapolation of the obtained density values over the entire study area; and (4) a *hybrid density model*.

density profile: the *Bouguer gravity* profile over a topographic feature having a substantial relief which is not related to the known near-surface geological *structure* in a *gravity survey* area. Several *density* profiles are constructed in the *Nettleton method* in order to obtain the Bouguer gravity profiles by application of *simple Bouguer correction* calculated with different *Bouguer density* values ranging from 1.6 g · cm^{-3} to 2.6–2.8 g · cm^{-3} with an increment of 0.1 g · cm^{-3}. The density value which minimizes the correlation between the Bouguer gravity *profile* and the topography profile is selected for calculation of the Bouguer correction for all observation points in a survey area. See also *Siegert method*.

density-profile method: See *Nettleton method*.

density surface: a surface obtained by *inverse calculation* of the equivalent distribution of *density* in the *equivalent-source method*. The density surface can be of any arbitrary shape, and its *elevation* is selected at or below the observation surface and above all true sources of the gravity anomalies.

density voxel: same as *voxel* with an emphasis on an elementary cube representing the unit volume *density* in a three-dimensional (3D) *voxel model* of the subsurface distribution of rock densities.

depocenter: a local- or regional-scale area of the largest accumulation of sediments. In the *Bouguer gravity* field, it has the *gravity signature* of a *gravity low*, whereas the *magnetic signature* over this area depends primarily on the presence, if any, of igneous intrusions and *volcanics*, magnetic properties (*susceptibility* and direction of *magnetization*) of the *basement* rocks below the depocenter, and to a lesser degree on the basement depth structure, i.e., a local depression in the basement can have the signature of a *magnetic high*. See also *igneous intrusion*.

depositional remanent magnetization: See *detrital remanent magnetization (DRM)*.

depth-ensemble filtering: *matched-filtering method* and *pseudodepth slicing* in the context of the filtering-based enhancement of the *wavelength* range associated with an assemblage of sources at a particular *grid*-averaged depth determined from a *power spectrum*.

depth estimation methods: methods of calculating the depth to: (a) an isolated *source body* the actual geometrical shape of which is approximated by one of the *ideal source* shapes; (b) a structural discontinuity, such as a *fault* or lateral lithological boundary, approximated by a *contact*; and (c) a depositional boundary, such as *basement*, obtained by *structural inversion*. There are *graphical source depth estimation* methods and *automated source depth estimation* methods. Graphical, or manual, methods are based on *depth rules* and measurements of the horizontal distances between the manually selected characteristic points on a *profile* or *contour grid* of the magnetic and gravity anomalies. The most common automated line- and *grid*-based methods represent a special group of *inversion* methods, such as *Euler deconvolution*, *Werner deconvolution*, and *local wavenumber methods*, that use simple source geometries (*thin sheet*, *contact*, *sphere*, *cylinder*, etc.) in order to linearize the *inverse problem* and obtain *source depth solutions*.

depth of compensation: 1. a depth level at which the topographic or bathymetric load is isostatically compensated and below which lateral *density* variations are expected to be insignificant for keeping the balance, or equilibrium, between the *crust* and upper *mantle*. **2.** a parameter in the *Pratt-Heyford model* of *isostasy*. Sometimes, the top of the *asthenosphere* is considered as the depth of compensation. See also *isostatic compensation*.

depth rules: empirical rules relating the depths of isolated source bodies to the *profile* parameters of their anomalies, such as *half-width*, *half-amplitude width*, *Peters length*, *Sokolov distance*, *straight-slope length*, and *inflection point*. Depth rules are based on the observation that the source depth is related to the *slope* and width of an anomaly on its *principal profile* orthogonal to the long axis of the *anomaly*. Anomaly slope and width depend on: (a) the geometrical shape of the source: slopes and widths of anomalies from sources of different shapes at the same depth are different; and (b) source depth: the shallower the source, the steeper the slope and the narrower the width. Depth rules are implemented in the graphical, or manual, methods of estimating source depth that include but are not limited to *ASA-derived source depth*, *half-width method*, *Hannel method*, *NSS method*, *Peters half-slope method*, *Sokolov method*, *straight-slope method*, *Thalen method*, *Tiburg method*. See also *Smith depth rules*.

depth-separation filter: same as *Jacobsen filter* with an emphasis on its operating principle based on the separation of anomalies from sources at different depth levels.

depth-slice filter: *matched filter* in the context of enhancing the *wavelength* components associated with a depth level "sliced" by the selection of a particular slope in the *power spectrum*.

depth slicing: See *pseudodepth slicing*.

depth solution: 1. a sedimentary horizon or *basement* depth structure obtained by *structural inversion*, i.e., depth values obtained by solving the *inverse problem*. **2.** *source depth solutions*.

derivative: the rate of change of the continuous function f per unit distance r in a given direction. It is expressed as

$$\partial f/\partial \mathrm{r} = \lim \Delta f/\Delta \mathrm{r},$$

where $\Delta \mathrm{r} \rightarrow 0$. Its absolute numerical value is equal to the *calculated gradient* at the observation point. The absolute value of the derivative of the *gravitational potential* or *magnetic potential* defines the *magnitude* of the *vector* component of a gravity or magnetic *field* in the direction of calculation. The first-order *horizontal derivative* describes the anomaly *slope* quantified by a corresponding *gradient* value. The second-order horizontal derivative describes the slope of the first-order derivative, or gradient of the gradient, and approximates the *curvature* of the anomalous gravity or magnetic field. According to the *Poisson theorem*, the *vertical derivative* of a *magnetic anomaly* is equivalent to the next higher order vertical derivative of a *gravity anomaly* from the same source, so that the *first vertical derivative* of the magnetic anomaly is equivalent to the *second vertical derivative* of the gravity anomaly. In the *Fourier domain*, the horizontal and vertical derivatives of the field F are related via the *Hilbert transform* as

$$\mathcal{F}(\partial F/\partial z) = H_x\mathcal{F}(\partial F/\partial x) + H_y\mathcal{F}(\partial F/\partial y),$$

where $\mathcal{F}(\partial F/\partial z)$, $\mathcal{F}(\partial F/\partial x)$, $\mathcal{F}(\partial F/\partial y)$ are Fourier transforms of the vertical and horizontal derivatives, respectively, and H_x and H_y are x- and y-components of the *3D Hilbert transform*. See also *directional derivative, partial derivative*, and *Cauchy-Riemann equations*.

derivative Euler deconvolution: *Euler deconvolution* applied to the vertical or horizontal derivatives of line-oriented or gridded anomalous gravity and magnetic fields. Because the regional background is suppressed in the field of derivatives, an overdetermined system of the Euler equations for each position of a *moving window* along a *survey* line or *grid* can be solved for the local source coordinates (x_0, y_0, z_0) and its *Euler structural index (SI)* value (Cooper, 2004; Florio et al., 2006). See also *derivative, gridding, Euler equation*, and *analytic signal Euler deconvolution*.

derivative filter: an *enhanced processing* algorithm that calculates the *derivative* of the *anomalous field*. Derivative filters are efficient in enhancing mid- and, especially, short-*wavelength* anomalies and are capable of detecting very small changes in the gravity and magnetic fields associated with elements of a subsurface geological *structure*. Both the first- and second-order *vertical derivative* filters are often used in the analysis of gridded data, whereas *horizontal derivative* calculations are commonly restricted to the first-order derivatives only. The horizontal derivative has its maximum over a *density contrast* or *magnetization contrast*, and it is used for mapping high-offset faults, lithological contacts, and source edges. The second-order vertical derivative filter emphasizes the local structures, and its zero contours in the *residual field* closely correlate with structural edges. Data errors and *noise*, both random and regular, are enhanced with the increasing order of derivatives, so the derivative filters are usually cascaded with *upward continuation* and other noise-suppressing techniques. Resolution of anomalies by application of the derivative filter depends on the *survey* parameters, such as *survey line spacing, sample interval*, measurement *bandwidth*, and velocity of a survey aircraft or ship. See also *cascading, edge, grid, lateral resolution, residual field*, and *integral filter*.

derivative map: a digital or printed map of one of the derivatives of an anomalous gravity or magnetic field. Its primary objective is to highlight the short- and mid-*wavelength* anomalies associated with elements of a subsurface geological *structure*. See also *anomalous field, derivative*, and *derivative filter*.

derived gradient: same as *calculated gradient* representing a *scalar* quantity equal to the absolute value of a *derivative* at a particular line or *grid* point.

despiking: *spike filtering* in the context of eliminating, or filtering out, the *spike*-like high-*amplitude* components of gravity and magnetic data.

destriping: same as *decorrugation* designed to eliminate residual *leveling* errors, or stripes, oriented along the *survey* lines in gridded gravity or magnetic data.

detailed survey: 1. a high-*resolution* airborne or marine (shipboard) gravity or magnetic *survey* with closely spaced survey lines providing the most detailed information about geological structures in the *exploration* interest area which has been identified by the *interpretation* of the *regional survey* data. The lowest flight *altitude* for a detailed fixed-wing airborne survey is limited by safety regulations to 80–150 m. The ratio of the flight altitude h to the *survey line spacing* is commonly 1:1 or 1:2, but not less, to avoid *aliasing* of the short-*wavelength* components in the measured field. For aeromagnetic *total field* surveys, the minimum ratio is 0.5, i.e., the line spacing should not exceed $2h$, whereas for aeromagnetic *gradient* surveys the optimal ratio is 1:1 (Reid, 1980). Flight line spacing is usually constant over a detailed survey area to maintain consistency in the gravity and magnetic signatures over a similar geology. **2.** a high-resolution *ground survey* conducted with the objective of a comprehensive study of the *prospect*, such as an *ore body* or other specific *target* of mineral exploration, identified by a detailed *airborne survey*. See also *gravity signature, magnetic signature, regional survey, reconnaissance survey, high-resolution aeromagnetic survey, high-resolution ground survey, drone survey, fixed-wing survey*, and *survey design aliasing*.

determinant: See *FTG determinant*.

detrending: 1. removing the regional *trend*, or regional-scale anomalous *gradient*, from an *observed field* to separate the low-amplitude *target* anomalies from the superposition of the high-*amplitude* long-*wavelength* anomalies associated with the regional-scale structures in sedimentary basins. The first-order polynomial surface is often used to approximate the regional trend in a study area. **2.** removing the linear trend of values from gridded or line-oriented data prior to the application of a *Fourier transform* in order to minimize *edge effects*. The removed trend is added back after *inverse Fourier transform* by a procedure called *trend replacement*. Detrending is also referred to as *trend removal*. See also *basin, polynomial fitting*, and *regional-residual separation*.

detrital magnetite: *magnetite* formed by chemical alterations in the process of rock weathering and accumulated in clastic sedimentary rocks (sandstones, shales, limestones, conglomerates) and within ancient fluvial systems, sand channels, beach sands, and alluvial placers. Detrital magnetite is a major source of *detrital magnetization*. See also *diagenetic magnetite* and *placer*.

detrital magnetization: *magnetization* of sedimentary rock formations by accumulations of magnetic minerals, primarily *detrital magnetite*, in clastic sedimentary rocks, ancient fluvial and submarine paleochannels, sand dunes and sand channels, alluvial fans, glacial till, and ancient beaches. Sandstones of a marine origin often contain significant amounts of detrital magnetite. Rocks with detrital magnetization generate the localized assemblages, or clusters, of low-*amplitude* short-*wavelength* magnetic anomalies. When their sources are in the upper part of a sedimentary section, these anomalies can be detected by a *high-resolution aeromagnetic survey.* See also *clustering, paleochannel, regolith,* and *SRM anomaly.*

detrital remanent magnetization (DRM): a permanent *magnetization* acquired by the detritus-like components of rocks during their slow settling in water. Clays can have a relatively strong magnetization of this type. Also known as *depositional remanent magnetization.* See also *remanent magnetization* and *detrital magnetite.*

DGPS: See *differential Global Positioning System (GPS).*

DGRF: See *Definitive Geomagnetic Reference Field.*

DGRF model: a mathematical model of the *Definitive Geomagnetic Reference Field (DGRF)* obtained from the retrospectively updated *IGRF model* for each past 5-year period, or *epoch.* Each update accounts for differences between the predicted and actual, or true, *geomagnetic field parameters* measured during the preceding epoch. For example, DGRF2005 represents the model for the epoch 2005 and describes the actual parameters of the *main geomagnetic field* during 2000 to 2005. See also *IGRF-12.*

diagenetic magnetic anomaly: a *sedimentary magnetic anomaly* generated by the accumulation of *diagenetic magnetite* or other *diagenetic magnetic minerals* in buried alluvial deposits and mineral placers, sand channels, depositional fairways, and other ancient fluvial systems. These anomalies primarily have low-*amplitude* and short-*wavelength* content identified in the field of *residual magnetics,* except for the high-amplitude *clinker anomaly* and other magnetic anomalies associated with *combustion magnetism.* See also *fairway, placer, regolith,* and *SRM anomaly.*

diagenetic magnetic minerals: high-*susceptibility* rock-forming minerals, such as *magnetite,* maghemite, greigite, and pyrrhotite, that are formed by oxidation and other chemical changes in the mineral composition of rocks and often associated with (a) recurrent upward migration of the highly mineralized and high-temperature fluids of the *juvenile water* type along *fault* or fracture planes, (b) microseepage of fluids from hydrocarbon reservoirs, and (c) conversion of low-susceptibility minerals by *combustion magnetism.* See also *magnetic susceptibility, hydrocarbon seepage effect, clinker anomaly,* and *SRM anomaly.*

diagenetic magnetite: *magnetite* formed by the sedimentary diagenetic processes, such as hydrothermal alteration, dolomitization, hydrocarbon seepage, *combustion magnetism,* and other natural processes resulting in changes of the mineral composition and *magnetization* of rocks. Its local accumulations contribute to *detrital magnetization* and generate, mostly, the low-*amplitude* diagenetic magnetic anomalies, except for the high-amplitude *clinker anomaly* and other anomalies associated with the combustion magnetism of organic-rich sedimentary formations. See also *detrital magnetite, hydrocarbon seepage effect, diagenetic magnetic anomaly,* and *SRM anomaly.*

diagonalized tensor matrix: the 3×3 *tensor matrix* where all six off-diagonal elements are zeroes. Tensor gradient elements change their values with the rotation of the Cartesian coordinate frame, and there exists the rotation position at which the nine-element tensor matrix becomes diagonalized. The *FTG matrix* is diagonalized to obtain *FTG eigenvalues* paired with *FTG eigenvectors* which show the orientation of axes of the rotated *Cartesian coordinates* at each observation point for the calculation of *true curvature attributes.* See also *matrix.*

diamagnetic: weakly magnetic with a small negative *magnetic susceptibility* and *magnetic permeability* less than that of a source-free space (vacuum). Diamagnetic rock-forming minerals, such as rock *salt* (halite), anhydrite, quartz, and others, cannot acquire a *remanent magnetization.* See also *diamagnetic susceptibility* and *diamagnetism.*

diamagnetic susceptibility: a small negative *magnetic susceptibility* of rock-forming minerals, such as halite (rock *salt*), anhydrite, quartz, gypsum, feldspar, calcite, and graphite, where the direction of *magnetization* is opposite, or antiparallel, to the ambient *Earth magnetic field.* The typical value of diamagnetic susceptibility is -10^{-6} in *cgs units* or -10^{-5} in *SI units.* Negative susceptibility values are used in *modeling* the diamagnetic effects of salt bodies and, sometimes, *remanent magnetization* of rocks with unknown orientation of the *magnetization vector.* See also *ambient field* and *paramagnetic susceptibility.*

diamagnetism: a weak negative *induced magnetization* which is antiparallel to an externally applied, or inducing, magnetic field, such as the *Earth magnetic field.* Diamagnetic rock-forming minerals have a low negative *magnetic susceptibility* at the order of -10^{-5} in *SI units* that is referred to as *diamagnetic susceptibility,* and their *permeability* is less than the permeability of a source-free space (vacuum). Diamagnetic effects usually do not exceed 1.0–1.5 *nanotesla (nT),* and they disappear with the removal of the external magnetic field, i.e., there is no *remanence* in the diamagnetism. Detectable diamagnetic anomalies are generated, primarily, by large salt domes and other massive *salt* bodies. A significant majority of the rock-forming minerals in the Earth's *crust,* including the copper, silver, and gold, are either *diamagnetic* or *paramagnetic.* See also *paramagnetism, ferromagnetism,* and *ferrimagnetism.*

diapir: a teardrop-shaped rock body that moves toward shallower depths because its *density* is lower than the densities of the surrounding rock formations. Sedimentary diapirs usually have a low-density *salt* or shale core. High-density *igneous intrusion,* which moves upward along a *fault* or fracture zone under the pressure of a deep *magma* chamber, also creates a diapiric structure. The gravity or magnetic *signature* of a diapir is defined by the content of its core: salt diapirs generate gravity and magnetic lows in a *residual field,* whereas igneous intrusions generate gravity and magnetic highs in both the total and residual fields. Shale diapirs can contain increased concentrations of *detrital magnetite* or *diagenetic magnetite* sufficient for generating a low-amplitude *magnetic high.* See also *fault, gravity high, gravity low, residual gravity,* and *residual magnetics.*

diatreme: the funnel-shaped brecciated upper portion of a *kimberlite intrusion*. It is formed by hydrovolcanic fragmentation and wall-rock collapse resulting from the explosion-like *extrusion*, or eruption, of a fast-raising pipe-shaped *igneous body* onto the Earth's surface. Diatremes can be up to 1.5–3.0 km in diameter and up to 2.0–2.5 km in depth, and they are usually buried under unconsolidated sediments or beneath shallow lakes. Their *gravity signature* is produced by the *interference of anomalies* generated by the localized high-*density* kimberlite body and broader depression composed of the reduced-density brecciated igneous rocks. Density and magnetization contrasts of the brecciated diatreme are usually 2–3 times lower than those of a high-density and high-*magnetization* kimberlite body. See also *density contrast* and *magnetization contrast*.

difference gravity: a gravity *modeling* term for the difference between the *observed gravity*, such as *Bouguer gravity* or *free-air gravity* used in modeling, and the *calculated gravity*, which is the gravity field generated by a two-dimensional (2D) or three-dimensional (3D) *model*. Difference gravity represents the observed gravity components associated with those elements of a geological *structure* that are not included into a *model* or modeled incorrectly using an increased or reduced *density* or depth and shape of the structural elements. Within some range of the difference gravity values, the iterative changes are made, manually or by an *inversion* algorithm, to the model parameters in order to reduce this difference to a user-specified level of *tolerance*. Referred to with the same meaning as *model misfit* for gravity models.

differential acceleration: 1. the difference between two values of the *Earth gravity field* (i.e., values of *gravitational acceleration*) measured by two gravity sensors mounted on the same reference frame and separated by a distance called the *baseline*. Differential acceleration per unit distance quantifies the *measured gradient* of the Earth's gravity field in the direction of the sensor separation. **2.** the difference between accelerations measured by pairs of identical and synchronized accelerometers along each *gradiometer arm* in a *gravity gradiometer* system. Sometimes referred to as *differential-mode acceleration*. See also *acceleration*, *accelerometer*, *common-mode acceleration*, and *inertial acceleration*.

differential accelerometer: same as *gravity gradiometer* with an emphasis on the use of paired and synchronized accelerometers mounted on the same reference frame and separated by a distance called the *baseline*.

differential continuation: 1. a method of recalculating the observed gravity or magnetic field from the *altitude* of measurements to a higher or lower altitude. Among the space-domain and wavenumber-domain methods, the *equivalent source* methods are often applied and referred to as *equivalent-source continuation*. Land gravity data are compared with *aerogravity* data using the differential continuation method. **2.** an *enhanced processing* technique based on the calculation of a difference between two upward or downward continuations of an anomalous gravity or magnetic field. See also *anomalous field*, *differential upward continuation*, *differential downward continuation*, and *chessboard method*.

differential curvature: a quantitative characteristic, or *attribute*, of the gravitational *equipotential surface* defined by the difference between the *maximum curvature* and *minimum curvature*. In the general form, it is expressed as

$$K_D = (1/r_{max} - 1/r_{min})g_z,$$

where g_z is *vertical gravity* representing the gravity field strength and r_{max} and r_{min} are radii of the maximum and minimum curvatures, respectively, at each point on the equipotential surface along the orthogonal x- and y-directions in *Cartesian coordinates*. In terms of gradients of the anomalous gravity field, the differential curvature K_D is expressed as

$$K_D = K_{max} - K_{min} = [(g_{xx} - g_{yy})^2 + 4g_{xy}^2]^{1/2}/g_z$$
$$= 2(g_{UV}^2 + g_{NE}^2)^{1/2}/g_z,$$

where K_{max} is maximum curvature; K_{min} is minimum curvature; $g_{xx} = \partial^2U/\partial x^2$, $g_{yy} = \partial^2U/\partial y^2$, $g_{xy} = \partial^2U/\partial x\partial y$ are *curvature gradients*; U is *gravitational potential*; and g_{NE} and g_{UV} are *differential curvature gradients* related to the curvature gradients as

$$g_{NE} = g_{xy}, \quad g_{UV} = (g_{xx} - g_{yy})/2.$$

The most accurate differential curvature values are calculated from the diagonal components of a *rotated gradient tensor* as

$$K_D = \pm(T_{xx} - T_{yy})/g_z,$$

where the *gradient tensor* components $T_{xx} = g_{xx}$, $T_{yy} = g_{yy}$, and the curvature sign depends on which of the x- and y-axes is aligned with the maximum curvature direction (Li, X., 2015; Li, Y., 2018). The differential curvature is zero for point and spherical sources, wherein the nonzero values quantify deviations of the equipotential surface from a spherical shape. The K_D reaches its maxima over high-*offset* faults and high-angle edges of source bodies, where deviations of the *density* distribution and source shape from a homogeneous *sphere* are the largest. The absolute value, or *magnitude*, of the differential curvature $|K_D|$ is calculated without g_z and referred to as the *horizontal directive tendency (HDT)* or *differential curvature amplitude*. See also *curvature*, *curvature attributes*, *edge*, *gradient*, and *point source*.

differential curvature amplitude: the *horizontal directive tendency (HDT)* expressed as

$$|K_D| = [(U_{xx} - U_{yy})^2 + 4U_{xy}^2]^{1/2},$$

where U is *gravitational potential* and the subscripts refer to its differentiation with respect to the Cartesian axes, so that $U_{xx} = g_{xx}$, $U_{yy} = g_{yy}$, and $U_{xy} = g_{xy}$. Also referred to as *differential curvature magnitude*.

differential curvature gradients: second-order horizontal derivatives of the *gravitational potential* implemented in the calculation of the *differential curvature* of its *equipotential surface*. They are often denoted by g_{NE} and g_{UV} and defined as, respectively,

$$g_{NE} = g_{xy} = \partial^2 U / \partial x \partial y,$$

$$g_{UV} = (g_{xx} - g_{yy})/2 = (\partial^2 U / \partial x^2 - \partial^2 U / \partial y^2)/2,$$

where U is gravitational potential and g_{xx}, g_{xy}, g_{yy} are horizontal gradients of the horizontal components of the *Earth gravity field* that are known as *curvature gradients*. Previously obtained from measurements by a *torsion-balance gradiometer*, now the differential curvature gradients g_{NE} and g_{UV} are measured by a *partial tensor gravity gradiometer* or obtained from measurements by a *full tensor gravity gradiometer*. See also *derivative* and *gradient*.

differential curvature magnitude: a numerical value of the *differential curvature* K_D calculated without the scaling factor g_z and referred to as the *horizontal directive tendency (HDT)* or *differential curvature amplitude*.

differential downward continuation: 1. an *enhanced processing* technique based on the calculation of the difference between two downward continuations of an anomalous gravity or magnetic field. **2.** a method of recalculating an observed gravity or magnetic field from the *altitude* of measurements to a lower altitude, such as *equivalent-source continuation*. See also *anomalous field*, *differential continuation*, and *downward continuation*.

differential GPS (DGPS): an advanced version of the *Global Positioning System (GPS)* providing the most accurate calculation of the *three-dimensional (3D)* coordinates of gravity or magnetic observation points: latitude, longitude, and *elevation*. Satellite-transmitted signals acquired by a moving GPS receiver in a *survey* area are compared with those at a stationary GPS receiver, or *GPS base station*, whose 3D coordinates are known with high *accuracy*. Microwave-range *phase* differences between the moving and stationary GPS receiver signals are used to correct the moving GPS receiver data. Achievable accuracy is less than 1 cm for the horizontal position and a few centimeters for the vertical position, or elevation, above a *reference ellipsoid*.

differential median filtering: a method of *leveling* and *microleveling* without *tie lines*. It is based on (a) the application of the one-dimensional (1D), or line-oriented, *median filter* and 2D median filter with a circular *moving window*, and (b) the calculation of the difference between their values for an *observation point* at the center of both moving windows. The difference between the 1D and 2D median filter values comprises the leveling misfit which is added to or subtracted from the data value. Moving window lengths for both median filters are determined empirically, wherein the radius for the 2D filter is larger than the maximum distance between three adjacent flight lines. Prior to application, the *regional trend* is removed from the data. Level-corrected data are smoothed before *gridding* (Mauring and Kihle, 2006). See also *smoothing* and *detrending*.

differential-mode acceleration: same as *differential acceleration* with an emphasis on its difference from *common-mode acceleration* of a moving *observation platform*, such as a *survey* ship or airplane and planet-orbiting satellite.

differential pseudogravity: a *pseudogravity field* obtained by *pseudogravity transformation* of the differentially reduced to pole *total magnetic intensity (TMI)* field. It is used in the regional-scale *interpretation* of the major magneto-tectonic provinces, because their anomalous magnetic signatures are enhanced by vertical integration in the process of the pseudogravity transformation. See also *differential reduction-to-pole (DRTP)*, *enhanced processing*, *magnetic signature*, and *first vertical integral (FVI)*.

differential reduction-to-pole (DRTP): the *grid*-based *reduction-to-pole (RTP)* applied to the aeromagnetic data in regional-scale *survey* areas extending over 6°–10° of latitude or longitude. Differing from standard RTP algorithms, which use the constant values of *magnetic inclination* and *magnetic declination* for the whole survey area, the DRTP algorithm accounts for inclination and declination variations at each grid point. DRTP assumes the *induced magnetization* of rocks, and it is implemented (a) in the *wavenumber domain* via transformation of the *total magnetic intensity (TMI)* data into an *equivalent layer*, calculation of its *magnetic potential*, and point-by-point conversion of the potential values into the RTP magnetic anomaly (Arkani-Hamed, 1988; Blakely, 1996), or (b) in the *space domain* using the *Taylor series* expansion of the RTP dataset obtained with the constant area-averaged inclination and declination values, wherein variations of the inclination and declination values over the survey area are defined by their deviations from a corresponding area-averaged value (Cooper and Cowan, 2005). Differences between the DRTP and RTP anomalies are the largest at the most distant ends of an *aeromagnetic survey* area.

differential upward continuation: 1. an *enhanced processing* technique based on calculation of the difference between two upward continuations of an anomalous gravity or magnetic field. A *Jacobsen filter* algorithm implements the differential upward continuation. **2.** a method of recalculating the observed gravity or magnetic field from the *altitude* of measurements to a higher altitude, such as for *merging* gravity and magnetic grids obtained from airborne data acquired at different flight altitudes. See also *acquisition*, *anomalous field*, *differential continuation*, *upward continuation*, *equivalent-source continuation*, and *grid*.

differential upward continuation filter: same as *Jacobsen filter* with an emphasis on its operational principle based on the use of *differential upward continuation*.

diffraction correction: a correction applied to *absolute gravity* measurements with the use of a laser interferometer and free-falling corner-cube reflector to compensate for the effect of diffraction on the interference of the reflected and reference wavefronts in the recombined laser beam of a *free-fall absolute gravimeter*. The diffraction effect is a source of systematic errors in high-*accuracy* measurements that makes the measured absolute gravity value lower than its actual value at an *observation point*. Diffraction correction depends on the laser beam diameter, and it is estimated at the order of 1–5 *microgal (μGal)*. The correction value is added to the measured absolute gravity value. See also *atmospheric pressure correction*, *polar motion correction*, *self-attraction correction*, *speed-of-light correction*, and *tilt correction*.

Digital Elevation Model (DEM): same as *Digital Terrain Model (DTM)* that represents a *three-dimensional (3D)* model of the Earth's surface described in values of *elevation*, or height, above a *survey datum*: sea level or *reference ellipsoid*.

Digital Terrain Model (DTM): a three-dimensional (3D) *model* of the Earth's surface constructed from airborne and spaceborne (satellite) radar and *LIDAR* measurements. In exploration applications, DTM is used for the calculation of *terrain correction* and pre-programming of the aircraft *drape flight* for airborne gravity and magnetic surveys. Also known as *Digital Elevation Model (DEM)*. See also *radar altimeter*.

dike: an *igneous intrusion* which moves upward through the previously deposited rock formations along the structural weaknesses created by faults and fracture zones in the sedimentary section and *basement*. It has the tabular shape of a finite thickness, or width, which is much shorter than its vertical and another lateral *dimension*. Due to its high *density* and *magnetization*, the dike is among the main sources of observed gravity and magnetic anomalies and one of the basic 2D *source geometry* approximations in the magnetic and gravity *inversion* methods, including *Euler deconvolution* and *Werner deconvolution*. There are *thin dike* and *thick dike* source geometries. Magnetic and gravity signatures of dikes are controlled, primarily, by their depth and width: the shallower and the narrower the dike, the higher the *amplitude* and the shorter the *wavelength* of a corresponding gravity or magnetic *anomaly*. Occasionally spelled as *dyke*. See also *diapir*, *fault*, *infinite dike*, *gravity signature*, *magnetic signature*, and *tabular body*.

dike swarm: a local- or regional-*scale* assemblage of dikes of a similar orientation *pattern*: parallel, subparallel, and radiating. Depending on the direction of the *main geomagnetic field* at the time of their intrusion into the sedimentary section and amount of *remanent magnetization* acquired, dike swarms have either positive or negative anomalous magnetic signatures in the *total field*, and they are enhanced by calculation of the *residual field*. In some areas, the high-*amplitude* linear and curvilinear gravity and magnetic anomalies generated by dike swarms can be correlated over hundreds of kilometers. See also *anomaly*, *dike*, *igneous intrusion*, *gravity signature*, and *magnetic signature*.

dilation: See *wavelet dilation*.

dimension: 1. a measurable property by which other properties are measured or calculated, e.g., *acceleration* is a physical dimension of a *gravitational field*. **2.** a fundamental measure which defines the number of variables in a given dataset, i.e., a N-dimensional dataset will have N variables described by their spatial parameters, such as *one-dimensional (1D)* length, two-*dimensional (2D)* area, and *three-dimensional (3D)* volume, or by their physical property characteristics, such as *mass*, distance, and time. Two-dimensional *modeling* describes a magnetic or gravity *model* in one dimension of the depth z and one dimension of the horizontal distance x, whereas 3D modeling describes a *model* in the vertical dimension of depth z and two horizontal dimensions along the mutually orthogonal x- and y-directions in *Cartesian coordinates*. Time is the fourth dimension in *4D gravity* measurements. **3.** the lateral or vertical extent of a *source body* or geological *structure*. See also *2.5D model* and *2.75D model*.

dimensionality grid: a *color-coded grid* of the *invariant ratio* values calculated for each observation point from *full tensor gradient (FTG)* data. Invariant ratio values are close to 0 for a two-dimensional (2D) *source body* approximation and approach 1 for the compact three-dimensional (3D) sources approximated by a *sphere*. See also *dimension*, *gridding*, and *2D source*.

dimensionality indicator: same as *invariant ratio* representing a spatial characteristic of a source shape approximated by a two-dimensional (2D) or three-dimensional (3D) *source body*.

dip: 1. the vertical angle between a structural surface, such as a fault plane or continuous surface of a gravity and magnetic *grid*, and the horizontal plane. Dip is measured in the vertical plane along the direction perpendicular to the *strike* of a given surface. **2.** *horizontal gradient* representing the *slope* of a two-dimensional (2D) *profile* or dip of a three-dimensional (3D) surface, such as a gridded gravity or magnetic field. See also *gradient-dip estimation* and *gridding*.

dip equator: same as *magnetic equator* representing an imaginary line on the Earth's surface along which the *main geomagnetic field* is horizontal and *magnetic inclination*, also known as *magnetic dip*, is zero for a particular *epoch*.

dip poles: locations on the Earth's surface where the *main geomagnetic field* is presently vertical. Dip poles are not coincident with the geographic poles defined by the Earth's axis of rotation. North and south dip poles are referred to as the *magnetic north pole* and *magnetic south pole*, respectively. Also referred to as *magnetic dip poles*.

dipolar anomaly: 1. a *magnetic anomaly* of the sources approximated by a *magnetic dipole* which is also referred to as a *dipolar source*. A dipolar anomaly has an asymmetric *shape* with non-equal positive and negative parts associated with the *dipolar magnetization* of its source. The asymmetric shape of a dipolar anomaly is controlled, primarily, by *magnetic inclination* and the amount of *remanent magnetization* in the source. Its *amplitude* depends on the *magnetization contrast* and source depth: the higher the contrast, the higher the amplitude, and the deeper the source, the lower the amplitude. For sources with positive magnetization contrasts, the maximum (positive) values of a dipolar anomaly are shifted from the source center toward the *magnetic equator*, whereas the minimum (negative) values are shifted toward the magnetic poles. The amount of shift depends on the inclination of the Earth's magnetic field, source depth, and source shape. The asymmetry and shift of a dipolar magnetic anomaly are eliminated by *reduction-to-pole (RTP).* All magnetic anomalies are source-dependent versions of a dipolar anomaly. **2.** a *free-air gravity* anomaly caused by *free-air edge effect.* See also *anomaly, anomaly shape (mag.),* and *monopolar anomaly.*

dipolar magnetization: same as *magnetization* with an emphasis on its dipolar nature, wherein all magnetized materials and bodies of the anomalous magnetization have two magnetic poles: a north, or positive, pole and a south, or negative, pole.

dipolar source: a magnetic *source body* approximated by a *magnetic dipole.* All magnetic materials and bodies of anomalous *magnetization* represent the assemblages of sources for which the fundamental building blocks are magnetic dipoles. See also *dipole field.*

dipole: a *point source* approximation of the source of a *magnetic anomaly.* Commonly referred to as *magnetic dipole.* See also *monopole.*

dipole field: 1. *magnetic dipole field.* **2.** a major portion of the *main geomagnetic field* that is also referred to as *eccentric dipole field.* See also *nondipole field.*

dipole moment: See *magnetic dipole moment.*

dipole strength: See *magnetic dipole strength.*

dipole surface: a spread of magnetic dipoles on the surface obtained by *inverse calculation* of the equivalent distribution of *magnetization* in the *equivalent-source method.* The dipole surface can be of any shape and located at or below the observation surface, but it is always above all magnetic sources. See also *magnetic dipole.*

direct analytic signal (DAS): a *complex variable* for which the real part is the function f representing an *anomalous field,* such as *Bouguer anomaly* or *RTP anomaly,* and the imaginary part consists of two Hilbert transforms of this function in the orthogonal x- and y-directions. Differing from the standard *analytic signal (AS)* calculated from the horizontal and vertical derivatives of an *anomalous field,* DAS is calculated directly from the gridded anomalous field. The DAS at the *grid* point (x, y) is expressed as

$$\mathrm{DAS}(x,y) = f\mathbf{e}_z + i\{H_x f\mathbf{e}_x + H_y f\mathbf{e}_y\},$$

where $i = \sqrt{-1}$ is a complex unit number indicating orthogonality, H_x and H_y are x- and y-components of the *3D Hilbert transform,* and $\mathbf{e}_x$, $\mathbf{e}_y$, $\mathbf{e}_z$ are unit vectors parallel to the mutually orthogonal x-, y-, z-directions, respectively, in *Cartesian coordinates.* The DAS is equivalent to the *monogenic signal (MS)* and analogous to the analytic signal (AS) of the *vertical integral* of a gravity or magnetic *anomaly,* i.e., it is analogous to the AS of the *anomalous potential.* The DAS is used for calculation of the four attributes of anomalous fields: *direct analytic signal amplitude (DASA), horizontal DAS (HDAS), improved tilt angle (ITA),* and *improved theta map* (Luo et al., 2011). See also *directional analytic signal.*

direct analytic signal amplitude (DASA): the absolute value, or *magnitude,* of a *direct analytic signal (DAS)* expressed as

$$\mathrm{DASA} = |\mathrm{DAS}| = \{f^2 + [H_x f]^2 + [H_y f]^2\}^{1/2},$$

where f is the anomalous gravity or magnetic field, such as *Bouguer anomaly* or *RTP anomaly* and H_x and H_y are x- and y-components of the *3D Hilbert transform* of the field f along the mutually orthogonal x- and y-directions, respectively, in *Cartesian coordinates.* The DASA is equivalent to (a) the *zero-order ASA* (ASA_0) and (b) the *local amplitude* of the *monogenic signal (MS),* and it is analogous to the *analytic signal amplitude (ASA)* of the *vertical integral* of a magnetic or gravity *anomaly,* i.e., it is analogous to the ASA of *anomalous potential.*

direct Fourier transform: See *forward Fourier transform.*

direct modeling: calculation of the gravity or magnetic field of a *model* representing an isolated *source body* or geological *structure* with a given or assumed depth, geometry, and *density contrast* or *magnetization contrast.* It is opposite to *inverse modeling* and referred to with the same meaning as *forward calculation* and *forward problem solution.*

direct problem: a *forward problem* in the context of calculating the gravity or magnetic effect, or *field,* of a particular *model,* as opposed to the *inverse problem* of calculating the model parameters from the observed gravity or magnetic field.

directional acceleration: See *inertial acceleration.*

directional analytic signal: an *analytic signal (AS)* composed of the second-order directional derivatives of the *gravitational potential* or *magnetic potential.* Amplitudes of three directional analytic signals are calculated along the x-, y-, z-directions in *Cartesian coordinates* using the three rows in the nine-element *matrix* of the *full tensor gradient (FTG)* as

$$|A_x| = (P_{xx}^2 + P_{xy}^2 + P_{xz}^2)^{1/2} = (T_{xx}^2 + T_{xy}^2 + T_{xz}^2)^{1/2},$$

$$|A_y| = (P_{xy}^2 + P_{yy}^2 + P_{yz}^2)^{1/2} = (T_{xy}^2 + T_{yy}^2 + T_{yz}^2)^{1/2},$$

$$|A_z| = (P_{xz}^2 + P_{yz}^2 + P_{zz}^2)^{1/2} = (T_{xz}^2 + T_{yz}^2 + T_{zz}^2)^{1/2},$$

where P is the gravity or magnetic *potential* and T_{ij} are FTG components. Directional analytic signal amplitudes are very sensitive to different azimuthal directions of their calculation and, therefore, can be used for mapping of *fault*-related structural discontinuities, high-angle edges of isolated source bodies, and other lithological contacts (Beiki, 2010). See also *azimuth, contact, directional derivative, edge-detector attribute,* and *direct analytic signal (DAS).*

directional ASA: the absolute value, or *magnitude,* of a *directional analytic signal* calculated from one of three rows in the nine-element *matrix* of the *full tensor gradient (FTG)* of a anomalous gravity or magnetic field. In the general form, the directional ASA is expressed as

$$|A_j| = [(T_{jx})^2 + (T_{jy})^2 + (T_{jz})^2]^{1/2},$$

where T_{jx}, T_{jy}, T_{jz} are FTG components in *Cartesian coordinates,* and $j = x, y, z$. Directional analytic signal amplitudes satisfy the *Euler homogeneity equation* and, therefore, can be used in high-resolution *Euler deconvolution* applications (Beiki, 2010). See also *analytic signal amplitude (ASA)* and *direct analytic signal amplitude (DASA).*

directional cosine filter: a *directional filter* with the cosine transition curve providing a smooth decrease of *grid* values to zero beyond the grid edges in order to minimize *edge effects* in the process of *microleveling*. See also *cosine taper filter*.

directional derivative: the rate of change of the gravity or magnetic *field* per unit distance in a particular direction. Directional derivative of the field F in the direction r is expressed as

$$\partial \mathrm{F}/\partial \mathrm{r} = \nabla \mathrm{F} \cdot \mathbf{r}_1 = \nabla \mathrm{F} \cos\theta,$$

where ∇ is vector *gradient* operator *del*, $\mathbf{r}_1$ is a unit vector parallel to the direction r, and θ is the angle between the original field *vector* and unit vector $\mathbf{r}_1$. See also *derivative* and *partial derivative*.

directional filter: a *grid*-based spectral-domain *filter* which either passes or rejects the *wavenumber* components of the gravity or magnetic anomalies that are trending, or striking, along a user-specified range of the azimuthal directions. There are two modes of its operation: *strike*-pass and strike-reject. A common side effect in the application of this filter is stretching the anomalies in the pass direction. The option of using the strike-reject mode with a narrow range of azimuthal directions creates much less distortions than the strike-pass mode. The selection of parameters for the directional filter is based on the use of a *map power spectrum*, where directions of the linear trends in the *spectral domain* are orthogonal to the directions of the original anomalous trends in the *space domain*. Sometimes referred to as *pie-slice filter* and *strike filter*. See also *azimuth*, *spectral domain*, *trend*, and *directional cosine filter*.

directional filtering: a *grid*-based spectral-domain *filtering* technique applied in order to (a) pass, or retain, the anomalous components of a gravity or magnetic *field* that are trending along a particular range of the azimuthal directions or (b) suppress *regular noise* components oriented in a user-defined azimuthal direction or along the *acquisition* lines. Both the suppression of regular noise trends and the enhancement of anomalous alignments improve the visualization of the low-*amplitude* anomalies of an *exploration* interest. A conceptual analogue of directional filtering in the space-domain processing is *sunshading*, also known as *shaded relief*, which highlights anomalies associated with the structural-tectonic trends in a study area. See also *azimuth*, *space domain*, *spectral domain*, *trend*, and *principal component analysis (PCA)*.

directional Hilbert attribute: a secondary *complex attribute* of the *monogenic signal (MS)*. It is composed of two imaginary parts obtained from the real part of the MS by the first-order *Riesz transform* operators R_1 and R_2 and projected on the direction of the monogenic signal orientation. It is expressed as

$$H_\theta = R_1 f(x,y)\cos\theta + R_2 f(x,y)\sin\theta,$$

where $f(x,y)$ is the real part representing the anomalous gravity or magnetic field value at the *grid* point (x,y) and θ is the angle of *local orientation* which is the orientation of the monogenic signal with respect to the mutually orthogonal imaginary axes on the complex planes. Calculation of the directional Hilbert attribute transforms an *anomalous field* in a manner similar to the *FTG rotation*, and it is used in *enhanced processing* of *gradient* and high-resolution *total field* data, such as *full tensor gravity gradient (FTGG)* and *high-resolution aeromagnetic survey* data. See also *complex plane*, *lateral resolution*, and *Riesz transform attribute*.

directional tilt angle: a quantitative characteristic, or *attribute*, of an *anomalous field* defined by the arctangent of the ratio of the *horizontal derivative* in a preselected direction to the square root of the sum of the squared horizontal derivative orthogonal to the selected direction and the squared *vertical derivative* of the anomalous magnetic or gravity field F:

$$T_x = \arctan\{(\partial \mathrm{F}/\partial x)/[(\partial \mathrm{F}/\partial y)^2 + (\partial \mathrm{F}/\partial z)^2]^{1/2}\},$$

$$T_y = \arctan\{(\partial \mathrm{F}/\partial y)/[(\partial \mathrm{F}/\partial x)^2 + (\partial \mathrm{F}/\partial z)^2]^{1/2}\}.$$

The directional tilt angle is the *amplitude*-normalized horizontal derivative that enhances the anomalous gravity or magnetic field components in the user-specified x- and y-directions, where x is any chosen direction and y is the direction orthogonal to x (Cooper and Cowan, 2006). See also *horizontal tilt angle (TDX)* and *normalized derivatives*.

Dirichlet conditions: two necessary and sufficient mathematical conditions for *Fourier analysis* of gravity and magnetic data represented by the time- or space-domain variable f: (1) f is a single-valued and continuous function, except for the finite number of its discontinuities; and (2) f has only a finite number of the maxima and minima and a finite energy. See also *space domain* and *time domain*.

discrete Fourier transform (DFT): *Fourier transform* of discretely sampled data with limitations on both the longest and shortest wavelengths. The DFT is periodic with a *period* Δk, which is inversely proportional to *sample interval* Δx and defined as

$$\Delta k = k_s = 2\pi/\Delta x,$$

where k_s is called *sampling wavenumber*. A set of the space-domain data sampled at the interval of Δx is converted by DFT into a set of equivalent values at the regular interval of $2\pi/\Delta x$ in the *Fourier domain*. Half of the sampling wavenumber is called the *Nyquist wavenumber* k_N, i.e., $k_N = \pi/\Delta x$, which represents the largest *wavenumber* in the DFT that corresponds to the shortest *wavelength* of the sampled data transformed adequately by the DFT, so it also can be expressed as

$$k_N = 2\pi/\lambda_N,$$

where $\lambda_N = 2\Delta x$ is the *Nyquist wavelength*. The smallest wavenumber available in the DFT is called the *fundamental wavenumber*, and it corresponds to the longest wavelength that can be transformed into the Fourier domain (Blakely, 1996). See also *sampling* and *fast Fourier transform (FFT)*.

discrete wavelet transform (DWT): the *wavelet transform (WT)* method that involves three basic procedures: (1) decomposition of the original line-oriented or gridded data into the successive ranges, or scales, of *wavelength resolution*, wherein the data are divided into high-pass (HP) and low-pass (LP) parts by *half-band filtering*, and each subsequent *scale* is represented by the *wavelength* components within the twice shorter *bandwidth* than at a previous scale; (2) *convolution* of the dilated and translated *mother wavelet* with the HP and LP parts of the data at each scale of decomposition, wherein the scale-corresponding *wavelet dilation* and *wavelet translation* are restricted to powers of 2, so the operating wavelet length is either doubled or halved, and it is shifted twice or one-half the increment at a previous scale of the wavelength resolution; and (3) calculation of the convolution-derived *wavelet coefficients* at all *wavelet* positions along the data points. Because the division into the HP- and LP-filtered parts doubles the dataset size, both parts are downsampled at each scale by discarding every second data point. At the second and all successive DWT scales, the data part containing the majority of energy (usually it is the LP-filtered part) is decomposed repeatedly into the HP- and LP-filtered parts. Because the DWT uses the scales that are powers of 2, the dataset length defines the number of scales into which the dataset can be decomposed: for the length $N = 2^m$, there are *m* possible scales of the DWT decomposition with two data points remaining at the last scale. Locations of the individual wavelength components of the data are determined by the line or *grid* intervals, where a particular wavelength component and operating wavelet closely match each other and the wavelet coefficients reach their maximum values. Obtained wavelet coefficients are (a) displayed on the DWT space-scale plane for visualization of scales contaminated by *random noise* and *coherent noise* and (b) arranged in the format of a *matrix* for subsequent processing of the transformed data (Hornby et al., 1999; Ridsdill-Smith, 2000). See also *continuous wavelet transform (CWT)*.

displacement vector: a *vector* directed from a targeted *source body* to a particular observation point in a gravity or magnetic *survey* area. See also *target*.

distant relief effect: the gravity effect of *topography* and *bathymetry* beyond the distance of 166.735 km within which *Bouguer correction* and *terrain correction* are commonly calculated. This effect is compensated in the process of *gravity reduction* of the data obtained from continental- and global-scale gravity studies.

disturbing potential: *gravitational potential* T defined by the difference between the *Earth gravity potential* W and *normal gravitational potential* U at the same point on the Earth's surface, i.e.,

$$T = W - U.$$

Because a *reference ellipsoid* has the same *angular velocity* of rotation as the Earth, its *centrifugal potential* Φ_{RE} is approximately the same as that of the Earth, i.e., $\Phi_{RE} \cong \Phi_E$, so the disturbing potential T can be expressed as

$$T = W - U \cong V_E - V_{RE},$$

where V_E and V_{RE} are self-gravitational potentials of the Earth and reference ellipsoid, respectively. Disturbing potential is the *potential* of the anomalous gravitational attractions that are not accounted by the ellipsoidal shape of the Earth and its rotation. It is used in *satellite gravity* studies and geodetic applications for the analysis of *geoid undulation* and mapping of the surface of a *geoid*. Also referred to as *anomalous potential* and *residual potential*. See also *Bruns equation*, *gravitational attraction*, and *self-gravitational potential*.

diurnal correction: a correction applied to *observed magnetics* to compensate for the time-variable irregular changes, or fluctuations, of *external magnetic field* of the Earth that are called *diurnals*. Its application is based on the point-for-point subtraction of records made by a synchronized stationary magnetometer, or *base-station magnetometer*, from the recorded ground or airborne *magnetic survey* data. For marine (shipboard) magnetic surveys, the application of the diurnal correction is often based on the *decorrelation* of acquired magnetic data with data recorded at the nearest *geomagnetic observatory*. See also *aeromagnetic survey*.

diurnal drift: a continuous record of *diurnals* by the stationary *base magnetometer* synchronized with a moving *magnetometer* during a ground or airborne *magnetic survey*. Diurnal drift records are examined on a *station*-by-station or line-by-line basis to determine those stations or survey lines that were acquired with the level of diurnals exceeding their *tolerance* defined by *survey specifications*. Airborne survey lines for which the diurnals do not satisfy the planned tolerance are re-flown. See also *acquisition*.

diurnal variation: *diurnals* in the context of the time-variable irregular changes of the *external magnetic field* of the Earth.

diurnals: 1. daily time-variable irregular changes, or fluctuations, of the *external magnetic field* of the Earth caused by the combined effects of: (a) solar radiation, i.e., *solar wind*; (b) the Earth's rotation within its *magnetosphere*; and (c) movements of the ionized molecules which release electrons and produce the transient electric currents in the *ionosphere*. Diurnals also include spatially variable disturbances, known as *micropulsations* and *bay*, and secondary magnetic effects, such as the *coastline effect*. The magnitudes of diurnals are generally on the order of 10–100 *nanotesla (nT)*, and they depend on the latitude of the measurements, season of the year, and time of the day. Diurnals increase in magnitude with the increase of their periods ranging widely from several seconds to several hours and days. Diurnals are recorded by a *base-station magnetometer* and removed from *observed magnetics* by the application of *diurnal correction* prior to *leveling* and *enhanced processing*. Also referred to as *diurnal variation* and *solar daily variation*. **2.** daily temperature cycles monitored during *ground gravity* surveys for the calculation of *temperature correction*. See also *equatorial electrojets* and *K-index*.

divergence: a *scalar* quantity representing the three-dimensional (3D) spatial *derivative* of *vector field* **A** and denoted by $\nabla\mathbf{A}$, where ∇ is *gradient* operator *del*. Divergence describes the outward field flow, or *flux*, per unit volume that is used to estimate a decrease in field strength due to geometrical spreading. When the flux of vector field **A** entering the volume is equal to the flux leaving the volume, the divergence is zero, and this field is said to be a divergenceless vector field, or *solenoidal field*. For example, *magnetic induction* **B**, also referred to as *magnetic field*, is a divergenceless field. In *Cartesian coordinates*, the divergence is expressed as

$$\nabla\mathbf{A} = \partial A_x/\partial x + \partial A_y/\partial y + \partial A_z/\partial z,$$

where A_x, A_y, A_z are components of vector field **A** along the *x*-, *y*-, *z*-directions, respectively. See also *divergence theorem* and *Gauss law*.

divergence theorem: the theorem postulating that the integral of the normal, or orthogonal, component of *vector* **A** over the closed surface *S* is equal to the *divergence* of this vector integrated over the volume *V* contained by this surface

$$\int_S \mathbf{A}\cdot\check{\mathbf{n}}\, dS = \int_V \nabla\mathbf{A} dv,$$

where $\check{\mathbf{n}}$ is a unit *vector* orthogonal to surface *S*, $\mathbf{A}\cdot\check{\mathbf{n}}$ is a *dot product* that defines the normal component of vector **A**, and ∇ is *gradient* operator *del*. Divergence theorem is implemented in the *Gauss law* to describe the properties of a *magnetic field*. Also referred to as *Gauss theorem*.

dividing factor: a denominator in the formula of a source depth determined by the *Peters half-slope method*, *Sokolov method*, and other graphical methods based on the use of *depth rules* which relate the characteristic points on profiles of isolated, or non-interfering, anomalies to the depth of their respective sources. See also *profile* and *index factor*.

DLC methods: a group of magnetic methods for the detection, location, and classification (DLC) of magnetic objects approximated by *point dipole* sources. The DLC methods are implemented in the search for naval mines, *unexploded ordnance (UXO)*, sunken ships, buried fuel drums, archaeological artefacts, etc.

DNSC08: the global model of the *Earth gravity field* released by the Danish National Space Centre (DNSC) in 2008 in the format of a *grid* with 1×1 *arc-minute* ($\sim2\times2$ km) *grid cell* size. It is based on integration of *satellite altimetry*-derived *free-air gravity* data over oceans and seas with the *field* of the continental gravity anomalies derived from the geopotential *Earth Gravitational Model 2008 (EGM08)* of the U.S. National Geospatial Intelligence Agency (NGA). The estimated *rms error* in calculation of anomalies is approximately 2–3 *milligal (mGal)* for wavelengths longer than 20 km. The DNSC08 presents improved *lateral resolution* of anomalies in the coastal and polar regions, and it is the first global model of the Earth's gravity field without a polar gap, as it covers the whole Earth from 90° North to 90° South. See also *gravity anomaly*, *geopotential model*, and *wavelength*.

domain: **1.** a mathematical set of variables that defines *dimension* as an independent variable and the corresponding values of a particular function as the dependent variables. For example, the *space domain* defines a gravity or magnetic *field* as a function which varies depending on three independent variables *x*, *y*, *z* in *Cartesian coordinates*. **2.** a microscopic-scale volume in a rock-forming mineral with a uniform *magnetization* that is known as *magnetic domain*. See also *Fourier domain* and *wavelet domain*.

dominant frequency: **1.** a characteristic of an anomalous gravity or magnetic field that describes the *spatial frequency* which is observed the most frequently within a particular area of a *survey*. **2.** a *wavelet* parameter representing the reciprocal, or inverse, of a peak-to-peak interval in the wavelet. **3.** the reciprocal of the *dominant period*. See also *anomalous field* and *frequency*.

dominant period: the time or distance interval between the principal adjacent peaks (highest values) or troughs (lowest values) in a non-periodic *waveform*. See also *period*.

dominant wavelength: **1.** a characteristic of an *anomalous field* that describes the *wavelength* that is observed the most frequently within a particular area of a *survey*. **2.** the horizontal distance between the principal adjacent peaks (highest values) or troughs (lowest values) in a non-periodic *waveform*.

dot product: same as *scalar product* with an emphasis on the dot "·" used as a mathematical symbol for multiplying the magnitudes of two vectors. See also *magnitude*.

double Bouguer correction: **1.** *Bouguer correction* applied to gravity measurements made by an *underwater gravimeter* to compensate for (a) the upward *gravitational attraction* of the water mass above the gravimeter and (b) the difference between the seawater *density* and *replacement density*. **2.** Bouguer correction applied to gravity measurements made above and below a *target* layer in a mine or at two successive depths in a drilled well (borehole). See also *borehole gravimetry* and *lake gravity*.

downsampling: decreasing the original *sampling* of (a) line-oriented data by averaging the observed values within a pre-selected *grid interval* before *gridding* or (b) gridded data by using the reduced number of the *grid* values, e.g., every second grid value, every third, fourth, etc. Downsampling by the power of 2 is implemented in a *discrete wavelet transform (DWT)* to represent the low-pass and high-pass parts of the transformed data, where at each level, or *scale*, of transformation the original data sampling is successively reduced by one-half: every second data point is used at the first level, every fourth at the second level, every eighth at the third level, etc. Referred to with the same meaning as *subsampling*. See also *undersampling*, *upsampling*, and *resampling*.

downward continuation: calculation of the gravity or magnetic field at an *elevation* that is lower than the elevation at which it was measured, i.e., closer to all sources in the subsurface. In the *Fourier domain*, its mathematical expression is

$$\mathcal{F}(k,h) = \mathcal{F}(k,0)e^{hk},$$

where $\mathcal{F}(k,h)$ is the *Fourier transform* of the field at elevation $z=h$, $\mathcal{F}(k,0)$ is the Fourier transform of the field at elevation $z=0$, *k* is *wavenumber*, and *h* is continuation distance. Downward continuation amplifies all components of an *anomalous field*, including the high-frequency *noise*. The degree of amplification increases exponentially with increasing k and h, thereby emphasizing the short-*wavelength* field components, because the shorter the wavelength, the higher its wavenumber. There are no *phase* changes in the downward-continued field when it is calculated by a level-to-level continuation. At some downward continuation distance, the process becomes unstable, as the high-wavenumber components from shallow and near-surface sources are amplified much faster than those from the deeper sources. Additional stability can be obtained by *upsampling*, which reduces the original *sample interval* and minimizes *Gibbs errors*. Downward continuation is analogous to a *high-pass filter*, and its application can help to improve *vertical resolution* of the short-*wavelength* anomalies of an exploration or research interest. See also *stabilized downward continuation*, *upward continuation*, and *differential continuation*.

drape correction: **1.** a correction applied to the aeromagnetic and, sometimes, airborne *gravity gradient* data to compensate for the *amplitude attenuation* of anomalies and other *drape effects* due to the variable *altitude* of an *airborne survey* flight. **2.** a procedure that re-calculates airborne magnetic or gravity data to a constant altitude in order to minimize the mis-ties at airborne survey line crossings. Drape correction has no effect on the *spatial frequency* content of measured data, because the *bandwidth* of measurements is not changed. Correction algorithms are based, primarily, on the *equivalent-source method* and *Taylor series* expansion. Referred to with the same meaning as *elevation correction* for aeromagnetic data. See also *mis-tie*.

drape effects: gravity and magnetic effects caused by variations of flight *altitude* during an *airborne survey*. Large flight-related changes in the distance between a *sensor* and *source body* result in variable *amplitude* attenuation of a measured field, spurious anomalies oriented along the flight lines, and non-geological variations of the high-*wavenumber* data content, especially in the airborne *gravity gradient* and aeromagnetic surveys for mineral exploration. Drape effects are reduced by *drape flight* and minimized by the application of *drape correction*, but they cannot be eliminated completely. See also *inertial acceleration* and *corrugations*.

drape flight: a *survey* aircraft flight which follows a smoothed topography at approximately the same *ground clearance* in order to maximize the lateral and vertical *resolution* of the short- and mid-*wavelength* anomalies and minimize *drape effects*. A drape flight is pre-programmed for the aircraft computer based on a *digital elevation model (DEM)* of the *airborne survey* area, flight characteristics of the aircraft, and safety regulations.

drape survey: an *airborne survey* during which measurements of the Earth's gravity or magnetic field and its gradients are made in the *drape flight* mode. A drape survey is common for *magnetic survey* measurements, whereas for gravity and *gravity gradient* airborne measurements a nearly constant *elevation* flight is preferred in order to minimize the non-gravitational inertial accelerations of a survey aircraft. See also *gradient*, *gravitational acceleration*, *inertial acceleration*, and *drone survey*.

draped image: a computer display or printed copy in which a *grid* image in *pseudocolor* is overlaid, or draped, on another grid image created in shades of gray or vice versa. See also *raster image*.

drift: an instrument characteristic describing the gradual change in its reference value, or zero setting, relative to which measurements are made, i.e., drift is one reason for an instrument tendency to change its readings as a function of time and temperature. Drift is related to the mechanical design of an instrument, thermoelastic properties of materials utilized in its measuring system, mechanical instabilities, and excessive vibrations during transportation. Drift is a particularly important consideration in gravity measurements using a *spring gravimeter* or spring-based *accelerometer*, and it is also present in certain types of magnetometers, such as a *fluxgate magnetometer*. Drift is different from *tare*, which is a sudden jump, or step-like change, in a sequence of measured values. Often referred to as *mechanical drift* and *instrument drift*. See also *residual drift*, *gravimeter drift*, *magnetometer drift*, *temperature drift*, *reading*, and *static test*.

drift correction: a time-varying correction applied to measured values to compensate for *gravimeter drift* and, if necessary, *magnetometer drift*. Drift correction is calculated with the use of a *drift curve* obtained from repeated measurements, or *reoccupations*, at the repeat stations and primary or secondary *base station* with time intervals of 2–3 hours in order to approximate *drift* by a linear function of time. Obtained drift curve values are distributed proportionally, or prorated, on a point-by-point basis between all values measured during the period of time used for calculation of the drift curve. A separately calculated gravity effect of the *Earth tides* is often included into drift correction. See also *automatic drift correction*, *Earth tide correction*, *loop drift rate*, *repeat station*, and *static correction*.

drift curve: a graphical plot obtained from gravity or magnetic measurements at the repeat stations and primary or secondary *base station* to determine *drift* as a function of time during each day of a *survey*. Intervals between the repeated measurements, or *reoccupations*, are minimized (usually, once every 2–3 hours, depending on the type of *gravimeter* or *magnetometer* and survey type) in order to approximate the drift curve by a linear function of time. Often, the *gravimeter drift* curve includes changes in the measured gravity values caused by the *Earth tides*, wherein the alternative approach is based on application of the *Earth tide correction* at each *station* before calculation of the drift curve. Obtained drift curve values are used for computing a time-varying *drift correction*. Drift curve is calculated with the assumption that a gravimeter had no tares in the process of observations. See also *primary gravity base*, *secondary gravity base*, *repeat station*, and *tare*.

drift rate: a change in the *gravimeter drift* value as a function of time during each day of a ground, marine (shipboard), or airborne *gravity survey*. The drift rate is commonly approximated by a linear function, and the amount of accumulated drift is determined by calculation of the difference between the repeated measurements, or *reoccupations*, at the primary or secondary base stations and other control points in a gravity survey area. The obtained difference value is distributed proportionally, or prorated, over readings at each observation point. See also *drift curve*, *loop drift rate*, *primary gravity base*, *secondary gravity base*, *reading*, and *gravimeter drift compensation*.

drone: 1. an *unmanned aerial vehicle (UAV)*, i.e., pilotless mini-aircraft, often a rotorcraft, equipped with the *GPS* antenna and attached magnetic sensor, such as a light-weight *fluxgate magnetometer*, *potassium magnetometer*, or *microfabricated atomic magnetometer (MFAM)*. Drone flies under remote control or autonomously following a software-controlled pre-programmed flight plan. **2.** an *autonomous underwater vehicle (AUV)* also known as a *sea-floor drone*.

drone magnetometer: a high-sensitivity *UAV magnetometer*, such as a *microfabricated atomic magnetometer (MFAM)* or light-weight *potassium magnetometer*, with its *sensitivity* up to 0.0002 *nanotesla (nT)*, *absolute accuracy* of ±0.1 nT, and *heading error* of ±0.05 nT.

drone survey: high-*resolution* airborne measurements of the *Earth magnetic field*, its gradients, and *gradient tensor* components with the use of a pre-programmed or remote-controlled *drone* and high-accuracy *positioning* system, such as *differential GPS (DGPS)*. A drone can fly at a very low *altitude* with a close *survey line spacing* to provide the highest possible *lateral resolution* of magnetic anomalies, especially those generated by sources in the upper sedimentary section and shallow subsurface, in various *exploration* and research applications, including shale hydrocarbon and mineral explorations. See also *gradient*, *fixed-wing survey*, *helicopter survey*, and *drone magnetometer*.

DRTP: See *differential reduction-to-pole (DRTP)*.

DTM: See *Digital Terrain Model (DTM)*.

dyke: See *dike*.

dynamic corrections: corrections applied to *dynamic gravity* measurements to compensate for the motion-related effects and inertial accelerations of a moving *observation platform*, such as a *survey* ship or aircraft. In accordance with the *Agerlin principle*, the calculated dynamic corrections are low-pass filtered with the same *low-pass filter* used for suppression of *dynamic noise* prior to application to the data. Referred to with the same meaning as *moving-platform corrections*. See also *cross-coupling corrections*, *Coriolis effect*, *Eötvös effect*, *inertial acceleration*, *turbulence effect*, and *wave-motion effect*.

dynamic gravimeter: a *gravimeter* the operational principle of which is based on measuring the parameters of a motion of its *sensor* elements, such as (a) the free-fall *acceleration* of a reflective object or streams of cold atoms in a *free-fall absolute gravimeter*, (b) the swing period of a *reversible pendulum* in a *pendulum gravimeter*, and (c) the resonant frequency of a vertically suspended and mass-loaded ribbon, or string, in a *vibrating-string gravimeter*. See also *static gravimeter*.

dynamic gravity: *relative gravity* and *gravity gradient* data acquired on a moving *observation platform*. Relative gravity measurements are made either by a *spring gravimeter* with a damped gravity *sensor* or by *accelerometer*-based instruments. Gravity gradients are measured by *gravity gradiometer* systems of various designs ranging from a simple assemblage of two vertically or horizontally separated gravimeters to the airborne *rotating-disk gravity gradiometer* and space-borne *GOCE gravity gradiometer*. See also *damping*.

dynamic noise: *noise* in gravity and magnetic measurements on a moving *observation platform*, such as a ship, aircraft, or satellite. The primary sources of dynamic noise in gravity measurements are the inertial accelerations produced by the *wave-motion effect* and *turbulence effect*, whereas for magnetic measurements these are temporal changes in the *external magnetic field* of the Earth and variations of the distance between shallow sources and a *magnetic sensor* that are caused by changes in aircraft *altitude* during a flight over rugged terrains. Dynamic gravity noise is minimized by: (a) a *stabilized platform* (b) specially designed instruments, such as a *rotating-disk gravity gradiometer*; and (c) application of *time averaging*, *time filter*, *cross-coupling corrections*, and other noise-suppressing procedures. Remaining, or residual, dynamic noise is commonly averaged over the operational *bandwidth* expressed in *hertz (Hz)*, so its root-mean-square (rms) estimates are presented in units of $1/\sqrt{Hz}$. Also known as *acquisition noise*. See also *Coriolis effect*, *Eötvös effect*, *inertial acceleration*, *temporal magnetic variations*, and *wavelet denoising*.

dynamic range: 1. an instrument characteristic described by the ratio of the maximum measured value, or maximum *reading*, to the minimum reading (often, *noise* level) which can be recorded by a particular *gravimeter* or *magnetometer* without changing its measurement scale. Commonly presented in units of *decibel (dB)*. **2.** the ratio of the highest mapped value to the lowest value on a *color-coded grid* or *contour grid*.

E

E: See *Eötvös (E)*.

Earth core field: same as *main geomagnetic field* with an emphasis on its origin from the aggregated electromagnetic effects of the *geomagnetic dynamo* process in the liquid outer Earth's core.

Earth curvature correction: *Bullard B correction* for the Earth's surface *curvature* which is not compensated by *Bouguer correction*.

Earth ellipsoid: *reference ellipsoid* in the context of the best-fitting mathematical approximation of the actual Earth shape and its heterogeneous internal composition by a homogeneous and rotating *ellipsoid*.

Earth Gravitational Model 1996 (EGM96): the *geopotential model* representing a global mathematical approximation of the *gravitational potential* of the Earth released by the U.S. National Geospatial Intelligence Agency (NGA) in 1996. The model is referenced to the *WGS84 ellipsoid* and completed to the spherical harmonic degree and order 360 corresponding to a *wavelength resolution* of 110 km in the mapping of global-scale geoid undulations. Also referred to as *EGM96 geoid*. See also *geoid*, *geoid undulation*, and *spherical harmonics*.

Earth Gravitational Model 2008 (EGM08): the *geopotential model* representing a global mathematical approximation of the *gravitational potential* of the Earth released by the U.S. National Geospatial Intelligence Agency (NGA) in 2008. The model is referenced to the *WGS84 ellipsoid* and based on data from worldwide satellite, airborne, marine (shipboard) and ground gravity observations. The EGM08 is expanded to the spherical harmonic degree and order 2159 corresponding to a *wavelength resolution* of gravity anomalies up to 20–30 km on the Earth's surface in areas of model-constraining gravity observations. Also referred to as *EGM2008 geoid*. See also *geoid*, *spherical harmonics*, and *constraints*.

Earth gravity field: a *vector field* of the force of *gravitational attraction*, or *gravitational force*, generated by the Earth's mass. Gravitational attraction at the point of measurement is defined by the gravitational force per unit *mass*. In accordance with the *Newton law of motion*, the force per unit mass describes a resulting quantity in the units of *acceleration*, so the physical *dimension* of the Earth's gravity field and measure of its strength is acceleration of a free-falling object. In a reference frame moving with the same *angular velocity* of rotation as the Earth, the Earth's gravity field values on or above the Earth's surface are defined by the *vector* sum of *gravitational acceleration* due to the attraction of the Earth's mass and *centrifugal acceleration* due to the Earth's rotation. The absolute value of the Earth's gravity field, or *absolute gravity*, is denoted by g and determined by direct measurements of the free-fall acceleration of cold atom streams or reflective object in a vacuum chamber, whereas the *relative gravity* Δg is determined by measurements of changes in the gravitational force acting on a spring-suspended or levitating *proof mass*. The average value of the Earth's gravity field is 9.8 m · s^{-2} = 980 *Gal* = 9.8 · 10^5 *milligal (mGal)*, and the *average vertical gravity gradient* is 0.3086 mGal per meter. There are two major parts in the observed Earth's gravity field: (1) the permanent part defined by the Earth's size, shape, heterogeneous mass, and rotational characteristics, and (2) the time-varying part defined by *temporal gravity variations*. The Earth's gravity field is a *potential field*, and its vector components are first-order spatial derivatives of the scalar *gravitational potential* of the Earth, also referred to as *Earth gravity potential*. Effects of the centrifugal acceleration and Earth's oblateness, or *polar flattening*, are integrated into the theoretical Earth gravity field represented by the gravity field of a homogeneous rotating *reference ellipsoid* and referred to as *theoretical gravity* or *normal gravity*. See also *derivative*, *electrostatic levitation*, *magnetic levitation*, *scalar*, and *observed gravity*.

Earth gravity potential: **1.** a mathematical function that describes through its spatial derivatives the *gravitational field* of the Earth, or *Earth gravity field*, at any point on or above the Earth's surface. The direction and *magnitude* of the *Earth gravity vector* are defined by the *gradient* of the *scalar* Earth gravity potential. **2.** a scalar quantity W defined by the sum of the *self-gravitational potential* U and *centrifugal potential* Φ at any point on or above the Earth's surface as

$$\mathrm{W} = \mathrm{U} + \Phi.$$

Its first-order spatial derivatives represent the *vector* components of the Earth's gravity field within a reference frame moving with the same *angular velocity* of rotation as the Earth. In geodetic applications, it is referred to as *total gravitational potential*. See also *derivative*, *potential*, and *disturbing potential*.

Earth gravity vector: **1.** a *gravity vector* with three mutually orthogonal components that are spatial derivatives of the *gravitational potential* of the Earth, i.e., spatial derivatives of *Earth gravity potential*. **2.** a three-component *vector* $\mathbf{g} = (g_x, g_y, g_z)$ which describes the strength and direction of the *Earth gravity field*. Vector direction and length, or *magnitude*, anywhere on or above the Earth's surface are defined by the *gradient* of the *scalar* Earth gravity potential. The field of the gravity vector values is the Earth's gravity field, the physical *dimension* of which and measure of its strength is *acceleration* of a free-falling object. **3.** a *vector* quantity defined by the vector sum of *gravitational acceleration* due to the attraction of Earth's mass and *centrifugal acceleration* due to the Earth's rotation within a reference frame moving with the same *angular velocity* of rotation as the Earth. At ground observation points, the direction of the

Earth's gravity vector is indicated by a *plumb line* that is orthogonal to the surface of a *geoid*. First-order spatial derivatives of the Earth's gravity potential W along three mutually orthogonal directions define the Earth's gravity vector as

$$g = \nabla W,$$

where ∇ is the vector gradient operator *del*. In *Cartesian coordinates*, the horizontal and vertical components of the Earth's gravity vector are defined as

$$g_x = \partial W/\partial x, \quad g_y = \partial W/\partial y, \quad g_z = \partial W/\partial z.$$

See also *derivative* and *normal gravity vector*.

Earth Magnetic Anomaly Grid-2 (EMAG2): a global *grid* of the *lithospheric magnetic field* compiled from satellite, airborne, and marine (shipboard) magnetic measurements. It shows *total magnetic intensity (TMI)* values with a *grid interval* of 2 arc-minutes corresponding to a map, or planar, *grid cell* of 3.68 × 3.70 km at the equator. Original line-oriented magnetic data and existing TMI grids have been upward continued for *grid merging* at the *elevation* of 4 km above *mean sea level (MSL)*. Because the long-*wavelength* components of the acquired data are limited by the area extent of each *survey*, the wavelengths longer than 330 km are represented in the final merged grid by magnetic data acquired during the *CHAMP mission*. See also *acquisition*, *arc-minute*, *upward continuation*, and *World Digital Magnetic Anomaly Map (WDMAM)*.

Earth magnetic field: a *vector field* representing the superposition and interference of three major magnetic fields within a wide range of spatial and temporal scales: (1) the slow-varying *main geomagnetic field* generated in the liquid outer Earth core and comprising approximately 90%–95% of the magnetic field values measured on the Earth's surface; (2) a highly variable *external magnetic field* generated in the *magnetosphere* by interactions between the Earth's rotational, tidal, and thermal effects, main geomagnetic field, *solar wind*, and magnetic fields of the Sun and other planets; (3) essentially a constant in time and location, or static, *lithospheric magnetic field* generated by the cumulative effect of *induced magnetization* and *remanent magnetization* of the sedimentary, metamorphic, and igneous rocks within the *crust* and upper *mantle*. Earth's magnetic field is a *potential field*, and its *vector* components are defined by the first-order spatial derivatives of the scalar *magnetic potential* of the Earth, also known as *Earth magnetic potential*. At any point on the Earth's surface, the main geomagnetic field is described by *geomagnetic field parameters*. The dominant component of the Earth's magnetic field, i.e., main geomagnetic field, is vertical at *Earth magnetic poles* and horizontal at the *magnetic equator*. By convention, the Earth's magnetic field is inclined downward from a horizontal level in the northern geomagnetic hemisphere and upward in the southern hemisphere, so the *magnetic inclination* $I = 90°$ at the *north magnetic pole* and $I = -90°$ at the *south magnetic pole*. Time-variable external space boundaries of the Earth's magnetic field are defined by a magnetosphere. Often referred to as *geomagnetic field*. See also *derivative*, *observed magnetics*, *temporal magnetic variations*, and *van Allen belts*.

Earth magnetic poles: two locations on the Earth's surface where the *main geomagnetic field* is vertical, its strength is close to maximum, *magnetic inclination* is ±90°, and *magnetic declination* is zero. By convention, magnetic inclination is +90° at the *magnetic north pole*, or *magnetic north*, whereas at the *magnetic south pole*, or *magnetic south*, magnetic inclination is −90°. For each *epoch*, their predicted locations are defined by a corresponding *IGRF model*. Earth's magnetic poles are not coincident with the geographic poles and *geomagnetic poles*, because the main geomagnetic field (which comprises 90%–95% of the magnetic field values measured on the Earth's surface) is composed of both *dipole*-approximated and nondipole parts. Sometimes referred to as *magnetic dip poles*. See also *dipole field*, *eccentric dipole field*, and *nondipole field*.

Earth magnetic potential: a mathematical function that describes through its spatial derivatives the *magnetic field* of the Earth, or *Earth magnetic field*, at any point on or above the Earth's surface. The direction and *magnitude* of the Earth's magnetic field *vector* **B** are defined by the *gradient* of the *scalar* Earth magnetic potential. Its first-order derivatives along the axes in *Cartesian coordinates* represent the vertical and two horizontal components of the Earth's magnetic field vector **B** as

$$\boldsymbol{B}_x = \partial \mathcal{V}/\partial x, \quad \boldsymbol{B}_y = \partial \mathcal{V}/\partial y, \quad \boldsymbol{B}_z = \partial \mathcal{V}/\partial z,$$

where $\mathcal{V}$ is the Earth's magnetic potential. See also *geomagnetic field parameters*, *geomagnetic vector*, and *magnetic potential*.

Earth magnetic vector: a three-component *vector* $\mathbf{B} = (\boldsymbol{B}_x, \boldsymbol{B}_y, \boldsymbol{B}_z)$ which describes the strength and direction of the *Earth magnetic field* at each *observation point*. Vector direction and length, or *magnitude*, anywhere on or above the Earth's surface are defined by the *gradient* of the scalar *Earth magnetic potential*. The Earth magnetic vector is different from the *geomagnetic vector* which is the vector of the *main geomagnetic field*. See also *scalar*.

Earth normal field: **1.** the theoretical *gravity field* of the Earth, or *theoretical gravity*, defined by the *International Gravity Formula (IGF)*. **2.** the theoretical *magnetic field* of the Earth, or *theoretical geomagnetic field*, defined by the *International Geomagnetic Reference Field (IGRF)*. The normal field is computed at any point on the Earth's surface using its geographic coordinates and, in case of *Earth magnetic field*, the dates of a particular *magnetic survey* (year and average *Julian day*) to select the appropriate *IGRF model*. In *exploration* applications, the Earth normal field is subtracted from the measured gravity or magnetic values on a point-by-point basis in the process of *data reduction*.

Earth tide correction: a correction applied to *observed gravity* to compensate for the time-varying and latitude-dependent *Earth tide effect*. In regular *ground gravity* surveys, it is often considered jointly with *gravimeter drift*, and their total effect is compensated by *drift correction* applied to the *raw gravity*. In marine (shipboard) and ground *microgravity* surveys, the Earth tide correction is calculated using the *Longman algorithm* and referred to as *tidal correction*. Depending on time of the day and location on the Earth's surface, the correction value is added to or subtracted from observed gravity values. In automated *relative gravimeter* systems used in ground gravity surveys, it is integrated into *automatic drift correction* and referred to as *automatic Earth tide correction*. See also *Earth tides*.

Earth tide effect: the time-varying gravity effect of *Earth tides* on measurements of the *Earth gravity field*. Its *magnitude* is a function of latitude, time of the day, and month of the year, and it interferes with ocean-bottom, marine (shipboard), submarine, and high-*accuracy* ground gravity measurements. There are three main contributors to the Earth tide effect: (1) direct *gravitational attraction* of the Sun and Moon; (2) change in the Earth's gravity field due to the tidal displacement of the Earth's center of mass; and (3) change in the gravity value measured by a *gravimeter* caused by the radial tidal displacement of the Earth's surface at an *observation point* – when the Earth's surface is pulled radially outward, the measured gravity values are decreased, and when it is pulled radially inward, the measured gravity values are increased relative to their true values. The Earth tide effect has a dominant periodic component of 12 hours, and it is greatest at low latitudes. The maximum tide effect variation within 24 hours is up to $\pm$0.20–0.35 *milligal (mGal)*. The rate of change is variable with its temporal maximum being 0.05–0.07 mGal per hour. The Earth tide effect is compensated by the application of the *Earth tide correction*, also known as *tidal correction*. Often referred to as *tidal effect*. See also *ocean loading effect*.

Earth tides: elastic deformations of the solid Earth caused by the combined *gravitational attraction* of the Sun and Moon. Earth tides are periodic: semi-diurnal (12 hours), diurnal, semi-monthly, monthly, semi-annual (6 months), and annual. The tidal gravity effect of the Moon is approximately three times the tidal effect of the Sun. Due to a different attraction of the Earth's center of mass and Earth's surface by these celestial bodies, the deformation of the solid Earth is different for the different areas: the surface areas which are the nearest and farthest from the Moon and Sun are both pulled radially outward from the Earth's center, whereas the surface areas orthogonal to the Earth–Moon and Earth–Sun directions are pulled radially inward. Earth tides result in changes in measured gravity values that are compensated by the application of the *Earth tide correction* or included into integrated *drift correction*. Sometimes referred to as *solid Earth tides*. See also *Earth tide effect*, *ocean tides*, and *permanent tide*.

earthquake noise: the effect of large earthquakes at any place around the world on *gravimeter* readings during a *ground gravity* survey. It is observed as unstable readings tending to oscillate repeatedly within a few seconds. The duration and intensity of earthquake noise depend on the geological structure in a *survey* area, and they are greatest in sedimentary basins. See also *basin*, *noise*, and *reading*.

easting: 1. the X-coordinate of a particular *grid* or map point that represents its distance in meters along the *x*-axis from a grid or map reference point called a *false easting*. **2.** the longitude difference between *survey* lines. See also *northing* and *XY coordinates*.

eccentric dipole field: the best-fit mathematical approximation of the dipole component of the *main geomagnetic field* by the field of a *magnetic dipole* located at approximately 400 km above the Earth's geometric center. See also *geocentric dipole field* and *nondipole field*.

eddy currents: electrical currents induced in electrically conductive materials by the time-varying, or fluctuating, externally applied *magnetic field*. Temporal eddy currents are generated in the conductive parts of a survey aircraft frame during its flight in the variable *Earth magnetic field*. The magnetic interfering effects of eddy currents are eliminated by an onboard magnetic compensation system. See also *aeromagnetic compensation*.

edge: 1. an abrupt discontinuity in the geological *structure* created by a *fault* with the vertical or horizontal rock displacement, known as *offset*, or by a lateral change in the rock *density* and *magnetization* or *magnetization vector* direction caused by changes in rock composition. **2.** a lateral lithological boundary of a *source body*, such as an *igneous intrusion* or *salt dome*. **3.** a fault-offset boundary between rock units of a similar geological origin. Edges are identified in an *anomalous field* as local areas of rapidly changing amplitudes of anomalies, wherein the amount of change is quantified by the *horizontal gradient* reaching its maximum values above the vertical or high-angle edge. In *modeling* and automatic source depth estimates, the edges are approximated by *contact* source geometries. See also *edge-detection methods*, *edge detectors*, *edge-detector attribute*, and *edge lineaments*.

edge detection: an *enhanced processing* methodology based on the use of *edge-detection methods* applied to highlight the gravity and magnetic anomalies associated with *source body* edges, high- and low-*offset* faults, boundaries of geological structures, and other structural and lithological discontinuities in the subsurface. See also *edge*, *edge detectors*, *fault*, and *structure*.

edge-detection methods: *enhanced processing* techniques applied to gravity or magnetic *total field*, *gradient*, and *full tensor gradient (FTG)* data to map the locations of *density* and *magnetization* contacts associated with structural and lithological edges, such as high- and low-*offset* faults, boundaries of relatively homogeneous lithological units, edges of igneous intrusions and shale or salt diapirs, edges of local-scale folded structures (anticlines, synclines, thrusts), boundaries of regional-scale tectonic provinces (*basin*, *magnetic terrane*, *rift*), edges of *allochthonous salt*, depositional edges, and boundaries of mineral alterations within ore bodies. Methods are based on the calculation of attributes of the anomalous gravity and magnetic fields that are known as *edge detectors*. Resolution of the edge-detection methods depends, primarily, on *survey line spacing* for marine (shipboard) and airborne surveys and *survey height* for airborne surveys. See also *anomalous field*, *attribute*, *contact*, *continental-oceanic boundary*, *density contrast, magnetization contrast, edge, edge lineaments, fault, igneous intrusion, lateral resolution,* and *ore body.*

edge-detector attribute: a quantitative characteristic of an anomalous gravity or magnetic field defined by the square root of the sum of the squared first vertical derivatives of the amplitudes of a *directional analytic signal* calculated along the mutually orthogonal *x*-, *y*-directions in *Cartesian coordinates* as

$$|\text{EDA}| = [(\partial|A_x|/\partial z)^2 + (\partial|A_y|/\partial z)^2]^{1/2},$$

where $|A_x|$ and $|A_y|$ are amplitudes of the directional analytic signal expressed as

$$|A_x| = (T_{xx}^2 + T_{xy}^2 + T_{xz}^2)^{1/2} \quad \text{and}$$

$$|A_y| = (T_{yx}^2 + T_{yy}^2 + T_{yz}^2)^{1/2},$$

where T_{ij} are derivatives of the *i*th component of the gravity or magnetic *vector* in the *j*th direction that comprise the *full tensor gradient (FTG)* components (Beiki, 2010). Because the calculation of the edge-detector attributes involves the third-order derivatives of the gravity or magnetic *potential*, the edge-detector attribute has a higher *resolution* of laterally interfering anomalies than a regular *horizontal gradient* defined by the second-order derivative of the potential. This high-resolution property is used in mapping the structurally complex source edges, such as those created by salt or shale overhangs, *sill*-like igneous intrusions, mineral alterations within ore bodies, near-surface *karst*, etc. See also *anomalous field, first vertical derivative, horizontal invariants, horizontal tilt angle (TDX), horizontal DAS,* and *ore body.*

edge detectors: attributes of the gravity or magnetic field implemented in *edge-detection methods* for analysis and mapping of *edge lineaments* and other structural and lithological edges. They include but are not limited to: *analytic signal amplitude (ASA), ASTA filter, curvature attributes, directional ASA, edge-detector attribute, enhanced horizontal derivative (EHD), horizontal gradient amplitude (HGA), horizontal local wavenumber, instantaneous frequency, horizontal invariants, local phase, local wavenumber (LW), modified-phase eigenvalue, monogenic signal attributes, theta angle, tilt angle (TA), tilt angle of horizontal gradient (TAHG), tilt ASA, second-order tilt ASA, directional tilt angle, horizontal tilt angle (TDX), tilt derivative,* and *total horizontal derivative of tilt.* See also *attribute* and *edge.*

edge-effect gravity anomaly: same as *shelf-edge gravity anomaly* that originates from the combined gravity effect of three types of a lateral *density contrast* produced by (1) increased seawater depth across a shelf edge, (2) increased thickness of *rift*-accumulated sediments, and (3) decreased depth of the *Moho boundary* in the transition zone from a thick continental *crust* to a much thinner oceanic crust.

edge effects: high-*amplitude* linear stripes and other distortions of the gridded gravity and magnetic data along the edges of a *grid.* They are generated in the process of spectral-domain *filtering* and other *data enhancement* procedures based on the *Fourier transform.* Fourier-domain grid edge effects are caused, primarily, by *Gibbs phenomenon,* and they are minimized by *grid expansion* and *edge-smoothing filters.* See also *gridding, padding,* and *spectral domain.*

edge lineaments: linear and curvilinear alignments of the *gradient* and gradient-derived anomalies calculated from a gridded *anomalous field* or obtained from *measured gradient* and *full tensor gradient (FTG)* data. They are generated by the structural and lithological edges within a sedimentary section and at the top of the *basement* and include, but are not limited to, high-*offset* faults, high-*contrast* lithological contacts (e.g., edges of igneous intrusions and *salt* bodies), depositional edges, boundaries of rock formations with a similar lithology but different *density* and *magnetization,* etc. See also *contact, edge, fault, gridding, igneous intrusion, lineaments, line lineaments,* and *strike lineaments.*

edge mismatch: mid- and long-wavelength artificial anomalies produced in the process of *grid merging.* They are caused by differences in the data values and *wavelength* content at the edges of the separately processed datasets acquired in the different years, such as recently acquired data and *vintage data.* See also *acquisition.*

edge-preserving smoothing filters: same as *edge-sharpening filters* that increase the *sharpness* of *edge*-generated anomalies and enhance the edges of *inversion*-derived structural elements while smoothing the data values elsewhere.

edge-sharpening filters: a group of *grid*-based space-domain filters that increase the *sharpness* of *edge*-generated anomalies and enhance the edges of structural elements obtained by *structural inversion.* As an option in *terracing* the gravity and magnetic fields, they can be applied repeatedly, until the preferred terraced effect is achieved. Filters include but are not limited to the *Kuwahara filter, Laplace filter, mean-of-least-variance (MLV) filter, median filter,* and *symmetric nearest neighbor (SNN) filter.* Sometimes referred to as *edge-preserving smoothing filters.* See also *edge-smoothing filters* and *sharpening filter.*

edge-smoothing filters: space- and spectral-*domain* filters applied to line-oriented and gridded data to provide their smooth transition to zero at the line ends and *grid* edges in order to minimize *edge effects* and errors in calculation of the data *spectrum* and application of *FFT filters.* See also *filter, gridding, cosine taper filter,* and *taper filter.*

effective inclination: an *inclination* of the ambient field *vector* or source body *magnetization vector* projected onto the plane normal, or orthogonal, to the *strike* of a two-dimensional (2D) *source body.* Effective inclination of the projected vector of an *ambient field* is defined as

$$I_{AF}^E = \arctan[\tan I_{AF}/\cos(D_{AF} - \theta)].$$

Effective inclination of the projected vector of a source body magnetization is defined as

$$I_{SB}^E = \arctan[\tan I_{SB}/\cos(D_{SB} - \theta)],$$

where I_{AF} and D_{AF} are *magnetic inclination* and *magnetic declination,* respectively, of the ambient magnetic field, i.e., *Earth magnetic field;* I_{SB} and D_{SB} are inclination and declination, respectively, of the source body magnetization vector; and θ is *azimuth* of the *profile* direction. Both effective inclinations are always equal to or higher than the true inclinations. When a source body strike is parallel to the ambient field declination or source body magnetization vector declination, the corresponding effective inclination is vertical (Blakely, 1996). See also *ambient field* and *dimensionality indicator.*

effective induced magnetization: 1. the *vector* quantity $\mathbf{M}_{\text{EI}}$ which describes *induced magnetization* of an isolated *source body* and accounts for its internal *demagnetization* properties. It is expressed as

$$\mathbf{M}_{\text{EI}} = k\mathbf{F}_i/(1 + Nk) = k_E\mathbf{F}_i,$$

where k is *bulk susceptibility,* $\mathbf{F}_i$ is *inducing magnetic field, N* is *demagnetization factor,* and k_E is *effective susceptibility.* **2.** the vector quantity $\mathbf{M}_{\text{EI}}$ defined by the product of the bulk susceptibility k and *internal magnetic field* $\mathbf{F}_i$ of a source body as

$$\mathbf{M}_{\text{EI}} = k\mathbf{F}_i.$$

See also *effective remanent magnetization.*

effective magnetization: *total magnetization* of a *source body* that accounts for its internal *demagnetization* properties in addition to *induced magnetization* and *remanent magnetization* within a source volume. Effective magnetization $\mathbf{M}_{\text{E}}$ is defined by a *vector* sum of *effective induced magnetization* $\mathbf{M}_{\text{EI}}$ and *effective remanent magnetization* $\mathbf{M}_{\text{ER}}$ as

$$\begin{aligned}\mathbf{M}_{\text{E}} &= \mathbf{M}_{\text{EI}} + \mathbf{M}_{\text{ER}} \\ &= (k\mathbf{F}_{\text{I}} + \mathbf{M}_{\text{R}})\,/\,(1 + Nk) = k_E\mathbf{F}_{\text{I}} + \mathbf{M}_{\text{ER}},\end{aligned}$$

where k is *bulk susceptibility*; $\mathbf{F}_{\text{I}}$ is the inducing *Earth magnetic field*; $\mathbf{M}_{\text{R}}$ is the original, or inherent, remanent magnetization; N is *demagnetization factor*; and k_E is *effective susceptibility* Also referred to as *effective total magnetization.* See also *inducing magnetic field.*

effective remanent magnetization: the *vector* quantity $\mathbf{M}_{\text{ERM}}$ which describes *remanent magnetization* of an isolated *source body* and accounts for its internal *demagnetization* properties. It is expressed as

$$\mathbf{M}_{\text{ERM}} = \mathbf{M}_{\text{R}}/(1 + Nk),$$

where $\mathbf{M}_{\mathbf{R}}$ is the original, or inherent, remanent magnetization; N is *demagnetization factor*; and k is *bulk susceptibility.* See also *effective induced magnetization* and *effective magnetization.*

effective susceptibility: 1. a ratio of the *magnitude* of the *total magnetization* vector to the inducing *magnetic field strength.* **2.** the *magnetic susceptibility* k_E which accounts for a decrease in *bulk susceptibility* due to the internal *demagnetization* of a highly magnetic rock body and defined as

$$k_E = k/(1 + Nk),$$

where k is *bulk susceptibility* and N is *demagnetization factor.* **3.** the magnetic susceptibility k_E associated with rock *magnetization* which is estimated by a *scalar* sum of *induced magnetization* and *remanent magnetization.* It is determined by the relation

$$k_E = k/(1 + Q),$$

where k is bulk susceptibility and Q is *Koenigsberger ratio.* See also *inducing magnetic field* and *volume susceptibility.*

effective total magnetization: same as *effective magnetization* represented as a *vector* sum of the *effective remanent magnetization* and *effective induced magnetization.*

EGM96 geoid height: the *geoid undulation* map of the Earth compiled by the U.S. National Geospatial-Intelligence Agency (NGA), for which *geoid* undulations have been calculated with the use of the *Earth Gravitational Model 1996 (EGM96).*

EGM1996 geoid: See *Earth Gravitational Model 1996 (EGM96).*

EGM2008 geoid: See *Earth Gravitational Model 2008 (EGM08).*

eigenimage: an image of the anomalous gravity or magnetic field in a new set of Cartesian coordinate axes aligned with *eigenvectors* determined by *principal component analysis (PCA).* Eigenimages show how much of a covariant similarity, or *coherence*, between data parameters is represented by the corresponding eigenvectors in every *dimension.* Eigenimages are sorted in descending order of data coherence: the first eigenimage represents the largest part of the coherent trends in the entire dataset, the second eigenimage represents a part of the coherent trends that was not represented by the first eigenimage, and so on. Linear combinations of the obtained eigenimages are used in composite images, such as an *RGB composite image*, for visualization of the PCA results. See also *covariance* and *eigenvalues.*

eigensystem: See *FTG eigensystem.*

eigenvalue decomposition: diagonalization of a symmetric 3×3 *tensor matrix*, such as the *FTG matrix* or *Hessian matrix*, at each observation point in a study area and calculation of *eigenvalues* λ_1, λ_2, λ_3, which are roots of the cubic characteristic equation

$$\lambda^3 + I_1\lambda - I_2 = 0,$$

where I_1 and I_2 are non-zero *principal invariants* of the original FTG matrix that are constrained as

$$(I_2/2)^2 + (I_1/3)^3 \leq 0,$$

where $I_1 \leq 0$ (Pedersen and Rasmussen, 1990). Obtained eigenvalues are three diagonal elements of a diagonalized nine-element matrix, where all six off-diagonal elements are zeros. Diagonalization of the FTG or Hessian matrix is equivalent to rotation of the original Cartesian coordinate frame, where the obtained eigenvalue-paired *eigenvectors* define the orientation of the *x*-, *y*-, *z*-axes of the rotated *Cartesian coordinates.* Arithmetic combinations of FTG and Hessian eigenvalues are invariant under any coordinate transformation and used in *enhanced processing* of gravity and magnetic data for high-*resolution* mapping of *lineaments* and evaluation of isolated *source body* parameters. See also *diagonalized tensor matrix*, *FTG attributes*, and *lateral resolution*.

eigenvalue ellipsoid: a representation of *gradient tensor* components by the three-dimensional (3D) shape of an *ellipsoid* which has three mutually orthogonal axial radii that are absolute values of *eigenvalues*. Spatial orientation of the ellipsoidal axes is defined by the orientation of the eigenvalue-paired *eigenvectors* at each *observation point*. See also *eigenvalue wheel*, *FTG anisotropic ellipsoid*, and *tidal tensor*.

eigenvalue magnitude: 1. the invariant *scalar* quantity $2\sqrt{Q}$ which is used, jointly with two other invariants $R = I_2/2$ and $\theta = (1/3)\cos^{-1}(R/Q^{3/2})$, to compute three *FTG eigenvalues* λ_1, λ_2, λ_3 as

$$\lambda_1 = 2\sqrt{Q}\cos\theta,$$

$$\lambda_2 = 2\sqrt{Q}\cos\theta\left(\theta - \frac{2\pi}{3}\right),$$

$$\lambda_3 = 2\sqrt{Q}\cos\left(\theta + \frac{2\pi}{3}\right),$$

where $Q = -I_1/3$ and I_1 and I_2 are non-zero *principal invariants* of the original *FTG matrix* with $I_1 \leq 0$. **2.** the length of a spoke, or radius, in a three-spoke *eigenvalue wheel*. When interpreted in the context of *curvature attributes*, the eigenvalue magnitude has the physical meaning of a measure of the amount of *curvature* of a gravitational or magnetic *equipotential surface* (Kindlmann, 2007; Cevallos, 2017). See also *magnitude* and *eigenvalue decomposition*.

eigenvalue phase: an imaginary angle between the *eigenvalue magnitude* and each of three *eigenvalues* λ_1, λ_2, λ_3 in their geometric representation by an *eigenvalue wheel*, where the *phase* of the largest eigenvalue λ_1 is defined by the invariant quantity θ as

$$\theta = (1/3)\cos^{-1}(R/Q^{3/2}),$$

where $R = I_2/2$, $Q = -I_1/3$, and I_1 and I_2 are non-zero *principal invariants* of the original *FTG matrix*, with $I_1 \leq 0$. The condition $\lambda_1 \geq \lambda_2 \geq \lambda_3$ limits the range of eigenvalue λ_1 phase variations to $0 \leq \theta \leq \pi/3$. Eigenvalues λ_2 and λ_3 have their phases defined as $\theta - 2\pi/3$ and $\theta + 2\pi/3$, respectively (Kindlmann, 2007). By analogy with the *invariant ratio* parameter of the *FTG matrix*, the eigenvalue phase is related to the dimensionality of a *source body*. See also *modified-phase eigenvalue* and *dimensionality indicator*.

eigenvalue wheel: the geometric representation of *eigenvalues* by a wheel which has three spokes, or radii, equally placed at $2\pi/3$ apart and centered at a reference point on the real number axis. Spoke length is equal to the invariant quantity $2\sqrt{Q}$ which is referred to as the *eigenvalue magnitude*. Eigenvalues are projections of three spokes of the same length but different orientations relative to the real number axis. Orientation of each spoke in the eigenvalue wheel is referred to as the *eigenvalue phase*, which is denoted by θ and quantified by the angle between a particular spoke and its corresponding eigenvalue. Eigenvalues λ_1, λ_2, λ_3 have their respective phases defined as θ, $\theta - 2\pi/3$, $\theta + 2\pi/3$, where the *eigenvalue phase* θ is an invariant quantity for each *FTG matrix* (Kindlmann, 2007). See also *invariants*.

eigenvalues: 1. three *scalar* quantities λ_1, λ_2, λ_3 obtained by transformation of a nine-element *gradient tensor* from its original *format* of a square 3×3 symmetric *matrix* into a diagonal matrix, where all six off-diagonal elements are zeros:

$$\begin{matrix} \lambda_1 & 0 & 0 \\ 0 & \lambda_2 & 0 \\ 0 & 0 & \lambda_3. \end{matrix}$$

These three eigenvalues are related as $\lambda_1 \geq \lambda_2 \geq \lambda_3$. The transformation is equivalent to the rotation of the axes of the Cartesian coordinate system to their orientation position at which the gradient tensor matrix becomes diagonalized. Under any Cartesian coordinate transformation, the gradient tensor will have three *invariants* related to eigenvalues as

$$I_0 = \lambda_1 + \lambda_2 + \lambda_3,$$
$$I_1 = \lambda_1\lambda_2 + \lambda_2\lambda_3 + \lambda_1\lambda_3,$$
$$I_2 = \lambda_1\lambda_2\lambda_3,$$

where $I_0 = 0$ in a source-free space. Eigenvalues are paired with the corresponding *eigenvectors*, which define the orientation of the *x*-, *y*-, *z*-axes of the rotated *Cartesian coordinates* at each *observation point* for the obtained eigenvalues. **2.** roots of the cubic characteristic equation relating the eigenvalues and invariants I_0, I_1, I_2 as

$$\lambda^3 - I_0\lambda^2 + I_1\lambda - I_2 = \lambda^3 + I_1\lambda - I_2 = 0,$$

where $I_0 = 0$ and $I_1 \leq 0$ (Pedersen and Rasmussen, 1990). **3.** diagonal elements of a locally diagonalized gradient tensor matrix. **4.** parameters in a *principal component analysis (PCA)* of a multidimensional dataset that quantify the *covariance* along the directions defined by eigenvectors for each *dimension* of the data: from the largest eigenvalue representing the highest covariance to the smallest eigenvalue representing the lowest covariance, including *random noise*. See also *diagonalized tensor matrix*, *eigenvalue decomposition*, *eigenvalue magnitude*, *eigenvalue phase*, *eigenvalue wheel*, and *curvature eigenvalues*.

eigenvectors: 1. three mutually orthogonal unit vectors $\mathbf{v}_1$, $\mathbf{v}_2$, $\mathbf{v}_3$ obtained by transformation of a nine-element *gradient tensor* from its original 3×3 square symmetric *matrix* format into a *diagonalized tensor matrix*. For each *observation point*, the eigenvectors show the orientation of *x*-, *y*-, *z*-axes in the new *Cartesian coordinates*, where all six off-diagonal elements of the gradient tensor matrix are zeros and three non-zero diagonal elements are *eigenvalues*. **2.** three unit vectors oriented in the mutually orthogonal *x*-, *y*-, *z*-directions and representing a *local reference frame* for the calculation of *true curvature attributes* of the gravitational *equipotential surface* for a local gravity field. **3.** unit vectors computed from a *covariance matrix* in the process of a *principal component analysis (PCA)* of a multidimensional dataset, where the number of eigenvectors is equal to the number of dimensions in the dataset. See also *apparent curvature attributes*, *FTG eigenvectors*, *FTG eigenvalues*, and *FTG matrix*.

elastic hysteresis: See *spring hysteresis*.

electrojets: See *auroral electrojets* and *equatorial electrojets*.

electromagnetic accelerometer: an *accelerometer* for which the *proof mass* is wrapped in two loops of an electric coil between two permanent magnets. The proof mass is constrained to move along the vertical axis and kept at the null position, or *null point*, by a balance between the *gravitational force* and the electromagnetic force induced by an electric current in the primary loop. Change in the gravitational force causes the proof mass to move up or down, and this movement generates the electric current in the secondary loop that results in a change of the electromagnetic balance which is restored by the primary loop current and, thereby, moves the proof mass back to its null position. Changes in the secondary coil current are linearly related to changes of the gravitational force and, therefore, can be calibrated in units of *gravitational acceleration*. See also *electrostatic accelerometer*.

electromagnetic pulsations: See *micropulsations*.

electrostatic accelerometer: **1.** an *accelerometer* for which the *proof mass* is held in a balanced suspended position sensed by capacitive pick-off rings. Change in the *gravitational force* results in displacement of the proof mass and, thereby, triggers the pick-off system to generate a capacitance signal, which is converted into the electric current driving the proof mass back into its balanced position, or *null point*. The generated electric current is directly proportional to the change in gravitational force at an *observation point*. **2.** a *virtual spring* accelerometer for which the proof mass is kept in a controlled balance position by *electrostatic levitation*. Arrangements of the paired and synchronized electrostatic accelerometers are used to measure gradients of *gravitational acceleration* by an *electrostatic gradiometer*. Also known as *capacitive accelerometer* and *capacitive levitating accelerometer*. See also *gradient*, *electromagnetic accelerometer*, and *GOCE gravity gradiometer*.

electrostatic gradiometer: an arrangement, or *array*, of paired and synchronized electrostatic accelerometers mounted on a *sensor*-controlled measurement platform. Values of *gravitational acceleration*, measured by pairs of accelerometers of the same *gradiometer arm*, are subtracted from each other, thereby suppressing the common mode *dynamic noise* which affects both accelerometers. The obtained difference is divided by a *baseline* value to obtain the *gravity gradient* at each observation point. See also *electrostatic accelerometer*, *GOCE gravity gradiometer*, and *common-mode rejection*.

electrostatic levitation: the operational principle of an *electrostatic accelerometer*, for which the *proof mass* is maintained in a balanced suspended position, or levitates, by the electric force of the stationary electric charges generated by a set of electrodes and controlled by the capacitive bridge circuits. Gravity-related change in the proof mass position is detected by measuring the capacitance difference between the levitating proof mass and electrodes. An amplified differential capacitance signal produces the electric voltage which generates a restoring force acting on the proof mass. Voltage value, which is required to return the proof mass back to its balanced position, is directly proportional to *gravitational acceleration* at the point of measurement. See also *magnetic levitation*.

elevation: **1.** the vertical distance, or height, above a *vertical datum* which is either sea level or *reference ellipsoid*. **2.** the *topographic height*, i.e., Earth surface height above sea level, variations of which represent the *topography* in a *survey* area. During *microgravity* surveys, it is measured by high-*precision* geodetic instruments. **3.** the water depth in sea-bottom gravity measurements by an *underwater gravimeter*. **4.** a regional-scale structurally high subsurface area in the upper *crust*, also known as *uplift*. Ground geodetic measurements of Earth surface elevations generally have been replaced by satellite-based *ellipsoidal height* measurements with the use of *Global Positioning System (GPS)*. In airborne gravity and magnetic surveys, the surface elevation data are obtained by subtracting the *radar altimeter* height, or *altitude*, from the GPS-derived ellipsoidal height of a survey aircraft. A correct elevation estimate is a major factor in the *accuracy* of measured gravity anomalies: a ground *gravity survey* accuracy of 1 *microgal* (μ*Gal*) requires the surface elevation to be determined with accuracy better than 0.3 cm. See also *datum* and *indirect effect*.

elevation base station: See *benchmark*.

elevation correction: **1.** the sum of *free-air correction* and *simple Bouguer correction*. Elevation correction is applied to *observed gravity* data acquired by a *ground gravity* survey to compensate for the gravity effect of the *elevation* of observation points above a *survey datum*: sea level or *reference ellipsoid*. Two corrections are combined, because they are functions of the same height difference between the survey datum and ground observation point, also known as *station*. Correction is calculated by multiplying the station elevation h and *elevation correction factor* as

$$\Delta g_{EC} = h \cdot \mathrm{ECF} = h(0.3086 - 0.04193\rho),$$

where Δg_{EC} is correction value in units of *milligal (mGal)*, h is station elevation in meters, and ρ is *density* of rocks between the survey datum and station in grams per cubic centimeter ($g \cdot cm^{-3}$). Sometimes, it is referred to as *combined elevation correction*. **2.** a *drape correction* applied to aeromagnetic data to compensate for large variations of *terrain clearance* during a *drape flight* over rugged terrain with reduced terrain clearance over topographic peaks and increased clearance over valleys. Large uncompensated variations of terrain clearance result in distortions of the amplitudes and shapes of magnetic anomalies, large *leveling* errors, and, eventually, incorrect estimates of *source body* parameters (depth, shape, etc.). See also *free-air gradient*, *height correction*, and *magnetic terrain correction*.

elevation correction constant: See *elevation correction factor*.

elevation correction factor: the gravity correction parameter which combines a *free-air factor* and a *flat-plate Bouguer factor*. It is defined in units of *milligal (mGal)* per meter ($mGal \cdot m^{-1}$) as

$$ECF = 0.3086 - 0.04193\rho,$$

where ρ is rock *density* in grams per cubic centimeter ($g \cdot cm^{-3}$) between a *survey datum* and ground observation point, also known as *station*. For the average crustal density 2.67 $g \cdot cm^{-3}$, ECF = 0.1976 $mGal \cdot m^{-1}$. Its multiplication by the height h of each station above a survey datum (sea level or *reference ellipsoid*) provides a value of *elevation correction* at any given station of a *ground gravity* survey. Sometimes referred to as *elevation correction constant*. See also *free-air gradient* and *Siegert method*.

elevation datum: a *vertical datum* in the context of measurements of an observation point height, or *elevation*, above sea level or a *reference ellipsoid*.

elevation effect: the effect of a decrease of measured gravity values with an increase in *elevation* of observation points. For the spherical Earth shape approximation, the rate of gravity decrease is defined as 0.3086 *milligal (mGal)* per meter ($mGal \cdot m^{-1}$) and referred to as the *free-air factor* or *average vertical gravity gradient*. For the ellipsoidal Earth shape approximation, its latitude-dependent values are defined by the more accurate *free-air gradient*. The elevation effect on the measured gravity values is compensated by *elevation correction*. See also *reference ellipsoid*.

ellipsoid: **1.** a *three-dimensional (3D)* geometrical shape with planar cross-sections that are ellipses. **2.** a spheroid that is flattened at the poles, or oblate. A rotating ellipsoid, or ellipsoid of revolution, with two semi-major axes is the best-fitting mathematical approximation of the actual Earth shape. See also *reference ellipsoid*, *eigenvalue ellipsoid*, and *FTG anisotropic ellipsoid*.

ellipsoidal anomaly: a *gravity anomaly*, such as a *free-air anomaly* or *Bouguer anomaly*, obtained by application of corrections calculated with the use of *ellipsoidal height*, which is the height above a *reference ellipsoid*. The term is used to distinguish these anomalies from the gravity anomalies obtained by the application of corrections calculated with the use of the height above sea level. Satellite-based *Global Positioning System (GPS)* and its high-*accuracy* versions, where the height calculations are referenced to the *WGS84 ellipsoid*, make this type of gravity anomalies a new standard in the *data reduction* process.

ellipsoidal Bouguer anomaly: same as *Bouguer anomaly* with an emphasis on a *reference ellipsoid* as the *vertical datum* for the calculation of *gravity corrections*.

ellipsoidal flattening: *polar flattening* in the context of a rotating *ellipsoid* as the best-fitting mathematical approximation of the actual Earth shape.

ellipsoidal gravity: same as *theoretical gravity* with an emphasis on approximation of the *gravitational field* of the heterogeneous Earth mass by the gravitational field of a uniformly dense, or homogeneous, mass of a rotating *ellipsoid* called a *reference ellipsoid*.

ellipsoidal gravity formula: same as *International Gravity Formula (IGF)*, which is the mathematical formula of the theoretical *gravitational field* of the Earth, where the heterogeneous Earth mass is approximated by a homogeneous mass of a rotating *ellipsoid* called a *reference ellipsoid*.

ellipsoidal height: **1.** the height above a *reference ellipsoid* determined by *Global Positioning System (GPS)*. Also known as *GPS height*. **2.** the height of a ground *station* relative to a reference ellipsoid as opposite to its height relative to sea level. Ellipsoidal height is recommended as a new standard for a *vertical datum*, because the gravity observation *positioning* is commonly determined from GPS data. Even so, this recommendation comes with the caveat that ellipsoidal heights are generally non-zero at shorelines and in oceans and seas due to the vertical separation between the actual sea level, *geoid*, and reference ellipsoid. Accordingly, *free-air anomaly* and *simple Bouguer anomaly* are not equivalent at the shorelines as they were, when the sea level was used as the vertical datum. See also *height correction* and *orthometric height*.

ellipsoidal theoretical gravity: same as *theoretical gravity* in the context of the *gravitational field* of the homogeneous rotating *reference ellipsoid* approximating the actual shape and heterogeneous mass of the Earth.

EMAG2: See *Earth Magnetic Anomaly Grid-2*.

ELW method: See *enhanced local wavenumber method*.

emu: the system of the electromagnetic units equivalent to *cgs units* of measurements. Used in vintage gravity and magnetic data, presently it is being replaced by standard cgs units and *SI units*. See also *vintage data*.

energy-density spectrum: a characteristic of the total energy, or power, of an anomalous gravity or magnetic field described by the squared *amplitude spectrum* and denoted by $|A(k)|^2$, where k is *wavenumber*. A *two-dimensional (2D)* graph of the energy-density spectrum shows a distribution of the squared amplitudes of Fourier-transformed data as a function of wavenumber. Also referred to as *power-density spectrum*, *power spectrum*, and *spectral density*. See also *complex spectrum* and *Fourier transform*.

energy envelope: an encompassing curve for *analytic signal amplitude (ASA)* that bounds, or envelopes, all *phase* shifts of the ASA anomaly above a two-dimensional (2D) *source body* in response to all possible directions of the *main geomagnetic field* and *total magnetization* of a source body. See also *envelope amplitude*.

energy leakage: See *spectral leakage*.

energy rosette: See *map power spectrum*.

enhanced analytic signal (eAS): the *analytic signal (AS)* of the Nth order *vertical derivative* of an anomalous magnetic or gravity field. It is expressed as

$$\begin{aligned} eAS(x, y) = {} & [\partial(\partial^N F/\partial z^N)/\partial_x]\mathbf{i} \\ & + [\partial(\partial^N F/\partial z^N)/\partial_y]\mathbf{j} \\ & + \sqrt{-1}[\partial(\partial^N F/\partial z^N)/\partial z]\mathbf{k}, \end{aligned}$$

where F is a field value at the observation point (x, y) on or above the XY plane; $\partial^N F/\partial z^N$ is the Nth order vertical derivative of the field; and **i**, **j**, **k** are unit vectors parallel to the mutually orthogonal x-, y-, z-directions in *Cartesian coordinates*. Calculation of vertical derivatives with the orders higher than two is often *noise*-prohibitive, so the eAS is commonly calculated as: (a) *first-order enhanced AS (eAS_1)*, which is the AS of the *first vertical derivative*, and (b) *second-order enhanced analytic signal (eAS_2)*, which is the AS of the *second vertical derivative* of an anomalous field (Hsu et al., 1996). See also *anomalous field* and *enhanced analytic signal amplitude (eASA)*.

enhanced analytic signal amplitude (eASA): a quantitative characteristic of an *enhanced analytic signal (eAS)* that defines its *magnitude* at the observation point (x,y) as

$$\begin{aligned}|\mathrm{eAS}(x,y)| &= \mathrm{eASA}\\ &= [(\partial\nabla^N F/\partial x)^2 + (\partial\nabla^N F/\partial y)^2\\ &\quad + (\partial\nabla^N F/\partial z)^2]^{1/2},\end{aligned}$$

where ∇ is *gradient* operator *del* and $\nabla^N F = \partial^N F/\partial z^N$ is the Nth order *vertical derivative* of the anomalous gravity or magnetic field F. The eASA components are the second-order (for $N=1$) and third-order (for $N=2$) derivatives of the field F, and they can significantly improve the *lateral resolution* of interfering anomalies generated by closely spaced sources. Prior to computations, the magnetic or gravity field is recomputed at the lowest empirically determined level of *upward continuation* to suppress the high-frequency *noise* which otherwise will be amplified by the calculation of the second- and third-order derivatives. See also *first-order ASA (ASA_1)*, *second-order ASA*, and *fractional-order ASA*.

enhanced horizontal derivative (EHD): the *total horizontal derivative* of the weighted sum of the different order vertical derivatives of an anomalous gravity or magnetic field F. For the *grid* point (x,y), its calculation formula is expressed as

$$\mathrm{EHD}(x,y) = [(\partial\Phi(x,y)/\partial x)^2 + (\partial\Phi(x,y)/\partial y)^2]^{1/2},$$

where $\Phi(x,y) = w_0 F^{(0)}(x,y) + w_1 F^{(1)}(x,y) + w_2 F^{(2)}(x,y) + \ldots + w_m F^{(m)}(x,y)$; $w_0, \ldots, w_m$ are weighting coefficients; $F^{(1)}, \ldots, F^{(m)}$ are vertical derivatives, from the first-order to the m-order *derivative*; and $w_i = C^n$; where $n = 0, \ldots, m$, C is constant, and m typically ranges from 1 to 3. Selection of different weights and derivatives for the summation controls *noise* and *lateral resolution* of the EHD grid. Prior to computations, the *TMI field* is reduced to the pole, and the magnetic or gravity field is recomputed at the lowest empirically determined level of *upward continuation* to suppress high-frequency noise. Correlation of the EHD maxima is implemented in the mapping of lithological contacts, faults, and other structural discontinuities (Fedi and Florio, 2001). See also *anomalous field*, *contact*, *fault*, *horizontal derivative*, *vertical derivative*, and *reduction-to-pole (RTP)*.

enhanced local wavenumber (ELW) method: a *profile*-based method of estimating the depth of *two-dimensional (2D)* magnetic sources approximated by *magnetic contact*, *thin dike*, and *horizontal cylinder* source geometries. It is based on the use of two complementary complex attributes: *horizontal local wavenumber k_x* and *vertical local wavenumber k_z*, which is the 90° *phase*-rotated version of k_x. Both k_x and k_z can be expressed in terms of the horizontal position of the source and its depth and *Euler structural index (SI)* as

$$\begin{aligned}k_x &= [(N+1)(z-z_0)]/[(x-x_0)^2 + (z-z_0)^2],\\ k_z &= -[(N+1)(x-x_0)]/[(x-x_0)^2 + (z-z_0)^2],\end{aligned}$$

where N is the Euler structural index with $N=0$ for the magnetic contact, $N=1$ for the thin dike, $N=2$ for the horizontal cylinder; x_0 is source position along the 2D profile; and z_0 is source depth. Rearranging the ratio k_x/k_z yields the linear equation without the Euler structural index:

$$k_x x + k_z z = k_x x_0 + k_z z_0.$$

The obtained linear equation is solved for x_0 and z_0 using the groups of data windows centered at the locations of the calculated k_x maxima. The number of data points in the window defines the overdetermined system of equations, and it depends on the data quality and degree of *interference of anomalies*. Once the source location parameters x_0 and z_0 have been determined, then a value of the structural index N for each source can be calculated from a corresponding local wavenumber expression to make the estimate of its 2D nature: contact-like, dike-like, or cylinder-like source (Salem et al., 2005). See also *local wavenumber (LW)*, *source parameter imaging (SPI™) method*, *improved SPI™ method*, and *advanced source-parameter imaging method*.

enhanced processing: a sequence of *data processing* operations applied to gravity and magnetic data in order to make visible their anomalous components associated with elements of the subsurface geological *structure*. Usually, these components are small-*amplitude*, short- and mid-*wavelength* anomalies hidden by superposition of the high-amplitude, long-wavelength regional anomalies and by interference with anomalies from nearby sources and background *noise*. Enhanced processing involves various data transformations, including the application of filters and the calculation of derivatives, gradients, residuals, and other attributes of the *anomalous field* that help to improve, or enhance, the detection and analysis of the anomalies of an exploration or research interest and their *interpretation*. Also referred to as *data enhancement*. See also *attribute*, *derivative*, *filter*, *gradient*, *residual*, and *wavelet transform*.

Enhanced World Magnetic Model: the *NGDC720 magnetic model* that has an increased *lateral resolution* of anomalies with short wavelengths at the order of 60–70 km, as compared to the shortest *wavelength* resolved by the *World Magnetic Model (WMM)* being approximately 3000 km.

enhancement: **1.** any transformation of an *anomalous field* that helps to improve the detection and visual representation of the *target* anomalies and other anomalous components of an exploration and research interest. **2.**

increasing the *signal-to-noise ratio (S/N)* in an observed gravity or magnetic field and its information content by *filtering* or calculation of attributes, derivatives, gradients, residuals, and application of other *enhanced processing* techniques. See also *anomaly*, *attribute*, *derivative*, *gradient*, and *residual*.

envelope: two low-frequency curves encompassing, or bounding, all deflections (peaks and troughs) of the relatively high *spatial frequency* components in obtained data from their average level. The envelope is drawn by a smooth connection of adjacent peaks and adjacent troughs in a two-dimensional (2D) data *profile*. See also *energy envelope*.

envelope amplitude: *2D analytic signal amplitude (2DASA)* that has the property of bounding, or enveloping, all *phase* shifts of a *magnetic anomaly* above a two-dimensional (2D) *source body* in response to all possible directions of the *main geomagnetic field* and *magnetization* of a source.

environmental magnetism: surface and near-surface magnetic properties of soils and rocks that are utilized in the research and studies of the environmental impact of heavy industry contamination, transportation pollution, and other commercial activity related to the accumulation and alteration of the magnetic materials. See also *forensic magnetism* and *magnetic archaeology*.

Eötvös (E): a unit of *gravity gradient* measurements. It quantifies the rate of change of *gravitational acceleration* relative to a horizontal or vertical distance as

$$\begin{aligned} 1\text{E} &= 0.1\text{mGal}\cdot\text{km}^{-1} = 10^{-4}\,\text{mGal}\cdot\text{m}^{-1} \\ &= 10^{-9}\,\text{Gal}\cdot\text{cm}^{-1} = 10^{-6}\,\text{mGal}\cdot\text{cm}^{-1} \\ &= 10^{-7}\,\text{mGal}\cdot\text{mm}^{-1} = 0.1\,\mu\text{Gal}\cdot\text{m}^{-1} \\ &= 10^{-3}\,\mu\text{Gal}\cdot\text{cm}^{-1} = 10^{-9}\text{s}^{-2} \\ &= 1\ \text{nanosec}^{-2}(\text{ns}^{-2}). \end{aligned}$$

Eötvös is the most commonly used unit of gravity gradients. Sometimes denoted by Eö. Also referred to as *Eötvös unit (Eu)*. See also *Gal*, *milligal (mGal)*, and *microgal (μGal)*.

Eötvös acceleration: the combined *inertial acceleration* of a moving *observation platform* produced by the *Eötvös effect*. It is defined by the sum of (a) the vertical component of the direction-dependent *Coriolis acceleration* due to platform movement within a coordinate frame fixed on the rotating Earth and (b) the direction-independent vertically oriented *centrifugal acceleration* due to platform movement over the curved Earth surface. Eötvös acceleration depends on the latitude of measurements, velocity of the observation platform, and its movement relative to the north–south or east–west directions. Its *magnitude* can be expressed as

$$|\alpha_{EA}| = 2\text{V}\omega\cos\theta\sin\alpha + \text{V}^2/\text{R},$$

where V is velocity of a *survey* ship or aircraft, ω is *angular velocity* of the Earth's rotation, θ is geographic latitude, α is *heading*, and R is the Earth's radius at the observation point. The Eötvös acceleration effect on marine (shipboard) and airborne gravity measurements is compensated by *Eötvös correction*. Also referred to as *total Eötvös acceleration*.

Eötvös acceleration correction: same as *Eötvös correction* with an emphasis on compensating for *Eötvös acceleration*, which represents the sum of the vertical component of *Coriolis acceleration* and the vertically directed *centrifugal acceleration* of an *observation platform* moving over the curved surface of the rotating Earth.

Eötvös correction: 1. the direction-, velocity-, and latitude-dependent correction applied to *observed gravity* values obtained from measurements on a moving *observation platform* in order to compensate for the vertical component of the direction-dependent *Coriolis acceleration* and direction-independent vertically oriented *centrifugal acceleration* of the observation platform moving over the curved surface of the rotating Earth. Its formula is expressed as

$$\Delta g_E = 2\text{V}\omega\cos\theta\sin\alpha + \text{V}^2/\text{R} \quad \text{or}$$
$$\Delta g_E = 2\text{V}\omega\cos\theta\cos\beta + \text{V}^2/\text{R},$$

where V is velocity of a *survey* ship or aircraft, ω is *angular velocity* of the Earth's rotation, θ is geographic latitude, α is *heading*, $\beta = 90° - \alpha$ is the angle of the movement direction relative to the east, and R is the Earth's radius at the measurement point. The first component in the formula is the *Coriolis term* and the second component is the *centrifugal term*. **2.** correction for the *Eötvös effect*. For the Earth average radius of 6371 km, the Eötvös correction in units of *milligal (mGal)* is expressed as

$$\Delta g_E = 4.040\,\text{V}\cos\theta\sin\alpha + 0.00121\text{V}^2$$

for velocity V in kilometers per hour ($\text{km}\cdot\text{h}^{-1}$) and

$$\Delta g_E = 7.487\,\text{V}\cos\theta\sin\alpha + 0.004154\text{V}^2$$

for velocity V in knots, i.e., nautical miles per hour, where 1 knot = 1.852 $\text{km}\cdot\text{h}^{-1}$. Navigation *accuracy* is one of the major factors in the measurement accuracy and *lateral resolution* of acquired data: 0.08 $\text{m}\cdot\text{s}^{-1}$ error in east–west velocity will result in 1 mGal error in the Eötvös correction. Velocity and heading errors have their greatest effects in the east–west and north–south movement directions, respectively. The Coriolis term is subtracted from the observed gravity values obtained in the westward direction and added to values obtained in the eastward direction, whereas the centrifugal term V^2/R is added to the observed gravity values regardless of the movement direction. Also referred to as *Eötvös acceleration correction* and *total Eötvös correction*. Often spelled as Eotvos correction. See also *Browne correction* and *centripetal acceleration correction*.

Eötvös effect: 1. the combined effect of *centrifugal acceleration* and *Coriolis acceleration*. **2.** the motion-related effect in gravity measurements on a moving *observation platform* that causes *gravimeter* readings to deviate from true values of *gravitational acceleration*: readings of a westward moving gravimeter are higher than the true values, whereas readings of an eastward moving gravimeter are lower than the true values. The amount of deviation depends on (a) the velocity of the observation platform, (b) platform movement relative to the north–south or east–west directions, and (c) the latitude of measurements. Marine (shipboard) measurements commonly have reading deviations up to tens of milligals. Deviations during airborne surveys can be within hundreds and thousands of milligals. The Eötvös effect on measurements of *vertical gravity* is compensated by *Eötvös correction*. Measurements of *gravity gradient* are essentially not affected by the Eötvös effect, because it is canceled out by the *common-mode rejection* of the *gradiometer arm* sensors. See also *Eötvös acceleration*, *equivalence principle*, and *milligal (mGal)*.

Eötvös torsion balance: same as *torsion balance* with an emphasis on the name of its inventor, Roland von Eötvös.

Eötvös unit (Eu): See *Eötvös (E)*.

epoch: the five-year interval adopted as a reference period for geomagnetic studies of the time-varying components of the *main geomagnetic field* measured in the past and their prediction for future years. Each past epoch within the year range of 1900–2015 is described by a corresponding historical *IGRF model*, whereas each future epoch for the following five years is described by the predictive IGRF model. For example, *geomagnetic field parameters* for the epoch 2015 are predicted by the IGRF2015 model for the period of 2015–2020. See also *International Geomagnetic Reference Field (IGRF)* and *IGRF-12*.

equatorial electrojets: time-variable, or temporal, components of the *external magnetic field* of the Earth produced by the interaction between the *main geomagnetic field* of the rotating Earth and *solar wind*. This interaction generates the charged particles which maintain their continuous flow as a ring electric current system above and along the equator in the *ionosphere*, where the instabilities in the ring current give rise to fluctuations of the external magnetic field up to tens and hundreds of *nanotesla (nT)* on and above the Earth's surface. In exploration applications, the magnetic effects of the equatorial electrojets are recorded by a *base-station magnetometer* and incorporated into *diurnal correction*. In *satellite magnetometry*, the localized sharp drop of several tens of nanoteslas in daytime satellite measurements of the magnetic field intensity above the ionosphere is used to determine the spatial position of the *magnetic equator* in order to solve the *Backus problem* in *modeling* of the main geomagnetic field. See also *auroral electrojets*, *temporal magnetic variations*, and *van Allen belts*.

equatorial gravity: a *theoretical gravity* value at the geographic equator. For *GRS80 ellipsoid*, it is equal to 9.7803267715 $m \cdot s^{-2}$ or 978032.67715 *milligal (mGal)*. See also *polar gravity*.

equipotential ellipsoid: a uniformly dense, or homogeneous, *ellipsoid* for which its surface is the *equipotential surface* for its own *gravity field*. A rotating equipotential ellipsoid, which has the same mass, center of mass, and *angular velocity* of rotation as those of the Earth, is referred to as a *reference ellipsoid* or *normal ellipsoid*, and it represents the equipotential surface of *normal gravitational potential*.

equipotential surface: a geometrical surface of the equal values of the gravitational or magnetic *potential*. It is everywhere perpendicular to the *vector* direction of a *potential field*, and no work is done in moving a particle along the equipotential surface, because the potential remains constant on it. Positive or negative departures from the uniform *density* or *magnetization* distribution below the equipotential surface will warp it upward or downward, respectively. The sea-level equipotential surface for the *Earth gravity field* is known as the *geoid* to which the direction of the Earth's gravity field is perpendicular everywhere. See also *gravitational potential* and *magnetic potential*.

equivalence principle: the physical rule postulating that it is not possible to distinguish between *acceleration* due to *gravitational attraction* caused by geological bodies and structures (i.e., true *gravitational acceleration*) and *inertial acceleration* due to the motion of an *observation platform*, such as a survey ship or aircraft. Accordingly, the measured accelerations in the dynamic environment of a moving ship or aircraft are defined by the *vector* sum of (a) true gravitational acceleration, and (b) combined horizontal and vertical inertial accelerations of the observation platform. See also *moving-platform corrections* and *superposition principle*.

equivalent layer: *Green's equivalent layer* in the context of the data transformation concept based on a hypothetical layer of *density* or *magnetization* that does not necessarily have any resemblance to a true *source body* but duplicates its gravity or magnetic field.

equivalent-layer gridding: See *equivalent-source gridding*.

equivalent source: the distribution of gravity or magnetic sources within a hypothetical layer of uniform thickness that is equivalent to the actual distribution of the *density* or *magnetization* in the subsurface and generates the same gravity or magnetic field as the *observed field*. The hypothetical layer is positioned at or below the original observation surface but above all real sources. Equivalent source construction involves (a) setting up a layer composed of the elementary cells and (b) computing its density or *susceptibility* distribution by *inversion* of the observed field. Then, any linear transformation of the gravity or magnetic data can be performed using the obtained physical property (density or susceptibility) distribution. Among other applications, it is implemented in the algorithms of *reduction-to-pole* (*RTP*) at low magnetic latitudes and *differential reduction-to-pole (DRTP)*. Referred to with the same meaning as *equivalent-source layer* and *Green's equivalent layer*. See also *equivalent-source method* and *magnetic latitude*.

equivalent-source continuation: 1. a version of the *equivalent-source method* applied for *upward continuation* or *downward continuation* of gravity and magnetic fields. **2.** a *three-dimensional (3D)* method of airborne magnetic or gravity data transformation to the level of a hypothetical layer of uniform thickness used in *equivalent-source gridding* and other applications.

equivalent-source gridding: a method of *gridding* where the interpolation between data points is based on the use of the *equivalent-source method.* Differing from *minimum-curvature gridding,* it can accommodate anomalously high gradients in the observed data and account for large deviations of the flight *altitude* from *survey specifications* that result in a *herringbone effect* and other gridding artifacts. Also known as *equivalent-layer gridding.* See also *artifact* and *gradient.*

equivalent-source layer: same as *Green's equivalent layer* or *equivalent source* which is a hypothetical layer of uniform thickness that generates the same gravity or magnetic field as the *observed field.*

equivalent-source method: a method of data transformation based on the theory of *Green's equivalent layer,* which represents the *density* or *magnetization* distribution in a single hypothetical layer of uniform thickness composed of multiple elementary cells of contrasting densities or susceptibilities, whose combined gravity or magnetic field is the same as the *observed field.* Equivalent source distribution does not necessarily represent a physical reality or resemble the true source distributions, but it is used as a tool in various applications, including (a) recalculation of the observed field and its derivatives and integrals at any spatial coordinates, (b) suppression of *acquisition noise* in *airborne gravity* and airborne *gravity gradient* data, (c) interpolation for *gridding,* (d) *regridding* at a constant or variable *elevation,* (e) *reduction-to-pole* (*RTP*) at low magnetic latitudes, where traditional methods are unstable, and (f) *differential reduction-to-pole (DRTP)* over regional-scale areas covering several zones with different magnetic inclinations (Dampney, 1969; Davis and Li, 2007). See also *equivalent-source continuation, equivalent-source gridding, low-latitude reduction-to-pole (RTP-LL),* and *magnetic inclination.*

equivalent-source RTP: a version of the *equivalent-source method* applied for the calculation of *reduction-to-pole (RTP)* at low magnetic latitudes and *differential reduction-to-pole (DRTP)* for regional-scale aeromagnetic *survey* areas covering several zones with different magnetic inclinations (Silva, 1986; Arkani-Hamed, 1988). See also *magnetic inclination, magnetic latitude,* and *low-latitude reduction-to-pole (RTP-LL).*

Euler deconvolution: a *profile-* or *grid-*based automated method of estimating the depth and other parameters of magnetic and gravity sources approximated by the *ideal source* geometries. It is based on solving the *Euler deconvolution equation* for each *observation point* within a *moving window* with the assumption that the *observed field* values within the window are due to a single *source body* with the known *Euler structural index (SI).* User-defined calculation parameters are (a) different window lengths for different depth intervals of sources, (b) different structural indices for different ideal source geometries, and (c) different increments for moving the calculation window along the survey line or grid. The moving window usually includes five to seven line data points and nine data points for the square grid window. For each window position, the overdetermined system of the Euler deconvolution equation is solved for estimates of the source depth, horizontal position, and other parameters (Thompson, 1982; Reid et al., 1990). Euler deconvolution can be applied to the *first vertical derivative* (FVD) and the *second vertical derivative* (SVD) of the gravity or magnetic field with the structural indices

$$N_{FVD} = N + 1 \quad \text{and} \quad N_{SVD} = N + 2,$$

where N is the Euler structural index in the standard version for *total field.* An excessive number of *noise* solutions is reduced by *clustering* and *acceptance criteria.* Euler deconvolution also can be applied to *analytic signal amplitude (ASA),* its derivatives, and *gradient tensor* components. Euler deconvolution is referred to as *2D Euler deconvolution* for line-oriented data or arbitrary-oriented profiles extracted from a grid and *3D Euler deconvolution* for gridded data. The obtained results are more conclusive for RTP magnetic data than TMI data, and they are more accurate in mapping the locations of structural and lithological boundaries than their depths, because the source depth and structural index N are linearly dependent and, therefore, the choice of N directly influences depth estimates. The same effect occurs for the calculation window length: the larger the window, the deeper the source depth estimate. When *reduction-to-pole (RTP)* is problematic due to low magnetic latitudes or *remanent magnetization* effects, the Euler deconvolution is applied to ASA data with the SI values increased by 1. Sometimes referred to as *Euler inversion, Euler method,* and *Euler modeling.* See also *analytic signal Euler deconvolution, derivative Euler deconvolution, extended Euler deconvolution, generalized Euler deconvolution, located Euler deconvolution, multiridge Euler deconvolution, reduced Euler deconvolution, tensor Euler deconvolution, wavelet Euler deconvolution,* and *weighted Euler deconvolution.*

Euler deconvolution equation: the first-order differential equation which is a theoretical basis for *Euler deconvolution.* It is obtained from the *Euler homogeneity equation* approximating the anomalous magnetic or gravity field F by (a) *homogeneous function* of the different homogeneity degrees and variable coordinates of its *source body* and (b) *vertical derivative* of the gravity or magnetic *potential* P, i.e., $\mathrm{F} = \partial\mathrm{P}/\partial z = P_z$. For the *three-dimensional (3D)* case, the Euler deconvolution equation is expressed as

$$(x - x_0)P_{zx} + (y - y_0)P_{zy} + (z - z_0)P_{zz} = -N(P_z - \mathrm{B}) = N(\mathrm{B} - P_z)$$

$$(x - x_0)\partial\mathrm{F}/\partial x + (y - y_0)\partial\mathrm{F}/\partial y + (z - z_0)\partial\mathrm{F}/\partial z = N(\mathrm{B} - \mathrm{F}),$$

where (x_0, y_0, z_0) is the spatial position of the source, for which the field F is measured at the observation points with coordinates (x, y, z); $P_{zx} = \partial\mathrm{F}/\partial x$, $P_{zy} = \partial\mathrm{F}/\partial y$, $P_{zz} = \partial\mathrm{F}/\partial z$ are the second-order derivatives of the potential P or first-order derivatives of the field F along the mutually orthogonal x-, y-, z-directions in *Cartesian coordinates*; B is a constant field value which accounts for variations of a regional component of the field F; and N is opposite, or negative, of a degree of field homogeneity n ($\mathrm{N} = -n$), called the *Euler structural index*

(SI), the value of which is a function of the source type. For the *two-dimensional (2D)* case of the *profile* or line-oriented data, the *y*-component of this equation is excluded. The equation can be written for each observation point, so that four unknown values x_0, y_0, z_0, and B, or three unknowns in the 2D case, are obtained by solving the overdetermined system of equations for each position of the *moving window* along the survey line, *grid*, or extracted *profile*. The number of equations is the same as the number of points in the moving window. Sometimes, it is referred to as the *3D Euler deconvolution equation* to highlight its application in the standard 3D Cartesian coordinates. See also *anomalous field*, *derivative*, *dimension*, and *gridding*.

Euler depth estimates: *source depth solutions* obtained within a *moving window* by the application of the *Euler deconvolution* to line-oriented or gridded magnetic and gravity data. For the same *anomaly*, Euler depth estimates tend to increase with the increasing value of the *Euler structural index (SI)* and calculation window length that are used in solving the overdetermined system of Euler equations for each window position along a survey line or *grid*. Depth estimates are calculated without regard for the actual *magnetization direction* in the source bodies and their actual shapes approximated by *ideal source* geometries, such as *sphere*, *vertical cylinder*, *horizontal cylinder*, vertical and dipping *thin dike* or *contact*, vertical *thick contact*, and horizontal *thin sheet*. Generally, the obtained depth solutions are much more accurate in mapping the locations of structural discontinuities, such as faults, *source body* edges, and other lithological boundaries, rather than their true depths. See also *dip*, *edge*, *fault*, *gridding*, and *Werner depth estimates*.

Euler differential equation: 1. See *Euler deconvolution equation*. **2.** See *Euler homogeneity equation*.

Euler equation: 1. See *Euler deconvolution equation*. **2.** See *Euler homogeneity equation*.

Euler homogeneity equation: a differential equation that describes the relationship between the first-order derivatives of the *homogeneous function* F(*x*,*y*,*z*) and its degree of homogeneity *n* as

$$x\partial \mathrm{F}/\partial x + y\partial \mathrm{F}/\partial y + z\partial \mathrm{F}/\partial z = n\mathrm{F}.$$

In application to observed gravity and magnetic fields, the modified Euler homogeneity equation is referred to as the *Euler deconvolution equation*, where the gravity or magnetic field is approximated locally by a *homogeneous field* within a variable-width *moving window* for calculation of the parameters of a field-generating source via *Euler deconvolution*. The negative, or opposite, of the homogeneity degree *n* of the approximated gravity or magnetic field is called the *Euler structural index (SI)* and denoted by *N*, i.e., $N = -n$. Also referred to as *Euler equation* and *Euler differential equation*.

Euler identity: a mathematical formula establishing the relationship between the complex exponential function of the variable θ and its orthogonal sine and cosine components as

$$e^{\pm i\theta} = \cos\theta \pm i\sin\theta,$$

where $i = \sqrt{-1}$ is a complex unit number indicating the orthogonality. See also *analytic signal (AS)*, *complex function*, *complex number*, *Hilbert transform*, and *Fourier transform*.

Euler index: See *Euler structural index (SI)*.

Euler inversion: same as *Euler deconvolution* used as a *two-dimensional (2D)* or *three-dimensional (3D)* method of *inversion* of line-oriented or gridded gravity and magnetic data to obtain the depth, location, and *Euler structural index (SI)* of isolated source bodies approximated by *ideal source* geometries.

Euler method: same as *Euler deconvolution* which is a semi-automated method of estimating depth and other parameters of isolated source bodies approximated by *ideal source* geometries.

Euler modeling: same as *Euler deconvolution* in the context of a mathematical algorithm that is based on the approximation of actual *source body* shapes by *ideal source* geometry models with different values of *Euler structural index (SI)*.

Euler structural index (SI): 1. a source-dependent parameter *N* in the *Euler deconvolution equation*. It is an integer value representing the negative, or opposite, of the homogeneity degree *n* of the gravity or magnetic field, which is considered a *homogeneous function* of the source coordinates and different homogeneity degrees that satisfies the *Euler homogeneity equation*, i.e.,

$$\mathrm{SI} = N = -n.$$

2. a *source body* parameter that specifies a type of *ideal source* approximation in *Euler deconvolution* for estimates of source depth and horizontal position from the derivatives and other attributes of an *anomalous field*. **3.** a *scale*-invariant *inverse power law* index, or scaling exponent, of the field generated by a *single-point source*. For the magnetic sources approximated by a *sphere* (*point source*), SI = 3. For the pipe-like *igneous intrusion*, such as *kimberlite pipe*, and the horizontal line of point sources approximated, respectively, by a *vertical cylinder* and a *horizontal cylinder*, SI = 2. For *thin dike* and other tabular-shaped magnetic bodies, edges of *sill* or *lava* flow, and small-*offset* faults approximated by a *thin sheet* edge, SI = 1.0. For the structural edge or deep-rooted high-offset *fault* approximated by *infinite contact*, SI = 0.0. The relationship between the gravity and magnetic structural indices is defined as

$$\mathrm{SI}_{grav} = \mathrm{SI}_{mag} - 1.$$

For the gravity sources approximated by the two-dimensional *finite step*, SI = −1, which is referred to as the *extended negative structural index* (Stavrev and Reid, 2010; Reid and Thurston, 2014). Structural indices for the *first vertical derivative* (FVD) and *second vertical derivative* (SVD) of the gravity or magnetic field are related as

$$N_{FVD} = N + 1 \quad \text{and} \quad N_{SVD} = N + 2,$$

where *N* is the Euler structural index (SI) in the standard version for *total field*. Because the source depth and structural index in the Euler deconvolution equation are linearly dependent, the choice of the SI directly influences *Euler depth estimates*: the higher the SI, the deeper the source depth estimate. Commonly referred to as *structural index (SI)*. See also *fractional structural index (fSI)*.

evaporites: sedimentary rocks, such as *salt*, gypsum, and anhydrite, formed by precipitation from saline waters in the process of their extensive or total evaporation. Salt has a small, near-zero negative *magnetic susceptibility*, and large salt volumes can generate negative magnetic anomalies up to several *nanotesla (nT)*. Densities of evaporites vary from 2.10–2.15 $g \cdot cm^{-3}$ for pure rock salt to approximately 2.70–2.85 $g \cdot cm^{-3}$ for anhydrites.

even function: a function $f(x)$ for which the basic property is expressed as $f(-x)=f(x)$, i.e., it has the same value when a sign of the variable x is changed. Any function can be split into even (symmetric) and odd (asymmetric) parts. Even and odd parts of a function represent its two independent averages. The *Fourier transform* of the even part of a function is called a *cosine transform*. Also known as the symmetric function. See also *odd function* and *Naudy method*.

even-odd grid comparison: a *grid*-based method of *quality control (QC)* and estimate of *survey accuracy* for *airborne gravity* measurements. It is based on the calculation of the difference between two grids obtained by a separate *gridding* of the even-numbered and odd-numbered traverse flight lines, wherein all *data processing*, including *terrain correction*, is performed separately for each grid. Because these grids cover the same area, the geological signal on the difference grid will be canceled, thereby, leaving only the accumulated *noise* of two grids that is twice the noise level of the combined grid. The method assumes that an airborne *gravity survey* has been flown at a traverse *line spacing* which is less than the length of the *low-pass filter* applied for the suppression of *dynamic noise*. See also *control lines*, *traverse lines*, and *test line*.

excess density structure: a local- or regional-scale geological *structure* composed of rocks having a *density* higher than the densities of the surrounding rock formations, e.g., *anticline*, *horst*, *igneous intrusion*, and *thrust*. Its *gravity signature* is a local or regional *gravity high*. See also *deficient density structure*.

exploration: application of geological and geophysical methods for the assessment of local, regional, and planetary geological structures in the search for commercial oil-and-gas accumulations, deposits of economically useful minerals, geothermal and water resources, etc. Sometimes referred to with the same meaning as *prospecting*. See also *gravity exploration*, *magnetic exploration*, and *structure*.

exploration leads: 1. local areas where *structural interpretation* of the gravity and magnetic data has identified geological structures favorable for the accumulation of hydrocarbons or the occurrence of mineral deposits. **2.** recommendations for future oil-and-gas, mineral, or geothermal *exploration* programs in a *survey* area that are based on the results of *integrated interpretation* of gravity, magnetic, seismic, and other geophysical and geological data. See also *prospect* and *structure*.

extended Euler deconvolution: *Euler deconvolution* extended by two differential equations obtained from a regular *Euler deconvolution equation* with the addition of the rotational invariant equation and formulated for the calculation of *dip* and *susceptibility contrast* of the *two-dimensional (2D)* sources approximated by the *magnetic contact* source geometries as

$$(x-x_0)\partial M/\partial x+(z-z_0)\partial M/\partial z=\alpha\cdot\sin\beta,$$

$$(x-x_0)\partial M/\partial z+-(z-z_0)\partial M/\partial x=\alpha\cdot\cos\beta,$$

where M is *total magnetic intensity (TMI)*, $\alpha=2Fkc\sin d$, $\beta=2I-d-90°$, F is *geomagnetic field strength*, k is susceptibility contrast, $c=1-\cos^2 i\cdot\sin^2\theta$, d is source dip, $\tan I=\tan i/\cos\theta$, i is *magnetic inclination*, and θ is the angle between the direction to *magnetic north* and the x-axis in *Cartesian coordinates*. Because the magnetic field M of a *thin dike* is described by the *horizontal derivative* of the magnetic contact field, the same equations are applicable for the calculation of thin dike dip and the product of the susceptibility contrast and thin dike thickness, wherein the corresponding *vertical derivative* is determined from the rotational invariant equation, $\alpha=2Fkct\sin d$ (where t is dike thickness), and source location x_0 and depth z_0 are obtained from the regular Euler deconvolution equation (Mushayandebvu et al., 2001). See also *invariants*, *Cauchy-Riemann equations*, and *extended 3D Euler deconvolution*.

extended negative structural index: the negative value of the *Euler structural index (SI)* which indicates a specific source type and represents an addition to the standard structural indices in the *Euler deconvolution* theory for gravity sources approximated by a horizontal layer with a large thickness relative to its depth and referred to as a *finite step*, whose *field* is described with the negative SI = −1 (Stavrev and Reid, 2010). See also *homogeneous function*, *homogeneous source*, and *Euler homogeneity equation*.

extended 3D Euler deconvolution: *generalized Euler deconvolution* in the context of a generalization of the two-dimensional (2D) *extended Euler deconvolution* to three dimensions via *3D Hilbert transform*. See also *dimension*.

external magnetic field: the external, or space, component of the *Earth magnetic field*. It is generated and maintained by interactions between the *main geomagnetic field*, cosmic radiation, interplanetary magnetic fields, electric currents in the *ionosphere*, and *solar wind* that are coupled with the Earth's rotation, tidal effects of the Moon and Sun, and various thermal effects. The external magnetic field is characterized by fast time-varying changes, which are partly cyclic and partly random and related, primarily, to the radiation activity of the Sun. The space around the Earth, where the external magnetic field exists, is called the *magnetosphere*. The external magnetic field is the source of *diurnals*, *magnetic storm*, and electric currents in the shallow subsurface that occasionally generate secondary magnetic fields on the Earth's surface. With periods from seconds to days and magnitudes from a few *nanotesla (nT)* up to hundreds of nanoteslas, the external magnetic field variations interfere with magnetic measurements, and episodes of their high intensities are prohibitive for a *magnetic survey*. Contributions of the external components of the Earth's magnetic field are recorded by a *base-station magnetometer* and removed from measured magnetic values by *diurnal correction*. See also *van Allen belts* and *internal magnetic field*.

extraneous acceleration: *inertial acceleration* in the context of a motion-related *acceleration* of a moving *observation platform* that is extraneous to the *gravitational acceleration*.

extrusion: the outflow of molten igneous rocks, or *lava*, onto the Earth's surface. Local accumulations of the relatively low-thickness igneous extrusions are referred to as *volcanics*. Basalt lava flows often cover large areas of thousands of square kilometers, and they can have thicknesses up to several kilometers, high-amplitude *magnetization*, and highly irregular anomalous signatures masking the local-scale magnetic anomalies from deeper sources. See also *intrusion* and *signature*.

F

fabric: *grain* in the context of a spatial configuration of anomalies, their *spatial frequency* content, and trends that are characteristic for a particular local- or regional-scale area.

fairway: a local-scale depositional or mineralization *trend* within which the hydrocarbon-bearing reservoir sands or *placer* deposits are likely to be found. Similar to buried drainage channels, fairways generate the low-*amplitude* short-*wavelength* negative gravity anomalies highlighted by the calculation of *residual gravity*. Sand transportation fairways can accumulate sufficient amounts of *detrital magnetite* and acquire *detrital magnetization* capable of generating localized clusters of low-amplitude short-wavelength positive magnetic anomalies. For relatively shallow fairways, such anomalies can be detected and mapped by a *high-resolution aeromagnetic survey*.

falling-atom gravimeter: same as *absolute atomic gravimeter* with an emphasis on its operating principle based on measurements of the free fall *acceleration* of cold atom streams in a vacuum chamber.

falling corner-cube gravimeter: See *falling-mirror gravimeter*.

falling-mirror gravimeter: an *absolute gravimeter* utilizing the laser interferometer and atomic clock to measure the *gravitational acceleration* of a reflective object, or mirror, in its free fall inside a vacuum chamber. Laser light beams are separated by the interferometer into two paths, where one beam reflects from the free-falling mirror and other beam passes straight through the system, so both beams interfere with one another and produce the light and dark optical fringes at each half of the laser *wavelength*. High-*precision* distance measurements by optical fringes and timing by the atomic clock are used to calculate the gravitational acceleration of a free-falling mirror, i.e., *absolute gravity*, for each free-fall cycle. Calculations are based on the relationship between the free-fall time t, travel distance d, and gravitational acceleration g that is expressed as

$$t = (2d/g)^{1/2}.$$

Falling-mirror gravimeter systems have neither *drift* nor *tare* over time. Achievable *accuracy* is approximately 1–2 *microgal (μGal)*, which depends on the integration time of multiple free-fall cycles: the longer the integration time, the higher the accuracy of the measured gravitational acceleration. Also referred to as *optical free-fall gravimeter*, *falling corner-cube gravimeter*, *laser gravimeter*, and *laser interferometer gravimeter*. See also *absolute atomic gravimeter*.

fall-off: a decrease in the *amplitude* of the gravity or magnetic *anomaly* with distance from its source. Fall-off is related to the geometrical shape and *dimension* (1D, 2D, 3D) of a *source body* and the direction of its *magnetization*. For *single-point source* geometries, it is described by the *inverse power law* as

$$A(r) = C/r^N,$$

where A(r) is *anomaly amplitude*, C is a constant specific for such factors as *anomalous mass* or *anomalous magnetization* and *geomagnetic field* parameters, r is distance from the source, and N is *fall-off rate*. For sources approximated by a *point source*, the gravity anomaly falls off as C/r^2, whereas the magnetic anomaly falls off as C/r^3, thereby showing that the *magnetic method* provides the higher lateral and vertical resolutions of anomalies than the *gravity method*. For *gradient* anomalies, the fall-off can be described as

$$A(r) = C/r^{N+1},$$

i.e., it is higher than the fall-off of the *total field* anomalies by one power of distance from the source. Also referred to as *amplitude attenuation* and *shape factor*.

fall-off factor: a dimensionless function which describes the decrease, or *fall-off*, of a gravity or magnetic anomaly with distance from an *ideal source* using the ratio x/z, where x is the offset distance, i.e., distance from the surface position of a source to the observation point on a *principal profile*, and z is the source depth. The fall-off factor and *central amplitude* are variables in the analytical expressions for calculating the gravity and magnetic fields of ideal sources. Also known as *attenuation factor*. See also *wavelength attenuation*.

fall-off rate: the rate of an *amplitude* decrease, or *attenuation*, of a gravity or magnetic *anomaly* with distance from a *source body*. For *single point source* geometries, it is quantified by the *inverse power law* exponent, or index, N. The fall-off rate is inversely proportional to the source body depth, and it depends on the source shape and *dimension* (1D, 2D, 3D). For *gradient* anomalies, the fall-off rate is higher by one power of distance from the source and expressed as

$$N_{grad} = N + 1,$$

where N is fall-off rate for the *total field* anomalies. Also referred to as *attenuation rate* and *decay rate*. See also *fall-off* and *wavelength attenuation*.

false color: *pseudocolor* in the context of using the color shades to display a range of gravity or magnetic field values and other measurable characteristics that do not have natural, or true, colors. See also *color-coded grid*.

false easting: a parameter of the *Universal Transverse Mercator (UTM)* coordinate system that represents the value of 50,0000 m assigned to the central meridian of each UTM zone to ensure that values of the X-coordinate within the UTM zone are always positive. It is a standard reference parameter required for *gridding* and mapping. See also *easting*, *false northing*, and *XY coordinates*.

false northing: a parameter of the *Universal Transverse Mercator (UTM)* coordinate system that represents either (a) the value of 0 m assigned to the equator for each UTM zone in the northern hemisphere or (b) the value of 10,000,000 m assigned to the equator for each UTM zone in the southern hemisphere to ensure that values of the Y-coordinate within the UTM zone are always positive. It is a standard reference parameter required for *gridding* and mapping. See also *northing*, *false easting*, and *XY coordinates*.

false-sun illumination: *sunshading* in the context of viewing a gridded *anomalous field* as a topographic surface of hills and valleys illuminated by an artificial source of light, or false sun, from different directions.

far-zone terrain correction: See *outer-zone terrain correction*.

fast Fourier transform (FFT): an optimized *data processing* algorithm executing the high-speed *Fourier transform*. The FFT assumes that a dataset is periodic or, equivalently, circular with the last point next to the first point. In order to eliminate first-order discontinuities at data edges and minimize *Gibbs errors*, the following requirements should be satisfied prior to FFT: (a) *trend removal* to minimize level shifts between the opposite edges of line-oriented or gridded data; (b) data extension, or *padding*, to provide the integer number of 2^N points and a smooth transition between the neighboring repetitions in a dataset; and (c) application of an *alias filter* to prevent *aliasing* of wavenumbers outside the range defined by the *Nyquist wavenumber*. All spectral-domain transformations of gravity and magnetic data are based on the use of FFT. See also *FFT filters*, *gridding*, and *spectral domain*.

fault: the structural discontinuity represented by a discrete surface or zone of discrete surfaces across which the rock units slide along each other horizontally and/or vertically. Vertical and dipping faults separate layers of sedimentary rock formations and blocks of metamorphic and igneous rocks within the *basement*. Depending on the type of sliding displacement, or *offset*, of rocks along a fault and the *scale* of the structural deformations, faults are categorized into normal, reverse, strike-slip, thrust, wrench (transform), listric, and other fault types. Faults can be acting as either conduits for the migration of water, hydrocarbons, and highly mineralized fluids or seals for their local accumulation. Fractured reservoirs with hydrocarbon accumulations and mineral ore deposits are often located at the intersections of faults and within their fracture zones. Faults commonly have distinct gravity and magnetic signatures among *total field* and *gradient* anomalies. Hydrocarbon or *juvenile water* migration within intrasedimentary faults can result in their *magnetization* sufficient for generating low-*amplitude* short-*wavelength* magnetic anomalies. Fault-related displacement of rock formations and their fracturing create the lateral *density* contrasts that generate gravity and *gravity gradient* anomalies of various amplitudes and wavelength content. Faults often create the structurally weak zones that act as natural conduits for igneous intrusions, which are the sources of various kinds of magnetic and gravity anomalies. For source depth estimates and *modeling*, a fault is often approximated by a vertical or dipping *contact*, *thin sheet (thin dike)*, or *faulted slab* source geometry. In the detection of faults, the higher *resolution* of the *magnetic method* allows it to be more efficient than the *gravity method*. See also *contrast*, *dip*, *igneous intrusion*, *magnetized fault*, *gravity signature*, *magnetic signature*, *shear zone*, and *strike-slip fault*.

fault zone: a zone of fracturing around a *fault* that can vary in its width from several centimeters to several hundred meters. It is a frequent source of low-*amplitude* short-*wavelength* negative gravity anomalies and positive and negative magnetic anomalies. See also *magnetized fault*.

faulted slab: the horizontal plate of a uniform thickness terminated by a vertical plane along one side that approximates an actual *fault*. Its calculated gravity or magnetic effects are used in *modeling* of faulted sedimentary layers. A relatively thin faulted slab is referred to with the same meaning as a faulted horizontal *thin sheet*.

Faye anomaly: See *free-air anomaly*.

Faye correction: **1.** the sum of a *free-air correction* and *terrain correction* used in processing the geodetic measurements of a *geoid*. **2.** *free-air correction*.

feasibility modeling: **1.** the two-dimensional (2D) or three-dimensional (3D) *forward modeling* of a targeted *structure* performed prior to a *detailed survey* in order to (a) estimate the *amplitude* and *wavelength* of a structure-generated *anomaly* and their change with increasing or decreasing *target* depth and (b) select the optimal parameters for an *airborne survey*, such as *survey height*, *survey layout*, and *survey line spacing* for *traverse lines* and *tie lines*, to ensure that the *target anomaly* will be resolved both laterally and vertically under cost-effective terms. **2.** *feasibility study*. See also *lateral resolution* and *vertical resolution*.

feasibility study: **1.** an assessment of (a) opportunities and limitations in the application of a specific gravity or magnetic method for the detection of *total field* and *gradient* anomalies generated by an exploration *target*; (b) *accuracy* and *resolution* requirements in the application of *time-lapse gravity* or *time-lapse gravity gradiometry* for *reservoir monitoring*. **2.** computer-based simulation of a pre-planned gravity or magnetic *survey* performed to ensure that the selected method and *survey specifications* are appropriate for the geological environment of a particular exploration target. It is based on (a) construction of a three-dimensional (3D) *model* of the target using the available seismic and well data, (b) *forward calculation* of the model-generated *amplitude* and *wavelength*, and (c) analysis of the *amplitude response* to changes in the *density* or *magnetization* contrast, target depth, and fluid content of a targeted reservoir. **3.** *feasibility modeling*. Also referred to as *feasibility test*. See also *density contrast* and *magnetization contrast*.

feasibility test: See *feasibility study*.

femtotesla (fT): a unit of *sensitivity* of advanced magnetometers, such as a *helium magnetometer* and *SQUID magnetometer*, defined as

$$1\,\mathrm{fT} = 10^{-15}\,Tesla = 10^{-3}\,picotesla\,(pT) = 10^{-6}\,nanotesla\,(nT).$$

ferrimagnetic: a characteristic of rock-forming magnetic minerals with a predominantly moderate *magnetization* (except for *magnetite*) and positive *magnetic susceptibility* within a wide range from approximately 10^{-6} to 0.7–1.5 *cgs units*. Ferrimagnetic minerals differ from ferromagnetic minerals by their combined parallel and antiparallel alignments of the atomic magnetic moments and, often, by *Curie temperature*, which is 560°–580°C for ferrimagnetic magnetite and 1130°C for ferromagnetic cobalt. Ferrimagnetic minerals can acquire *remanent magnetization*. See also *ferrimagnetism* and *magnetic dipole moment*.

ferrimagnetism: a subdivision of *ferromagnetism* representing the rock *magnetization* produced by the combined parallel and antiparallel alignments of the adjacent atomic magnetic moments that result in the total orientation of the elementary magnetic dipoles parallel to the externally applied magnetic field, such as *Earth magnetic field*. Ferrimagnetic minerals, such as *magnetite*, titanomagnetite, maghemite, and ilmenite, are accessory to the dominant rock-forming minerals, but they are major sources of magnetization of rocks in the upper *crust*, wherein magnetite (FeO-Fe_2O_3) is a principal naturally occurring *ferrimagnetic* mineral, whose volume percentage defines the magnetic properties of rock formations. See also *ferromagnetism*, *diamagnetism*, *paramagnetism*, *magnetic dipole moment*, and *Weiss theory*.

ferromagnetic: a characteristic of the moderately-to-strongly magnetic rock-forming minerals with a wide range of positive *magnetic susceptibility* values varying from 0.5–1.0 *cgs units* in iron-based minerals, such as hematite and pyrrotite, to approximately 10^6 cgs units in ferrous metals, such as iron, cobalt, and nickel. Their magnetic properties include *hysteresis* and *remanent magnetization*. See also *ferrimagnetic* and *paramagnetic*.

ferromagnetism: *magnetization* of the rock-forming minerals by interactions between the adjacent atomic magnetic moments and alignments of groups of atoms, or magnetic domains, in the direction of the externally applied magnetic field, such as *Earth magnetic field*. Ferromagnetic minerals, such as the iron-titanium oxides and iron sulfides, have a positive and commonly large *magnetic susceptibility* and strong *remanent magnetization*. Magnetization of the ferromagnetic minerals is highly variable, but it is generally much stronger than magnetization of the diamagnetic or paramagnetic minerals. Depending on the type of alignment, or coupling, of the atomic magnetic moments and magnetic domains, there are several groups of the *ferromagnetic* rock-forming minerals, including: (a) ferromagnetic minerals, where the atomic magnetic moments are aligned parallel to each other, and their total magnetization is the strongest; (b) antiferromagnetic minerals, where the atomic magnetic moments are aligned antiparallel and cancel one another, so their total magnetization is nearly zero; and (c) minerals, where the atomic magnetic moments are aligned both parallel and antiparallel but do not cancel each other, so the resulting *magnetization vector* is oriented in the direction of the dominant magnetic moments. Above the *Curie temperature*, ferromagnetic minerals can keep some degree of paramagnetic properties, but they become essentially non-magnetic. Rare earth elements (REE), such as neodymium, can greatly intensify the ferromagnetism of their host rocks. See also *diamagnetism*, *paramagnetism*, *ferrimagnetism*, *magnetic domain*, *magnetic dipole moment*, *rare earth elements (REE)*, and *Weiss theory*.

ferry: the distance from the nearest airport facility to an airborne gravity or magnetic *survey* area.

FFT: See *Fast Fourier Transform (FFT)*.

FFT filters: a large group of spectral-domain filters, such as *low-pass filter*, *high-pass filter*, and *band-pass filter*, with algorithms that implement a *fast Fourier Transform (FFT)* for mathematical transformations of the data. Application of FFT filters is based on the assumption that the transformed dataset is periodic or, equivalently, circular beyond the dataset area, i.e., line or *grid*, and it requires *data conditioning*, including (a) *detrending* to minimize level shifts between the opposite edges of a dataset area and (b) *padding* to provide a smooth extrapolation and transition from the original to the padded data values. The majority of filters used in the *enhanced processing* of gravity and magnetic data are FFT filters. See also *filter*, *spectral domain*, and *2D FFT filters*.

fiducial time: the absolute chronological time of gravity and magnetic measurements along *acquisition* lines during marine (shipboard) and airborne surveys.

fiducials: 1. fixed reference points along airborne and marine (shipboard) *survey* lines that are used for the *positioning* and synchronization of measurements. **2.** survey pickets of ground gravity and magnetic measurements. See also *airborne survey*, *marine survey*, and *ground survey*.

field: 1. the space in which a physical force exists and throughout which the continuity and variations of this force are described by a measurable quantity, such as *acceleration* of a free-falling body of *mass* under the action of *gravitational force*, and expressed mathematically by functions of the spatial coordinates and time. The field of a gravitational or magnetic force is called *gravitational field* or *magnetic field*, respectively. A field has only one *scalar* or *vector* value of the measured quantity at every point within the space at a given time. **2.** an outdoor area away from the office or laboratory, where a geophysicist conducts gravity or magnetic measurement operations. See also *scalar field*, *vector field*, and *potential field*.

field continuation: See *downward continuation* and *upward continuation*.

field intensity: 1. a quantitative characteristic of the gravity or magnetic *field* described by the *gravitational force* per unit *mass* (i.e., by *gravitational acceleration*) or *magnetic force* per unit *magnetic pole* (magnetic pole of a unit strength), respectively. **2.** a quantitative characteristic of the *potential field* defined at any point in space by the *magnitude* and orientation of a *gradient vector* which represents a spatial rate of change of the scalar *potential*. Also known as *field strength* and *flux density*. See also *magnetic pole strength*, *scalar*, and *total magnetic intensity (TMI)*.

field strength: See *field intensity*.

field terrain correction: same as *innermost terrain correction* with an emphasis on high-*accuracy* geodetic or *drone*-assisted measurements of *elevation* in a *ground gravity* survey area prior to calculation of *terrain correction* values in the office.

field-to-potential transformation: 1. calculation of the *vertical integral* of a gravity or magnetic *anomaly*, i.e., calculation of an *anomalous potential*. **2.** Fourier-domain transformation of an anomalous *gravity field* expressed as

$$\mathcal{F}(U) = \mathcal{F}\{g_z\}/|k|,$$

where $\mathcal{F}(U)$ is *Fourier transform* of *gravitational potential* U, $\mathcal{F}\{g_z\}$ is the Fourier transform of the *vertical gravity* g_z, and $|k| \neq 0$ is a *wavenumber* defined as

$$|k| = \left(k_x^2 + k_y^2\right)^{1/2},$$

where $k_x = 2\pi/\lambda_x$, $k_y = 2\pi/\lambda_y$, and λ_x, λ_y are *wavelength* components in the mutually orthogonal x- and y-directions, respectively, in *Cartesian coordinates*. For an anomalous *magnetic field*, the transformation is expressed as

$$\mathcal{F}\{\mathcal{V}\} = -\mathcal{F}\{\Delta T\}/|k|\theta_f,$$

where $\mathcal{F}\{\mathcal{V}\}$ is the Fourier transform of *magnetic potential* $\mathcal{V}$, $\mathcal{F}\{\Delta T\}$ is the Fourier transform of *magnetic anomaly* ΔT, and θ_f is a *complex function* of k_x and k_y that depends on the direction of the *ambient field*, such as *Earth magnetic field* (Blakely, 1996). See also *anomalous field*, *Fourier domain*, and *scalar-to-vector transformation*.

figure-of-merit (FOM): a measure of the effectiveness of *aeromagnetic compensation* for the magnetic interference effects produced by aircraft maneuvers in the variable *Earth magnetic field*. The FOM value is determined by a pre-survey *calibration flight* during a period of steady *temporal magnetic variations* and at a relatively high (2–3 km) *altitude* over the area of low-*amplitude* magnetic anomalies, so any recorded variations over the same area can be attributed to the magnetic effects of the *heading* and maneuvers. In each of the cardinal compass directions (east, west, north, south), the aircraft performs the maneuvers of ±5° *pitch*, ±5° *yaw*, and ±10° *roll*, and the sum of the *reading* variations, recorded over 4–5 second intervals regardless of sign for each of these 12 maneuvers, represents the FOM value. For an active, or adaptive, *real-time compensation* system, the acceptable FOM value is less than 1 *nanotesla (nT)*. A FOM test is flown prior to airborne *magnetic survey* measurements and after all major changes of aircraft equipment and configuration. The test is required for wing-tip and *stinger*-mounted airborne magnetometers but not for towed-*bird* or *slung sensor* measurements. See also *heading effect* and *pre-survey flight tests*.

filter: 1. a computer algorithm which changes the *amplitude*, *phase*, and *wavelength* components of input data by dividing them into two parts: passed and rejected. The passed part, or *passband*, is retained without *attenuation* and modified in accordance with a specified *transfer function*, whereas the rejected part, or *stopband*, is suppressed or attenuated substantially. The length of a transition zone between the passband and stopband parts is defined by the *filter order*: the higher the order, the shorter the transition zone length. Effective passband limits are defined by a *filter cutoff*. **2.** an algorithm for calculation of a particular *attribute* of an anomalous gravity or magnetic field, such as the first- or second-order *derivative*, *gradient*, *tilt angle (TA)*, *vesselness (VS)*, and *curvature attributes*. **3.** any processing operator that transforms input data in accordance with the properties of a mathematical function, or transfer function, which defines this operator. See also *band-pass filter*, *high-pass filter*, *low-pass filter*, *matched filter*, and *notch filter*.

filter cutoff: a *filter* parameter that defines the effective limit of a retained part of data, or *passband*, by a specific space-domain *wavelength* or spectral-domain *wavenumber* at which the data amplitudes are approximately 50%–60% of their original values. For example, a *high-pass filter* with a cutoff of 1600 m passes without *attenuation* all data components with wavelengths shorter than 1600 m and rejects all components with wavelengths longer than 1600 m. The filter cutoff value is within the transition zone between the passband and *stopband* part of the data, wherein the length of this zone is defined by the *filter order*. See also *space domain* and *spectral domain*.

filter fall-off: See *filter order*.

filter length: a *time filter* parameter that defines the low-pass *filter cutoff* value in units of time, usually in seconds. Multiplying the filter length by the velocity of a survey ship or airplane in meters per second ($m \cdot sec^{-1}$) yields a filter cutoff in meters. For example, the filter length of 80 seconds at an aircraft velocity of 40 $m \cdot sec^{-1}$ (144 km per hour) is equivalent to a *low-pass filter* with a filter cutoff of 3200 m. In marine (shipboard) and airborne gravity and *gravity gradient* surveys, the filter length defines the limit of the achievable *half-wavelength resolution* of the acquired gravity and gravity gradient data. See also *full-wavelength resolution*.

filter montage: See *filter panel*.

filter order: a *filter* parameter that defines the slope of the transition from the *passband* to *stopband* parts of the filtered data: the higher the order, the steeper the slope. The highest orders can produce a *filter response* similar to the response of a *boxcar filter*, but they distort the data by various space-domain side effects and by *Gibbs ringing* in the spectral-domain filters. Also referred to as *filter fall-off*. See also *Butterworth filter*, *space domain*, and *spectral domain*.

filter panel: a printed or on-screen display that shows side-by-side *grid* images obtained by the application of different filters. They are used for a comparative analysis of filters in *enhanced processing*. Also known as *filter montage* and *filter scan*. See also *filter*.

filter response: same as *transfer function* with an emphasis on the characteristic of a *filter* that describes a particular type of changes which will be made to the data by application of this filter.

filter scan: See *filter panel*.

filter spectrum: an *amplitude*-versus-*wavenumber* graphical plot of a *filter response* that shows the range of wavenumbers separated by a spectral-domain *filter* from the rest of the data in the pass or reject mode of its operation. See also *spectral domain.*

filtering: 1. any numerical transformation of obtained data that reveals the anomalies of an exploration and research interest. **2.** application of space-invariant filters for separating an *anomalous field* into groups of residual anomalies for their subsequent analysis and *interpretation.* Because the anomalous field represents a result of the interference and superposition of multiple anomalies generated by various sources at different depths and distances around each *observation point,* the analysis of such mixed-source data without filtering can produce the interpretation solutions that are low-resolving structurally and irrelevant geologically. The purpose of filtering is to (a) separate anomalies from each other, (b) remove a regional component of the observed field, (c) suppress *noise,* and (d) isolate anomalies of the exploration or research interest. Depending on the nature of the *anomaly,* the filtering can be based on amplitudes, anomalous trends, gradients, wavelengths, and wavenumbers. Anomaly separation by *wavelength* criteria is performed in the *space domain,* whereas the separation by *wavenumber* criteria is performed in the *spectral domain.* **3.** *space-varying filtering.* See also *regional-residual separation, residual anomaly,* and *trend.*

finite contact: a *model* approximation of a *source body* by the *edge* of a horizontal plate, or slab, of finite thickness and *infinite length* in both horizontal dimensions. Also referred to as *finite fault, finite step, thick plate,* and *thick step.* See also *dimension* and *infinite contact.*

finite-difference method: a space-domain method of computing the horizontal and vertical derivatives of an anomalous gravity or magnetic *field.* The horizontal *derivative* is obtained by calculating the difference between two data points at the same *elevation* and dividing it by the distance between two points. The vertical derivative is obtained by (a) *upward continuation* of data to the levels h and $2h$, where h is less than the shallowest source depth or *grid interval* in the gridded data, or (b) calculating the difference between the data values at the observation level and elevation $2h$ and dividing the obtained value by the upward continuation distance $2h$ to obtain the *vertical derivative* in the middle of the upward continuation range, i.e., at elevation h. Absolute values, or magnitudes, of the obtained derivatives represent calculated gradients of a gravity or magnetic field in the direction of their calculation. See also *anomalous field, gridding, space domain, magnitude, calculated gradient,* and *measured gradient.*

finite fault: See *finite contact.*

finite step: See *finite contact.*

first-derivative filter: an *enhanced processing* operator for calculating the first-order *horizontal derivative* or first-order *vertical derivative* of an anomalous gravity or magnetic field. Because its spectral-domain *frequency response* is directly proportional to the *wavenumber,* this filter enhances the relatively high-frequency (short-*wavelength*) field components generated, primarily, by shallow and moderately deep elements of subsurface geological structures. See also *anomalous field, first-derivative map, spectral domain,* and *structure.*

first-derivative map: a digital or printed map of the *first horizontal derivative (FHD)* or *first vertical derivative (FVD)* of an anomalous gravity or magnetic field. Because the attenuation, or *fall-off,* of anomalies is inversely proportional to the distance from their sources, the anomalous field components generated by shallow and mid-depth geological structures will be enhanced much more than anomalies from deep structures. The FVD map of *Bouguer gravity* or marine *free-air gravity* shows positive values over structurally elevated areas (uplifts, anticlines, thrusts, horsts, etc.) and negative values over structural lows (troughs, synclines, grabens, etc.). Trends of discontinuity, or truncation, of FVD-enhanced anomalies often indicate a presence of cross-cutting faults or shear zones related to the strike-slip faulting. See also *anticline, horst, thrust, uplift, shear zone, strike-slip fault,* and *second-derivative map.*

first harmonic: same as *fundamental frequency* for which the simple multiples represent *harmonics* of variable orders in a *Fourier series.*

first horizontal derivative (FHD): the first-order *horizontal derivative,* also denoted by 1HD.

first-order analytic signal (AS_1): the *analytic signal (AS)* of the first-order *vertical derivative* of an anomalous gravity or magnetic field. Referred to with the same meaning as *first-order enhanced AS (eAS_1)* and, occasionally, *second-order analytic signal (AS_2).*

first-order ASA (ASA_1): the *analytic signal amplitude (ASA)* of the first-order *vertical derivative* of an anomalous gravity or magnetic field. The ASA_1 of *total magnetic intensity (TMI)* is implemented in the *AN-EUL method* and *ASA-ratio method.* Referred to with the same meaning as *first-order enhanced ASA ($eASA_1$)* and, occasionally, *second-order ASA (ASA_2).*

first-order enhanced AS (eAS_1): the *analytic signal (AS)* of the first-order *vertical derivative* of an anomalous gravity or magnetic field. Anomalies of the eAS_1 field are represented by the second-order derivatives of an *anomalous field* that can improve *lateral resolution* of the interfering anomalies generated by closely spaced shallow and moderately deep sources (Hsu et al., 1996). Equivalent to *first-order analytic signal (AS_1).* See also *derivative.*

first-order enhanced ASA ($eASA_1$): 1. *amplitude* of *first-order enhanced AS (eAS_1).* **2.** *analytic signal amplitude (ASA)* of the first-order *vertical derivative* of an anomalous gravity or magnetic field expressed as

$$eASA_1 = \left|eAS_1\right| = \{\partial[(\partial M/\partial z)/\partial x]^2 + \{\partial[\partial M/\partial z)/\partial y]^2 + \{\partial[(\partial M/\partial z)/\partial z]^2\}^{1/2},$$

where M is the *magnitude* of the *anomalous field.* Prior to computations, the magnetic and gravity fields are recomputed at the lowest empirically determined level of *upward continuation* to suppress high-frequency *noise* (Hsu et al., 1996). It is equivalent to *first-order ASA (ASA_1)* and *second-order ASA (ASA_2),* which was so named to highlight the use of second-order derivatives of the anomalous field in $eASA_1$ calculations. See also *derivative* and *second-order enhanced ASA ($eASA_2$).*

first-order free-air correction: *free-air correction* calculated with the use of the first-order approximation formula and expressed in units of *milligal (mGal)* as

$$\Delta g_{FA} = 0.3086h,$$

where 0.3086 is a *free-air factor* in units of milligal per meter (mGal · m^{-1}) representing the theoretical value of the *free-air gradient* at 45° latitude and *h* is *elevation* in meters. The difference between the first- and second-order free-air corrections is directly proportional to the observation point elevation: the higher the elevation, the larger the difference. See also *second-order free-air correction.*

first-order local wavenumber: same as *horizontal local wavenumber* that is the *attribute* of an anomalous gravity or magnetic field defined by the first-order *horizontal derivative* of the *local phase.* See also *second-order local wavenumber.*

first-order tilt angle: same as *tilt angle (TA)* that is the original *attribute* of an anomalous magnetic or gravity field from which the *second-order tilt angle* can be calculated.

first vertical derivative (FVD): the first-order *vertical derivative* of anomalous gravity or magnetic field F that quantifies its rate of change in the vertical direction *z* and denoted by ∂F/∂z. The FVD *magnitude* is much higher over shallow and mid-depth sources than over deep sources, because the rate of the field change, or *fall-off rate,* is inversely proportional to the distance from its sources: the larger the distance, the lower the fall-off rate. The FVD can be calculated by: (a) *Hilbert transform* of the *horizontal derivative,* (b) multiplying the Fourier-transformed data by |*k*|, where *k* is *wavenumber*; (c) *vertical integral* of the *second vertical derivative (SVD)* obtained from *Laplace equation*; and (d) *finite-difference method.* Often referred to simply as *vertical derivative.* Also denoted by 1VD. See also *anomalous field, Fourier transform, fractional vertical derivative,* and *variable order vertical derivative.*

first vertical integral (FVI): 1. the *vertical integral* for N = 1, where N is the vertical integration order which can be either an integer or fractional number. **2.** an algorithm for the vertical integration of the *RTP magnetic field* scaled by the ratio of *density* to *magnetization* of rocks. It is used in *pseudogravity transformation* to convert the reduced-to-pole (RTP) magnetic field values into the equivalent *pseudogravity* values in accordance with *Poisson theorem.* **3.** an algorithm for transformation of an *anomalous field* into *anomalous potential.* See also *reduction-to-pole (RTP)* and *fractional vertical integral.*

fish: a watertight and streamlined non-magnetic container, or housing, where a high-*sensitivity* magnetometer, usually an *Overhauser magnetometer,* is mounted. During a marine (shipboard) *magnetic survey,* the fish is towered at several ship lengths behind the *survey* ship at the depth of 3–5 m below the sea surface on a cable with electrical connections to power and operate the fish-mounted magnetometer. See also *bird* and *stinger.*

5C inversion: the *joint inversion* of five independent components of the *full gradient tensor (FTG)* obtained from measurements by an *airborne gravity gradiometer (AGG)* system. For the same *signal-to-noise ratio (S/N),* the increased number of *constraints* in the multicomponent *inversion* can provide a higher *accuracy* in estimates of structural parameters, such as *density contrast,* depth, *dip,* and shape of *target,* compared to those obtained by a single-component T_{zz} *inversion* (Paoletti et al., 2016).

fixed-wing survey: airborne magnetic or gravity measurements with the use of an airplane and high-accuracy *positioning* system, such as *differential GPS (DGPS).* Detailed fixed-wing surveys are flown in the *drape flight* mode or at a constant low-altitude *elevation* determined by the airplane flight characteristics and safety regulations. See also *altitude, helicopter survey,* and *drone survey.*

flat-Earth approximation: approximation of rock formations above a *survey datum* (sea level or *reference ellipsoid*) by the infinite horizontal *Bouguer plate,* usually for the purpose of calculating the *simple Bouguer correction.*

flat-plate Bouguer factor: *Bouguer correction factor (BCF)* in the context of approximating rock formations above a *survey datum* by the infinite horizontal *Bouguer plate.*

flattening: See *polar flattening* and *gravitational flattening.*

flexural isostasy: the *Vening Meinesz hypothesis* in the theory of *isostasy* based on the assumption that the Earth's *crust* strength is sufficient for its regional bending, or flexing, in response to rock mass loading and unloading due to large-scale topographic and bathymetric relief variations, as opposed to the crust breaking into rigid blocks bounded by the high-angle faults that is assumed by the *Airy hypothesis* and *Pratt hypothesis.* Also referred to as *regional isostasy.*

flexural isostatic anomaly: the *isostatic anomaly* compensated for the strength and elastic, or flexural, properties of the *lithosphere* quantified by its effective elastic thickness in the *Vening Meinesz model* of *isostasy.* It represents the *residual gravity* field obtained by the application of *flexural isostatic correction* to a *Bouguer anomaly* or *free-air anomaly.* Gravity effects of crustal thickness variations and bending of the lithosphere under a given topographic or bathymetric load are calculated with different elastic thickness estimates in order to obtain several flexural isostatic anomaly maps for subsequent analysis and *interpretation.*

flexural isostatic correction: a correction for the gravity effect of regional, or flexural, *isostatic compensation.* Its calculation is based on parameters of the *Vening Meinesz model* which incorporates the flexural strength of the *lithosphere* quantified by its effective elastic thickness. Estimates of lithospheric elastic thickness involve calculations of *gravitational admittance* and *gravitational coherence.* At the rifted continental margins offshore, this correction includes the gravity effects of the accumulated sediments, bathymetry, and crustal thinning. See also *Jachens-Roberts method.*

flight elevation correction: 1. a correction which subtracts the flight *elevation* value from the calculated magnetic source depth in order to adjust the obtained depth values to a *survey datum*, because the standard methods of estimating the source depth are based on the use of the total height of a *magnetic sensor* above a *source body*. Also referred to as *sensor elevation correction* or *sensel correction*. **2.** an optional correction applied to aeromagnetic data to compensate for the flight elevation of a survey aircraft above the survey datum: sea level or *reference ellipsoid*. **3.** *drape correction* for high-*resolution* aeromagnetic data. In *airborne gravity* surveys, the gravity effect of the flight elevation is included into *free-air correction*.

flower structure: the local or regional-scale geological *structure* created by a convergent (for a positive flower structure) or divergent (for a negative flower structure) movement of rock formations along the *fault* splays within a *strike-slip fault* zone and their block-folding upward or downward, respectively. Flower structures generate the linear-trending positive and negative gravity and magnetic anomalies highlighted by *regional-residual separation*. Both types of the flower structure often generate distinct *gravity gradient* and *magnetic gradient* anomalies. See also *horst, pull-apart basin, residual gravity, residual magnetics, trend*, and *uplift*.

flux: lines of force of the *magnetic field* in the form of an imaginary fluid-like flow, where the number of lines of force is directly proportional to *field intensity*, also known as *field strength*. See also *flux density, Gauss flux theorem*, and *magnetic flux*.

flux density: a *vector field* parameter defined by the number of theoretical field lines, or *flux* lines, per unit volume or per unit area perpendicular to the direction of the flux *vector*. Flux density is directly proportional to *field strength*. See also *divergence* and *magnetic flux density*.

fluxgate gradiometer: a *gradiometer* system composed of paired and synchronized fluxgate magnetometers mounted on the same frame and separated by the distance called *baseline* to measure the vertical or horizontal gradients of the *Earth magnetic field*. Light-weight versions can be attached to a *drone* for a detailed aeromagnetic survey in *exploration*, environmental, and archaeological applications. See also *drone survey, detailed survey, horizontal gradient, vertical gradient*, and *fluxgate magnetometer*.

fluxgate magnetometer: the *magnetometer* which can be used in the *scalar* or *vector* mode of operation. In the scalar mode, it operates as a *scalar magnetometer* measuring the total magnetic field by orienting the axis of its *magnetic sensor* in the approximate direction of the *Earth magnetic field*. In the vector mode, it operates as a *vector magnetometer* measuring the vector component of the Earth's magnetic field relative to the spatial orientation of the sensor axis and reference, or *baseline*, value which is calibrated periodically by a non-drifting scalar magnetometer. A three-sensor, or three-component, version of a fluxgate magnetometer with sensors oriented along the mutually orthogonal *x*-, *y*-, *z*-directions is referred to as a 3C or triaxial fluxgate magnetometer. Its magnetic sensor is composed of two identical parallel cylindrical bars or one ring-shaped cylindrical bar composed of a high-permeability *ferromagnetic* material and wrapped by two windings: a primary, or excitation, coil and a secondary, or pick-up, coil which detects the changes in *magnetic flux* passing through the sensor. A fluxgate magnetometer is also used to measure the magnetic effects produced by the aircraft frame and electronic equipment for their compensation in the *real-time* mode by an *aeromagnetic compensation* system. Light-weight versions can be attached to a *drone* for a low-*altitude* pre-programmed or remote-controlled *high-resolution aeromagnetic survey*. A fluxgate magnetometer is directionally sensitive and subjected to *drift* caused, primarily, by temperature effects. Modern versions have a temperature coefficient of 0.10–0.12 *nanotesla (nT)* per each °C and a few nanoteslas of drift per year. The best reported *sensitivity* is 0.05–0.10 nT (Gavazzi et al., 2016). See also *magnetic permeability, magnetometer drift, 3C magnetometer*, and *drone survey*.

folding frequency: *Nyquist frequency* in the context of *sampling*, where a symmetry of the aliased data samples about the Nyquist frequency value is referred to as folding. See also *aliasing*.

folding wavenumber: same as *Nyquist wavenumber* representing the largest *wavenumber* corresponding to the shortest *wavelength* of the sampled data that can be transformed without *aliasing* by a *discrete Fourier transform (DFT)*.

FOM test: a flight test performed to determine *figure-of-merit (FOM)* prior to airborne *magnetic survey* measurements. See also *pre-survey flight tests*.

force of gravity: the force of the mutual attraction between all bodies of *mass*, wherein the source of this force is their masses. On the Earth's surface, the force of gravity is defined by the combined effect of *gravitational attraction*, or pull, by the Earth's mass and repulsion, or push, by *centrifugal force* resulting from the Earth's rotation about its axis. Referred to interchangeably with the same meaning as *gravity* and *gravitational force*. Unlike *magnetic force*, the force of gravity cannot be shielded. See also *centrifugal acceleration* and *Newton law of gravitational attraction*.

forensic magnetism: measurements and study of the surface and near-surface magnetic properties of disturbed soils in the search for buried objects and shallow graves. Digging a pit destroys the mostly homogeneous *bulk magnetization* of the soil by randomizing the original orientation of magnetic dipoles in the excavated soil, thereby resulting in a weak (less than 2–3 *nanotesla*) but detectable residual *magnetic low* over a grave location. Measurements are made along closely spaced *acquisition* lines to ensure the subsequent *gridding* of the acquired data at 0.20–0.25 m *grid cell size*, often with the use of a *fluxgate magnetometer* attached to a low-flying *drone*. See also *drone survey, magnetic dipole, residual field*, and *archaeomagnetism*.

format: **1.** the digital arrangement of data that permits their identification and loading into computer applications. **2.** a specific form of arrangement, presentation, display, and mathematical description of gravity and magnetic data for their multiple use, transformation, visualization, and analysis. The principal standard format in digital *data processing* and presentation of its results is *grid*. See also *ASCII format*.

format correction: *scale correction* in the context of converting continuous marine (shipboard) and airborne gravity or magnetic measurements from the original time-based *format* into the distance-based format of standard *data processing* in accordance with the instantaneously measured speed of a survey ship or aircraft.

formation density: *apparent density* in the context of rock formation *density* computed from the geophysical measurements in drilled wells (boreholes) and seismic interval velocities or obtained by two-dimensional (2D) or three-dimensional (3D) *gravity inversion*.

formation susceptibility: *apparent susceptibility* in the context of *magnetic susceptibility* computed from magnetic measurements in boreholes or obtained by two-dimensional (2D) or three-dimensional (3D) *magnetic inversion*.

forward anomaly: the gravity or magnetic field generated by the constructed *model*, also known as a *forward model*, of a particular geological *structure* or other *target* in a study area. See also *feasibility modeling* and *synthetic anomaly*.

forward calculation: 1. calculation of the gravity or magnetic field generated by the constructed *model* of a particular geological *structure* or any other *exploration* target. **2.** calculation of the gravity or magnetic field of an isolated *source body* with a known or assumed *source geometry* and *density contrast* or *susceptibility contrast*. In accordance with the *forward problem* principle, the forward calculation is always unique, as opposed to the *inverse calculation* which is always nonunique. Referred to with the same meaning as *forward problem solution*. See also *contrast*, *density*, *susceptibility*, and *magnetization contrast*.

forward Fourier transform: *Fourier transform* applied to decompose gravity and magnetic data from their original space-domain waveforms into the Fourier-domain series of sinusoids (sine and cosine wave functions) of different spatial frequencies, each having its own *amplitude*, *phase*, and *wavenumber*. The forward Fourier transform is opposite to the reconstitution, or synthesis, of the original waveforms of data performed by *inverse Fourier transform*. Sometimes referred to as *direct Fourier transform*. See also *spatial frequency* and *waveform*.

forward method: a method of gravity and magnetic data *interpretation* based on the multi-step *iterative modeling* process which involves: (1) construction of an initial *model* from available geological and geophysical data; (2) calculation of the gravity or magnetic field generated by the constructed model; (3) comparison of the calculated field with the observed gravity or magnetic field, such as *Bouguer gravity* or *RTP magnetic field*; and (4) adjustment of the model parameters to reduce *model misfit*, i.e., difference between the model-calculated and observed fields. The last three steps are repeated until the model misfit falls into a margin of error quantified by *tolerance*. The forward method is implemented primarily in two-dimensional (2D) *modeling* and sometimes in the 3D modeling. Also referred to as *forward modeling* and *forward calculation*. See also *inverse method*.

forward model: a gravity or magnetic *model* constructed from available geological and geophysical data, where the subsurface geological *structure* and its particular elements, such as the *basement* depth, intrasedimentary *target*, or shallow ore body, are described by their known parameters or by conceptual forms derived from publications and experience. A forward model is built with the use of elementary model blocks, such as the seismic-derived horizons, *lamina*, polygon, *sphere*, *horizontal cylinder*, *vertical cylinder*, *prism*, *polyhedron*, etc., whose gravity or magnetic effects are defined by the corresponding closed-form analytical expressions (formulas). Analytically expressed effects of all individual building blocks are numerically integrated at each model data point to obtain the gravity and magnetic fields of the arbitrary-shaped model components that comprise the calculated, or predicted, field of the constructed model. The model-generated field is compared with the *observed field* of gravity or magnetic anomalies in the process of *forward modeling*. See also *Talwani modeling*.

forward modeling: 1. *forward method*. **2.** a *modeling* technique based on the iterative comparison of the field generated by a constructed *forward model* with the observed gravity and magnetic fields. Trial-and-error adjustments of the initial *model* parameters are made until the calculated field of the iteratively adjusted model matches the *observed field* within a specified margin of error quantified by *tolerance*. **3.** calculation of the gravity and magnetic field of a *2D model* or *3D model* of a subsurface geological *structure* or isolated *ore body* for *feasibility modeling*. The field calculated in the process of forward modeling is referred to as the *model response* or *forward response*. See also *inverse modeling*.

forward problem: a fundamental principle of *modeling* stating that the calculation of the gravity or magnetic field generated by a particular *source body* or geological *structure* of a given or assumed depth, geometry, and physical properties (*density* or *susceptibility* contrast), i.e., the result of *forward modeling*, is always unique. Sometimes referred to as *direct problem*. See also *density contrast*, *magnetization contrast*, and *inverse problem*.

forward problem solution: 1. calculation of the gravity or magnetic field generated by a *source body* of a given or assumed depth, shape, and density or magnetization *contrast*. **2.** calculation of the gravity or magnetic field generated by a constructed *2D model* or *3D model*. Unlike the *inverse problem solution*, the solution of a *forward problem* is always unique and stable. Also referred to as *forward calculation* and *forward solution*.

forward projection: conversion of acquired gravity and magnetic data from the latitude-longitude coordinate pairs provided by *Global Positioning System (GPS)* into the XY coordinate pairs of the *Universal Transverse Mercator (UTM)* coordinate system. See also *XY coordinates* and *inverse projection*.

forward response: *model response* in the context of the gravity or magnetic field generated by a constructed *model* in the process of *forward modeling*.

forward solution: See *forward problem solution.*

4D gravity: same as *time-lapse gravity* with an emphasis on time as the fourth *dimension* of high-*accuracy* repeated measurements of changes in the *Earth gravity field* at the same coordinate points.

4D gravity gradiometry: same as *time-lapse gravity gradiometry* with an emphasis on time as the fourth *dimension* of high-*accuracy* repeated measurements of changes in the *gradient* or *full tensor gradient (FTG)* components of the *Earth gravity field* at the same coordinate points.

4D gravity monitoring: *time-lapse gravity* measurements of *microgravity* changes in the *absolute gravity* and *relative gravity* values caused by fluid movements and redistribution of fluid masses in the subsurface, such as those associated with groundwater-level changes in natural aquifers, hot fluid withdrawal from geothermal reservoirs, waterflooding of oil-and-gas reservoirs under enhanced development, steam injection in heavy oil recovery, and CO_2 sequestration. High-*accuracy* gravity measurements are made over a network of ground observation points at the timescales of months, years, and decades, and their processing and *interpretation* are integrated with time-lapse high-*resolution* surface *elevation* measurements of the ground surface subsidence or uplift due to changes in reservoir fluid saturation (Pearson et al., 2015). See also *aquifer gravity, CO_2 sequestration gravity, geothermal gravity, volcano gravity,* and *InSAR imaging.*

4D gravity survey: See *time-lapse gravity survey.*

Fourier amplitude spectrum: same as *amplitude spectrum* with an emphasis on its calculation in the *Fourier domain.*

Fourier analysis: **1.** an analytical representation of space-domain gravity and magnetic data in the *Fourier domain* by an equivalent series of mutually orthogonal sinusoids (90° phase-shifted sine and cosine wave functions) of different spatial frequencies, or wavenumbers, wherein each sinusoid has its own *amplitude, phase,* and *wavenumber.* Data decomposition into the constituent sine and cosine components is performed by *forward Fourier transform.* As a subdivision of *Fourier transform,* Fourier analysis is the opposite of a *Fourier synthesis* performed by *inverse Fourier transform.* **2.** *spectral analysis* in the context of evaluating the amplitude and phase components of the *spectrum* computed from Fourier-transformed gravity or magnetic data. See also *space domain, spectral domain,* and *spatial frequency.*

Fourier continuation method: a spectral-domain method of the upward or downward continuation of Fourier-transformed gravity and magnetic fields. Its mathematical expression can be written as

$$\mathcal{F}(k,h_2) = \mathcal{F}(k,h_1)e^{\Delta hk},$$

where $\mathcal{F}(k,h_2)$ is the *Fourier transform* of *anomalous field* values at *elevation* h_2, $\mathcal{F}(k,h_1)$ is the Fourier transform of anomalous field values at elevation h_1, $\Delta h = h_2 - h_1$ is a continuation distance, $k = 2\pi/\lambda$ is the *wavenumber* for one-dimensional (1D) line-oriented data, and λ is *wavelength.* For 2D gridded data, the wavenumber is expressed as

$$k = \left(k_x^2 + k_y^2\right)^{1/2} = (u^2 + v^2)^{1/2},$$

where u and v are wavenumbers in the mutually orthogonal x- and y-directions, respectively, and related to their respective wavelengths as: $u = k_x = 2\pi/\lambda_x$ and $v = k_y = 2\pi/\lambda_y$. When the z-axis in *Cartesian coordinates* is positive down, such as in a *NED reference frame,* and Δh is positive, the process is *downward continuation,* whereas for the negative Δh the process is *upward continuation.* See also *dimension, gridding,* and *spectral domain.*

Fourier convolution theorem: the theorem stating that *convolution* of two functions in the *space domain* is equivalent to multiplying their *amplitude* spectra and summing their *phase* spectra in the *Fourier domain.* See also *convolution filter.*

Fourier decomposition: See *forward Fourier transform.*

Fourier domain: a *domain* where gravity and magnetic data are dependent variables represented by a series of the mutually orthogonal sinusoids (90° phase-shifted sine and cosine wave functions), each having its own *amplitude, phase,* and *wavenumber* defined by a *Fourier transform.* Directional trends in the Fourier domain, such as those shown on a *map power spectrum,* differ from their original orientation in the *space domain* by 90°. Also referred to as *frequency domain, spectral domain,* and *wavenumber domain.* See also *Fourier series.*

Fourier phase spectrum: same as *phase spectrum* with an emphasis on its calculation in the *Fourier domain.*

Fourier power spectrum: same as *power spectrum* with an emphasis on its calculation in the *Fourier domain.*

Fourier series: a mathematical representation of *periodic function* by a linear combination, or sum, of mutually orthogonal sinusoids (sine and cosine wave functions), whose frequencies are multiples of *fundamental frequency.* For band-limited and non-periodic gravity and magnetic data, the *Fourier transform* assumes that the data are equivalently circular beyond the dataset area (i.e., the last data point is next to the first one), so the Fourier series represents the Fourier-domain decomposition of the original space-domain waveforms into the constituent sinusoidal *harmonics* of *fundamental wavelength.* See also *band-limited function, Fourier theorem, waveform,* and *Taylor series.*

Fourier spectrum: same as *spectrum* with an emphasis on its calculation in the *Fourier domain* from results of the *forward Fourier transform* of gravity or magnetic data.

Fourier synthesis: **1.** transformation, or conversion, of Fourier-transformed gravity and magnetic data from the *Fourier domain* back to their original *space domain* by the application of *inverse Fourier transform,* i.e., it is a synthesis of space-domain waveforms from the Fourier-domain sine and cosine wave functions. **2.** superimposing the Fourier-domain sine and cosine wave functions of the proper amplitudes and phases to reconstruct the space-domain waveforms. See also *phase, waveform,* and *Fourier analysis.*

Fourier terrain correction: *terrain correction* obtained by the Fourier-domain calculation of the *gravitational attraction* of a two-dimensional (2D) layer with an irregular top surface, such as that of *topography* or *bathymetry*, where the exponential in the transform integral is expanded into a *Taylor series* to produce a series of Fourier transforms in powers of this surface (Parker, 1973). The method is upgraded for terrain correction calculations in shallow waters and on top of uneven variable-thickness layers (Parker, 1995; 1996). See also *Parker formula*.

Fourier theorem: the theorem stating that *periodic function* can be synthesized by summation of the weighted orthogonal sinusoids: sine and cosine wave functions. The Fourier theorem is the basis for the *Fourier transform* of non-periodical gravity and magnetic data, because (a) their non-periodicity has no influence on the sinusoidal *Fourier series* expression within a given interval of *fundamental wavelength*, (b) all components outside the fundamental wavelength are ignored, and (c) the dataset is assumed to be circular beyond the data area with the last point next to the first point. See also *circular convolution*.

Fourier transform: 1. a mathematical operation that decomposes the continuous space-domain waveforms of a gravity or magnetic field into the wavenumber-domain series of orthogonal sinusoids (90° phase-shifted sine and cosine wave functions) of different spatial frequencies, or wavenumbers, wherein each sinusoid has its own *amplitude*, *phase*, and *wavenumber*. **2.** representation of the space-domain function $f(x)$ by a linear combination of complex exponential functions of the variable $\theta(k)$, where k is wavenumber, and each complex exponential is expressed as

$$e^{i\theta(k)} = \cos\theta(k) + i\sin\theta(k),$$

where $i = \sqrt{-1}$ is a complex unit number indicating orthogonality. The Fourier transform $\mathcal{F}(k)$ of the one-dimensional (1D) non-periodic function $f(x)$ representing the *profile* of an anomalous gravity or magnetic field is expressed as

$$\mathcal{F}(k) = \int_{-\infty}^{\infty} f(x)e^{-ikx}dx,$$

where k is inversely related to *wavelength* λ as $k = 2\pi/\lambda$. For the 2D function $f(x, y)$ representing the *grid* of an *anomalous field*, the Fourier transform formula is expressed as

$$\mathcal{F}(u,v) = \int\int_{-\infty}^{\infty} f(x,y)e^{-i(ux+vy)}dxdy,$$

where u and v are wavenumbers k_x and k_y, respectively, in the orthogonal x- and y-directions and related to their respective wavelengths as $u = k_x = 2\pi/\lambda_x$ and $v = k_y = 2\pi/\lambda_y$. For an arbitrary direction, the wavenumber k_a is expressed as

$$k_a = (k_x^2 + k_y^2)^{1/2} = (u^2 + v^2)^{1/2}.$$

Power, amplitude, and phase characteristics of an anomalous field are described by *power spectrum*, *amplitude spectrum*, and *phase spectrum*, respectively, calculated from the respective components of the Fourier-transformed data. **3.** a *complex function* for which the real part Re is the *cosine transform* and imaginary part Im is the *sine transform*, and it can be expressed as

$$\mathcal{F}(k) = \mathrm{Re}[\mathrm{F}(k)] + i\mathrm{Im}[\mathrm{F}(k)] = |\mathrm{A}(k)|\, e^{i\theta(k)},$$

where $|\mathrm{A}(k)| = (\mathrm{Re}^2 + \mathrm{Im}^2)^{1/2}$ is *amplitude spectrum* and $\theta(k) = \arctan(\mathrm{Im/Re})$ is *phase spectrum*. **4.** a mathematical operation involving two opposite types of data transformation: (1) *forward Fourier transform*, which converts the space-domain data into a series of sine and cosine wave functions of different spatial frequencies, amplitudes and phases in the *Fourier domain*; and (2) *inverse Fourier transform*, which converts the Fourier-transformed data back to their original *space domain*. Sometimes referred to as *spectral transform*. See also *discrete Fourier transform (DFT)*, *window Fourier transform (WFT)*, *Hilbert transform*, *Riesz transform*, and *Walsh transform*.

Fourier transform pair: a relationship between original gravity or magnetic data in the *space domain* and Fourier-transformed data in the *wavenumber domain*, i.e., a relationship between different representations, or formats, of the same data. See also *format* and *Hilbert transform pair*.

fourth difference: a quantitative characteristic of *acquisition noise* in line-oriented gravity or magnetic data defined as

$$4^{\text{th}}\,\text{diff} = (R_{-2} - 4R_{-1} + 6R_0 - 4R_{+1} + R_{+2})/16,$$

where R_{-2}, R_{-1}, R_0, R_{+1}, R_{+2} are five consecutive readings centered at the R_0 *reading*. Its values set a limit of detectability of low-*amplitude* anomalies in the process of measurements on a moving *observation platform*. The fourth difference is calculated in the *real-time* mode, and its digital *channel* records are used for monitoring the acquisition noise level, spikes, and other data-contaminating effects. See also *spike* and *noise envelope*.

fractal dimension: a parameter of the spatial distribution of measured values, such as *density* or *magnetization*, that has its *power spectrum* proportional to some power of the *spatial frequency*, or *wavenumber*, described as

$$\Phi(k) = \mathrm{A}k^{\mathrm{D}},$$

where $\Phi(k)$ is power spectrum, k is wavenumber, A is a constant, and D is the fractal dimension. When $\mathrm{D} = 0$, the distribution is uncorrelated, or random, as assumed in the statistical models for magnetic source depth estimates; when $\mathrm{D} < 0$, the distribution is correlated to a degree which increases with the decreasing fractal dimension D. See also *dimension*.

fractal magnetization: rock *magnetization* which is not completely uncorrelated, or random, but rather has some degree of a *scale*-invariant spatial correlation, or self-similarity, in the distribution of magnetization. The degree of correlation is quantified by the *power spectrum* parameter called *fractal dimension* (Maus and Dimri, 1994; Pilkington et al., 1994).

fractals: 1. geometric shapes that are self-similar at different scales, such as the shapes of snow crystals or mineral concentrations in some ore deposits. **2.** structurally complex objects exhibiting the similar patterns, such as the specific properties or visual characteristics, that are *scale*-invariant, i.e., they are similar at different levels, or scales, of *lateral resolution*. See also *fractal dimension, fractal magnetization, ore body*, and *pattern*.

fractional anisotropy attribute (FAA): the rotationally invariant dimensionless *scalar* quantity obtained by the combination of magnitudes of three *FTG eigenvalues* calculated at each *grid* point. It is expressed as

$$\mathrm{FAA} = \left[\left(|\lambda_1|-|\lambda_2|\right)^2 + \left(|\lambda_2|-|\lambda_3|\right)^2 + \left(|\lambda_3|-|\lambda_1|\right)^2\right]^{1/2} \Big/ \left[2\left(\lambda_1^2+\lambda_2^2+\lambda_3^2\right)\right]^{1/2},$$

where $|\lambda_1|$, $|\lambda_2|$, $|\lambda_3|$ are magnitudes of the FTG eigenvalues. The FAA values range between 0 and 1 for cases of *FTG isotropy* ($|\lambda_1| = |\lambda_2| = |\lambda_3|$) and cylindrically symmetric *FTG linear anisotropy* ($|\lambda_1| \gg |\lambda_2| = |\lambda_3|$), respectively (Alexander et al., 2000). The FAA is highly sensitive to variations of structural composition and distribution of the *density* and *magnetization* contrasts in rock formations and ore bodies. Calculated values are gridded, and the obtained FAA grid is integrated into the analysis and *interpretation* of high-*resolution* airborne gravity and magnetic data (Hassan and Goussev, 2020). See also *attribute, contrast, FTG anisotropy, gridding, invariants*, and *magnitude*.

fractional calculus: application of the horizontal and vertical derivatives of the non-integer orders and selection of the highest order which provides the most resolving images of the *structure*-generated anomalies with an acceptable level of *noise*. See also *anomaly, derivative*, and *resolution*.

fractional-order ASA: *enhanced analytic signal amplitude (eASA)* calculated with a non-integer, or fractional, value of the *vertical derivative* order NI and expressed as

$$eASA_{NI} = \left\{\left[\partial^{\mathrm{NI}}F/\partial z^{\mathrm{NI}}\right]^2 + \left[H_x\left(\partial^{\mathrm{NI}}F/\partial z^{\mathrm{NI}}\right)\right]^2 + \left[H_y\left(\partial^{\mathrm{NI}}F/\partial z^{\mathrm{NI}}\right)\right]^2\right\}^{1/2},$$

where F is the gridded gravity or magnetic field; H_x and H_y are components of the *3D Hilbert transform* operator in the mutually orthogonal x- and y-directions, respectively; and $\partial^{\mathrm{NI}}F/\partial z^{\mathrm{NI}}$ is *fractional vertical derivative*. The larger NI values increase the *sharpness* of anomalies with a reduced amplification of the high-frequency *noise* (Cooper and Cowan, 2003). See also *gridding*.

fractional structural index (fSI): 1. an *inverse power law* index, or scaling exponent, which is not an integer *source geometry* constant, such as the *scale*-invariant integer *Euler structural index (SI)*, but rather a spatially variable function of the source-observation distance that depends on the source depth and source distribution type, such as the three-dimensional (3D) *sphere* which is finite in all three standard x, y, z dimensions and two-dimensional (2D) *dike* infinite in one or two of three standard dimensions. Non-integer, or fractional, structural indices are by-products of the *AN-EUL method, improved SPI™ method*, and some other *automated source depth estimation* techniques (Reid and Thurston, 2014). **2.** the opposite, or negative, of a fractional degree of homogeneity in the local approximation and *multiscale analysis* of the nonhomogeneous fields of geological sources in the concept of *multihomogeneity* (Fedi et al., 2015). See also *dimension, Euler homogeneity equation, homogeneous field, nonhomogeneous field*, and *infinite length*.

fractional vertical derivative: the *vertical derivative* with a non-integer value of the derivative order NI. It is denoted by $\partial^{\mathrm{NI}}F/\partial z^{\mathrm{NI}}$, where F is the gridded gravity or magnetic *anomalous field*. A fractional vertical derivative, such as 1.5 or 1.75 vertical derivative, is calculated with the objective of finding the optimal *resolution* of anomalies between the lower resolving Nth-order vertical derivative and the higher resolving but noisier $(N+1)$th-order vertical derivative. Often, it is calculated as a less noisy alternative to the *second vertical derivative (SVD)* with the reduced amplification of high-frequency *noise*. Fractional vertical derivatives do not correspond to any of the traditionally measured parameters, but their calculation can provide additional *lateral resolution* of the anomalous features of an exploration or research interest and assist in their *interpretation* (Cooper and Cowan, 2003). See also *gridding, first vertical derivative (FVD), fractional-order ASA*, and *fractional vertical integral*.

fractional vertical integral: a Fourier-domain data transformation operator which performs a numerical vertical integration of the *anomalous field* F with a non-integer, or fractional, order of integration. It can be expressed as

$$\mathcal{F}(k) = \mathcal{F}(u,v) = (1/|k|^{\mathrm{NI}})\mathcal{F}(\mathrm{F}),$$

where $\mathcal{F}(\mathrm{F})$ is *Fourier transform* of field F; k is *wavenumber* for line-oriented data; $|k| = (k_x^2 + k_y^2)^{1/2}$ for gridded data; $u = k_x$ and $v = k_y$ are wavenumbers in the mutually orthogonal x- and y-directions, respectively; and NI is the non-integer vertical integration order (Cooper and Cowan, 2003). See also *vertical integral*.

free-air admittance: *gravitational admittance* calculated with the use of *free-air anomaly* data. See also *Bouguer admittance* and *free-air coherence*.

free-air anomaly: the *field* of gravity anomalies, or anomalous gravity field, obtained from the *observed gravity* by the application of the following corrections: (1) *theoretical gravity correction*, which accounts for the rotating homogeneous Earth mass and change in the observed gravity values with the latitude, and (2) *free-air correction* or *height correction*, which accounts for the height of each observation point above sea level or *reference ellipsoid*, respectively. In high-*accuracy* measurements, the observed gravity is also compensated for (3) the gravity effect of the atmospheric air mass by the application of *atmospheric mass correction* and (4) the gravity effect resulting from the difference in calculation of the observation point height relative to different datums, i.e., sea level or reference ellipsoid, by the application of *indirect effect correction*. The general calculation formula can be written as

$$\Delta g_{\mathrm{FAA}} = g_{\mathrm{obs}} - \Delta g_{\mathrm{TG}} + \Delta g_{\mathrm{FA}} + \Delta g_{\mathrm{AM}} + \Delta g_{\mathrm{IE}},$$

where g_{obs} is observed gravity value, Δg_{TG} is theoretical gravity correction, Δg_{FA} is free-air correction, Δg_{AM} is atmospheric mass correction, and Δg_{IE} is indirect effect correction. The free-air anomaly strongly correlates with *topography* and *bathymetry* as a reflection of changes in the ground surface *relief* and thickness of the water column. The ridge-and-trough ocean floor relief of oceanic fracture zones and spreading centers is clearly visible on free-air anomaly maps. Neogene deltaic fans, where the sea water is replaced by river-transported sediments, are often shown as prominent free-air anomaly highs. Marine (shipboard) gravity and *satellite gravity* data are commonly presented in the *format* of a free-air anomaly. Often referred to as *free-air gravity* and, occasionally, *Faye anomaly*. See also *Bouguer anomaly*, *gravity high*, *observed gravity components*, and *free-water anomaly*.

free-air coherence: *gravitational coherence* calculated with the use of *free-air anomaly data*. See also *Bouguer coherence* and *free-air admittance*.

free-air correction: 1. a correction applied to *observed gravity* to compensate for the *free-air effect* of a decrease in gravity values with the increasing height, or *elevation*, of observation points that is caused by the additional distance (relative to sea level) from the center of the Earth's mass. For the average Earth radius, it is expressed in units of *milligal (mGal)* as

$$\Delta g_{FA} = (0.3086 - 0.00023\cos 2\theta + 2\cdot 10^{-7})h,$$

where θ is geographic latitude in degrees and h is elevation above sea level in meters (LaFehr and Nabighian, 2012). **2.** adjustment of *theoretical gravity* that accounts for the theoretical *free-air gradient* at the observation point above or below sea level. For example, it reduces the theoretical gravity value at the top of Mt. Everest by 2731 mGal and increases the theoretical gravity value at the bottom of the Challenger Deep by 3368 mGal. The correction value is dependent on latitude, but in regular surveys this dependence is often disregarded, and the free-air correction Δg_{FA} is calculated using the first-order approximation formula

$$\Delta g_{FA} = 0.3086h,$$

where 0.3086 is *free-air factor* in milligals per meter ($\mathrm{mGal\cdot m^{-1}}$). The calculated correction value is added to the observed gravity value when the *observation point* is above sea level, and it is subtracted from the observed gravity when the observation point is below sea level. For marine (shipboard) surveys, the correction value is zero when the effect of water tides is ignored. Both definitions of this correction do not imply that its application changes the observation point positions, i.e., observation points are not moved to *survey datum*. After the release of *Global Positioning System (GPS)* into commercial applications, the *height correction* has been recommended as a new standard in compensating for the free-air effect (Hinze et al., 2005), because it is based on the GPS-determined height above *reference ellipsoid* and more accurate second-order approximation formula. Also, its name is more appropriate for the conceptually equivalent gravity correction applicable to planets without an air atmosphere. In regular ground surveys, it is sometimes referred to as *first-order free-air correction* and *Faye correction*. See also *Bouguer correction*, *elevation correction*, *free-water correction*, and *second-order free-air correction*.

free-air edge effect: the regional-scale anomalous gravity effect observed in shelf edge areas within the rifted continental margins, where the gravity effects of (a) rapid change in sea water depth from a shallow shelf to a deep continental slope, (b) increased thickness of rift-accumulated sediments, and (c) crustal thinning due to a decrease of the *Moho* depth in the transition from the thick continental *crust* to much thinner oceanic crust are combined in producing the characteristic positive-to-negative paired, or dipolar, *gravity signature* referred to as *shelf-edge gravity anomaly*.

free-air effect: a decrease in observed gravity values with increasing *elevation* in the ground measurements and height of a *survey* aircraft or orbiting satellite above the Earth's surface. It is a physical effect of non-geological origin, and it is caused by the additional "free-air" distance (relative to a *survey datum*: sea level or *reference ellipsoid*) between an *observation point* and the center of the Earth's mass. The free-air effect is a function of the geographic latitude and inverse square of the distance increment, and its theoretical, or predicted, estimate is compensated by the application of a *free-air correction* or more accurate *height correction*. See also *Bouguer effect*, *terrain effect*, and *free-air edge effect*.

free-air factor: 1. a parameter in the first-order approximation formula of the *free-air correction* that is expressed in units of *milligal (mGal)* per meter and defined as $0.3086\ \mathrm{mGal\cdot m^{-1}}$. Its multiplication by the height h of an *observation point* above sea level comprises the free-air correction value at this point. **2.** the absolute value of the *free-air gradient* at 45° latitude referred to as the *average vertical gravity gradient* of the theoretical gravity field of the Earth. See also *flat-plate Bouguer factor* and *elevation correction factor*.

free-air gradient: 1. the vertical rate of change in *theoretical gravity* due to a change in the distance between an observation point and the center of mass of a homogeneous *reference ellipsoid* without accounting for local and regional geological structures. **2.** an absolute value of the *vertical derivative* of *gravitational acceleration g*. The free-air gradient depends on the latitude of a *gravity survey*, because both gravitational acceleration and distance to the Earth's or ellipsoidal center of mass vary with latitude. The free-air gradient at the latitude θ and height, or *elevation*, h above the *GRS80 ellipsoid* is expressed in milligals per meter ($\mathrm{mGal\cdot m^{-1}}$) as

$$\left|\partial g/\partial h\right| = 0.3087691 - 0.0004398\sin^2\theta - 0.00000014425h.$$

Its value at 45° latitude is referred to as the *average vertical gravity gradient* of the theoretical gravity field of the Earth, or *free-air factor*, and it is closely approximated as $0.3086\ \mathrm{mGal\cdot m^{-1}} = 0.3086\ \mu\mathrm{Gal\cdot mm^{-1}} = 3086$ E. The height increase of 1 m results in a decrease of the theoretical gravity value of approximately 0.3086 mGal or 308.6 *microgal (μGal)*; therefore, for a planned ground survey *accuracy* of 0.001 mGal (1 μGal), the height of each *station* above sea level or reference ellipsoid should be determined with the accuracy 2–3 mm.

Also referred to as *free-air gravity gradient, nominal vertical gravity gradient, normal gravity gradient,* and *standard vertical gravity gradient.* See also *Eötvös (E)* and *vertical gravity gradient.*

free-air gravity: See *free-air anomaly.*

free-air gravity anomaly (FGA): the *field* of anomalous gravity values obtained from airborne or satellite gravity measurements in the geodetic studies of a *geoid,* defined as

$$\Delta g_{FGA} = g_{obs} - \Delta g_{FAH} - g_0,$$

where g_{obs} is *observed gravity* in *absolute gravity* values, Δg_{FAH} is *second-order free-air correction* calculated with the use of *orthometric height H,* and g_0 is *normal gravity* on *reference ellipsoid.* See also *free-air gravity disturbance (FGD)* and *free-air gravity residual (FAR).*

free-air gravity disturbance (FGD): the *field* of anomalous gravity values obtained from airborne or satellite gravity measurements in the geodetic studies of a *geoid,* defined as

$$\Delta g_{FGD} = g_{obs} - \Delta g_{FAh} - g_0,$$

where g_{obs} is *observed gravity* in *absolute gravity* values, Δg_{FAh} is *second-order free-air correction* calculated with the use of *ellipsoidal height h,* and g_0 is *normal gravity* on *reference ellipsoid.* See also *free-air gravity anomaly (FGA)* and *free-air gravity residual (FAR).*

free-air gravity gradient: same as *free-air gradient* that represents the *vertical gradient* of the theoretical *Earth gravity field* in a free-air, or source-free, space.

free-air gravity residual (FAR): the *field* of anomalous gravity values obtained from airborne or satellite gravity measurements in the geodetic studies of a *geoid,* defined as

$$\Delta g_{FAR} = g_{obs} - \Delta g_{FAH} - g_{geoid},$$

where g_{obs} is *observed gravity* in *absolute gravity* values, Δg_{FAH} is *second-order free-air correction* calculated with the use of *orthometric height H,* and g_{geoid} is *normal gravity* on the geoid instead of *reference ellipsoid.* See also *free-air gravity anomaly (FGA)* and *free-air gravity disturbance (FGD).*

free-fall absolute gravimeter: 1. an *absolute gravimeter* for high-*accuracy* absolute gravity measurements by the *free-fall method.* **2.** an instrument which measures the *acceleration* of a reflective object or stream of cold atoms in a vacuum free fall, wherein a numerical value of the measured acceleration represents the absolute value of *Earth gravity field* known as *absolute gravity.* Two operational concepts are presently in use: (1) measurements of a free-fall time and dropping distance of the corner-cube reflector, or "falling mirror," inside a vacuum tube by the atomic clock and laser interferometer and (2) measurements of the *phase* differences between the optical interferometer fringes, formed by streams of free-falling cold atoms, by the atom interferometer. The best estimated *precision* is $0.5 \cdot 10^{-8}$ *g,* where *g* is *nominal gravity* defined as $g = 9.8$ m · s^{-1}. Achievable *accuracy* is approximately 1–2 *microgal (μGal),* and it depends on the integration time for the repeated measurement cycles: the longer the integration time, the higher the accuracy. Free-fall absolute gravimeters are *drift*-free. Also referred to as *absolute free-fall gravimeter, ballistic absolute gravimeter,* and *free-fall gravimeter.* See also *absolute atomic gravimeter* and *falling-mirror gravimeter.*

free-fall gravimeter: See *free-fall absolute gravimeter.*

free-fall gravimetry: instruments and methodology for measurements and processing of the absolute values of the *Earth gravity field,* or *absolute gravity,* using the *free-fall method.* See also *free-fall absolute gravimeter* and *absolute gravity corrections.*

free-fall method: a method of direct measurements of *absolute gravity,* where its numerical value is defined by the free-fall *acceleration* of a test object in a vacuum. Depending on the test object type, the free-fall method is based on high-*precision* measurements of (a) free-fall time and distance of a reflective object, such as the corner-cube reflector dropped inside a vacuum tube, determined by the atomic clock and laser interferometer, or (b) *phase* differences between the optical interferometer fringes, which are formed by streams of free-falling cold atoms, by the atom interferometer. Achievable *accuracy* is up to 1–2 *microgal (μGal),* and it is a function of the integration time for the repeated free-fall measurement cycles: the longer the integration time, the higher the accuracy of measurements. Corrections applied to the absolute gravity values measured by this method include: *atmospheric pressure correction, datum transfer correction, diffraction correction, ocean loading correction, polar motion correction, self-attraction correction, speed-of-light correction, tidal correction,* and *tilt correction.* The International Bureau of Weights and Measures (BIPM) has proclaimed the free-fall method a primary method for absolute gravity measurements. See also *absolute atomic gravimeter* and *falling-mirror gravimeter.*

free-water anomaly: the *field* of the gravity anomalies obtained from measurements of *relative gravity* by an *underwater gravimeter* or *autonomous underwater vehicle (AUV).* Measured values are corrected for the relevant *acquisition* and non-geological effects and referenced to sea level or *reference ellipsoid* by the application of *free-water correction.* Also referred to as *free-water gravity.* See also *raw gravity corrections* and *observed gravity corrections.*

free-water correction: a correction applied to sea floor *relative gravity* measurements by an *underwater gravimeter* and submarine measurements by an *autonomous underwater vehicle (AUV)* in order to reference the measured gravity values to sea level or *reference ellipsoid.* The calculation formula is expressed in units of *milligal (mGal)* as

$$\Delta g_{FW} = \text{FWG}\,\Delta z,$$

where FWG is *free-water gradient* and Δz is seawater depth at the observation point in meters. The obtained correction value is subtracted from the *observed gravity* values. See also *AUV gravity, free-water anomaly, raw gravity corrections,* and *observed gravity corrections.*

free-water gradient: an analogue of the *free-air gradient* in the processing of *underwater gravity* data. It accounts for the *density* of sea water and is expressed in units of *milligal (mGal)* per meter ($mGal \cdot m^{-1}$) as

$$\mathrm{FWG} = \mathrm{FAG} - 4\pi G\rho,$$

where FAG is free-air gradient approximated by the *average vertical gravity gradient* of 0.3086 $mGal \cdot m^{-1}$, G is *universal gravitational constant* in *cgs units*, and ρ is seawater density in grams per cubic centimeter ($g \cdot cm^{-3}$). For the seawater density of 1.028 $g \cdot cm^{-3}$, the free-water gradient is 0.2224 $mGal \cdot m^{-1}$. See also *free-water correction*.

free-water gravity: See *free-water anomaly*.

frequency: 1. the time-domain parameter *f* which describes the repetition rate (i.e., number of the wave cycles per unit time or per regular time interval) of a periodic *waveform*. Frequency is measured in units of *hertz (Hz)*, so the frequency $f = 5$ Hz corresponds to 5 cycles per second and $f = 0.55$ Hz corresponds to 55 cycles per 100 seconds. **2.** the reciprocal, or inverse, of *period T*, i.e., $f = 1/T$. The space-domain equivalent of the time-domain frequency is called *spatial frequency*. See also *angular frequency*, *dominant frequency*, *space domain*, and *time domain*.

frequency content: 1. amplitudes of the specific sinusoids into which all space-domain waveforms of an *anomalous field* are decomposed by *forward Fourier transform* and referred to as the *frequency spectrum*. **2.** a characteristic of an anomalous gravity or magnetic *field* which describes its spatial frequencies that are dominant in a qualitative visual evaluation or in the quantitative estimate by an *amplitude spectrum*. See also *space domain*, *Fourier domain*, *spatial frequency*, and *waveform*.

frequency-depth rule: a general rule of *qualitative interpretation* that is based on the assumption that the higher *spatial frequency* (shorter *wavelength*) components of gravity and magnetic fields are generated by shallower sources, whereas the lower frequency (longer wavelength) components are generated by deeper sources. However, the *frequency content* of detected anomalies depends not only on the *source body* depth but also on its shape and lateral extent (width). Accordingly, the anomalies from narrower sources will exhibit higher spatial frequencies than those of wider sources at the same or shallower depth. See also *wavenumber-depth relation*.

frequency domain: *Fourier domain* in the context of the *frequency content* of Fourier-transformed time- or space-domain data. See also *domain* and *Fourier transform*.

frequency matching: 1. the *data processing* approach that is based on a low-pass *filtering* of calculated terrain corrections before their application to *airborne gravity* and airborne *gravity gradient* data in order to avoid compensating for the *terrain effect* components suppressed by *acquisition filtering*. **2.** the *modeling* concept that is based on the *Agerlin principle* stating that a comparison between the observed and *model*-generated fields should be made for the observed and model data processed by the same techniques. See also *airborne terrain correction* and *low-pass filter*.

frequency response: 1. a characteristic of a space-domain *filter* that describes the filter output as a function of *spatial frequency*. **2.** a characteristic of a spectral-domain *filter* that describes the filter output as a function of *wavenumber*. See also *frequency spectrum*, *space domain*, *spectral domain*, and *transfer function*.

frequency spectrum: 1. *amplitude* and *phase* characteristics of a space-domain *waveform* described as functions of the *wavenumber* in the *Fourier domain*. **2.** *frequency content*.

FTG: See *full tensor gradient (FTG)*.

FTG amplitude: the *FTG attribute* that represents the *magnitude* of the gravity or magnetic *full tensor gradient (FTG)* by the square root of the sum of its squared components. In accordance with the *Laplace equation* and symmetry of the *FTG matrix*, the FTG amplitude can be expressed as

$$\begin{aligned}\mathrm{FTGA} &= \left[\left(T_{xx}^2 + T_{yy}^2 + T_{zz}^2\right) + 2\left(T_{xy}^2 + T_{xz}^2 + T_{yz}^2\right)\right]^{1/2} \\ &= \left[2\left(T_{xy}^2 + T_{xz}^2 + T_{yz}^2\right)\right]^{1/2}.\end{aligned}$$

The FTGA is robust in the presence of *random noise* and, therefore, used in various applications, including the high-*resolution* mapping of *gradient* anomalies generated by compact *three-dimensional (3D)* sources, such as *ore body*, and sources approximated by *point source* geometries, such as *kimberlite intrusion* and *unexploded ordnance (UXO)*.

FTG analytic signal amplitude: *analytic signal amplitude (ASA)* calculated along one of the *x*-, *y*-, *z*-directions in *Cartesian coordinates* using a corresponding row in the nine-element *matrix* of the *full tensor gradient (FTG)* of an anomalous gravity or magnetic field. Also known as *directional ASA*. See also *anomalous field*.

FTG anisotropic ellipsoid: a *three-dimensional (3D)* representation of the directional variations of the *full tensor gradient (FTG)* components by an *ellipsoid* wherein the three mutually orthogonal axial radii are oriented along the *FTG eigenvectors* and their lengths are defined by the magnitudes of the eigenvectors-paired *FTG eigenvalues*. Geometrical parameters (shape, size, spatial orientation) of the FTG anisotropic ellipsoid are functions of three FTG eigenvalues and three FTG eigenvectors calculated at each observation point. In the case of *isotropy*, the FTG anisotropic ellipsoid becomes a sphere. The degree of deviation of an FTG anisotropic ellipsoid from its spherical, or isotropic, shape is a measure of *FTG anisotropy* quantified by *FTG anisotropic indices*. See also *anisotropy* and *magnitude*.

FTG anisotropic indices: rotationally invariant normalized *scalar* values representing measures of *FTG anisotropy*. They are calculated by the combination of magnitudes of three *FTG eigenvalues* to quantify the degree of FTG anisotropy by a single index value at each *observation point*. There are three primary anisotropic indices: linear, planar, and spherical (isotropic). Indices are normalized to range between 0 and 1 for isotropic and most anisotropic cases, respectively, and to make their sum equal to 1. Due to *normalization*, the FTG anisotropic indices are dimensionless. The distribution of their values over a study area describes the relationship between the linear, planar, and spherical parameters of the FTG anisotropy, and it is visualized by a monochromatic or color-coded *barycentric image* (Westin et al., 1999; Alexander et al., 2000). See also *invariants*, *linear anisotropic index*, *planar anisotropic index*, *spherical anisotropic index*, and *fractional anisotropy attribute*.

FTG anisotropy: a characteristic of directionally dependent *three-dimensional (3D)* variations of the *full gradient tensor (FTG)* components described by an *FTG anisotropic ellipsoid*. There are three basic types of FTG anisotropy: *FTG linear anisotropy*, *FTG planar anisotropy*, and *FTG spherical anisotropy*. Each type is represented by a particular shape, size, and orientation of the FTG anisotropic ellipsoid, wherein the mutually orthogonal axial radii are oriented along the *FTG eigenvectors* and their lengths are defined by the magnitudes of the eigenvectors-paired *FTG eigenvalues* at each observation point. The degree of FTG anisotropy is quantified by the rotationally invariant *FTG anisotropic indices* which are highly sensitive to changes in the structural composition and physical properties of rocks (*density* or *magnetization*) and their *anomaly-generating* contrasts. The spatial relationship between the calculated anisotropic indices is visualized by a color-coded *barycentric image* which shows the variations of the FTG anisotropy parameters over a study area (Hassan and Goussev, 2020). See also *contrast*, *magnitude*, *sensitivity*, and *tidal tensor*.

FTG attributes: quantitative characteristics of gravity and magnetic *gradient* fields obtained from *full tensor gradient (FTG)* data. They include but are not limited to *curvature attributes*, *directional ASA*, *edge-detector attribute*, *fractional anisotropy attribute*, *FTG amplitude*, *FTG anisotropic indices*, *FTG invariants*, and *normalized source strength (NSS)*.

FTG-derived gravity: *relative gravity* Δg_z computed by numerical transformations of the *full tensor gradient (FTG)* gravity data. The FTG-derived relative gravity differs from the relative gravity measured with the use of a regular, or conventional, *gravimeter* by the amount of a regional long-*wavelength* component in the relative gravity values, because this component is suppressed by *gradient* measurements. See also *regional-residual separation*.

FTG determinant (det FTG): a *scalar* quantity obtained by the combination of *FTG eigenvalues* or elements of 3 × 3 *FTG matrix*. It represents one of three *principal invariants* of the FTG matrix, namely the invariant I_2, and is expressed as

$$\begin{aligned}\det \mathrm{FTG} &= I_2 = \lambda_1\lambda_2\lambda_3 \\ &= T_{xx}(T_{yy}T_{zz} - T_{yz}^2) + T_{xy}(T_{yz}T_{xz} - T_{xy}T_{zz}) \\ &\quad + T_{xz}(T_{xy}T_{yz} - T_{xz}T_{yy}),\end{aligned}$$

where λ_1, λ_2, λ_3 are FTG eigenvalues obtained from *eigenvalue decomposition* of the original FTG matrix and T_{ij} are FTG matrix elements, each representing the measured *gradient* or calculated *derivative* of a particular component of the *anomalous field* and the axis in *Cartesian coordinates* along which it is measured or calculated. Calculation of the FTG determinant transforms five independent sets of *full tensor gradient (FTG)* data into a single dataset for subsequent *gridding* and analysis. Its calculation is also used for the high-*resolution* mapping of anomalies from sources approximated by *point source* geometry, such as *kimberlite intrusion* and *unexploded ordnance (UXO)*. See also *FTG trace*, *FTG invariants*, *lateral resolution*, and *source geometry*.

FTG eigensystem: a set of three mutually orthogonal *FTG eigenvectors* paired with three *FTG eigenvalues* obtained by *eigenvalue decomposition* of gravity or magnetic *FTG matrix*. Eigenvectors are oriented in the directions of the largest changes of the gravity or magnetic field components representing the first-order directional derivatives of the gravity or magnetic *potential*, respectively. Eigenvalues are magnitudes of these changes quantified by the second-order directional derivatives of the respective gravity and magnetic potentials. See also *directional derivative*, *magnitude*, *gravitational potential*, and *magnetic potential*.

FTG eigenvalues: three *eigenvalues* λ_1, λ_2, λ_3 obtained by *eigenvalue decomposition* of the gravity or magnetic *FTG matrix* and related as $\lambda_1 \geq \lambda_2 \geq \lambda_3$. The relationship between the FTG eigenvalues, FTG matrix elements, and *principal invariants* I_0, I_1, I_2 is expressed as

$$\lambda_1 + \lambda_2 + \lambda_3 = T_{xx} + T_{yy} + T_{zz} = I_0 = 0,$$

$$\begin{aligned}\lambda_1\lambda_2 + \lambda_2\lambda_3 + \lambda_1\lambda_3 &= T_{xx}T_{yy} + T_{yy}T_{zz} + T_{xx}T_{zz} - T_{xy}^2 \\ &\quad - T_{yz}^2 - T_{xz}^2 = I_1,\end{aligned}$$

$$\begin{aligned}\lambda_1\lambda_2\lambda_3 &= T_{xx}\left(T_{yy}T_{zz} - T_{yz}^2\right) + T_{xy}\left(T_{yz}T_{xz} - T_{xy}T_{zz}\right) \\ &\quad + T_{xz}\left(T_{xy}T_{yz} - T_{xz}T_{yy}\right) = I_2,\end{aligned}$$

where T_{ij} are FTG matrix elements, each representing the measured *gradient* or calculated *directional derivative* of the *i*th component of the gravity or magnetic field along the *j*th axial direction in the *Cartesian coordinates of i, j* = (*x*, *y*, *z*). In *enhanced processing*, FTG eigenvalues are determined at every observation or *grid* point and used for calculation of *FTG invariants* and other *FTG attributes*. Gridded FTG attributes are integrated into the high-*resolution* mapping of *lineaments* and *inversion*-based estimates of isolated *source body* parameters. See also *gridding*, *lateral resolution*, and *normalized source strength (NSS)*.

FTG eigenvectors: three mutually orthogonal *eigenvectors* obtained by *eigenvalue decomposition*, which transforms the nine-element *FTG matrix* from its original 3 × 3 symmetric *matrix* into a *diagonalized tensor matrix*. At each observation point, FTG eigenvectors define the orientation of the new *x*-, *y*-, *z*-axes of *Cartesian coordinates*, where six off-diagonal elements of the diagonalized FTG matrix are zeros and three non-zero diagonal elements are *FTG eigenvalues*. The FTG eigenvectors are paired with the corresponding FTG eigenvalues to compose the *FTG eigensystem*. See also *full tensor gradient (FTG)*.

FTG ellipsoid: 1. the *eigenvalue ellipsoid* in which three mutually orthogonal axial radii are absolute values, or magnitudes, of *FTG eigenvalues* that are calculated at each observation point with a spatial orientation of the ellipsoidal axes defined by the eigenvalue-paired *FTG eigenvectors*. **2.** *FTG anisotropic ellipsoid*.

FTG field: a multi-component *field* of measured or calculated gradients, where each coordinate point is represented by a 3 × 3 symmetric *FTG matrix* composed of nine second-order spatial derivatives of the gravity or magnetic *potential* in three mutually orthogonal *x*-, *y*-, *z*-directions. Five of nine calculated second-order derivatives are independent, and they can be combined into *FTG invariants* and other *FTG attributes* implemented in the *interpretation* of the gravity and magnetic *total field* and *gradient* data. For the sources approximated by *point mass* and *point dipole* geometries, the gravity and magnetic FTG fields fall off as $1/r^3$ and $1/r^4$, respectively. See also *derivative, gradient*, and *fall-off.*

FTG gravity: 1. *gravity gradient* anomalies obtained from measurements by the multi-*accelerometer* system of an *FTG gravity gradiometer*, where the measured *scalar* values of *gravitational acceleration* are converted into the spatial derivatives, or gradients, of three mutually orthogonal *vector* components g_x, g_y, g_z of the *Earth gravity field*. In theory, these gradients represent five independent components of the *full tensor gradient (FTG)*: g_{xy}, g_{xz}, g_{yz}, and any two of the three diagonal elements in the *FTG matrix*, namely g_{xx}, g_{yy}, g_{zz}. In practice, the different amount of *acquisition noise* in the measured gradients makes all three diagonal elements independent. A *three-dimensional (3D)* source, generating a simple bell-shaped *vertical gravity* anomaly, is represented in FTG gravity by (a) the single g_{zz} anomaly, (b) the doublet of g_{xz} and g_{yz} anomalies, (c) the triplet of g_{xx} and g_{yy} anomalies, and (d) the quadruplet of g_{xy} anomalies. Complexity of the anomalous FTG gravity representation by individual FTG components is moderated by their specific mathematical combinations, such as *FTG amplitude, FTG anisotropic indices, FTG invariants, FTG eigenvalues, directional ASA*, and *edge-detector attribute*. **2.** the *field* of FTG components calculated from gridded *relative gravity* data. **3.** *FTG-derived gravity*. Also referred to as *multi-component gravity* and *3D gravity*. See also *derivative, gradient, gridding, FTG attributes, GOCE gravity gradiometer*, and *rotating-disk gravity gradiometer.*

FTG gravity gradiometer: See *full tensor gravity gradiometer.*

FTG invariants: five *invariants* implemented in the *enhanced processing* and *interpretation* of gravity and magnetic data: three invariants I_0, I_1, I_2 computed from all independent *full tensor gradient (FTG)* components and two *horizontal invariants* HI_1 and HI_2 computed from the horizontal FTG components only. Any arithmetic combination of the FTG invariants or *FTG eigenvalues* is also invariant. Grids of the FTG invariants are integrated into the high-*resolution* mapping of *lineaments* and *inversion*-based estimates of isolated *source body* parameters. See also *eigenvalue decomposition, grid, lateral resolution*, and *normalized source strength (NSS).*

FTG isotropy: See *FTG spherical anisotropy.*

FTG linear anisotropy: a characteristic of the deviation of the *FTG anisotropic ellipsoid* from its spherical, or isotropic, shape. It describes the case in which the ellipsoidal radius defined by the *magnitude* of the FTG eigenvalue λ_1 is much larger than any of other two radii defined by the magnitudes of the eigenvalues λ_2 and λ_3, i.e., $|\lambda_1| \gg |\lambda_2| \approx |\lambda_3|$, so the FTG anisotropic ellipsoid has the shape of a cigar. The degree of FTG linear anisotropy is quantified by the normalized *linear anisotropic index*, for which values close to 1 indicate that the largest changes in the rock *density* and *magnetization* contrasts are orthogonal to the plane whose orientation is defined by *FTG eigenvectors* paired with the eigenvalues λ_2 and λ_3 and related to the *lateral contrast* changes associated with a *fault*, lithological *contact*, and other structural discontinuities. See also *FTG anisotropy, FTG eigenvalues, FTG planar anisotropy, FTG spherical anisotropy*, and *FTG anisotropic indices.*

FTG magnetic gradiometer: See *SQUID magnetic tensor gradiometer.*

FTG magnitude: See *FTG amplitude.*

FTG matrix: a symmetric square 3 × 3 *tensor matrix* wherein the nine elements represent the components of the *full tensor gradient (FTG)* of the anomalous gravity or magnetic *field* at each observation point. Individual FTG matrix elements are identified by their subscripts *i* and *j* that indicate a particular component of the field *vector* and one of the mutually orthogonal axes in *Cartesian coordinates* along which the *gradient* is measured or *derivative* is calculated. For example, $T_{xy} = g_{xy}$ and $T_{zz} = g_{zz}$ are gradients of the g_x and g_z components of the *Earth gravity field* vector that are measured along the *y*- and *z*-axes, respectively. In accordance with the *Laplace equation*, the FTG matrix is traceless: the *FTG trace* equals zero. See also *FTG attributes, FTG determinant, FTG eigensystem*, and *FTG gravity.*

FTG planar anisotropy: a characteristic of the deviation of the *FTG anisotropic ellipsoid* from its spherical, or isotropic, shape. It describes the case in which two ellipsoidal radii defined by the magnitudes of the *FTG eigenvalues* λ_1 and λ_2 are much larger than the third radius defined by the *magnitude* of the eigenvalue λ_3, i.e., $|\lambda_1| \approx |\lambda_2| \gg |\lambda_3|$, so the FTG anisotropic ellipsoid has the shape of a disk. The degree of FTG planar anisotropy is quantified by the normalized *planar anisotropic index*, for which values close to 1 indicate that the dominant lateral changes in the rock *density* and *magnetization* contrasts are within the plane orthogonal to orientation of the eigenvector paired with the smallest eigenvalue λ_3, such as within the *shear zone* and fractured shale layers. See also *FTG anisotropy, FTG anisotropic indices, FTG linear anisotropy*, and *FTG spherical anisotropy.*

FTG rotation: recomputing the *full tensor gradient (FTG)* components into a different reference frame by rotation about the *x*-, *y*-, *z*-axes in order to align these axes with the orientation of the subsurface structural elements or their dominant structural trend to obtain a better image of the *target*-generated *gradient* anomalies. For example, T_{xz} and T_{yz} components can be recomputed as

$$T^R_{xz} = T_{xz}\cos\theta + T_{yz}\sin\theta;$$

$$T^R_{yz} = -T_{xz}\sin\theta + T_{yz}\cos\theta,$$

where θ is angle of rotation about the *z*-axis (Dickinson et al., 2009). See also *rotational invariants* and *rotated gradient tensor.*

FTG spherical anisotropy: a characteristic of the spatial variations of the *full gradient tensor (FTG)* components, when all three axial radii of the *FTG anisotropic ellipsoid* have the same length, and it has the shape of a sphere. Because the radii lengths are defined by the absolute values, or magnitudes, of the corresponding *FTG eigenvalues*, the FTG spherical anisotropy represents the case of $|\lambda_1| = |\lambda_2| = |\lambda_3|$, where λ_1, λ_2, λ_3 are FTG eigenvalues computed at each observation or *grid* point. The degree of FTG spherical anisotropy is quantified by the normalized *spherical anisotropic index*, for which values close to 1 indicate very small lateral changes in the rock *density* or *magnetization* contrasts that are below the *resolution* of the FTG measurements. Also referred to as *FTG isotropy*. See also *FTG anisotropy*, *FTG anisotropic indices*, *FTG linear anisotropy*, *FTG planar anisotropy*, *density contrast*, and *magnetization contrast*.

FTG survey: airborne, marine (shipboard), and submarine measurements of the *full tensor gradient (FTG)* components of the *Earth gravity field* or *Earth magnetic field*. Its design is based on the use of a high-*accuracy* positioning system, such as *differential GPS (DGPS)*, short *sampling interval*, low *survey height*, and close *survey line spacing* to ensure the proper *acquisition* parameters for utilizing the high-*resolution* properties of the *gradient* measurements. See also *survey design* and *survey design aliasing*.

FTG technology: the geophysical exploration technology based on measurements of the *full tensor gradient (FTG)* components of the Earth's gravity or magnetic field. Extra-high *sensitivity* of FTG components to small lateral changes in the *density* and *magnetization* of rocks is the most efficient in application for mineral and oil-and-gas explorations at relatively shallow *prospect* levels. See also *FTG anisotropy*.

FTG trace: **1.** the sum of three diagonal elements in the symmetric 3 × 3 square *FTG matrix* that are related by the *Laplace equation* as $T_{xx} + T_{yy} + T_{zz} = 0$. **2.** the sum of three *FTG eigenvalues* λ_1, λ_2, λ_3 that defines the invariant I_0 as $I_0 = \lambda_1 + \lambda_2 + \lambda_3 = 0$. See also *FTG determinant (det FTG)* and *invariants*.

full-Earth model: a regional-scale *profile*-based *2D model* or *grid*-based *3D model* which includes the boundary between the Earth's lower *crust* and upper *mantle*, i.e., *Moho boundary*. See also *2.5D model* and *2.75D model*.

full-field gravity: same as *absolute gravity* that represents a full numerical value of the *Earth gravity field* measured in a reference frame rotating with the Earth.

full gradient: See *total gradient (TG)* and *analytic signal amplitude (ASA)*.

full-sine resolution: *full-wavelength resolution* in the context of representing the *wavelength* components of an observed gravity or magnetic field by the sinusoidal waveforms of different spatial frequencies. Full-sine resolution is twice *half-sine resolution*.

full spectrum gravity (FSG): **1.** the airborne and marine (shipboard) *acquisition* methodology for simultaneous measurements of the *full tensor gravity gradient (FTGG)* components by a *full tensor gravity gradiometer* and long-*wavelength* gravity field components (which are suppressed in the *gradient* measurements) by a *relative gravimeter* on the same *survey* aircraft or ship. **2.** an *anomalous gravity* dataset which combines the FTGG-derived gravity data and long-wavelength gravity data from a *ground gravity* survey, *airborne gravity*, and *satellite-derived gravity* data. See also *airborne gravimeter*, *marine gravimeter*, *FTG gravity*, and *GOCE gravity gradiometer*.

full tensor gradient (FTG): a nine-element *gradient tensor* representing the spatial rates of change of three mutually orthogonal *vector* components of the gravity or magnetic field along the *x*-, *y*-, *z*-directions in *Cartesian coordinates*. Because every vector component of the *potential field* is defined by the *directional derivative* of its *potential*, the FTG components are second-order derivatives of the *gravitational potential* or *magnetic potential*. In Cartesian coordinates, nine FTG components are denoted by T_{xx}, T_{xy}, T_{xz}, T_{yx}, T_{yy}, T_{yz}, T_{zx}, T_{zy}, T_{zz} and arranged in the *format* of a 3 × 3 square symmetric *matrix* for each observation point:

$$\mathrm{FTG} = \begin{matrix} T_{xx} & T_{xy} & T_{xz} \\ T_{yx} & T_{yy} & T_{yz} \\ T_{zx} & T_{zy} & T_{zz} \end{matrix} = \begin{matrix} \partial P_x/\partial x & \partial P_x/\partial y & \partial P_x/\partial z \\ \partial P_y/\partial x & \partial P_y/\partial y & \partial P_y/\partial z \\ \partial P_z/\partial x & \partial P_z/\partial y & \partial P_z/\partial z \end{matrix}$$

$$= \begin{matrix} \partial^2 P/\partial x^2 & \partial^2 P/\partial x\partial y & \partial^2 P/\partial x\partial z \\ \partial^2 P/\partial y\partial x & \partial^2 P/\partial y^2 & \partial^2 P/\partial y\partial z \\ \partial^2 P/\partial z\partial x & \partial^2 P/\partial z\partial y & \partial^2 P/\partial z^2 \end{matrix}$$

where P is the gravitational or magnetic potential and P_x, P_y, P_z are vector components of the gravity or magnetic field defined as $P_x = \partial P/\partial x$, $P_y = \partial P/\partial y$, and $P_z = \partial P/\partial z$. There are three pairs of symmetric elements: $T_{xy} = T_{yx}$, $T_{xz} = T_{zx}$, and $T_{zy} = T_{yz}$. Three diagonal elements comprise the *FTG trace* and are related by the *Laplace equation* as $T_{xx} + T_{yy} + T_{zz} = 0$. There are five independent elements T_{xy}, T_{xz}, T_{yz} and any two of three diagonal elements T_{xx}, T_{yy}, T_{zz}. The highest *resolution* of the *gradient* anomalies is obtained with the FTG components measured by a *full tensor gradiometer*. A less resolving option is calculation of the FTG-corresponding first-order derivatives of the gridded *total field* data in the *Fourier domain* with the following issues properly addressed: (a) a reliable interpolation algorithm for *gridding* the line-oriented data, (b) a data *sampling* sufficient to avoid *undersampling* and eliminate high-frequency *grid aliasing* across the survey lines, (c) *tapering* and *detrending* of gridded data prior to transformation into the Fourier domain to minimize *Gibbs errors* of the forward and inverse parts of the *Fourier transform*, and (d) compensation for *drape effects* and *magnetic terrain effects* in the *airborne survey* data. See also *FTG attributes*, *FTG field*, *FTG invariants*, and *circular convolution*.

full tensor gradiometer: a *gradiometer* system for measurements of the *full tensor gradient (FTG)* components of the Earth's gravity or magnetic field on a moving *observation platform*, such as a ship, airplane, or satellite. See also *full tensor gravity gradiometer*, *full tensor magnetic gradiometer*, and *GOCE gravity gradiometer*.

full tensor gradiometry: instruments to measure and record the *full tensor gradient (FTG)* components of the Earth's gravity or magnetic field and the methodology for processing and *interpretation* of the acquired *gradient tensor* data. Tensor mathematics offers a wide range of data processing algorithms for computing the attributes of gravity and magnetic *gradient* fields, including *curvature attributes*, *curvature gravity gradient tensor*, *directional analytic signal*, *edge-detector attribute*, *eigenvalues*, *eigenvectors*, *FTG attributes*, *FTG anisotropic indices*, *invariants*, *horizontal invariants*, *invariant ratio*, *normalized source strength (NSS)*, and *strike-angle invariant* as well as for the development of new *inversion* methods, such as *tensor Euler deconvolution* and *tensor deconvolution*, implemented in high-*resolution* exploration and research applications. See also *attribute* and *total field gradiometry*.

full tensor gravity: *vertical gravity* g_z and *gravity gradient* data obtained from measurements of the *full tensor gradient (FTG)* components of the planetary *gravity field*.

full tensor gravity gradient (FTGG): the *full tensor gradient (FTG)* in which the components represent the measured or calculated spatial rates of change, or gradients, of three mutually orthogonal *vector* components of the *Earth gravity field* g_x, g_y, g_z along the *x*-, *y*-, *z*-axes, respectively, in *Cartesian coordinates*. See also *full tensor magnetic gradient (FTMG)*.

full tensor gravity gradiometer: 1. *rotating-disk gravity gradiometer* system mounted on a *stabilized platform*. Its design is based on the use of three tilted rotating disks with a total of 12 accelerometers. Each rotating disk holds two pairs of equally spaced and orthogonally mounted accelerometers with their sensitive axes tangential to the disk plane and orthogonal to the disk rotation axis. Each pair of synchronized accelerometers forms a *gradiometer arm* along one of three mutually orthogonal axes of the measurements. Three disk assemblages are tilted toward each other, and their axes of rotation are mutually orthogonal. The assembly of three rotating disks is rotated at a constant speed about the vertical axis. Calculation of the difference between the gravitational accelerations measured by each pair of accelerometers compensates for most of the aircraft turbulence and motion-related inertial accelerations in marine (shipboard) and submarine observations. In total, six *gradient tensor* elements are measured in three separate coordinate frames from which all independent *full tensor gradient (FTG)* components are reconstructed in the standard *Cartesian coordinates*. The obtained g_{zz} component is transformed by vertical integration into the *vertical gravity* Δg_z, which differs from the vertical gravity measured with the use of a regular *relative gravimeter* by the amount of a regional long-*wavelength* component in the *gravity survey* area. The best reported *full-wavelength resolution* in airborne applications is approximately 0.3–1.0 km with an operational *accuracy* within the range of 3–7 *Eötvös* (*E*). **2.** a *GOCE gravity gradiometer*, wherein 18 electrostatic accelerometers are used for independent measurements of all nine FTG components. Commonly referred to as *FTG gravity gradiometer*. See also *accelerometer*, *common-mode rejection*, *electrostatic accelerometer*, *gravitational acceleration*, *inertial acceleration*, *vertical integral*, and *partial tensor gravity gradiometer*.

full tensor gravity gradiometry: instruments for measurements of the *full tensor gravity gradient (FTGG)* components and methodology for processing and interpretation of acquired *gravity gradient* data. The resolving power, i.e., *lateral resolution* and *vertical resolution*, of gravity gradient anomalies achievable with the use of measured FTGG components is superior to the resolution of FTGG components calculated from gridded *free-air gravity* or *Bouguer gravity* data, primarily due to the inherent spatial averaging of the *gridding* process and Fourier-domain calculations. Sometimes referred to as *3D gravity*. See also *Fourier domain*, *gravity gradiometry*, and *full tensor gravity gradiometer*.

full tensor magnetic gradient (FTMG): the *full tensor gradient (FTG)* in which the components represent the measured or calculated spatial rates of change, or gradients, of the three mutually orthogonal *vector* components of *Earth magnetic field* B_x, B_y, B_z along the *x*-, *y*-, *z*-axes, respectively, in *Cartesian coordinates*. See also *full tensor gravity gradient (FTGG)*.

full tensor magnetic gradiometer: See *SQUID magnetic tensor gradiometer*.

full tensor magnetic gradiometry: instruments for measurements of the *full tensor magnetic gradient (FTMG)* components and methodology for processing and *interpretation* of the acquired *magnetic gradient* data. Achievable *lateral resolution* of the magnetic *gradient* anomalies by the measured FTMG components is superior to the lateral resolution of the FTMG components calculated from gridded TMI or RTP magnetic data, primarily due to the inherent spatial averaging of the *gridding* and Fourier-domain calculations. Sometimes referred to as *3D magnetics*. See also *Fourier domain*, *magnetic gradiometry*, *total magnetic intensity (TMI)*, *RTP magnetic field*, and *SQUID magnetic tensor gradiometer*.

full-wavelength resolution: the shortest *wavelength* of anomalies that can be detected, or resolved, in the line-oriented data acquired by a marine (shipboard) *gravity survey* and an *airborne gravimeter (AG)* or *airborne gravity gradiometer (AGG)* survey. Its quantitative estimate is defined by the length, or time duration, of a low-pass *time filter* applied to suppress the high-frequency *noise* due to uncompensated inertial accelerations of the flying aircraft and wave-like motion of the survey ship. The space equivalent of the required time filter length is determined by multiplying the aircraft or ship speed and time duration of the filter operation, e.g., for the aircraft speed of 50 m · s^{-1} (180 km per hour) and 80-second low-pass time filter, the effective filter length and corresponding full-wavelength resolution are 4.0 km. Sometimes referred to as *full-sine resolution*. See also *acquisition*, *inertial acceleration*, *half-wavelength resolution*, *low-pass filter*, *turbulence effect*, and *wave-motion effect*.

Fuller filter: a space-domain *convolution filter* the algorithm of which is based on the computation of an average value within the *moving window* of a user-specified length. It is applied in *detrending*, *microleveling*, and *smoothing* of high-frequency *noise* components of line-oriented or gridded data. See also *space domain* and *gridding*.

fundamental frequency: 1. the lowest *frequency* of the sinusoids (sine and cosine wave functions) in

Fourier-transformed gravity or magnetic data. **2.** the lowest frequency of a periodic *waveform*, wherein its *period* is the inverse of the fundamental frequency. **3.** *spatial frequency* representing the reciprocal, or inverse, of the *sample interval* which controls the *lateral resolution* of acquired data. Simple (whole number) multiples of the fundamental frequency are called *harmonics* and implemented in a *Fourier series*, so the fundamental frequency is also referred to as the *first harmonic*. See also *acquisition* and *Fourier transform*.

fundamental period: the shortest *period* of the sinusoids (sine and cosine wave functions) in Fourier-transformed gravity or magnetic data. Minimization of the spectral-domain *Gibbs errors* and *wavenumber leakage* requires the length of the gridded or line-oriented dataset to be an integer multiple of the fundamental period. See also *Fourier transform*, *gridding*, *spectral domain*, and *dominant period*.

fundamental wavelength: 1. the longest *wavelength* which defines the range of wavelengths outside of which the gravity and magnetic anomalies are not included into the *Fourier transform*. It is a reference value for decomposition of space-domain waveforms into sinusoidal *harmonics* by the Fourier transform. **2.** the longest wavelength recorded in the *space domain* that can be transformed into the *Fourier domain*. **3.** a *period* of a periodic *waveform*. See also *Fourier series*, *space domain*, and *waveform*.

fundamental wavenumber: the smallest *wavenumber* in Fourier-transformed gravity or magnetic data, defined as

$$k_{fw} = 2\pi/\lambda_L,$$

where λ_L is the longest space-domain *wavelength* of the data transformed into the *Fourier domain* that is referred to as the *fundamental wavelength*. See also *Fourier transform*.

fusion: an *enhanced processing* technique based on the application of a *discrete wavelet transform (DWT)* or *machine learning (ML)* algorithm in order to (a) decompose line-oriented or gridded gravity and magnetic data into the successive levels, or scales, of *wavelength resolution* by the DWT or into *structure*-related characteristic components by the ML algorithms, and (b) to integrate, or fuse, the obtained data into a combined gravity-magnetic *grid* for a better visualization of the *total field* and *gradient* anomalies associated with elements of subsurface geological structures (Erkan et al., 2012; Hassan and Goussev, 2019b). See also *cross-gradient method* and *wavelet decomposition*.

G

G: See *universal gravitational constant.*

***g*: 1.** a *vector* quantity the *magnitude* and direction of which represent the *gravitational acceleration* of a free-falling body with the mass *m* due to attraction of the Earth's mass *M*. At any observation point *P*, it is defined by the *gravitational force* per unit *mass* and expressed as

$$\boldsymbol{g}(P) = F/m = -\hat{\mathbf{r}}\,GM/R^2,$$

where F is gravitational force between masses *M* and *m* formulated by the *Newton law of universal gravitation*, $\hat{\mathbf{r}}$ is a unit vector directed from the center of the Earth's mass *M* to the mass *m*, G is the *universal gravitational constant*, and R is a distance between the center of the Earth's mass *M* and observation point *P*. By convention, the negative sign indicates that the force of attraction (gravitational force) is directed opposite to the unit vector $\hat{\mathbf{r}}$. As a *three-dimensional (3D)* vector quantity, ***g*** is described by the directional derivatives of *gravitational potential* U as

$$\boldsymbol{g} = \nabla U = \mathbf{i}\partial U/\partial x + \mathbf{j}\partial U/\partial y + \mathbf{k}\partial U/\partial z,$$

where ∇ is vector *gradient* operator *del*; **i**, **j**, **k** are unit vectors parallel to the *x*-, *y*-, *z*-axes, respectively, in *Cartesian coordinates*; and $\partial U/\partial x = g_x$, $\partial U/\partial y = g_y$, $\partial U/\partial z = g_z$ are three components of the *gravity vector*. **2.** *absolute gravity* for which the numerical value is defined by the *magnitude* of gravitational acceleration at a given location. **3.** *nominal gravity,* which is a globally averaged theoretical value of the gravitational acceleration *g* at the Earth's surface, defined as

$$g \cong 9.8 \text{ m} \cdot \text{s}^{-2} = 980\ \textit{Gal} = 9.8 \cdot 10^5\ \textit{milligal (mGal)} = 9.8 \cdot 10^8\ \textit{microgal}\ (\mu \textit{Gal}).$$

Its actual value varies from approximately $9.78 \text{ m} \cdot \text{s}^{-2} = 978$ Gal at the equator to $9.83 \text{ m} \cdot \text{s}^{-2} = 983$ Gal at the geographic poles. Nominal gravity on the Earth is also used as a reference value for estimates of the gravity fields of other planets, such as ~0.38 *g* for Mars and ~0.17 *g* for the Moon. See also *directional derivative*.

Gal: a unit of the *gravitational acceleration* adopted as a basic unit in gravity measurements. An abbreviation of Galileo, the Gal represents *acceleration* of 1 centimeter per second squared:

$$1 \text{ Gal} = 1 \text{ cm} \cdot \text{s}^{-2} = 10^{-2} \text{ m} \cdot \text{s}^{-2} = 10^{-2} \text{newton} \cdot \text{kg}^{-1} = 10^4\ \mu\text{m} \cdot \text{s}^{-2} \cong 10^{-3} g,$$

where *g* is *nominal gravity* defined as $g = 980$ Gal $= 9.8 \text{ m} \cdot \text{s}^{-2}$ in *SI units* and $g = 980 \text{ cm} \cdot \text{s}^{-2}$ in *cgs units*. In exploration applications, the standard units are *milligal (mGal), microgal (μGal),* and micrometer per second squared (μms^{-2}), whereby 1 mGal = 10 μm s^{-2} = 10^3 μGal = 10^{-3} Gal = 10^{-5} m · s^{-2} = ~10^{-6} *g* and 1 μGal = 10^{-2} μms^{-2} = 10^{-3} mGal = 10^{-6} Gal = 10^{-8} m · s^{-2} ≅ 10^{-9} *g*.

gamma (γ): an obsolete unit of magnetic measurements. Previously, magnetic survey maps were contoured in gammas, but that has been replaced by the equivalent SI unit of *nanotesla (nT)* as the preferred name, whereby 1 nT = 1γ = 10^{-5} *gauss* = 10^{-9} *tesla (T).* See also *SI units* and *cgs units*.

gamma-gamma log: same as *density log* with an emphasis on its gamma-ray source and detector of the secondary gamma-ray radiation from rock formations.

Gardner equation: an empirical relationship of proportionality between *apparent density* and velocity V_p of the seismic compressional wave, or P-wave, that is valid for the majority of sedimentary rocks (not including *salt*) within the velocity range of 1.5–6.1 km · s^{-1}. The Gardner equation is expressed as

$$\rho = \alpha \cdot V_P^{\beta},$$

where ρ is apparent density in grams per cubic centimeter (g · cm^{-3}), and α and β are proportionality coefficients calibrated by the local *density log* data. For seismic velocity V_P in meters per second (m · s^{-1}) and feet per second (ft · s^{-1}), it is often approximated as $\rho = 0.31\,V_P^{0.25}$, and $\rho = 0.23\,V_P^{0.25}$, respectively. Sometimes referred to as *Gardner rule* and *Gardner velocity-density relationship*. See also *Adams-Williamson equation, Brocher formula*, and *Nafe-Drake curve*.

Gardner rule: *Gardner equation* in the context of an empirical rule that describes the relationship between *apparent density* and velocity of the seismic compressional wave, or P-wave, in sedimentary rock formations (not including *salt*).

Gardner velocity-density relationship: same as *Gardner equation* with an emphasis on the empirical relationship between *apparent density* and velocity of the seismic compressional P-wave.

gate: See *window*.

gauss: a unit of *magnetic induction*, or *magnetic flux density*, in the cgs system. It is defined by the *magnetic force*, or the number of imaginary magnetic lines of force, per unit area. Also used as the unit of *magnetization* in terms of the *magnetic dipole moment* per unit volume, whereby 1 gauss = 10^{-4} *tesla (T)* = 10^5 *nanotesla (nT)* = 1 *maxwell* · cm^{-2} = 10^{-4} *weber* · m^{-2}. See also *cgs units* and *flux*.

Gauss coefficients: time-varying coefficients in the *spherical harmonic expansion* of the global *magnetic potential* that specify the relative importance, or weights, of *spherical harmonics* in the mathematical model of the *main geomagnetic field*. The first three coefficients represent the components approximated by the *dipole field*. The coefficient values are revised every 5 years and included in predictive mathematical models of the *International Geomagnetic Reference Field (IGRF)* with the assumption that their variations are linear during the 5-year period called an *epoch*. Also referred to as *IGRF coefficients*. See also *IGRF model* and *Stokes coefficients*.

Gauss flux theorem: the theorem relating the *vector field* flow, or *flux*, through a closed surface to properties of the vector field enclosed by this surface. It postulates that the total flux Φ through any closed surface is equal to the constant $4\pi k$ multiplied by the hypothetical strength M of a source enclosed by this surface, i.e., $\Phi = 4\pi k\text{M}$, where constant k depends on the field type and units of measurement. Here, Φ represents (a) *gravitational flux*, when M is *mass* and $k = -\text{G}$, where G is a *universal gravitational constant*, or (b) *magnetic flux*, when M is the theoretical *magnetic pole strength*. The theorem validates the *nonuniqueness problem* inherent for the potential fields: if the field is known on a bounding surface, then there are infinitely many source distributions below this surface that can generate the same *observed field*. Also referred to as *Gauss theorem*. See also *Gauss law* and *potential field*.

Gauss law: 1. for the *gravity field*, it postulates that the total *mass* in a region bounded by a closed surface is directly proportional to the vertical component of *gravitational attraction* integrated over the bounding surface S of the region and expressed as

$$\int_S g_z dS = -4\pi G \int_V \rho dV = -4\pi G M_T,$$

where g_z is the vertical, or normal, component of the gravity field; G is a *universal gravitational constant*; ρ is rock *density*; and M_T is total *mass* inside the surface S. The Gauss law provides a way to estimate uniquely the total anomalous mass ΔM, or *mass differential*, below the horizontal surface of the Earth by numerical integration of the gravity anomaly Δg_z over its entire area extent as

$$\Delta \text{M} \cong (1/2\pi \text{G}) \int_S \Delta g_z dS.$$

Anomalous mass calculation does not require any assumptions about the mass shape, depth, or *density* distribution, if (a) the measured anomaly is isolated from the gravity effects of other sources and (b) the anomalous mass is small relative to the *gravity survey* area. **2.** for the *magnetic field*, the Gauss law is based on the *divergence theorem* and expressed as

$$\int_S \mathbf{B} \cdot \breve{\mathbf{n}}\, dS = \int_R \nabla \mathbf{B}\, dv = 0,$$

where **B** is *magnetic induction*, $\breve{\mathbf{n}}$ is unit *vector* orthogonal to the surface S, $\mathbf{B} \cdot \breve{\mathbf{n}}$ is the *dot product* representing the normal component of the magnetic induction **B**, and ∇ is *gradient* operator *del*. The Gauss law postulates that the normal component of the magnetic induction **B** entering any region enclosed by the surface S is equal to the normal component of **B** leaving the region, i.e., the normal component of **B** integrates to zero over any closed surface and, hence, the net normal *magnetic flux* through any region is zero. The Gauss law implies that: (a) the magnetic field has no *divergence*, i.e., it is a divergenceless or *solenoidal field*; (b) the source of the magnetic field is a *magnetic dipole* for which the imaginary field lines, or magnetic *flux* lines, exist in the form of closed loops directed from a positive pole to a negative pole of the magnetic dipole; (c) magnetic monopoles are hypothetical entities, at least macroscopically; and (d) once the regional-scale magnetic anomalies have been removed from the field of *total magnetic intensity (TMI)* in a survey for which the lateral extent is large compared to sizes of the individual magnetic sources, then the TMI *residual field* will contain approximately equal amounts of the positive and negative magnetic *anomaly energy*.

Gauss theorem: 1. *Gauss flux theorem*. **2.** *divergence theorem*. **3.** a theorem of the arithmetic mean stating that the value of a *harmonic function* at any point is the average of the harmonic function over any sphere concentric about that point, when the function is harmonic throughout the sphere (Blakely, 1996). See also *Gauss law*.

Gaussian: 1. *Gaussian filter*. **2.** *Gaussian function*.

Gaussian bell: same as *Gaussian function* with an emphasis on its symmetric bell shape.

Gaussian convolution filter: same as *Gaussian filter* that acts as a space-domain *filter* of the *convolution* type. See also *space domain*.

Gaussian curvature: an *attribute*, or parameter, of *curvature* of the gravitational *equipotential surface* calculated by multiplying the *maximum curvature* K_{max} and *minimum curvature* K_{min} and expressed as

$$K_G = K_{max} K_{min} = (g_{xx} g_{yy} - g_{xy}^2)/g_z^2,$$

where K_G is the Gaussian curvature; g_{xx}, g_{yy}, g_{xy} are *curvature gradients*; and g_z is *vertical gravity* (Li, 2015). Zero contour lines of the gridded Gaussian curvature values are indicators of structural edges, such as high-*offset* faults and lithological boundaries associated with the strong *density* contrasts within sedimentary formations, including those formed by *igneous intrusion*, high-density carbonate buildup (*reef*), and low-density reservoir-grade turbidite flows. Sometimes referred to as *total curvature*. See also *density contrast* and *gridding*.

Gaussian distribution: a bell-shaped distribution of a large number of repetitive observation values around a central value which is the arithmetic mean, or average, of all values in a dataset. The occurrence frequency is much larger for values near the arithmetic mean. Values that depart widely from the average are the low-occurring values called outliers. A Gaussian distribution is typical for unpredictable, or random, measurement errors. Also referred to as *normal distribution*.

Gaussian filter: 1. a *derivative*-based *convolution filter* the *impulse response* of which is a *Gaussian function*. It is applied to improve the continuity of anomalous alignments and detect *line lineaments*. **2.** a space-domain *smoothing filter* for the suppression of high-frequency *noise*. The degree of smoothing is defined by a preselected *standard deviation (SD)* of the filter *kernel* that is used as a weighting function for each *grid cell*, wherein the central grid cell in the kernel has the highest SD weight among the surrounding grid cells. Differing from an *averaging filter* with uniform weighting for all grid cells, the SD-based weighted average provides a more gentle data *smoothing*, and it can be applied to increase the *sharpness* of *edge*-generated anomalies when used as a residual *sharpening filter*.

Gaussian function: a bell-shaped function which has the property of being localized in both the space and spectral domains, because its *Fourier transform* produces a *power spectrum* with the shape of a scaled Gaussian function in the *space domain*. Among various applications, it is used as a *window function* in a *window Fourier transform (WFT)*, and its *horizontal derivative* can be used as a *mother wavelet* in the *wavelet transform*. Often referred to simply as *Gaussian*. See also *spectral domain* and *Gaussian wavelet*.

Gaussian noise: a pseudo-random *noise* with the zero average, or zero mean, and a user-specified *standard deviation (SD)*. It is implemented in *synthetic modeling* in lieu of the actual measurement noise. See also *random noise*.

Gaussian smoothing: *smoothing* of gridded gravity or magnetic data with the use of a *Gaussian filter*. See also *gridding*.

Gaussian wavelet: a *mother wavelet* obtained by calculation of the *horizontal derivative* of a *Gaussian function*. See also *Poisson wavelet* and *wavelet transform*.

Gaussian window: a space-domain *window* that has the shape of a *Gaussian function*. Its *Fourier transform* is also a Gaussian function in the *spectral domain*. See also *space domain* and *windowing*.

general source: an arbitrary-shaped *source body* generating a gravity or magnetic *anomaly*. For the purposes of *forward modeling* and *inversion*, the general source is approximated by assemblages of the elementary ideal sources of simple geometrical forms which fill the general source volume. Analytically expressed gravity and magnetic fields of all individual ideal sources are numerically integrated at each *observation point* to obtain a corresponding gravity or magnetic *anomaly*. See also *ideal source*.

generalized Euler deconvolution: a generalization of the two-dimensional (2D) *extended Euler deconvolution* to three dimensions with the use of the *3D Hilbert transform*. Modified Euler deconvolution equations in three dimensions for the arbitrary *Euler structural index (SI)* are obtained at each observation point. Three Euler equations for the *anomalous field* F instead of one *Euler deconvolution equation* stabilize calculations of the source position and structural index. Its formula can be expressed as

$$(x-x_0)\partial\mathbf{H}(\mathrm{F})/\partial x+(y-y_0)\partial\mathbf{H}(\mathrm{F})/\partial y$$
$$+(z-z_0)\partial\mathbf{H}(\mathrm{F})/\partial z+N\mathbf{H}(\mathrm{F})=\beta,$$

where x_0, y_0, z_0 are coordinates of the source location and depth; N is the Euler structural index; β is a constant which is usually zero, except for the case of $N=0$; $\mathbf{H}(\mathrm{F})=\boldsymbol{i}H_x(\mathrm{F})+\boldsymbol{j}H_y(\mathrm{F})$, where $H_x(\mathrm{F})$ and $H_y(\mathrm{F})$ are x- and y-components, respectively, of the 3D Hilbert transform; and $\boldsymbol{i}$ and $\boldsymbol{j}$ are unit vectors parallel to the x- and y-directions, respectively, in *Cartesian coordinates* (Nabighian and Hansen, 2001). Also referred to as *extended 3D Euler deconvolution*.

generalized Hilbert transform: 1. *Riesz transform* in the context of a generalization of the *Hilbert transform* for dimensions $\mathrm{D}\geq 2$. **2.** *3D Hilbert transform*. See also *dimension*.

geocentric dipole field: a first-order approximation of the dominant long-*wavelength* component of the *Earth magnetic field* by the field of a *geocentric magnetic dipole* located at the geometric center of the Earth. A better approximation is provided by the *eccentric dipole field* which represents the field of a *geomagnetic dipole* located at approximately 400 km above the Earth's center.

geocentric equipotential ellipsoid: same as *reference ellipsoid* with an emphasis on its center of mass coincident with the center of mass of the Earth, and for which the surface of this ellipsoid is an *equipotential surface* for its own *gravity field*.

geocentric gravitational constant (GM): a *reference ellipsoid* parameter obtained by multiplying the *universal gravitational constant* G and the Earth mass M which includes the mass of the atmosphere. Recently updated, the value is defined as $\mathrm{GM}=3.986004418\cdot 10^{14}\ \mathrm{m^3s^{-2}}$.

geocentric magnetic dipole: the hypothetical *magnetic dipole* located at the geometric center of the Earth. Its magnetic field is considered the first-order approximation of the dominant long-*wavelength* component of the *Earth magnetic field* and referred to as the *geocentric dipole field*.

geocentric reference ellipsoid: See *geocentric equipotential ellipsoid*.

geodetic datum: a reference surface representing a specific *reference ellipsoid*, its applicable regional extent, position relative to the Earth's center, and orientation of its coordinate axes. Different nations use the different geodetic datums of the *Geodetic Reference System* to determine the 3D geographic coordinates (latitude, longitude, and *elevation*) of observation points. Referred to with the same meaning as *regional geodetic datum* and *geographic datum*.

geodetic ellipsoid: See *geoid*.

Geodetic Reference System: a terrestrial *coordinate system* for the geographic *positioning* on or above the Earth's surface, where the *three-dimensional (3D)* coordinates of any point (latitude, longitude and height, or *elevation*) are determined relative to a specific *reference ellipsoid* known as a *geodetic datum*. Regional and national geodetic reference systems in different parts of the world implement the different reference ellipsoids which closely approximate, or conform to, the surface of a *geoid* over a particular region rather than the whole Earth surface. The *International Terrestrial Reference Frame (ITRF)* and *World Geodetic System 1984 (WGS84)* are the most recent internationally accepted global geodetic reference systems. Differences between the ITRF and WGS84 coordinates are negligible for exploration applications. See also *Universal Transverse Mercator (UTM)* and *GRS80 ellipsoid.*

geodynamo: See *geomagnetic dynamo.*

geographic datum: *geodetic datum* in the context of a reference surface for the geographic *positioning* of observation points on or above the Earth's surface.

geoid: a sea-level *equipotential surface* for the *Earth gravity field.* It approximates the most closely, in the least-squares sense, *mean sea level (MSL)* in the oceans and seas without tides and currents, and it is everywhere perpendicular to the direction of the Earth's gravity field defined by the spatial orientation of its *gravity vector.* A geoid is a fundamental reference surface, or *datum*, for geodetic and gravity measurements. A geoid coincides with a surface which would be attained by the ocean water, if it could flow in the imaginary channels through continents and settle at its own level across the entire globe in order to accommodate the combined effect of the inward-directed *gravitational force* of attraction produced by the Earth's mass and outward-directed *centrifugal force* resulting from the Earth's rotation. Deviation of the geoidal surface from the ellipsoidal shape is called *geoid undulation*, and it is controlled by the distribution of rock masses inside the Earth: the geoid warps upward over mass excesses, such as on the continents, and it warps downward over mass deficiencies, such as in the oceans. Geoid highs and lows are directly related to anomalous gravity highs and lows: a geoid slope of 1 microradian (µrad) corresponds to a *free-air anomaly* of approximately 1 *milligal (mGal).* As a standard reference surface, the geoid is now being replaced in exploration and research applications by the *reference ellipsoid*, also known as the *equipotential ellipsoid*, which is the reference surface for *Global Positioning System (GPS).* Sometimes referred to as *nominal sea level* and *geodetic ellipsoid.* See also *marine geoid, mass excess, mass deficiency, EGM2008 geoid*, and *indirect effect.*

geoid-ellipsoid separation: See *geoid height* and *geoid undulation.*

geoid height: 1. the vertical distance between the mathematical surfaces of *geoid* and *reference ellipsoid.* Geoid height varies slowly because of the small gradients of the geoidal surface, and it is positive over the continents, where the geoid is above the reference ellipsoid, and it is negative over the oceans, where the geoid is below the reference ellipsoid. Referred to with the same meaning as *geoid undulation* in the context of the mathematical representation of variations in the geoidal surface by its deviations from the surface of a particular reference ellipsoid. **2.** the height above the geoid which is referred to as *orthometric height* in geodetic applications. Also known as *geoid-ellipsoid separation.* See also *EGM96 geoid height.*

geoid-to-gravity method: a method of computing marine gravity anomalies, known as *marine free-air gravity*, from *satellite altimetry* data over oceans and seas. Because *mean sea level (MSL)* is equivalent to the *geoid*, which is the *equipotential surface* for the *Earth gravity field*, the method is based on the construction of a *grid* of the sea surface and calculation of its *first vertical derivative (FVD)*, i.e., vertical derivative of the equipotential surface, that represents the vertical component of the Earth's gravity field *vector* and known as *vertical gravity.* Satellite altimetry data along satellite orbital tracks are corrected for all measurement effects prior to the cross-track *leveling* and *microleveling* applied to minimize the temporal sea surface disturbances (waves, currents, tides). See also *gravity vector* and *slope-to-gravity method.*

geoid undulation: deviation of the surface of a *geoid* from a *reference ellipsoid.* Because the geoidal shape is influenced by the underlying mass deficiencies and excesses, the geoid warps downward below the reference ellipsoid in the oceans, and it warps upward above the reference ellipsoid on the continents. By convention, the sign of the geoid undulation is positive over continents and negative over oceans. On the global scale, geoid undulation varies from +85 m on the continents to −107 m in the oceans relative to the *WGS84 ellipsoid.* Dominant global-scale temporal variations of the geoidal surface are assumed to be caused by widespread convection within the *mantle.* Tabulated values are available in the *format* of continuously refined global, regional, and local geoid undulation maps. For *two-dimensional (2D)* approximations, the geoid undulation *amplitude* A and *wavelength* λ are related to the gravity anomaly Δg as

$$\Delta g = 2\pi A g_N / \lambda,$$

where $g_N = 980000$ mGal is the average value of *theoretical gravity* on the Earth's surface known as *nominal gravity.* Also referred to as *geoid height* and *geoid-ellipsoid separation.* See also *EGM96 geoid height, indirect effect*, and *milligal (mGal).*

geological basement: same as *basement* that represents the geological foundation on which all sedimentary and, sometimes, metamorphic rocks are deposited.

geological correction: a correction applied to the *Bouguer anomaly* or *free-air anomaly* to compensate for the gravity effects of the known or approximated elements of a subsurface geological structure, such as the upper part of the sedimentary section, whose depth structure and formation densities are known from seismic and well (borehole) data. The total amount of the geological correction is defined by the summation of the calculated gravity effects of all known local and regional geological structures in a *gravity survey* area. Calculation and application of this correction provides a basis for the *gravity stripping* method. Also referred to as *stripping correction.* See also *geological gravity anomaly* and *stripped gravity.*

geological gravity anomaly: the *field* of gravity anomalies compensated for the gravity effects, or *gravitational attraction*, of all known and predicted geological sources in a *gravity survey* area, including those associated with elements of subsurface structures known from seismic and well (borehole) data. As a final result of the *gravity reduction* process, it includes only the gravity effects of the unknown *density* and structural variations in the upper *crust.* Examples of the geological gravity anomaly are the *isostatic anomaly* and *stripped gravity* anomalous fields. See also *anomalous field*, *geological correction*, and *gravity anomaly*.

geomagnetic correction: *IGRF correction* in the context of compensation of the *observed magnetics* for the contribution of the *main geomagnetic field*.

geomagnetic declination: same as *magnetic declination* which is a parameter of the *main geomagnetic field*.

geomagnetic dipole: a slow-rotating *magnetic dipole* located at about 400 km from the geometric center of the Earth with its axis of rotation presently tilted by about 11° with respect to the Earth's rotational axis. Its magnetic field is called the *eccentric dipole field*, and it closely approximates 90% of the dominant long-*wavelength* component of the *Earth magnetic field* that is known as the *main geomagnetic field*. Sometimes referred to as *virtual geomagnetic dipole*. See also *geocentric dipole field*.

geomagnetic dynamo: the physical process of converting the kinetic energy of moving masses in the inner and outer parts of the Earth's core into a magnetic energy. It forms the basis for the theory which explains the origin of the *main geomagnetic field* by the aggregated mechanical and electromagnetic effects of the convective mass movements in the liquid outer Earth's core, rotation of the solid (iron) inner core, and rotation of the Earth. All planets in the solar system appear either to have or previously have had the dynamo-driven magnetic fields. Often referred to as *geodynamo*. See also *magnetohydrodynamics*.

geomagnetic equator: the intersection of a geocentric plane perpendicular to the axis of a hypothetical *geomagnetic dipole* with the Earth's surface. Because the *main geomagnetic field* is composed of both the dipole and non-dipole magnetic fields, the geomagnetic equator is not coincident with the *magnetic equator*. See also *geomagnetic poles*.

geomagnetic excursions: 1. periods of the unusually high-amplitude *secular variation* when the north and south poles of the *geomagnetic dipole* deviate, or wander away, from the geographic poles for more than 40°–45°. **2.** large changes in the location of the ancient *geomagnetic poles* without *geomagnetic polarity reversals*, when the geomagnetic poles moved to the equatorial latitudes and, after 10^3–10^4 years, they returned to their initial location near the Earth's rotation axis (geographic poles). The origin of geomagnetic excursions is assumed to be related to the *geomagnetic dynamo* processes in the liquid outer core and solid inner core of the Earth. Also referred to as *polarity excursions*. See also *polar wandering*.

geomagnetic field: 1. *Earth magnetic field*. **2.** *main geomagnetic field*.

geomagnetic field parameters: characteristic quantities that describe seven parameters of the *main geomagnetic field* at any point on or above the Earth's surface: (1) *magnetic inclination I*; (2) *magnetic declination D*; (3) vertical component B_z, known as *vertical intensity*; (4) horizontal north component B_x; (5) horizontal east component B_y; (6) *horizontal intensity H*, also known as *total horizontal component*; and (7) *geomagnetic field strength F*, also known as *total absolute magnetic intensity (TAMI)*. Geomagnetic field parameters are obtained from the worldwide measurements by vector magnetometers stationed at geomagnetic observatories and mounted on Earth-orbiting satellites with star-imaging cameras for orienting each *vector magnetometer*. The direction of the main geomagnetic field is defined by the declination D and inclination I calculated from the measured components of the *geomagnetic vector* **B** as

$$D = \arctan(B_y/B_x), \quad I = \arctan[B_z/(B_x^2 + B_y^2)^{1/2}],$$

where B_x, B_y B_z are mutually orthogonal horizontal and vertical components of the geomagnetic vector **B** in *Cartesian coordinates* with the x-axis directed to geographic north, the y-axis directed to geographic east, and the z-axis is positive down. Horizontal intensity H defines the *magnitude* of the total horizontal component B_H, which is oriented to *magnetic north* and calculated as

$$H = B_H = (B_x^2 + B_y^2)^{1/2}.$$

Geomagnetic field strength F defines the magnitude of the geomagnetic vector **B**. It is also denoted by B_F and calculated as

$$F = |\mathbf{B}| = B_F = (B_x^2 + B_y^2 + B_z^2)^{1/2}.$$

The relationship between the geomagnetic field parameters is expressed as

$$B_H = B_F \cos \mathrm{I}, \quad B_z = B_F \sin \mathrm{I} = B_H \tan \mathrm{I},$$
$$B_x = B_H \cos \mathrm{D}, \quad B_y = B_D \sin \mathrm{D}.$$

See also *geomagnetic observatory*, *IGRF model*, and *DGRF model*.

geomagnetic field strength: 1. a *magnitude* of the *main geomagnetic field* obtained from the *IGRF model*. **2.** a *scalar* quantity which defines the magnitude of the *geomagnetic vector* **B** calculated from measurements at geomagnetic observatories and expressed as

$$|\mathbf{B}| = (B_x^2 + B_y^2 + B_z^2)^{1/2},$$

where B_x, B_y B_z are mutually orthogonal horizontal and vertical components of the geomagnetic vector **B** measured by a *vector magnetometer*. Measurements by the worldwide network of geomagnetic observatories and Earth-orbiting satellites are integrated with marine (shipboard) and airborne *magnetic survey* data and summarized in *isodynamic maps*. Referred to with the same meaning as *total absolute magnetic intensity (TAMI)*. See also *geomagnetic observatory* and *satellite magnetometry*.

geomagnetic field vector: See *geomagnetic vector*.

geomagnetic inclination: same as *magnetic inclination* which is a parameter of the *main geomagnetic field*.

geomagnetic indices: the summary of the short- and mid-term *temporal magnetic variations* obtained from monitoring the changes in the *Earth magnetic field* by geomagnetic observatories. Geomagnetic indices are calculated for different intervals of time and different components: from tens of seconds for *micropulsations* to hourly and daily ranges for *diurnals* and three orthogonal components of the *main geomagnetic field*, including the different frequency ranges at the global and local scales. See also *geomagnetic observatory* and *K-index*.

Geomagnetic Information Nodes: points of collecting the real-time magnetic data from the international network of geomagnetic observatories called *INTERMAGNET*. They are connected to the INTERMAGNET observatories by satellite, computer, and telephone networks, so the minute-averaged observations of the *Earth magnetic field* and its *vector* components can be relayed within 72 hours of their recording. See also *geomagnetic observatory* and *real-time mode*.

geomagnetic jerks: *secular variation* components with variable rates of relatively fast changes. They are identified by the second-order time derivatives of the observed *Earth magnetic field* at the timescale of several months. Also referred to as *secular accelerations*.

geomagnetic latitude: *magnetic latitude* in the context of analogy with the geographic latitude and associated variations of the *geomagnetic field parameters*.

geomagnetic north pole: the northern intersection of the *geomagnetic dipole* axis with the Earth's surface. Because the dipole axis is inclined, or tilted, by approximately 11° relative to the rotational axis of the Earth, the geomagnetic north pole is not coincident with the geographic north pole. Due to the presence of the dipole and non-dipole components in the *main geomagnetic field*, the geomagnetic north pole is not coincident with the actual, or true, *magnetic north pole*. See also *geomagnetic south pole*.

geomagnetic observatory: a ground-based observatory for the stationary high-*precision* continuous measurements of the direction (*inclination* and *declination*) and intensity (vertical, horizontal, total) of the *Earth magnetic field*, also known as the *geomagnetic field*. Measurements by geomagnetic observatories at fixed locations across the world make possible the monitoring of both the slow periodic *secular variation* and rapid irregular *magnetic disturbances* for the long-term and short-term global-scale mapping of the Earth's magnetic field variations. Instruments include (a) a *vector magnetometer* of the *fluxgate magnetometer* type for measurements of three mutually orthogonal *vector* components of the geomagnetic field, (b) a *scalar magnetometer* from the group of *quantum magnetometers* for measurements of the *magnitude* of the *geomagnetic vector* and calibration of the *baseline* for vector magnetometers, and (c) a non-magnetic theodolite for directional, or angular, measurements. Worldwide observations are summarized in the format of *isoclinic map*, *isogonic map*, and *geomagnetic indices*. Often referred to simply as *magnetic observatory*. See also *absolute magnetic direction* and *declination-inclination magnetometer*.

geomagnetic polarity: a characteristic of the *main geomagnetic field* that describes its predominantly dipolar nature represented by the positive *geomagnetic north pole* and negative *geomagnetic south pole*. See also *magnetic dipole*, *magnetic north pole*, and *magnetic south pole*.

geomagnetic polarity reversals: hypothetical changes in the orientation, or *geomagnetic polarity*, of the dipole component of the *main geomagnetic field* in the geological past, when the positions of the *geomagnetic north pole* and *geomagnetic south pole* were reversed several times. The transition time from one polarity to the opposite is estimated at approximately 10^3–10^4 years, i.e., much shorter than the periods of stable orientation known as *polarity chrons*. Reversals are assumed to be accompanied by a significant decrease in the total intensity of the main geomagnetic field with a relative increase of its non-dipole component. See also *geomagnetic excursions*, *geocentric dipole field*, and *nondipole field*.

geomagnetic poles: two intersections of the *geomagnetic dipole* axis with the Earth's surface. Locations of these intersections are known as the *geomagnetic north pole* and *geomagnetic south pole*. Because the *main geomagnetic field* is composed of the dipole and non-dipole magnetic fields, the geomagnetic poles are not coincident with the actual, or true, magnetic poles of the Earth known as the *magnetic north pole* and *magnetic south pole*. See also *polar wandering*.

Geomagnetic Reference Field (GRF): the *magnetic field* generated by the Earth's core and known as the *main geomagnetic field*. Commonly, it is referred to as the *International Geomagnetic Reference Field (IGRF)*.

geomagnetic reversals: See *geomagnetic polarity reversals*.

geomagnetic secular variation: See *secular variation*.

geomagnetic south pole: the southern intersection of the *geomagnetic dipole* axis with the Earth's surface. Because the geomagnetic dipole axis is inclined, or tilted, by approximately 11° relative to the rotational axis of the Earth, the geomagnetic south pole is not coincident with the geographic south pole. Due to the presence of the dipole and non-dipole components in the *main geomagnetic field*, the geomagnetic south pole is not coincident with the actual, or true, *magnetic south pole*. See also *geomagnetic north pole*.

geomagnetic striping: global-scale high-*amplitude* oceanic magnetic anomalies running in parallel and subparallel linear stripes within the differently oriented lithospheric plates and with an apparent symmetry relative to the ancient and present mid-ocean ridges. Their origin is attributed to the formation of new oceanic *crust* by *sea-floor spreading* and the destruction of old oceanic crust in the subduction zones. In the hypothesis of *geomagnetic polarity reversals*, these linear magnetic anomalies are magnetic signatures of the normal and reversed polarities in *magnetization* of the newly formed oceanic crust. Geomagnetic striping is the dominant long-*wavelength* component of the *Earth magnetic field* over oceans, and it is in contrast to the much less regular *pattern* of the magnetic anomalies over continents. Also referred to as *ocean floor striping* and *oceanic magnetic lineations*. See also *magnetic anomaly*.

geomagnetic variation method: a method of the *mantle* convection studies by simultaneous measurements of the lateral variations of the three components of a *geomagnetic vector* at different geomagnetic observatories.

See also *geomagnetic observatory* and *geomagnetic field parameters*.

geomagnetic vector: a three-component *vector* $\mathbf{B} = (\boldsymbol{B}_x, \boldsymbol{B}_y, \boldsymbol{B}_z)$ which describes the strength and direction of the *main geomagnetic field*. Its orientation and *magnitude* anywhere on or above the Earth's surface are defined by the *gradient* of the scalar *Earth magnetic potential*. Mutually orthogonal vertical and two horizontal vector components $\boldsymbol{B}_z$, $\boldsymbol{B}_x$, $\boldsymbol{B}_y$ are measured by vector magnetometers stationed at geomagnetic observatories and mounted on Earth-orbiting satellites with star-imaging cameras for orienting the onboard *vector magnetometer*. Measured geomagnetic vector components are used in the calculation of *geomagnetic field parameters*. Geomagnetic vector and geomagnetic field parameters can be obtained for any point on and above the Earth's surface by selecting the *three-dimensional (3D)* coordinates of an observation point (latitude, longitude, *elevation*) and the *IGRF model* appropriate for the specified day, month, and year. Also referred to as *geomagnetic field vector* and *total magnetic vector*. See also *geomagnetic observatory* and *IGRF-12*.

geomagnetic vector magnitude: a *magnitude*, or length, of the three-dimensional (3D) *geomagnetic vector* **B** that quantifies *geomagnetic field strength* and expressed as

$$|\mathbf{B}| = (B_x^2 + B_y^2 + B_z^2)^{1/2},$$

where B_z is the vertical component, also known as *vertical intensity*; B_x is the horizontal north component; and B_y is the horizontal east component of the geomagnetic vector **B**.

geomagnetic vector phase: a parameter of a spatial orientation of the three-dimensional (3D) *geomagnetic vector* defined by *magnetic inclination* at an observation point. See also *geomagnetic field parameters*.

geometric flattening: same as *polar flattening* which is a geometric parameter of the oblate, or flattened, spheroid known as the *ellipsoid* that approximates the actual Earth shape.

geometric inversion: *structural inversion* in the context of *inversion* of the observed gravity or magnetic field for the geometric parameters of a modeled exploration *target*, such as the shape of an *ore body*, base of *salt*, and depth to *basement*.

geopotential field: *Earth gravity field* or *Earth magnetic field* spatial variations of which are described by the directional derivatives of *gravitational potential* or *magnetic potential*, respectively. See also *directional derivative*.

geopotential model: a global mathematical *model* of the *gravitational potential* of the Earth, such as the *Earth Gravitational Model 2008 (EGM2008)*, components of which are represented by *spherical harmonics*. See also *potential coefficients*.

geopotential number: the difference between the calculated *gravitational potential* on a *geoid* and the gravitational potential determined at a particular *observation point* on the Earth's surface. It is used primarily in the geodetic applications.

georeferencing: 1. assigning *XY coordinates* to the corners of raster images or scanned copies of printed images, such as those of the surface geology map, structural-tectonic map, or seismic section, in order to scale them to the extent of a gravity or magnetic *survey* area for the integrated *structural interpretation*. Also referred to as *image registration*. **2.** vertical and horizontal settings (*model depth*, height of *sensor* above topography, *XY coordinates*) for construction of a *2D model*, *2.5D model*, or *2.75 model*. See also *raster image* and *coordinate registration*.

geotherm: a graph of increasing temperature as a function of depth within the Earth. A continuous line, or *isoline*, connecting the depths of the same geotherm values within the *lithosphere* is called an *isotherm*. A regional-scale negative *thermal gravity anomaly* is observed over the large areas of laterally elevated temperatures and respective isotherms, where the thermal expansion of rocks in the lower *crust* and upper *mantle* results in a lateral decrease of rock densities. See also *Curie temperature*, *density*, and *thermal density*.

geothermal gravity: *time-lapse gravity* data from high-*accuracy* measurements of *absolute gravity* and *relative gravity* for monitoring of the geothermal reservoir development. Time-lapse *microgravity* data are referenced to the *baseline* gravity data obtained from ground measurements within and beyond reservoir boundaries prior to development and integrated with high-*accuracy* surface *elevation* and groundwater level data. The magnitude of the observed gravity changes depends on the reservoir depth, volume, and temperature, and their main causes are (a) hot water withdrawal, (b) rock subsidence due to withdrawal, (c) liquid zone pressure decline with the creation of a steam zone and subsequent fluid saturation changes, (d) liquid zone *density* changes due to temperature and pressure declines, (e) *water-table effect* due to shallow groundwater level changes, (f) fluid recharge by cold groundwater invading the steam and liquid zones from surrounding areas, and (g) pore compaction due to rock subsidence. Absolute gravity observations are made primarily in the central production area to measure directly the gravity changes caused by withdrawal of hot fluids and rock subsidence and to provide the reference points for more widely expanded relative gravity measurements for the delineation of the spatial distribution of the short- and long-term fluid mass changes. The obtained *hybrid gravity* data are used as *constraints* in refining a reservoir development *model* in accordance with the well-by-well production data (Allis and Hunt, 1986; Sugihara and Ishido, 2008). See also *aquifer gravity*, *CO_2 sequestration gravity*, *volcano gravity*, and *4D gravity monitoring*.

Gibbs errors: *Gibbs phenomenon* in the context of the accumulated errors in the forward and inverse parts of a *Fourier transform* of the data from one *domain* to another: from *space domain* to *spectral domain* and from the spectral domain back to the original space domain.

Gibbs phenomenon: high-frequency oscillations at data margins and around data spikes, discontinuities, and other abrupt changes in the data values that represent the approximation errors accumulated during the forward and inverse numerical transformations of the data from one *domain* to another. In gridded gravity and magnetic data, its effects are associated with the use of the *Fourier transform* and presented primarily by various forms of *edge effects* involving several rows and columns along the *grid* margins. Gibbs phenomenon also can be observed around the *cutoff frequency* of spectral-domain *FFT filters*. Distortions of the data are minimized by (a) *grid expansion* and data *padding* to provide a smooth transition of the data values to zero, (b) avoiding the use of narrow band-pass filters and other spectral-domain filters with a steep transition between the passed and rejected *wavenumber* ranges, and (c) removal of the regional long-*wavelength* component before transforming the gridded data into the *Fourier domain*. Also referred to as *Gibbs errors* and *Gibbs ringing*. See also *band-pass filter*, *gridding*, and *spectral filtering*.

Gibbs ringing: *Gibbs phenomenon* in the context of the high-frequency oscillations, or ringing, at the *grid* margins that represent the accumulated errors of the forward and inverse parts of a *Fourier transform*.

glacial isostatic adjustment: deformation of the Earth's surface (commonly, slow uplifting) that results from the removal of a glacial load due to its melting. Contribution of the glacial isostatic adjustment to the regional component of high-*accuracy* repeated gravity measurements, such as *time-lapse gravity*, is compensated by procedures analogous to calculation and application of the *Earth tide correction*. See also *isostatic compensation* and *isostasy*.

global bathymetric correction: a correction applied to the global-scale crustal *gravity model* data to compensate for the reduced *gravitational attraction* of a seawater mass (mean *density* 1.03 g · cm^{-3}) between *reference ellipsoid* and sea bottom due to its negative *density contrast* of −1.64 g · cm^{-3} relative to the reference crustal density of 2.67 g · cm^{-3}. It is a long-*wavelength* equivalent of a *simple Bouguer correction* in the oceans and seas with a *replacement density* of 1.64 g · cm^{-3} (Tenzer et al., 2009). See also *crust* and *global topographic correction*.

global crustal correction: a four-step correction applied to the global-scale crustal *gravity model* data. Its calculation involves (1) *global topographic correction*, (2) *global bathymetric correction*, (3) compensation for density contrasts of the ice, sedimentary and crystalline layers in the global *CRUST 1.0* model relative to the reference crustal density 2.67 g · cm^{-3}, and (4) compensation for the negative *density contrast* of the entire *crust* (approximated by rock masses with constant density of 2.67 g · cm^{-3} between the *reference ellipsoid* and *Moho boundary*) relative to the average upper *mantle* density which varies between 3.19 g · cm^{-3} and 3.40 g · cm^{-3}. Long-*wavelength* components of the anomalous gravity field remaining after application of this correction are used in *structural inversion* for the Moho boundary depth (Tenzer et al., 2009).

Global Gravitational Model: a mathematical *model* of *gravitational potential* of the attracting planetary body of mass, such as the Earth. See also *Earth Gravitational Model 2008 (EGM08)*.

global gravity datum: See *International Gravity Standardization Network 1971 (IGSN71)*.

Global Navigation Satellite System (GNSS): the global navigation and *positioning* system based on the use of Earth-orbiting satellites for *real time* calculations of (a) *three-dimensional (3D)* geocentric coordinates (latitude, longitude, and height above *reference ellipsoid*) at any point on or above the Earth's surface and (b) 3D velocity of a moving *observation platform*. At present, there are four operational global navigation satellite systems: U.S. Department of Defense *Global Positioning System (GPS)*, Chinese BEIDOU, Russian GLONASS, and European GALILEO. There are also the Japanese and Indian satellites launched for regional navigation projects.

Global Positioning System (GPS): the satellite-based navigation and *positioning* system which provides in *real-time mode* the *three-dimensional (3D)* coordinates (latitude, longitude, height above *reference ellipsoid*) of observation points and 3D velocity of a moving *observation platform*, such as a ship or aircraft. Position coordinates, velocity, and standard time increments are calculated by the GPS receiver based on measurements of differences between the arrival times and phases of radio signals transmitted by, at least, four satellites. The standard reference ellipsoid for GPS is the *WGS84 ellipsoid*. At present, the basic satellite network, or constellation, consists of 31 satellites orbiting the Earth at altitudes of approximately 20,000 km (12,500 miles) in six equally spaced orbiting planes ~60° apart with four satellites per orbiting plane, wherein five to eight satellites are continuously observable from any point on the Earth. The latest *gravimeter* and *magnetometer* systems are equipped with built-in *GNSS* receivers and software modules for the automatic correction of differences in the determined height above the reference ellipsoid and *elevation* above sea level. See also *altitude*, *GPS accuracy*, and *differential GPS (DGPS)*.

global spectrum: amplitudes and phases of all sinusoidal components, or *harmonics*, into which an anomalous gravity or magnetic field is decomposed by a standard *Fourier transform* as opposed to the *local spectra* in a *window Fourier transform (WFT)*. See also *amplitude* and *phase*.

global topographic correction: a correction applied to global-scale crustal *gravity model* data to compensate for the density contrasts of topographic rock masses (relative to the reference crustal *density* of 2.67 g · cm^{-3}) above *reference ellipsoid*. It is a long-*wavelength* equivalent of a *simple Bouguer correction* on the continents (Tenzer et al., 2009). See also *crust*, *density contrast*, *global bathymetric correction*, and *global crustal correction*.

GLONASS: the Russian *Global Navigation Satellite System (GNSS)*.

glyph: a parameterized icon for the visualization of relationships between different attributes of an anomalous gravity and magnetic field by the icon shape, color, texture, etc. See also *attribute*, *barycentric image*, and *RGB composite image*.

GNSS: See *Global Navigation Satellite System (GNSS)*.

GOCE gravity: the global-scale representation of the long-*wavelength* components of the *Earth gravity field* reconstructed from the *full tensor gradient (FTG)* measurements by *GOCE gravity gradiometer* and *satellite-to-satellite tracking* by *GNSS* satellites. The estimated *lateral resolution* of anomalies in the integrated GOCE gravity field is 80–100 km at the Earth's surface with the highest achievable *accuracy* of 1–2 *milligal (mGal)*.

GOCE gravity gradiometer: a spaceborne *gravity gradiometer* system for measurements of the *full tensor gradient (FTG)* components of the *Earth gravity field* during the *GOCE mission*. It has six identical *electrostatic accelerometer* units arranged in three mutually orthogonal gradiometer arms with the same *baseline* of 50 cm. The gradiometer reference frame co-rotates with the host satellite, so that the x-axis is aligned along the satellite orbital trajectory, the y-axis is perpendicular to the orbital plane, and the z-axis is pointed to the Earth's center. Extra-high *sensitivity* of $2 \cdot 10^{-12}$ m · s^{-2} is achieved by the *electrostatic levitation* design. Basic observables are control voltages applied to eight electrode pairs for each of the six accelerometer units to keep the *proof mass* levitating at the center of an ultra-stable carbon cage. Forty-eight voltage values, corrected for measurement effects, are transformed into 36 *acceleration* values from which nine *common-mode acceleration* and nine *differential-mode acceleration* datasets are formed by calculating the sum and difference of accelerations measured by pairs of the accelerometer units in three mutually orthogonal directions. Acquired datasets are used to (a) separate *gravitational acceleration* from components of *angular acceleration* and *centrifugal acceleration* and (b) make real-time adjustments to the ion propulsion engine to keep the satellite in a free-fall motion. The estimated *accuracy* of measuring the T_{xx}, T_{yy}, T_{zz}, and T_{xz} components is 0.03–0.05 *Eötvös* (*E*) per *hertz (Hz)* within the measurement *bandwidth* of 0.005–0.1 Hz (Fehringer et al., 2008). See also *gradiometer arm* and *real-time mode*.

GOCE mission: the Gravity field and steady-state Ocean Circulation Explorer (GOCE) satellite mission of the European Space Agency (ESA), launched in March 2009 and completed in November 2013, for mapping the global gravity gradients of the Earth at an *altitude* of approximately 250 km with the use of an *accelerometer*-based *gravity gradiometer* system. See also *CHAMP mission*, *GRACE mission*, *GRAIL mission*, *Swarm mission*, *GOCE gravity*, and *GOCE gravity gradiometer*.

Goussev filter: a high-*sensitivity* and high-*resolution* line- or *grid*-based space-domain *filter* for mapping of *gradient* anomalies by calculating the difference between the total and horizontal gradients of the *RTP magnetic field* or *Bouguer gravity*. This difference has its maxima over sources approximated by the vertical *thin sheet* and *horizontal cylinder* source geometries above which the *total gradient (TG)* has its maximum values and the *horizontal gradient (HG)* has its minimum values. Both the maximum-versus-minimum disparity in *amplitude response* of the TG and HG and suppression of *noise* by the gradient differencing amplify the sensitivity of this filter and, thereby, allow the detection of very small gradient anomalies, the amplitudes of which can be 100–150 times less than the amplitudes of the regular *total field* gradients (Goussev et al., 2003). Over structural discontinuities, such as a magnetized normal *fault*, low-offset *strike-slip fault*, narrow *salt* dissolution or fracture zone, buried *astrobleme* boundary, vertical or high-angle tabular *igneous intrusion*, fault-related alluvial *placer* deposit, etc., it responds with a single-*peak* anomaly above the location of a positive or negative lateral *density contrast* or *magnetization contrast* created by a particular discontinuity. The highest *lateral resolution* and improved visualization of anomalies are obtained in the application to a *residual field* and by *cascading* with the *Jacobsen filter*, *matched filter*, medium-order *band-pass filter*, and other *data enhancement* techniques. Over the *contact*-approximated boundaries, such as a high-*offset* fault, the *filter response* is the anomalous triplet of a minimum bracketed by two artificial maxima, wherein the minimum value is immediately above the vertical contact. In the RTP magnetic and Bouguer gravity fields over a deep sedimentary or *basement* grabens and horsts, the filter responds with its maxima above the long axes of these structures, as they are both consistent with the same vertical thin sheet or horizontal cylinder *source geometry* approximation. See also *graben*, *horst*, *magnetized fault*, *tabular body*, and *filter order*.

GPS: *Global Positioning System (GPS)*. Historically, it is used as a general reference to any *Global Navigation Satellite System (GNSS)*.

GPS accuracy: *accuracy* in determining the *three-dimensional (3D)* geocentric coordinates (latitude, longitude, and height above *reference ellipsoid*) of ground, marine, and airborne observation points by the satellite-based *Global Positioning System (GPS)*. The GPS accuracy depends on the GPS equipment type, configuration of satellites (the more available satellites, the higher the *positioning* accuracy) and varies from tens of meters to several millimeters. For regular applications, the achievable accuracy is 1.5–3.0 m for the horizontal positioning and 4.0–7.0 m for the vertical positioning. In the *differential GPS (DGPS)* mode with a stationary reference GPS receiver (*GPS base station*), the horizontal accuracy is better than 1 centimeter with vertical accuracy of a few centimeters. Recent commercial upgrades are capable of providing a sub-centimeter order of the positioning accuracy. The GPS accuracy is directly related to the *gravity survey* accuracy: a vertical GPS positioning error of 10 cm results in a measured gravity value error of 0.03 *milligal (mGal)*, wherein the horizontal positioning error has the largest effect along the north–south direction with the latitude error of 100 m resulting in the measured gravity value error of 0.05 mGal.

GPS base station: a stationary multi-channel *GPS* receiver, or *reference station*, for which the three-dimensional (3D) coordinates (latitude, longitude, and *elevation*) are known with a high *accuracy*. Its continuous records are used to correct differentially the GPS data acquired by a moving, or roving, GPS receiver on the ground or in the air and provide the most accurate coordinates of observation points along *survey* lines. The distance between a GPS base station and a roving GPS receiver is called the *baseline*, and it is limited to 20–25 km for ground gravity and magnetic surveys and to 80–100 km for the GPS receiver onboard a survey aircraft. See also *differential GPS (DGPS)*.

GPS height: height, or *elevation*, above *reference ellipsoid* determined with the use of the satellite-based *Global Positioning System (GPS)*. Sometimes referred to as *ellipsoidal height*. See also *orthometric height*.

graben: an intrasedimentary or upper *basement* structure created under the extensional tectonic regime. A graben is an elongated and relatively narrow sedimentary depression or down-dropped intrasedimentary or basement block bounded on both sides by high-angle normal faults which generate *gradient* anomalies. In *Bouguer gravity*, a graben commonly has the *gravity signature* of an elongated *gravity low* or negative *residual gravity* anomaly, except for a high-*density* down-dropped basement block. In the *RTP magnetic field*, the graben *signature* is less dependent on its *structural low* parameters and more affected by the lithological composition of the down-dropped block. Accordingly, it can have the *magnetic signature* of an elongated residual *magnetic high* generated by *volcanics* and other igneous rocks intruded into a graben during its creation by the divergent movement of rock formations within a *strike-slip fault* zone. Under all structural environments, the anomalous gravity and magnetic signatures of a graben are enhanced by calculation of the *residual field*. Amplitudes of the gravity and magnetic anomalies depend on the graben depth and lateral *density* and *magnetization* contrast with adjacent rocks. See also *contrast*, *igneous intrusion*, *residual magnetics*, and *horst*.

GRACE gravity: the global-scale representation of the long-*wavelength* components of the *Earth gravity field* reconstructed from *satellite-to-satellite tracking* with the use of the *GPS* satellites and microwave ranging system between two *GRACE mission* satellites. The estimated *half-wavelength resolution* of the gravity anomalies recovered from the 30-day averaged tracking measurements is 170–200 km, which is adequate for studies of the large-scale basins and upper *mantle* (Braitenberg and Ebbing, 2009). See also *basin* and *mantle gravity anomaly*.

GRACE mission: the Gravity Recovery and Climate Experiment (GRACE) satellite mission launched in 2002 and completed in October 2017. Two satellites were orbiting the Earth in tandem 220 km apart at the initial *altitude* of 485 km to observe the time-variable components of the *Earth gravity field*. Its *temporal gravity variations* are determined by high-*accuracy* measurements of the changes in distance between two free-falling satellites using the microwave ranging systems with *precision* of 0.002 mm. High *sensitivity* of the satellite separation distance to variations of the Earth's gravity field allows the recovery of the gravity anomalies to the *spherical harmonics* degree of 120, which is equivalent to *full-wavelength resolution* of 340 km. Estimates of the temporal gravity variations are made by accumulating the observations on a monthly basis. The GRACE mission has provided *time-lapse gravity* data for a wide range of planetary studies, including regional changes of atmospheric mass, melting ice sheets, ocean-bottom pressure, groundwater depletion in aquifer basins, and upper *mantle* mass variations. It also enabled the measurement of temporal variations of the surface of a *geoid* with *accuracy* up to 2–3 mm at the *spatial resolution* of 600 km (Wahr et al., 2004; Hinze et al., 2013). See also *CHAMP mission*, *GOCE mission*, *GRAIL mission*, *Swarm mission*, and *satellite-to-satellite tracking*.

gradient: **1.** a *scalar* quantity representing the spatial rate of change of an *observed field*, i.e., rate of change per unit distance in the horizontal or vertical direction, defined by the first-order *derivative* of this field in the direction of calculation. **2.** a *vector* quantity representing a vector which is orthogonal to the lines of the equal field values, or *contour* lines, and pointing in the direction of a maximum change of the *vector field*. Horizontal and vertical components of the three-dimensional (3D) *gradient vector* are *horizontal gradient* and *vertical gradient*, respectively. **3.** a scalar quantity measured by a *gradiometer* or calculated as a difference between two *total field* measurements divided by their separation distance and, sometimes, referred to as *magnitude gradient*. **4.** a quantitative measure of *curvature* of the *equipotential surface* for gravity and magnetic fields that is closely approximated by the second-order derivative of the gravitational or magnetic *potential*. The gradient of the field F is commonly denoted by ∇F, where ∇ is a mathematical symbol of the gradient operator *del*. For the 3D case, the gradient is expressed as

$$\nabla F = \mathbf{i}\partial F/\partial x + \mathbf{j}\partial F/\partial y + \mathbf{k}\partial F/\partial z,$$

where $\partial F/\partial x$, $\partial F/\partial y$, $\partial F/\partial z$ are first-order derivatives of the field F along the mutually orthogonal *x*-, *y*-, *z*-directions, respectively, in *Cartesian coordinates* and **i**, **j**, **k** are unit vectors parallel to these directions. Sometimes, the gradient is called *slope* or *dip*, because its value quantifies a graphical slope of the *profile* and is equivalent to a structural dip of the 3D surface. Gradients highlight the short- and mid-*wavelength* components of a gravity or magnetic field and suppress the long-wavelength regional components. For gravity fields, the gradients are measured in units of the inverse second squared (s^{-2}) and inverse nanosecond squared (ns^{-2}), where $1\ ns^{-2} = 10^{-9}\ s^{-2}$, or in units of *milligal (mGal)* per kilometer ($mGal \cdot km^{-1}$) and *Eötvös (E)*, where $1E = 0.1\ mGal \cdot km^{-1} = 10^{-4}\ mGal \cdot m^{-1} = 10^{-9}\ s^{-2} = 1\ ns^{-2}$. For magnetic fields, the common measurement units are *nanotesla (nT)* per kilometer ($nT \cdot km^{-1}$) and nanotesla per meter ($nT \cdot m^{-1}$). See also *gravity gradient*, *magnetic gradient*, *longitudinal gradient*, *transverse gradient*, *cross-line gradient*, *in-line gradient*, *full tensor gradient (FTG)*, and *magnitude gradient*.

gradient array: an assemblage of synchronized accelerometers or magnetometers mounted on the same platform and separated horizontally and/or vertically for measurements of gravity or magnetic gradients and *gradient tensor* components. Referred to with the same meaning as *gradiometer*. See also *absolute gravity gradiometer*, *rotating-disk gravity gradiometer*, *magnetic gradiometer*, and *accelerometer*.

gradient-depth method: **1.** a graphical, or manual, *profile*-based method of estimating the depth of gravity sources approximated by a *point source* or *sphere*. It is based on an empirical formula relating the source depth and horizontal distance Δx between the maximum values of a *gravity anomaly* and its *horizontal gradient* as

$$D_{SC} \approx 2\Delta x,$$

where D_{sc} is depth to center of the sphere. Any *3D source* can be approximated by point source geometry, when its anomaly is measured at sufficient distance (in practice, 8–10 times the largest *dimension* of the 3D source). **2.** an automated method of estimating the depth to a *magnetic contact* by the two-dimensional (2D) *Werner deconvolution* applied to the *horizontal derivative* of line-oriented *TMI field* or *RTP magnetic field* data. See also *Euler deconvolution.*

gradient-dip estimation: a graphical, or manual, estimation of the *dip* of a magnetic or gravity *contact.* It is based on the spatial relationship between the maximum values of the *total gradient (TG)* and *horizontal gradient (HG),* wherein the TG maxima tend to be located above the top of the contact and HG maxima are displaced downdip. The formula is

$$\Delta x = D \text{ cotan } \beta = D/\tan \beta,$$

where Δx is horizontal distance between the maxima of the total and horizontal gradients, D is depth to the top of the contact, and β is dip angle of the contact (Grauch and Cordell, 1987). Locations of the *gradient* maxima can be determined and mapped by automated methods, such as the *Blakely-Simpson method.*

gradient-enhanced gridding: a *gridding* technique where the interpolation of the airborne *gradient* data between *survey* flight lines is based on the measured *lateral gradient* rather than on a purely mathematical interpolation of the *observed field* values. Its application improves the *lateral resolution* and mapping of anomalies, especially those with the wavelengths shorter than a distance between the flight lines. See also *Hardwick method* and *Nelson method.*

gradient-enhanced minimum-curvature: *minimum-curvature gridding* based on the use of the measured *longitudinal gradient* and *transverse gradient* in a truncated *Taylor series* for interpolation of the *observed field* values between *survey* lines. Only the first- and second-order Taylor terms are implemented, wherein the first-order terms are the measured longitudinal and transverse gradients and the second-order terms are calculated by the *finite-difference method.* The method improves a continuity of high-*amplitude* short-*wavelength* anomalies and prevents *grid aliasing* (O'Connell et al., 2005). See also *boudinage artifact.*

gradient filtering: calculation of the gradients, derivatives, and their combinations and transformations in order to detect the *gradient* anomalies generated by elements of subsurface geological structures. See also *analytic signal amplitude (ASA), ASTA filter, directional analytic signal, Goussev filter, monogenic signal, theta angle, tilt angle (TA), tilt ASA, tilt THD, directional tilt, hyperbolic tilt,* and *tilt derivative.*

gradient survey: ground, marine (shipboard), and airborne measurements of the horizontal and vertical gradients or *full tensor gradient (FTG)* components of the Earth's gravity or magnetic field over areas of *exploration* or research interest. Subtraction of the simultaneous readings of two sensors in a *gradiometer arm* from one another effectively suppresses the common-mode *dynamic noise* in the *gravity gradient* measurements and, essentially, eliminates *diurnals* in the *magnetic gradient* measurements. Suppression of *noise* and long-wavelength *regional field* components by *gradient* measurements provides a much higher *lateral resolution* of the *target* anomalies, such as those generated by relatively shallow and small sources, compared to the lateral resolution achievable with *total field* measurements, especially in geologically complex areas. See also *common-mode rejection, interference of anomalies, reading, sensor,* and *UXO detection.*

gradient tensor: a nine-element symmetric *tensor* the elements of which represent the spatial rates of change, or gradients, of the three *vector* components of gravity or magnetic field F. In *Cartesian coordinates,* these vector components are denoted by F_x, F_y, F_z, each varying in three mutually orthogonal x-, y-, z-directions, so the gradient tensor elements can be displayed in the *format* of a 3×3 symmetric *matrix* as

$$\begin{matrix} F_{xx} & F_{xy} & F_{xz} \\ F_{yx} & F_{yy} & F_{yz} \\ F_{zx} & F_{zy} & F_{zz} \end{matrix}.$$

Three pairs of the gradient tensor elements are symmetric: $F_{xy} = F_{yx}$, $F_{xz} = F_{zx}$, and $F_{zy} = F_{yz}$. There are five independent elements: F_{xy}, F_{xz}, F_{yz}, and any two of the three diagonal elements F_{xx}, F_{yy}, F_{zz}. Gradient tensor elements have a different *sensitivity* to the property (*density* or *magnetization* contrast) and structural variations, and they can be used as additional *constraints* in a single element *inversion* (primarily, F_{zz}) or *joint inversion* for the density or magnetization *contrast* and shapes of source bodies. For example, F_{xx} can constrain the east–west edges of a three-dimensional (3D) *source body*, and F_{yy} can constrain the north–south edges. Both *curvature gradients* F_{xx} and F_{yy} can be used to estimate source body thickness and overall shape, whereas F_{zz} constrains its density or *susceptibility* contrast. Individual gradient tensor elements are combined into *invariants* and their modifications for the detailed *structural interpretation.* Referred to with the same meaning as *full tensor gradient (FTG).* See also *contrast, property inversion, structural inversion, FTG invariants,* and *tidal tensor.*

gradient-tensor gridding: *gradient-enhanced gridding* that uses the measured components of the *full gradient tensor (FTG).*

gradient tensor operator: a mathematical operator relating the *gradient tensor* T to the *gravitational potential* U or *magnetic potential* $\mathcal{V}$ and expressed as

$$\mathrm{T} = \nabla\nabla^{\mathrm{T}}\mathrm{U} \quad \text{or} \quad \mathrm{T} = \nabla\nabla^{\mathrm{T}}\mathcal{V},$$

where ∇ is *gradient* operator *del.*

gradient tensor trace: the sum of three diagonal elements T_{xx}, T_{yy}, T_{zz} of the *gradient tensor* that represent three mutually orthogonal components of the *full tensor* *gradient (FTG)* of the Earth's gravity or magnetic field and related to each other via the *Laplace equation* as

$$T_{xx} + T_{yy} + T_{zz} = 0,$$

i.e., the gradient tensor is traceless.

gradient terrain correction: *terrain correction* applied to airborne *gravity gradient* data to compensate for *terrain*-related gravity-gradient effects. Dransfield and Zeng (2009) describe its calculation and application procedures for data measured by an *airborne gravity gradiometer (AGG)*, including: (1) construction of *a Digital Terrain Model (DTM)* of the gradient *survey* and surrounding areas; (2) DTM discretization into square vertical prisms with widths that are zone-dependent and heights that are averages of the terrain *elevation* values within each *prism*; (3) assignment of a uniform *density* for all prisms; (4) calculation of the gravity-gradient effect of the DTM at each observation point to obtain the correction values; (5) low-pass filtering of the calculated corrections to match *acquisition filtering*; and (6) application of the low-pass filtered corrections to the measured gradient values. Because the gradients are influenced, primarily, by the local topography, the DTM zone boundaries expand laterally to 1000 m, 2000 m, and 6000 m from each observation point, where the prism widths are equal to one DTM *grid cell*, four DTM grid cells, and 16 DTM grid cells, respectively. Another approach (Li et al., 2010a) implements the space-domain algorithm based on an *adaptive quadtree discretization* of the DTM and calculation of the gradient terrain corrections around each observation point to the distances at which the combined gravity-gradient effect of all individual prisms becomes less than the estimated level of *noise* in the acquired gradient data. See also *acquisition*, *Agerlin principle*, *fourth difference*, *noise envelope*, and *mesh discretization*.

gradient tolerance: an instrument characteristic that quantifies the maximum *gradient* value of the *Earth magnetic field* at which the *magnetometer* is able to make coherent (mutually consistent) measurements. Gradient tolerance describes the ability of a magnetometer to obtain reliable measurements under extreme variations of *magnetic field strength* and defines the operational limit for each magnetometer type, not necessarily within its specifications regarding *sensitivity* and *accuracy*. See also *tracking speed*.

gradient vector: 1. a three-component *vector* which defines the spatial orientation and *magnitude* of the highest rate of change of a gravity or magnetic field. At each *grid* point, its mutually orthogonal components are represented by the magnitudes of one vertical and two horizontal derivatives of the *anomalous field*. The mathematical expression of a length, or *magnitude*, of a *three-dimensional (3D)* gradient vector is identical to those of the *total gradient (TG)* and 3D *analytic signal amplitude (ASA)* and defined as

$$|\nabla \mathbf{F}| = \mathrm{TG} = |\mathrm{ASA}| = [(\partial F/\partial x)^2 + (\partial F/\partial y)^2 + (\partial F/\partial z)^2]^{1/2},$$

where ∇ is a mathematical symbol of the gradient operator *del*, and $\partial F/\partial x$, $\partial F/\partial y$, $\partial F/\partial z$ are first-order derivatives of the field F along the mutually orthogonal *x*-, *y*-, *z*-directions, respectively, in *Cartesian coordinates*. Gradient vector *phase*, or vertical angle relative to the XY plane, is defined by the *tilt angle (TA)*. On the *contour grid* of an anomalous field, the gradient vector points to a local positive *anomaly*, and it is everywhere orthogonal to the contour lines. **2.** a three-component vector for which the direction and length quantify the orientation and strength, or intensity, of a gravity or magnetic field at any point in space by a spatial rate of change of the *scalar* gravity or magnetic *potential*, respectively, i.e., the gravity or magnetic *vector field* can be described by the *field* of the gradient vector of the corresponding potential. Also referred to as *3D gradient vector*. See also *derivative*, *gradient*, *gravitational potential*, *magnetic potential*, and *2D gradient vector*.

gradient vector phase: a *gradient vector* angle relative to the XY plane defined by a numerical value of *tilt angle (TA)*.

gradiometer: an instrument for measuring the horizontal and vertical gradients of the planetary gravity or magnetic field. Except for a *torsion-balance gravity gradiometer*, a gradiometer for the ground measurements is composed of two or three gravimeters or magnetometers separated horizontally and/or vertically. Marine (shipboard), airborne, and satellite gravity and magnetic gradient measurements are made by the assemblage, or *gradient array*, of the paired and synchronized accelerometers and magnetometers, respectively. The difference between the gravity or magnetic *sensor* readings divided by their separation distance, or *baseline*, defines the value of the *gradient* of a gravity or magnetic *field* along the separation direction. Baseline value commonly varies from several decimeters to several meters, and it depends on the gradiometer design and objectives of the *gradient survey*. See also *accelerometer*, *full tensor gradiometer*, *gravity gradiometer*, *magnetic gradiometer*, *GOCE gravity gradiometer*, *rotating-disk gravity gradiometer*, and *rotating-sensor magnetic gradiometer*.

gradiometer arm: a pair of matched gravity or magnetic sensors, such as the identical and synchronized accelerometers or magnetometers, that are mounted on the same frame and separated by a specific distance (which is called *baseline*) to measure the *gradient* of a gravity or magnetic *field* in the separation direction. See also *accelerometer* and *sensor*.

gradiometry: methods and instruments for measuring the *gradient* and *full tensor gradient (FTG)* components of the *Earth gravity field* and the *Earth magnetic field*, i.e., components of the three-dimensional (3D) *gradient vector*. Measured values represent the first-order derivatives of the gravity or magnetic field and second-order derivatives of *gravitational potential* or *magnetic potential*, respectively. In addition to the advantages of suppressing *diurnals* in magnetic surveys and minimizing *dynamic noise* during airborne and marine (shipboard) gravity surveys, the gradient and FTG measurements are more sensitive to the *density* and *magnetization* contrasts associated with the boundaries of local geological structures than measurements of the *total field* due to the higher *lateral resolution* of the gradient anomalies and suppression of the regional long-*wavelength* field components. See also *contrast*, *derivative*, *gravity gradiometry*, *magnetic gradiometry*, *interference of anomalies*, and *total field gradiometry*.

GRAIL mission: the Gravity Recovery and Interior Laboratory (GRAIL) satellite mission of two satellites orbiting the Moon in tandem to map its *gravity field*. Its mean *altitude* was 55 km during the primary *acquisition* phase (March–May 2012) and 23 km during its extension (August–December 2012). Satellite separation ranged from 70 to 200 km. Gravity field variations were determined by high-*accuracy* measurements of changes in the distance between two free-falling satellites using the Lunar Gravity Ranging System with a *precision* of 0.002 mm. Combined data from *satellite-to-satellite tracking* and radio tracking of the satellite lunar orbits by the Deep Space Network on Earth allowed to model *gravitational potential* of the Moon to the *lateral harmonics* degree and order 900 with *lateral resolution* of the lunar surface blocks up to 6 km (Zuber et al., 2012; Lemoine et al., 2014). See also *CHAMP mission*, *GOCE mission*, *GRACE mission*, and *Swarm mission*.

grain: an arrangement, or *pattern*, of the amplitudes, spatial frequencies, shapes, and dominant alignments of anomalies that is specific for a particular region or local area, thereby making it distinctly different from the surrounding and adjacent areas. Grain characteristics are used in the qualitative analysis and comparative evaluation of anomalous gravity and magnetic fields for (a) mapping of structural boundaries between adjacent areas and regional tectonic provinces and (b) *skeletonization*- and *streerable filter*-based analysis of patterns in the distribution of faults and fracture zones in the upper sedimentary section for the oil-and-gas exploration of shale formations. Also referred to as *structural grain*, *fabric*, and *texture*. See also *pattern*, *spatial frequency*, and *magnetic terrane*.

graphical source depth estimation: 1. estimation of an isolated *source body* depth based on the use of *depth rules* and measurements of the horizontal distances between the manually selected characteristic points on a *profile* or *contour grid* of the magnetic and gravity anomalies. Also referred to as *manual source depth estimation*. **2.** *multiridge geometric method*. See also *half-width method*, *slope methods*, and *automated source depth estimation*.

gravimeter: an instrument for measuring the planetary *gravitational field*, such as the *Earth gravity field*. There are two basic types of gravimeters: (1) *absolute gravimeter,* which measures a full numerical value of the gravitational field called *absolute gravity*, and (2) *relative gravimeter,* which measures changes in the gravitational field relative to a *base station* or between two observation points, i.e., *relative gravity*. Among various designs, the most often used are *free-fall gravimeter* and *pendulum gravimeter* for absolute gravity measurements and *spring gravimeter*, *vibrating-string gravimeter*, and *accelerometer*-based *airborne gravimeter (AG)* for relative gravity measurements. The majority of gravimeters for sea floor, marine (shipboard), ground, and borehole gravity surveys are adaptations of the spring gravimeter design. Sometimes referred to as *gravity meter*. See also *absolute atomic gravimeter* and *superconducting gravimeter*.

gravimeter drift: a gradual change in measured *relative gravity* values caused by a shift of the zero setting of a *relative gravimeter*. As a measurement-related effect, it can be considered a time-dependent and variable change of the gravimeter *scale factor*. Drift is an inherent property of all *spring gravimeter* types and spring-based accelerometers, and it is related primarily to the stress relaxation of a quartz or metal spring, its thermal expansion and elastic aging (fatigue), shock or vibration during transportation, heat stress, etc. For correction purposes, *drift* is approximated by a linear function of time, and it is determined by *reoccupations* in ground surveys and by repeated measurements at the primary or secondary *base station* in marine (shipboard) and airborne gravity surveys. Depending on the gravimeter design, the drift values range from a few *microgal* (μ*Gal)* per year in a *superconducting gravimeter* to 100–200 microgals per day in quartz-spring gravimeters. Gravimeters with a *drift rate* exceeding 0.05 *milligal (mGal)* per hour require maintenance and scale factor adjustments. Drift-related measurement errors are compensated by the application of a *drift correction*. See also *accelerometer*, *base-check method*, *cross-check method*, *looping method*, and *residual drift*.

gravimeter drift compensation: minimization of *drift*-related gravity measurement errors by the application of a *drift correction*. Because *gravimeter drift* is controlled by *reoccupations* at coordinate points with known gravity values and repeat stations that comprise only approximately 10% of all observation points in a regular *ground gravity* survey area, the drift correction estimate is commonly based on approximation of the *drift rate* between the control points by a linear function of time. Total accumulated drift is determined by calculation of the difference between repeated measurements. The calculated difference value is distributed proportionally, or prorated, over readings at each observation point in ground, marine (shipboard), and airborne gravity surveys. See also *open-end loop*, *reading*, *repeat station*, and *automatic drift compensation*.

gravimeter gradiometry: stationary ground measurements of the *vertical gradient* or *horizontal gradient* of the *Earth gravity field* with the use of two conventional, or regular, gravimeters separated vertically or horizontally. Measurements are conducted with the objective of utilizing the higher *lateral resolution* and *sensitivity* of the gravity gradients than those of the *vertical gravity* measurements by a single *gravimeter*. See also *magnetometer gradiometry*.

gravimeter precision: an instrument characteristic that defines a range of gravimeter *reading* repeatability, i.e., the scattering of measured values at the same observation points regardless of *gravimeter drift* and external non-instrumental conditions. Modern *relative gravimeter* types have a reading *precision* of approximately 0.001 *milligal (mGal)*.

gravimetry: methods and instruments to acquire, process, and interpret measurements of the planetary *gravitational field*, such as the *Earth gravity field*, its gradients, and *full tensor gradient (FTG)* and *vector* components. See also *acquisition*, *data processing*, *interpretation*, *scalar gravimetry*, and *vector gravimetry*.

gravitation: 1. *gravitational attraction*. **2.** *gravity* in exploration and research applications. **3.** *Earth gravity potential* in geodetic applications, where it is defined by the sum of the *self-gravitational potential* due to the stationary, or non-rotating, Earth mass and *centrifugal potential* due to the Earth's rotation about its axis. See also *Newton law of universal gravitation*.

gravitational acceleration: 1. *acceleration* of a body of *mass* due to the *gravitational attraction* of another, much larger body of mass, such as the Earth or any other planet. **2.** a *vector* quantity $\mathbf{g}$ representing the value and direction of acceleration of a free-falling body of mass m_0 under the action of *gravitational force* **F** generated by the attracting mass M. According to the *Newton law of motion*, gravitational acceleration is defined by the ratio of the gravitational force **F** to the body mass m_0 and expressed as

$$\mathbf{g} = \mathbf{F}/m_0 = -\mathrm{G}M\hat{\mathbf{r}}/r^2,$$

where **F** is the gravitational force vector whose *magnitude* is defined by the *Newton law of gravitational attraction*, G is a *universal gravitational constant*, $\hat{\mathbf{r}}$ is the unit vector directed from mass M to mass m_0, and r is distance between the body m_0 and attracting mass M. A negative sign indicates that the force of attraction is directed opposite to the unit vector $\hat{\mathbf{r}}$. The formula shows that, in the absence of other forces, the gravitational acceleration of a free-falling body is independent of its mass and directly proportional to the mass of the attracting body. **3.** a vector quantity defined by the *gradient vector* of *gravitational potential* U and expressed as

$$\mathbf{g} = \nabla \mathrm{U},$$

where ∇ is the vector gradient operator *del*. The vector *field* of the gravitational acceleration values represents the *field* of the gravitational attraction known as the *gravity field*. The vector components of the gravity field along the mutually orthogonal x-, y-, z-directions are defined by first-order derivatives of the gravitational potential U as

$$g_x = \partial \mathrm{U}/\partial x, \quad g_y = \partial \mathrm{U}/\partial y, \quad g_z = \partial \mathrm{U}/\partial z.$$

Gravitational acceleration is directly related to a particular attracting mass which defines it, so that measurements of the spatial variations of the gravitational acceleration and its gradients contain information about the distribution of such masses, including those of the exploration interest. In the SI system, the gravitational acceleration is measured in meters per second squared ($\mathrm{m \cdot s^{-2}}$) and in micrometers per second squared ($\mathrm{\mu m \cdot s^{-2}}$), where $1\ \mathrm{\mu m \cdot s^{-2}} = 10^{-6}\ \mathrm{m \cdot s^{-2}}$. In the cgs system, the gravitational acceleration unit is *Gal*: $1\ \mathrm{Gal} = 1\ \mathrm{cm \cdot s^{-2}} = 10^{-2}\ \mathrm{m \cdot s^{-2}}$. This unit is commonly used with the prefixes milli, such as *milligal (mGal)*, where $1\ \mathrm{mGal} = 10^{-3}\ \mathrm{Gal} = 10^{-5}\ \mathrm{m \cdot s^{-2}} = 10\ \mathrm{\mu m \cdot s^{-2}}$, and micro, such as *microgal (μGal)*, where $1\ \mathrm{\mu Gal} = 10^{-6}\ \mathrm{Gal} = 10^{-3}\ \mathrm{mGal} = 10^{-8}\ \mathrm{m \cdot s^{-2}} = 10^{-2}\ \mathrm{\mu m \cdot s^{-2}}$. Previously, *gravity unit (g.u.)* also has been used, where $1\ \mathrm{g.u.} = 1\ \mathrm{\mu m \cdot s^{-2}} = 0.1\ \mathrm{mGal} = 100\ \mathrm{\mu Gal}$. Sometimes referred to as *Newtonian gravitational acceleration*, *acceleration of free fall*, and *acceleration of gravity*. See also *centrifugal acceleration*, *inertial acceleration*, and *total acceleration*.

gravitational admittance: 1. a *wavenumber*-dependent function that describes the relationship between a regional-scale *gravity anomaly* due to the *isostatic equilibrium* within the *lithosphere* and *topography* or *bathymetry*. **2.** a wavenumber parameter that depends on the flexural properties of the lithosphere and represents the function which relates the topography or bathymetry to a *Bouguer anomaly* or *free-air anomaly*. Gravitational admittance is denoted by $Z(k)$ and expressed as

$$Z(k) = F_{GA}(k)/F_{TB}(k),$$

where $F_{GA}(k)$ is the *Fourier transform* of the observed *gravity anomaly*, $F_{TB}(k)$ is the Fourier transform of the topography or bathymetry, and k is the wavenumber defined as $k = 2\pi/\lambda$, where λ is *wavelength*. The unit of measurement is *milligal (mGal)* per kilometer ($\mathrm{mGal \cdot km^{-1}}$). The admittance function is calculated for quantitative estimates of *isostatic compensation* via *inverse modeling* in order to (a) examine a response of the lithosphere to the topographic or bathymetric loads of different wavelengths and (b) constrain the isostatic model parameters, such as the *Moho boundary* depth in the *Airy-Heiskanen model*, the depth of isostatic compensation in the *Pratt-Hayford model*, and the equivalent elastic thickness of the lithosphere in the *Vening Meinesz model*. Gravitational admittance is also used for calculation of gravity anomalies and corresponding admittance curves for different isostatic models using *forward modeling* to compare them with their observed equivalents and find the best fitting isostatic model for a study area. Also referred to as *admittance function* and *isostatic response function*. See also *anomaly*, *gravitational coherence*, and *isostasy*.

gravitational attraction: a natural phenomenon of the mutual attraction between all bodies of mass, wherein the source of this attraction is their masses. The force of the gravitational attraction, or *gravitational force*, is described by the *Newton law of gravitational attraction*. Mathematically, the gravitational attraction is represented by a *vector* which is directed toward the attracting body, wherein the length, or *magnitude*, of this vector is defined by the gravitational force per unit mass. In accordance with the *Newton law of motion*, the ratio of the unbalanced force F to the moving mass m is expressed in units of *acceleration* and, therefore, the quantitative measure, or physical *dimension*, of the gravitational attraction is *gravitational acceleration*. A gravitational attraction vector is always orthogonal to the gravitational *equipotential surface*. Denoted by $\mathbf{g}$, the gravitational attraction is also referred to as *Newtonian gravitational attraction*. For the unit mass at observation point P, the gravitational attraction by the mass M is expressed as

$$\mathbf{g}(P) = -\mathrm{G}M \cdot \hat{\mathbf{r}}/r^2,$$

where G is the *universal gravitational constant*, $\hat{\mathbf{r}}$ is the unit vector directed from mass M to observation point P, and r is the distance between the observation point and attracting mass M. A negative sign indicates that the force of attraction is directed opposite to the unit vector $\hat{\mathbf{r}}$. The vector *field* of gravitational attraction is referred to as the *gravitational field* or *gravity field*. Used interchangeably with the same meaning as *gravitation*. See also *scalar field* and *vector field*.

gravitational coherence: **1.** a cross-spectral parameter which is a measure of correlation between an observed *gravity anomaly*, such as a *Bouguer anomaly* or *free-air anomaly*, and *topography* or *bathymetry* in studies of the isostatic response of the *lithosphere* to the topographic and bathymetric loads of different wavelengths. **2.** the degree of correlation between an observed gravity anomaly and topography or bathymetry as a function of *wavelength*. In the *spectral domain*, it is defined by the ratio of the squared cross-correlation *power spectrum* to the product of the auto-correlation power spectra of the gravity anomaly and topography or bathymetry as

$$\gamma^2(k) = |P_{GTB}(k)|^2 / P_{GA}(k) \cdot P_{TB}(k),$$

where $P_{GTB}(k)$ is the cross-correlation power spectrum of the gravity anomaly and topography or bathymetry, $P_{GA}(k)$ is the auto-correlation power spectrum of the gravity anomaly, $P_{TB}(k)$ is the auto-correlation power spectrum of the topography or bathymetry, and k is *wavenumber* defined as $k = 2\pi/\lambda$, where λ is *wavelength*. Calculated coherence values are compared with the coherence curves obtained with different values of the equivalent elastic thickness of the lithosphere in order to find the best fitting value for a study area. See also *Bouguer coherence*, *free-air coherence*, and *gravitational admittance*.

gravitational constant: **1.** *universal gravitational constant (G)*. **2.** *geocentric gravitational constant (GM)*.

gravitational field: a *vector field* of *gravitational attraction* between all bodies of *mass*, wherein the source of this field is the mass of each body. It is described by the *Newton law of gravitational attraction* and expressed in terms of *gravitational force* per unit mass. In accordance with the *Newton law of motion*, the force per unit mass defines the *acceleration* of a body of mass, so the planetary gravitational field is equivalent to and described by the field of *gravitational acceleration* of a free-falling body under the action of the gravitational force generated by the attracting planetary mass. The measured value of the planetary gravitational field, such as the *Earth gravity field*, is a *vector* sum of the gravitational attractions of all individual masses within the planet. The direction and *magnitude* of the Earth's gravity field $\boldsymbol{g}$ at an arbitrary point P are described by the *gradient* of *gravitational potential* U as

$$\mathbf{g}(P) = \nabla \mathrm{U} = \mathbf{i}\partial \mathrm{U}/\partial x + \mathbf{j}\partial \mathrm{U}/\partial y + \mathbf{k}\partial \mathrm{U}/\partial z,$$

where ∇ is gradient operator *del*; $\partial \mathrm{U}/\partial x$, $\partial \mathrm{U}/\partial y$, $\partial \mathrm{U}/\partial z$ are first-order derivatives of the *scalar* gravitational potential U along the mutually orthogonal x-, y-, z-directions, respectively, in *Cartesian coordinates*; and **i**, **j**, **k** are unit vectors parallel to these directions. The vector component of the gravitational field in any direction is defined by a *derivative* of the gravitational potential in that direction. First-order derivatives of the gravitational potential define the vertical component $g_z = \partial \mathrm{U}/\partial z$ and two horizontal components $g_x = \partial \mathrm{U}/\partial x$ and $g_y = \partial \mathrm{U}/\partial y$ of the gravitational field vector $\boldsymbol{g}$. Second-order derivatives of the gravitational potential represent the first-order derivatives of each vector component of the gravitational field in three orthogonal directions, so that three derivatives of three vector components of a gravitational field comprise the nine components of a *full tensor gradient (FTG)*. Unlike magnetic fields, the gravitational field cannot be shielded. Commonly referred to as *gravity field*.

gravitational flattening: the gravitational analogue of the geometrical *polar flattening*. It quantifies a percentage of change in the *theoretical gravity* value from the Earth's equator to the geographic poles, and it is defined as

$$f_{gr} = (g_p - g_E)/g_E,$$

where g_P is theoretical gravity at the geographic poles and g_E is theoretical gravity at the equator. For $g_P = 9.832186\ \mathrm{m \cdot s^{-2}}$ and $g_E = 9.780327\ \mathrm{m \cdot s^{-2}}$, the gravitational flattening is 0.00530, i.e., the gravity field of the *reference ellipsoid* decreases over its surface by 0.53% from the equator to the geographic poles.

gravitational flux: lines of the *gravitational field* in the form of imaginary fluid-like flow, or *flux*, wherein the density of this flow, or *flux density*, is directly proportional to the strength of the *gravitational force* and, hence, to the field intensity. Used as a theoretical measure of the gravitational field passing through a given surface or volume. See also *Gauss flux theorem*.

gravitational force: a *vector* force **F** of the mutual attraction between all bodies of *mass*, or force of *gravitational attraction*, wherein the source of this force is their masses. The *Newton law of universal gravitation* postulates that the *magnitude* of the gravitational force between two bodies is directly proportional to the *mass* of each body and inversely proportional to the square of distance between them. Gravitational force cannot be determined independently of the mass it is acting upon, and, hence, the gravitational force per unit mass is a measured quantity. According to the *Newton law of motion*, the force per unit mass describes this quantity as *acceleration*, so that the ratio of the gravitational force **F** to the body mass m_0 defines the *gravitational acceleration* $\boldsymbol{g}$ under the action of the gravitational force **F**, i.e., $\boldsymbol{g} = \mathbf{F}/m_0$. The direction and magnitude, or vector length, of the gravitational force vector on the Earth's surface is defined by the *gradient* of the scalar *gravitational potential* and represented by a vector sum of the gravitational force due to the stationary, or non-rotating, Earth's mass and *centrifugal force* due to the Earth's rotation about its axis. At any space point, the gravitational force vector is orthogonal to the *equipotential surface* passing through this point. The vector *field* of gravitational force is known as *gravitational field* and *gravity field*. See also *scalar field* and *vector field*.

gravitational mass: the *mass* of a gas, water, rock, or planetary body which is the source of the *gravitational force* whose *vector field* is referred to as the *gravitational field*. The larger the gravitational mass, the stronger the gravitational field generated by this body. In gravity *exploration*, variations in the subsurface distribution of gravitational masses are determined by measurements of *relative gravity* and its gradients and interpreted in terms of elements of the subsurface geological *structure* and their physical properties. See also *gravimeter*, *gradient*, and *interpretation*.

gravitational potential: 1. a mathematical function which describes through its derivatives the spatial parameters of a *gravitational field*. **2.** a *scalar* quantity representing the amount of energy, or work, required to move a unit mass from one point to another within the gravitational field acting upon the unit mass. **3.** the negative of the amount of work required to move a unit mass from infinity to a given point against the *gravitational force*. For the *point mass m*, its gravitational potential U at the distance r is expressed as

$$U = Gm/r,$$

where G is the *universal gravitational constant*. For measurements on and above the Earth's surface without accounting for atmospheric mass, i.e., in a mass-free space, the gravitational potential obeys the *Laplace equation*, whereas in the mass-filled space it obeys the *Poisson equation*. Gravitational potential complies with the *superposition principle*. At any point in space, the direction and length, or *magnitude*, of the gravitational field *vector* $\boldsymbol{g}$ are described by the *gradient* of the *scalar* gravitational potential. First-order derivatives of the gravitational potential in *Cartesian coordinates* define the horizontal components g_x and g_y and the vertical component g_z of the gravitational field vector $\boldsymbol{g}$ as

$$g_x = \partial U/\partial x, \quad g_y = \partial U/\partial y, \quad g_z = \partial U/\partial z,$$

where U is the gravitational potential. Second-order derivatives of the gravitational potential represent the first-order derivatives of each vector component of the gravity field in three orthogonal directions, so that three derivatives of three vector components of the gravity field comprise the nine elements of a *full tensor gradient (FTG)*. The relationship between the gravitational potential and *magnetic potential* is described by the *Poisson theorem*. Also referred to as *Newtonian potential*. See also *derivative*, *normal gravitational potential*, and *total gravitational potential*.

gravity: 1. a *vector* force of the mutual attraction between all bodies of *mass* that is known as *gravitational force* whose *magnitude* is defined by the *Newton law of gravitation*. The physical *dimension* of gravity and measure of its strength is *acceleration*. **2.** *gravitational attraction* of the Earth's mass measured by a free-fall acceleration of the reflective object or streams of cold atoms in an *absolute gravimeter*. Scalar values and vector components of the *field* of gravitational attraction of the Earth, or *Earth gravity field*, are quantities measured and interpreted in gravity exploration. **3.** a *scalar* quantity representing the magnitude of *total acceleration* of a free-falling body of mass at the Earth's surface. **4.** a term in geophysical exploration that is used to describe the type of *survey*, measuring instruments, measured data, data corrections and processing techniques, detected anomalies, their forward and inverse *modeling*, and *interpretation* concepts. Referred to with the same meaning as *gravitation* and *force of gravity*. See also *absolute gravity*, *relative gravity*, and *vertical gravity*.

gravity acceleration vector: 1. *vector* $\boldsymbol{g}$ representing the *total acceleration* of a free-falling body of mass at the Earth's surface. Its length, or *magnitude*, $g = |\boldsymbol{g}|$ and orientation in the space are defined by the vector sum of *gravitational acceleration* due to attraction of the Earth's mass and *centrifugal acceleration* due to the Earth's rotation about its axis. A gravity acceleration vector at the Earth's surface is oriented in the direction of a *plumb line*. Also known as *Earth gravity vector* and *gravity vector*. **2.** a vector representing the direction and numerical value (magnitude) of *acceleration* of a free-falling body of *mass* under the action of the *gravitational force* of attraction generated by the planetary mass. The magnitude of this vector is commonly described in units of the *nominal gravity g*, such as $\sim 0.38\ g$ for Mars and $\sim 0.17\ g$ for Moon, where $1\ g = 9.8\ \mathrm{m \cdot s^{-2}}$. See also *gravitational attraction*.

gravity anomaly: 1. the difference Δg between *observed gravity* and *predicted gravity* which accounts for all known planetary and non-geological gravity effects contributing to and interfering with measured gravity values, i.e., it is the difference between observed and *model*-predicted gravity values. **2.** the *field* of gravity values obtained by the application of all relevant *observed gravity corrections* and referred to as *free-air anomaly*, *Bouguer anomaly*, and *isostatic anomaly*. **3.** a local area where the *amplitude*, *shape*, and *spatial frequency* characteristics of the observed gravity field are different from the same characteristics in the surrounding *gravity survey* area. **4.** the local variation of the measured *Earth gravity field* or its gradients associated with a particular geological *structure* or lateral change in *density contrast* related to the rock composition. **5.** the gravity field of a single geological body calculated by *forward modeling*. **6.** *anomalous gravity*. In order to generate an observable gravity anomaly, the *source body* must satisfy the following requirements: (a) sufficient density *contrast* with the surrounding rocks, (b) sufficient volume, and (c) sufficient proximity to the observation points. Depending on the sign of a lateral *mass differential*, gravity anomalies are either positive or negative. Unlike magnetic anomalies, anomalous gravity highs and lows often represent the structural highs and lows, such as anticlines, synclines, horsts, grabens, etc., within a sedimentary section and at the top of a *basement*. See also *anomaly*, *observed gravity components*, *residual gravity anomaly*, *gravity high*, *gravity low*, and *magnetic anomaly*.

gravity anomaly amplitude: a quantitative measure of a *gravity anomaly* controlled by three primary factors: (1) *mass differential*: the larger the source volume and its *density contrast*, the higher the amplitude; (2) lateral density *contrast*: the larger the contrast, the higher the amplitude; and (3) *source body* depth: the deeper the source, the lower the amplitude. Additional factors include the distance to observation points, *background noise*, interference with anomalies generated by closely spaced sources, and superposition of regional anomalies. See also *amplitude*, *anomaly*, *regional anomaly*, *interference of anomalies*, and *superposition principle*.

gravity anomaly shape: See *anomaly shape (grav.).*

gravity anomaly vector: **1.** *anomalous gravity vector.* **2.** a three-component *vector* $\Delta\mathbf{g}$ for which the length, or *magnitude*, $|\Delta\mathbf{g}|$ is defined as

$$|\Delta\mathbf{g}| = (\Delta g_x^2 + \Delta g_y^2 + \Delta g_z^2)^{1/2},$$

where Δg_x, Δg_y, Δg_z are mutually orthogonal horizontal and vertical vector components of the *gravity anomaly* Δg obtained by *scalar-to-vector transformation* of the gridded *residual gravity* field. See also *gridding, residual field*, and *magnetic anomaly vector.*

gravity base: a gravity *base station* in the context of a ground location with known three-dimensional (3D) coordinates (latitude, longitude, *elevation*) and an *absolute gravity* or *relative gravity* value which is used as a reference for gravity measurements in a *gravity survey* area.

gravity basement: **1.** *density basement.* **2.** a *basement* for which the depth structure is obtained by two-dimensional (2D) gravity *modeling* or three-dimensional (3D) gravity *inversion.* See also *magnetic basement* and *acoustic basement.*

gravity-constrained velocity model: a three-dimensional (3D) *inverse model* which is built with the use of available seismic and gravity data to constrain the seismic velocities for the prestack and poststack time or depth migration of the seismic CMP reflection data. Its velocity- and structure-constraining properties are based on the results of 3D *structural inversion* of either *Bouguer gravity* or *full tensor gravity gradient (FTGG)* data. The model is built, primarily, for improved subsalt seismic imaging, such as the base of *allochthonous salt* and salt overhangs in areas of extensive *salt* tectonic deformations. See also *constraints, Gardner velocity-density relationship*, and *gravity-derived static corrections.*

gravity contact: a two-dimensional (2D) *source body* approximation in the gravity *modeling* of structural and lithological boundaries that assumes an "infinite" depth and horizontal extent of the lateral *contact* between rocks with different *density* values. See also *infinite bottom, infinite length, density contrast*, and *magnetic contact.*

gravity corrections: See *gravity data corrections.*

gravity curvature: See *gravity field curvature.*

gravity data: digital records of the measured *Earth gravity field*, its gradients, *full tensor gradient (FTG)*, and *vector* components that are made during ground, underwater, marine (shipboard), and airborne gravity surveys and space satellite missions. There are four major types of gravity data: (1) *absolute gravity* data obtained from measurements by an *absolute gravimeter* and integrated into the global *International Gravity Standardization Network 1971 (IGSN71)*; (2) *relative gravity* data obtained from measurements by a *relative gravimeter* and presented as *free-air anomaly, Bouguer anomaly*, and *isostatic anomaly*; (3) *gravity gradient* data obtained from measurements of vertical and horizontal gradients of the Earth's gravity field; and (4) *gradient tensor* data obtained from measurements of the *gravity gradient tensor (GGT)* components. By the status of *data processing*, the gravity data are divided into three categories: *raw gravity, observed gravity*, and *anomalous gravity.* See also *gradient, full tensor gravity gradient (FTGG), gravity gradiometer*, and *satellite gravimetry.*

gravity data corrections: **1.** *absolute gravity corrections.* **2.** corrections applied to *relative gravity* data in order to compensate for (a) measurement effects in the process of *acquisition*, such as *cross-coupling effect, Eötvös effect, turbulence effect, wave-motion effect gravimeter drift*, and (b) planetary, regional, and local non-geological effects, such as *atmospheric effect, Bouguer effect, curvature effect, free-air effect, isostatic compensation effect, terrane effect, theoretical gravity*, and *tidal effect.* These effects interfere with the measured gravity values and conceal the gravity anomalies generated by elements of the geological structures in the upper *crust.* There are two basic types of gravity corrections: *raw gravity corrections* and *observed gravity corrections.* Compensation for all known and predicted gravity effects of the non-geological origin transforms the measured gravity values into a *field* of gravity anomalies, or *anomalous gravity*, represented by *free-air anomaly, Bouguer anomaly*, and *isostatic anomaly.* The process of applying gravity data corrections is referred to as *gravity reduction.* See also *gradient terrain correction, self-gradient correction*, and *magnetic data corrections.*

gravity datum: the *absolute gravity* or *relative gravity* value at a *base station* that is used as a reference value, or *datum*, relative to which all measurements are made during a *gravity survey.* Local gravity datums are integrated into the regional or national network of absolute gravity values that is tied to the global absolute gravity network known as the *International Gravity Standardization Network 1971 (IGSN71).* Ties between the different datums are completed with the use of portable absolute gravimeters or high-*precision* relative gravimeters. Depending on the absolute or relative gravity value known at a base station, the gravity datum is referred to as *absolute gravity datum* or *relative gravity datum.* See also *magnetic datum.*

gravity depth: a depth to the isolated source of a *gravity anomaly* or continuous structural boundary, such as *basement* or *Moho boundary*, determined with the use of depth estimation methods, such as the automated *Euler deconvolution*, manual (graphic) techniques, and any method of *structural inversion* of gravity and *gravity gradient* data. Generally, any source depth estimate obtained from gravity data.

gravity-derived static corrections: seismic static corrections calculated from a *gravity model* of the near-surface low-velocity weathering layer and shallow *karst.* The model is based on high-accuracy *ground gravity* measurements with *amplitude resolution* of 0.02–0.05 mGal and *wavelength resolution* of 5–10 m. See also *milligal (mGal)* and *gravity-constrained velocity model.*

gravity disturbance: **1.** the difference between the observed *Earth gravity field* referenced to a *geoid* and *normal gravity* at the same point on a *reference ellipsoid.* **2.** a *scalar* quantity defined by the first-order *radial derivative* of *disturbing potential* T and expressed as

$$\partial g = \partial T / \partial \mathrm{r},$$

where r is direction along the Earth's radius on the reference ellipsoid. In *gravity inversion* with the use of global-scale crustal models, gravity anomalies are sometimes referred to as gravity disturbances. See also *anomalous field*, *crust*, and *gravity anomaly*.

gravity exploration: application of the *gravity method* for assessment of local, regional, or planetary geological *structure* in the search for commercial accumulations of hydrocarbons, economically useful minerals, and geothermal and water resources. Sometimes referred to as *gravity prospecting*. See also *magnetic exploration*.

gravity field: See *gravitational field*.

gravity field curvature: a measure of deviation, or bending, of the *gravitational field* from a planar, or flat, surface. Its estimates are commonly based on the spatial rate of change of a *gradient* of the gravity field and obtained from (a) two separate measurements at different spatial locations with the use of an *absolute gravity gradiometer* or (b) simultaneous measurements at three equally spaced positions by a set of three synchronized *absolute gravimeter* systems. Because the *gravity gradient* represents the first-order *derivative* of a gravity field or second-order derivative of *gravitational potential*, the gravity field curvature is approximated by the second-order directional derivatives of the *anomalous gravity* field or third-order directional derivatives of *gravitational potential*. Gravity field curvature is a quantitative characteristic of the *three-dimensional (3D)* deformation of the gravity field at a particular observation point: the more deformed the gravity field, the larger its curvature value. See also *curvature*, *curvature attributes*, *directional derivative*, and *geopotential model*.

gravity field line curvature: the rate of change of a direction, or bending, of the vertical component of the *Earth gravity field* due to a heterogeneous distribution of rock masses. It is defined by the ratio of *horizontal gradient amplitude (HGA)* to *vertical gravity* g_z and expressed as

$$K_f = (g_{zx}^2 + g_{zy}^2)^{1/2} / g_z,$$

where g_{zx} and g_{zy} are horizontal gravity gradients along the mutually orthogonal *x*- and *y*-directions, respectively, in *Cartesian coordinates*. Here, K_f has its maximum values above the structural edges, such as high-*offset* faults and lithological boundaries. The direction of the gravity field at a ground observation point is determined by a *plumb line*, so the gravity field line curvature is also referred to as *plumb line curvature*. See also *deflection of the vertical*, *edge*, and *fault*.

gravity field strength: 1. a quantitative characteristic of a *gravitational field* generated by a *source body* or planetary body of *mass*. Its estimate is based on the *Newton law of gravitational attraction* and defined by the amount of *gravitational force* F that will be exerted upon a unit mass *m* at a distance r from the center of the attracting mass M:

$$\mathrm{F} = \mathrm{GM}m/\mathrm{r}^2,$$

where G is *universal gravitational constant*. Relative changes in the strength of the *Earth gravity field* between observation and reference points are quantities measured by a *relative gravimeter*. **2.** the length, or *magnitude*, of a *gravity vector* at an observation point. See also *absolute gravimeter*.

gravity field units: units of measurement and analysis of a gravity field that are expressed in units of *gravitational acceleration*, because the physical *dimension*, or quantitative measure, of a *gravitational field* is the mass-independent *acceleration*, such as the acceleration of a free-falling body of *mass* under the action of *gravitational force*. Basic gravity field units in exploration and research applications are *milligal (mGal)* and *microgal (μGal)*. See also *Gal*, *inertial acceleration*, and *Newton law of motion*.

gravity field vector: See *gravity vector*.

gravity force vector: See *gravity vector*.

gravity gradient: 1. a *scalar* quantity, such as *horizontal gradient* or *vertical gradient*, representing the rate of change of a *gravitational field* in the direction of a horizontal or vertical separation of two or three sensors in a *gravity gradiometer*, i.e., it is a measured rate of change of the gravity field per unit of separation distance. **2.** a scalar quantity representing the horizontal or vertical rate of change of a gravity field calculated from measurements by a *relative gravimeter*, i.e., it is a calculated rate of change of a gravity field. **3.** a *vector* quantity defined by a three-component *gravity gradient vector*. For gridded data, the horizontal gravity gradient is defined by the absolute value of the *horizontal derivative*, and the vertical gravity gradient is equal to the *first vertical derivative (FVD)* calculated with the *z*-axis positive down. Decay, or *attenuation*, of a gravity gradient *anomaly* is inversely proportional to the cube of the distance from the sources approximated by the *point mass* geometries. In *SI units*, the gravity gradient is measured in inverse seconds squared (s^{-2}) or inverse nanoseconds squared (ns^{-2}), where $1\ ns^{-2} = 10^{-9}\ s^{-2}$. In *cgs units*, it is measured in units of *milligal (mGal)* per kilometer ($mGal \cdot km^{-1}$) or per meter ($mGal \cdot m^{-1}$). Gravity gradient measurements are often presented in units of *Eötvös (E)*, where $1\ E = 1\ ns^{-2} = 10^{-9}\ s^{-2} = 0.1\ mGal \cdot km^{-1} = 10^{-4}\ mGal \cdot m^{-1}$. See also *gridding* and *gravity gradiometry*.

gravity gradient tensor (GGT): 1. a nine-element symmetric *tensor* the elements of which are the spatial derivatives of three mutually orthogonal *vector* components of the *gradient vector* of *gravitational potential* along three mutually orthogonal directions in space. **2.** the nine-element *gradient tensor T* representing the derivatives of each of the g_x, g_y, g_z components of *gravity vector* **g** along the *x*-, *y*-, *z*-directions, respectively, in *Cartesian coordinates* and displayed in the format of a 3 × 3 *matrix* as

$$T = \nabla \mathbf{g} = \nabla\nabla U = \begin{matrix} g_{xx} & g_{xy} & g_{xz} \\ g_{yx} & g_{yy} & g_{yz} \\ g_{zx} & g_{zy} & g_{zz} \end{matrix} = \begin{matrix} \partial g_x/\partial x & \partial g_x/\partial y & \partial g_x/\partial z \\ \partial g_y/\partial x & \partial g_y/\partial y & \partial g_y/\partial z \\ \partial g_z/\partial x & \partial g_z/\partial y & \partial g_z/\partial z \end{matrix},$$

where ∇ is the mathematical symbol of the gradient operator *del*, U is the gravitational potential, $g_{xy} = g_{yx}$, $g_{xz} = g_{zx}$,

$g_{yz} = g_{zy}$, and $g_x = \partial U/\partial x$, $g_y = \partial U/\partial y$, $g_z = \partial U/\partial z$. In accordance with the *Laplace equation*, $g_{xx} + g_{yy} + g_{zz} = 0$, i.e., the sum of the diagonal gradient tensor elements, also known as the trace of the gradient tensor, is zero. Nine tensor elements comprise the *full tensor gradient (FTG)* of the anomalous gravity field at each observation point. There are five independent elements: g_{xy}, g_{xz}, g_{yz}, and any two of three diagonal elements g_{xx}, g_{yy}, g_{zz}. In airborne and marine (shipboard) gradient surveys, the independent GGT components are obtained from measurements by a *rotating-disk gravity gradiometer.* An assembly of electrostatic accelerometers was used for spaceborne *full tensor gravity gradient (FTGG)* measurements by the *GOCE gravity gradiometer.* An alternative, but a less laterally resolving, option is calculation of the GGT components from gridded gravity data, such as *Bouguer gravity* (Pilkington, 2014). Sometimes referred to as *gravity tensor* and *tidal tensor.* See also *derivative, gridding, electrostatic accelerometer, lateral resolution,* and *rotated gradient tensor.*

gravity gradient vector: a three-component *vector* defined by the vector sum of the first-order vertical and horizontal derivatives of an anomalous gravity field or the second-order vertical and horizontal derivatives of *gravitational potential* along the mutually orthogonal *x*-, *y*-, *z*-directions in *Cartesian coordinates.* It is orthogonal to the gravity field isolines and pointing in the direction of the maximum change of an anomalous gravity field. Its length, or *magnitude*, is equal to a numerical value of the *3D analytic signal amplitude (3DASA),* and its *phase*, or angle relative to the XY plane, is defined by the *tilt angle (TA)* at each observation point. See also *anomalous field, isoline, horizontal derivative,* and *vertical derivative.*

gravity gradiometer: an instrument for measurements of a *gravity gradient.* Gravity gradiometer designs vary from a simple assembly of two gravimeters, separated vertically or horizontally by a distance called the *baseline*, to complex assemblages with twelve synchronized accelerometers mounted on three tilted rotating disks for measuring the components of the *full tensor gradient (FTG).* The gradient component in the direction of separation of the paired accelerometers is obtained from the difference between their measurements divided by the baseline value. Differencing the synchronized measurements of *gravitational acceleration* essentially eliminates the non-geological inertial accelerations by *common mode rejection.* Standard attributes of a gravity field, such as *first vertical derivative, horizontal gradient,* and *analytic signal amplitude (ASA),* can be calculated from the measured gradient components and gridded for subsequent *enhanced processing* and *interpretation.* Measured *gravity gradient tensor (GGT)* components are transformed into *relative gravity* values, which differ from the relative gravity values measured with the use of a conventional, or regular, *gravimeter* by the amount of the regional long-*wavelength* component in a *survey* area, because this component is suppressed by *gradient* measurements. Sometimes, the *accelerometer*-based gravity gradiometer is referred to as *differential accelerometer.* See also *inertial acceleration, GOCE gravity gradiometer, rotating-disk gravity gradiometer,* and *magnetic gradiometer.*

gravity gradiometry: methods and instruments for measuring the gradients and *gradient tensor* components of a *gravitational field.* Gravity gradiometry has several advantages over measurements of the vertical component of the *Earth gravity field* by a conventional *gravimeter,* especially for airborne and shipboard (marine) gravity surveys, including: (a) increased *lateral resolution* of anomalies generated by closely spaced sources, (b) reduced *noise* due to *common mode rejection* of the motion-related inertial accelerations of a *survey* aircraft or ship, (c) improved detection of short-*wavelength* anomalies due to suppression of regional gravity field components and anomalies from deep sources, and (d) improved interpolation between survey lines for *gridding* with the use of *lateral gradient.* See also *anomaly, full tensor gravity gradiometry, gradient-enhanced gridding, gradient-enhanced minimum-curvature,* and *inertial acceleration.*

gravity high: the *gravity anomaly* over a local- or regional-scale area where the *observed gravity* field values are higher than those in surrounding areas. It is commonly associated with a *structural high* or lateral increase in the *density* of rock formations, such as the one created by an *igneous intrusion*, shallow shale *diapir,* and high-density lithological unit within the *basement.* In deep basement areas, gravity highs in the *residual field* often, but not always, correspond to intrasedimentary geological structures, such as *anticline, horst, reef, thrust,* and *uplift.* See also *gravity low, magnetic low,* and *magnetic high.*

gravity interpretation methods: qualitative and quantitative methods of gravity data *interpretation* that are based on the analysis of *total field, gradient,* and *full tensor gradient (FTG)* anomalies and their mathematical transformations by *enhanced processing* techniques. Methods include but are not limited to: (1) mapping of the anomalous alignments, or *lineaments,* generated by structural discontinuities, such as faults, fracture and shear zones, local lithological edges, and regional tectonic boundaries; (2) mapping of subsurface geological structures, such as igneous intrusions, *salt* and shale diapirs, carbonate buildups (reefs), uplifts, depressions, horsts, grabens, thrusts, and other structural highs and lows; (3) source depth estimates with the use of *automated source depth estimation* and *manual source depth estimation* techniques; (4) 2D, 2.5D, and 2.75D *forward modeling*; (5) 2D and 3D *gravity inversion* or *joint inversion* for the depth and physical properties (*density contrast* or *magnetization contrast*) of a targeted subsurface *structure,* parameters of a discovered *ore body, volcanics,* and other exploration targets; (6) *pattern recognition*; and (7) construction of a *gravity-constrained velocity model.* Due to a much narrower range of *density* variations relative to variations of *magnetic susceptibility* in rock formations, gravity interpretation methods are more efficient in mapping of structural highs and lows than *magnetic interpretation methods,* which are the most efficient in mapping of structural discontinuities, especially faults. The inherent *interpretation ambiguity* of the individual gravity interpretation methods is minimized by integration with magnetic data and other geophysical and geological information in a manner of *integrated interpretation*. See also *gravity anomaly, target, attribute analysis, boundary analysis, curvature analysis, multiscale analysis, principal component analysis (PCA), source depth analysis, spectral analysis, trend analysis*, and *wavelet analysis*.

gravity inversion: *inversion*, or *inverse modeling*, for structural parameters (depth, shape, etc.) or distribution of *density* contrasts as a function of *two-dimensional (2D)* or *three-dimensional (3D)* subsurface position with the use of line-oriented *total field* and *gravity gradient* data or gridded *residual gravity* and *full tensor gradient (FTG)* data with the use of a *voxel model*, respectively. Advances in developing of the 3D gravity inversion methods have made possible the simultaneous inversion for both the density contrasts and depth *structure* of targeted rock formations (Li and Oldenburg, 1998a; Rossi, 2017). See also *density contrast, gridding, target*, and *magnetic inversion*.

gravity loop network: a set of interlocking closed loops of gravity measurements in a *ground survey* area. It includes the repeat and tie stations and relative or absolute *base station*. See also *loop, network, relative gravity base, repeat station*, and *tie station*.

gravity low: the *gravity anomaly* over a local- or regional-scale area where the *observed gravity* field values are lower than those in the surrounding areas. It can be associated with a *structural low* or caused by a lateral decrease in the *density* of rock formations, such as the one created by *allochthonous salt*, salt *diapir, salt dome*, and low-density lithological unit within the *basement*. In deep basement areas, the gravity lows in the *residual field* often correspond to intrasedimentary geological structures, such as *syncline, graben*, local *pull-apart basin*, etc. See also *gravity high, magnetic high*, and *magnetic low*.

gravity method: the geophysical method based on ground, submarine, marine (shipboard), airborne, and satellite measurements of the planetary gravity field, such as the *Earth gravity field*, its gradients, *full tensor gradient (FTG)*, and *vector* components. The majority of gravity measurements are made for the relative changes in the Earth's gravity field, or *relative gravity*, with the use of a *relative gravimeter*. Gradients of the Earth's gravity field are measured by *gravity gradiometer* systems or calculated from line-oriented or gridded *anomalous gravity* data. Components of the *gravity gradient tensor (GGT)* are measured by a *rotating-disk gravity gradiometer* or by an *array* of electrostatic accelerometers, such as the one implemented in the *GOCE gravity gradiometer*. As a less resolving but still very informative option, the gradient tensor components can be calculated from gridded anomalous gravity data. In exploration applications, the gravity method is applied for detection and mapping of the *total field* and *gradient* anomalies generated by: (a) elements of subsurface geological structures, such as *anticline, carbonate platform* or *reef, graben, horst, syncline, thrust, uplift*, and other intrasedimentary and upper *basement* structures targeted in oil-and-gas exploration; (b) high-*density* igneous rock bodies, such as *carbonatite, ore body, kimberlite intrusion*, and other targets of mineral exploration; (c) structural discontinuities, such as faults, fracture and shear zones, and *igneous body* contacts; (d) low-density *allochthonous salt, salt dome*, and other *salt*-related structures in areas of salt tectonic deformations; (e) depositional fairways along which hydrocarbon-bearing reservoir sands and *placer* deposits can be accumulated; (f) changes in fluid volume and content caused by oil-and-gas or geothermal reservoir development and depletion of underground aquifers; and (g) changes in the igneous rock volume in active volcanos. After application of all relevant *gravity data corrections, leveling*, and, if necessary, *microleveling*, the *gravity survey* data are gridded and presented in the standard *format* of grids and maps of *free-air anomaly* and *Bouguer anomaly* with the addition of *isostatic anomaly* for regional-scale areas. Unlike the rock *magnetization* with variations reaching two to three orders of magnitude, i.e., up to 100–1000 times, the crustal rock densities vary within 2.0–2.5 times of their lowest value, so the anomalous gravity field is equally sensitive to the presence of high- and low-density source bodies and to changes of their depth, especially within the sedimentary section. For the same *source geometry*, the *fall-off rate* of gravity anomalies is less than that of magnetic anomalies, and for *point source* or *sphere* it is defined as $1/r^2$ versus $1/r^3$, where r is distance to the point source. Accordingly, gravity anomalies are broader than magnetic anomalies from the same source, more influenced by the *interference of anomalies*, and less responsive to source depth variations, thereby making the gravity method less resolving, both laterally and vertically, than the *magnetic method*. Achievable *accuracy* of gravity measurements is approximately 1 *microgal (μGal)*, which is one billionth, or 10^{-9}, part of the average *theoretical gravity* value on the Earth's surface. Airborne and marine (shipboard) gravity and *gravity gradient* surveys are the principal sources of data for *gravity exploration*. The gravity method is a *potential field* method, because its basic variables are the first- and second-order spatial derivatives of *gravitational potential*. See also *anomalous field, crust, derivative, fairway, fault, gravity anomaly, gravity-constrained velocity model, grid, gridding, lateral resolution, sensitivity, shear zone*, and *source body*.

gravity model: 1. a local-scale *2D model* or *3D model* of a particular exploration *target* that is obtained by *gravity inversion* with the use of *constraints* provided by the seismic, magnetic, surface geology, borehole, and other geological-geophysical data. **2.** a regional-scale 2D or 3D model based on the regional structural-tectonic concepts and constrained by available geological and geophysical information. Subsurface geology is modeled with the use of layers and separate bodies of either the constant or vertically and laterally variable densities. Allochthonous *salt* models are built to constrain the prestack and poststack migration velocities for improved subsalt seismic imaging. **3.** a 3D *property inversion*-based *voxel model* built for monitoring of (a) waterflooding in enhanced oil or gas reservoir development, (b) enhanced heavy oil recovery with steam injection, (c) depletion of groundwater aquifers,

(d) withdrawal of hot fluids from geothermal reservoirs, (e) CO_2 sequestration, and (f) growth of active volcanos with the use of *relative gravity* and *hybrid gravity* data. See also *hybrid density model*, *2.5D model*, *2.75D model*, *gravity-constrained velocity model*, and *4D gravity monitoring*.

gravity monitoring: See *4D gravity monitoring*.

gravity prospecting: See *gravity exploration*.

gravity reduction: a multi-step process of the application of *gravity data corrections* to measured gravity values in order to obtain *free-air anomaly*, *Bouguer anomaly*, and *isostatic anomaly*. Application and relevant corrections depend on the *survey* objectives, survey type (ground, marine, submarine, airborne), survey scale (local or regional), and measurement *accuracy* requirements. Corrected gravity values represent the *field* of the gravity anomalies, or *anomalous gravity*, associated with local and regional geological structures in an *exploration* project area. See also *anomaly*, *gravity anomaly*, and *observed gravity components*.

Gravity Reference Field (GRF): the *gravity field* generated by the uniformly dense, or homogeneous, rotating *reference ellipsoid* and defined at each observation point by the *International Gravity Formula (IGF)*. Positive and negative deviations from this field represent the local- and regional-scale gravity anomalies resulting from the Earth's heterogeneity due to the lateral changes in *structure* and *density* of rock formations within the *crust* and upper *mantle*. Referred to with the same meaning as *theoretical gravity* and *normal gravity field*.

gravity resolution: the *amplitude resolution* and *wavelength resolution* of gravity anomalies achievable with the use of a particular observation technique, measuring instrument, and *gravity survey*. Gravity resolution requirements for *survey* and *target* types are generalized in tables by Johnson (1998) and Sheriff (2002).

gravity repeats: *reoccupations* in the context of gravity measurements repeated at selected stations during a *ground survey* or at a fixed airfield location during an *airborne survey* in order to determine (a) *drift* of a *spring gravimeter* and (b) *static correction* for *accelerometer*-based *airborne gravity* measurements. See also *station*.

gravity sensor: a device with the natural and mechanical components implemented in the design of a *gravimeter* or *gravity gradiometer* for measurements of *absolute gravity*, *relative gravity*, and gradients of the *Earth gravity field*. There are seven main types of gravity sensors: (1) free-falling streams of cold atoms or free-falling reflective object in a *free-fall absolute gravimeter*, (2) *reversible pendulum* in an absolute *pendulum gravimeter*, (3) *proof mass* suspended by a metal or quartz spring in a *spring gravimeter*, (4) proof masses suspended by a fine torsion fiber or wire in a *torsion-balance gravity gradiometer*, (5) superconducting sphere levitating in the electromagnetic field of a *virtual-spring gravimeter*, (6) mass-loaded conducting ribbon, or string, in a *vibrating-string gravimeter*, and (7) proof mass of the electromagnetic or electrostatic *accelerometer* in *airborne gravimeter*, *rotating-disk gravity gradiometer*, and *GOCE gravity gradiometer* systems. See also *electromagnetic accelerometer*, *electrostatic accelerometer*, *electrostatic levitation*, *magnetic levitation*, and *equivalence principle*.

gravity signature: **1.** the specific *shape* and sign (positive or negative) of a *gravity anomaly* that (a) make this *anomaly* distinctly different from other anomalies in the *observed gravity* and *residual gravity* fields, and (b) it is known to be associated with a particular *source body* or geological *structure*. Gravity signatures are used as guides in the *structural interpretation* of *anomalous gravity* data. **2.** a three-dimensional (3D) *forward model* of the gravity field generated by (a) the mass of an *absolute gravimeter* and near-field components of its stationary mounting platform or (b) the mass of a survey ship or aircraft and movable components of the *stabilized platform*. This 3D forward model is used in calculation of the *self-attraction correction* for the *absolute gravity* measurements and *self-gradient correction* for airborne and marine *gravity gradient* and *full tensor gravity gradient (FTGG)* measurements. See also *magnetic signature*.

gravity standard: **1.** an *absolute gravity* value used as a reference, or standard, for measurements of *relative gravity* in a *gravity survey* area. **2.** *International Gravity Standardization Network 1971 (IGSN71)*.

gravity stripping: **1.** a method of removing, or stripping-off, the gravity effects of the subsurface structural elements whose size, shape, and *density contrast* are known from the available information or approximated for *modeling* purposes. **2.** a method of compensation for the positive or negative density contrasts (relative to *background density* $0 \text{ g} \cdot \text{cm}^{-3}$ or $2.67 \text{ g} \cdot \text{cm}^{-3}$, respectively) of the sedimentary layers, whose depth structure and rock densities are known from seismic and well (borehole) data. Their calculated gravity effect, or *gravitational attraction*, is removed from *Bouguer gravity* or *free-air gravity* in order to emphasize the gravity anomalies generated by the geological structures of an exploration or research interest in the lower sedimentary section, upper *basement*, and lower parts of the *crust*. Obtained gravity field, or *stripped gravity*, is used in 2D *forward modeling*, 3D structural and property *inversion*, and *integrated interpretation*. Stripping is analogous to application of gravity corrections, where *Bouguer correction* and *isostatic correction* are examples of compensating for the *observed gravity* components of the shallowest and deepest origins, respectively. As in the case of gravity corrections, the observation points remain at their original positions. Sometimes referred to as *backstripping* and *layer stripping* (Hammer, 1963). See also *geological correction*, *stripping boundary*, *inversion-based stripping*, *property inversion*, and *structural inversion*.

gravity survey: ground, marine (shipboard), submarine, sea-bottom, and airborne measurements of the *Earth gravity field*, its gradients and *full tensor gradient (FTG)* components over areas of exploration or research interest. Occasionally, *absolute gravity* is measured, but usually it is *relative gravity* and *gravity gradient* measurements. Measurements are made at regularly distributed observation points along pre-planned *survey* lines and along accessible ground routes. Coordinates of observation points are provided by the satellite-based *Global Positioning System (GPS)* or by more accurate *differential GPS (DGPS)*. There are three basic types of gravity surveys which differ by *sample interval*, *survey line spacing*, and *accuracy* of measurements: (1) large-scale *regional survey* for tectonic and geodynamic studies, where the observation points, or stations, can be up to several kilometers from each other and measurement accuracy is approximately 1 *milligal (mGal)*; (2) medium-scale *reconnaissance survey* for delineation of promising areas for hydrocarbon or mineral exploration, with *station spacing* of 500–1000 m and measurement accuracy of 0.1–0.5 mGal; and (3) small-scale *detailed survey* for mapping of the gravity, gravity gradient, and *gravity gradient tensor (GGT)* anomalies generated by *target* structures, where the required accuracy of measurements is 0.01 mGal, station spacing is several meters, and survey line spacing varies from tens to a few hundred meters. At present, regional and reconnaissance *ground gravity* surveys mostly have been replaced by *airborne gravity* surveys. Estimates of the expected *gravity survey accuracy* and *lateral resolution* of the detectable anomalies are based on *survey specifications*. See also *time-lapse gravity survey* and *gravity method*.

gravity survey accuracy: an estimate of the ability of a *gravity survey* to obtain the true values of the *Earth gravity field* and its components. Projected *accuracy* requirements are defined by the expected *target* parameters, such as depth, size, and *density contrast*. Measurement accuracy depends, primarily, on five types of accumulated errors at each observation point: (1) measured gravity value, or *reading*, error; (2) horizontal position, or latitude and longitude, error; (3) vertical position, or *elevation*, error; (4) *data reduction* errors, especially in calculation of *terrain correction*; and (5) residual *mis-tie* errors. For *ground gravity* surveys, it can be estimated in terms of *gravity survey precision*. For marine (shipboard) and *airborne gravity* surveys, survey accuracy estimates are based on *intersection statistics* and averaged *rms error* at the mis-ties after *leveling* and *microleveling*. Airborne gravity and *gravity gradient* data are compared to ground gravity data, if they are available in a survey area, using a *ground truth test*.

gravity survey precision: a quantitative estimate of the *gravity survey accuracy* and consistency in the performance characteristics of a *spring gravimeter*. It is based on the use of randomly distributed repeated observations, or *reoccupations*, in *ground gravity* measurements and defined by (a) the percentage of reoccupations in the total number of stations in a *gravity survey* area, (b) *repeatability* of the reoccupations, and (c) errors of individual measurements expressed by the mean square error m_i^2 at the *i*th *station* as

$$m_i^2 = m_G^2 + (\partial\Delta g/\partial\theta)^2 m_L^2 + (\partial\Delta g/\partial h)^2 m_h^2 + m_{TC}^2,$$

where m_G^2 is *variance*, or uncertainty, of the measured gravity value which is a function of *gravimeter precision*, *gravimeter drift*, and soil and weather conditions; $\partial\Delta g/\partial\theta$ is *latitudinal gradient*; m_L^2 is uncertainty in the *station* location controlled by the latitude θ term in the *theoretical gravity* formula; $\partial\Delta g/\partial h$ is *free-air gradient*; m_h^2 is uncertainty in elevation h estimated from free-air and Bouguer corrections; and m_{TC}^2 is variance of the calculated *terrain correction* controlled by *lateral resolution* of the *Digital Terrain Model* (Long and Kaufmann, 2013). Optimally, 10%–15% of all stations in a *ground gravity* survey area should be reoccupied at random times and with different base station loops. A high-precision *microgravity* survey requires the repeatability of ±0.005 *milligal (mGal)*. For regional-scale and local surveys other than a *detailed survey*, the required repeatability is ±0.05 mGal. See also *base station loop* and *loop drift rate*.

gravity survey resolution: *lateral resolution* of gravity anomalies defined, primarily, by two parameters of the *survey specifications*: (1) *sample interval* and (2) *survey line spacing*. Marine (shipboard) and airborne gravity surveys use gravimeters and gravity gradiometers with high sampling rates, so the actual lateral resolution of the line-oriented data is defined by the *cutoff* value of a *low-pass filter* applied to suppress *dynamic noise*. For gridded data, the lateral resolution depends on the survey line spacing: the closer the line spacing, the smaller *grid cell*, and the higher the lateral resolution of anomalies. See also *gravimeter*, *gravity gradiometer*, *sampling rate*, and *gridding*.

gravity tensor: See *gravity gradient tensor (GGT)*.

gravity tensor field: a field of the *gravity gradient* anomalies obtained from one of the independent components of *gravity gradient tensor (GGT)* representing (a) five horizontal derivatives of three mutually orthogonal *vector* components of the *Earth gravity field*, defined as

$$g_{xx} = \partial g_x/\partial x = \partial^2 U/\partial x^2,$$

$$g_{xy} = \partial g_x/\partial y = \partial^2 U/\partial x \partial y,$$

$$g_{zx} = \partial g_z/\partial z = \partial^2 U/\partial z \partial x,$$

$$g_{zy} = \partial g_z/\partial y = \partial^2 U/\partial z \partial y,$$

$$g_{yy} = \partial g_y/\partial y = \partial^2 U/\partial y^2;$$

and (b) the *vertical derivative* of vertical component g_z, defined as

$$g_{zz} = \partial g_z/\partial z = \partial^2 U/\partial z^2,$$

where U is *gravitational potential* and g_x, g_y, g_z are mutually orthogonal vector components of the Earth's gravity field. See also *horizontal derivative* and *invariants*.

gravity tie: high-*precision* measurements by a *relative gravimeter* along a *traverse* connecting the ground or airborne *gravity survey* area with an *absolute base station* in order to present the survey data in both *relative gravity* and *absolute gravity* values. See also *base-tie value*.

gravity unit (g.u.): a unit of gravity field measurements in *vintage data*. At present, the standard gravity measurement units are *milligal (mGal)* and *microgal (μGal)*, where 1 g.u. = 0.1 mGal = 10^2 μGal = 10^{-6} m · s^{-2} = 1 μm · s^{-2}.

gravity vector: a three-component *vector* for which the direction and *magnitude* at every point in space are defined by the spatial derivatives of a scalar *gravitational potential* U. The field of gravity vector values is a *vector field* known as the *gravitational field*, and its physical *dimension*, or quantitative measure of its strength, is *acceleration*. For the *Earth gravity field*, the gravity vector is expressed as

$$\boldsymbol{g} = \nabla \mathrm{U} = (\boldsymbol{g}_x, \boldsymbol{g}_y, \boldsymbol{g}_z),$$

where ∇ is gradient operator *del*. In *Cartesian coordinates*, the gravity vector components are defined as

$$\boldsymbol{g}_x = \partial \mathrm{U}/\partial x, \quad \boldsymbol{g}_y = \partial \mathrm{U}/\partial y, \quad \boldsymbol{g}_z = \partial \mathrm{U}/\partial z,$$

where $\partial \mathrm{U}/\partial x$, $\partial \mathrm{U}/\partial y$, $\partial \mathrm{U}/\partial z$ are first-order spatial derivatives of the gravitational potential U that represent two horizontal components and one vertical component of the *Earth gravity vector*, respectively. Its length, or magnitude, is defined as

$$g = |\boldsymbol{g}| = (g_x^2 + g_y^2 + g_z^2)^{1/2},$$

and its *phase*, or angle relative to the XY plane, is defined by the *tilt angle (TA)*. Also known as *3D gravity vector*, *gravity acceleration vector*, *gravity field vector*, and *gravity force vector*. See also *derivative*, *scalar*, *anomalous gravity vector*, and *magnetic vector*.

GRF correction: 1. same as *theoretical gravity correction* used to correct *observed gravity* values for the *Gravity Reference Field (GRF)* defined at each observation point by the *International Gravity Formula (IGF)*. **2.** *IGRF correction* for the long-*wavelength* components of the magnetic field generated by the Earth's core that sometimes is referred to as the *Geomagnetic Reference Field (GRF)*, whereas the internationally accepted standard term is the *International Geomagnetic Reference Field (IGRF)*.

Green's equivalent layer: 1. a thin hypothetical layer of mass or magnetic material for which the *gravitational potential* or *magnetic potential* is identical to the *potential* caused by the actual three-dimensional (3D) *density* or *magnetization* distribution. This layer is spread over any of its equipotential surfaces at or below an observation surface and above all gravity or magnetic sources. Total mass or total magnetization of the actual 3D distribution of sources is equivalent to that of the Green's equivalent layer. **2.** a *surface density* or *magnetostatic charge* distribution which produces the gravity or magnetic field outside the closed surface that is identical to the field produced by the density or magnetization distribution within a volume contained by the closed surface. As a data transformation concept, the Green's equivalent layer is used for *low-latitude reduction-to-pole (RTP-LL)*, *differential reduction-to-pole (DRTP)*, *gridding* of irregularly spaced ground data, *vertical continuation* between variable surfaces, stable *downward continuation*, and other applications. Green's equivalent layer validates the *nonuniqueness problem*, also known as the *inverse problem*: any observed gravity or magnetic *anomaly* can be generated by an infinite variety of density or magnetization distributions. Often referred to simply as *equivalent layer*. See also *equipotential surface*, *equivalent source*, and *Gauss law*.

grid: 1. a continuous mathematical surface through all data points in a *survey* area that is obtained by *gridding*. The grid represents the equally spaced gravity or magnetic data values arranged in rows (lines of data values oriented east–west) and columns (lines of data values oriented north–south). Intersections of the rows and columns are called *grid nodes* or *mesh points*. The horizontal distance between the grid nodes is called the *grid interval*. Each grid value is assigned to the center of a *grid cell*, the size of which is defined by the grid interval. **2.** the standard digital *format* of gravity and magnetic data for visualization of the obtained *anomalous field*, results of *enhanced processing*, and comparative analysis and *interpretation* of the detected anomalies of an exploration and research interest. See also *mesh*.

grid aliasing: errors in the *gridding* of irregularly spaced ground gravity or magnetic data and data acquired along the regularly spaced marine (shipboard) and airborne *survey* lines. Due to a large spacing between stations and survey lines, the data interpolation algorithm of the gridding cannot accurately represent the actual, or true, *field* values within the large data gaps, thereby altering the real shapes and gradients of anomalies, *wavelength* content, and *spectrum* of the gridded field. When the *grid interval* is too large to represent the short-*wavelength* components of line-oriented data, then (a) artificial, or aliased, long-wavelength components are included into the gridding and (b) a laterally continuous short-wavelength anomaly is gridded as a sequence of isolated circular anomalies resembling a "string of beads" or "bicycle chain" and referred to as a *boudinage artifact*. In a *detailed survey* with relatively close *survey line spacing*, the mid- and long-wavelength components of the *anomalous field* are essentially not affected by grid aliasing. See also *aliasing*, *bull's-eye effect*, *station*, *spatial aliasing*, and *wavenumber aliasing*.

grid cell: a square or, occasionally, rectangular elementary unit of a *grid*. Equally spaced grid cells of the same size comprise the continuous mathematical surface of a grid. Grid cell size is defined by the distance between *grid nodes*, i.e., it is equal to a pre-selected *grid interval*. Grid cell value is calculated by averaging the grid values of four grids nodes at the corners of each cell and assigning the obtained average value to the grid cell center. See also *gridding*.

grid expansion: 1. addition of several rows and columns of artificial data along the *grid* margins to provide a smooth transition of the grid values to zero in order to minimize *edge effects* and *Gibbs errors*. **2.** assigning a *roll-off widow*, or *taper*, to the edges of gridded data or ends of *survey* lines prior to *gridding*. Also known as *grid extension* and *padding*.

grid extension: See *grid expansion*.

grid fill-in: an auxiliary post-*gridding* procedure that fills the data gaps in a *grid* with the equally spaced values calculated by interpolation between the valid grid values around the *grid gap*.

grid filter: a *filter* applied to gridded data, as opposed to a *line filter* which is applied to line-oriented data. Grid filters are the primary tools for the *enhancement* of the gravity and magnetic anomalies of an exploration interest, except for the *low-amplitude* short-*wavelength* micromagnetic anomalies, which are often detected by the application of line filters only. See also *micromagnetic anomaly.*

grid gap: a *grid* area without data values. It represents a local area of missing observation points that is much larger than the *grid interval* and radius of the data interpolation in a regular *gridding* algorithm. Large grid gaps result in *grid aliasing.* See also *grid fill-in.*

grid interval: 1. the distance between equally spaced values in a gravity or magnetic *grid.* **2.** the distance between *grid nodes* in gridded data. **3.** the *sample interval* of gridded data. Selection of the grid interval Δx defines the *Nyquist wavelength* λ_N, which is the shortest *wavelength* resolvable in gridded data, expressed as $\lambda_N = 2\Delta x$. According to the *Nyquist theorem*, the grid interval should not be larger than half of the shortest wavelength in the pre-gridded data. In practice, the grid interval is selected within the range from 1/3 to 1/5 of the distance between *traverse lines* of a marine (shipboard) or airborne *survey.* For ground surveys with irregularly spaced observation points, the grid interval *GI* is determined from the empirical relationship:

$$GI = (S^2/n)^{1/2},$$

where S is the survey area extent and *n* is number of observation points. For airborne and marine gravity and *gravity gradient* surveys, the grid interval is commonly defined by a half-length of a *low-pass filter* applied to suppress *dynamic noise.* Also referred to as *grid spacing* and *grid sampling.* See also *filter length, half-wavelength resolution*, and *sampling bandwidth.*

grid merging: the process of combining the overlapping or geographically adjacent grids into a united, or merged, *grid.* Preparation involves (a) *regridding* one of the grids to make the same *grid interval* for both grids and (b) selection of the merge zone length and weighting coefficients for the gridded data within the merge zone or *DC shift* for one of the grids to provide a smooth transition of the grid values between the merging grids. Also known as *grid stitching.*

grid nodes: intersections of the data rows (lines of data values oriented east–west) and columns (lines of data values oriented north–south) that correspond to the points on a *grid*, where rows and columns have the same data values. Grid nodes represent the equally spaced coordinate points on the continuous mathematical surface computed by a *gridding* algorithm. The distance between grid nodes is called the *grid interval.* Also known as *grid points* and *mesh points.*

grid Nyquist wavelength: an analogue of the *Nyquist wavelength* for gridded gravity and magnetic data. It is equal to the doubled *grid interval* Δx, i.e.,

$$\lambda_N = 2\Delta x,$$

and it represents the shortest unaliased *wavelength* and, hence, the highest achievable *lateral resolution* of anomalies in the gridded *anomalous field.* Also referred to as *grid resolution.* See also *gridding* and *grid aliasing.*

grid points: See *grid nodes.*

grid resampling: *regridding* in the context of changing the original *grid interval* (which is also known as *grid sampling*) of gridded gravity or magnetic data.

grid residual: 1. short- and mid-*wavelength* components of the gridded *anomalous field* remaining after subtraction of its long-wavelength component, which is assumed to represent the *regional field* in a *survey* area. **2.** a method of *regional-residual separation* of anomalous gravity and magnetic fields based on the approximation of a regional component of the gridded data and subtraction of this component from the original *grid.* The regional component can be approximated with the use of (a) *polynomial fitting*, (b) *upward continuation*, (c) *low-pass filtering*, and (d) averaging techniques analogous to the *Griffin method.* See also *polynomial residual, residual field*, and *inversion-based separation.*

grid resolution: *grid Nyquist wavelength* in the context of the shortest *wavelength* which can be resolved in the gravity or magnetic data gridded with a selected *grid interval.* See also *gridding* and *lateral resolution.*

grid sampling: See *grid interval.*

grid smoothing: suppression of high-frequency *noise* in gridded data by averaging the *grid cell* values by a processing operator whose length and shape are defined by the number of cells surrounding each particular grid cell at the center of a *running window.* Generally, the operator with nine grid cells in a 3×3 square *kernel* provides sufficient *smoothing*, although the larger square and rectangular kernels, such as 5×5 and 5×7, can be used too. See also *gridding, smoothing filter*, and *convolution grid filter.*

grid spacing: same as *grid interval* that represents the distance between equally spaced *grid nodes* in gridded gravity or magnetic data.

grid stitching: See *grid merging.*

gridding: 1. conversion of the separate gravity or magnetic values measured along regular-spaced *survey* lines or at irregularly spaced ground observation points into a continuous mathematical surface of equally spaced values that is called a *grid.* With the same user-selected *grid interval*, the gridding algorithm performs (a) *downsampling* by averaging the values measured at very short intervals along the survey lines and (b) *upsampling* by interpolation within much larger intervals between the survey lines in order to compute a continuous surface providing the *least-squares fit* with all observed values in the survey area. Gridding is applied after *leveling* and other *data conditioning* procedures, and it is often combined with an *alias filter* to prevent data contamination by *aliasing.* The most commonly used method of gridding is *minimum-curvature gridding.* Any gridding method acts similar to a *low-pass filter* in suppressing the high-frequency (short-*wavelength*) components of the *observed field* that have their wavelengths less than a doubled grid interval. **2.** spatial reconstruction of a continuous gravity or magnetic *field* and its components from the discrete measurements at observation points in a survey area. Gridding limitations in preserving the original slopes, or gradients, of line-oriented data can result in errors in the process of *inversion* of the gridded data and incorrect source depth estimates. See also *bi-directional gridding, equivalent-source gridding, gradient-enhanced gridding, gradient-enhanced minimum-curvature, Hardwick method, Nelson method*, and *grid aliasing.*

Griffin method: a *grid*-based method of *regional-residual separation* of anomalous gravity and magnetic fields by calculating the average of the gridded values along the perimeter of a circle with a user-specified radius and assigning the obtained average value to the center of this circle as its regional value. The residual value is obtained by computing the difference between the observed value at the circle center and the obtained perimeter-averaged value. A regional or residual grid is named by a selected averaging radius, e.g., for a 10-km radius circle, the grid is named a 10-km regional grid or a 10-km residual grid. Digital version is based on the use of a square kernels of variable sizes centered at each *observation point*. See also *kernel*.

ground clearance: same as *terrain clearance*, or aircraft *altitude*, representing the aircraft height above the ground during an *airborne survey*.

ground gravimeter calibration: a standard *quality control (QC)* procedure in a *ground gravity* survey for monitoring the *scale factor*, *repeatability*, and other performance characteristics of a *relative gravimeter* before and after a survey. It involves measurements at two coordinate points with known gravity values which differ by 40–60 mGal and represent the largest range of *relative gravity* values measured during a survey. See also *airborne gravimeter calibration*, *calibration range*, and *milligal (mGal)*.

ground gravity: *relative gravity* and, occasionally, *absolute gravity* data acquired by stationary measurements of the *Earth gravity field* on the ground. The standard method of relative gravity measurements is *looping*. Coordinates of each observation point, or *station*, are provided by *Global Positioning System (GPS)* or more accurate *differential GPS (DGPS)* and any other *GNSS*. The distance between stations, or *station spacing*, depends on the accessible pathways and assumed depth and size of the geological structures generating the gravity anomalies. In *microgravity* surveys, the *GPS* observations are supplemented with geodetic *surveying* to ensure the highest-possible *accuracy* in measurements of *elevation* at each station and calculation of adequate *gravity data corrections*. Measurement accuracy of 1 *microgal (μGal)* requires the surface elevation to be determined with an accuracy of 2–3 mm. Also referred to as *land gravity*. See also *airborne gravity*, *marine gravity*, *submarine gravity*, *satellite gravity*, and *time-lapse gravity*.

ground magnetic intensity: measurements of the *Earth magnetic field* recorded by a *base-station magnetometer* during an *aeromagnetic survey*. The obtained data are used for monitoring the *external magnetic field* of the Earth and calculation of the *diurnal correction*.

ground magnetics: magnetic data acquired by a *ground survey*, as opposed to *airborne magnetics*, *marine magnetics*, and *satellite magnetics*. The three-dimensional coordinates of the observation points (latitude, longitude, and *elevation*) are provided by *Global Positioning System (GPS)* or more accurate *differential GPS (DGPS)* and any other *GNSS*. See also *forensic magnetism* and *magnetic archaeology*.

ground magnetometer: a *base-station magnetometer* in the context of the stationary measurements of *diurnals* on the ground during an *aeromagnetic survey*.

ground reference dataset: See *ground truth*.

ground survey: measurements of the *Earth gravity field* or *Earth magnetic field* and its gradients on the Earth's surface along the regularly spaced acquisition lines or at irregular observation points, or stations, over the area of exploration or research interest. See also *acquisition line*, *gradient*, *station*, *airborne survey*, *marine survey*, and *submarine survey*.

ground truth: 1. gravity data acquired along a single line, which is known as a *test line*, or several lines on the ground with relatively close spacing between observation points. It is used as a reference dataset for evaluation of the *accuracy* and *lateral resolution* of *airborne gravity* measurements and control over performance characteristics of *accelerometer*-based *airborne gravimeter (AG)* and *airborne gravity gradiometer (AGG)* systems. Also referred to as *ground reference dataset*. **2.** coordinate points on the ground with known *relative gravity* values which can be upward continued to an airborne survey *altitude* and filtered to match the parameters of airborne measurements for a *ground truth test*. See also *static test* and *upward continuation*.

ground truth test: a *quality control (QC)* procedure for *airborne gravity* measurements that is based on comparison of the acquired *airborne gravity* or *gravity gradient* data with the *ground truth* data available in an *airborne survey* area. Flying over the ground truth line is used for the estimates of *accuracy* and *resolution* of the new methods and instruments for the gravity and gravity gradient measurements on an airborne *observation platform*. See also *even-odd grid comparison*, *static test*, and *test line*.

GRS80: the *Geodetic Reference System* where the *GRS80 ellipsoid* is used as a *geodetic datum*. See also *World Geodetic System 1984 (WGS84)*.

GRS80 datum: same as *GRS80 ellipsoid* that is used as a reference surface, or *geodetic datum*, for local and regional coordinate systems.

GRS80 ellipsoid: the *reference ellipsoid* with four principal parameters (equatorial radius, *geometrical flattening*, angular velocity of rotation, and *geocentric gravitational constant*) which have been adopted by the Geodetic Reference System 1980 (GRS80). The GRS80 ellipsoid is a reference surface, or *geodetic datum*, for the *International Terrestrial Reference Frame (ITRF)* representing the global network of *three-dimensional (3D)* coordinates for observation points: latitude, longitude, and *elevation*, which is also known as *ellipsoidal height*. The GRS80 ellipsoid and its updated version, the *WGS84 ellipsoid*, have insignificant differences for exploration applications. See also *Geodetic Reference System*.

gyration radius: the perpendicular distance between an *observation point* on the Earth's surface and the axis of the Earth's rotation that is defined as

$$r_g = \mathrm{R} \cdot \cos\theta,$$

where R is the Earth radius at a given point and θ is the geographic latitude. Its value varies from the maximum equal to the Earth's radius at the equator to zero at the geographic poles. Gyration radius is a parameter of *centrifugal acceleration*.

gyromagnetic constant: same as *gyromagnetic ratio* that represents a constant specific for the subatomic particles, such as protons and electrons, whose properties are utilized in *quantum magnetometers*.

gyromagnetic ratio: a proportionality constant which relates the *Larmor frequency* in *quantum magnetometers* to the strength, or intensity, of the *Earth magnetic field*. Each subatomic particle, such as a proton or electron, has its own gyromagnetic ratio which is directly proportional to the charge and inversely proportional to the mass of a particle. The gyromagnetic ratio of electrons is two to three orders of magnitude higher than that of protons and, therefore, the design of high-*sensitivity* magnetometers is based on measurements of the Larmor frequency of electrons. The unit of the gyromagnetic ratio is *hertz (Hz)* per *nanotesla (nT)*.

H

half-amplitude width: the horizontal distance between *principal profile* points, where a gravity or magnetic *anomaly* is one-half of its maximum or minimum value for the positive or negative anomaly, respectively. It is implemented in source depth estimates with the use of principal profiles of the original gravity or magnetic anomalies and anomaly-derived *analytic signal amplitude (ASA)* and *horizontal derivative*. Also referred to as *anomaly half-width* and *half-maximum width*. See also *ASA-derived depth*, *half-width method*, and *Peters half-slope method*.

half-band filtering: application of low-pass or high-pass filters to separate half of the data *bandwidth*. For example, for the data with the highest *frequency* component of 20 Hz, a low-pass half-band filter removes all frequencies above 10 Hz, whereas a high-pass half-band filter removes all frequencies below 10 Hz from the data. Half-band filtering is implemented in a *discrete wavelet transform (DWT)* for the iterative decomposition of line-oriented gravity and magnetic data into the successively larger scales of *wavelength resolution*, where each following *scale* is represented by the *wavelength* components within the twice shorter bandwidth than at the previous scale. See also *high-pass filter* and *low-pass filter*.

half-maximum method: See *Tiburg method*.

half-maximum width: See *half-amplitude width*.

half-sine resolution: *half-wavelength resolution* in the context of representing the *wavelength* components of the anomalous gravity or magnetic field by sinusoids of different spatial frequencies. See also *spatial frequency*.

half-slope length: same as *Peters length* that represents the horizontal distance between two *profile* points known as *half-slope points*, where the *slope* of the magnetic anomaly is one-half of its maximum slope value.

half-slope points: two points at both sides of the *inflection point* on a *principal profile* of the magnetic anomaly that define the *half-slope length*, which is also known as the *Peters length*. Their positions are determined by projections of two points corresponding to the *half-amplitude width* of the *horizontal derivative* profile for the same *magnetic anomaly*.

half-space: a *model* approximation of the homogeneous rock volume which has an "infinite" extent in the x-, y-, z-dimensions below a bounding horizontal plane. Also referred to as *infinite half-space*. See also *dimension* and *infinite strike length*.

half-wavelength resolution: **1.** the width of the narrowest *anomaly* which can be identified, or resolved, in marine (shipboard), *airborne gravimeter (AG)*, and *airborne gravity gradiometer (AGG)* data. As one-half of the full *waveform* cycle, the anomaly width is measured between zero values on both sides of the positive or negative *amplitude*. Half-wavelength resolution is often estimated by the half-length of a *low-pass filter* applied to suppress the uncompensated inertial accelerations of the flying aircraft and wave-like motions of the survey ship. Filter length is determined by multiplying the aircraft or ship velocity and time duration of the filter work: for an aircraft velocity of 50 m · s^{-1} (180 km per hour) and a 60-second filter, the low-pass filter length is 3.0 km corresponding to a half-wavelength resolution of 1.5 km. **2.** a visual estimate of *lateral resolution* defined by approximate widths of the narrowest anomalies in the observed gravity or magnetic field that are coherent, or repeatable, over adjacent *survey* lines. Sometimes referred to as *half-sine resolution*. See also *inertial acceleration* and *full-wavelength resolution*.

half-width: a parameter of the *principal profile* measured in the vertical plane perpendicular to the source body *strike*. Depending on the assumed *source geometry*, the half-width is (a) the *half-amplitude width* above a *sphere* or *horizontal cylinder*, (b) the horizontal distance between the points where a *gravity anomaly* is three-quarters and one-quarter of its maximum value on the profile across the edge of a *horizontal sheet* or *fault*, and (c) half of the horizontal distance between the maximum and minimum values of *magnetic anomaly* on the profile across the edge of a horizontal sheet or fault. See also *half-width half-maximum* and *half-width method*.

half-width depth: a depth to the source of a gravity or magnetic *anomaly* estimated with the use of the *half-width method*.

half-width depth rules: empirical relationships between the *half-width* parameter of the *principal profile* of a magnetic or gravity *anomaly* and the depth to its source that are used in the *half-width method*. Rules are applied for the source approximations by simple *ideal source* geometries, such as *sphere*, *horizontal cylinder*, *vertical cylinder*, vertical *thin dike*, and *horizontal sheet*. For other source geometries, the half-width rules are less reliable and depend on their *dip*, width, and depth extent. See also *Smith depth rules*.

half-width half-maximum: a parameter of the *principal profile* of an *anomaly* defined by one-half of the horizontal distance at half of the maximum *amplitude* value of a particular anomaly in the *normalized source strength (NSS)* field. It is used in depth estimates of magnetic sources by the graphical version of the *NSS method*.

half-width method: a graphical method of estimating the depth h to a source of a gravity or magnetic *anomaly* by measuring the *half-width* distance on its *principal profile* and multiplying the measured value by the *index factor*. Gravity source depth estimates are made for: (a) the center of a *sphere*: $h_{sph} \leq 1.3\ x_{1/2}$, where $x_{1/2}$ is the half-width distance and 1.3 is the index factor; (b) the center of a *horizontal cylinder*: $h_{hc} \leq 1.0\ x_{1/2}$; (c) the center of a thin semi-infinite slab or *horizontal sheet*: $h_{hs} \leq x_{1/2} = 0.5(x_{3/4} - x_{1/4})$, where $x_{3/4}$ and $x_{1/4}$ are profile points where the anomaly value is three-quarters and one-quarter, respectively, of its *amplitude*; (d) the top of a *vertical cylinder*: $h_{vc} \leq 0.58\ x_{1/2}$; and (e) a vertical *thin dike*: $h_{thd} \leq 0.75\ x_{1/2}$. Magnetic source depth estimates are made using the anomalies of the *RTP magnetic field* for: (a) the center of sphere: $h_{sph} \leq 2.1\ x_{1/2}$; (b) the center of a horizontal cylinder: $h_{hc} \leq 2.0\ x_{1/2}$; (c) the center of a thin semi-infinite slab or horizontal sheet: $h_{hs} \leq 1.73\ x_{1/2}$; (d) the top of *vertical cylinder*: $h_{vc} \leq 1.3\ x_{1/2}$; and (e) a vertical *thin dike*: $h_{thd} \leq x_{1/2}$ (Sheriff, 2002; Hinze et al., 2013). See also *straight-slope method* and *Peters half-slope method*.

halo effect: **1.** an artificial ring of the opposite sign values around an *anomaly* that represents a side effect of the calculation of a high-order *vertical derivative* or *polynomial residual*. **2.** a micromagnetic ring-shaped anomaly associated with the *hydrocarbon seepage effect*. See also *micromagnetic anomalies*.

Hammer chart: a transparent template for the manual calculations of *terrain correction* by the *Hammer method*. The template has 13 concentric ring-shaped zones from A to M, each divided into equidimensional compartments. Compartment dimensions expand with the increasing zone bounding radii. The total number of compartments is 132. Bounding radii vary from 2.0 to 170.1 m for the first four zones and up to 21.94 km for the last zone. The template is prepared at the *scale* of a topographic map in a *gravity survey* area, and it is centered at the location of each *station* to calculate the difference between the station *elevation* and averaged elevation within each compartment. Obtained elevation difference values are used to determine the terrain corrections corresponding to each compartment around each station from the pre-calculated tables by Hammer (1939).

Hammer method: a method of the *terrain correction* calculation. It is based on dividing the *topography* around each observation point, or *station*, into a set of concentric ring-shaped zones segmented into equidimensional compartments whose dimensions increase with the increasing zone distance from the station. The difference (regardless of its sign) between the average *elevation* within a particular compartment and station elevation corresponds to a specific terrain correction value in the tables pre-calculated with a constant *density* of 2.0 g · cm^{-3}. All individual corrections are summed to obtain the total terrain correction which represents the *terrain effect* of all compartments around each station. The procedure is repeated for all stations in the *gravity survey* area. For any density other than 2.0 g · cm^{-3}, the obtained total terrain corrections are multiplied by the factor $\rho/2$, where ρ is the assumed rock density. Terrain corrections are calculated with the formula for *gravitational attraction* of a vertical hollow cylinder and expressed in units of *milligal (mGal)* as

$$g_{vhc} = 2\pi G\rho\,[R_2 - R_1 + (R_1^2 + h^2)^{1/2} - (R_2^2 + h^2)^{1/2}],$$

where G is *universal gravitational constant* in *cgs units*; R_1 and R_2 are the inner and outer radii, respectively; and h is height of the vertical cylinder defined by the difference between the station elevation and averaged elevation of each compartment. Bounding radii and terrain corrections corresponding to the elevation differences are tabulated by Hammer (1939) to the distance of 21.94 km from each station.

Hammer zones: concentric annular, or ring-shaped, zones segmented by radial lines into equidimensional compartments whose dimensions increase with the increasing distance between the zone and *station*. Hammer zones are modifications of the *Hayford-Bowie zones*, and their equidimensional compartments are obtained by relating the ratio of the outer and inner radii of each zone to a compartment width within the zone in order to maximize the *accuracy* of calculations and minimize the total number of compartments. There are 13 Hammer zones, from A to M, to the distance of 21.94 km for the last zone, beyond which the gravity data should be corrected for *curvature* of the Earth's surface. They are used for calculation of *terrain correction* by the *Hammer method*. See also *Hammer chart* and *Bullard B correction*.

Hannel method: a graphical method of estimating the depth to magnetic sources approximated by a *point dipole*. Source depth is equal to one-half of the horizontal distance at one-third of the *amplitude* on the *principal profile* of a *magnetic anomaly*. Also referred to as *Hannel rule*.

Hannel rule: See *Hannel method*.

Hardwick method: a method of *gradient-enhanced gridding* of aeromagnetic data. It is based on the use of measured lateral gradients for extrapolation of *total magnetic intensity (TMI)* values to the imaginary "pseudo-lines" between the actual flight lines. Extrapolation to the pseudo-line TMI_{PL} value is calculated as

$$TMI_{PL} = TMI_{AFL} + G_L(+/- \Delta Y),$$

where TMI_{AFL} is TMI value on the actual flight line, G_L is a measured *lateral gradient*, and $\pm\Delta Y$ is the iteratively determined distance to the pseudo-line on either ($\pm$) side of a particular flight line. Obtained TMI values along the pseudo-lines are included into the original dataset for subsequent *gridding*. Sometimes referred to as *pseudo-lines method* (Hardwick, 1999; 2004). See also *Nelson method*.

harmonic function: a mathematical function which satisfies three main criteria: (1) it has the continuous single-valued first-order derivatives, (2) it has the second-order derivatives, and (3) it obeys the second-order differential equation known as the *Laplace equation*. Outside their sources, the gravity and magnetic fields are harmonic. See also *derivative* and *homogeneous function*.

harmonics: **1.** elementary sine and cosine wave functions, or sinusoids, that are simple (whole number) multiples of *fundamental frequency* or simple fractions of *fundamental wavelength*. For example, the third and fourth harmonics of Fourier-transformed gravity and magnetic waveforms are their constituent sinusoids, whose *spatial frequency* values are equal to three and four times the value of the fundamental frequency or one-third and one-fourth of the fundamental wavelength, respectively. Two frequencies or wavelengths are harmonically related when they are harmonics of the same fundamental *frequency* or *wavelength*. **2.** all components of a *Fourier series* except for the fundamental frequency or fundamental wavelength, which is also known as the *first harmonic*. See also *spherical harmonics*.

Harrison correction: 1. a correction applied to marine (shipboard) and airborne gravity measurements to account for deviation of the sensing axis of a *shipboard gravimeter* or *airborne gravimeter* from the vertical orientation due to *tilt*, or deflection from the horizontal, of the inertial *stabilized platform* caused by strong horizontal inertial accelerations. For small angles of deflection, the correction value is expressed as

$$\Delta g_{Har} \cong -a_x\alpha - a_y\beta - \Delta g_z(\alpha^2 + \beta^2)/2,$$

where a_x is the x-component of horizontal *inertial acceleration*; a_y is the y-component of horizontal inertial acceleration; α and β are angles of deflection in the XZ and YZ planes, respectively, derived from measurements of the platform horizontal inertial accelerations by independent accelerometers; and Δg_z is true *relative gravity* value measured with the vertical sensing axis orientation. **2.** correction of *dynamic gravity* values for their artificial increase due to *off-level errors*. Also referred to as *horizontal acceleration correction*, *platform tilt correction*, and *off-level correction*. See also *Browne correction*.

Hayford-Bowie zones: concentric annular, or ring-shaped, zones segmented by radial lines into equidimensional compartments whose dimensions increase with the increasing zone distance from each *station*. There are 15 zones in the Hayford-Bowie system, from A to O, to the distance of 166.7 km. *Hammer zones* are derived from the Hayford-Bowie zones. Both Hayford-Bowie and Hammer zones are used interchangeably in the algorithms for calculation of three types of terrain corrections: (1) *innermost terrain correction*, (2) *inner-zone terrain correction*, and (3) *outer-zone terrain correction*. See also *terrain correction*.

heading: 1. the direction of an aircraft or ship movement as indicated by the navigation instruments during gravity or magnetic airborne and marine (shipboard) surveys. **2.** *azimuth*.

heading correction: a correction applied to marine (shipboard) magnetic and aeromagnetic data to compensate for the aggregated effects of incomplete *aeromagnetic compensation*, *positioning* errors, *lag error*, and other non-geological contributions to the measured magnetic values that are associated with the alternating movement directions of a *survey* ship or aircraft relative to the direction of the *Earth magnetic field*.

heading effect: the *aeromagnetic survey* effect resulting in the different *observed magnetics* values at the same (*x*,*y*) positions during flights in opposite directions. Referred to with the same meaning as *heading error*.

heading error: 1. a systematic error in *aeromagnetic survey* measurements that is associated with the alternating directions of an aircraft flight relative to the direction of the *Earth magnetic field* and caused by the aggregated effects of incomplete *aeromagnetic compensation*, *positioning* errors, and *lag error*. Accumulated heading errors can completely obscure the low-*amplitude* short-*wavelength* anomalies of an exploration interest and produce a *herringbone effect* on the aeromagnetic *grid*. Its total amount is estimated by a *heading error test* and compensated by a *heading correction*. **2.** a *reading* error in certain types of *quantum magnetometers*, such as a *potassium magnetometer*, that is related to changes in the orientation of a magnetometer *sensor* relative to the Earth's magnetic field direction. Its value depends on the amount of operational misalignment of the sensor axis or displacement from its original setting by the manufacturer due to mechanical shifts in the process of the *magnetometer* mounting.

heading error test: the aeromagnetic performance test conducted prior to regular *aeromagnetic survey* flights in order to estimate *heading error*. The test is flown in a clover-leaf pattern, which involves flights along the four principal directions of survey lines (east, west, north, south) over the area of low-*amplitude* magnetic anomalies or at a high *altitude*. Heading errors are determined by comparing the measured values obtained during the flights in the opposite directions. For an airborne *vector magnetometer* survey, the test is flown along eight different directions, or headings: 0°, 45°, 90°, 135°, 180°, 225°, 270°, and 315° (Christensen and Dransfield, 2002). Also known as *clover-leaf test*.

height correction: 1. a correction applied to *observed gravity* to compensate for the predicted, or theoretical, *free-air effect*, of *ellipsoidal height* which is the *GPS*-determined height of each observation point above a *reference ellipsoid*. **2.** a correction applied to *theoretical gravity* to compensate for the observation point height above a reference ellipsoid. Both definitions of this correction do not imply that its application changes the observation point positions, i.e., observation points are not moved to *survey datum*. Height correction is introduced as a more accurate version of the regular *free-air correction* calculated by the first-order approximation formula with the height h_{SL} above sea level and expressed as

$$\Delta g_{FA} \cong 0.3086 \cdot h_{SL}.$$

For the *GRS80 ellipsoid*, the second-order approximation formula for calculation of the height correction is expressed in units of *milligal (mGal)* as

$$\Delta g_h \cong (0.3087691 - 0.0004398 \sin^2\theta)h - 7.2125 \cdot 10^{-8}h^2,$$

where h is the ellipsoidal height in meters and θ is the geographic latitude in degrees. The first-order term in this formula represents the regular free-air correction, which is also referred to as *first-order free-air correction*. The difference between the height correction and regular free-air correction is a function of the geographic latitude and topography in a survey area, and it may be as large as several milligals at heights of several thousand meters. Height correction is subtracted from the theoretical gravity values, when observation points are above the reference ellipsoid. In geodetic applications, the analogue of the height correction is calculated with the use of *orthometric height* and the *Heiskanen-Moritz formula*. Also referred to as *second-order free-air correction*.

height effect: a decrease in measured gravity and magnetic field values with increasing height, or *elevation*, of observation points above a *survey datum*: sea level or *reference ellipsoid*. For gravity data, this effect is compensated by *free-air correction* or more accurate *height correction*. For aeromagnetic data, this effect is minimized, if necessary, by *drape correction* or *flight elevation correction*.

Heiskanen-Moritz formula: a formula for the most accurate calculation of gravity correction for the predicted, or theoretical, *free-air effect*, i.e., the gravity effect of height h of each observation point above a *reference ellipsoid*. It is expressed in units of *milligal (mGal)* as

$$\Delta g_{\mathrm{HM}} = (2g_{\mathrm{E}}/a)\{1 + f + m + [2.5m - 3f]\sin^2\theta\}h - 3g_{\mathrm{E}}h^2/a^2,$$

where g_E is *equatorial gravity*; f is *polar flattening* with $f=(a-b)/a$, where a is equatorial radius of the reference ellipsoid in meters and b is polar radius of the reference ellipsoid; $m=\omega^2a^2b^2/\mathrm{GM}$, where ω is *angular velocity* of the Earth's rotation in radians per second; GM is *geocentric gravitational constant* in *cgs units*; θ is latitude in degrees; and h is height of the observation point above a particular reference ellipsoid in meters. First-order approximation of this formula is *free-air correction* and second-order approximation is *height correction*.

Helbig integrals: integrals that define the first-order *integral magnetic moments* of a localized, or compact, magnetic *source body*. Zero-order integral moments become zeros, or vanish, over the range of integration from "$-\infty$" to "$+\infty$" in accordance with the *Gauss law*. Nonvanishing first-order integral moments relate the mutually orthogonal *vector* components of the magnetic field generated by a compact source body to their respective m_x, m_y, m_z components of *vector magnetic moment* **m** as

$$m_x = -(1/2\pi)\int\int_{-\infty}^{\infty} x\Delta Z\partial x\partial y,$$

$$m_y = -(1/2\pi)\int\int_{-\infty}^{\infty} y\Delta Z\partial x\partial y,$$

$$m_z = -(1/2\pi)\int\int_{-\infty}^{\infty} x\Delta X\partial x\partial y = -(1/2\pi)\int\int_{-\infty}^{\infty} y\Delta Y\partial x\partial y,$$

where ΔX, ΔY, ΔZ are vector components of the *magnetic anomaly* generated by the source body and $\mathbf{m}=(m_x, m_y, m_z)$ is the vector magnetic moment of the source body (Helbig, 1963). Vector magnetic moment components are calculated by the *Helbig method* and other techniques to determine *magnitude* |**m**| and spatial orientation (*magnetic inclination* and *magnetic declination*) of the *magnetization vector* of compact magnetic source bodies, such as a small *pluton* and mineral *ore body* (Phillips, 2005).

Helbig method: the method of estimating the *magnitude* and direction of the *magnetization vector* of a localized, or compact, magnetic *source body* that is based on calculation of the truncated first-order *Helbig integrals* by numerical integration of the gridded *vector* components of the *TMI field* within an empirically selected *moving window*. Vector components of the TMI field are determined with the use of a Fourier-domain *component filter*. The best-fit first-order, or planar, polynomial surface is removed from the component data within the moving window prior to the calculation of Helbig integrals. Magnitude |**m**|, *magnetic inclination* I_{csb}, and *magnetic declination* D_{csb} of the magnetization vector within a compact source body are defined as

$$|\mathbf{m}| = (m_x^2 + m_y^2 + m_z^2)^{1/2},$$

$$I_{csb} = \arcsin(m_z/|\mathbf{m}|),$$

$$D_{csb} = \arctan(m_y/m_x) \quad \text{for} \quad m_x \geq 0,$$

$$D_{csb} = \arctan(m_y/m_x) + 180^\circ \quad \text{for} \quad m_x < 0, m_y \geq 0,$$

$$D_{csb} = \arctan(m_y/m_x) - 180^\circ \quad \text{for} \quad m_x < 0, m_y < 0,$$

where m_x, m_y, m_z are components of the *vector magnetic moment* of the compact source body. Obtained results are presented as grids of the inclination and declination values determined for each localized magnetic source in a study area (Phillips, 2005). See also *Fourier domain*, *grid*, *gridding*, and *polynomial fitting*.

helicopter survey: **1.** high-*resolution* airborne measurements of the *Earth magnetic field* and its gradients with the use of a helicopter and high-accuracy *positioning* system, such as *differential GPS (DGPS)*. Better control over the flight path positioning and *altitude* makes possible the flying at the very low altitudes and close *survey line spacing* (20–50 m) to provide a much higher *lateral resolution* of anomalies than that of a *fixed-wing survey*. Measurements of *relative gravity* are conducted with the use of a *helicopter-suspended gravimeter* and mostly in the rugged terrains. **2.** helicopter-based high-resolution *gravity gradient* measurements of the *full tensor gradient (FTG)* components of the *Earth gravity field*. See also *drone survey*.

helicopter-suspended gravimeter: a fully automated *relative gravity* measuring system, where the *gravimeter* module is installed within a self-leveling tripod assembly suspended on a cable. When the tripod is on the ground, the ground-contact sensor generates a signal for servomotors to initiate the self-leveling process, which is followed by automatic gravity measurement at the *observation point*. The best reported *precision* of measurements is approximately 25 *microgal* (μ*Gal*).

helium magnetometer: an ultra-high *sensitivity* type of *optically pumped magnetometer* utilizing the helium isotopes He^3 or He^4 as its *magnetic sensor* and properties of *magnetic resonance* amplified by a laser pumping process. Achievable sensitivity is up to 50 *femtotesta (fT)*, where 1 fT $=10^{-6}$ *nanotesla (nT)*. A helium magnetometer is used as a *scalar magnetometer* to measure the *Earth magnetic field* intensity or as a *vector magnetometer* to measure its *vector* components with the same sensor. In the vector mode of operation, the *relative accuracy* of measurements is approximately 0.1% of the measured values, whereas in the *scalar* mode its relative accuracy is up to 0.001%. It is used primarily in *satellite magnetometry*. See also *absolute scalar magnetometer (ASM)*.

helium vector magnetometer: an *optically pumped magnetometer* with a helium isotope He^4 *magnetic sensor* for both the triaxial *vector* and omnidirectional *scalar* measurements of the magnetic fields of the Earth and other planets. In the vector mode of operation, the magnetic measurements of the *ambient field* are superposed with three mutually orthogonal low-frequency modulations of the auxiliary electromagnetic fields generated by the triaxial coil system. Vector components are reconstructed by processing the modulated scalar output signal containing the principal *harmonics* of the modulation from which the vector amplitudes and their respective orientations are derived. The three obtained vector components are used to calculate the *total magnetic vector* and determine its direction at each *observation point*. The achievable *accuracy* of measurements is about 50 *picotesla (pT)*. See also *absolute scalar magnetometer (ASM)*.

Helmholtz coils: an assembly of two identical coaxial coils separated by a distance equal to their radius, where the electric current flows of the same direction and strength in both coils generates a constant and uniform *magnetic field* between the coils that can be measured with high *accuracy*. It is used for the calibration of magnetometers and shielding the ambient magnetic field for measurements of the magnetic properties of rock samples. In the latter case, three pairs of square or circular coils are usually implemented. See also *ambient field* and *magnetometer*.

Helmholtz theorem: the theorem stating that any *vector field* **F**, which is continuous throughout the space and zero at infinity, can be expressed by the *gradient* of its *scalar* potential and *curl* of its *vector* potential as

$$\mathbf{F} = \nabla \mathrm{U} + \nabla \times \mathbf{A},$$

where ∇ is the gradient operator *del*, U is the scalar *potential* of vector field **F**, and **A** is its vector potential. In regions outside the mass and magnetic material and in the absence of electric currents (in the case of a magnetic field), both gravity and magnetic fields are irrotational, i.e., their curl is zero, and they are defined by gradients of their respective scalar potentials. See also *irrotational field* and *magnetic vector potential*.

henry: a unit of electrical inductance in the SI system of measurements. Inductance of one henry is equal to one volt induced into the conductive circuit by an electric current change of one *ampere* per second. The magnetic constant of proportionality is expressed in *SI units* as

$$C_m = \mu_0/4\pi = 10^{-7}\text{henry} \cdot \text{meter}^{-1},$$

where μ_0 is *magnetic permeability* of a source-free space. See also *Coulomb law* and *magnetic induction*.

herringbone effect: the systematic offset, or deviation, of map contours or color-coded values on a gravity or magnetic *grid* along the *survey* lines. Herringbone effects are produced by (a) displacement of one or several survey lines from their true coordinate locations, (b) accumulated *heading error* and *lag error* in the aeromagnetic measurements, (c) systematic mislocation of data recorded at adjacent lines due to accumulated *positioning* errors, and (d) shifts of *datum* in the process of *acquisition*. Subtle herringbone effects are more visible over the high-*gradient* areas of a gravity or magnetic survey. See also *corrugations* and *acquisition footprint*.

hertz (Hz): a unit of *frequency* in the SI measurement system. One hertz is equal to one cycle per second. In observations on a moving *observation platform*, it is used to describe the acquisition *sample rate* by the number of measurements per second or per 100 seconds: sample rate of 5 Hz is equal to 5 measurements per second, and sample rate of 0.18 Hz is equal to 18 measurements per 100 seconds. Estimates of the airborne *acquisition noise* and *accuracy* of the *airborne gravimeter (AG)* and *airborne gravity gradiometer (AGG)* systems are averaged over a measurement *bandwidth* expressed in the Hz values, so their respective root-mean-square (rms) estimates are presented in units of $1/\sqrt{Hz}$. See also *rms accuracy*.

Hessian: 1. *Hessian matrix*. **2.** *curvature matrix*.

Hessian eigensystem: a set of mutually orthogonal *eigenvectors* paired with *eigenvalues* that is obtained by *eigenvalue decomposition* of a two-dimensional (2D) or three-dimensional (3D) *Hessian matrix*. See also *FTG eigensystem*.

Hessian eigenvalues: *eigenvalues* obtained by *eigenvalue decomposition* of a two-dimensional (2D) or three-dimensional (3D) *Hessian matrix*. Hessian eigenvalues are paired with *eigenvectors* and implemented in quantitative estimates of the *curvature* parameters of an anomalous gravity or magnetic field and calculation of its *curvature attributes*. Combinations of 2D Hessian eigenvalues closely approximate *Gaussian curvature*, *maximum curvature*, and *mean curvature* of an anomalous gravity or magnetic field at each observation point. See also *anomalous field* and *FTG eigenvalues*.

Hessian matrix: a square symmetric 2 × 2 or 3 × 3 *matrix* whose elements are second-order derivatives of a mathematical function, such as the one which describes *potential field* or its *equipotential surface*. For the gravity or magnetic field F, the Hessian matrix is expressed as

$$H(x,y) = \begin{matrix} \partial^2 F/\partial x^2 & \partial^2 F/\partial x\partial y \\ \partial^2 F/\partial y\partial x & \partial^2 F/\partial y^2 \end{matrix}$$

$$H(x,y,z) = \begin{matrix} \partial^2 F/\partial x^2 & \partial^2 F/\partial x\partial y & \partial^2 F/\partial x\partial z \\ \partial^2 F/\partial y\partial x & \partial^2 F/\partial y^2 & \partial^2 F/\partial y\partial z \\ \partial^2 F/\partial z\partial x & \partial^2 F/\partial z\partial y & \partial^2 F/\partial z^2 \end{matrix}$$

Four elements of the two-dimensional (2D), or 2 × 2, Hessian matrix represent three different measures of *curvature* of an *anomalous field*, wherein $\partial^2 F/\partial x\partial y = \partial^2 F/\partial y\partial x$ due to the matrix symmetry. *Gaussian curvature* K_G is approximated by a determinant of the Hessian matrix and expressed as

$$K_G \cong \lambda_1\lambda_2,$$

maximum curvature is approximated as

$$K_{max} \approx (\lambda_1 + \lambda_2) + (\lambda_1^2 + \lambda_1\lambda_2 + \lambda_2^2)^{1/2},$$

and *mean curvature* is approximated by a trace of the Hessian matrix as

$$K_m \cong \lambda_1 + \lambda_2,$$

where λ_1 and λ_2 are *Hessian eigenvalues* paired with *eigenvectors* on the XY plane. Also referred to as 2D *Hessian* or 3D Hessian, respectively. See *FTG matrix, FTG determinant (det FTG), FTG trace*, and *vesselness.*

high-cut filter: *low-pass filter* in the context of its property to suppress, or cut off, all components of an *observed field* with spatial frequencies higher than *cutoff frequency.*

high-density basement: same as *density basement* or *gravity basement* that represents a major structural boundary in the subsurface with a high-*density contrast* across this boundary.

high-frequency magnetics: short-*wavelength* (high *spatial frequency*) magnetic anomalies of the sub-*nanotesla (nT)* range and of near-surface or shallow sedimentary origin. Their possible sources include local accumulations of *detrital magnetite* and *diagenetic magnetite*, mineral alterations due to the *hydrocarbon seepage effect*, salt or anhydrite *diamagnetism*, dissolution of *salt*, and along-the-*fault* mineralization by recurrent flows of *juvenile water.* Referred to interchangeably with the same meaning as *micromagnetic anomalies.*

high-line effect: a magnetic interference effect, often in the form of periodic *spike*-like anomalies, caused by electric transmission lines with a power frequency of 50 or 60 *hertz (Hz)*, and 16 2/3 Hz for some electric railways. High-line anomalies are eliminated from the obtained magnetic data by application of a *notch filter.* Sometimes referred to as *periodic noise.*

high-pass filter: 1. a spectral-domain *filter* which passes without suppression all components of an *observed field* with spatial frequencies higher than its *cutoff wavenumber.* **2.** a space-domain filter which passes without suppression all components of a gravity or magnetic field that have wavelengths shorter than its *cutoff wavelength.* Sometimes referred to as *low-cut filter* and *residual wavelength filter.* See also *low-pass filter, spatial frequency*, and *wavelength.*

high-pass roll-off filter: a spectral-domain *filter* which retains, or passes, the high-*wavenumber* part of a *power spectrum* using a smooth curve in its *roll-off window* in order to minimize *ringing.* See also *cosine roll-off filter* and *spectral domain.*

high-resolution aeromagnetic survey: 1. a detailed *aeromagnetic survey* flown by airplane at the lowest *altitude* permitted by safety regulations (usually 80–150 m) with a distance between flight lines 100–300 m and high-accuracy *positioning* by *differential GPS (DGPS)* or other analogous *GNSS.* To prevent *aliasing* of the short-*wavelength* components of the *Earth magnetic field*, the ratio of *survey height* to *survey line spacing* should be 1:1 or 1:2 (Reid, 1980). Measurements are made by a high-*sensitivity* (0.001–0.005 nT) *magnetometer* and recorded at a *sample rate* providing the data *sampling interval* of 5–7 m on the ground. **2.** *drone survey* or *helicopter survey.* Often referred to as *HRAM survey.* See also *detailed survey, nanotesla (nT)*, and *survey design aliasing.*

high-resolution aeromagnetics (HRAM): high-*resolution* aeromagnetic data acquired along the closely spaced *airborne survey* lines in an area of exploration or research interest during one of three types of airborne magnetic surveys: *drone survey, helicopter survey*, and *fixed-wing survey.* See also *acquisition* and *spatial resolution.*

high-resolution ground survey: a ground magnetic or gravity *survey* with *station* intervals within the range of 0.1–5.0 m and *survey line spacing* of 0.2–10.0 m. It is used in detailed studies of mineral prospects, such as *placer* deposits, and in the *UXO detection*, archaeological and forensic applications. See also *forensic magnetism, magnetic archaeology*, and *prospect.*

Hilbert transform: 1. a data transformation the basic property of which is the *phase* shift by 90° without a change of *amplitude spectrum*: it keeps the amplitudes of anomalies unchanged but changes their phases by ±90° ($\pm\pi/2$), depending on the sign the *anomaly amplitude.* The Hilbert transform is used for calculation of the *vertical derivative* of an anomalous gravity or magnetic *field* from its *horizontal derivative* (with the assumption that both have the same amplitude) and for other *enhanced processing* applications. For the one-dimensional function $f(x)$ representing line-oriented *anomalous field* data, its *analytic signal (AS)* and Hilbert transform are related as

$$\mathrm{AS}(x) = f(x) + i\mathrm{H}f(x) = \mathrm{A}(x)e^{i\varphi(x)},$$

where $i = \sqrt{-1}$ is a complex unit number indicating the orthogonality, $\mathrm{H}f(x)$ is the Hilbert transform of $f(x)$, $\mathrm{A}(x)$ is *analytic signal amplitude (ASA)* defined as

$$\mathrm{A}(x) = |\mathrm{AS}| = \{[f(x)]^2 + [\mathrm{H}f(x)]^2\}^{1/2},$$

and $\varphi(x)$ is *analytic signal phase* defined as

$$\varphi(x) = \arctan[\mathrm{H}f(x)/f(x)].$$

2. a *transform* which converts original data into complex variables composed of the "real" and "imaginary" parts, where the real part represents the original data and the imaginary part is the 90° phase-shifted real part. The imaginary part of the *complex variable* obtained by the Hilbert transform is also referred to as *quadrature* or *quadrature signal.* See also *Riesz transform, 3D Hilbert transform*, and *direct analytic signal (DAS).*

Hilbert transform pair: the relationship between components of a *complex variable*, where the real and imaginary parts have the same *amplitude spectrum* but differ in *phase* by 90°. Horizontal and vertical derivatives of a gravity or magnetic field represent the Hilbert transform pair in an *analytic signal (AS).* Also referred to as *quadrature pair.* See also *derivative.*

hillshading: *sunshading* in the context of visualizing gridded gravity or magnetic anomalies as a topographic surface composed of hills and valleys illuminated by an artificial source of light from different directions.

histogram equalization: the image enhancement algorithm which finds the maximum and minimum values of *color-coded grid* and extends, or stretches, its color shade range to assign an equal number of contour lines or equal areas of a *grid* to each color shade. Small anomalous zones of low and high values can become better visible by the color equalization of the gridded gravity and magnetic fields. See also *amplitude balancing.*

historical IGRF model: See *IGRF model.*

homogeneous field: 1. a gravity or magnetic *field* described by the *homogeneous function* of the degree *n* which is the same integer value regardless of the source-observation distance or level of *upward continuation* of the field. A homogeneous field is used as a local approximation of an *anomalous field* within a pre-selected width of a *moving window* for calculation of the source parameters in *Euler deconvolution.* It satisfies the *Euler deconvolution equation* with *Euler structural index (SI)* denoted by *N* and defined as

$$N = -n,$$

where structural index *N* is the integer value varying from 0 to 3 for magnetic sources and from –1 to 2 for gravity sources. **2.** the gravity or magnetic field generated by one of the four *ideal source* geometries: *point source* (*sphere*), line of the point sources (*horizontal cylinder* or *vertical cylinder*), *thin sheet*, and *infinite contact.* See also *homogeneous source*, *multihomogeneous field*, *nonhomogeneous field*, and *fractional structural index (fSI).*

homogeneous function: a mathematical function of a set of variables that satisfies the equation

$$\mathrm{F}(kx, ky, \ldots ku) = k^n \mathrm{F}(x, y, \ldots u).$$

When all variables are multiplied by the same coefficient *k*, the function F(*x*,*y*, … *u*) is multiplied by the power *n* of this coefficient, and the function is said to be homogeneous of the degree *n*. Functions of the type $F(r) = C/r^N$ (where C is a constant) that describe *amplitude attenuation*, or *fall-off*, of the gravity or magnetic *anomaly* with the distance r from its source, are homogeneous of the degree $N = -n$. Gravity and magnetic fields generated by the *ideal source* geometries of *sphere*, *cylinder*, *thin sheet*, and *infinite contact* are homogeneous functions that satisfy the *Euler homogeneity equation*, and they are used as the local approximations of the *anomalous field* for calculation of real source parameters by means of the *Euler deconvolution equation* within a *moving window.* See also *homogeneous field* and *nonhomogeneous field.*

homogeneous source: a type of *ideal source* that approximates real sources in the theory of *Euler deconvolution* and generates the *homogeneous field.* There are four basic homogeneous sources, each having its own integer *Euler structural index (SI)*: *sphere* or *point source* with SI = 3 and SI = 2 for magnetic and gravity sources, respectively; line of point sources approximated by a *horizontal cylinder* or *vertical cylinder* (SI = 2 and 1); *thin sheet* or *thin dike* (SI = 1 and 0); and *infinite contact* (SI = 0 and –1). The SI value of the homogeneous source is constant regardless of the source-observation distance and level of *upward continuation* of the *anomalous field.* There are also more complex homogeneous source distributions generating homogeneous fields and characterized by the *fractional structural index (fSI)*, whose value remains constant regardless of the source-observation distance (Fedi et al., 2015). See also *nonhomogeneous source.*

Honkasalo correction: a correction for the latitude-dependent gravity effect of *permanent tide* representing the average value of the combined *gravitational attraction* of the Sun and Moon. It is defined in units of *milligal (mGal)* as

$$\Delta g_H = 0.0371(1 - 3\sin^2\theta),$$

where θ is the geographic latitude of a *station* in the northern or southern hemisphere in degrees. Honkasalo correction values vary from +0.037 mGal at the equator to –0.019 mGal at the 45° latitude and to –0.074 mGal at the geographic poles. This correction was initially included into the *absolute gravity* values in the database of the *International Gravity Standardization Network 1971 (IGSN71).* Subsequently, the International Association of Geodesy (IAG) recommended the removal of the Honkasalo correction from the IGSN71 database. Also known as *Honkasalo term* and *Honkasalo tidal correction.* See also *Earth tide correction.*

Honkasalo term: *Honkasalo correction* in the context of its application as the additional member, or term, in corrections applied to *absolute gravity* data.

Honkasalo tidal correction: same as *Honkasalo correction* that is used as a correction for the latitude-dependent gravity effect of *permanent tide* representing the average value of the combined *gravitational attraction* of the Sun and Moon.

horizontal acceleration correction: a correction applied to gravity data acquired on a moving *observation platform* to compensate for the interfering effects of the inertial accelerations resulting from the horizontal motion of a *survey* ship or aircraft. There are two horizontal acceleration corrections: *Browne correction* and *Harrison correction.* See also *inertial acceleration* and *inertial acceleration corrections.*

horizontal cylinder: 1. one of the *ideal source* geometries equivalent to the horizontal line of point sources, or spheres, that generates a *homogeneous field* which is used as a local approximation of the *anomalous field* in *Euler deconvolution.* Its *Euler structural index (SI)* is the integer value of 2. **2.** a two-dimensional (2D) *source body* approximation for anomalies elongated in one of the horizontal directions and having approximately the same shape of their profiles orthogonal to this direction. Its gravity or magnetic anomalous *amplitude* at the observation point is proportional to the product of the density or magnetization *contrast* and cross-section of the horizontal cylinder. It is used in a *3D model* for approximation of anticlines, synclines, paleochannels, engineering tunnels, and other subsurface objects, where one of the two orthogonal horizontal dimensions is at least 5–6 times longer than the other. See also *anticline*, *syncline*, *paleochannel*, *point source*, and *vertical cylinder.*

horizontal DAS (HDAS): an edge-detection *attribute* of the *anomalous field* F, such as *Bouguer anomaly* or *RTP anomaly*, that represents the horizontal component of the *direct analytic signal (DAS)*, expressed as

$$\mathrm{HDAS} = \{[H_x(\mathrm{F})]^2 + [H_y(\mathrm{F})]^2\}^{1/2},$$

where H_x and H_y are x- and y-components of the *3D Hilbert transform* of the anomalous field F in the orthogonal x- and y-directions, respectively (Luo et al., 2011). The HDAS is equivalent to the *Riesz transform attribute* of a *monogenic signal (MS)*. See also *edge*, *edge-detector attribute*, and *edge detectors*.

horizontal datum: a *reference ellipsoid* relative to which the geographic position (latitude and longitude) of observation points is determined. Regional and national geographic datums implement different reference ellipsoids approximating the most closely the surface of the Earth over particular regions rather than the whole Earth surface. The *International Terrestrial Reference Frame (ITRF)*, which is referenced to the *GRS80 ellipsoid*, is recommended as the standard global ground-based horizontal datum. Average differences between the high-*accuracy* horizontal coordinates in the satellite-based *Global Positioning System (GPS)*, which is referenced to the *WGS84 ellipsoid*, and ITRF coordinates are within 10 cm. Unlike vertical *positioning*, such small differences in horizontal position are disregarded in gravity and magnetic explorations. Also referred to as *geographic datum*. See also *regional geodetic datum* and *vertical datum*.

horizontal derivative: the calculated rate of change of the gravity or magnetic field F per unit of horizontal distance. It is denoted by $\partial F/\partial x$ or $\partial F/\partial y$, where x and y are mutually orthogonal horizontal directions in *Cartesian coordinates*. The horizontal derivative is a *vector* quantity, but its *scalar* value, calculated by the *finite-difference method*, is used the most often. The first-order horizontal derivative is zero at the maxima and minima of an *anomalous field*, and its *principal profile* over a *source body* approximated by a *thin sheet* geometry has a characteristic *peak*-to-*trough* or trough-to-peak shape depending on the positive or negative *magnetization contrast* or *density contrast* created by the source body, wherein the transitional zero value is above the source center. Over the *contact* source geometries, the horizontal derivative profile has the shape of either a peak or trough depending on the positive or negative lateral magnetization or density *contrast*, wherein the respective maximum or minimum value marks the location of a high-offset *fault*, lithological boundary, or other vertical contact. The first-order horizontal derivative is a measure of the slope, or *dip*, of an anomalous gravity or magnetic field, whereas the second-order horizontal derivative is a measure of its *curvature*. See also *horizontal gradient (HG)* and *vertical derivative*.

horizontal directive tendency (HDT): the *magnitude*, or absolute value, of *differential curvature* K_D of the gravitational *equipotential surface*. The HDT is calculated without the scaling *vertical gravity* g_z value in the original K_D formula and expressed as

$$|K_D| = [(g_{xx} - g_{yy})^2 + (2g_{xy})^2]^{1/2} = 2(g_{NE}^2 + g_{UV}^2)^{1/2},$$

where $g_{xx} = \partial^2 U/\partial x^2$, $g_{yy} = \partial^2 U/\partial y^2$, $g_{xy} = \partial^2 U/\partial x \partial y$ are *curvature gradients*; U is *gravitational potential*; and $g_{NE} = g_{xy}$ and $g_{UV} = (g_{xx} - g_{yy})/2$ are *differential curvature gradients*. Sometimes referred to as *differential curvature amplitude*. See also *curvature*.

horizontal gradient (HG): 1. the measured rate of change of the gravity or magnetic field F per unit of horizontal distance. **2.** a *scalar* quantity representing the absolute value of the *horizontal derivative*. **3.** a *vector* quantity representing the horizontal component of the *gradient vector*. The HG can be measured by a *gradiometer* or calculated from the gridded field as the square root of the sum of the squared horizontal derivatives $\partial F/\partial x$ and $\partial F/\partial y$ in the orthogonal x- and y-directions:

$$HG = [(\partial F/\partial x)^2 + (\partial F/\partial y)^2]^{1/2},$$

where F is the anomalous gravity or magnetic field. Grid-calculated HG is also referred to as *total horizontal derivative* and *total horizontal gradient*. **4.** the horizontal rate of change of the vertical component of *gravitational acceleration* g_z, wherein its numerical value and azimuthal direction of the largest variation α_H are calculated from the components g_{zx} and g_{zy} of the *full tensor gravity gradient (FTGG)* as:

$$HG = (g_{zx}^2 + g_{zy}^2)^{1/2} = HI_1, \quad \alpha_H = \arctan(g_{zy}/g_{zx}),$$

where $g_{zx} = \partial^2 U/\partial z \partial x$, $g_{zx} = \partial^2 U/\partial z \partial y$, U is *gravitational potential*, and HI_1 is one of two *horizontal invariants* showing that the HG is independent of, or invariant to, the orientation of the x- and y-axes in *Cartesian coordinates*. HG tends to place its maxima over the abrupt lateral changes in *density* and *magnetization* associated with structural and lithological contrasts, such as high-*offset* faults or edges of igneous intrusions and *salt* bodies. The higher the *contrast*, the higher the *horizontal gradient amplitude (HGA)*. Its *principal profile* over the *thin sheet* (*thin dike*) and *horizontal cylinder* source geometries has a characteristic trough-between-peaks shape with the *trough* positioned immediately above the horizontal cylinder and vertical or high-angle thin sheet. The HG has its maximum value above the vertical *contact*, whereas for a non-vertical case its maximum is displaced down-*dip* from the contact top, and this property is used in *gradient-dip estimation*. Horizontal gradients of deep-sourced long-*wavelength* anomalies are commonly low and, hence, an HG *grid* highlights the gradient anomalies associated with the short- and mid-wavelength components of an *anomalous field*. The HG property to acquire its maximum values over the boundaries of laterally contrasting rock units is implemented in *edge-detection methods*. See also *edge*, *fault*, *igneous intrusion*, and *vertical gradient (VG)*.

horizontal gradient amplitude (HGA): 1. a numerical value of the *horizontal gradient (HG)*. For measurements by *gradiometer* and calculations from line-oriented *total field* data, HGA values are (a) positive for an increasing field with the *peak* above a laterally positive *density contrast* or *magnetization contrast*, (b) negative for a decreasing field with the *trough* over a laterally negative *contrast*, and (c) zero at the total field maxima. For gridded data, the calculated HGA is always positive regardless of the positive or negative sign of the gravity or magnetic *anomaly*, and for that reason it is also referred to as *horizontal gradient magnitude (HGM)*. **2.** the *magnitude*, or length, of a *2D gradient vector*. The HGA grids highlight structural discontinuities, such as local large-*offset* faults, regional tectonic boundaries, and *source body* edges. For magnetic fields, the HGA is commonly calculated from the reduced-to-pole (RTP) magnetic data. When applied to *anomalous potential*, HGA defines the horizontal component of the anomalous gravity or magnetic field. The HGA of *tilt angle* represents the absolute value of the *horizontal local wavenumber* and is referred to as *total horizontal derivative of tilt* and *horizontal gradient of tilt*. Decay, or *fall-off*, of the gravity or magnetic HGA with the distance r from a *source body* is proportional to $1/r^2$ or $1/r^3$ for two-dimensional (2D) sources and to $1/r^3$ or $1/r^4$ for a *point source*, respectively. See also *contact*, *fault*, *grid*, and *reduction-to-pole (RTP)*.

horizontal gradient azimuth: the *phase*, or orientation, of a *2D gradient vector* on the XY plane. It is defined as

$$\phi = \arctan[(\partial F/\partial y)/(\partial F/\partial x)],$$

where $\partial F/\partial y$ and $\partial F/\partial x$ are horizontal derivatives of the gridded *anomalous field* F along the mutually orthogonal x- and y-axes, respectively, in *Cartesian coordinates*. For magnetic fields, it defines *magnetic declination* at the observation point. See also *derivative*, *gradient*, and *gridding*.

horizontal gradient magnitude (HGM): same as *horizontal gradient amplitude (HGA)* with an emphasis on its positive value regardless of the positive or negative sign of the gravity or magnetic *anomaly* in the gridded data.

horizontal gradient of tilt: See *total horizontal derivative of tilt*.

horizontal gradient vector: same as *2D gradient vector* with an emphasis on its horizontal direction in the three-dimensional (3D) *vector* space of the gravity and magnetic fields.

horizontal invariants: *invariants* calculated from the *full tensor gradient (FTG)* components which represent derivatives of the horizontal components of an anomalous gravity or magnetic field. There are two horizontal invariants HI_1 and HI_2, expressed as

$$HI_1 = (T_{xz}^2 + T_{yz}^2)^{1/2}, \quad HI_2 = \{T_{xy}^2 + [(T_{yy} - T_{xx})/2]^2\}^{1/2}.$$

Grids of horizontal invariants are interpreted in integration with gridded T_{zz} values which represent the *vertical derivative* of the vertical component T_z of an *anomalous field*. Horizontal invariants highlight *lineaments* associated with *fault*-bounded structural trends, lithological contacts, and *source body* edges, whereas the T_{zz} component provides information about the location, size, and depth of isolated source bodies and other elements of the geological *structure* (Murphy, 2007). Referred to with the same meaning as *lineament invariants*. See also *contact*, *derivative*, and *gridding*.

horizontal local wavenumber: a *horizontal derivative* of the *local phase* that defines the rate of change of the local phase θ along the horizontal direction, expressed as

$$k_x = \partial\theta/\partial x = \partial\{\arctan[(\partial M/\partial z)/(\partial M/\partial x)]\}/\partial x$$

$$= |ASA|^{-2}[(\partial^2 M/\partial x\partial z)(\partial M/\partial x) - (\partial^2 M/\partial x^2)(\partial M/\partial z)];$$

where M is *magnitude* of the gravity or magnetic field and |ASA| is the two-dimensional (2D) *analytic signal amplitude (ASA)*. Because the local phase is equivalent to the *analytic signal phase*, the horizontal local wavenumber is also expressed as

$$k_x = (\partial|ASA|/\partial z)/|ASA|.$$

The horizontal local wavenumber is equivalent to (a) the normalized *vertical derivative* of the ASA and (b) the *tilt derivative* calculated in the horizontal direction, wherein its components are represented by the second-order derivatives. Combinations of the horizontal local wavenumber and *vertical local wavenumber* are implemented in the *source parameter imaging (SPI™) method*, *improved SPI™ method*, *advanced source-parameter imaging method*, and *enhanced local wavenumber method* for estimates of depth, *Euler structural index (SI)*, and other parameters of the 2D magnetic sources approximated by the *magnetic contact*, *thin dike*, and *horizontal cylinder* source geometries (Thurston and Smith, 1997; Smith and Salem, 2005; Salem et al., 2005). Also known as *first-order local wavenumber*. See also *derivative*, *local wavenumber*, *second-order local wavenumber*, and *vertical local wavenumber*.

horizontal magnetic intensity: 1. the *vector* quantity **H** defined by the vector sum of two horizontal components of a three-dimensional (3D) *geomagnetic vector* measured by a triaxial *vector magnetometer*. **2.** the *scalar* quantity H which defines the absolute value, or *magnitude*, of the *total horizontal component* of a geomagnetic vector calculated from measurements by a triaxial vector magnetometer. See also *vertical intensity*, *total intensity*, and *geomagnetic field parameters*.

horizontal sheet: a *model* approximation of (a) a faulted rock formation wherein the *offset*, i.e., vertical displacement of rocks along a deep-rooted *fault*, is much less than its depth; (b) a predominantly horizontal *source body*, such as a suprabasement or intrasedimentary *sill* and any other relatively thin flat-laying source body of finite thickness and much larger length. See also *vertical sheet* and *finite contact*.

horizontal slab effect: *Bouguer effect* in the context of the *gravitation attraction*, or gravity effect, of an infinite horizontal plate (slab).

horizontal spatial resolution: same as *lateral resolution* that represents the horizontal extent, or width, of the narrowest *anomaly* which can be mapped by a particular gravity or magnetic *survey* or identified in the gridded data. See also *gridding*.

horizontal tensor gradiometer: same as *partial tensor gravity gradiometer* with an emphasis on its use of the horizontal rotating disk(s) with paired and synchronized accelerometers.

horizontal tilt angle (TDX): an edge-detection *attribute* defined by the arctangent of the ratio of the *total horizontal derivative (THDR)*, also known as *horizontal gradient magnitude (HGM)*, to the absolute value of the *first vertical derivative (FVD)* of the *anomalous field* F:

$$TDX = \arctan\{[(\partial F/\partial x)^2 + (\partial F/\partial y)^2]^{1/2}/|\partial F/\partial z|\}$$

$$= \arctan\{HGM/|FVD|\}.$$

In terms of spatial geometry, TDX represents the angle between the opposite side and hypotenuse of a right-angled triangle, where the adjacent side is the THDR, the opposite side is the FVD, and the hypotenuse is *analytic signal amplitude (ASA)*. The TDX values vary from 0° to 90°, and it is complementary to the *tilt angle (TA)*, i.e., TDX = $\pi/2$ – TA. The THDR *normalization* by the absolute value of the FVD has an *amplitude*-equalizing effect analogous to *automatic gain control (AGC)*, so the TDX can resolve both shallow- and deep-sourced anomalies, and it has a maximum value of 90° above *source body* edges. The TDX is different from the *total horizontal derivative of tilt*, because the TDX calculation does not involve the second-order derivatives and, hence, it is less sensitive to *noise*. Also referred to as *TDX attribute* (Cooper and Cowan, 2006). See also *edge-detection methods*, *directional tilt angle*, and *normalized derivatives*.

horizontal variometer: a *gravity gradiometer* for measurements of the horizontal gradients of the vertical component g_z of the Earth's gravity field *vector* **g** that are defined as

$$g_{zx} = \partial g_z/\partial x = \partial^2 \mathrm{U}/\partial z \partial x$$

$$g_{zy} = \partial g_z/\partial y = \partial^2 \mathrm{U}/\partial z \partial y,$$

where U is *gravitational potential*. See also *curvature variometer* and *horizontal gradient*.

horst: the elongated and elevated crustal *block structure* bounded on both sides by normal faults which dip away from one another and generate *gradient* anomalies. In *Bouguer gravity*, the horst has the *gravity signature* of an elongated *gravity high*. In the *RTP magnetic field*, the intrasedimentary horst can have no *magnetic signature* due to the low *magnetization* contrasts between sedimentary rocks, whereas an upper-*basement* horst has the *signature* of a positive or negative elongated *magnetic anomaly*, whose sign depends on the dominant type of magnetization of its igneous or metamorphic rocks: *induced magnetization* or *remanent magnetization*, respectively. Amplitudes of a horst-generated gravity and magnetic *total field* and gradient anomalies are controlled, primarily, by the amount of vertical rock displacement, or *offset*, and lateral *density contrast* or *magnetization contrast* with the adjacent rock formations. Horst signatures are highlighted by calculation of *residual gravity* and *residual magnetics*. See also *dip*, *fault*, *flower structure*, and *graben*.

hot spot: 1. a volcano which has been active for millions of years. Hot spots are surface expressions of the *mantle* plumes forming the nearly linear chains of the extinct volcanoes on the lithospheric plates that have passed over mantle plumes. Hot spots commonly have the strong anomalous signatures of circular- or oval-shaped positive gravity anomalies and positive or negative magnetic anomalies, whose sign depends on the dominant type of *magnetization* of the constituent igneous rocks: *induced magnetization* or *remanent magnetization*, respectively. **2.** a zone of increased fracturing at the intersection of faults mapped by a *high-resolution aeromagnetic survey* for oil-and-gas exploration of fractured reservoirs. See also *fault*, *lithosphere*, *signature*, and *bright spot*.

hot spot track: a regional-scale chain of extinct volcanoes or a ridge of volcanic rocks formed when the lithospheric plate was moving over a *mantle* plume. It is identified by a linear or curvilinear *trend* of the high-*amplitude* gravity and magnetic anomalies, primarily in the oceans. See also *gravity anomaly*, *magnetic anomaly*, and *lithosphere*.

HRAM data: See *high-resolution aeromagnetics (HRAM)*.

HRAM survey: See *high-resolution aeromagnetic survey*.

Hubbert method: a method of calculating the anomalous gravity effect Δg of a two-dimensional (2D) *source body* of an arbitrary-shaped cross-section and constant *density contrast* with the use of the line integrals and two formulas as

$$\Delta g = 2\mathrm{G}\Delta\rho \oint \theta \partial z \quad \text{or} \quad \Delta g = 2\mathrm{G}\Delta\rho \oint z \partial\theta,$$

where G is *universal gravitational constant* in *cgs units*, $\Delta\rho$ is density contrast, z is vertical coordinate on the XZ plane, and θ is angle in radians between the horizontal level $z = 0$ and radial line connecting the calculation point with a point on the boundary of the cross-section, which is orthogonal to the long axis of the source body. Calculations are repeated by moving the calculation point over the horizontal surface, until the calculated anomalous gravity effect becomes negligible at the distant points (Hubbert, 1947). See also *forward modeling*, *radian*, and *Talwani method*.

hybrid density model: a three-dimensional (3D) *gravity model* that combines a *voxel model* of the laterally and vertically variable rock densities in a targeted *source body*, such as *allochthonous salt* or mineral *ore body*, with the constant rock densities in other model layers.

hybrid gravity: *gravity data* obtained from measurements of both *relative gravity* and *absolute gravity*, such as those conducted for monitoring the active volcanos, hot water withdrawal from geothermal reservoirs, enhanced oil-and-gas recovery, and depletion of groundwater aquifers. See also *4D gravity monitoring*.

hybrid inversion: 1. three-dimensional (3D) *gravity inversion* with the use of a *hybrid density model*. **2.** *property inversion* with the use of *hybrid gravity* for upgrading a 3D *density model* built for waterflood surveillance of oil or gas reservoir development, enhanced heavy oil recovery with steam injection, and monitoring of geothermal reservoirs, groundwater aquifers, and active volcanos. Hybrid gravity inversion for the base of *salt* is performed to obtain *constraints* for the initial seismic velocity *model* prior to the seismic prestack or poststack migration. See also *joint inversion*, *gravity-constrained velocity model*, and *4D gravity monitoring*.

hybrid time-lapse gravity: *time-lapse gravity* obtained from the *relative gravity* measurements during a *time-lapse gravity survey* and complemented by *absolute gravity* measurements at the same or selected reference, or benchmark, observation points.

hydrocarbon seepage effect: the secondary magnetic effect produced by changes in rock composition caused by migration of hydrocarbon fluids along faults, fractures, joints, and bedding planes within a sedimentary section. Plumes of rising hydrocarbons above oil-and-gas accumulations can create favorable conditions for the growth of *magnetite*-producing bacteria, precipitation of ferrous iron oxides and sulfides, and microbial alteration of weakly magnetic hematite into a strongly magnetic *magnetite* (Reynolds et al., 1993). Hydrocarbon seepage effects can enhance *magnetization* of rocks altered by the hydrocarbons and related fluids and produce the low-amplitude *micromagnetic anomalies* detectable by a *high-resolution aeromagnetic survey*. Burning of the hydrocarbon seepage fluids within the near-surface parts of faults and fracture zones can generate short-*wavelength* high-*amplitude* magnetic anomalies up to hundreds of *nanotesla (nT)*. Also referred to as *micromagnetic seep effect*. See also *chimney*, *combustion magnetism*, *clinker anomaly*, *magnetic bright spot*, *magnetized fault*, and *micromagnetic aureoles*.

hydromagnetics: See *magnetohydrodynamics*.

hysteresis: **1.** a property of *ferrimagnetic* and *ferromagnetic* minerals to retain a portion of their initial *magnetization* after the removal of an externally applied magnetic field. Retained, or remaining, magnetization is called *remanent magnetization*. **2.** a property of the *spring gravimeter* known as *spring hysteresis*.

hysteresis loop: a graphical representation of a full cycle of *magnetic saturation* of rock samples. It describes a non-linear relationship between the intensity of *magnetization* of a rock sample and variations of the external magnetizing field. The cycle begins with an increase in the intensity of the external field that causes a non-linear increase in magnetization of the initially demagnetized rock sample up to its saturation level. When the magnetizing field is decreased back to zero, the rock sample retains a portion of its magnetization that is called *remanent magnetization* or *remanence*. Reversing the polarity of the external magnetic field and increasing its intensity will reduce the rock sample remanence to zero. Further increase of the reversed magnetizing field intensity results in a negative saturation magnetization of the rock sample at the point that is antipodal (same value but opposite sign) to the positive saturation point. The second half of the rock sample magnetization cycle repeats this sequence in the reverse order, so the hysteresis loop is closed along the curve which is antipodal to the non-linear curve describing the transition from positive to negative saturation magnetization.

I

ice-cap gravity: *field* of the gravity anomalies obtained from measurements on top of the variable thickness ice cap or ice sheet. For the ice bottom above sea level, the *complete Bouguer anomaly* Δg_B value at each *observation point* is expressed in units of *milligal (mGal)* as

$$\Delta g_B = g_{obs} - \Delta g_{TG} + 0.3086h - 2\pi G[\rho_{ice}d + \rho_r(h - d)] + \Delta g_T,$$

where g_{obs} is *observed gravity*, Δg_{TG} is *theoretical gravity correction*, h is *elevation* of the ice sheet or cap above sea level, G is *universal gravitational constant*, d is ice thickness, ρ_{ice} is ice *density*, ρ_r is assumed density of the underlying rocks, and Δg_T is *terrain correction* (Hinze et al., 2013). See also *gravity anomaly* and *lake gravity*.

ideal source: 1. a *source body* having the shape of a simple geometric form, such as *sphere*, *plate*, *prism*, *vertical sheet*, *horizontal sheet*, *horizontal cylinder*, and *vertical cylinder*. Gravity and magnetic fields of ideal sources are defined by the closed-form analytical expressions, or formulas, implemented in the algorithms for *forward modeling* and *inversion*. Ideal sources are used as the elementary fill-in blocks the analytically expressed gravity or magnetic fields of which are numerically integrated at each observation point to obtain the corresponding gravity or magnetic *anomaly* of the arbitrary-shaped *general source*. **2.** a *homogeneous source*.

ideal spring: same as *virtual spring* with an emphasis on the absence of the mechanical deficiencies of metal and quartz springs, such as *creep* and *tare*.

ideal-spring gravimeter: same as *superconducting gravimeter* with an emphasis on its operating principle based on the use of *electrostatic levitation* or *magnetic levitation* instead of a metal or quartz spring.

IGF: See *International Gravity Formula (IGF)*.

IGF1930: See *International Gravity Formula 1930 (IGF1930)*.

IGF1967: See *International Gravity Formula 1967 (IGF1967)*.

IGF1980: See *International Gravity Formula 1980 (IGF1980)*.

IGF1987: See *International Gravity Formula 1987 (IGF1987)*.

IGF1999: See *International Gravity Formula 1999 (IGF1999)*.

igneous body: igneous rock solidified in the process of cooling of a molten *magma* intruded into the *basement* or sedimentary section. Igneous body *density* and *magnetization* are among the highest of all rock types. Due to the high *density contrast* and *magnetization contrast* with surrounding rock formations, the igneous bodies are the most common sources of high-*amplitude* magnetic and gravity anomalies. Pre-depositional (i.e., buried under the subsequent deposition of sediments) igneous bodies and their brecciated alterations can create the structural and stratigraphic traps for accumulation of hydrocarbons. Post-depositional igneous body intrusions often damage and destroy the hydrocarbon accumulations in their immediate vicinity. Gravity *total field* and *magnetic gradient* signatures of isolated and locally extended intrasedimentary igneous bodies can be similar to those of *reef* and *horst*, respectively. Igneous bodies are commonly associated with a wide range of the mineral deposits. Alkaline igneous bodies, such as *carbonatite*, are quite rare, and they are unusually enriched in mineral elements such as niobium, lithium, and *rare earth elements (REE)*. For depth estimates, the igneous body is approximated by simple *two-dimensional (2D)* and *three-dimensional (3D)* geometrical shapes of *thin dike*, *thick dike*, *vertical cylinder*, *horizontal cylinder*, and *sphere*. Mafic and ultramafic igneous bodies have a strong *remanent magnetization* component. Igneous bodies are prolific sources of the thermal energy and highly mineralized hydrothermal fluid known as *juvenile water*. Also referred to as *igneous intrusion*, *magmatic body*, and *plutonic body*. See also *intrusion*, *signature*, and *trap*.

igneous extrusion: flow-out of molten igneous rocks, or *lava*, onto the planetary surface. Basalt lava flows can cover regional-scale areas and form a relatively thick layer of highly magnetized igneous (volcanic) rocks. For the purposes of *modeling* and depth estimation, basalt flows and *sill*-like igneous bodies are approximated by the *horizontal sheet* source geometries. See also *igneous body* and *volcanics*.

igneous intrusion: same as *intrusive body* composed of the igneous rock and originating from *magma* that has intruded into the sedimentary section, *basement*, or other parts of the *crust* along faults, fracture zones, and other structural weaknesses.

IGRF: See *International Geomagnetic Reference Field (IGRF)*.

IGRF-12: the *IGRF model* of the 12th generation that includes (a) historical, or definitive, IGRF models for each past *epoch* in the range of years from 1900 to 2015, (b) predictive IGRF model for the epoch 2015 and 2015–2020 years, and (c) maps of *magnetic inclination*, *magnetic declination*, and *total absolute magnetic intensity (TAMI)* for the epoch 2015 and their predicted rates of change for 2015–2020 years. The IGRF-12 is used for calculation of *IGRF correction* and *geomagnetic field parameters* for a specified day, month, and year at any point on or above the Earth's surface with known *three-dimensional (3D)* geographic coordinates: latitude, longitude, and *elevation* (Thébault et al., 2015). See also *International Geomagnetic Reference Field (IGRF)*.

IGRF coefficients: See *Gauss coefficients*.

IGRF correction: a correction applied to *observed magnetics* to compensate for the latitude- and longitude-dependent magnetic field generated by the Earth's core and referred to as the *International Geomagnetic Reference Field (IGRF)*. It is applied on a point-by-point basis and subtracted from measured magnetic field values along *magnetic survey* lines. Correction for each observation point is defined by a *geomagnetic field strength* value determined from the retrospectively updated *IGRF model* appropriate for the day, month, and year of the actual measurement using *three-dimensional (3D)* geographic coordinates of each observation point: latitude, longitude, and *elevation*. After subtraction of the IGRF correction, the remaining components of the *Earth magnetic field* represent (a) the essentially static magnetic field of the upper *crust* and (b) the dynamic, or time-varying, *external magnetic field* of the Earth. The average of the calculated IGRF values in a survey area is often added back to the individual observed values in a manner of a *baseline shift*. The IGRF correction is applied prior to *merging* of the data acquired in different years in order to account for changes in geomagnetic field strength between the calendar dates of the merged surveys. Also referred to as *magnetic latitude correction*, *core-field correction*, *geomagnetic correction*, *GRF correction*, and *theoretical magnetic correction*. See also *lithospheric magnetic field*.

IGRF model: a mathematical description of the *main geomagnetic field* of the Earth that is known as the *International Geomagnetic Reference Field (IGRF)*. There are predictive and historical IGRF models. A predictive IGRF model represents parameters of the main geomagnetic field and its *secular variation* for the dates of the next five-year period, or *epoch*, e.g., the IGRF2015 model describes the predicted *geomagnetic field parameters* for 2015–2020 years. Historical IGRF models represent the retrospectively updated predictive IGRF models for each of the past epochs from 1900 to 2015 with the subdivision into annual IGRF models used for calculation of *IGRF correction* and geomagnetic field parameters at any point on or above the Earth's surface. Predictive and historical IGRF models are incorporated into the continuously revised and updated generation models. The IGRF model of the 12th generation is referred to as *IGRF-12*, and it includes the IGRF parameters for the years from 1900 to 2020. The latest IGRF model is IGRF-13 released in October 2019. See also *DGRF model*.

IGSN base value: the *absolute gravity* value at a local or regional *base station* which is tied to the *International Gravity Standardization Network 1971 (IGSN71)*.

IGSN71: See *International Gravity Standardization Network 1971*.

IGSN71 datum: the *International Gravity Standardization Network 1971 (IGSN71)* that is used as a reference dataset, or *gravity datum*, to which worldwide measurements of *absolute gravity* are tied.

image registration: *georeferencing* in the context of assigning project *XY coordinates* to the corners of raster images or scanned copies of printed images in order to scale them to the study area size and use in the process of *integrated interpretation*. See also *coordinate registration* and *raster image*.

imaginary component: **1.** the component of a *complex variable*, such as the *analytic signal (AS)* or *monogenic signal (MS)*, that is *phase*-shifted by 90° relative to the real component. It is denoted by the attached letter *i* or *j* which (a) indicates the orthogonality of a given component relative to the real component, and (b) represents the complex unit number expressed as

$$i = j = \sqrt{-1} = (-1)^{1/2}.$$

In a *complex spectrum*, the imaginary components are defined by *sine transform*, whereas the real components are defined by *cosine transform*. The AS of an *anomalous field* is composed of the field derivatives in the mutually orthogonal directions, where two horizontal derivatives represent the real components and the *vertical derivative* represents the imaginary component. **2.** a phase-orthogonal, or 90° phase-shifted, component of an anomalous gravity or magnetic *field*. Referred to with the same meaning as *quadrature*.

impact crater: See *astrobleme*.

impact structure: same as *astrobleme* that represents a structural depression formed by the blow, or impact, of a hypervelocity cosmic body, such as a meteorite or asteroid, on the ancient or present planetary surface.

improved SPI™ method: an upgraded *grid*-based *source parameter imaging (SPI™) method* for the automatic estimation of the parameters of *two-dimensional (2D)* magnetic sources approximated by a dipping *magnetic contact*, dipping *thin dike*, and *horizontal cylinder*. The source depth and *structural index (SI)* are determined with the use of *multimodel local wavenumber* k_{mm}, which is the same for all three source geometries and defined as

$$k_{mm} = k_2 - k_1,$$

where k_2 is *second-order local wavenumber* and k_1 is *first-order local wavenumber*. The source location is determined by the maximum value of either or both k_2 and k_1. The source depth h and actual value of the *SPI™ structural index* N_{SPI} are defined at the source location as, respectively,

$$h = 1/k_{mm} = 1/(k_2 - k_1), \quad N_{SPI} = k_1/(k_2 - k_1) - 1,$$

where $N_{SPI} = 0$ for the magnetic contact, $N_{SPI} = 1$ for the thin dike, and $N_{SPI} = 2$ for the horizontal cylinder. Both local wavenumbers are independent of *inclination*, *declination*, intensity of the *main geomagnetic field*, *susceptibility contrast*, and *dip* of sources. This method is used in *structural interpretation* applications, including those in the low *magnetic latitude* and equatorial areas (Smith et al., 1998). See also *advanced source-parameter imaging method*, *AS-LW relationship*, *AN-EUL method*, *enhanced local wavenumber method*, and *multimodel method*.

improved theta map: **1.** an upgraded *theta map* technique for improved *edge detection*, where *horizontal DAS (HDAS)* and *direct analytic signal amplitude (DASA)* are used instead of *total horizontal derivative (THDR)* and *analytic signal amplitude (ASA)* in the calculation algorithm. The cosine of *theta angle* θ_i is calculated as

$$\cos\theta_i = \{[H_x(F)]^2 + [H_y(F)]^2\}^{1/2} / \{F^2 + [H_x(F)]^2 + [H_y(F)]^2\}^{1/2},$$

where H_x and H_y are components of the *3D Hilbert transform* of *anomalous field* F in the mutually orthogonal x- and y-directions in *Cartesian coordinates* (Luo et al., 2011). **2.** a theta map technique modification based on the use of the second-order horizontal and vertical derivatives of an anomalous gravity or magnetic field (Chen et al., 2017). See also *derivative*, *edge detectors*, and *second-order theta map*.

improved tilt angle (ITA): an *attribute* of the *anomalous field F* that represents the modified *tilt angle (TA)*, where the field F and *horizontal DAS (HDAS)* are used instead of the *vertical derivative* and *horizontal gradient amplitude (HGA)* in the calculation formula:

$$\text{ITA} = \arctan\{F/\{H_x[(F)]^2 + [H_y(F)]^2\}^{1/2}\},$$

where H_x and H_y are x- and y-components of the *3D Hilbert transform* of the anomalous field in the mutually orthogonal x- and y-directions in *Cartesian coordinates* (Luo et al., 2011). See also *tilt ASA* and *improved theta map*.

impulse response: a characteristic of a *convolution* operator that describes its output, when the input signal is a single impulse. The response of a *convolution filter* to a series of impulses is a linear combination of the impulse responses controlled by the convolution operator length and the number of sampling points in the operator. See also *transfer function* and *amplitude response*.

incised valley: a buried *paleochannel* or valley formed by an ancient fluvial system which eroded deeply into the underlying sedimentary strata. It can be up to hundreds of meters deep with its width from one-half kilometer to several kilometers, especially when it has developed over a deep-rooted *fault*. Sand-filled incised valleys often create the stratigraphic traps with the *gravity signature* of a low-*amplitude* negative *residual gravity* anomaly. Incised valleys within the upper sedimentary sections can have the detectable *magnetic signature* of the assembled, or clustered, low-*amplitude* short-*wavelength* positive magnetic anomalies generated by local accumulations of *detrital magnetite*. See also *trap*.

inclination: **1.** *magnetic inclination*. **2.** *analytic signal inclination*. **3.** the vertical angle between a *vector* direction and a horizontal plane, i.e., vector *phase*.

incomplete Bouguer correction: same as *simple Bouguer correction* with an emphasis on its distinction from a *complete Bouguer correction* that compensates not only for the *Bouguer effect* but also for *terrain effect*, and, if necessary, *curvature effect*.

index factor: the multiplying constant in a source depth formula used by a particular *graphical source depth estimation* method. Index factors are related to the specific source shapes and source dimensions in the *half-width method*, *straight-slope method*, and *Smith depth rules*. See also *dividing factor*.

index map: See *track plot*.

indirect effect: a non-geological gravity effect on the amplitudes of *observed gravity* anomalies that is a result of differences in the *elevation* values determined at observation points with the use of the different vertical datums: sea level or *reference ellipsoid*. Indirect effect combines the gravity effect of the elevation in *free-air correction* and gravity effect of the rock mass between different datums as defined by *simple Bouguer correction*. Compensation for the indirect effect follows the historical convention to measure the observation point elevations relative to sea level approximated by *geoid*. Indirect effect has the same sign as *geoid height*: it is positive over the continents, where the geoid is above the reference ellipsoid, and it is negative over the oceans, where the geoid is below the reference ellipsoid. In local-scale gravity surveys, the indirect effect is too small to be considered separately from the regional gravity component, which is commonly removed by *regional-residual separation*. For regional-scale surveys, this effect can be significant in *interpretation* of long-*wavelength* gravity anomalies. Indirect effect is compensated by application of *indirect effect correction*. See also *geoid undulation* and *vertical datum*.

indirect effect correction: a correction applied to *free-air anomaly* and *Bouguer anomaly* to compensate for *indirect effect*. Using the horizontal layer with its thickness equal to *geoid height*, the indirect effect correction for ground observation points is expressed in units of *milligal (mGal)* as

$$\Delta g_{IE} = (0.3086 - 2\pi G\rho)\Delta h_g,$$

where G is *universal gravitational constant* in *cgs units*, ρ is rock *density* in grams per cubic centimeter ($g \cdot cm^{-3}$), and Δh_g is geoid height in meters. For the average upper crust density of 2.67 $g \cdot cm^{-3}$, the indirect effect correction value is defined as

$$\Delta g_{IE} = (0.3086 - 0.1119)\Delta h_g = 0.1967\Delta h_g.$$

For offshore observations, the correction value is

$$\Delta g_{IE} = 0.2655\,\Delta h_g$$

For local surveys, where *geoid* wavelengths are less than 10 km, the geoid height is less than 10 cm and the maximum change in the calculated Bouguer anomaly is 0.02 mGal. For surveys where geoid wavelengths are less than 100 km, the geoidal height variations are commonly less than 1 m, and the Bouguer anomaly change is 0.2 mGal or less. For regional surveys where geoid wavelengths are above 100 km, the geoidal height variations can reach several meters per 100–150 km *wavelength*. On the global scale, the geoid height varies from +87 m on the continents to –107 m in the oceans relative to the *WGS84 ellipsoid*, and it interferes with the measurement *accuracy* of regional long-*wavelength* gravity anomalies. Indirect effect correction for the Bouguer anomaly has its maximum values up to +17 mGal in the high mountains and –28 mGal in the deep ocean regions, whereas for the free-air anomaly the maximum indirect effect correction values are up to +27 mGal onshore and –33 mGal offshore. Indirect effect correction is unnecessary when the elevations of observation points are determined relative to the *reference ellipsoid*. See also *elevation*.

induced magnetic field: a *vector* component of rock *magnetization* that is oriented in the direction of the present *magnetic field*, such as the *Earth magnetic field*. Vector length, or *magnitude*, is defined by the product of the *magnetic susceptibility* of a particular rock unit and the *magnetic field strength* at the point of measurement. See also *magnetic induction* and *inducing magnetic field*.

induced magnetization: *magnetization* acquired by rocks in the direction of the ambient *magnetic field*, such as the *Earth magnetic field*. It is a *vector* quantity and its *magnitude*, or vector length, is directly proportional to (a) the intensity, or strength, of the planetary magnetic field at a measurement point, and (b) the capacity of rocks to be magnetized as defined by *magnetic susceptibility*. An induced *magnetization vector* is parallel to the direction of an externally applied magnetic field. The general expression is written as

$$\mathbf{M}_i = k\mathbf{B} = k\mathbf{H},$$

where k is magnetic susceptibility and $\mathbf{B}$ is the external magnetic field, which is referred to with the same meaning as *magnetizing force* and can be denoted by $\mathbf{H}$. Induced magnetization lines up, or polarizes, the originally random distributed magnetic dipoles within the rock-forming minerals and, therefore, it is also called *magnetic polarization*. Induced magnetization is dominant in granitic and metamorphic rocks, whereas *remanent magnetization* often dominates in basaltic and pyrrhotite-bearing fine-grained igneous rocks. See also *ambient field*, *diamagnetism*, and *magnetic field strength*.

inducing magnetic field: any external local or planetary-scale magnetic field, such as the *Earth magnetic field*, which creates a *magnetic field* inside, or induces *magnetization* into, a rock body by lining up, or polarizing, the originally random distributed magnetic dipoles within the rock-forming minerals. See also *magnetic induction* and *magnetic dipole*.

induction: See *magnetic induction*.

inertial acceleration: *acceleration* resulting from the motion of an *observation platform*, such as a *survey* ship or aircraft. Horizontal and vertical inertial accelerations of a moving observation platform are monitored separately by onboard accelerometers or determined by high-accuracy *differential GPS (DGPS)* data. Motion-related accelerations can have magnitudes of more than 100 *milligal (mGal)* on a shipboard platform and up to thousands of milligals on an airborne platform. Their *vector* components along the sensing axis of a *gravity sensor* interfere with measurements of *gravitational acceleration* and result in various degree errors of measured gravity values. Inertial accelerations are minimized by a *stabilized platform*, while their residual interfering components are eliminated by *inertial acceleration corrections*. High-frequency inertial accelerations are suppressed by a *low-pass filter* and by averaging the measured values over the empirically determined period of time or interval of the horizontal distance. Also referred to as *directional acceleration*, *extraneous acceleration*, *kinematic acceleration*, and *translational acceleration*. See also *equivalence principle* and *cross-coupling effect*.

inertial acceleration corrections: corrections applied to *relative gravity* data acquired on a moving *observation platform* to compensate for the interfering effects of the inertial accelerations resulting from the horizontal and vertical motions of a *survey* ship or aircraft. They include *Browne correction*, *Harrison correction*, *cross-coupling corrections*, and *vertical acceleration correction*. See also *inertial acceleration* and *moving-platform corrections*.

inertial accelerometer: See *inertial-grade accelerometer*.

inertial-grade accelerometer: *accelerometer* with the performance characteristics of industry-standard accelerometers for measuring inertial accelerations and adapted to operate as a *gravity sensor* in *airborne gravimeter (AG)* and *airborne gravity gradiometer (AGG)* systems. Also known as *mass-spring accelerometer* and *inertial accelerometer*. See also *inertial acceleration* and *vector accelerometer*.

inertial platform: See *stabilized platform*.

infield processing: preliminary *data processing* of measured data at the gravity or magnetic *survey* location in order to identify the *acquisition* errors and monitor *drift* or *diurnals* as a part of the *quality control (QC)* procedures.

infinite bottom: the *model* approximation of a *source body* bottom that has a depth of approximately 8–10 times the depth of the source body top or its largest horizontal *dimension*, so the gravity or magnetic effect of the infinite bottom can be disregarded. Also referred to as *infinite depth extent*. See also *infinite length*.

infinite contact: the *model* approximation of a *source body* by the *edge* of a horizontal plate, or slab, which has *infinite length* in its horizontal and vertical dimensions. Sometimes referred to as *infinite fault*. See also *dimension* and *finite contact*.

infinite depth extent: See *infinite bottom*.

infinite dike: **1.** *dike* with a horizontal top, parallel sides, *infinite strike length*, and *infinite bottom*. **2.** model approximation of a *tabular body* having the lateral extent, or width, that is much greater (in practice, 8–10 times) than its thickness. See also *infinite length*.

infinite fault: See *infinite contact*.

infinite half-space: same as *half-space* with an emphasis on the *infinite depth extent* of a rock volume below the deepest bounding horizontal plane in a *2D model* or *3D model*.

infinite length: the length of a *source body*, such as *horizontal cylinder* or vertical *prism*, that is so long (in practice, 8–10 times its horizontal or vertical *dimension*) that the gravity and magnetic effects of its lateral ends or bottom can be disregarded. See also *infinite bottom*.

infinite strike length: the distance along the *strike* of a structural layer or *source body* that is so long that the gravity and magnetic effects of their ends are negligible for the purposes of *modeling* and source depth estimates. Body and layer strikes are assumed to be perpendicular to the vertical plane of a *2D model* and have no changes in the property (*density* or *magnetization*) and structural composition along the strike. In practice, the infinite strike length requirement is satisfied for distances 8–10 times longer than the orthogonal horizontal *dimension* of a particular layer or source body.

inflection point: a characteristic point on a *profile* or map of an *anomaly* where the *slope* changes its type of *curvature* from convex to concave or vice versa. For anomalies from isolated sources, the inflection points correspond to (a) maxima or minima of the first-order *horizontal derivative* and (b) zeroes of the second-order horizontal derivative of the anomaly. Inflection points closely correlate with the *source body* edges and vertical rock formation contacts associated with high-*offset* faults and edges of geological structures, such as *horst* and *graben*. The width of an anomaly between its inflection points is measured for estimating the source depth from profiles of the *analytic signal amplitude (ASA)* and by other *graphical source depth estimation* methods. See also *ASA-derived depth, contact, edge, fault*, and *straight-slope method.*

inhomogeneous field: See *nonhomogeneous field.*

initial gridding: a *data conditioning* procedure applied prior to *gridding* of the gravity or magnetic data measured at irregularly spaced observation points during a *ground survey* in order to fill-in data gaps and, sometimes, data margins by interpolation and extrapolation of the available data.

injection effect: a change in *observed gravity* values caused by changes in the *density* of a reservoir formation resulting from water flooding and injection of a gas or steam in the process of enhanced oil-and-gas recovery and CO_2 sequestration. The magnitude of the gravity changes can reach tens and hundreds of microgals depending on the volume of the accumulated injection fluids and reservoir depth: the larger the injected fluid volume and the shallower the reservoir, the stronger the injection effect. See also *microgal (mGal), microgravity*, and *4D gravity monitoring.*

in-line gradient: 1. same as *longitudinal gradient* that represents a *gradient* value measured along the *survey* line. **2.** a *full tensor gradient (FTG)* component representing the rate of change of a particular *vector* component of a gravity or magnetic field along the spatial orientation of this vector component. There are three in-line gradients in the symmetrical nine-element *FTG matrix* defined for the *Earth gravity field* as

$$g_{xx} = \partial g_x/\partial x = \partial^2 U/\partial x^2,$$
$$g_{yy} = \partial g_y/\partial y = \partial^2 U/\partial y^2,$$
$$g_{zz} = \partial g_z/\partial z = \partial^2 U/\partial z^2,$$

where g_x, g_y, g_z are mutually orthogonal components of *gravity vector* **g** in *Cartesian coordinates* and U is *gravitational potential.* According to the *Laplace equation*, the sum of the in-line gradients of the FTG matrix is zero in a source-free space. See also *longitudinal gradient, transverse gradient*, and *cross-line gradient.*

inner-zone terrain correction: *terrain correction* applied to high-accuracy *ground gravity* and *microgravity* measurements to compensate for the gravity effects of topographic *relief* variations within either (a) the four *Hayford-Bowie zones* C through F, for the distances from 68 to 2290 m, or (b) the five *Hammer zones* D through H, for the distances from 53 to 2610 m, around each *station.* See also *innermost terrain correction, near-zone terrain correction, outer-zone terrain correction*, and *airborne terrain correction.*

innermost terrain correction: *terrain correction* applied to high-*accuracy* ground gravity and *microgravity* measurements to compensate for the gravity effects of topographic *relief* variations within either (a) the first two *Hayford-Bowie zones* A and B to the distance 68 m, or (b) the first three *Hammer zones* A, B, C to the distance 53 m, around each *station*, where the *terrain effect* is the largest. Correction values are calculated using the *elevation* data from one of the following: (a) high-*precision* ground measurements by geodetic instruments, (b) high-precision airborne measurements by a *drone*-attached *altitude*-measuring device, (c) *Digital Terrain Model (DTM)* with *vertical resolution* of 2–3 cm, or (d) high-precision measurements by satellite-mounted Interferometric Synthetic Aperture Radar (InSAR). Also referred to as *field terrain correction.* See also *InSAR imaging, sloping-wedge method*, and *terrain surveying.*

in-situ density: 1. rock *density* determined by laboratory measurements of core samples retrieved from drilled wells (boreholes). **2.** *apparent density* in the context of its measurements in boreholes with the use of *borehole gravity meter (BHGM), gamma-gamma log*, and *sonic log.*

in-situ modeling: *modeling* of an anomalous gravity or magnetic field, where the *model* points are at the same spatial locations as those of the actual observation points. See also *anomalous field.*

InSAR imaging: the satellite-based Interferometric Synthetic Aperture Radar (InSAR) imaging of the Earth's surface via the transmission of successive pulses of microwaves and recording the amplitudes and phases of the returned signals. Time-lapse InSAR observations are carried out for high-*precision* measurements of surface deformations, such as a surface subsidence caused by compaction of a depleted hydrocarbon reservoir and surface uplift due to a steam chamber growth in a geothermal reservoir, with *vertical resolution* up to 3–5 mm (Rahmoune et al., 2021). The InSAR-derived high-resolution *elevation* data also can be used for monitoring ground surface changes, caused by seasonal variations of groundwater levels in shallow aquifers, and integrated into *4D gravity monitoring* for calculation of gravity corrections for surface deformation effects. See also *amplitude, aquifer, phase*, and *time-lapse gravity.*

instantaneous amplitude: same as *analytic signal amplitude (ASA)* and *local amplitude* with an emphasis on its calculation at each observation point to determine the direct, or instant, *amplitude response* to variations of the subsurface *structure* and lithological composition.

instantaneous frequency: an *attribute* of the *analytic signal (AS)* defined by the rate of change of the *analytic signal phase* along the mutually orthogonal x- and y-directions in *Cartesian coordinates.* For gridded data, its numerical value represents the *total horizontal derivative* of the analytic signal phase Φ, expressed as

$$\varphi(x,y) = [(\partial\Phi/\partial x)^2 + (\partial\Phi/\partial y)^2]^{1/2}.$$

Instantaneous frequency is equivalent to *local frequency* and *total horizontal derivative of tilt.* See also *gridding.*

instantaneous phase: same as *analytic signal phase* and *local phase* with an emphasis on its calculation at each observation point to determine the direct, or instant, *phase response* to variations of the subsurface *structure* and lithological composition.

instantaneous wavenumber: See *local wavenumber.*

instrument corrections: corrections applied to *raw gravity* and *raw magnetics* to compensate for the instrument and measurement effects accumulated in the process of *acquisition* and obtain *observed gravity* and *observed magnetics*. There are two types of the instrument corrections: *raw gravity corrections* and *raw magnetic corrections*.

instrument drift: See *gravimeter drift* and *magnetometer drift*.

instrument lag: error in the coordinate position of an instrument *reading* caused by the combined effects of *antenna offset* and time delay in the electronic response of the *positioning* and measuring systems during marine (shipboard) and airborne surveys. Instrument lag is eliminated by the application of *lag correction.* See also *lag test* and *positioning*.

instrument noise: unpredictable errors in gravity and magnetic measurements caused by the nonlinear components of *gravimeter drift* or uncompensated *magnetometer drift* and fragments of *diurnals*. See also *noise*.

integral filter: a *data processing* operator for calculating the integral of a function, such as a particular *derivative* or measured *gradient* of a gravity or magnetic field. The first-order vertical derivative, or *first vertical derivative*, can be obtained by the application of this filter to the *second vertical derivative* calculated with the use of the *Laplace equation.* In accordance with the *Poisson theorem*, a *pseudogravity anomaly* is calculated by the vertical integration of reduced-to-pole (RTP) magnetic data. The integral filter amplifies the long-*wavelength* components of an *anomalous field*, whereas the short-wavelength components are suppressed, i.e., its performance is analogous to a *low-pass filter.* See also *reduction-to-pole (RTP).*

integral magnetic moments: integral relations between the mutually orthogonal *vector* components of the anomalous *magnetic field* of a localized, or compact, *source body*, its *magnetization* parameters, and its position coordinates. Integral magnetic moments are expressed as

$$I_i^{jk} = \iint_{-\infty}^{\infty} \Delta B_i x^j y^k \partial x \partial y,$$

where $\Delta\mathbf{B}(\Delta x, \Delta y, \Delta z)$ is the anomalous magnetic field and index i specifies a particular component of the anomalous magnetic field and marks its respective integral moment. The sum of the indices $j=(0, 1)$ and $k=(0, 1)$ defines the order of the integral moment. Zero-order integral moments ($j=k=0$) vanish, i.e., become zeros, over the range of integration from $-\infty$ to $+\infty$ in accordance with the *Gauss law*, whereas the non-vanishing first-order magnetic moments are calculated by numerical integration and interpreted for (a) the m_x, m_y, m_z components of the *magnetic moment* and (b) the center of the magnetization distribution within the compact magnetic sources, such as a small *pluton* and mineral *ore body* (Phillips, 2005; Caratori Tontini and Pedersen, 2008). See also *Helbig integrals*, *Helbig method*, and *magnetic moment*.

integrated imaging: 1. *inversion* of several datasets for estimates of a subsurface *structure* or distribution of the complementary physical properties of rock formations, such as *density* and *magnetization* contrasts. **2.** construction of a subsurface *model* with the use of several datasets of the different modalities and different physical properties, such as the gravity, magnetic, seismic, electromagnetic, well formation top, and well log data. **3.** *fusion*. See also *contrast, joint inversion, multimodal inversion*, and *RGB composite image*.

integrated interpretation: *interpretation* which is based on the integrated analysis of gravity, magnetic, seismic, and other available geophysical, geological, and topographic data jointly with the information from geophysical and geological publications related to the study area and its region. Integration of multiple sources of relevant information helps to guide the selection of geologically realistic interpretation scenarios and necessary *constraints* for two-dimensional (2D) or three-dimensional (3D) *forward modeling* and *inversion*. Such integration is the most reliable approach to minimizing the inherent ambiguity, or nonuniqueness, of the gravity and magnetic interpretation and obtaining the most accurate description of the detected geological *structure*. See also *ACLAS* and *nonuniqueness problem*.

interface inversion: See *structural inversion.*

interference of anomalies: a constructive or destructive summation of the positive and negative anomalies generated by the individual sources of different shapes, sizes, and *density* or *magnetization* contrasts at different depths. The term "interference" is commonly used to describe a partial overlaying of two or more anomalies from the closely spaced sources, whereas "superposition" describes a complete overlapping of the low-*amplitude* short-to-mid-*wavelength* local anomalies by the high-amplitude long-wavelength deep-sourced and regional anomalies. Interference-related changes of the original shapes of anomalies result in the various degree errors in depth estimates by *manual source depth estimation* and *automated source depth estimation* methods. Interference and superposition define the composite nature of observed anomalies which represent the total volumetric effect of all subsurface sources. The application of filters and calculation of derivatives and attributes of an *anomalous field* attempt to recover, or isolate, its constituent components of the exploration or research interest. See also *anomaly*, *attribute*, *contrast*, *derivative*, *wavelength filter*, and *wavelet filtering*.

INTERMAGNET: the International Real-time Magnetic Network, which is a global network of cooperating geomagnetic observatories which monitor the *Earth magnetic field* and provide minute-averaged digital observation records of its *vector* and *scalar* components in the nearly *real-time* mode of operation via *Geomagnetic Information Nodes*. See also *geomagnetic observatory*.

internal demagnetization: same as *self-demagnetization* which can reduce *magnetization* and change the direction of an *inducing magnetic field* within a localized, or compact, *three-dimensional (3D)* rock body.

internal magnetic field: 1. the primary and dominant part of the *Earth magnetic field* that includes (a) the time-varying magnetic field generated by the Earth's core, known as the *main geomagnetic field*, and (b) the essentially static magnetic field generated by magnetized rocks within the *lithosphere*, known as the *lithospheric magnetic field*. The main geomagnetic field dominates at wavelengths longer than 1,500–2,000 km, whereas the lithospheric magnetic field dominates at wavelengths shorter than 1,000–1,200 km. **2.** the *vector* magnetic field within a three-dimensional (3D) *source body* quantified by the vector sum of the ambient magnetic field $\mathbf{F}_{AMF}$, such as the Earth's magnetic field, and internal *demagnetizing field* $\mathbf{F}_{DF}$ generated by the magnetic poles at the opposite boundaries of a source body. It is expressed as

$$\mathbf{F}_{IMF} = \mathbf{F}_{AMF} - \mathbf{F}_{DF} = \mathbf{F}_{AMF} - \mathrm{N}\mathbf{M}_{E},$$

where N is the *demagnetization factor* and $\mathbf{M}_E$ is the *effective magnetization*. See also *ambient field*, *magnetic pole*, *wavelength*, *self-demagnetization*, and *external magnetic field*.

International Geomagnetic Reference Field (IGRF): **1.** the internationally accepted best-fitting mathematical model that describes the *main geomagnetic field* by a series of *spherical harmonics* of increasing degree and predicts the variations of its components in space and time for the next 5-year period, or *epoch*. At the end of each epoch, the International Committee of the International Association of Geomagnetism and Aeronomy (IAGA) adopts a new *IGRF model* that predicts *geomagnetic field parameters* based on a linear extrapolation of the spherical harmonic coefficients. Each IGRF model is built from measurements by the worldwide network of geomagnetic observatories and satellite-mounted magnetometers with the incorporation of data acquired by marine (shipboard) and airborne magnetic surveys. For calculation of the *total magnetic intensity (TMI)* values, the IGRF is subtracted from *observed magnetics* by application of *IGRF correction* determined from the IGRF model appropriate for the year, months, and days of a particular *magnetic survey*. **2.** a mathematical set of three mathematical models: (1) the provisional, or temporary, IGRF model predicting the main geomagnetic field for each epoch; (2) the provisional *secular variation* model predicting the annual variation rate for each epoch; and (3) the *Definitive Geomagnetic Reference Field (DGRF)* model representing a revision of the provisional IGRF model for the previous epoch. The revised provisional model is designated as the definitive model which is not revised in the subsequent generations of IGRF models. Sometimes referred to as *Geomagnetic Reference Field (GRF)*, *normal magnetic field*, and *theoretical geomagnetic field*. See also *IGRF-12* and *total absolute magnetic intensity (TAMI)*.

International Gravity Formula (IGF): **1.** the mathematical formula that defines *theoretical gravity* as a function of the geographic latitude on the surface of a *reference ellipsoid*. **2.** the *Somigliana-Pizzetti formula*. In accordance with the sequentially refined reference ellipsoid parameters, five standard IGF versions have been published in 1930, 1967, 1980, 1984, and 1999: *International Gravity Formula 1930 (IGF1930)*, *International Gravity Formula 1967 (IGF1967)*, *International Gravity Formula 1980 (IGF1980)*, *International Gravity Formula 1984 (IGF1984)*, and *International Gravity Formula 1999 (IGF1999)*. In *exploration* applications, the differences between formulas are generally disregarded, because all analyzed and interpreted gravity anomalies are relative quantities. For *merging* the anomalous gravity fields obtained in several areas, these differences should be accounted, unless the same formula is used for each of the merged areas. Also referred to as *ellipsoidal gravity formula*, *normal gravity formula*, *Somigliana-Pizzetti gravity*, *standard gravity formula*, and *theoretical gravity formula*.

International Gravity Formula 1930 (IGF1930): the mathematical formula for calculation of *theoretical gravity* at any point on the Earth's surface. The IGF1930 defines the theoretical gravity field of the homogeneous and rotating *reference ellipsoid* with parameters adopted by the Geodetic Reference System 1930 (GRS30) as

$$g_0 = 9.78049(1 + 0.0052884 \sin^2 \vartheta - 0.0000059 \sin^2 2\theta),$$

where g_0 is the theoretical gravity value in units of meter per second squared ($\mathrm{m \cdot s^{-2}}$) and θ is geographic latitude in degrees. The first term in the formula is the theoretical gravity value at the equator, i.e., *equatorial gravity*.

International Gravity Formula 1967 (IGF1967): the mathematical formula for calculation of *theoretical gravity* at any point on the Earth's surface. The IGF1967 defines the theoretical gravity field of the homogeneous and rotating *reference ellipsoid* with parameters adopted by the Geodetic Reference System 1967 (GRS67) as

$$g_0 = 9.78031846(1 + 0.0053024 \sin^2 \theta - 0.0000058 \sin^2 2\theta),$$

where g_0 is the theoretical gravity value in units of meter per second squared ($\mathrm{m \cdot s^{-2}}$) and θ is geographic latitude in degrees. The first term in the formula is the theoretical gravity value at the equator, i.e., *equatorial gravity*.

International Gravity Formula 1980 (IGF1980): the mathematical formula for calculation of *theoretical gravity* at any point on the Earth's surface. The IGF1980 defines the theoretical gravity field of the homogeneous and rotating *reference ellipsoid* (referred to as the *GRS80 ellipsoid*) with parameters adopted by the Geodetic Reference System 1980 (GRS80) as

$$g_0 = 9.7803267715(1 + 0.001913185135 \sin^2 \theta) / (1 - 0.0066943800229 \sin^2 \theta)^{1/2},$$

where g_0 is the theoretical gravity value in units of meter per second squared ($\mathrm{m \cdot s^{-2}}$) and θ is geographic latitude in degrees. The gravity effect of the Earth's atmosphere mass, or *atmospheric effect*, is included into the formula with the assumption that this mass is within the GRS80 ellipsoid.

International Gravity Formula 1984 (IGF1984): the mathematical formula representing an upgrade of the *International Gravity Formula 1980 (IGF1980)* for calculation of the *theoretical gravity* at any point on the Earth's surface. The IGF1984 defines the theoretical gravity field of the homogeneous and rotating *reference* *ellipsoid* (referred to as the *WGS84 ellipsoid*) with parameters adopted by the World Geodetic System 1984 (WGS84) as

$$g_0 = 9.7803267714(1 + 0.00193185138639 \sin^2\theta)/ (1 - 0.00669437999013 \sin^2\theta)^{1/2},$$

where g_0 is the theoretical gravity value in units of meter per second squared ($m \cdot s^{-2}$) and θ is geographic latitude in degrees. The gravity effect of the Earth's atmosphere mass, or *atmospheric effect*, is included into the formula with the assumption that this mass is within the WGS84 ellipsoid.

International Gravity Formula 1999 (IGF1999): the mathematical formula representing an upgrade of the *International Gravity Formula 1984 (IGF1984)*. It is published by the U.S. National Imagery and Mapping Agency (NIMA) for calculation of the *theoretical gravity* at any point on the Earth's surface with the updated parameters of the *WGS84 ellipsoid* as

$$g_0 = 9.7803253359(1 + 0.00193185265241 \sin^2\theta)/ (1 - 0.00669437999014 \sin^2\theta)^{1/2},$$

where g_0 is the theoretical gravity value in units of meter per second squared ($m \cdot s^{-2}$) and θ is geographic latitude in degrees. The gravity effect of the Earth's atmosphere mass, or *atmospheric effect*, is included into the formula with the assumption that this mass is within the WGS84 ellipsoid.

International Gravity Standardization Network 1971 (IGSN71): the worldwide network of *absolute gravity* values established in 1971 as the *gravity datum*, or *gravity standard*, to which all absolute gravity measurements are referenced and tied. The IGSN71 supersedes the *Potsdam Datum*, or *Potsdam gravity*, determined from measurements by a *pendulum gravimeter*. The Potsdam absolute gravity value has been reduced by approximately 14 *milligal (mGal)* to make it compatible with the IGNS71. In the past, the IGSN71 absolute gravity values have included *Honkasalo correction*. The International Association of Geodesy (IAG) has recommended the removal of the amount of this correction from the absolute gravity values. Also referred to as *IGSN71 datum* and *global gravity datum*. See also *absolute gravimeter*.

International Terrestrial Reference Frame (ITRF): the *Geodetic Reference System* representing the continuously updated global network of 3D terrestrial coordinates: latitude, longitude, and *elevation*. Coordinates are determined by high-*accuracy* ground geodetic measurements at different locations across the world using the *GRS80 ellipsoid* as a reference surface, or *geodetic datum*. Estimated differences between the ITRF coordinates and coordinates determined with the use of the *World Geodetic System 1984 (WGS84)*, which is a standard reference frame for satellite-based *Global Positioning System (GPS)*, are insignificant for *exploration* applications. See also *horizontal datum* and *vertical datum*.

interpretation: the process of obtaining new information about the subsurface *structure* and rock properties in a study area that can be summarized in a plausible geological *model* consistent with the detected gravity and magnetic anomalies and available geological and geophysical data. There are qualitative and quantitative parts in the process of interpretation. Qualitative interpretation is based on (a) analysis of the gravity and magnetic anomalies revealed by calculation of the attributes, derivatives, gradients, *invariants*, application of filters, and other *enhanced processing* techniques, and (b) identification of the specific elements of the geological structure that have generated these anomalies. Quantitative interpretation is based on the results of the qualitative part and can be subdivided into seven main objectives which are selected and prioritized in accordance with a particular *exploration* or research program: (1) mapping of *lineaments* representing the structural discontinuities within the sedimentary section and upper *basement*, such as faults, fracture and shear zones, lithological contacts, etc.; (2) mapping of the intrasedimentary structures favorable for the accumulation of hydrocarbons and other targets of oil-and-gas exploration; (3) mapping of the igneous intrusions, zones of anomalous *density* and *magnetization*, depositional fairways, and other targets of mineral exploration; (4) mapping of the basement depth structure with the use of *manual source depth estimation*, *automated source depth estimation*, *pattern recognition*, and *inversion*-based methods; (5) estimates of depth and other parameters of the selected source bodies; (6) 2D, 2.5D, or 2.75D *modeling* along the regional transects, available 2D seismic sections, and profiles of anomalies of the exploration or research interest; and (7) 3D *structural inversion*, *property inversion*, or *stochastic inversion* for parameters of the targeted geological objects. Interpretation results provide a basis for the preparation of recommendations for future exploration in the survey or regional study area that are referred to as *exploration leads*. See also *gravity interpretation methods*, *magnetic interpretation methods*, *target*, and *integrated interpretation*.

interpretation ambiguity: the nonuniqueness of *interpretation* as a consequence of the general *nonuniqueness problem* inherent for the potential fields. Limitations on the alternative choices in the process of interpretation and minimization of the interpretation ambiguity depend, primarily, on two key factors: (1) the quality of the gravity and magnetic data defined by the *accuracy* of measurements and *lateral resolution* of detected anomalies and (2) the availability of independent geophysical data and geological information that can be used as *constraints* in the selection of different interpretation scenarios. See also *potential field* and *integrated interpretation*.

interpreter: a person in charge of *interpretation* and its preparation procedures, including but not limited to (a) assembling a *database* from the reduced data, i.e., data obtained in the process of *data reduction* which is usually performed by a *survey* contractor, (b) *gridding* the line-oriented or irregularly spaced data values using the assembled database, (c) *enhanced processing* of the gridded data for detection of *structure*-related anomalies, (d) analysis of the detected gravity and magnetic anomalies, identification and mapping of the targeted elements of the subsurface geological structures in integration with the available geological and geophysical data, (e) 2D, 2.5D, 2.75D *modeling* and 3D *inversion*, and (f) preparation of the exploration report and *exploration leads*. See also *anomaly, integrated interpretation, target*, and *observer.*

interruption trend: the direction on a gravity or magnetic *grid* or map along which the observed anomalies are interrupted, i.e., displaced or truncated. It is used as the identification criterion in mapping of structural discontinuities, such as the local-scale intrasedimentary and *basement* faults (especially strike-slip faults and shear zones) and the regional-scale tectonic boundaries and wrench fault (regional *strike-slip fault*) systems. Also known as *offset trend* and *truncation trend.* See also *fault, shear zone*, and *trend.*

intersection statistics: airborne and marine (shipboard) *quality control (QC)* data representing the results of the statistical analysis of differences between the measured values of the Earth's gravity or magnetic field at the intersections of the traverse and control lines. Calculation of the *standard deviation (SD)* of all intersection differences is used for estimating the internal consistency of the measurements and *survey accuracy*. See also *control lines, traverse lines*, and *crossover error.*

interval density: *apparent density* in the context of (a) borehole *density* measurements and (b) conversion of seismic interval velocities into rock densities via the *Gardner equation* and other velocity-*density* relationships.

interval factor: a manufacturer-established *gravimeter* constant for converting the original gravimeter readings from dial units into milligals or microgals within the *operating range* of a particular gravimeter. See also *milligal (mGal), microgal (μGal)*, and *reading.*

intrabasement anomalies: gravity and magnetic anomalies generated by lateral changes in the *density* and *magnetization* of the igneous and metamorphic rocks associated with elements of *basement* structures. In two-dimensional (2D) and three-dimensional (3D) *modeling*, their sources are commonly approximated by the vertical or high-angle *thin dike, thick dike*, and *contact* source geometries, wherein their boundaries are defined by the interpreted and mapped *lineaments* (basement faults and lithological contacts) and assumed to have *infinite bottom*, i.e., depth extent which is significantly greater than the basement depth. See also *anomaly, source geometry, structure*, and *suprabasement anomalies.*

intrasedimentary anomalies: gravity and magnetic anomalies generated by lateral changes in the *density* and *magnetization* of the rock formations associated with elements of sedimentary structures, including anticlines and synclines, horsts and grabens, uplifts and depressions, thrusts, igneous intrusions, lava flows (*volcanics*), faults, fracture and shear zones, reservoir-grade sand channels, isolated ore bodies, shallow mineral deposits, as well as diapirs and other low-density bodies created by *salt*- or shale-driven structural deformations. Modeling of intrasedimentary anomalies is constrained by the seismic and borehole data with approximation of the individual source shapes by *ideal source* assemblages. Intrasedimentary anomalies are made better visible, or enhanced, by *regional-residual separation* and other *enhanced processing* techniques. See also *anticline, constraints, diapir, fault, graben, horst, igneous intrusion, ore body, syncline, thrust, uplift*, and *intrabasement anomalies.*

intrinsic magnetization: rock *magnetization* resulting from magnetic properties of the rock-forming minerals without regard for the shape-dependent *self-demagnetization* effects within a rock volume.

intrinsic susceptibility: *magnetic susceptibility* which is equivalent to *bulk susceptibility*, also known as *true susceptibility.*

intrusion: 1. emplacement, or injection, of an *igneous body, allochthonous salt*, shale or salt *diapir* into the pre-existing sedimentary environment as a result of the magmatic activity or *salt*- or shale-driven structural deformations under the abnormal pressure of accumulated sediments. **2.** a sedimentary *structure*, such as *igneous intrusion* and *salt* or shale *diapir*, formed by the process of intrusion. **3.** an *intrusive body*. Due to the high *density* and *magnetization* contrasts of intrusive bodies with the surrounding rock formations, they often have strong gravity and magnetic anomalous signatures of various amplitudes and shapes. See also *anomaly, density contrast, magnetization contrast, gravity signature, magnetic signature*, and *magma.*

intrusive body: an *igneous body, allochthonous salt*, shale or salt *diapir* which has been emplaced, or injected, into a pre-existing sedimentary structural environment by the process of *intrusion.*

invariant ratio: a dimensionless quantity defined by the scaled ratio of *invariants* I_2 and I_1 calculated from components of the *full tensor gradient (FTG)*, expressed as

$$I = -(I_2/2)^2/(I_1/3)^3 = -27I_2^2/4I_1^3,$$

where $0 \le I \le 1$ and $I_1 \le 0$. The invariant ratio is zero when the gravity or magnetic field is the same, or invariant, along some direction, such as the *strike* of a *two-dimensional (2D)* source, and it is equal to 1 for the radially symmetric fields generated by a *point source* or *sphere*. Invariant ratio values can be calculated for each observation point and gridded to create a *dimensionality grid*. Also referred to as *dimensionality indicator* (Pedersen and Rasmussen, 1990). See also *gridding* and *2D source.*

invariants: *scalar* quantities obtained by the combination of *full tensor gradient (FTG)* components or *FTG eigenvalues*. Invariants and their linear combinations have the property to remain unchanged by, or invariant to, any spatial orientation of the axes in *Cartesian coordinates*. There are three *principal invariants* denoted by I_0, I_1, I_2 and defined as

$$I_0 = T_{xx} + T_{yy} + T_{zz} = \lambda_1 + \lambda_2 + \lambda_3 = 0,$$

$$I_1 = T_{xx}T_{yy} + T_{yy}T_{zz} + T_{xx}T_{zz} - T_{xy}^2 - T_{yz}^2 - T_{xz}^2 = \lambda_1\lambda_2 + \lambda_2\lambda_3 + \lambda_1\lambda_3,$$

$$I_2 = T_{xx}(T_{yy}T_{zz} - T_{yz}^2) + T_{xy}(T_{yz}T_{xz} - T_{xy}T_{zz}) + T_{xz}(T_{xy}T_{yz} - T_{xz}T_{yy}) = \lambda_1\lambda_2\lambda_3,$$

where T_{ij} are the FTG components and λ_1, λ_2, λ_3 are the FTG eigenvalues. For a gravity *point source*, the amplitude attenuation, or *fall-off*, of invariants I_1 and I_2 is a function of the inverse sixth and nineth powers of the distance r from the source, i.e., $1/r^6$ and $1/r^9$, respectively. Invariants can be considered the non-linear filters which transform the original *gradient* field into a new representation with the increased *lateral resolution* of the gradient anomalies. The high *sensitivity* of invariants I_1 and I_2 to lateral changes in the *density* and *magnetization* of rocks is used in high-*resolution* exploration and engineering applications, including: (a) detection and mapping of *lineaments*; (b) mapping of low-density contrasts between the sediments and *salt* or shale overhangs; (c) separation of the interfering gradient anomalies generated by localized volumetric sources, such as isolated ore bodies within an igneous feeder-*dike* system or *banded iron formation (BIF)*, and by magnetic heterogeneity within mineral alteration zones; (d) mapping of near-surface dissolution cavities, or *karst*; (e) detection of *unexploded ordnance (UXO)* and archaeological objects; and (f) kimberlite exploration. Invariants are also referred to as *rotational invariants*, so the invariants I_1 and I_2 can be denoted by R_1 and R_2, respectively. See also *kimberlite intrusion, ore body, FTG invariants, horizontal invariants, invariant ratio, strike-angle invariant*, and *normalized source strength (NSS)*.

inverse calculation: 1. calculation of the *source body* parameters (depth, shape, *dip*, density or magnetization *contrast*) or depth *structure* of a targeted rock formation from the observed *anomaly* generated by this source body or from the obtained *anomalous field*, respectively. **2.** *inversion*. Automated anomaly-by-anomaly inverse calculations over a gravity or magnetic *survey* area are performed by inversion algorithms applied to line-oriented and gridded data. See also *gridding* and *forward calculation*.

inverse Fourier transform: a mathematical operation that converts Fourier-transformed gravity or magnetic data from the *spectral domain* back to their original *space domain*. Reconstitution, or synthesis, of the space-domain waveforms of gravity or magnetic data by the inverse Fourier transform is the opposite of data decomposition by the *forward Fourier transform*. For a one-dimensional (1D) function *f*(*x*) representing a *profile-* or line-oriented *anomalous field* data, the inverse Fourier transform is expressed as

$$f(x) = (1/2\pi)\int_{-\infty}^{\infty} \mathrm{F}(k)e^{ikx}dx,$$

where k is the *wavenumber* related to the *wavelength* λ as $k=2\pi/\lambda$, F(k) is the *Fourier transform*, and $i=\sqrt{-1}$ is a complex unit number indicating the orthogonality. For a 2D function *f*(*x,y*) representing a *grid* of the anomalous gravity or magnetic field, the inverse Fourier transform is expressed as

$$f(x,y) = (1/4\pi^2)\iint_{-\infty}^{\infty} F(u,v)e^{i(ux+vy)}dxdy,$$

where F(u,v) is the Fourier transform and u and v are wavenumbers k_x and k_y in the mutually orthogonal x- and y-directions in *Cartesian coordinates* related to their respective wavelengths as $u=k_x=2\pi/\lambda_x$ and $v=k_y=2\pi/\lambda_y$. For arbitrary directions, the wavenumber k_a is expressed as $k_a=(k_x^2+k_y^2)^{1/2}$ (Blakely, 1996).

inverse magnetostriction: a change in the *magnetization* parameters, such as *susceptibility, remanence*, and *permeability*, of certain types of rock-forming minerals in response to the application of external stress or deformation. Referred to with the same meaning as *piezomagnetism*. See also *magnetostriction*.

inverse method: same as *inversion* with an emphasis on it as an automatic or semi-automatic method of *modeling* and, therefore, *interpretation*, wherein the unknown parameters of an exploration *target* are determined from its gravity and magnetic anomalies.

inverse model: the two-dimensional (2D) or three-dimensional (3D) gravity or magnetic *model* of a targeted geological *structure* obtained by *inversion*. Residual and *gradient* gravity or magnetic fields are inverted for (a) a selected model component, such as a sedimentary horizon and *source body* structure (depth, *dip*, shape) or physical property (*density* or *susceptibility* contrast) using standard *structural inversion* or *property inversion*, respectively, or (b) structure and physical properties using *stochastic inversion* or 3D inversion algorithms based on minimization of the model *objective function*. Anomalous gravity or magnetic field can be divided by *filtering* into two or more parts for the separate cycles of property and structural inversions with the subsequent integration of the independently obtained results into a final inverse model. See also *contrast, residual field, joint inversion*, and *target*.

inverse modeling: 1. a *quantitative interpretation* technique applied to obtain a *2D model* or *3D model* of a targeted geological *structure* from the residual and *gradient* fields. Referred to interchangeably with the same meaning as *inversion*. **2.** an *inverse problem solution* derived from *model*-based calculations of *target* parameters which are geologically plausible, consistent with the available *constraints*, and generating the gravity or magnetic field matching the *observed field* within a specified margin of error called *tolerance*. See also *residual field* and *forward modeling*.

inverse power law: a mathematical rule which describes the *amplitude attenuation*, or *fall-off*, of a gravity or magnetic *anomaly* A with the distance from its source approximated by *single-point source* geometry. It is expressed as

$$\mathrm{A(r) = C/r^N},$$

where C is a constant specific for the *anomalous mass* of the gravity source or *anomalous magnetization* and *geomagnetic field* parameters for the magnetic source, r is distance between the source and observation point, and N is *fall-off rate*. For *point source* geometries, the gravity anomaly falls off as C/r^2, whereas the magnetic anomaly falls off as C/r^3 and, thereby, demonstrates that the *magnetic method* provides a higher resolution laterally and vertically than the *gravity method*. For *gradient* fields, the inverse power law is expressed as

$$\mathrm{A(r) = C/r^{N+1}}.$$

See also *lateral resolution*, *vertical resolution*, and *inverse-square law*.

inverse problem: 1. the estimation of *source body* parameters from its gravity or magnetic *anomaly*, i.e., determining the cause from measurements of its effect. **2.** the fundamental principle of *inversion* postulating that a solution of the inverse problem, i.e., estimation of the subsurface *structure* parameters (*density contrast* or *magnetization contrast*, depth, etc.) from an isolated anomaly or observed gravity and magnetic fields, is not unique, because there are infinitely many individual source parameters and source distributions in the subsurface that can generate the anomalies which fit acceptably the observed *anomalous field*. Moreover, the inverse problem is ill-conditioned, or unstable, especially for deep targets, because the large changes in a targeted structure *model* can result in small changes in the calculated gravity or magnetic field, whereas small errors in the observed data can produce large errors in the calculated parameters of a deep *target*. Also referred to as *nonuniqueness problem*. See also *Gauss flux theorem*, *observed field*, and *forward problem*.

inverse problem solution: 1. *model*-based reconstruction of the *density* or *magnetization* distribution and depth *structure* of a targeted rock formation, such as the top of *basement*, base of *allochthonous salt*, surface of *subcrop*, isolated *ore body*, etc., with the use of *inversion* algorithms applied to residual and *gradient* fields. **2.** *modeling* of the physical property (density or magnetization *contrast*) of an exploration *target* by *property inversion* or its structure (depth, *dip*, shape) by *structural inversion*. **3.** *voxel model*-based reconstruction of both the rock properties and depth structure by a three-dimensional (3D) *stochastic inversion* or 3D inversion with minimization of the model *objective function*. Unlike the unique solution of a *forward problem* and its integrating effect on the model field, the solution of an *inverse problem* is not unique, and it is based on the iterative field discretization which is inherently unstable. Inverse problem solution is optimized by *numerical regularization*, wherein the nonuniqueness of the solution is minimized by available *constraints*, which also provide additional information for the construction of an initial model and selection of the optimization criteria for inversion algorithms. See also *residual field* and *forward problem solution*.

inverse projection: conversion of gravity and magnetic coordinates from the XY coordinate pairs of the *Universal Transverse Mercator (UTM)* coordinate system into the original latitude-longitude coordinate pairs of the *Global Positioning System (GPS)*. See also *forward projection* and *XY coordinates*.

inverse-square law: the physical law stating that the *magnitude* of *gravitational force* or *magnetic force* varies in inverse proportion to the square of the distance between two bodies of *mass* or two magnetic dipoles, respectively. The inverse-square law is a partial case of the *Newton law of universal gravitation* for gravity fields and the *Coulomb law* for magnetic fields. See also *magnetic pole* and *inverse power law*.

inverse wavelet transform: *wavelet synthesis* in the context of the space-domain reconstruction of gravity and magnetic data after *wavelet denoising* and other *wavelet transform* operations performed in the *wavelet domain*.

inversion: 1. a method of the *quantitative interpretation* of gravity and magnetic data that is based on the *inverse problem solution* for estimating the parameters of a subsurface geological *structure*. **2.** an automated or semi-automated interactive technique for the calculation of *source body* parameters (*density contrast*, *magnetization contrast*, depth, shape, etc.) in a constructed *2D model* or *3D model* from the residual, *gradient*, and *gradient tensor* fields using available *constraints* and mathematical optimization criteria. **3.** solution of the *inverse problem* by calculating the parameters of a *causative body* from its measured *anomaly*, i.e., determining the cause from measurements of its effect. **4.** calculation of the depth, *dip*, and other parameters of the isolated rock bodies approximated by *ideal source* geometries. The principal objectives of inversion are: (a) to determine the *target* parameters which are consistent with the available constraints, geologically plausible, and reproducing the *observed field* within a specified *tolerance*, and (b) to develop a reference *model* for the analytical testing of a response of the target parameters to the mathematical transformations of an *observed field*. There are four basic types of inversion: (1) *property inversion* for the physical properties, such as *density* and *susceptibility* contrasts within a targeted rock formation or density and magnetization distribution within a constrained *ore body* shape; (2) *structural inversion* for a targeted structure: ore body shape, depth of a particular sedimentary formation, *basement* depth, and base of *allochthonous salt*; (3) inversion of a 3D *voxel model* for both the physical properties and structure by iterative minimization of the model *objective function* (Li and Oldenburg, 1996; 1998a); and (4) 3D voxel model-based *stochastic inversion* for the structure and properties of an exploration or research target (Rossi, 2017). The 2D *profile*-based inversion and 3D *grid*-based inversion for parameters of ideal sources are implemented in *Euler deconvolution*, *Werner deconvolution*, *local wavenumber methods*, *tilt-derivative method*, and other inversion techniques. Inversion results are strongly dependent on the initial model, available constraints, and assumptions about the sources of anomalies. Magnetic inversion and, especially, *gravity inversion* are value-adding *interpretation* techniques complementary to the seismic exploration in areas where seismic data are sparse or their *lateral resolution* is low. See also *Agerlin principle*, *residual field*, *cross-gradient joint inversion*, *5C inversion*, *gravity inversion*, *hybrid inversion*, *joint inversion*, *magnetic inversion*, *multimodal inversion*, *vector inversion*, and *window inversion*.

inversion-based separation: the *regional-residual separation* method based on the *inversion* of a three-dimensional (3D) *voxel model*. The gridded anomalous gravity or magnetic field is, first, upward-continued to suppress *noise* and shallow short-*wavelength* components and, then, it is inverted for the 3D distribution of the densities or susceptibilities using a coarse *mesh* with voxels 2–5 times larger than those of the original mesh. In the constructed coarse-mesh voxel model, the local-source volume (which includes local sources of the targeted anomalies) is defined by an *interpreter*, and its *density* or *susceptibility* voxel values are set to zero. The remaining part of the coarse-mesh voxel model is assumed to represent only the regional sources and, therefore, its *forward modeling* produces a *regional field* in the targeted area. Subtraction of the forward-calculated regional field from the original *anomalous field* yields a *residual field* of the targeted anomalies. Also referred to as *inversion-based stripping* (Li and Oldenburg, 1998b). See also *gridding*, *forward calculation*, and *upward continuation*.

inversion-based stripping: same as *inversion-based separation* with an emphasis on removing, or stripping-off, the regional component of an *anomalous field* by subtracting the forward-calculated gravity or magnetic field of the 3D distribution of densities or susceptibilities obtained by *property inversion* of a course-mesh *voxel model*.

inversion errors: errors in the *inversion*-obtained parameters of a targeted *structure* that are caused, primarily, by: (1) insufficient geological and geophysical *constraints*; (2) distortion of amplitudes and shapes of anomalies in the process of *gridding* the line-oriented gravity and magnetic data; (3) consequences of the ill-conditioned, or unstable, *inverse problem*, whereby large changes in the depth and shape or physical properties (*density contrast* or *susceptibility contrast*) of a deep *target* result in small changes of a *model*-calculated field, while small *data processing* errors produce large errors in the deep target parameters; (4) conceptual errors associated with discrepancies between the mathematical representation of a *target* in the constructed model and real geological structure, including the approximation of a three-dimensional (3D) structure by *2D model*, *2.5D model*, or *2.75D model*; and (5) linearization of a nonlinear inverse problem.

ionosphere: the uppermost part of the atmosphere and innermost part of the *magnetosphere* at altitudes between 60–100 km and 1000–1200 km, where the interaction between the Earth's rotation, tidal and thermal effects, and ionization of the moving atmospheric particles by *solar wind* generate the electric currents. These currents produce their own temporal magnetic fields with highly variable periods of duration and magnitudes up to hundreds of *nanotesla (nT)* and, thereby, contribute to the spatial and temporal variations of the *external magnetic field* of the Earth. See also *altitude*, *magnitude*, *temporal magnetic variations*, and *van Allen belts*.

irrotational field: *vector field* **F** whose *curl* is zero at each space point, i.e.,

$$\nabla \times \mathbf{F} = 0,$$

where ∇ is *vector* gradient operator *del* and $\times$ denotes *vector product*. Such fields have no vortex components: if a fluid flow can be presented as the irrotational field, then a paddlewheel placed within this fluid flow with its axis of rotation parallel to the flow direction would not rotate. In accordance with the *Helmholtz theorem*, the irrotational field is described by the *gradient* of the scalar *potential*. In the space outside the mass and magnetic material and in the absence of electric currents, gravity and magnetic fields are irrotational. See also *solenoidal field*, *conservative field*, and *potential field*.

isoanomaly: a continuous line on a *contour map* connecting the coordinate points of equal values of an anomalous gravity or magnetic *field*. See also *anomalous field*.

isoclinal map: a global-scale *contour map* where the contours represent lines of equal values, or isolines, of the *magnetic inclination* of the *main geomagnetic field*. The map is compiled from worldwide *magnetic vector* measurements by vector magnetometers stationed at geomagnetic observatories and mounted on Earth-orbiting satellites with high-*accuracy* star-imaging cameras for orienting their magnetic sensors. Also referred to as *main field inclination map*. See also *geomagnetic observatory*, *magnetic sensor*, and *vector magnetometer*.

isocline: a continuous line, or contour, of equal *magnetic inclination* values, or inclination *isoline*, on an *isoclinal map*. See also *isogon*.

isodynamic maps: global-scale contour maps where the contours represent lines of equal values of *total intensity*, *vertical intensity*, and *horizontal intensity* of the *geomagnetic field* and its *vector* components. The maps are compiled from worldwide *magnetic vector* measurements by vector magnetometers stationed at geomagnetic observatories and mounted on Earth-orbiting satellites that are integrated with the measurements by scalar magnetometers during ground, marine (shipboard), and airborne magnetic surveys. See also *geomagnetic observatory*, *scalar magnetometer*, and *vector magnetometer*.

isogal: a line of equal *gravity field* values on a *contour map* where the anomalous gravity field is shown in units of *milligal (mGal)* or *microgal (μGal)*. See also *anomalous field* and *isogam*.

isogam: a line of equal *total magnetic intensity (TMI)* or *RTP magnetic field* values on a contour map complied from *vintage data*, where the anomalous magnetic field values are shown in units of *gamma*. See also *anomalous field* and *isogal*.

isogon: a continuous line, or contour, of equal *magnetic declination* values, or declination *isoline*, on an *isogonic map*. See also *isocline*.

isogonic map: a global-scale *contour map* where contours represent lines of equal values, or isolines, of the *magnetic declination* of the *main geomagnetic field*. The map is compiled from the worldwide *magnetic vector* measurements by vector magnetometers stationed at geomagnetic observatories and mounted on Earth-orbiting satellites with high-*accuracy* star-imaging cameras for orienting their magnetic sensors. Also referred to as *main field declination map*. See also *geomagnetic observatory*, *magnetic sensor*, and *vector magnetometer*.

isoline: a continuous line on a *contour map* connecting the coordinate points of equal values of (a) the *anomalous field*, (b) calculated derivatives and attributes, (c) interpreted *basement* depth, (d) *inversion*-obtained parameters of a targeted *structure*, and (e) *geomagnetic field parameters*. See also *isocline*, *isogal*, *isogon*, *isoanomaly*, and *target*.

isomagnetic charts: See *isomagnetic maps*.

isomagnetic maps: global-scale contour maps where contours represent lines of equal values, or isolines, of the *geomagnetic field parameters*. The magnitudes of the *vector* components of the *Earth magnetic field* and its strength, or dynamic, parameters are shown on *isodynamic maps*. Values of *magnetic inclination* are shown on an *isoclinal map* and values of *magnetic declination* are shown on an *isogonic map*. Maps are compiled from the worldwide measurements at geomagnetic observatories and by Earth-orbiting satellites in integration with data acquired by ground, marine (shipboard), and airborne magnetic surveys. Also referred to as *isomagnetic charts*. See also *contour map*, *geomagnetic observatory*, and *magnitude*.

isometric display: See *3D view*.

isoporic map: a global-scale *contour map* where contours represent lines of equal values, or isolines, of the rate of change of the *secular variation* in units of *nanotesla (nT)* per year or in degrees per year. The maps are compiled with the use of *DGRF* and *IGRF* data. See also *DGRF model* and *IGRF model*.

isostasy: the concept of the geodynamic balance, or equilibrium, between the *crust* and upper *mantle*. By analogy with icebergs in sea water, the large crustal blocks are assumed to be floating in a denser underlying layer of partially melted rocks in the *asthenosphere*, so the less dense regions of the Earth's crust have a higher topographic *elevation* than the more dense regions. Mountains and large topographic elevations above sea level are balanced, or compensated, by the underlying lateral mass deficiencies known as *isostatic root*, whereas the oceanic and sea basins and large topographic depressions below sea level are balanced by the underlying lateral mass excesses known as *isostatic antiroot*. There are three basic theories of isostasy: *Airy hypothesis*, *Pratt hypothesis*, and *Vening Meinesz hypothesis*. In the Airy and Vening Meinesz hypotheses, the rock densities in the crustal blocks and upper mantle are constant, so the load of a laterally varying topographic or bathymetric relief is balanced either locally (Airy) or regionally (Vening Meinesz) by the matching increase or decrease of the crustal block thickness. In the Pratt hypothesis, both the topographic and bathymetric relief variations are compensated within the *lithosphere* at the constant depth approximately 100 km below sea level by the lateral changes in rock densities, so the lowest densities are under the highest topographic elevations, and the highest rock densities are under the deepest parts in oceans and seas. The Airy hypothesis and Vening Meinesz hypothesis are implemented in regional gravity studies, except for the mid-ocean ridge areas where the Pratt hypothesis is preferred. See also *local isostasy*, *mass deficiency*, *mass excess*, *regional isostasy*, *thermal isostasy*, and *isostatic rebound*.

isostatic anomaly: **1.** the remaining, or residual, part of *observed gravity* after application of *observed gravity corrections* relevant to the *survey* type (ground, marine, airborne), including the *isostatic correction*. **2.** the *field* of the residual gravity anomalies obtained by application of isostatic correction to *free-air anomaly* or *Bouguer anomaly*. The isostatic anomaly is determined over the large-scale, high-relief and mountainous areas onshore and rifted continental margins offshore as a part of the *regional-residual separation* process. In general, the relatively short- and mid-*wavelength* isostatic anomalies are associated with lateral variations in the geological structures and densities within the sedimentary section and upper *basement*. Long-wavelength components are assumed to be related to regional-scale overcompensated or undercompensated areas and regions of the deep contrasting densities in the *lithosphere*, wherein the near-zero long-wavelength isostatic anomalies are indicative of the isostatically compensated regions. Sometimes, the approximate offshore *Airy isostatic anomaly* is calculated by subtracting the forward-calculated gravity effect of the gridded *Moho boundary* from the Bouguer anomaly *grid*, and the obtained isostatic anomaly mimics the *free-air anomaly* scaled by the ratio of the ocean bottom and Moho density contrasts. More accurate Airy isostatic anomaly is obtained when the isostatic correction for the gravity effect of the Moho depth variations is calculated by the *Jachens-Roberts method*. Also referred to as *isostatic residual anomaly*. See also *density contrast*, *forward calculation*, *gridding*, and *flexural isostatic anomaly*.

isostatic antiroot: the lateral *mass excess* under the oceans, seas, and regional-scale continental depressions. In accordance with the *Airy-Heiskanen model* and *Vening Meinesz model* of *isostatic compensation*, this rock mass maintains the isostatic balance, or *isostatic equilibrium*, between the *crust* and upper *mantle*. The isostatic antiroot approximates the *Moho depth* in the regions of greatly reduced crustal thickness (6–9 km under the ocean floor versus 25–40 km under major continental areas and 60–75 km under high mountain ranges) that correspond to the structural highs in the global *Moho boundary* depth structure. Because the crustal densities are lower than densities in the upper mantle, these structural highs create the lateral mass excesses producing the long-*wavelength* regional trends of the positive values in the offshore *Bouguer anomaly* that can be removed, or accounted for, by application of *isostatic correction* to obtain *isostatic anomaly*. Also referred to as *Airy antiroot*, *compensating antiroot*, and *crustal antiroot*. See also *structural high* and *isostatic root*.

isostatic antiroot depth: a depth to the *isostatic antiroot* predicted by the *Airy-Heiskanen model* of *isostatic compensation*. Its calculation formula is based on the *Airy-Heiskanen equations* and expressed as

$$D_{IAR} \cong D_{Moho} = T_{SL} - h_w(\rho_{ao} - \rho_w)/\Delta\rho,$$

where D_{IAR} is the isostatic antiroot depth below sea level that approximates the *Moho depth* D_{Moho} in the oceans and seas, T_{SL} is the *reference crustal thickness* at the shoreline (zero) *elevation* assumed to be 30–40 km, h_w is ocean/sea water depth at the *observation point*, ρ_{ao} is average oceanic crust *density*, ρ_w is ocean/sea water density, and $\Delta\rho$ is *density contrast* between the lower *crust* and upper *mantle* estimated at 0.40–0.45 g·cm^{-3}. Isostatic root depth values are gridded and used for calculation of *Airy isostatic correction* in the oceans and seas (Chapin, 1996a). Sometimes referred to as *offshore Airy depth*. See also *AIRYROOT*, *Jachens-Roberts method*, and *isostatic root depth*.

isostatic compensation: balancing, or compensation, of the regional-scale elevated areas and depressions on the Earth's surface by deep lateral mass deficiencies and mass excesses in accordance with the theory of *isostasy*. The extra load of rock masses above sea level on continents is compensated by the underlying *mass deficiency*, which is created either by the laterally increased thickness of the *crust* with constant rock densities in the *Airy-Heiskanen model* and *Vening Meinesz model* or by a lateral decrease in rock densities above the constant *depth of compensation* in the *Pratt-Hayford model*. Similarly, the reduced mass loads of large continental depressions and water in oceans and seas are compensated by the underlying *mass excess* created either by the laterally decreased thickness of the crust with constant rock densities or by a lateral increase in rock densities above the constant depth of compensation in the *lithosphere*. The gravity effect of the isostatic compensation is referred to as *isostatic compensation effect*. See also *isostatic anomaly*.

isostatic compensation correction: same as *isostatic correction* applied to *observed gravity* to account for the regional-scale *isostatic compensation effect*.

isostatic compensation effect: the regional-scale gravity effect of *isostatic compensation* which interferes with the long-*wavelength* component of *observed gravity*. The isostatic compensation effect is negative on the continents and positive in the oceans and seas, thereby resulting in the inverse correlation between the long-wavelength components of *Bouguer anomaly* with *topography* and direct correlation with *bathymetry*: the onshore Bouguer anomaly becomes increasingly negative (in some areas, down to –400 mGal and greater) with increasing *elevation*, whereas the offshore Bouguer gravity becomes increasingly positive (up to 300–320 mGal) with the deepening ocean. The isostatic compensation effect does not include the local-scale gravity effect of rock masses above a survey *datum* known as the *Bouguer effect*. When the *Airy-Heiskanen model* of isostatic compensation is used, the isostatic compensation effect is referred to as the *root isostatic effect* onshore and the *antiroot isostatic effect* offshore. The numerical value of the isostatic compensation effect comprises the amount of *isostatic correction* at each *observation point*. Due to its regional-scale extent, the calculated isostatic compensation effect is often referred to with the same meaning as *isostatic regional anomaly*. Also referred to as *isostatic effect* and *isostatic gravity effect*.

isostatic compensation level: the depth level at which an *isostatic compensation* is maintained. In the *Airy-Heiskanen model*, the isostatic compensation level is predicted at the *Moho boundary*, whereas in the *Pratt-Hayford model* it is predicted within the lower part of the *lithosphere* at the depth of approximately 100 km below sea level, called *depth of compensation*, i.e., it is below the Moho boundary.

isostatic compensation model: a mathematical model that describes *isostatic compensation* in accordance with one of three basic theories of *isostasy*: *Airy hypothesis*, *Pratt hypothesis*, and *Vening Meinesz hypothesis*. There are three main isostatic compensation models known as the *Airy-Heiskanen model*, the *Pratt-Hayford model*, and the *Vening Meinesz model*.

isostatic correction: 1. a correction for the long-*wavelength* gravity effect of *isostatic compensation* caused by the regional-scale lateral variations in the thickness or *density* of the Earth's *crust* in accordance with a particular *isostatic compensation model*. **2.** gravity correction for the *isostatic compensation effect*. In exploration applications, the *Airy-Heiskanen model* is commonly used for the calculation of isostatic correction. The *Pratt-Hayford model* is preferred in the mid-ocean ridge areas, and the *Vening Meinesz model* is used primarily in scientific research projects. The correction value is calculated with the use of a *Digital Terrain Model (DTM)* and assumed isostatic parameters, including *density contrast* at the *Moho boundary* and its depth, which is called *isostatic root depth* onshore and *isostatic antiroot depth* offshore in the Airy-Heiskanen model. Isostatic correction removes the dominant regional trend of negative *Bouguer gravity* values on continents and positive Bouguer gravity values in oceans and seas. In terms of *regional-residual separation*, the isostatic correction represents a numerical value of the regional-scale isostatic compensation effect at each observation point and, therefore, a *grid* of the calculated isostatic correction values is referred to as an *isostatic regional anomaly* grid. See also *AIRYROOT*, *Jachens-Roberts method*, and *flexural isostatic correction*.

isostatic effect: See *isostatic compensation effect*.

isostatic equilibrium: a state of the geodynamic balance in which the large crustal blocks have a tendency to settle after the addition or removal of mass loads, such as those related to the growth and decay of volcanoes, deposition of sediments, erosion of mountains, and melting of ice sheets. Crustal loading results in isostatic depression, or downwarping, whereas crustal unloading results in *isostatic rebound*, or upwarping. A region or tectonic province is considered to be in isostatic equilibrium, or isostatically compensated, when it is not subjected to vertical buoyancy forces. Often referred to with the same meaning as *isostatic compensation*. See also *isostasy*.

isostatic gravity effect: See *isostatic compensation effect*.

isostatic rebound: the isostatic adjustment by a slow and continuous increase in the *elevation* of the Earth's surface that follows the removal of a regional-scale mass load as a result of the erosion of mountains and melting of ice sheets. Melting of the Pleistocene continental glaciers has initiated the isostatic rebound whose relaxation time, i.e., time for restoration of *isostatic equilibrium*, has been estimated at the order of 4000 years. Also referred to as *isostatic uplift*.

isostatic regional anomaly: 1. a long-*wavelength* component of *observed gravity* that represents the regional-scale *isostatic compensation effect* whose numerical value is defined by *isostatic correction* at each observation point. **2.** a *grid* of isostatic correction values in a regional-scale *gravity survey* area. Removing the isostatic regional anomaly from *Bouguer anomaly* eliminates the dominant regional trend of the negative values on the continents and analogous trend of the positive values in oceans and seas. The obtained *residual gravity* field represents the *isostatic residual anomaly* which is commonly referred to as *isostatic anomaly.*

isostatic residual anomaly: same as *isostatic anomaly* that represents the residual part of *observed gravity* remaining after the application of *observed gravity corrections* relevant to the *survey* type (ground, marine, airborne), including *isostatic correction.*

isostatic response function: *gravitational admittance* in the context of the *wavenumber*-dependent function which describes the relationship between a gravity anomaly caused by the isostatic deformation of the *lithosphere* and topographic or bathymetric mass loads of different wavelengths.

isostatic root: the lateral *mass deficiency* under the mountains and regional-scale topographic elevations on the continents. In accordance with the *Airy-Heiskanen model* and *Vening Meinesz model* of *isostatic compensation,* this crustal rock mass maintains the isostatic balance, or *isostatic equilibrium,* between the *crust* and upper *mantle.* The isostatic root bottom approximates the *Moho depth* in the regions of increased crustal thickness (25–40 km under major continental areas and 60–75 km under high mountain ranges) that correspond to the structural lows in the global *Moho boundary* depth structure. Because the crustal densities are lower than densities in the upper mantle, these structural lows create the lateral mass deficiencies producing the long-*wavelength* regional trends of the negative values in the onshore *Bouguer anomaly* that can be removed, or accounted for, by application of *isostatic correction* to obtain *isostatic anomaly.* Also referred to as *Airy root, compensating root,* and *crustal root.* See also *structural low* and *isostatic antiriot.*

isostatic root depth: a depth to the *isostatic root* predicted by the *Airy-Heiskanen model* of *isostatic compensation.* Its calculation formula is based on the *Airy-Heiskanen equations* and expressed as

$$D_{IR} \cong D_{Moho} = T_{SL} + h\rho_{ac}/\Delta\rho,$$

where D_{IR} is the isostatic root depth below sea level that approximates the *Moho depth* D_{Moho} on the continents; T_{SL} is the *reference crustal thickness* at the shoreline (zero) *elevation* assumed to be 30–40 km; h is elevation, or topographic height above sea level, at the *observation point*; ρ_{ac} is average continental crust *density*; and $\Delta\rho$ is *density contrast* between the lower *crust* and upper *mantle* estimated at 0.40–0.45 $g \cdot cm^{-3}$. Isostatic root depth values are gridded and used for calculation of *Airy isostatic correction* on the continents (Chapin, 1996a). Sometimes referred to as *onshore Airy depth.* See also *AIRYROOT, Jachens-Roberts method,* and *isostatic antiroot depth.*

isostatic uplift: See *isostatic rebound.*

isotherm: a continuous line, or *isoline,* connecting points with the same heat flow values or depths of the same temperature on a *contour map* and in the lithospheric isothermal *model.* See also *Curie isotherm depth, geotherm,* and *lithosphere.*

isothermal remanent magnetization (IRM): *remanent magnetization* retained by surface rocks after short-term exposure to a strong externally applied magnetic field, such as that generated by a lightning strike. See also *combustion magnetism* and *thermoremanent magnetization (TRM).*

isotropic index: See *spherical anisotropic index.*

isotropic sensor: See *omnidirectional sensor.*

isotropy: uniform variations of a physical property, such as *density* or *magnetization,* without regard for the direction of its measurement. The isotropy of a particular property parameter, such as the *gradient* of an anomalous gravity or magnetic field, can be described by a sphere whose radius is equal to the gradient value at an *observation point,* whereas deviations from the uniform variations, i.e., *anisotropy,* can be described by an *ellipsoid* whose axial radii are equal to the gradient values measured or calculated along the mutually orthogonal *x-, y-, z-*directions in *Cartesian coordinates.* Sometimes referred to as *spherical anisotropy.* See also *FTG isotropy* and *principal component analysis (PCA).*

iSPI™ method: See *improved SPI™ method.*

iterative modeling: same as *modeling* with an emphasis on multiple cycles of repeated calculations, or iterations, resulting in a progressively improved *model misfit* that is performed automatically by a computer algorithm or interactively by an *interpreter* in order to solve either the *forward problem* or *inverse problem.*

iterative rescaling (ITRESC) method: a *grid*-based semi-automated method of mapping the depth to *basement* with the use of *residual gravity* (*isostatic anomaly* or residual *Bouguer anomaly*) and basement depth estimates obtained from independent data (borehole, seismic, EM, etc.). Residual gravity values are iteratively transformed, or rescaled, into basement depths via the following procedures: (1) building a plot of the known basement depths versus residual gravity values; (2) polynomial fitting of the depth-gravity plot and using the obtained coefficients to build a first-approximation depth model independent from the density contrasts; (3) estimation of the density-depth function by building a plot of the observed versus unit-density predicted gravity that is, first, approximated by a suitable polynomial and, then, divided into several linear segments whose slopes represent the density contrasts at different depths; (4) recalculating the gravity field of the first-approximation depth model with the obtained depth-varying density contrasts and subtraction from the observed residual gravity field; and (5) rescaling the obtained *model misfit* into the basement depths and adding the first-approximation depth model. Iterative depth model adjustments are repeated until the model misfit reaches the *stopping criterion* (Florio, 2018; 2020). See also *Bott method.*

J

Jachens-Roberts method: a method of calculating the *isostatic correction* for the regional-scale *isostatic compensation effect* produced by the varying depth, or thickness, of the *isostatic root* or *isostatic antiroot* in accordance with the *Airy-Heiskanen model* of *isostasy*. It is based on the *Airy-Heiskanen equations* and similar to the process implemented in the calculation of *terrain correction*. Here, a *Digital Terrain Model (DTM)* is divided into 3 × 3 *arc-minute* compartments and the average elevations of all individual compartments are determined. The gravity effect, or *gravitational attraction*, of the isostatic root or antiroot below each compartment is computed without accounting for the *curvature* of the Earth's surface around all observation points. For areas above sea level, the isostatic root effect is expressed in units of *milligal (mGal)* as

$$\Delta g_{IR} = -\mathrm{G}\Delta\rho \mathrm{A}\{1/[r^2 + (T_{SL} + h_t)^2]^{1/2} - 1/[r^2 + (T_{SL} + h_t + T_{IR})^2]^{1/2}\},$$

where G is a *universal gravitational constant* in *cgs units*; $\Delta\rho$ is the *density contrast* between the lower *crust* and upper *mantle*; A is the compartment area; r is the horizontal distance between the center of a particular compartment and the *observation point* in meters; T_{SL} is the *reference crustal thickness* at the shoreline (zero) elevation assumed to be 30–40 km; h_t is topographic height, or *elevation*, above sea level; T_{IR} is thickness of the isostatic root below the T_{SL} level defined as

$$T_{IR} = h_t\rho_{ac}/\Delta\rho,$$

where ρ_{ac} is average crustal *density* and $\Delta\rho$ is assumed *density contrast* between the lower *crust* and upper *mantle*. For oceans, seas, and regional-scale depressions below sea level, the isostatic antiroot effect is expressed as

$$\Delta g_{IAR} = \mathrm{G}\Delta\rho \mathrm{A}\{1/[r^2 + (T_{SL} - h_w - T_{AR})^2]^{1/2} - 1/[r^2 + (T_{SL} - h_w)^2]^{1/2}\},$$

where h_w is ocean/sea water depth and T_{AR} is thickness of the isostatic antiroot above the T_{SL} level defined as

$$T_{AR} = h_w(\rho_{ac} - \rho_w)/\Delta\rho,$$

where ρ_w is ocean/sea water density. Summing the calculated values for all compartments comprises the amount of isostatic correction at each observation point (Jachens and Roberts, 1981). See also *AIRYROOT* and *Parker formula*.

Jacobsen filter: a *grid*- or line-based spectral-domain operator calculating the difference between two upward continuations of a gravity or magnetic field to separate the anomalous components generated by sources within a user-specified depth interval. For separation of components from the interval between depths h_1 and h_2 ($h_2 > h_1$), the Jacobsen filter *transfer function* is expressed as

$$\mathrm{F}(k) = e^{-2|k|h_1} - e^{-2|k|h_2},$$

where k is the *wavenumber* defined as

$$|k| = (k_x^2 + k_y^2)^{1/2},$$

and k_x and k_y are wavenumbers in the mutually orthogonal x- and y-directions, respectively, in *Cartesian coordinates* related to their respective wavelengths as $k_x = 2\pi/\lambda_x$ and $k_y = 2\pi/\lambda_y$. Because this procedure is known as *differential upward continuation*, the Jacobsen filter is sometimes referred to as a *differential upward continuation filter*. The Jacobsen filter simultaneously suppresses both high-frequency *noise* and the *regional field*: the former by *upward continuation* and the latter by differencing the two upward-continued fields. Also known as *separation filter*. See also *spectral domain*.

jerks: See *geomagnetic jerks*.

joint inversion: *inversion* for the parameters of an exploration *target* with the use of two or more types of the data which have different *sensitivity*, i.e., *amplitude response* and *phase response*, to variations of the targeted parameters, such as a different sensitivity of the anomalous gravity and magnetic fields or independent components of the *full tensor gradient (FTG)*. Joint inversion can be more efficient than the regular inversion of a single dataset, because it provides additional *constraints* for the inversion process and, hence, more accurate estimates of the *target* parameters. Joint *property inversion* of the gravity and magnetic *total field* and *gradient* data with constraints by petrophysical data from available wells is applied for the construction of a three-dimensional (3D) *pseudolithology model*, where the inversion-obtained ranges of the physical properties are used for differentiation of the mapped *ore body* into the specific lithological units associated with the mineralization zones, mineral alteration types, and localized concentrations of the high-*density* and high-*magnetization* rock minerals within the ore body (Li et al., 2019). Referred to with the same meaning as *multimodal inversion*. See also *cross-gradient joint inversion*, *5C inversion*, and *integrated imaging*.

joint modeling: **1.** *forward modeling* of an exploration *target* with the use of all available geophysical and geological data: gravity, magnetic, seismic, surface geology, well logs, well formation tops, etc. **2.** joint *inverse modeling* commonly referred to as *joint inversion*.

Josephson junction: a thin non-conductive layer, or barrier, between two superconductive layers that is too thin to prevent completely the flow of electrons and, thereby, sustains a very small resistance to their flow in a superconducting loop. The property of a through-barrier-flow of electrons to oscillate in response to an externally applied voltage at the oscillation frequency proportional to the ambient magnetic field is used for measurements of the *Earth magnetic field*. Josephson junctions are included into the superconducting loop of a *SQUID sensor* to measure the *magnitude* of the total magnetic field and its *gradient* along the direction of the sensor separation. See also *ambient field* and *magnetic sensor*.

Julian date: the four-digit number representing the calendar date of observations that is recorded in the digital *format* of measured magnetic or gravity field values. It is composed of the *Julian day* and the last two digits of a current year, e.g., the Julian date of February 2, 2017 is denoted as 3317. Sometimes, the sequence of the Julian day and current year is reversed, e.g., February 2, 2017 will be denoted as 1733.

Julian day: the calendar day number counted from the first of January in a current year, e.g., February 2 is Julian day 33. The Julian day is a standard parameter in the description of digital records of magnetic and gravity *survey* data and the calculation of *IGRF correction*. See also *Julian date*.

Julians: days of a gravity or magnetic *survey* counted from the first of January of each calendar year. See also *Julian date* and *Julian day*.

juvenile water: **1.** water that has originated directly from *magma* and entered into groundwater circulation for the first time. **2.** highly mineralized, gas-saturated, hydrothermal and potentially ore-forming fluid released in large quantities by *igneous intrusion* during its transition from the molten state into a solidified *igneous body*. The abnormally high pressure and temperature of an igneous body environment are capable of driving the juvenile water into sedimentary formations to open and fill-in faulted and fractured rocks and precipitate the magnetic minerals, such as *diagenetic magnetite*, within the faults and fracture zones in quantities sufficient for the generation of detectable magnetic anomalies (Peirce et al., 1998; Williams et al., 1999). Juvenile water, released in the process of transformation of the "magma ocean" into the Archean protocrust of the Earth, is a primary source of the original, or primordial, water in the ancient oceans (Genda, 2016; Schaefer and Elkins-Tanton, 2018). Also known as *magmatic water* and *plutonic water*. See also *crust* and *magnetized fault*.

K

K-index: a measure of variability of the *external magnetic field* of the Earth over 3-hour time intervals that is determined by stationary observations at a *geomagnetic observatory*. The K-index has 10 basic levels varying from the lowest variability of the zero level to the nineth level of the highest variability, wherein each level represents an increase in the variability by 10 *nanotesla (nT)*. The average of the K-values from 14–20 observatories across the Earth describes the global variations, excluding the *diurnals*. The K-values of 5 and higher are indicators of a *magnetic storm*.

karst: a near-surface or subsurface cavity, such as a sinkhole, cave, or underground drainage channel, formed by dissolution of limestones, dolomites, gypsum, and other soluble rocks. Karst has the *gravity signature* of a residual *gravity low* in the *microgravity* range and, occasionally, the *magnetic signature* of the residual clustered low-*amplitude* magnetic anomalies generated by localized accumulations of *detrital magnetite*. See also *clustering, residual gravity*, and *residual magnetics*.

karst gravity: *microgravity* and *gravity gradient* data from high-*accuracy* and high-*resolution* ground gravity measurements for the detection and delineation of shallow subsurface cavities, such as the natural air- or water-filled dissolution cavities in limestones, dolomites, and *evaporites*, or man-made empty objects, such as tunnels, shafts, and abandoned mines. Karst gravity surveys require *station* intervals of 2–5 m and *survey line spacing* of 4–10 m to detect short-*wavelength* anomalies and *elevation* measurement accuracies of 2–3 mm to enable adequate *gravity data corrections*. Typical amplitudes of karst-generated anomalies are within the range of 10–100 microgals (Butler, 1984; 2005). See also *gravity anomaly, microgal (μGal)*, and *InSAR imaging*.

Keating method: a *grid*-based *pattern recognition* technique for locating the magnetic anomalies generated by kimberlite intrusions that have shapes approximated by a *vertical cylinder*. Several grids represent the magnetic field of the vertical cylinder calculated with different cylinder parameters (depth, diameter, length) that are used with a *moving window*. Correlation coefficients between the calculated and observed fields are determined for each *grid node*. Local areas with correlation coefficients exceeding a pre-selected threshold value are assumed to be locations of kimberlite intrusions (Keating, 1995). See also *kimberlite intrusion*.

kernel: 1. the principal component of a *data processing* operator that defines a main property of a data transformation, such as the size and shape of a *moving window*, mathematical formula of transformation, conversion *matrix*, and linear transform function. For example, the kernel of a *grid*-based *convolution filter*, such as an *averaging filter*, defines the shape of its moving window and number of grid cells within it, so a 3×3 or 5×5 square filter kernel represents a square array, or matrix, of 9 or 25 grid cells, respectively. Also referred to as *convolution matrix*. **2.** an algorithm for performing a particular mathematical operation.

kimberlite intrusion: an ultramafic igneous rock body of the deep magmatic origin and carrot-like or mushroom-like shape that often is a host rock for diamonds. Its diameter can vary from several hundred meters up to 1.5–3.0 km, and it is usually buried under unconsolidated sediments or beneath small lakes which fill a downward-curved, or concave, *diatreme*. Typical kimberlite intrusion is composed of a near-surface funnel-shaped diatreme and a much narrower deep-rooted crater pipe. Due to the high positive or negative *magnetization* and *density* contrasts with surrounding rocks, its gravity and magnetic signatures on residual grids are expressed as variations of the "bright spot" anomaly which is a high-*amplitude* positive gravity or magnetic anomaly of a circular or oval shape. In the case of high-density surrounding rocks and strong *remanent magnetization*, a kimberlite intrusion can have signatures of the "dim spot" and "dark spot" anomalies: low-amplitude positive or negative *gravity anomaly* and negative *magnetic anomaly*. See also *residual grid, gravity signature, magnetic signature*, and *magma*.

kimberlite pipe: a vertical or high-angle *kimberlite intrusion* with a crater pipe of up to several hundred meters in diameter and a much wider, funnel-shaped *diatreme* near or at the ground surface. Density and magnetization contrasts of a brecciated diatreme are usually 2–3 times lower than those of the high-*density* and high-*magnetization* crater pipe. See also *density contrast, magnetization contrast*, and *Keating method*.

kinematic acceleration: same as *inertial acceleration* that originates from the motion of an *observation platform*.

Koenigsberger ratio: the ratio Q of the absolute value, or *magnitude*, of *remanent magnetization* $|\mathbf{M}_r|$ to the magnitude of *induced magnetization* $|\mathbf{M}_i|$, expressed as

$$Q = |\mathbf{M}_r| / |\mathbf{M}_i| = |\mathbf{M}_r| / k|\mathbf{B}|,$$

where k is *magnetic susceptibility* and $|\mathbf{B}|$ is external *magnetic field strength* or *geomagnetic field strength* in the case of the *Earth magnetic field*. Its common variations are within the range of 0.1 to 1.0, but it can be up to 10.0 and higher in basalts, kimberlite pipes, pyrrhotite-bearing fine-grained rocks and ores, young volcanic (extrusive) rocks, and some iron or steel objects. Generally, the Koenigsberger ratio is much larger in oceanic *crust* rocks than in continental crust rocks. Also referred to as *Q-factor*.

kriging: a method of the geostatistical interpolation and *gridding* of irregular- and regular-spaced field values that is based on calculation of a data *variogram*, which quantifies the level of *noise* and correlation length of measurements. The obtained variogram model is adaptive to the spatial trends in the observed values and, thereby, makes kriging a data-driven technique superior to those constrained by purely mathematical interpolation and extrapolation. Kriging can be used for suppression of *random noise* in gridded *airborne gravity gradient (AGG)* data and estimates of the total amount of *dynamic noise* in AGG measurements by the "nugget" value of a variogram curve (Pilkington and Shamsipour, 2014). See also *cokriging*, *least-squares collocation*, *gridding*, and *minimum-curvature gridding*.

Kuwahara filter: a *grid*-based space-domain *edge-sharpening filter*. Its *kernel* is defined by a square *moving window* divided into four subwindows of the same shape and size that overlap the central row and central column of grid cells in the moving window. As a *default* option, 5×5 grid cells compose the moving window with 3×3 grid cells for each subwindow. The filter operator calculates the arithmetic mean, or average, value and its *variance* in each of four subwindows. The filter output is a value at the center *grid cell* in the moving window that represents the mean value in a subwindow with the smallest variance. Among other applications, such as enhancing the edges of structures obtained by *inversion*, it can be used in *terracing*. An upgraded version of the Kuwahara filter is referred to as *mean-of-least-variance (MLV) filter* (Li, 2016).

L

L1 norm: an *inversion* regularization parameter implemented in solving the gravity or magnetic *inverse problem* to prevent *overfitting* and amplification of *noise.* The L1 norm is defined by sum of the absolute values of differences per data point between the observed (residual) gravity or magnetic *anomaly* and the anomaly generated by the *model* of an isolated *source body* or geological *structure.* See also *model misfit, numerical regularization,* and *residual field.*

L2 norm: an *inversion* regularization parameter used in solving the gravity or magnetic *inverse problem* to prevent *overfitting* and amplification of *noise.* The L2 norm is defined by the sum of the squares of the differences per data point between the observed (residual) gravity or magnetic *anomaly* and the anomaly generated by the *model* of an isolated *source body* or geological *structure.* The square root of the sum of the squares of differences is also used as the L2 norm. See also *model misfit, numerical regularization,* and *residual field.*

LaFehr formula: a formula for calculation of *Bullard B correction* for the gravity effect of the Earth's surface *curvature.* It converts *simple Bouguer correction* values calculated for the infinite horizontal *Bouguer plate* into values corresponding to the finite *Bouguer spherical cap* of the same thickness with the arc length, or surface radius, of 166.735 km around each observation point. The LaFehr formula is applied to high-*accuracy* regional-scale gravity data and expressed in units of *milligal (mGal)* as

$$\Delta g_{BB} = 2\pi G\rho(\mu h - \lambda R),$$

where G is *universal gravitational constant* in *cgs units*; ρ is *density* of rocks in the Bouguer spherical cap; h is Bouguer cap thickness defined by height, or *elevation,* of the observation point above sea level in meters; R is the Earth's radius at the observation point defined as $R = (R_0 + h)$, where $R_0 \cong 6371$ km is the Earth's radius to the sea level at 45° latitude; and μ and λ are dimensionless coefficients (LaFehr, 1991). See also *Cogbill formula.*

lag correction: 1. a correction applied to aeromagnetic data to compensate for *lag error* in the horizontal position of an instrument *reading* that is caused by the time delay, or lag, due to (a) *antenna offset* which is a horizontal distance separating the measuring instrument, such as a *stinger*-mounted airborne *magnetometer,* and *GPS* antenna and (b) electronic delays between the acquisition and navigation systems. Lag correction is a function of aircraft velocity, so the lag value is determined in seconds from the results of a *lag test,* and the correction value is calculated by multiplying the lag value in seconds and the aircraft velocity in meters per second at the time of each measurement. Sometimes referred to as *parallax correction.* **2.** *cable correction* applied to marine (shipboard) magnetic data to compensate for *lag error* due to a horizontal distance between the GPS antenna and cable-towed *marine magnetometer.* See also *heading correction* and *lever arm correction.*

lag error: error in the positioning of an instrument *reading* due to *antenna offset* and electronic delays between the acquisition and navigation systems of a *survey* aircraft or ship. Lag error is compensated by *lag correction,* which is referred to as *cable correction* for marine (shipboard) magnetic data. See also *lag test* and *lever arm effect.*

lag test: one of three standard *pre-survey flight tests* performed to estimate errors in the horizontal position of an instrument *reading* that are caused by the time delay, or lag, between the electronic response of the onboard navigation system and the *stinger*-mounted *magnetometer.* The test is flown in the opposite directions along the flight line over the easily identified source of a *magnetic anomaly,* such as a metal bridge or other large metal object on the ground. Lag value in seconds is defined by one-half of the time shift required to match the horizontal position of the magnetic anomaly determined on flights in the opposite directions. Lag test results are used in the calculation of *lag correction.*

lake gravity: gravity data obtained from measurements on a lake surface with the use of a light stable boat, such as a catamaran. For small land terrain effects and without significant changes in the lake *bathymetry,* the *Bouguer anomaly* Δg_B value at each *observation point* is expressed in units of *milligal (mGal)* as

$$\Delta g_B = g_{obs} - \Delta g_{TG} + 0.3086h - 2\pi G[\rho_r(h - d) + \rho_W d] \pm \Delta g_{Eo},$$

where g_{obs} is *observed gravity,* Δg_{TG} is *theoretical gravity correction,* h is lake *elevation* above sea level, G is *universal gravitational constant*; d is lake depth, ρ_r is assumed rock *density,* ρ_w is lake water density, and Δg_{Eo} is *Eötvös correction* for measurements on a moving boat. For stationary measurements with an *underwater gravimeter* lowered to the lake bottom from a survey boat, the Bouguer anomaly value is expressed as

$$\Delta g_B = g_{obs} - \Delta g_{TG} + 0.3086(h - d) + 2\pi G\rho_w d - 2\pi G\rho_r(h - d),$$

where the gravity effect of the water mass above each observation point is accounted by *double Bouguer correction* (Peirce and Lipkov, 1988; Hinze et al., 2013).

lamina: a thin horizontal layer, or sheet, of constant *density* or *magnetization.* Laminas are used as the elementary building blocks for *modeling* of *three-dimensional (3D)* source bodies of arbitrary shapes. Lamina thickness is usually less than one-tenth of its depth. Lamina shape is defined by the N-sided polygon bounded by straight line segments. Summing the gravity or magnetic effects of all individual laminas in a modeled *source body* will produce the gravity or magnetic field of, or *anomaly* generated by, this body. See also *Talwani modeling* and *3D source*.

land gravity: See *ground gravity*.

Laplace equation: the second-order differential equation which is a special case of the *Poisson equation* for the empty, or source-free, space. The Laplace equation describes the gravity or magnetic field at every space point in terms of the second-order derivatives of the gravity or magnetic *potential*. In *Cartesian coordinates*, it is expressed as

$$\nabla^2 P = \partial^2 P/\partial x^2 + \partial^2 P/\partial y^2 + \partial^2 P/\partial z^2 = 0,$$

where ∇^2 is *Laplacian*, ∇ is *gradient* operator *del*, and P is *gravitational potential* or *magnetic potential*. Among various applications, the Laplace equation is used for calculation of the *vertical derivative* of an anomalous gravity or magnetic field from its horizontal derivatives. See also *anomalous field* and *horizontal derivative*.

Laplace filter: **1.** a *grid*-based *edge-sharpening filter* which highlights the local areas of rapid changes in gravity or magnetic field values associated with the boundaries between structures of the contrasting gravity and magnetic properties, such as edges of high-*offset* structural blocks or bodies of anomalous *density* and *magnetization*. Laplace filter output is obtained by summing the second-order derivatives of the anomalous gravity or magnetic field that are calculated along the mutually orthogonal x-, y-, z-directions in *Cartesian coordinates*. Its algorithm can be expressed as

$$\nabla^2 F = \nabla \cdot (\nabla F) = \partial^2 F/\partial x^2 + \partial^2 F/\partial y^2 + \partial^2 F/\partial z^2,$$

where F is *anomalous field*, ∇^2 is *Laplacian*, and ∇ is *gradient* operator *del*. Profiles of the filtered data over the orthogonally striking step-like edges have a characteristic peak-to-trough *signature*, where the zero value corresponds to the *inflection point* which marks a transition from the positive to negative values of the field *curvature* and location of a structural *edge*. **2.** a grid-based *convolution* operator for estimates of the local curvature of a gravity or magnetic field: high values indicate high curvatures and zero values represent their inflection points. Also referred to as *Laplacian filter*. See also *strike* and *gravity field curvature*.

Laplacian: a data processing operator denoted by ∇^2 and representing the product of two *gradient* operators, i.e., $\nabla^2 = \nabla \cdot \nabla$, where ∇ is gradient operator *del*. It is used for calculation of the *three-dimensional (3D)* second-order spatial derivatives of anomalous gravity and magnetic fields and their *potential*. The Laplacian can be considered as the "gradient-of-gradient" operator, expressed as

$$\nabla^2 F = \nabla \cdot (\nabla F),$$

where F is *anomalous field*. In *Cartesian coordinates*, the Laplacian is expressed as

$$\nabla^2 = \partial^2/\partial x^2 + \partial^2/\partial y^2 + \partial^2/\partial z^2.$$

Laplacian filter: See *Laplace filter*.

Larmor frequency: the *angular frequency* at which charged subatomic particles (positively charged protons and negatively charged electrons) precess, or rotate their axes of rotation, about the direction of a stable ambient magnetic field, such as the *Earth magnetic field*, in response to the application of a specific electromagnetic field. The Larmor frequency ω_L of *precession* is defined in units of *hertz (Hz)* as

$$\omega_L = \gamma H/2\pi,$$

where γ is *gyromagnetic ratio* and H is *magnetic field strength*, i.e., the Larmor frequency is directly proportional to the strength, or *total intensity*, of the ambient magnetic field. For the same magnetic field strength, the subatomic particles with different gyromagnetic ratios have different Larmor frequencies. For the Earth's magnetic field of 50,000 *nanotesla (nT)*, the Larmor frequency of protons is 2129 Hz. When the oscillating electromagnetic waves or monochromatic laser light waves are emitted with a frequency equal to the Larmor frequency and at right angles to the direction of a stable ambient magnetic field, the major part of the electromagnetic energy will be absorbed by charged subatomic particles. The energy absorption property of the precessing unpaired electrons in vapor cells is used to measure the Larmor frequency and determine the *magnitude* of the Earth's magnetic field by an *optically pumped magnetometer*. See also *ambient field* and *proton magnetometer*.

Larmor signal: an electric signal generated by the sensors of *quantum magnetometers* in a direct proportion to the strength, or *total intensity*, of a stable ambient magnetic field, such as the *Earth magnetic field*. The Larmor signal is converted into values of the measured Earth's magnetic field in units of *nanotesla (nT)* using the *gyromagnetic ratio* of the subatomic particles in a particular *magnetometer* type. See also *ambient field*, *Larmor frequency*, and *magnetic sensor*.

laser altimeter: **1.** the type of *altimeter* installed on Earth-orbiting satellites to measure the *topography* and ocean/sea surface undulations. **2.** *LIDAR*.

laser gravimeter: a *falling-mirror gravimeter* with an emphasis on the use of a laser-generated monochromatic light in its measurement system.

laser interferometer gravimeter: a *falling-mirror gravimeter* with an emphasis on the use of a laser interferometer in its measurement system.

laser magnetometer: See *optically pumped magnetometer*.

laser-pumped magnetometer: an *optically pumped magnetometer* with an emphasis on the use of the laser-generated monochromatic light waves in its measurement system.

laser scanner: See *LIDAR*.

lateral contrast: *contrast* in the context of a difference in *density* or *magnetization* between laterally adjacent rock units, as opposed to a vertical contrast (such as the one between homogeneous horizontal layers) which does not generate a gravity or magnetic *anomaly*.

lateral gradient: same as *transverse gradient* with an emphasis on the *gradient* measurements by two laterally separated *magnetometer* sensors, such as those mounted at the wingtips of a *fixed-wing survey* aircraft.

lateral resolution: 1. a characteristic parameter of data that describes their ability to identify, or resolve, anomalies generated by horizontally separated sources. For the *total field* data, it is defined by a width of the narrowest *anomaly* for which the *amplitude* exceeds the *accuracy* of measurements. The anomaly width is determined by the horizontal distance between the inflection points on its *profile* or between zero *grid* contours on both sides of the anomaly maximum, and this distance is called the *half-wavelength resolution* or *half-sine resolution*. Accordingly, *full-wavelength resolution* is a doubled width of the narrowest anomaly which can be detected and mapped. For the *gradient* field data, it is defined by the horizontal distance between the gradient maxima corresponding to the opposite edges of the smallest isolated *source body* or to the spatially closest structural discontinuities, such as faults and lithological contacts. **2.** the projected estimate of the shortest *wavelength* of anomalies that can be mapped by a gravity or magnetic *survey* with the selected *survey specifications*. In marine (shipboard) and airborne surveys, the full-wavelength resolution is defined primarily by (a) the length of a *low-pass filter* applied to suppress *dynamic noise* and (b) the *survey line spacing*. In airborne surveys, the lateral resolution is inversely proportional to the flight *altitude*: the higher the altitude, the lower the resolution. **3.** the smallest structural element, such as a *structural low* or *structural high*, of a mapped surface (sea surface, ground surface, sea bottom, sedimentary horizon, *basement* surface, etc.) that can be identified on a *grid* or map. Sometimes referred to as *horizontal spatial resolution*. See also *contact*, *edge*, *fault*, *inflection point*, and *vertical resolution*.

latitude correction: 1. a correction for the *latitude effect* of the latitude-dependent increase in *theoretical gravity* values from the equator to the geographic poles. **2.** *theoretical gravity correction*. **3.** equivalent of the theoretical gravity correction applied to *ground gravity* data acquired by *microgravity* and small-area regular surveys, where the locations of each *station* are not referenced to the global latitudes, and gravity measurements are made relative to a single *base station* with known latitude and theoretical gravity values. It is also referred to as *normal gravity gradient correction*, because it is based on the use of the *normal gravity gradient* and approximation of the theoretical gravity variations by a linear function of the north–south distance from a base station. Correction values are defined in units of *milligal (mGal)* and *microgal (μGal)* as

$$\Delta g_L(\text{mGal}) = \pm 0.8144 \cdot 10^{-3} D \cdot \sin 2\theta,$$
$$\Delta g_L(\mu\text{Gal}) = \pm 0.8144 D \cdot \sin 2\theta,$$

where θ is the base station latitude and D is the north–south distance in meters between the base station and each measurement station. In the northern hemisphere, the calculated correction values are added to the known theoretical gravity value for all stations positioned to the south of the base station, and they are subtracted from the theoretical gravity value for all stations positioned to the north of the base station. The calculation procedure is reversed for surveys in the southern hemisphere. The obtained theoretical gravity values for each station are subtracted from the *observed gravity* values in the process of *gravity reduction* (Yule et al., 1998; Hinze et al., 2013).

latitude effect: smooth global-scale changes in the dominant components of the Earth's magnetic and gravity fields. Magnetic latitude effect is caused by latitude-related changes in the *main geomagnetic field* described by the *International Geomagnetic Reference Field (IGRF)*. Gravity latitude effect of an increase in the *theoretical gravity* values from the equator to the geographic poles is caused by (a) a decrease of the Earth's radius by approximately 21 km and (b) a decrease of *centrifugal acceleration* from its maximum value at the equator to zero at the poles. The contribution of the latitude effect to measured gravity and magnetic values is compensated by the application of *theoretical gravity correction*, also known as *latitude correction*, and *IGRF correction*, respectively. Also referred to as *latitudinal variation*.

latitudinal gradient: the rate of change of the *theoretical gravity* per unit distance along a geographic meridian (latitude) estimated in units of *milligal* per kilometer ($\text{mGal} \cdot \text{km}^{-1}$). The average value is 0.8144 $\text{mGal} \cdot \text{km}^{-1}$ corresponding to the latitude of ±45°. See also *longitudinal gradient*.

latitudinal variation: *latitude effect* in the context of a smooth global-scale change, or variation, of the *normal gravity* and *normal magnetic field* values as a function of the geographic latitude.

lava: a molten igneous rock which flows out of volcanos onto the Earth's surface. Basalt lava flows can spread over areas of hundreds and thousands of square kilometers and accumulate in thickness up to 8–10 km. Due to the high *density* and *magnetization* of solidified lava rocks, the lava flows buried under thick sequences of younger sediments can form a local- or regional-scale *gravity basement* and *magnetic basement* that have no relation to the geological, or true, *basement* in a survey area. Lava flows generate the regional- and local-scale *magnetic high* and *gravity high* anomalies which in the latter case can be comparable to or exceed the *gravity anomaly* generated by a large *anticline* or *thrust*.

layer filtering: same as *separation filtering* with an emphasis on separating the gravity and magnetic field components generated by the layer-approximated assemblages of sources at sufficiently different depth levels.

layered-Earth model: a simplified *model* of the Earth's *crust* composed of homogeneous layers, each having a constant *density* or *susceptibility* of the corresponding rock formations, above a homogeneous *half-space* approximating the deep crustal part. See also *full-Earth model*.

layered stripping: *gravity stripping* in the context of a layer-by-layer removal of the gravity effects of sedimentary layers for which the depth structure and *density* are known from seismic and well (borehole) data.

leads: See *exploration leads*.

least-squares approximation: See *least-squares fit*.

least-squares collocation: a method of the data interpolation based on the use of a *covariance matrix* and weighted average of the available values surrounding each interpolation point, wherein the weighted averages are obtained in a minimum-*variance* sense. The method is applied in the *merging* of gravity and magnetic data with highly different *lateral resolution* and *wavelength* content, such as those acquired by ground, marine (shipboard), and airborne surveys and satellite missions, into the global-scale compilations, e.g., *Earth Magnetic Anomaly Grid-2 (EMAG2)* and *World Digital Magnetic Anomaly Map (WDMAM)*. See also *least-squares fit*, *kriging*, and *cokriging*.

least-squares filter: *optimum filter* in the context of the *least-squares minimization* of the differences between a desired result of *filtering* and the actual filter output.

least-squares fit: the best-fitting approximation of a given set of observed gravity or magnetic values. The best fit in the least-squares sense means that the approximating function minimizes the sum of the squared differences between the individual values in a set and its best-fitting approximation. Also referred to as *least-squares approximation* and *least-squares minimization*. See also *standard deviation (SD)*.

least-squares minimization: **1.** *least-squares fit*. **2.** *least-squares optimization*.

least-squares optimization: an *inversion* result or *filter* output optimized, or minimized, in the least-squares sense, which means that the sum of the squared differences between the observed and inverted *model* values or between the actual filter output and its desired output is minimized. Also referred to as *least-squares minimization*.

legacy data: See *vintage data*.

level correction: See *level shift*.

level shift: a *leveling* procedure applied to gravity and magnetic values measured along the individual lines of an airborne or marine (shipboard) *survey* in order to minimize differences between the measured values at the intersections of survey lines. The amount of level shift is determined by averaging the crossover errors between the traverse and tie lines. Level shift is applied after suppressing the *dynamic noise* components by low-pass *filtering* of all survey lines. Sometimes referred to as *level correction*. See also *low-pass filter*, *tie lines*, *traverse lines*, and *crossover error*.

leveling: **1.** minimization of the differences between the measured values, or minimization of mis-ties, at crossover points of line-oriented ground, marine (shipboard), and airborne *survey* data prior to *gridding*. Leveling iteratively adjusts the mis-tie values in the manner of *least-squares fit*, wherein the calculated mis-tie variations are approximated by a smooth function to obtain the values of *leveling corrections* at all observation points along the survey lines. Leveling compensates for (a) discrepancies in the measured *elevation* of each observation point during the ground surveys, (b) *altitude* variations and *positioning* errors during airborne and marine surveys, (c) non-linear components of *gravimeter drift*, and (d) residual *diurnals* and other *temporal magnetic variations* remaining after application of *diurnal correction*. Because the leveling procedure changes the local gradients and, hence, curvatures of the measured fields over the network of the line intersections and, thereby, changes the shapes and amplitudes of the actual anomalies, it can have a negative effect on results of the three-dimensional (3D) *inversion* of gridded data. Also referred to as *crossover leveling*, *line leveling*, *tie-line leveling*, and *network adjustment*. **2.** aligning the sensing axis of a *gravimeter* with the vertical direction at each *station* during a ground *gravity survey*. **3.** geodetic mapping of heights, or elevations, of the ground observation points above sea level. See also *crossover point*, *gradient*, *mis-tie*, and *microleveling*.

leveling corrections: correction, or adjustment, values for each observation point along the *survey* lines obtained by *least-squares minimization* of mis-ties at crossover points of the line-oriented magnetic and gravity data. See also *crossover point*, *level shift*, *leveling*, *mis-tie*, and *microleveling corrections*.

leveling errors: systematic residual errors remaining in gridded marine (shipboard) and airborne gravity or magnetic data after *leveling*. They are represented by the low-*amplitude* wave-like ridges, or stripes, along *survey* lines and commonly referred to as *corrugations*. Leveling errors are best visible on a *map power spectrum* as alignments of *power spectrum* values oriented orthogonally to the directions of corresponding survey lines. Leveling errors are reduced to the level of *random noise* by the application of *microleveling*. Also referred to as *line-level errors*. See also *gridding*.

lever arm: the distance, or instrument offset, between the *GPS* antenna and *gravity sensor* onboard a *survey* aircraft or ship. Its high-*accuracy* estimate is required for calculation of (a) *lever arm correction* applied to the *relative gravity* data acquired by airborne and marine gravity surveys and (b) *lag correction* applied to aeromagnetic data. Also referred to as *antenna offset*. See also *airborne gravity*, *marine gravity*, and *lever arm effect*.

lever arm correction: a correction for the motion-related *lever arm effect*. It is applied to *airborne gravity* and *marine gravity* data, and its calculation requires three types of the high-*accuracy* data: (1) *attitude* variations of a survey aircraft or ship by *differential GPS (DGPS)*, (2) inertial accelerations calculated from DGPS data or measured independently, and (3) *lever arm* length (de Saint-Jean et al., 2007). See also *inertial acceleration* and *moving-platform corrections*.

lever arm effect: a motion-related effect in *marine gravity* and *airborne gravity* surveys. Due to the different spatial positions of the GPS antenna and *gravity sensor*, the *inertial acceleration* components in a sensing *accelerometer* and inertial accelerations measured independently or calculated from DGPS measurements are different. Moreover, when the GPS antenna and gravity sensor are at the opposite sides from the center of gravity (COG) of a survey aircraft or ship, their sense of motion (and, therefore, response to the *cross-coupling effect*) would be opposite: GPS antenna ahead of the COG moves up with the aircraft or ship pitching up, whereas the gravity sensor behind the COG moves down for the same maneuver. The lever arm effect is the greatest in airborne gravity measurements, and its *magnitude* depends, primarily, on the *lever arm* length and aircraft *attitude* variations during a flight. Compensated by application of a *lever arm correction* (Jekeli, 2000; de Saint-Jean et al., 2007). See also *pitch*, *roll*, and *yaw*.

levitation: See *electrostatic levitation* and *magnetic levitation*.

LIDAR: the **Li**ght **D**etection **A**nd **R**anging airborne system for measurements of *altitude* of a *survey* aircraft by laser-generated monochromatic light waves. The achievable *accuracy* is 2–5 cm within the measurement range up to 400 m. In integration with *GPS* data, closely sampled high-*precision* altitude measurements are used for the three-dimensional (3D) mapping of *topography* and the construction of a *Digital Terrain Model (DTM)* of an *airborne survey* area. Also referred to as *laser altimeter* and *laser scanner*. See also *sampling*.

line averaging: averaging of *airborne gravity* data acquired along the closely spaced (50–100 m) flight lines of a *fixed-wing survey* in order to suppress *dynamic noise*. For *random noise*, line averaging increases the *signal-to-noise ratio (S/N)* by $\sqrt{N}$, where N is the number of averaged lines. Before the averaging, the original time-based line-oriented data are low-pass filtered by a *time filter* to eliminate the high frequencies beyond the *measurement bandwidth*. See also *low-pass filter* and *time averaging*.

line data: gravity and magnetic data acquired with a pre-selected *sample interval* along the *survey* lines, which are also referred to as *acquisition* lines. See also *fiducials*.

line filter: a *convolution filter* applied to *line data* before *gridding* in order to suppress the high-*frequency* components of *noise*, such as *cultural noise* and *spike*-like changes of the *external magnetic field* of the Earth. In marine (shipboard) and airborne gravity and *gravity gradient* surveys, the length of the applied line filter, such as a *time filter*, defines the achievable *half-wavelength resolution* of the acquired gravity data. See also *filter length*.

line-level errors: See *leveling errors*.

line leveling: *leveling* in the context of the minimization of mis-ties at the crossover points, i.e., at coordinate points where *traverse lines* and *tie lines* cross each other. See also *mis-tie*.

line lineaments: alignments of anomalies in gridded *total field*, *gradient*, *gradient tensor* data and their calculated attributes associated with the linear and curvilinear elements of a subsurface geological *structure* and structural discontinuities, such as high-angle and vertical thin dikes, *thrust*, *flower structure*, low-*offset* magnetized faults, strike-slip faults, buried drainage channels, *shear zone*, fracture and *salt* dissolution zones, etc. See also *anomaly*, *attribute*, *thin dike*, *fault*, *gridding*, *lineaments*, *magnetized fault*, *strike-slip fault*, *edge lineaments*, and *strike lineaments*.

line power spectrum: *power spectrum* that shows the energy of anomalies by their squared amplitudes measured along a particular *survey* line. It is used primarily for selecting the parameters of a *line filter*.

line resampling: a *data conditioning* procedure applied to ground gravity or magnetic *survey* data in order to (a) fill data gaps by interpolation of the measured values and (b) replace the original irregular sample intervals by the same *sample interval* for an entire survey line before the application of a *line filter*.

line source: a *source body* approximation by a horizontal or vertical line of point masses or point dipoles. A horizontal line source generates a *two-dimensional (2D)* gravity or magnetic *anomaly*, whereas a vertical line source generates a 3D anomaly. See also *point dipole* and *point mass*.

line spacing: See *survey line spacing*.

lineament interpretation: analysis and *interpretation* of *lineaments* obtained by *enhanced processing* of the gridded *total field*, *gradient*, *gradient tensor* data, and their attributes. It follows the principles of *integrated interpretation* for the identification and mapping of anomalous alignments in agreement with their structural sources and available geological and geophysical *constraints*. Different sources of *edge lineaments* and *line lineaments* are the main reasons for their different positions on a *grid*. Differences, or mismatch, between positions of lineaments of the same type are commonly related to: (a) different *acquisition* program parameters: lineaments may be obtained from the data acquired with different *sampling* and *lateral resolution* of anomalies; (b) differences in *density* and *magnetization* contrasts: a *source body* can have a *density contrast* sufficient for generating a detectable *gravity anomaly* and an insufficient *magnetization contrast* and vice versa; (c) *positioning* errors in the merged data acquired with positioning systems of different accuracies; and (d) different *resolution* capabilities of the *magnetic method* and *gravity method*: the former is more resolving, especially laterally, so the total amount of the detected magnetic lineaments is usually larger than that of the gravity lineaments in the same area. See also *accuracy*, *attribute*, *gridding*, *merging*, and *ACLAS*.

lineament invariants: See *horizontal invariants*.

lineaments: 1. linear and curvilinear alignments of anomalies in the gridded gravity and magnetic *total field*, *gradient*, *gradient tensor* data, and their calculated attributes that are associated with elements of a subsurface geological *structure*. There are two major types of lineaments: (1) *edge lineaments* representing the anomalies generated by the structural and lithological edges within a sedimentary section and at the top of a *basement*, e.g., high-*offset* faults, edges of igneous intrusions, depositional edges, and other high-*contrast* lateral contacts of rock formations; and (2) *line lineaments* representing the anomalies generated by linear and curvilinear elements of a subsurface structure and structural discontinuities, e.g., high-angle and vertical thin dikes, low-offset magnetized faults, strike-slip faults, buried drainage channels, shear zones, fracture and *salt* dissolution zones, etc. **2.** linear topographic features identified on airborne and satellite images as the long-extent scarps, stream courses, alignments of vegetation, and other surface features associated with fracture zones and faults that can have deep roots in a sedimentary section and *basement*. See also *contact*, *edge*, *fault*, *gridding*, *igneous intrusion*, *magnetized fault*, *shear zone*, *thin dike*, and *strike lineaments*.

linear anisotropic index: a quantitative measure of the degree of *FTG linear anisotropy* defined as

$$\mathrm{LAI} = \left(|\lambda_1| - |\lambda_2|\right)/|\lambda_1|,$$

where $|\lambda_1|$ and $|\lambda_2|$ are magnitudes of the two largest *FTG eigenvalues*. The index is normalized by $|\lambda_1|$ to range between 0 and 1 for isotropic and most anisotropic cases, respectively, and to make the sum of the linear, planar, and spherical indices to be equal to 1 at every *grid* point. The relationship between these three calculated anisotropic indices is visualized by a monochromatic or color-coded *barycentric image* to show variations of the *FTG anisotropy* parameters associated with the elements of a subsurface geological *structure* (Hassan and Goussev, 2020). See also *planar anisotropic index* and *spherical anisotropic index*.

linear anisotropy: See *FTG linear anisotropy*.

linear filter: a *filter* for which the output is a linear combination, or sum, of the line-oriented or gridded gravity and magnetic data convolved with the *transfer function* of this filter. See also *convolution*, *gridding*, and *convolution filter*.

lithosphere: the outer layer of the Earth that includes the entire *crust* and underlying uppermost part of the *mantle*. Its thickness ranges from 100–120 km to 150–250 km on the continents and from 40–50 km to 80–100 km in the oceans, whereas along the axes of the mid-ocean ridges it may be as thin as the oceanic crust (6–9 km). The lithosphere is segmented into six major and 14–15 minor lithospheric plates, whose movements and interactions have resulted in the global structural-tectonic deformations of the Earth's crust. The lithosphere overlays a much hotter and more mobile *asthenosphere*, the relative weakness of which makes possible the floating motion of the lithospheric plates. The lithospheric bottom is not a petrological boundary (i.e., boundary between different rock types), but rather the quantitatively estimated lower limit of a relatively strong outer layer of the Earth. At the long-*wavelength* scale of satellite measurements with *lateral resolution* of magnetic anomalies of approximately 350–400 km, the *magnetization* of the lithosphere is mostly induced on the continents and mostly remanent in the oceans. See also *induced magnetization*, *remanent magnetization*, and *isostasy*.

lithospheric magnetic field: the component of the *Earth magnetic field* generated by magnetized rocks within the *lithosphere*. At each observation point, its numerical, or *scalar*, value and *vector* orientation represent the combined magnetic effects of *induced magnetization* and *remanent magnetization* of the igneous, metamorphic, and sedimentary rocks comprising the local- and regional-scale geological structures within the *crust* and upper *mantle*. In exploration applications, the long-*wavelength* contributions from the iron-bearing magnetic rocks in the upper mantle are assumed to be negligible, and the lithospheric magnetic field is commonly referred to with the same meaning as *crustal magnetic field*.

lithospheric thermal gravity anomaly: *thermal gravity anomaly* with an emphasis on its origin from a lateral decrease in the *density* of rocks in the deep parts of the *lithosphere* due to their thermal expansion.

local amplitude: **1.** the primary *complex attribute* of a *monogenic signal (MS)*, defined as

$$\mathrm{LA}(x,y) = \{[f(x,y)]^2 + [R_1 f(x,y)]^2 + [R_2 f(x,y)]^2\}^{1/2},$$

where $f(x,y)$ is the real part representing the anomalous gravity or magnetic field, such as *RTP anomaly* or *Bouguer anomaly*, at *grid* point (x,y) and $R_1f(x,y)$ and $R_2f(x,y)$ are two imaginary, or 90° *phase*-shifted, parts obtained by *Riesz transform* of $f(x,y)$ in the orthogonal complex planes (Felsberg and Sommer, 2001). Because the Riesz transform operators R_1 and R_2 are equivalent to the x- and y-components H_x and H_y of the *3D Hilbert transform*, the local amplitude of the monogenic signal is equivalent to (1) the *direct analytic signal amplitude (DASA)*, (2) the *zero-order ASA* (ASA_0), and (3) the *analytic signal amplitude (ASA)* of the *vertical integral* of the magnetic or gravity *anomaly*, i.e., it is equivalent to the ASA of the *anomalous potential*. Also known as *monogenic signal amplitude* and *monogenic instantaneous amplitude*. **2.** *instantaneous amplitude* in the context of its characteristic as (a) a *space domain* parameter of an *anomalous field* calculated at each observation point to determine the direct, or instant, *amplitude response* to variations of a subsurface geological *structure*, and (b) a direct response to changes in the *amplitude* content of a *complex variable* known as *analytic signal (AS)*. See also *anomalous field*, *complex plane*, *local energy*, *local frequency*, *local orientation*, and *local phase*.

local energy: a measure of the *amplitude* content of a *complex variable*, such as *analytic signal (AS)* or *monogenic signal (MS)*, expressed as

$$\mathrm{LE}(x,y) = [\mathrm{Re}Pf(x,y)]^2 + [\mathrm{Im}Pf(x,y)]^2,$$

where $[\mathrm{Re}Pf(x,y)]$ is the real part representing the anomalous gravity or magnetic field f at *grid* point (x,y) and $[\mathrm{Im}Pf(x,y)]$ is the imaginary, or 90° *phase*-shifted, part obtained by *Riesz transform* or *3D Hilbert transform*. See also *anomalous field*, *local amplitude*, *local frequency*, *local orientation*, and *local phase*.

local frequency: **1.** the horizontal rate of change, or *horizontal derivative*, of a *local phase* that can be expressed as

$$f_L = \{\partial\{\arctan[(\partial F/\partial z)/(\partial F/\partial x)]/\partial x\}\}/2\pi,$$

where $\partial F/\partial x$ and $\partial F/\partial z$ are horizontal and vertical derivatives, respectively, of the anomalous gravity or magnetic field F. Local frequency is referred to with the same meaning as *instantaneous frequency* in the context of a direct, or instantaneous, response of an *anomalous field* to variations of the local phase. Often, *local wavenumber* k_L is used instead of the local frequency f_L based on their relationship: $k_L = 2\pi f_L$. **2.** a quantitative measure of the *spatial frequency* content of the gridded data described by the dominant *frequency* band in a particular local area. See also *gridding*, *local amplitude*, *local energy*, *local orientation*, and *local phase*.

local gravity: *residual gravity* in general and *residual Bouguer gravity* in particular with an emphasis on its association with a local-scale geological *structure*.

local isostasy: *Airy hypothesis* and *Pratt hypothesis* in the theory of *isostasy*. They assume a relative weakness of the Earth's *crust* and its isostatic response to rock mass loading and unloading due to topographic and bathymetric relief variations by local-scale block faulting without involving neighboring regions. See also *regional isostasy*.

local magnetic anomalies: 1. *residual magnetics*. **2.** magnetic anomalies of a restricted, or local, distribution generated by magnetic sources in the uppermost part of the *crust*.

local orientation: the primary *complex attribute* of the *monogenic signal (MS)* that describes the MS orientation in a *vector* space by the angle of rotation of its projection on the *complex plane* formed by mutually orthogonal imaginary axes. It is denoted by $\theta(x,y)$ and expressed as

$$\theta(x, y) = \arctan(|R_2 f(x, y)|/|R_1 f(x, y)|),$$

where $f(x,y)$ is the real part representing the anomalous gravity or magnetic total field f at *grid* point (x,y) and $R_1 f(x,y)$ and $R_2 f(x,y)$ are two imaginary, or 90° *phase*-shifted, parts obtained by *Riesz transform* of $f(x,y)$ in the orthogonal complex planes (Felsberg and Sommer, 2001). The local orientation is orthogonal to the *local phase* and *local amplitude*, and it tends to align along a continuity of anomalies, so their truncations by a cross-cutting *fault*, such as a *strike-slip fault*, create a distinct *truncation trend* in the gridded local orientation values and, therefore, provide high-*resolution* information for the analysis and *interpretation* of structural discontinuities, such as faults, fracture and shear zones, lithological contacts, etc. Also referred to as *monogenic signal orientation* and *monogenic instantaneous orientation*. See also *anomalous field*, *contact*, *gridding*, *local amplitude*, *local energy*, and *local frequency*.

local phase: 1. a *complex attribute* of the gravity or magnetic field defined in a *vector* space by the angle between the *complex variable* vector and real axis on a *complex plane*. It is denoted by ϕ and expressed as

$$\phi = \arctan(\mathrm{ImP}/\mathrm{ReP}),$$

where ImP is the imaginary, or 90° phase-shifted, part of the complex variable, and ReP is its real, or in-phase, part. In *enhanced processing*, the local phase ϕ is represented by two angular quantities: (1) the arctangent of the ratio of the vertical and horizontal derivatives of the *anomalous field* F:

$$\phi = \arctan[(\partial F/\partial z)/(\partial F/\partial h)],$$

where h is any horizontal direction on the XY plane; and (2) the arccosine of the ratio of the *total horizontal derivative* to the *analytic signal amplitude (ASA)*:

$$\phi = \arccos[(\partial F/\partial x)^2 + (\partial F/\partial y)^2]^{1/2}/|\mathrm{ASA}|.$$

The local phase is analogous to *tilt angle (TA)*, but it utilizes the sign of the *horizontal derivative* along a particular direction, whereas in calculation of the tilt angle the absolute value of the horizontal derivative is used. **2.** the primary complex attribute of a *monogenic signal (MS)* that describes the monogenic signal *phase* by the angle between the MS vector and the real axis on a complex plane. The MS local phase is expressed as

$$\Phi(x, y) = \arctan\{[R_1 f(x, y)]^2 + [R_2 f(x, y)]^2\}^{1/2} f/(x, y)\},$$

where $f(x,y)$ is the real part representing the anomalous gravity or magnetic field f at *grid* point (x,y) and $R_1 f(x,y)$ and $R_2 f(x,y)$ are two imaginary parts obtained by *Riesz transform* of $f(x,y)$ in orthogonal complex planes (Felsberg and Sommer, 2001). The MS local phase angle is complementary to the tilt angle (TA). Both angles range from –90° to +90°, but the TA varies continuously, whereas the MS local phase changes abruptly from –90° to +90° across the zero contour on the $f(x,y)$ *grid*. The MS local phase is orthogonal to the *local orientation*, and its values reach maxima above the structural edges, such as high-*offset* faults, lithological *density* and *magnetization* contacts, and *igneous body* edges. Also referred to as *monogenic signal phase* and *monogenic instantaneous phase*. See also *contact*, *edge*, *local amplitude*, *local energy*, *local frequency*, and *local orientation*.

local phase filter: 1. an algorithm for calculation of the *complex attribute* known as the *local phase*. **2.** an edge-detection algorithm having an operational principle based on the concept of *normalized derivatives*. The local phase filter converts a gravity or magnetic field into the *field* of the normalized angular, or *phase*, quantities that (a) represent a balanced *amplitude response* to sources from different depths and (b) reach their maximum values above the structural edges, such as high-*offset* faults, lithological *density* and *magnetization* contacts, and edges of isolated source bodies. Algorithms for calculation of *tilt angle (TA)*, *theta angle*, and their modifications are local phase filters. See also *edge*, *edge detection*, and *filter*.

local reference frame: a Cartesian coordinate frame for which the vertical z-axis is aligned with the direction of a local anomalous gravity field defined by the orientation of an *anomalous gravity vector* at each observation or *grid* point. Because *eigenvectors* can be used to determine the orientation of axes in local reference frames for point and spherical sources only, the anomalous gravity vector direction is estimated via the process of *scalar-to-vector transformation* of the local gravity field reconstructed from the measured *gravity gradient tensor (GGT)* components using *wavenumber*-domain operators (Li, Y. 2018). After the application of all relevant corrections to the measured *gradient* values, the point-by-point rotations of the GGT from the fixed Cartesian coordinate frame of its original measurements to every local reference frame are required prior to calculation of the *curvature attributes* of the gravitational *equipotential surface* to obtain their most accurate values referred to as *true curvature attributes*. See also *apparent curvature attributes*, *point source*, and *rotated gradient tensor*.

local spectra: amplitudes and phases of the sinusoidal components, or *harmonics*, obtained by *window Fourier transform (WFT)* in localized areas of an anomalous gravity or magnetic field, as opposed to the *global spectrum* obtained by the standard *Fourier transform* and representing the total of amplitudes and phases of all harmonics in a gridded *anomalous field.*

local wavenumber (LW): an *attribute* of an *anomalous field* implemented in the automatic estimation of parameters of magnetic sources approximated by the two-dimensional (2D) source geometries of a *magnetic contact, thin sheet* (*thin dike*), *finite step*, and *horizontal cylinder.* There are four local wavenumber types: (1) *horizontal local wavenumber* k_x defined by the first-order *horizontal derivative* of the *local phase* and also referred to as *first-order local wavenumber* k_1, (2) *vertical local wavenumber* k_z defined by the first-order *vertical derivative* of the local phase, (3) *second-order local wavenumber* k_2 defined by the first-order horizontal derivative of the local phase of the first-order vertical derivative of the anomalous magnetic field, and (4) a *multimodel wavenumber* determined from k_2 and k_1. The same as the *analytic signal amplitude (ASA)* or *total gradient (TG)* for 2D sources, the LW has its maxima over the structural edges regardless of the edge *dip*, geomagnetic field *inclination* and *declination*, direction of the source *magnetization*, and sign of *susceptibility contrast.* Combinations of the calculated local wavenumbers k_x and k_z are used for estimating the depth, *structural index (SI)*, and other *source body* parameters at all observation points and visualizing them on *grid* images (Smith and Salem, 2005). Sometimes referred to as *instantaneous wavenumber.* See also *AS-LW relationship* and *enhanced local wavenumber (ELW) method.*

local wavenumber methods: methods of automatic estimation of magnetic source parameters, such as horizontal position, depth, and *dip*, that are based on the use of the *local wavenumber (LW).* They include the *source parameter imaging (SPI™) method, improved SPI™ method, enhanced local wavenumber (ELW) method, multimodel method,* and *tilt-derivative method.*

located Euler deconvolution: the *grid*-based *Euler deconvolution* which differs from the standard *3D Euler deconvolution* by using only the grid values within a square window centered at locations of the *analytic signal amplitude (ASA)* maxima. The window size is defined by the distance between the inflection points of each *anomaly* selected by *interpreter* in the gridded *Bouguer gravity* or *RTP magnetic field.* The advantages over the standard version are: (a) variable-length calculation windows are tailored to the size of each *anomaly*; (b) depth solutions are obtained for the selected anomalies only; and (c) the number of solutions is much less than those of the standard version, where calculation results are overpopulated with the interference, *noise*, and other artificial depth solutions (Bainbridge et al., 2002). See also *gridding, inflection point,* and *weighted Euler deconvolution.*

log-energy spectrum: See *power spectrum.*

log-map power spectrum: See *map power spectrum.*

log-power plot: See *power spectrum.*

log-power spectrum: *power spectrum* with an emphasis on the logarithmic scale of the squared amplitudes representing the energy, or power, of the individual components of the Fourier-transformed data. See also *Fourier transform.*

longitudinal: a term applied to describe the direction along the major axis of a *structure* or along a *survey* or *modeling* line to highlight its difference from the transverse, or orthogonal, direction. See also *longitudinal gradient* and *transverse gradient.*

longitudinal gradient: *horizontal gradient* measured along the direction of the flight lines in an *airborne survey.* For *magnetic gradient* surveys with two wingtip-mounted sensors and one *sensor* at the tail *stinger,* it is measured by: (a) the difference between the average of two wingtip measurements and tail measurement of the *total magnetic field* divided by the distance between the tail sensor and the projection of the wingtip sensors to the aircraft axis and (b) the time *derivative* $\partial F/\partial t$ of the total magnetic field F between the successive tail sensor measurements divided by the instantaneous velocity, i.e., aircraft velocity at the time of measurements. Measured longitudinal gradients can be used in *gradient-enhanced gridding.* Also referred to as *in-line gradient.* See also *transverse gradient* and *gradient-enhanced minimum-curvature.*

Longman algorithm: an algorithm for calculation of the numerical value and sign (positive or negative) of the *Earth tide correction,* also known as *tidal correction,* at any time and any place on the Earth's surface. The latitude and longitude of each *observation point* and Universal Coordinated Time/Greenwich Mean Time (UTC/GMT) are inputs to this algorithm. Software modules of modern gravimeters include the Longman algorithm into the data processing program, automatically compensating for the gravity effect of *Earth tides* on a *station*-by-station basis.

Longman formulas: mathematical expressions for calculation of the *Earth tide correction* and *tidal accelerations* at any time and any place on the Earth's surface regardless of the *reference ellipsoid* version (Longman, 1959). The formulas are sufficiently accurate for regular gravity surveys, but they do not account for the *ocean loading effect* and variations of crustal elasticity that are reasons for (a) different amounts of tide-related crustal deformation at different locations and (b) variable delay, or phase lag, in response to the time-varying and latitude-dependent gravity effect of the combined *gravitational attraction* of the Sun and Moon. Also referred to as *Longman tidal equations.* See also *automatic Earth tide correction, Earth tide effect,* and *gravitational force.*

Longman tables: the original tabulated *format* of the *Longman formulas* used in vintage gravity surveys for calculation of the *Earth tide correction.* See also *vintage data.*

Longman tidal equations: See *Longman formulas.*

loop: a sequence of gravity and magnetic observations in which the first and last measurements are made at the same coordinate point, or *station,* during a *ground survey.* The objective of loop measurements, or measurements by the *looping method,* is control over *drift* and other systematic instrument errors in *gravimeter* and *magnetometer* readings. The time duration of each loop sequence should allow the approximation of *gravimeter drift* or *magnetometer drift* by a linear function of time. Also referred to as *traverse loop.* See also *gravity loop network, open-end loop,* and *reading.*

loop-base station: the coordinate point at the start and end of a particular *loop*. Such points are used for repeated measurements, or *reoccupations*, in several loops during the same *ground gravity* survey. See also *tie station* and *gravity loop network*.

loop closure: a *closure error* with an emphasis on the *reading* error accumulated during the *ground gravity* measurements along a single *loop*.

loop-closure correction: **1.** a correction of the measured *ground gravity* values for the *reading* errors accumulated in the process of measurements along a single *loop*. **2.** a correction applied to measurements of *elevation* at each observation point during a high-*accuracy* ground *gravity survey*. The correction procedure is based on the iterative adjustment of the gravity and elevation values at each *crossover point* and minimization of *loop closure* as a part of the *leveling* process and control over *drift* and measurement artifacts, such as *bull's-eye anomaly*. See also *artifact* and *observed gravity corrections*.

loop drift rate: a change of the *gravimeter drift* as a function of time during measurements along a single *loop* of a *ground gravity* survey. Its values Δg_{ldr} at each *station* are calculated as

$$\Delta g_{ldr} = [(g_{out} - g_{in})/(t_{out} - t_{in})]\Delta t_{st},$$

where g_{out} is the *reading* at the loop beginning, g_{in} is the reading at the loop end, t_{out} is the time at the loop beginning, t_{in} is the time at the loop end, and Δt_{st} is the time increment at each station relative to t_{out}. Commonly, *Earth tide correction* is applied to every station within the loop before calculation of the loop drift rate. Obtained results are used in calculation of *loop-closure correction*. See also *drift* and *drift correction*.

looping: magnetic and, primarily, gravity measurements conducted in closed *loop* sequences during a *ground survey*. Monitoring and control over *gravimeter drift* and, if necessary, *magnetometer drift* are based on the measurement results obtained during the looping. See also *reoccupations* and *looping method*.

looping method: a method of a control over *instrument drift* and other systematic instrument errors. It is based on calculation of the difference between the repeated gravity or magnetic measurements at the successively reoccupied stations during *ground gravity*, *microgravity*, and, if necessary, magnetic surveys. See also *base-check method*, *cross-check method*, *station*, and *reoccupations*.

low-cut filter: same as *high-pass filter* with an emphasis on the ability to suppress, or reject, all components of a gravity or magnetic field with spatial frequencies lower than its *cutoff frequency*. See also *spatial frequency*.

low-latitude reduction-to-pole (RTP-LL): *reduction-to-pole (RTP)* at low magnetic latitudes, where the *magnetic inclination* is less than ±15–20°. Here, calculations with the use of standard RTP algorithms become unstable, amplify all types of *noise*, and produce the *magnetic declination*-oriented artificial linear anomalies on the *RTP grid*. The RTP-LL is based on one of four concepts: (1) the *equivalent source* (Silva, 1986); (2) the optimum *Wiener filter* with the iterative refinement of the *misfit* between the observed data and RTP-LL data projected back to the original *inclination* (Keating and Zerbo, 1996); (3) the *inverse problem* in which the *discrete Fourier transform (DFT)* of the RTP field is specified as the *model* to be recovered and the *inverse problem solution* is found by minimizing the model *objective function*, so that the RTP field can be obtained at any inclination, including the *magnetic equator* (Li and Oldenburg, 2001); and (4) the *pseudo-inclination method* (Grant and Dodds, 1972; MacLeod et al., 1993). The main criteria of a correct RTP-LL application for gridded data are (a) a close correlation between the RTP-LL maxima and the *analytic signal amplitude (ASA)* maxima over the same source bodies and (b) the absence of noise stripes and stretching of anomalies in the magnetic declination direction. The same as the standard RTP, the RTP-LL calculations are made with the assumption of a negligible *remanent magnetization* of sources. See also *magnetic latitude*.

low-pass filter: **1.** a *filter* that passes without suppression all components of an *observed field* with spatial frequencies lower than its *cutoff frequency*. **2.** a filter that passes without suppression all components of an observed gravity or magnetic field with wavelengths longer than its *cutoff wavelength*. Sometimes referred to as *high-cut filter*. See also *high-pass filter*, *spatial frequency*, and *wavelength*.

low-pass roll-off filter: a line- or *grid*-based *filter* that retains, or passes, the low-*wavenumber* range of the *power spectrum* of an observed gravity or magnetic field using a smooth curve in a *roll-off window* in order to minimize *Gibbs ringing* in the process of forward or inverse *Fourier transform*.

luni-solar effect: the time-varying and latitude-dependent gravity effect of the combined *gravitational attraction* of the Moon and Sun that results in the crustal deformations called *Earth tides*. It is compensated by application of the *Earth tide correction* calculated with the use of the *Longman formulas*.

luni-solar gravity correction: same as *Earth tide correction* with an emphasis on its use to correct for the time-varying and latitude-dependent gravity effect of the combined *gravitational attraction* of the Moon and Sun.

M

machine learning (ML): data processing and analysis methodology based on the use of computer algorithms with the ability to adapt and improve their performance through automatic "learning" from the statistical analysis of multidimensional datasets with (supervised mode) or without (unsupervised mode) human assistance. Input geological and geophysical datasets, or layers, include the original data, their enhancements, and calculated attributes. There are "shallow learning" with a few layers and "deep learning" with tens and hundreds of layers in the format of grids over the same area, where the borehole data represent the validation coordinate points. In the process of learning with the use of training and validation data subsets, the ML algorithms develop the functions which find targeted patterns in the data and their relationships at different scales of *lateral resolution* to accommodate the multiscale characteristics of the input data (LeCun et al., 2015). These relationships are used for mineral prospectivity mapping (McMillan et al., 2021) and in the automatic *interpretation* of anomalous gravity and magnetic fields in a manner of *pattern recognition* for mapping faults, fracture zones, lithological boundaries, and other elements of subsurface geological structures (Hassan and Goussev, 2019a). The ML algorithms also can be implemented in optimization of the *inversion* process and solving the *inverse problem* in both the supervised and unsupervised modes of their operation (Kim and Nakata, 2018). See also *anomalous field*, *attribute*, *dimension*, *enhancement*, *grid*, *pattern*, and *target*.

magma: molten or partially molten rock material formed in the upper *mantle*. During the ascend into the *crust*, its chemical composition changes in response to variations of the pressure, temperature, and components of the surrounding rocks to produce a great diversity of rock-forming minerals. Alkaline magmas are enriched with rare elements such as lithium, niobium, zirconium, and other *rare earth elements (REE)*. Magma either rises to the Earth's surface to form *igneous extrusion*, such as *kimberlite pipe*, basalt *lava* flows, and *volcanics*, or it cools within the upper crust to form intrusive, or plutonic, igneous rock bodies, such as *batholith*, *dike*, *pluton*, *sill*, and *stock*. Both the extrusive and intrusive igneous rocks create the high lateral *density* and *magnetization* contrasts in a sedimentary section and *basement* and, therefore, generate the high-*amplitude* magnetic and gravity *total field* and *gradient* anomalies. Cooling and solidification of magma is accompanied by the release of large quantities of hydrothermal fluids, such as *juvenile water*, which are the main contributors to the reservoir-grade dolomitization of carbonate rocks, ore-forming mineralization, and magnetization of intrasedimentary faults. See also *contrast*, *igneous intrusion*, and *magnetized fault*.

magmatic body: an *igneous body* with an emphasis on its origin from *magma*.

magmatic water: See *juvenile water*.

magnetic accelerometer: *superconducting gravimeter* with an emphasis on *magnetic levitation* implemented in its design for the measurement of changes in *gravitational acceleration* which is a physical *dimension*, or quantitative measure, of a *gravity field*.

magnetic amplitude inversion: *inversion* of a *magnitude* of the calculated or measured three-component *magnetic anomaly vector* for the subsurface distribution of a *magnetization* from which the *effective susceptibility* of rock formations is derived. Obtained results are weakly dependent on the *remanent magnetization* and *self-demagnetization* effects, and they are used in the mapping of *volcanics* and imaging of kimberlite intrusions, magnetic ore bodies, and their contact mineralization zones, including the skarn deposits and iron-oxide-copper-gold deposits at low magnetic latitudes (Li et al., 2021). See also *apparent susceptibility*, *kimberlite intrusion*, *magnetic latitude*, and *ore body*.

magnetic anisotropy: directionally dependent *magnetization* of the rock-forming minerals related to the fundamental *anisotropy* of their crystalline structure and shape of the non-spherical *ferromagnetic* grains. Because the magnetization of minerals is produced by alignments of magnetic dipole moments of the constituent grains, the magnetic anisotropy of the individual grains defines how closely the magnetic moments of minerals align with the external magnetizing field. The most common cause of magnetic anisotropy in rock formations is the presence of a dominant distribution and orientation of rock-forming minerals. Among the ferromagnetic and *ferrimagnetic* minerals, hematite and pyrrhotite are the most anisotropic, whereas *magnetite* is one of the least anisotropic magnetic minerals. See also *magnetic dipole moment*, *magnetostriction*, and *FTG anisotropy*.

magnetic anomaly: **1.** the *diurnals*-corrected difference ΔB between *observed magnetics* and predictive or historical *IGRF model* values, i.e., the difference between the observed and *model*-predicted magnetic field values. **2.** the *field* of magnetic values obtained by application of *observed magnetic corrections* for all known planetary and non-geological magnetic effects contributing to and interfering with magnetic measurements. The obtained field is referred to as *total magnetic intensity (TMI)* or *total magnetic anomaly (TMA)*. **3.** a local area where the *amplitude*, *shape*, and *spatial frequency* characteristics of the observed magnetic field are different from the same characteristics in the surrounding *magnetic survey* area. **4.** a local variation of the measured *Earth magnetic field* or its gradients that is associated with (a) a particular geological *structure*, *UXO* and other iron-bearing objects, or (b) a lateral change in the *magnetization* of rocks related to their mineral composition. **5.** the magnetic field of a single geological body calculated by *forward modeling*. **6.** *anomalous magnetics*. In order to generate an observable magnetic anomaly, a *source body* must satisfy the following basic requirements: (a) sufficient *magnetization contrast* with surrounding rocks, (b) sufficient volume, and (c) sufficient proximity to observation points. Depending on the *scalar* and *vector* changes in the lateral magnetization contrasts, the magnetic anomalies are either positive or negative. Due to the very large differences in *magnetic susceptibility* between the sedimentary rocks and igneous intrusive and *basement* rocks, the anomalies generated by intrasedimentary igneous intrusions and sources within a basement are dominant in an observed magnetic field. Unlike gravity anomalies, the anomalous magnetic highs and lows more often represent variations of the lithology-related magnetization rather than real structural highs and lows in a sedimentary section or basement. For example, in the case of an *igneous body* intruded under an extensional tectonic regime into a local depression or *graben*-like structure, this *structural low* will have the *signature* of a *magnetic high* in the gridded *residual field* data. See also *gridding*, *magnetic low*, *observed magnetic components*, *residual magnetics*, and *gravity anomaly*.

magnetic anomaly amplitude: a quantitative characteristic of a *magnetic anomaly* defined by four major factors: (1) lateral *magnetization contrast*: the higher the contrast, the higher the amplitude; (2) *source body* volume: the larger the volume, the higher the amplitude; (3) source body depth: the shallower the source, the higher the amplitude; and (4) *magnetic inclination* and *magnetic declination* of the *main geomagnetic field* in the *survey* area. Among other factors are *background noise*, interference with anomalies from closely spaced sources, and superposition of the regional magnetic anomalies. See also *interference of anomalies* and *superposition principle*.

magnetic anomaly shape: See *anomaly shape (mag.)*.

magnetic anomaly vector: the three-component *vector* $\Delta\mathbf{B}$ having a length, or *magnitude*, $|\Delta\mathbf{B}|$ defined as

$$|\Delta\mathbf{B}| = (\Delta B_x^2 + \Delta B_y^2 + \Delta B_z^2)^{1/2},$$

where ΔB_x, ΔB_y, ΔB_z are mutually orthogonal horizontal and vertical vector components of the *magnetic anomaly* ΔB obtained by *scalar-to-vector transformation* of the gridded residual magnetic field. Vector magnitudes are calculated at each *grid* point to obtain a new *field* of anomalous magnetic amplitudes that are independent of the *magnetization direction* of the *two-dimensional (2D)* sources and weakly dependent on the magnetization direction of the *three-dimensional (3D)* sources, i.e., these amplitudes are much less influenced by *remanent magnetization* effects. An obtained magnetic *amplitude* dataset is inverted by a 3D *property inversion* algorithm for the distribution of *magnetization* as a function of the 3D subsurface position and mapping of *volcanics* (Li et al., 2012). See also *gridding*, *residual field*, and *gravity anomaly vector*.

magnetic archaeology: high-*resolution* measurements of the *Earth magnetic field* and, primarily, its gradients in search for buried remnants of burned and destroyed ancient cities, settlements, and fortifications. Fast thermal oxidation by fire converts a *paramagnetic* pyrite into the *ferromagnetic* magnetite or hematite and, thereby, increases significantly the *magnetization* of the burned objects. Combustion by surface fires with heating and cooling through the *Curie temperature* creates the magnetization contrasts between the burned buildings and surrounding soil that are sufficient for generating the *total field* anomalies with amplitudes ranging from several *nanotesla (nT)* to hundreds of nanoteslas. Magnetic gradients are very sensitive to differences in the magnetic properties of (a) ancient stone foundations and surrounding soils and (b) disturbed and undisturbed soils in areas of defensive trenches. See also *combustion magnetism*, *magnetization contrast*, *drone survey*, *high-resolution ground survey*, and *forensic magnetism*.

magnetic artifacts: See *artificial magnetic anomalies*.

magnetic azimuth: the *azimuth*, or horizontal angle, measured in degrees clockwise from a direction to the *magnetic north pole* through 360°.

magnetic basement: the upper surface of a highly heterogeneous, both lithologically and magnetically, crystalline rock complex composed of igneous and metamorphic rocks within which the *magnetization* is approximately two to three orders of magnitude higher than the magnetization of the sedimentary rocks. For this reason, the anomalies generated at the top and within the magnetic basement are dominant among the anomalies of a *total magnetic intensity (TMI)* and *RTP magnetic field*. Based on its characteristics and properties, the magnetic basement is often referred to with the same meaning as the geological, or true, *basement*. Sometimes, thick and extensive sequences of highly magnetic igneous rocks, such as basalt *lava* flows spreading over large areas, are considered as substitutes for the magnetic basement, because magnetic anomalies from sources below these sequences are poor or not resolvable. The magnetic basement depth structure is one of the primary objectives of magnetic data *interpretation*, and its mapping is based on the use of *automated source depth estimation*, *manual source depth estimation*, *pattern recognition*, and *inversion*-based methods. See also *acoustic basement*, *gravity basement*, and *resolution*.

magnetic bottom depth: the depth of a basal surface, or bottom, of a randomly magnetized horizontal layer in the lowest part of the *crust* or in the uppermost part of the *mantle*, where rocks can still retain their *magnetization*. It is used as the regional-scale approximation of the *Curie depth* in estimates of its variations. See also *Curie depth estimation*.

magnetic bright spot: **1.** a cluster of sedimentary residual magnetic (SRM) anomalies generated by diagenetically altered rock-forming minerals in the subsurface zone, or *chimney*, directly above a hydrocarbon accumulation and related to the *hydrocarbon seepage effect*. **2.** a high-*amplitude* circular-shaped positive *magnetic anomaly* generated by a *kimberlite intrusion* or other localized *igneous body* with a small amount of *remanent* *magnetization.* See also *SRM anomaly, detrital magnetization, diagenetic magnetic minerals,* and *regolith.*

magnetic chrons: See *polarity chrons.*

magnetic compensation: See *aeromagnetic compensation.*

magnetic contact: **1.** the location of a lateral change in the *magnetization* of rocks that is associated with a structural or lithological boundary. **2.** a *source body* approximation in the magnetic *modeling* of the structural and lithological boundaries that assumes an "infinite" lateral and depth extent of a *contact* between rocks with different *total magnetization.* **3.** the interface, often a *fault,* between rock formations with contrasting magnetizations. See also *contrast, finite contact, infinite contact, infinite bottom,* and *gravity contact.*

magnetic contrast: See *magnetization contrast* and *susceptibility contrast.*

magnetic corrections: See *magnetic data corrections.*

magnetic crust: a part of the Earth's *crust* and uppermost *mantle* where rocks are magnetized. Its variable thickness is bounded from below by the *Curie depth* and controlled by the regional heat flows and thermal processes within the upper mantle, e.g., the magnetic crust below the Tibetan plateau is thinner than the surrounding continental crust due to the increased heat flow. See also *magnetization.*

magnetic data: **1.** digital records of the measured *Earth magnetic field* and its components that are made during ground, submarine, marine (shipboard), and airborne surveys as well as space satellite missions. There are four basic types of magnetic data: (1) *scalar* data obtained by measurements of the strength, or intensity, of the Earth's magnetic field and displayed as the *total magnetic intensity (TMI)* and *RTP magnetic field*; (2) *vector* data obtained by measurements of the *magnetic vector* components; (3) *gradient* data obtained by measurements of the horizontal and vertical gradients of the Earth's magnetic field; and (4) *gradient tensor* data obtained by measurements of the *magnetic gradient tensor* components. By the status of *data processing,* magnetic data are divided into three categories: *raw magnetics, observed magnetics,* and *anomalous magnetics.* **2.** digital records of the measured planetary magnetic field made by planet-orbiting satellites and automated remote-controlled probes, or rovers, on a planet's surface. See also *scalar magnetometer, vector magnetometer, magnetic gradiometer, full tensor magnetic gradiometer,* and *satellite magnetometry.*

magnetic data corrections: corrections applied to *magnetic data* to compensate for (a) measurement effects in the process of *acquisition,* such as *heading error, lag error,* and, if necessary, *magnetometer drift*; and (b) planetary and non-geological effects, such as *main geomagnetic field, diurnals, micropulsations, secular variation,* and *cultural magnetic effects.* These effects contribute to measured magnetic values and interfere with the magnetic anomalies generated by elements of the geological structures in the Earth's *crust.* There are two basic types of corrections: *raw magnetic corrections* and *observed magnetic corrections.* Compensation for all known and predicted magnetic effects of planetary and local origin transforms *observed magnetics* into a *field* of the magnetic anomalies, or *anomalous magnetics,* that is referred to as either the *total magnetic intensity (TMI)* or *total magnetic anomaly (TMA).* Final magnetic correction compensates for *magnetic declination* and *magnetic inclination* in a survey area, and it is known as *reduction-to-pole (RTP).* The process of the application of magnetic data corrections is referred to as *magnetic reduction.* See also *aeromagnetic compensation, reduction-to-equator (RTE),* and *gravity data corrections.*

magnetic datum: **1.** a *background field* that is used as a reference level, or *datum,* relative to which *diurnals* are measured by a *base-station magnetometer.* **2.** a *baseline* value that is a reference value for the *magnetic vector* measurements by a *vector magnetometer.*

magnetic declination: **1.** a directional parameter of the *main geomagnetic field* denoted by D and defined by the horizontal angle between the geographic meridian and the *magnetic meridian,* i.e., it is the angle between directions to the geographic north or south poles and corresponding *magnetic north pole* or *magnetic south pole,* respectively. **2.** the *azimuth* of the magnetic meridian. **3.** the *phase angle* of the two-dimensional (2D) *geomagnetic vector* **B**. Declination value D is determined at a *geomagnetic observatory* as

$$D = \arctan(B_y/B_x) = \arcsin[B_y/(B_x^2 + B_y^2)^{1/2}],$$

where B_y and B_x are mutually orthogonal horizontal components of the 3D geomagnetic vector **B** in *Cartesian coordinates,* where the x-axis is directed to the north, the y-axis is to the east, and the z-axis is positive down. By convention, the declination is positive to the east of a geographic meridian and negative to the west of a geographic meridian. For the major part of the northern geomagnetic hemisphere, the declination varies within −30° to +30°. In the southern geomagnetic hemisphere, it varies within −80° to +80° except for the area around the magnetic south pole, where this range expands to nearly ±180°. Declination is zero at both *Earth magnetic poles.* Worldwide measurements by geomagnetic observatories and Earth-orbiting satellites are summarized for each *epoch* and its annual subdivisions in the format of an *isogonic map,* also known as a *main field declination map.* Declination is variable over time and location, and its value at any point on or above the Earth's surface for any date in the range of years from 1900 to 2015 is obtained from the *IGRF model* of the 12th generation, which is referred to as the *IGRF-12* and predicts the declination and other *geomagnetic field parameters* for the epoch 2015 corresponding to years 2015–2020. Sometimes referred to as *geomagnetic declination.* **4.** one of two directional parameters of a localized, or compact, magnetic *source body* determined by calculation of the *vector magnetic moment* components in order to estimate the spatial orientation of its *magnetization vector,* i.e., direction of its *total magnetization.* See also *magnetic inclination.*

magnetic depth: **1.** a depth to the source of a *magnetic anomaly.* **2.** a *magnetic basement* depth obtained with the use of magnetic source depth estimation methods, such as the automated *Werner deconvolution* and *Euler deconvolution,* manual (graphical) techniques, or any method of magnetic or *pseudogravity* data *inversion*. Generally, it is any structural depth estimate obtained from magnetic data. See also *automated source depth estimation* and *graphical source depth estimation*.

magnetic dip: *magnetic inclination* in the context of the *dip* of magnetic field lines that is defined by the vertical angle between the direction of a *geomagnetic vector* and horizontal plane.

magnetic dip poles: *Earth magnetic poles* identified as two locations on the Earth's surface, where the *main geomagnetic field* lines have vertical *dip* corresponding to the *magnetic inclination* of +90° at the *magnetic north pole* and −90° at the *magnetic south pole*.

magnetic dipole: **1.** a pair of hypothetical magnetic poles of equal *pole strength* and opposite sign, or *polarity*, separated by infinitely small distance. Its mathematically described equivalent is the infinitely small loop of an electric current. Elementary magnetic dipoles are formed at the atomic level by the orbiting and spinning (rotating about axes of rotation) motion of negatively charged electrons and positively charged spinning protons. Constructive and destructive summation of the vector fields of these magnetic dipoles define the magnetic properties of rock-forming minerals. The same as *point mass* for gravity sources, the magnetic dipole is a fundamental theoretical approximation and elementary building block of all magnetic sources. Its field lines exist in the form of closed loops, where each line is directed from the positive (north) pole to the negative (south) pole outside the magnetic dipole and from the south pole to the north pole within the magnetic dipole. Almost 90% of the *main geomagnetic field* is approximated by the *field* of a slow-rotating magnetic dipole placed at about 400 km from the Earth's center with its axis of rotation tilted by ~11° relative to the Earth's rotational axis. Intersections of the dipole axis with the Earth's surface are the locations of the *geomagnetic poles*. **2.** the *point source* approximation of a magnetic *source body*. The field of magnetic dipole satisfies the *Euler deconvolution equation* with the *Euler structural index* of SI = 3. See also *magnetic pole* and *Coulomb law*.

magnetic dipole field: the *vector field* **B** generated by a *magnetic dipole*. At all points in space other than the magnetic dipole itself, it can be expressed as

$$\mathbf{B} = C_m m[3(\hat{\mathbf{m}} \cdot \hat{\mathbf{r}})\hat{\mathbf{r}} - \hat{\mathbf{m}}]/\mathrm{r}^3 = C_m m(3\cos\theta\hat{\mathbf{r}} - \hat{\mathbf{m}})/\mathrm{r}^3,$$

where $m = |\mathbf{m}|$ is the *magnitude*, or vector length, of the *magnetic dipole moment* **m**; $\hat{\mathbf{m}}$ is a unit vector showing the orientation of the magnetic dipole moment; r is distance to the magnetic dipole; $\hat{\mathbf{r}}$ is a unit vector directed to the *observation point*; θ is the angle between $\hat{\mathbf{m}}$ and $\hat{\mathbf{r}}$; C_m is a magnetic constant of proportionality defined as $C_m = 1$ in the cgs system and $C_m = \mu_0/4\pi = 10^{-7}$ *henry*/meter in the SI system; and μ_0 is the *magnetic permeability* of a source-free space. The strength of a magnetic dipole field is quantified by its magnitude $B = |\mathbf{B}|$, which is directly proportional to the magnetic dipole moment **m** and inversely proportional to the cube of the distance r between the magnetic dipole and observation point (Blakely, 1996). Field lines, or *flux* lines, of the magnetic dipole exist in the form of closed loops directed from the *positive pole*, or north pole, to *negative pole*, or south pole, outside the magnetic dipole and from the negative to positive pole inside the magnetic dipole. Approximately 90% of the *main geomagnetic field* is approximated by the magnetic dipole field. See also *magnetic north* and *magnetic south*.

magnetic dipole moment: a *vector* quantity that describes the strength and orientation of a *magnetic dipole*. It is a measure of the magnetic dipole's ability to align with the externally applied magnetic field and contribute to the *magnetization* of a host rock at the micro and macro levels. The absolute value, or *magnitude*, of a magnetic dipole moment is called the *magnetic dipole strength*. Atomic magnetic dipole moments have their origin in the spinning-and-orbiting motion of negatively charged electrons and spinning of positively charged protons. The smallest, or fundamental, unit of a magnetic dipole moment is the *Bohr magneton*. Aggregated magnetic moments of elementary magnetic dipoles define the magnetic properties of rock-forming minerals at the atomic level. The theoretical magnetic dipole moment is expressed as

$$\mathbf{m_d} = \rho \cdot \Delta x \cdot \hat{\mathbf{r}},$$

where ρ is the *magnetic pole strength* of each *magnetic pole*, Δx is the distance between magnetic poles, and $\hat{\mathbf{r}}$ is a unit vector pointing from the *negative pole* of the magnetic dipole to its *positive pole*. The magnetic dipole moment has a physical *dimension* of either the electric current times the equivalent loop area of the electric current or energy divided by *magnetic flux density*. Units of the magnetic dipole moment are *gauss*·cm^3 in the cgs system and *ampere*·m^2 ($\mathrm{A}\cdot\mathrm{m}^2$) in the SI system, where $1\ \mathrm{A}\cdot\mathrm{m}^2 = 10^3$ gauss·cm^3. Often referred to as *dipole moment* and *magnetic moment*. See also *normalized magnetic dipole moment*.

magnetic dipole potential: a *scalar* quantity representing the *magnetic potential* $\mathcal{V}$ of a *magnetic dipole* at observation point P, expressed as

$$\mathcal{V}(\mathrm{P}) = C_m \mathbf{m} \cdot \hat{\mathbf{r}}/\mathrm{r}^2 = C_m |\mathbf{m}|/\mathrm{r}^2,$$

where $|\mathbf{m}|$ is the *magnitude* of *magnetic dipole moment* **m**; r is the distance between the magnetic dipole and observation point P; $\hat{\mathbf{r}}$ is a unit vector directed to the *observation point*; and C_m is a magnetic constant of proportionality whose value depends on the system of measurements: $C_m = 1$ in *cgs units* and $C_m = \mu_0/4\pi = 10^{-7}$ *henry*·meter^{-1} in *SI units*, where $\mu_0 = 4\pi \cdot 10^{-7}$ is the *magnetic permeability* of a source-free space. See also *magnetic potential*.

magnetic dipole strength: the absolute value, or *magnitude*, of a *magnetic dipole moment*. It is directly proportional to the maximum amount of torque, or twist, of a *magnetic dipole*, when it is oriented orthogonally to the external magnetic field direction. For each elementary rock unit having its *magnetization* approximated by a *magnetic dipole field*, the magnetic dipole strength $|\mathbf{m}|$ is defined by the *dot product* of the magnetization and volume of a rock unit:

$$|\mathbf{m}| = |\Delta\mathbf{M}| \cdot \Delta V,$$

where $|\Delta\mathbf{M}|$ is the magnitude of rock unit magnetization and ΔV is rock unit volume. See also *normalized source strength (NSS)*.

magnetic disturbances: fluctuations of *external magnetic field* of the Earth, that are highly variable in duration, amplitude, and *wavelength* content, and caused, primarily, by the interaction of *solar wind* and other solar radiation effects with the *magnetosphere*. They include *bay*, *diurnals*, *auroral electrojets*, *equatorial electrojets*, *magnetic storm*, *magnetic substorm*, and *micropulsations*. See also *transient disturbances* and *gravity disturbance*.

magnetic domain: the microscopic-scale volume in a rock-forming mineral, where the atomic magnetic dipole moments are aligned with one another to produce uniform *magnetization*. The size, shape, orientation, and stability of the magnetic domains depend on the rock composition, mineral grain size and shape, presence of microfractures and microinclusions, and ambient temperature. Alignments of the magnetic domains under exposure to the externally applied magnetic field, such as the *Earth magnetic field*, result in the *ferromagnetism*, *ferrimagnetism*, *diamagnetism*, and *paramagnetism* of rocks. See also *Barkhausen effect* and *Weiss theory*.

magnetic equator: the non-stationary line on the Earth's surface along which *magnetic inclination* is zero and the *main geomagnetic field* strength is close to its minimum for a particular *epoch*. Because the *geomagnetic vector* is horizontal here, the *magnetic anomaly* generated by a *sphere* with a positive *magnetization contrast* has a trough-like *magnetic signature*, and it is zero for a north–south striking *thin dike* in the absence of *remanent magnetization* components. The magnetic equator is not coincident with the geographic equator due to the differently tilted, or inclined, axis of the *main geomagnetic field*. By the same reason that locations of the *Earth magnetic poles* are different from the locations of the *geomagnetic poles* and geographic north and south poles, the magnetic equator location differs from the location of the *geomagnetic equator*. Sometimes referred to as *aclinic line* and *dip equator*. See also *magnetic meridian*.

magnetic exploration: application of the *magnetic method* for the assessment of local, regional, and planetary geological structures in the search for commercial accumulations of hydrocarbons, deposits of economically useful minerals, geothermal resources, etc. Sometimes referred to as *magnetic prospecting*. See also *gravity exploration*.

Magnetic Fault Interpretation Cube (MaFIC): a *voxel*-type *three-dimensional (3D)* volume for visualization of the magnetic *source depth solutions* obtained by *profile*-based 2D *Euler deconvolution* and 2D *Werner deconvolution* and presented in the standard SEGY *format* of seismic workstations. The scrolling through the volume of numerous depth solutions and depth slicing along imported seismic horizons are used to identify the spatially coherent alignments of the depth solutions associated with faults and other structural discontinuities. Original or filtered magnetic and gravity grids, seismic horizons, formation top data from wells, and other digital geological and geophysical information can be combined for the *workstation*-based comprehensive analysis and *integrated interpretation* of all available data (Rhodes and Peirce, 1999). See also *coherence*, *fault*, and *grid*.

magnetic field: the *vector field* of a *magnetic force* generated at the atomic level by motion and interaction of the charged subatomic particles: the orbiting and spinning negatively charged electrons and positively charged protons. Joining-up, or coupling, of the rotational spins between particular atoms defines the alignment, or *polarization*, of the motion-derived elementary magnetic dipoles, i.e., the fundamental source of the magnetic field is the *magnetic dipole* at the atomic level. At every space point, the characteristic magnetic field parameters are *magnetic flux density* and *magnetic induction*. The direction and *magnitude* of magnetic vector field **B** at an arbitrary point P within the field are described by the *gradient* of the scalar *magnetic potential* $\mathcal{V}$, expressed as

$$\mathbf{B}(P) = -\nabla\mathcal{V},$$

where ∇ is *vector* gradient operator *del* and the minus sign is a theoretical convention. The vector component of the magnetic field in any direction is defined by the *derivative* of the magnetic potential in that direction. Magnetic field lines, or *magnetic flux* lines, exist in the form of closed loops of various sizes, up to infinity, around a field-generating magnetic dipole. Flux lines are directed from a *positive pole*, or north pole, to a *negative pole*, or south pole, outside the magnetic dipole and from the south pole to the north pole within the magnetic dipole to close the loop. Unlike the *gravitational field*, the magnetic field can be shielded. See also *Ampere law*, *Coulomb law*, *Gauss law*, and *Earth magnetic field*.

magnetic field curvature: a measure of the deviation, or bending, of the *magnetic field* from a planar, or flat, surface. Its estimates are commonly based on the spatial rate of change of the *gradient* of the magnetic field and obtained from (a) two separate measurements at different spatial locations with the use of a *magnetic gradiometer* or (b) simultaneous measurements of the *Earth magnetic field* at three equally spaced positions by a set of three synchronized *magnetometer* systems. Because the *magnetic gradient* represents the first-order *derivative* of the magnetic field or second-order derivative of *magnetic potential*, the magnetic field curvature is approximated by the second-order directional derivatives of the *anomalous field* or third-order directional derivatives of the *gravitational potential*. Magnetic field curvature is a quantitative characteristic of the *three-dimensional (3D)* deformation of the magnetic field at a particular observation point: the more deformed, or disturbed, the magnetic field is, the larger its curvature value. See also *curvature*, *curvature attributes*, and *directional derivative*.

magnetic field intensity: See *magnetic field strength*.

magnetic field strength: 1. the strength, or intensity, of a *magnetic field* generated during the process in which the rocks are magnetized by an externally applied magnetic field. It is quantified by the *magnitude*, or absolute value, of *magnetic induction* **B** and denoted by |**B**|. In *SI units*, the magnetic field strength is expressed as

$$|\mathbf{B}| = \mu_0(|\mathbf{H}| + |\mathbf{M}|) = \mu_0(1+k)|\mathbf{H}| = \mu|\mathbf{H}|,$$

where **H** is *magnetizing force*, **M** is *induced magnetization*, and μ is *magnetic permeability* defined as

$$\mu = \mu_0(1+k),$$

where k is *magnetic susceptibility* and μ_0 is magnetic permeability of a source-free space. In *cgs units*, the magnetic field strength is expressed as

$$|\mathbf{B}| = |\mathbf{H}| + 4\pi|\mathbf{M}| = (1+4\pi k)|\mathbf{H}| = \mu|\mathbf{H}|,$$

where $\mu = (1+4\pi k)$. **2.** the magnitude |**H**| of magnetizing force **H** which is a quantitative measure of the externally applied, or inducing, magnetic field that defines the *induced magnetization* of rocks. **3.** a theoretical parameter defined by *magnetic force* per unit of *magnetic pole*. The strength, or intensity, of the *Earth magnetic field* is the quantity measured in *magnetic exploration* by a conventional, or regular, *magnetometer*. Also referred to as *magnetic field intensity*. See also *total magnetic intensity (TMI)* and *total absolute magnetic intensity (TAMI)*.

magnetic flux: a quantitative measure of the theoretical *magnetic field* lines passing through a given surface or volume. Magnetic flux through a closed surface is defined by the surface integral of the normal (perpendicular) component of the magnetic field *vector* over this surface. According to the *Gauss law*, the net normal magnetic flux through any region is zero, so the normal component of *flux* entering the region is equal to the normal component of the flux leaving the region. Magnetic flux units are *weber* in the SI system and *maxwell* in the cgs system, where 1 weber $= 10^8$ maxwell. See also *Gauss flux theorem*.

magnetic flux density: 1. a theoretical *vector* quantity that describes the intensity and direction of a *magnetic field*, such as the field around a *magnetic dipole*, magnet, or electric current. **2.** *magnetic induction* **B** described in terms of *magnetic flux* per unit volume or per unit area perpendicular to the direction of the magnetic flux vector. Magnetic flux density is directly proportional to the *magnetic field strength* which is quantified by the *magnitude* of magnetic induction **B**. In a source-free space, the magnetic flux density is defined as

$$\mathbf{B}_0 = \mu_0\mathbf{H},$$

where μ_0 is the *magnetic permeability* of a source-free space and **H** is the *magnetizing force*. In rock formations,

$$\mathbf{B} = \mu\mathbf{H},$$

where μ is the magnetic permeability of a particular rock, such that $\mu = \mu_0(1+k)$ in the SI system and $\mu = 1+4\pi k$ in the cgs system, and k is *magnetic susceptibility*. The difference between the flux density in rocks and the flux density in a vacuum represents the magnetic flux density $\Delta\mathbf{B}$ added by rocks and expressed in *SI units* as

$$\Delta\mathbf{B} = \mu\mathbf{H} - \mu_0\mathbf{H} = \mu_0 k\mathbf{H} = \mu_0\mathbf{M}_i,$$

where $\mathbf{M}_i = k\mathbf{H}$ is *magnetization* induced into rocks by an externally applied magnetic field, i.e., *induced magnetization*. The standard unit of magnetic flux density is the *nanotesla (nT)*.

magnetic force: 1. the physical force generated by the motion and interaction of charged subatomic particles: positively charged protons and negatively charged electrons. **2.** a *vector* force of interaction between the poles of magnetic dipoles from the atomic level throughout the planetary and interplanetary levels. Differing from the attractive *gravitational force*, the magnetic force is both attractive and repulsive: it acts as a force of attraction between magnetic poles of opposite signs and as a force of repulsion between magnetic poles of the same sign. Analogous to the gravitational force, the magnetic force is directly proportional to each pole strength and inversely proportional to the square of their separation distance as postulated by the *Coulomb law*. The vector *field* of the magnetic force acting upon a unit *magnetic pole* is termed the *magnetic field* whose fundamental source is a *magnetic dipole* at the atomic level. **2.** *magnetizing force (H)*.

magnetic gradient: 1. a *scalar* quantity, such as the *horizontal gradient* or *vertical gradient*, representing the rate of change of the *magnetic field* in the direction of a vertical or horizontal separation of two or three sensors in a *magnetic gradiometer*, i.e., it is a measured rate of change of the magnetic field per unit of separation distance. **2.** a scalar quantity representing the horizontal or vertical rate of change of the magnetic field calculated from measurements by a regular *magnetometer*, i.e., it is a calculated rate of change of the magnetic field. **3.** a *vector* quantity defined by a three-component *magnetic gradient vector*. For gridded data, the horizontal magnetic gradient is defined by the absolute value of the *horizontal derivative*, and the vertical magnetic gradient is equal to the *first vertical derivative (FVD)* with the z-axis positive down. The decay, or *attenuation*, of a magnetic gradient *anomaly* generated by a *point source* is inversely proportional to the fourth power ($1/r^4$) of the distance r from this source. The standard units of measurement are *nanotesla (nT)* per meter ($nT \cdot m^{-1}$) and nanotesla per kilometer ($nT \cdot km^{-1}$). See also *gridding* and *magnetic gradiometry*.

magnetic gradient tensor: 1. a nine-element symmetric *tensor* the elements of which represent the spatial rates of change of three mutually orthogonal *vector* components of the *gradient vector* of *magnetic potential* along three mutually orthogonal directions. **2.** a nine-element *gradient tensor* representing the derivatives of the B_x, B_y, B_z components of the *Earth magnetic field* vector $\boldsymbol{B}$ along the x-, y-, z-axes in *Cartesian coordinates* in the format of a 3×3 *matrix*:

$$\begin{matrix} B_{xx} & B_{xy} & B_{xz} \\ B_{yx} & B_{yy} & B_{yz} \\ B_{zx} & B_{zy} & B_{zz} \end{matrix} = \begin{matrix} \partial B_x/\partial x & \partial B_x/\partial y & \partial B_x/\partial z \\ \partial B_y/\partial x & \partial B_y/\partial y & \partial B_y/\partial z, \\ \partial B_z/\partial x & \partial B_z/\partial y & \partial B_z/\partial z \end{matrix}$$

where $B_{xy}=B_{yx}$, $B_{xz}=B_{zx}$, $B_{yz}=B_{zy}$, $B_x=\partial V/\partial x$, $B_y=\partial V/\partial y$, $B_z=\partial V/\partial z$, and V is the magnetic potential. In accordance with the *Laplace equation*, $B_{xx}+B_{yy}+B_{zz}=0$, i.e., the magnetic gradient tensor is traceless. Its nine elements comprise the *full tensor gradient (FTG)* of an anomalous magnetic field at each observation point. On a moving *observation platform*, five independent elements B_{xx}, B_{yy}, B_{zz}, B_{xy}, B_{xz} are obtained from measurements by a *SQUID magnetic tensor gradiometer* or calculated from the gradients measured by a regular *magnetic gradiometer*. A less laterally resolving option is the calculation of the magnetic gradient tensor elements from gridded magnetic data, such as an *RTP grid*. Measured gradient tensor elements are essentially independent from the direction and intensity of the *main geomagnetic field*, because the *gradient* measurements suppress the *regional field* components and respond, primarily, to the variations of a local-scale anomalous magnetic field. Often referred to as simply the *magnetic tensor*. See also *anomalous field*, *derivative*, *FTG trace*, *gridding*, *lateral resolution*, and *invariants*.

magnetic gradient vector: a three-component *vector* defined by the vector sum of the first-order vertical and horizontal derivatives of the *magnetic field* or the second-order vertical and horizontal derivatives of *magnetic potential* along the mutually orthogonal *x*-, *y*-, *z*-directions in *Cartesian coordinates*. It is orthogonal to the magnetic field isolines and points in the direction of the maximum change of an anomalous magnetic field. Its length, or *magnitude*, is equal to the numerical value of the *3D analytic signal amplitude (3DASA)*, and its *phase*, or angle relative to the XY plane, is defined by the *tilt angle (TA)* at each observation point. See also *anomalous field*, *isoline*, *horizontal derivative*, and *vertical derivative*.

magnetic gradiometer: an instrument for measurements of the *magnetic gradient*. Its design is based on the use of an assemblage, or *array*, of two or three identical and synchronized magnetometers mounted on the same frame and separated vertically and/or horizontally by a fixed distance called the *baseline*. In a *fixed-wing survey*, two magnetometers are installed in the opposite wingtips and one in the tail *stinger* of a survey airplane, so that measurements of the Earth's magnetic field are made simultaneously by all three magnetometers. The value of the *magnetic gradient* in the direction of separation of the magnetometers is obtained by a digital processor calculating the difference between their measurements divided by the baseline value with a proper adjustment for the airborne or towed marine array of magnetometers. The difference between the synchronized measurements of the Earth's magnetic field over relatively short baselines effectively eliminates *diurnals* and suppresses the regional field components, so the *gridding* of *horizontal gradient*, *first vertical derivative*, *tilt*, *analytic signal amplitude*, and other attributes of the Earth's magnetic field calculated from the measured gradients represents the low-*noise* and high-*resolution* magnetic gradient data. See also *attribute*, *SQUID magnetic tensor gradiometer*, and *gravity gradiometer*.

magnetic gradiometry: methods and instruments for measuring the gradients and *gradient tensor* components of the *magnetic field*. Magnetic gradiometry has several advantages over traditional measurements of the *magnitude* of the *Earth magnetic field*, including (a) increased *lateral resolution* of anomalies from closely spaced sources due to the suppression of regional long-*wavelength* components and higher *sensitivity* to changes in the local-scale *magnetization* of rocks, (b) reduced *noise* and elimination of *diurnals* via *common mode rejection* by paired and synchronized magnetometers or other sensors in *magnetic gradiometer* systems, (c) enhancement of anomalies from local structures and shallow sources by the suppression of deep-sourced anomalies, (d) improved interpolation between *gradient survey* lines for *gridding* with the use of a measured *lateral gradient*, and (e) the ability to conduct a *magnetic gradient* survey without a *base-station magnetometer*. See also *full tensor magnetic gradiometry*, *gradient-enhanced gridding*, and *SQUID magnetic tensor gradiometer*.

magnetic high: a *magnetic anomaly* over a local- or regional-scale area where the observed magnetic field values are higher than those in the surrounding areas. It is associated with a lateral increase in the *magnetization* of rock formations and can be generated by *igneous intrusion*, basalt *lava* flow, *banded iron formation (BIF)*, thick-skin *thrust*, and high-magnetization lithological units within a shallow or mid-depth *basement*. Variable-*amplitude* magnetic highs in a *residual field* are often generated by *volcanics* and local accumulations of *detrital magnetite* in buried drainage channels and depositional or mineralization fairways. In the case of an *igneous body* intruded into a *graben*, the magnetic high will be observed at the location of a residual *gravity low* produced by this graben. See also *observed field*, *fairway*, *residual field*, *magnetic low*, and *gravity high*.

magnetic inclination: **1.** a directional parameter of the *main geomagnetic field* denoted by I and defined by the vertical angle between the direction of the three-dimensional (3D) *geomagnetic vector* and the horizontal plane. **2.** the *phase angle* of a 3D geomagnetic vector. Inclination value I is determined at a *geomagnetic observatory* as

$$I = \arctan[B_z/(B_x^2 + B_y^2)^{1/2}],$$

where B_x, B_y, and B_z are mutually orthogonal horizontal and vertical components of the 3D geomagnetic vector in *Cartesian coordinates*, where the *x*-axis is directed to the north, the *y*-axis is to the east, and the *z*-axis is positive down. By convention, the inclination is positive, or inclined downwards, in the northern geomagnetic hemisphere, and it is negative, or inclined upwards, in the southern geomagnetic hemisphere. On the global scale, the inclination varies from –90° at the *magnetic south pole* to 0° at the *magnetic equator* and +90° at the *magnetic north pole*. Inclination has a profound effect on the *amplitude* and shape of an observed *magnetic anomaly*: a compact 3D source with a positive *magnetization contrast* and *induced magnetization* generates (a) a *peak*-like symmetric magnetic anomaly at the *Earth magnetic poles*, (b) a *trough*-like symmetric magnetic anomaly at the *magnetic equator*, and (c) an asymmetric *dipolar anomaly* with a peak-to-trough *signature* at intermediate inclinations. Worldwide measurements by geomagnetic observatories and Earth-orbiting satellites are summarized for each *epoch* and its annual subdivisions in the format of an *isoclinal map*, also known as a *main field inclination map*. Inclination is variable over time and location, and its value at any point on or above the Earth's surface for any date in the years 1900 to 2015 is obtained from the *IGRF model* of the 12th generation, which is referred to as *IGRF-12* and predicts the inclination and other *geomagnetic field parameters* for the epoch 2015 corresponding to the years 2015–2020. Also known as *geomagnetic inclination* and *magnetic dip*. **3.** one of two directional parameters of a localized, or compact, magnetic *source body* determined by calculation of the *vector magnetic moment* components in order to estimate the spatial orientation of its *magnetization vector*, i.e., direction of its *total magnetization*. See also *effective inclination* and *magnetic declination*.

magnetic induction: 1. the physical process by which rocks are magnetized and generate their own magnetic fields, when they are exposed to externally applied magnetic fields, such as the *Earth magnetic field*. **2.** a *vector* quantity that is a measure of the strength and direction of the *magnetic field* produced as a result of rock *magnetization* by an externally applied, or inducing, magnetic field. Denoted by **B**, the magnetic induction represents the vector sum of two vector fields: (1) the internal magnetic field, i.e., the magnetic field induced into a rock material, and (2) the external magnetic field for which the strength is quantified by *magnetizing force* **H**. Magnetic induction **B** and external magnetizing force **H** are related as

$$\mathbf{B} = \mu \mathrm{H},$$

where μ is *magnetic permeability* defined as $\mu = \mu_0(1+k)$ in the SI system and $\mu = (1+4\pi k)$ in the cgs system, k is *magnetic susceptibility*, and μ_0 is the magnetic permeability of a source-free space. The total magnetic induction vector is expressed as in *SI units* as

$$\mathbf{B} = \mu_0(\mathbf{H} + \mathbf{M})$$

and in *cgs units* as

$$\mathbf{B} = \mathbf{H} + 4\pi\mathbf{M},$$

where **M** is magnetization induced by an external magnetic field into the rock material, i.e., *induced magnetization*. **3.** a *magnetic vector* for which the direction and *magnitude* at any point in space outside the magnetic media are described by the *gradient* of the *scalar* potential as

$$\mathbf{B} = -\nabla \mathcal{V},$$

where ∇ is *gradient* operator *del*, $\mathcal{V}$ is *magnetic potential*, and the minus sign follows a theoretical convention. In terms of *magnetic anomaly*, the magnetic induction can be expressed as

$$\mathbf{B} = \mathbf{B}_0 + \Delta\mathbf{B},$$

where $\mathbf{B}_0$ is the planetary magnetic field, or *ambient field*, and $\Delta\mathbf{B}$ is the anomalous magnetic induction component generated by a *source body*. Magnetic induction has no *divergence*, so that *field* of magnetic induction values is divergenceless, and it is termed the *solenoidal field* and expressed as

$$\nabla\mathbf{B} = 0.$$

Nominal units of measurement are *gauss* in the cgs system and *tesla (T)* in the SI system. The unit of *nanotesla (nT)* is used much more often and related to those units as 1 nT $= 10^{-9}$ tesla $= 10^{-5}$ gauss. Magnetic induction is often referred to with the same meaning as *magnetic field* and *magnetic flux density*.

magnetic intensity: See *magnetic field strength*.

magnetic interpretation methods: qualitative and quantitative methods of magnetic data *interpretation* that are based on analysis of the *total field*, *gradient*, and *full tensor gradient (FTG)* anomalies and their mathematical transformations obtained by the application of *enhanced processing* techniques. Methods include but are not limited to: (1) identification and mapping of the anomalous alignments, or *lineaments*, generated by structural discontinuities, such as faults, fracture or shear zones, and lithological contacts; (2) mapping of the isolated igneous bodies, *volcanics*, and zones of anomalous *magnetization* associated with the hydrothermal mineral alterations at the intersections of faults and around igneous bodies; (3) source depth estimates with the use of *automated source depth estimation* and *manual source depth estimation* techniques; (4) mapping of the *magnetic basement* depth structure; (5) 2D, 2.5D, and 2.75D *forward modeling*; (6) 2D and 3D *magnetic inversion* or *joint inversion* for parameters of an *ore body* or other exploration *target*; and (7) *pattern recognition* for the detection of structures of exploration interest. High-*resolution* properties of the *magnetic method* form the basis of more detailed interpretation and mapping of the structural discontinuities and more accurate source depth estimates than the *gravity interpretation methods*. Inherent *interpretation ambiguity* of the individual magnetic interpretation methods is minimized by integration with gravity data and other geophysical and geological information through the process of *integrated interpretation*. See also *attribute analysis*, *boundary analysis*, *multiscale analysis*, *principal component analysis (PCA)*, *source depth analysis*, *spectral analysis*, *trend analysis*, *wavelet analysis*, and *igneous body*.

magnetic inversion: *inversion* for or *inverse modeling* of the structural parameters (depth, size, shape, etc.) and distribution of *magnetization* as a function of the two-dimensional (2D) or three-dimensional (3D) subsurface position with the use of line-oriented *total field* and *magnetic gradient* data or gridded *residual magnetics* and *full tensor gradient (FTG)* data with the use of a *voxel model*, respectively. Advances in the development of 3D magnetic inversion methods, including the inversion of magnetic *amplitude* and *vector* data strongly affected by *remanent magnetization* and *self-demagnetization*, have made it possible to invert for (a) *effective susceptibility* and geometric shapes of isolated source bodies, such as an *ore body*, (b) *magnetization contrast* and boundaries of anomalously magnetized geological units, such as *volcanics*, (c) 3D structure and distribution of the total magnetization *magnitude*, and (d) 3D structure and distribution of the *magnetization vector* (Li et al., 2012; 2021; Leão-Santos et al., 2015). Sometimes referred to as *magnetization inversion*. See also *gridding*, *magnetic moment*, and *pseudolithology model*.

magnetic latitude: **1.** *magnetic inclination* in the context of its global variation and difference from the standard frame of the geographic latitudes, especially in the southern hemisphere. **2.** a magnetic inclination value determined on a smooth regional basis rather than at a local point.

magnetic latitude correction: same as *IGRF correction* with an emphasis on its conceptual analogy with *latitude correction* applied to *observed gravity*.

magnetic levitation: a basic operating principle of a *superconducting gravimeter*, where *proof mass* is suspended, or levitates, in the auxiliary magnetic field generated by the electric currents running through the assemblage of coils. The voltage value, which is necessary to maintain the proof mass in a fixed suspended position corresponding to a *null point* of this type of *relative gravimeter*, is directly proportional to the amount of change in the *gravitational field* at each observation point. See also *Meissner effect* and *electrostatic levitation*.

magnetic log: See *magnetic susceptibility log*.

magnetic low: a *magnetic anomaly* over a local- or regional-scale area where the observed magnetic field values are lower than those in the surrounding areas. It is associated with a lateral decrease in *magnetization* of rock formations, and it can be generated by the thick accumulation of *allochthonous salt*, local *salt dome*, or salt *diapir*, and by relatively low-magnetization lithological units within the shallow and mid-depth *basement*. In the case of an intrasedimentary *igneous intrusion* with a strong *remanent magnetization* component, the residual magnetic low can be observed at the location of a *gravity high* produced by this igneous intrusion. See also *magnetic high*, *observed magnetics*, *residual magnetics*, and *gravity low*.

magnetic meridian: **1.** the imaginary line on the Earth's surface that connects the *Earth magnetic poles*. **2.** the vertical plane of a *geomagnetic field vector*. **3.** the direction of the *total horizontal component* of the *Earth magnetic field*, i.e., the direction to the *magnetic north pole* for a particular *epoch*. The horizontal angle between the directions to the geographic north or south poles and corresponding magnetic north pole or *magnetic south pole*, respectively, is referred to as *magnetic declination*, also known as the *azimuth* of the magnetic meridian. See also *magnetic equator*.

magnetic method: a geophysical method based on ground, submarine, marine (shipboard), airborne, and satellite measurements of a planetary magnetic field, such as the *Earth magnetic field*, its gradients, *full tensor gradient (FTG)* and *vector* components. The majority of the magnetic measurements are made of the *amplitude* of the magnetic field using high-*sensitivity* magnetometers. Gradients are measured by *magnetic gradiometer* systems or calculated from gridded *anomalous magnetics*. Components of a *magnetic gradient tensor* are measured by a *SQUID magnetic tensor gradiometer* or calculated from gridded anomalous magnetic data. Vector components are measured by a *vector magnetometer* which is used, primarily, in the detailed studies of a discovered *ore body*, for stationary observations at a *geomagnetic observatory*, and for *magnetic vector* measurements of the *external magnetic field* by planet-orbiting satellites. As a part of *enhanced processing*, the components of a *magnetic anomaly vector* can be calculated by *scalar-to-vector transformation*. In the *exploration*, infrastructure, and environmental applications, the magnetic method is applied for the detection and mapping of the magnetic anomalies generated by (a) faults, fault-related *salt* dissolution, fracture and shear zones, lithological heterogeneity of the *basement* and its faulted depth structure, buried impact structures, shallow- and mid-depth fracturing of shale formations, reservoir-grade *volcanics*, boundaries of the pull-apart basins, edges of *allochthonous salt* bodies, and other targets of oil-and-gas *exploration*; (b) igneous rock bodies, such as *carbonatite*, *ore body*, and *kimberlite intrusion*, zones of hydrothermal mineral alterations at the intersections of faults and around large igneous intrusions, *banded iron formation (BIF)*, accumulations of *detrital magnetite* in the shallow *paleochannel* and depositional or mineralization *fairway* within which *placer* deposits can be found, magnetic spherules on the deep-sea floor, and other targets of mineral exploration; (c) shallow- and mid-depth igneous intrusions in geothermal exploration; (d) abandoned wells, buried containers, and pipelines; and (e) *unexploded ordnance (UXO)*. After the application of all relevant *magnetic data corrections*, *leveling* and, if necessary, *microleveling*, the *magnetic survey* data are gridded and presented in the standard *format* of grids and maps of *total magnetic intensity (TMI)* and *RTP magnetic field*. Unlike the crustal rock densities varying mostly by a factor of two, the *magnetization* of rocks varies up to two or three orders of magnitude, or by factors of 100–1000, so the observed magnetic field is very sensitive to the presence and lateral heterogeneity of high-*susceptibility* rock formations, such as igneous intrusions and volcanics in a sedimentary section and metamorphic and igneous rocks in the basement. For the same *source geometry*, the *fall-off rate* of magnetic anomalies is higher than that of gravity anomalies, and for a *point source*, or *sphere*, it can be expressed as $1/r^3$ versus $1/r^2$, where r is distance to the point source. Accordingly, magnetic anomalies are narrower than gravity anomalies from the same source, more responsive to the source depth and, thereby, make the magnetic method more resolving laterally and vertically than the *gravity method*. Achievable *accuracy* of magnetic measurements is up to 0.005 *nanotesla (nT)*, which is 10^{-8} part of the average Earth magnetic field value on the Earth's surface.

Airborne magnetic surveys, such as a *high-resolution aeromagnetic survey* and *drone survey*, are the primary sources of data for *magnetic exploration*. The magnetic method is a *potential field* method, because its basic variables are first- and second-order spatial derivatives of *magnetic potential*. See also *astrobleme*, *gridding*, *igneous body*, *lateral resolution*, *magnetic anomaly*, *magnetic basement*, *magnetic archaeology*, *magnetic spherule*, *forensic magnetics*, *pull-apart basin*, and *UXO detection*.

magnetic model: 1. a local-scale *2D model* or *3D model* of a particular *target* of magnetic exploration, such as an *ore body* or *basement* depth, that is obtained by *magnetic inversion* with the use of *constraints* provided by borehole, gravity, and other geological-geophysical data. **2.** a regional-scale 2D model based on the regional structural-tectonic concepts and constrained by available geological and geophysical information. Usually, regional-scale magnetic *modeling* is combined with gravity modeling into an integrated 2D gravity and magnetic model. See also *gravity model*, *2.5D model*, *2.75D model*, and *voxel model*.

magnetic moment: 1. a *vector* quantity representing the *magnetization intensity* and orientation of the *magnetization vector* in a rock body. It is directly proportional to the *magnetic field* generated by this body and quantified by its effect on the resulting magnetic field, when the rock body is subjected to an externally applied magnetic field, such as the *Earth magnetic field*. For a rock volume unit, whose field is approximated by a *magnetic dipole field*, the individual magnetic moment $\mathbf{m_i}$ is defined by the *dot product* of the rock unit *magnetization* $\Delta\mathbf{M}$ and unit volume ΔV:

$$\mathbf{m_i} = \Delta\mathbf{M} \cdot \Delta\mathrm{V}.$$

The total magnetic moment $\mathbf{m_T}$ of a rock body is the vector sum of all individual magnetic moments $\mathbf{m_i}$ within the rock body volume:

$$\mathbf{m_T} = \Sigma\mathbf{m_i}.$$

In the same way that total *mass* per unit volume defines rock *density*, the total magnetic moment per unit volume defines rock *magnetization* and its vector **M** as

$$\mathbf{M} = \mathbf{m_T}/\mathrm{V} = \Sigma\Delta\mathbf{M} \cdot \Delta\mathrm{V}/\mathrm{V},$$

where V is rock body volume. **2.** *magnetic dipole moment*. See also *vector magnetic moment* and *integral magnetic moments*.

magnetic moment vector: See *vector magnetic moment*.

magnetic north: See *magnetic north pole*.

magnetic north pole: the coordinate point on the Earth's surface in the northern hemisphere where the *main geomagnetic field* is vertical, its strength is close to maximum, and *magnetic inclination* is +90°. For each *epoch*, the location of the magnetic north pole is determined by the corresponding *IGRF model*. Due to the presence of both dipole and non-dipole components in the main geomagnetic field, the magnetic north pole is not coincident with the *geomagnetic north pole* and geographic north pole. Also referred to as *magnetic north*. See also *magnetic pole*, *magnetic south pole*, and *magnetic equator*.

magnetic observatory: See *geomagnetic observatory*.

magnetic permeability: a measure of the easiness with which the *magnetic induction* can be established within a particular rock by an external, or inducing, magnetic field. It is defined by the ratio of the magnetic induction **B** to the inducing field strength, or *magnetizing force*, **H** and represented by a dimensionless constant of proportionality μ expressed as

$$\mu = \mathbf{B}/\mu_0\mathbf{H},$$

where μ_0 is the permeability of a source-free space with $\mu_0 = 4\pi \cdot 10^{-7}$ *weber* per *amper*·meter (*henry* per meter) in the SI system and $\mu_0 = 1$ *gauss* per *oersted* in the cgs system. Rock-forming minerals with $\mu < 1$ are called *diamagnetic*, and minerals with low $\mu \geq 1$ values are called *paramagnetic*. Minerals with high magnetic permeability values are called *ferromagnetic* and *ferrimagnetic*, and a great majority of rocks referred to as "magnetic" fall into these two categories. Sometimes, the quantity $\mu\mu_0 = |\mathbf{B}|/|\mathbf{H}|$ is used in lieu of the magnetic permeability. The relationship between magnetic permeability and *magnetic susceptibility* is different in the SI and cgs measurement systems and described by the *permeability-susceptibility relation*. See also *induced magnetization*.

magnetic polarity: 1. a characteristic of the spatial orientation of a *magnetic field* and the position of the north (positive) and south (negative) poles of a *magnetic dipole* approximating the source of this field. **2.** the positive or negative *amplitude* of a gravity or magnetic *anomaly* defined by the corresponding positive or negative *density contrast* or *magnetization contrast* of its source. **3.** the direction of the *remanent magnetization* vector in a rock sample or *magnetization vector* in a rock body. See also *dipolar anomaly*, *monopolar anomaly*, and *geomagnetic polarity*.

magnetic polarization: same as *magnetization* with an emphasis on its property to line up, or polarize, the originally randomly oriented magnetic dipoles within a rock volume. See also *magnetic dipole*.

magnetic pole: 1. one of two poles of a *magnetic dipole* toward which the imaginary lines of *magnetic force* are oriented and concentrated. According to the *Gauss law*, single magnetic poles, or magnetic monopoles, are hypothetical entities that exist only as a pair in a magnetic dipole. There are a *positive pole* and a *negative pole* in a magnetic dipole. The magnetic field lines of a magnetic dipole form the closed loops directed from the positive, or north, pole to the negative, or south, pole outside the magnetic dipole and from the south pole to the north pole within the magnetic dipole to close the loop. All magnetic sources have a dipolar nature, and they are called dipolar sources, whereas all gravity sources are monopolar sources. **2.** one of two areas at the opposite ends of a magnetized bar (magnet). When the magnet is allowed to rotate about its center, the pole which points in the direction to *magnetic north* is a positive, or north-seeking, pole, and the other pole is a negative, or south-seeking, pole. The magnetic poles of the *Earth magnetic field* are referred to as the *magnetic* *north pole* and *magnetic south pole*, and they are not coincident with *geomagnetic poles* and geographic north and south poles. See also *magnetic pole strength* and *magnetic dipole field*.

magnetic pole strength: a *magnetic dipole* parameter quantified by the strength of a hypothetical *magnetic pole*. Magnetic pole strength is equivalent to the *magnetic dipole moment* per unit length of the electric current loop which is equivalent to the elementary magnetic dipole. For a magnetized bar, or magnet, it is expressed as

$$p = \mathrm{MA},$$

where p is magnet pole strength, M is magnet *magnetization*, and A is the cross-section area orthogonal to the magnetization direction.

magnetic poles: **1.** two poles of a *magnetic dipole*. **2.** *Earth magnetic poles*.

magnetic potential: **1.** a mathematical function that describes through its spatial derivatives the parameters of a *magnetic field*. **2.** a *scalar* quantity defined by the amount of energy, or work, required to move a *point dipole* from one position to another within the magnetic field acting upon the point dipole. **3.** *anomalous potential* defined by the *vertical integral* of the *RTP magnetic field*. For an elementary *magnetic dipole*, or point dipole, the magnetic potential $\mathcal{V}$ at the distance r is expressed as

$$\mathcal{V} = C_m |\mathbf{m}| / \mathrm{r}^2,$$

where $|\mathbf{m}|$ is *magnitude* of *magnetic dipole moment* **m**; C_m is a proportionality constant which depends on the system of measurements: $C_m = 1$ in the cgs system and $C_m = \mu_0/4\pi = 10^{-7}$ *henry* per meter in the SI system with $\mu_0 = 4\pi \cdot 10^{-7}$ as the *magnetic permeability* of a source-free space. Magnetic potential complies with the *superposition principle*: the magnetic potential of an assemblage of magnetic dipoles is the sum of the magnetic potentials of all individual dipoles. At any point in space, the direction and magnitude of magnetic field *vector* **B** are defined by the *gradient* of the magnetic potential. In *Cartesian coordinates*, the first-order derivatives of the magnetic potential represent the horizontal components B_x and B_y and vertical component B_z of magnetic field vector **B** as

$$B_x = \partial \mathcal{V} / \partial x, \quad B_y = \partial \mathcal{V} / \partial y, \quad B_z = \partial \mathcal{V} / \partial z,$$

where $\mathcal{V}$ is magnetic potential. Second-order derivatives of the magnetic potential represent the first-order derivatives, or gradients, of each vector component of a magnetic field in three orthogonal directions, so that the three first-order derivatives of the three vector components of the magnetic field comprise the nine elements of the *magnetic gradient tensor*. The relationship between the magnetic and gravitational potentials is described by the *Poisson theorem*. The calculated *3D analytic signal amplitude (3DASA)* and *tilt angle (TA)* of the magnetic potential are equal to the *total magnetic intensity (TMI)* and *magnetic inclination*, respectively, of the anomalous magnetic field at each *grid* point. See also *anomalous field*, *magnetic vector potential*, and *gravitational potential*.

magnetic prospecting: See *magnetic exploration*.

magnetic pulsations: See *micropulsations*.

magnetic reciprocity: a physical principle postulating that the *magnetic anomaly* generated by a *source body* remains unchanged when the direction of the external, or inducing, magnetic field and the direction of the source body *magnetization* are interchanged. See also *magnetic induction*.

magnetic reduction: the process of application of *magnetic data corrections* to measured magnetic values in order to obtain the values of *total magnetic intensity (TMI)*, also known as *total magnetic anomaly (TMA)*, and its reduced-to-pole (RTP) version. Corrected values represent the *field* of the magnetic anomalies, or *anomalous magnetics*, generated by rock *magnetization* distributions associated with elements of the local- and regional-scale geological structures in a *magnetic survey* area. See also *anomaly*, *magnetic anomaly*, *reduction-to-pole (RTP)*, and *observed magnetic components*.

magnetic resonance: a quantum physics phenomenon of the interaction between magnetic dipole moments of charged subatomic particles (positively charged protons and negatively charged electrons) in response to an externally applied oscillating electromagnetic field or monochromatic pulses of laser light that results in the *precession* of the subatomic particles with an *angular frequency* directly proportional to the strength, or intensity, of a stable ambient magnetic field, such as the *Earth magnetic field*. Precession frequency is called the *Larmor frequency*. Magnetic resonance is the basic operating principle of a *proton magnetometer*, *Overhauser magnetometer*, *potassium magnetometer*, and other *quantum magnetometers*. See also *ambient field*, *magnetic dipole moment*, and *Zeeman effect*.

magnetic saturation: a physical phenomenon which is observed when the increase in intensity of an externally applied *magnetic field* does not result in an increase of *magnetization* induced into the ferromagnetic or ferrimagnetic rock material. See also *hysteresis loop*, *induced magnetization*, *ferromagnetism*, and *ferrimagnetism*.

magnetic sensor: the principal element in a *magnetometer* or *magnetic gradiometer* for measurements of the planetary *magnetic field*, its *gradient*, *gradient tensor* and *vector* components. There are four general types of magnetic sensors that utilize either the magnetic properties of the permalloy cores or the quantum properties of protons and electrons: (1) the magnetically saturated permalloy cores with their electrical coil windings in the opposite directions, such as the two parallel cylindrical cores in a *fluxgate magnetometer*; (2) the hydrogen-rich fluid in a *proton magnetometer* and *Overhauser magnetometer*; (3) vaporized chemical elements from the alkali group, such as cesium, potassium, rubidium, and the helium isotopes He^3 and He^4, in the *optically pumped magnetometer* types; and (4) the superconducting loop containing one or two Josephson junctions in a *SQUID sensor*. See also *Josephson junction* and *gravity sensor*.

magnetic signature: **1.** the specific *shape* and positive or negative sign of a *magnetic anomaly* that (a) make this *anomaly* distinctly different from other anomalies in the gridded *RTP magnetic field* and *residual magnetics*, and (b) it is known to be associated with a particular *source body* or geological *structure*. The magnetic signatures of an *anomalous field* are compared with the known and forward-modeled anomalies and used as guides in *structural interpretation*. They are also implemented in automated *pattern recognition* techniques, such as the *Keating method*. **2.** a three-dimensional (3D) *forward model* of the magnetic effects produced by *survey* aircraft components (frame, engine, electronics, power cables, etc.). The model-calculated magnetic field of aircraft components is integrated into the adaptive *aeromagnetic compensation* system. See also *gravity signature*.

magnetic south: See *magnetic south pole*.

magnetic south pole: the coordinate point on the Earth's surface in the southern hemisphere where the *main geomagnetic field* is vertical, its strength is close to maximum, and the *magnetic inclination* is −90°. For each *epoch*, the location of the magnetic south pole is defined by the corresponding *IGRF* model. Due to the presence of both dipole and non-dipole components in the main geomagnetic field, the magnetic south pole is not coincident with the *geomagnetic south pole* and geographic south pole. Also referred to as *magnetic south*. See also *magnetic pole*, *magnetic north pole*, and *magnetic equator*.

magnetic spherule: a small-diameter dark-colored spherical body composed mostly of *magnetite*, sometimes with a metal core, and deposited in deep ocean/sea sediments. Large accumulations of magnetic spherules generate detectable magnetic anomalies, and these accumulations are considered as commercial sources of high-grade iron ore. See also *mineral nodule* and *submarine magnetics*.

magnetic storm: rapid, high-*amplitude* irregular fluctuations of the *external magnetic field* of the Earth generated by the interaction between high-energy *solar wind* and electric currents in the *ionosphere*. With a duration from several hours to several days, magnetic storms are much higher in *magnitude*, more irregular, and often higher in their dominant frequencies than *diurnals*. Magnetic storms directly correlate with periods of unusually high radiation activity of the Sun, such as coronal mass ejections and solar flares. Rapid changes in solar radiation levels can result in magnetic storm magnitudes of several hundred nanoteslas and, sometimes, up to several thousands nanoteslas within a few hours. During magnetic storms, all exploration magnetic surveys are suspended. See also *magnetic substorm*, *magnetosphere*, *nanotesla (nT)*, and *transient disturbances*.

magnetic stratigraphy: See *magnetostratigraphy*.

magnetic striping: See *geomagnetic striping*.

magnetic substorm: a *magnetic storm* with a duration from one-half hour to several hours. The primary source of energy for magnetic storms and substorms is high-energy *solar wind*. See also *magnetic disturbances*.

magnetic survey: ground, marine (shipboard), submarine, and airborne measurements of the *Earth magnetic field*, its *gradient*, *gradient tensor*, and *vector* components in areas of exploration or research interest. Measurements are commonly made at regularly distributed observation points along the pre-planned *survey* lines with the use of high-*sensitivity* and high-*precision* magnetometers. The coordinates of observation points are provided by the satellite-based *Global Positioning System (GPS)* or more accurate *differential GPS (DGPS)*. There are two basic types of magnetic surveys which differ by *sample interval*, *survey line spacing*, flight *altitude* (*survey height*), and *accuracy* of measurements: (1) a large-scale *regional survey* for tectonic and geodynamic studies, where the observation points are tens and hundreds of meters apart, survey lines are up to 2–3 km from each other, and measurement accuracy is 0.5–1.0 *nanotesla (nT)*; (2) a small-scale *detailed survey* for mapping of the magnetic, *magnetic gradient*, and *magnetic gradient tensor* anomalies generated by targeted structures, where the required accuracy of measurements is 0.05–0.1 nT, the sample interval is 2–5 m, and the survey line spacing varies from 2–10 m to several tens or hundreds of meters. At present, most regional and detailed ground magnetic surveys have been replaced by airborne magnetic surveys, such as *high-resolution aeromagnetic survey* and *drone survey*, that are capable of detecting anomalies less than 0.05 nT. Estimates of the expected *magnetic survey accuracy* and *lateral resolution* of detectable anomalies are based on *survey specifications*. See also *fixed-wing survey*, *helicopter survey*, *high-resolution ground survey*, *target*, and *magnetic method*.

magnetic survey accuracy: an estimate of the capability of a *magnetic survey* to obtain the true values of the *Earth magnetic field*, its *gradient*, *gradient tensor*, and *vector* components. Projected *accuracy* requirements are defined by the expected *target* parameters, such as depth, size, and *magnetization contrast*. Measurement accuracy depends, primarily, on five types of accumulated errors: (1) measured magnetic value, or *reading*, error; (2) horizontal position, or latitude and longitude, error; (3) vertical position, or *elevation*, error; (4) aggregated effects of *heading error*, *lag error*, and incomplete *aeromagnetic compensation*; and (5) residual *mis-tie* errors. In measurements of the anomalous vector components, the required accuracy in the orientation of a *magnetic sensor* is higher than 0.01°. For marine (shipboard) and airborne magnetic and *magnetic gradient* surveys, survey accuracy estimates are based on *intersection statistics* and averaged *rms error* at the mis-ties after *leveling* and *microleveling*. See also *FOM test*, *positioning*, and *vector magnetometer*.

magnetic survey resolution: *amplitude resolution* and *wavelength resolution* of anomalies in the data acquired by a *magnetic survey*. The high *sensitivity*, *sampling rate*, and *accuracy* of modern magnetometers provide an amplitude, or vertical, resolution sufficient for detailed exploration and research applications. The wavelength, or lateral, resolution of a magnetic survey depends primarily on two parameters of the *survey specifications*: the along-the-line *sample interval* for ground surveys and the *survey line spacing*. For line-oriented marine (shipboard) and airborne data, the *lateral resolution* of anomalies is defined by the *cutoff wavelength* of a *low-pass filter* applied to eliminate short-wavelength *noise* due to the uncompensated magnetic interference effects of the survey aircraft and unsuppressed components of *diurnals*. For gridded data, the lateral resolution is defined by the pre-selected survey line spacing: the closer the line spacing, the higher the lateral resolution of anomalies. In comparison with *airborne gravity* and airborne *gravity gradient* measurements, the airborne magnetic and *magnetic gradient* measurements are, practically, not affected by *dynamic noise* and, thereby, make the aeromagnetic surveys very efficient in resolving, both laterally and vertically, the low-*amplitude* short-*wavelength* anomalies, especially those generated by faults, fracture and shear zones, lithological contacts, and other structural discontinuities mapped for oil-and-gas or mineral exploration. See also *fault, gridding, high-resolution aeromagnetic survey, drone survey, magnetometer, magnetic gradiometer,* and *shear zone.*

magnetic susceptibility: 1. the physical property of rocks and a measure of the degree to which the rocks can be magnetized by an externally applied magnetic field, such as the *Earth magnetic field.* **2.** the proportionality constant *k* in the relationship between *induced magnetization* $\mathbf{M}_{\mathrm{IM}}$ and *inducing magnetic field* $\mathbf{F}_{\mathrm{IMF}}$ expressed as

$$\mathbf{M}_{\mathrm{IM}} = k\mathbf{F}_{\mathrm{IMF}}.$$

The inducing magnetic field is also called *magnetizing force* **H**, with

$$\mathbf{M}_{\mathrm{IM}} = k\mathbf{F}_{\mathrm{IMF}} = k\mathbf{H}.$$

For each particular rock type, the magnetic susceptibility is directly proportional to a volume percentage of the highly magnetic accessory rock-forming minerals, primarily *magnetite*, i.e., magnetic susceptibility is not a bulk rock property such as *density*. Susceptibility is dimensionless in both the SI and cgs systems, but its value differs by 4π: $k_{SI} = 4\pi k_{cgs}$. Susceptibility k is related to *magnetic permeability* μ in the cgs system as

$$k = (\mu - 1)/4\pi$$

and in the SI system as

$$k = \mu - 1.$$

In the cgs system, the magnetic susceptibility is often reported in units of 10^{-6} called *microcgs (μcgs)* units. Generally, the susceptibility is described in positive values. Negative susceptibility values are used in the *modeling* of the diamagnetic effects of *salt* and *remanent magnetization* of rock units. Unlike rock density, the susceptibility varies up to two or three orders of magnitude, or 10^2–10^3, with high variations within the same rock type. On average, susceptibility values vary from 10–50 μcgs in sedimentary rocks, up to 7000–8000 μcgs in metamorphic rocks, and 10,000–14,000 μcgs in igneous rocks. The *SI units* and *cgs units* of magnetic susceptibility are related as 1 SI ≅ 12.56 cgs = 12560000 μcgs. See also *diamagnetic susceptibility, bulk susceptibility, effective susceptibility,* and *volume susceptibility.*

magnetic susceptibility log: a logging tool for direct measurements of *magnetic susceptibility* of borehole-penetrated rocks for stratigraphic differentiation and cross-well correlation of sedimentary formations. See also *magnetostratigraphy.*

magnetic tensor: See *magnetic gradient tensor.*

magnetic tensor field: the *field* of *magnetic gradient* anomalies obtained from the independent elements of a *magnetic gradient tensor* that represent: (a) five horizontal derivatives of three mutually orthogonal *vector* components of the *Earth magnetic field*, defined as $B_{xx} = \partial B_x/\partial x = \partial^2 V/\partial x^2$, $B_{xy} = \partial B_x/\partial y = \partial^2 V/\partial x\partial y$, $B_{zx} = \partial B_z/\partial x = \partial^2 V/\partial z\partial x$, $B_{zy} = \partial B_z/\partial y = \partial^2 V/\partial z\partial y$, $B_{yy} = \partial B_y/\partial y = \partial^2 V/\partial y^2$; and (b) one *vertical derivative* defined as $B_{zz} = \partial B_z/\partial z = \partial^2 V/\partial z^2$, where V is *magnetic potential* and B_x, B_y, B_z are mutually orthogonal vector components of the Earth's magnetic field. See also *horizontal derivative* and *invariants.*

magnetic terrain correction: a correction applied to aeromagnetic data acquired over high relief *terrain* in order to reduce *magnetic terrain effects*. Its calculation is based on a three-dimensional (3D) *model*, where (a) the *Digital Terrain Model (DTM)* is the model top, (b) the *magnetization* of rocks is a constant value determined from rock sample measurements, and (c) the model bottom is either flat with a constant *elevation* below the lowest point of the topographic surface or parallel to the smoothed topographic surface. The model-calculated magnetic field is subtracted from *observed magnetics*. Grauch (1987) describes the *grid*-based method of calculating the variable-magnetization terrain corrections. See also *gridding, reduction-to-pole (RTP),* and *observed magnetic corrections.*

magnetic terrain effects: magnetic effects produced by the *magnetization* of rocks and unconsolidated rock material that form the topographic *relief* in a *magnetic survey* area. In areas of high relief, the *magnitude* of the terrain effects can overwhelm and obscure the *target* anomalies from sources in the upper sedimentary section. Magnetic terrain effects are recognized by the strong correlation of the shapes of the observed magnetic anomalies with the corresponding features of the *topography*, and they are different from *drape effects* caused by the irregular flight *altitude* of an aeromagnetic survey. Magnetic terrain effects are minimized by application of *magnetic terrain correction*. Referred to with the same meaning as *topographic magnetic anomalies*. See also *regolith.*

magnetic terrane: a regional-scale area where the spatial characteristics (amplitudes, shapes, wavelengths) and dominant trends of the magnetic anomalies are distinctly different from those in the adjacent and surrounding areas. Magnetic terranes are assumed to represent different parts of the ancient tectonic plates, whose boundaries are mapped by correlating the maximum values of the *horizontal gradient* of a low-pass filtered *RTP magnetic field*. See also *grain, low-pass filter,* and *wavelength.*

magnetic theodolite: See *declination-inclination magnetometer.*

magnetic vector: 1. *geomagnetic vector.* **2.** a three-component *vector* $\boldsymbol{B} = (B_x, B_y, B_z)$ for which the direction and *magnitude* at every point in space are defined by the *gradient* of the scalar *magnetic potential* V. The magnetic vector is defined as

$$\mathbf{B} = \nabla V,$$

where ∇ is the vector gradient operator *del*. In *Cartesian coordinates*, three magnetic vector components are expressed as

$$B_x = \partial \mathcal{V} / \partial x,\ B_y = \partial \mathcal{V} / \partial y,\ B_z = \partial \mathcal{V} / \partial z,$$

where $\partial\mathcal{V}/\partial x$, $\partial\mathcal{V}/\partial y$, $\partial\mathcal{V}/\partial z$ are first-order spatial derivatives of the magnetic potential that represent two horizontal components and one vertical component of the *magnetic field*, respectively. Magnetic vector length, or magnitude, is defined as

$$|\boldsymbol{B}| = (B_x^2 + B_y^2 + B_z^2)^{1/2},$$

and its *phase*, or angle relative to the XY plane, is equal to the *tilt angle (TA)* value at each observation point. Also referred to as *3D magnetic vector*. See also *anomalous magnetic vector, residual magnetic vector, magnetization vector*, and *airborne vector magnetometry*.

magnetic vector attributes: quantitative characteristics of the *magnetic vector* components of individual magnetic source bodies. They include magnetic vector *inclination*, magnetic vector *declination*, magnetic vector horizontal component, magnetic vector vertical component, magnetic vector *amplitude*, and their residual components. See also *anomalous magnetic vector, magnetization vector, residual magnetic vector*, and *source body*.

magnetic vector calculation: the *grid*-based calculation of three *vector* components of an anomalous *magnetic field* from its measured single-component *scalar* values via *scalar-to-vector transformation*. Obtained data representation also can be considered in a manner of the local- or regional-scale *directional filtering*. The results of local-scale calculations of a *magnetic anomaly vector* are used as the inputs to three-dimensional (3D) *magnetic inversion* for estimates of isolated *source body* parameters and mapping of *volcanics*, including those strongly affected by *remanent magnetization*.

magnetic vector potential: *magnetic potential* in the presence of an electric current. Its *curl* defines the magnetic field which is external to the fields generated by the magnetic sources in the subsurface. Commonly, the electric currents are assumed to be negligible in areas where the *Earth magnetic field* is measured. See also *Helmholtz theorem*.

magnetic viscosity: a gradual change of *magnetization direction* in the individual grains of some rock-forming minerals in the direction of the ambient magnetic field, such as the *Earth magnetic field*. See also *ambient field* and *viscous remanent magnetization (VRM)*.

magnetism: a general term applied to various physical phenomena associated with (a) permanent and temporal magnetic effects in magnetic rock-forming minerals, (b) planetary magnetic fields, such as the *Earth magnetic field*, and (c) electric currents in conductive materials. See also *magnetosphere* and *van Allen belts*.

magnetite: a strongly magnetic mineral that has the highest *magnetic susceptibility* among the common rock-forming minerals, ranging from approximately 1200–19,200 *microcgs* (μcgs) with the average of 6000 µcgs. Its *Curie temperature* is 560°–580°C. Widely distributed, magnetite is the most important magnetic mineral because its volume percentage is a dominant contributor to the *magnetization* of rocks. Magnetite is a generic name for the iron-titanium oxide minerals within the chemical composition range between the iron oxide FeO-Fe_3O_4 and the ulvospinel Fe_2TiO_4. Igneous rocks are usually magnetite-rich. There are several types of sedimentary magnetite, including: (a) *diagenetic magnetite* formed by hydrothermal alterations, hydrocarbon seepage, dolomitization, and other chemical changes in the mineral composition of rocks; (b) diagenetic magnetite formed as a result of *combustion magnetism*; (c) *detrital magnetite* formed by rock weathering and accumulated in clastic rocks, such as sandstones, shales, and limestones, and in paleo-drainage systems and beach sands; and (d) magnetite formed in faults and fracture zones with a history of repeated vertical migration of highly mineralized hydrothermal fluids, such as *juvenile water*. Outcrops and upper sedimentary subcrops of magnetite-bearing layers generate high-*amplitude* alignments of magnetic anomalies which can dominate in *high-resolution aeromagnetic survey* data. See also *clinker anomaly, magnetized fault, outcrop, subcrop*, and *subcrop anomaly*.

magnetization: **1.** a measure of the magnetic effect of a rock on the resulting magnetic field, when this rock is subjected to *magnetizing force (H)* representing the strength, or intensity, of an external magnetic field, such as the *Earth magnetic field*. As a *vector* quantity, magnetization is denoted by **M** or **I**, and its relationship with *magnetic induction* **B** is expressed in *SI units* as

$$\mathbf{B} = \mu_0(\mathbf{H} + \mathbf{M})$$

and in *cgs units* as

$$\mathbf{B} = \mu_0(\mathbf{H} + 4\pi\mathbf{I}),$$

where μ_0 is *magnetic permeability* of a source-free space. **2.** vector quantity **M** defined by the total *magnetic dipole moment* per unit volume and expressed as

$$\mathbf{M} = \Sigma\mathbf{m_i}/\mathrm{V},$$

where $\Sigma\mathbf{m_i}$ is a vector sum of all individual magnetic dipole moments $\mathbf{m_i}$ of mineral grains in a rock body and V is rock body volume. **3.** vector quantity **M** defined by the product of the *magnetic field strength* **F** or magnetizing force **H** and *bulk susceptibility* k_b of a rock body and expressed as

$$\mathbf{M} = k_b\mathbf{F} = k_b\mathbf{H}.$$

There are two basic types of rock magnetization: *induced magnetization* $\mathbf{M_I}$ and *remanent magnetization* $\mathbf{M_R}$. Vector addition of the induced and remanent magnetization vectors defines *magnetization direction* of a rock body and the *magnitude* of its *total magnetization*. Magnetization of rock bodies depends on many parameters, including their size, shape, mineralogy, origin, mineral grains, temperature, etc., that can be generalized as: (a) the *magnetic susceptibility* of a particular rock; (b) the magnitude and direction of the

Earth's magnetic field, both past and present; (c) the volume percentage of *magnetite* and other *ferrimagnetic* and *ferromagnetic* minerals in the rock composition; (d) the vector direction of the remanent magnetization, if any present; and (e) the geological and thermal history of the rock. Unlike rock *density* varying 2.0–2.5 times its lowest value and directionally independent, the magnetization of rocks can vary up to 100–1000 times and it is directionally dependent. Because magnetization is controlled, primarily, by a few ferromagnetic and ferrimagnetic minerals which are accessory rock components, the most common rock types do not have diagnostic magnetic properties. Without sufficient *magnetic susceptibility log* data and susceptibility estimates from properly preserved rock samples, the prediction of rock magnetization for the purposes of *forward modeling* and *inversion* can be made by general approximations only. Units of measurement are *ampere* per meter ($A \cdot m^{-1}$) in the SI system and *gauss* in the cgs system, where $1\ A \cdot m^{-1} = 10^{-3}$ gauss. Also known as *intensity of magnetization*, *dipolar magnetization*, and *magnetic polarization*. See also *effective magnetization*.

magnetization contrast: 1. a quantitative characteristic of a lateral change in the *magnetization* of rocks defined by (a) the *scalar* difference in *magnetic susceptibility* values between two adjacent rock formations that is referred to as the *susceptibility contrast* and (b) the *vector* addition of the *remanent magnetization* and *induced magnetization* vectors in two adjacent rock formations. Lateral magnetization contrast is a fundamental reason for, or cause of, a *magnetic anomaly*: there is no magnetic anomaly in the absence of a lateral magnetization contrast. It is positive when the *source body* magnetization is higher than that of the adjacent rocks, and it is negative for the reverse case. The amount of the lateral magnetization contrast is one of the primary controls over the *amplitude* of a magnetic anomaly: the higher the contrast, the higher the amplitude of the anomaly. A change of the magnetization contrast sign reverses the sign of the magnetic anomaly. Referred to with the same meaning as *anomalous magnetization*. **2.** the difference in magnetization of rocks across a lateral lithological or structural boundary between two rock formations of different ages, such as the younger overlying rocks and the older underlying rocks. Magnetization contrast of this type is the largest, up to two or three orders of magnitude, at the *basement* top separating the low-magnetization sedimentary formations and high-magnetization metamorphic and igneous basement rocks. See also *unconformity*, *suprabasement anomalies*, and *density contrast*.

magnetization direction: the direction, or orientation, of the *magnetization vector* in a magnetic *source body*. Standard *reduction-to-pole (RTP)* is based on the assumption that the ambient *geomagnetic field* and source body have the same magnetization direction. Calculation of the *analytic signal amplitude (ASA)* of a gridded *TMI field* reduces the effect of the *remanent magnetization* direction in three-dimensional (3D) sources and essentially eliminates this effect in the case of 2D sources. The true direction of *total magnetization* provides the most accurate RTP for isolated source bodies and estimates of their parameters, and it can be determined by the *cross-correlation method*, *Helbig method*, and other techniques (Clark, 2014). See also *ambient field*, *analytic signal (AS)*, and *cross-correlation remanence*.

magnetization intensity: 1. a *scalar* quantity $|\mathbf{M}|$ that represents the *magnitude* of rock body *magnetization* **M**. **2.** a *vector* quantity **M** representing the strength, or intensity, of a *magnetic field* within a rock body defined by the vector sum of *induced magnetization* and *remanent magnetization*. **3.** the length, or magnitude, of a *magnetization vector*. See also *self-demagnetization*.

magnetization inversion: See *magnetic inversion*.

magnetization vector: a *source body* parameter that defines the direction, or orientation in the *vector* space, and *magnitude* of its *total magnetization*. In the presence of *remanent magnetization*, the resultant magnetization direction is defined by the vector sum of the ambient *geomagnetic field* vector and remanent magnetization vector of a particular source body. Knowledge of the magnetization vector orientation in a source body improves the *inversion*-based estimates of its physical properties (volume, shape, *susceptibility contrast*) and results of *reduction-to-pole (RTP)*, because the standard RTP assumes that the geomagnetic field and source body have the same *magnetization direction*. When the vector data from measurements by a *vector magnetometer* are not available, the magnitude and orientation of the magnetization vector of a source body can be determined by the *Helbig method* and other techniques (Clark, 2014). See also *ambient field*, *normalized magnetic moment vector*, and *vector magnetic moment*.

magnetized fault: a *fault* or fracture zone within a sedimentary section, where the recurrent upward migration of highly mineralized and high-temperature fluids of the *juvenile water* type or seepage of hydrocarbons has resulted in chemical alterations and hydrothermal mineralization of the host rocks and deposition of strongly magnetic minerals, such as *diagenetic magnetite*, along the fault or fracture plane. Magnetized vertical and high-angle faults and fracture zones generate the low-*amplitude*, short-*wavelength* magnetic anomalies that can be detected and mapped by a *high-resolution aeromagnetic survey* and *magnetic gradient* surveys. In *modeling*, the magnetized faults are approximated by *thin dike* source geometries (Peirce et al., 1998; Williams et al., 1999). See also *hydrocarbon seepage effect*.

magnetizing field intensity: *magnetizing force (H)* in the context of a quantitative measure of strength, or intensity, of an externally applied *magnetic field*, such as the *Earth magnetic field*.

magnetizing force (H): 1. a *vector* quantity that is a measure of strength, or intensity, of an external magnetic field that defines the *magnetization* of rocks. It has two components and can be expressed in *SI units* as

$$\mathbf{H} = (\mathbf{B} - \mathbf{M})/\mu_0$$

and in *cgs units* as

$$\mathbf{H} = \mathbf{B} - 4\pi\mathbf{M},$$

where **B** is *magnetic induction*, also referred to as *magnetic field*; **M** is *induced magnetization*; and μ_0 is *magnetic permeability* of a source-free space. **2.** a vector quantity defined by the *gradient* of *magnetic potential*. The magnetizing force is measured in units of *oersted* (Oe) in the cgs system and in units of *ampere* per meter ($A \cdot m^{-1}$) in the SI system, where $1\ A \cdot m^{-1} = 4\pi \cdot 10^{-3}$ Oe. Referred to with the same meaning as *magnetic force*. Also known as *magnetizing field intensity*.

magnetochronology: rock age dating based on the correlation of *vector* parameters of the ancient *Earth magnetic field* (determined from estimates of the *remanent magnetization* direction in rock samples) with vector parameters of the magnetic anomalies generated by rock formations of the known geological times. See also *magnetostratigraphy* and *paleomagnetism*.

magnetohydrodynamics: a general term for the physical phenomena associated with flows of electrically conducting fluids, such as a liquid metal or ionized gas, within the magnetic fields. Also referred to as *hydromagnetics*. See also *geomagnetic dynamo* and *van Allen belts*.

magnetometer: an instrument for measuring the planetary *magnetic field*, such as the *Earth magnetic field*. There are two basic magnetometer types: *scalar magnetometer* and *vector magnetometer*. A scalar magnetometer measures a numerical value, or *magnitude*, of the total magnetic field regardless of its *vector* orientation. Scalar magnetometers are dominant in exploration applications, and they include the *cesium magnetometer*, *helium magnetometer*, *Overhauser magnetometer*, *potassium magnetometer*, *proton magnetometer*, and *rubidium magnetometer*. Vector magnetometers measure the *vector* components of the magnetic field in the direction of orientation of their magnetic sensors, and they include the *helium magnetometer*, *fluxgate magnetometer*, and *SQUID magnetometer*. Depending on the objectives of a *magnetic survey*, the ground surveys use magnetometers measuring either the total magnetic field of the Earth or its vertical or horizontal vector components. Marine (shipboard) and airborne magnetic surveys use scalar magnetometers and their assemblages in *magnetic gradiometer* systems to measure the Earth's magnetic field, its *gradient*, and *full tensor gradient (FTG)* components. Satellite missions are commonly equipped with both scalar and vector magnetometers. See also *magnetic sensor* and *satellite magnetometry*.

magnetometer drift: a slow gradual change of *scalar magnetometer* readings over time caused by changes in its zero setting and related to the specific characteristics of the *magnetic sensor* and its measuring device, temperature effects, and the stability of the magnetometer mounting, such as in a *fluxgate magnetometer*. Drift of the *baseline* value in a *vector magnetometer* is monitored and calibrated periodically by scalar magnetometer. See also *reading* and *gravimeter drift*.

magnetometer gradiometry: stationary ground measurements of the *vertical gradient* or *horizontal gradient* of the *Earth magnetic field* with the use of two magnetometers separated vertically or horizontally. Measurements are made with the objectives of (a) eliminating the effect of *diurnals* and (b) utilizing the higher *lateral resolution* and *sensitivity* of magnetic gradients as compared to the *total field* measurements by a single *magnetometer*. See also *gravimeter gradiometry*.

magnetometry: methods and instruments to acquire, process, and interpret measurements of a planetary *magnetic field*, such as the *Earth magnetic field*, its *gradient*, *full tensor gradient (FTG)*, and *vector* components. See also *acquisition*, *data processing*, *interpretation*, and *airborne vector magnetometry*.

magneton: See *Bohr magneton*.

magnetosphere: the space around the Earth where the *main geomagnetic field* and the long-*wavelength* components of the *lithospheric magnetic field* interact with *solar wind* and magnetic fields of the Sun and other planets, and where this interaction produces a highly variable *external magnetic field* of the Earth. Continuous flow of the Sun-radiated charged particles (solar wind) compresses the external magnetic field toward the Earth on the sunlit side and extends it out into the space on the shadow side, so that the magnetosphere has a comet-like shape. On the sunlit side, the magnetosphere is nearly hemispherical and extends to approximately 64,000 km into space under regular conditions, but it can be compressed by a *magnetic storm* to nearly 40,000 km. On the shadow side, the magnetosphere has a long tail of more than 600,000 km. Changes in the interplanetary magnetic fields and intensity of the solar wind influence continuously the position and shape of the magnetospheric boundary. See also *ionosphere* and *van Allen belts*.

magnetospheric storm: a *magnetic storm* with an emphasis on its origin and development in the *magnetosphere*.

magnetostatic charge: a hypothetical magnetic charge whose surface and volume distributions are used to represent the *magnetization* of three-dimensional (3D) source bodies. If the magnetization of a *source body* is uniform, then the *magnetic potential* of the surface distribution of the magnetostatic charge is identical to the magnetic potential caused by magnetization of the whole source body. Magnetostatic charge representation is implemented in data transformation algorithms applied by the *equivalent-source method*. See also *Green's equivalent layer*.

magnetostratigraphic classification: arranging rock formations into groups of stratigraphic units based on the similarity of their magnetic characteristics, such as *susceptibility* and *magnetization direction*. See also *magnetostratigraphic unit*.

magnetostratigraphic unit: a rock formation described by specific magnetic properties, such as *magnitude*, *magnetization direction*, and *susceptibility*, that make it distinct from the surrounding rock formations.

magnetostratigraphy: **1.** identification of the stratigraphic rock units with different magnetic properties for the cross-correlation of sedimentary rock formations. It is based on direct measurements of *magnetic susceptibility* in boreholes and laboratory studies of magnetic properties of extracted core samples. **2.** measurements of the *remanent magnetization* direction in rock samples of a known geological age for the purposes of *magnetochronology*, construction of the *polar wandering* curves, and restoration of the paleolatitudes for plate tectonic analysis. **3.** a method of using the *magnetic polarity* patterns in sedimentary and volcanic rock outcrops and oriented rock samples for their correlation with the global magnetic timescale. See also *magnetic susceptibility log*, *outcrop*, and *pattern*.

magnetostriction: 1. a natural phenomenon of changes in shapes and/or dimensions of grains in the rock-forming *ferromagnetic* minerals by an externally applied *magnetic field*. **2.** a change in the elastic strain or deformation of rocks as a result of changes in their *magnetization*. See also *dimension* and *inverse magnetostriction*.

magnitude: 1. the numerical, or absolute, value of a *scalar* quantity regardless of its sign, such as the anomalous *amplitude* or field intensity, that is used as a reference for comparison with other quantities of the same type. **2.** the absolute value of a *vector* quantity, such as rock magnetization **M**, denoted by |**M**|. **3.** the length of the vector **A** defined as

$$|\mathbf{A}| = (A_x^2 + A_y^2 + A_z^2)^{1/2},$$

where A_x, A_y, A_z are orthogonal vector components defined by projections of vector **A** onto the mutually orthogonal x-, y-, z-directions in *Cartesian coordinates*. The magnitude of a *complex variable* in the vector space is called the *modulus*. For both scalar and vector quantities, the magnitude is used interchangeably and with the same meaning as *anomaly amplitude*. See also *anomaly*, *eigenvalue magnitude*, and *FTG magnitude*.

magnitude gradient: a *gradient* with an emphasis on a quantitative characteristic of the *scalar* and *vector* fields, where it represents the absolute value of the *derivative* and length, or *magnitude*, of the three-dimensional (3D) *gradient vector*, respectively.

main field declination: *magnetic declination* of the *main geomagnetic field*.

main field declination map: a global-scale *isogonic map* showing the *magnetic declination* values of the *main geomagnetic field*.

main field inclination: *magnetic inclination* of the *main geomagnetic field*.

main field inclination map: a global-scale *isoclinic map* showing the *magnetic inclination* values of the *main geomagnetic field*.

main geomagnetic field: the long-*wavelength* dominant part of the *Earth magnetic field* generated in the outer Earth core and explained by the theory of *geomagnetic dynamo*. Its components have wavelengths exceeding 2,000–2,500 km and amplitudes varying slowly over time and position on the Earth's surface from 25,000–40,000 *nanotesla (nT)* in the South Atlantic and equator areas to 60,000–65,000 nT in northern Canada, eastern Siberia, and near both magnetic poles. Except for a few anomalous areas, the main geomagnetic field comprises approximately 95% of the magnetic field values measured on the Earth's surface. Almost 90% of the main geomagnetic field is approximated by an *eccentric dipole field*, which is the magnetic field of a slow-rotating *magnetic dipole* located at about 400 km from the Earth's center with its axis of rotation presently tilted by 11° with respect to the Earth's rotational axis. Two intersections of the magnetic dipole axis with the Earth's surface are called *geomagnetic poles*. About 10% of the main geomagnetic field is approximated by a *nondipole field*. The best-fitting mathematical approximation of the main geomagnetic field is described by the *International Geomagnetic Reference Field (IGRF)*. Its average *vertical magnetic gradient* is 0.025 nT per meter ($\text{nT} \cdot \text{m}^{-1}$). The main geomagnetic field lines are nearly vertical over the geomagnetic poles and horizontal over the *magnetic equator*, elsewhere between those locations their *inclination* varies as a function of both latitude and longitude. The two locations on the Earth's surface, where the main geomagnetic field is vertical, are called the *magnetic north pole* and the *magnetic south pole*. Basic characteristics of the main geomagnetic field are referred to as *geomagnetic field parameters* and determined from measurements by the worldwide network of geomagnetic observatories and Earth-orbiting satellites. Global measurements are summarized in the IGRF models and displayed in various formats, including *isoclinic map*, *isogonic map*, *isodynamic maps*, and *geomagnetic indices*. The contribution of the main geomagnetic field to *observed magnetics* is compensated by *IGRF correction*. Also referred to as *Earth core field*, *core field*, *main magnetic field*, and, sometimes, *geomagnetic field*. See also *geomagnetic observatory*, *IGRF model*, and *INTERMAGNET*.

main gravity field: the latitude-dependent smooth dominant part of the *Earth gravity field* that varies from approximately $9.78 \cdot 10^5$ *milligal (mGal)* at the equator to $9.83 \cdot 10^5$ mGal at the geographic poles. Its values are functions of the Earth's mass, size, shape, and rotational characteristics, and its best-fitting mathematical approximation is the *gravitational field* of a rotating homogeneous ellipsoid called the *reference ellipsoid*, which has the same mass, center of mass, and *angular velocity* of rotation as those of the Earth. Main gravity field values at any point on the Earth's surface are defined by the *International Gravity Formula (IGF)* and referred to as *theoretical gravity*. The contribution of the main gravity field to *observed gravity* is compensated by *theoretical gravity correction* which is also known as *latitude correction*.

main magnetic field: 1. *main geomagnetic field*. **2.** the dominant long-*wavelength* part of the planetary *magnetic field* measured by magnetometers mounted on planet-orbiting satellites.

mantle: the Earth's interior layer below the *crust* and above the core. It is divided into the upper mantle and lower mantle with a transition zone between them. The upper mantle includes the lower part of the *lithosphere* and *asthenosphere*. The boundary between the crust and upper mantle is known as the *Mohorovičić discontinuity* or *Moho boundary*. Upper mantle *density* estimates vary between $3.19\ \text{g} \cdot \text{cm}^{-3}$ and $3.40\ \text{g} \cdot \text{cm}^{-3}$. Mantle rocks are assumed to be either low-magnetic or essentially non-magnetic, because rock temperatures in the mantle exceed the *Curie point*. Recent evidence suggests that the upper mantle can contribute significantly to the observed long-*wavelength* magnetic anomalies (Ferré et al., 2014). See also *magnetic anomaly*.

mantle gravity anomaly: the *field* of the global-scale *residual gravity* anomalies associated with the distribution of anomalous rock masses in the upper *mantle*. It is obtained by calculating the difference between *satellite gravity*, such *GRACE gravity* or *GOCE gravity*, and crustal gravity effect, including the oceans and seas. See also *crust* and *crustal correction*.

mantle residual anomaly: **1.** a global-scale *mantle gravity anomaly* obtained by application of *crustal correction* for the gravity effect of the Earth's *crust* calculated with the use of *satellite gravity* data, global crustal models, and average crustal density of 2.67 gcm^{-3}. Calculated residual anomalies are caused mainly by heterogeneous distribution of densities in the upper *mantle* (Herceg et al., 2016). **2.** *residual gravity* Δg representing the gravity effect of changes in the *Moho boundary* depth. It is calculated primarily in the oceanic basins and expressed as

$$\Delta g_{MRA} = \Delta g_{FAA} - \Delta g_{bath} - \Delta g_{TGA} - \Delta g_{sed},$$

where Δg_{FAA} is satellite-derived *free-air gravity*, Δg_{bath} is the gravity effect of lateral changes in *bathymetry*, Δg_{TGA} is the lithospheric *thermal gravity anomaly*, and Δg_{sed} is the *gravity anomaly* due to changes in sedimentary thickness and *density* (Chappell and Kusznir, 2008). See also *global bathymetric correction*, *gravity stripping*, and *Parker formula*.

manual source depth estimation: *graphical source depth estimation* in the context of the manual selection of characteristic points on a *profile* or *contour map* of an anomalous magnetic or gravity *field*.

map power spectrum: *power spectrum* of the gridded magnetic or gravity field in the *format* of a map showing the variations of its spectral components as a function of the *azimuth* of power spectrum calculations. The highest power spectrum values associated with the smallest wavenumbers are plotted at the map center with the successively lower power values descending to ±0.5 *Nyquist frequency* in standard algorithms. A map power spectrum is used for: (a) selection of *directional filter* parameters; (b) identification of dominant trends of anomalies; and (c) quality control of *data processing* operations, because *leveling* errors, *microleveling*, *cultural noise*, and *aliasing* effects are clearly visible as alignments of relatively high power spectrum values along the *survey* lines. Alignments of power spectrum values on a spectral-domain map are orthogonal to the corresponding anomalous alignments on a space-domain *grid*, so that *corrugations* and other leveling errors along the east–west oriented survey lines will be shown as the north–south alignments on the map power spectrum. Also referred to as *energy rosette*, *power rosette*, *log-map power spectrum*, and *power spectrum map*. See also *space domain*, *spectral domain*, and *radial power spectrum*.

map projection: **1.** a concept of transferring the actual positions of observation points on the three-dimensional (3D) curved Earth surface to the corresponding coordinate points on a flat, or planar, surface of a map based on a regional or national *reference ellipsoid*. **2.** a map parameter that describes the *geodetic datum* of a particular map. More than 250 different map projections are currently used worldwide for local, regional, and global mapping. See also *coordinate system*, *regional geodetic datum*, and *Universal Transverse Mercator (UTM)*.

marine Bouguer anomaly: See *marine Bouguer gravity*.

marine Bouguer correction: *complete Bouguer correction* applied to observed marine (shipboard) and submarine gravity data. Its application involves two data processing steps: (1) the addition of *simple Bouguer correction* calculated with *replacement density* to compensate for *mass deficiency* below sea level or *reference ellipsoid* and (2) compensation for the *bathymetric effect* produced by variations of *bathymetry* in the survey and surrounding areas by *marine terrain correction*. Some marine data processing contractors refer to it as *3D Bouguer correction*. See also *double Bouguer correction*, *observed gravity*, and *observed gravity corrections*.

marine Bouguer gravity: *field* of the *Bouguer gravity* anomalies obtained from marine (shipboard) and submarine gravity measurements by application of all relevant instrument and data corrections, including *marine Bouguer correction*. Often referred to as *marine Bouguer anomaly*. See also *raw gravity corrections* and *observed gravity corrections*.

marine free-air gravity: **1.** a global-scale *free-air gravity* field of anomalies computed from *satellite altimetry* data acquired over the oceans and seas. The achievable *lateral resolution* of the satellite-derived marine free-air gravity anomalies is 7–20 km with the *accuracy* of 3–5 mGal and *waveform* tracking in the shelf-area waters to 10–15 km off the coastline. **2.** a regional- or local-scale *gravity field* obtained from marine (shipboard) *relative gravity* measurements and application of all relevant instrument and data corrections, except for *marine Bouguer correction*. See also *raw gravity corrections*, *observed gravity corrections*, and *satellite-altimeter gravity*.

marine geoid: the marine part of a *geoid* closely approximated by *mean sea level (MSL)* in the oceans and seas. A geoid is a mathematical surface of the equal *gravitational potential* values, or *equipotential surface*, so the calculation of the *first vertical derivative* of a marine geoid (i.e., *derivative* of the gravitational potential in the vertical direction) will produce the vertical component of the *Earth gravity field* in the oceans and seas that is known as *marine free-air gravity*. The surface of the marine geoid is mapped by *satellite altimetry*. Currently available coverage by satellite altimetry tracks is used for the *track*-oriented calculations of marine free-air gravity and *gridding* of the obtained values. See also *geoid-to-gravity method* and *slope-to-gravity method*.

marine gravimeter: See *shipboard gravimeter*.

marine gravimetry: methods and instruments to record and process measurements of the relative changes in the *Earth gravity field*, or *relative gravity*, and its gradients on the surface of the oceans and seas. Marine gravity data are acquired along regularly spaced *acquisition* lines over a *marine survey* area under stable ocean/sea surface conditions. High-*precision* measurements with the use of *Global Positioning System (GPS)* are required to determine the *real-time* coordinates and velocity of a survey ship for the calculation of all necessary corrections applied to the *marine gravity* data. See also *airborne gravimetry* and *submarine gravimetry*.

marine gravity: 1. *relative gravity* data obtained from measurements by *shipboard gravimeter* during a marine (shipboard) *gravity survey* in the oceans and seas. The best reported *accuracy* of marine gravity data is 0.1–0.2 *milligal (mGal)* at *full-wavelength resolution* of 800–1200 m depending on the water depth: the shallower the water, the higher the resolution. With its low operating costs, simple logistics, and valuable geophysical information, marine gravity measurements are often included into regular 2D and 3D seismic programs on the same ship as an *add-on survey*. **2.** *gravity gradient* data, such as *full tensor gradient (FTG)* data, acquired by a *full tensor gravity gradiometer* or other *gradiometer* system during a marine gravity survey. Sometimes referred to as *shipboard gravity*. See also *acquisition*, *airborne gravity*, *ground gravity*, *satellite gravity*, and *submarine gravity*.

marine magnetics: 1. *total field* data obtained from measurements by an *Overhauser magnetometer* mounted inside a *fish* and towed behind a survey ship during a marine (shipboard) *magnetic survey* in the oceans and seas or behind an *autonomous underwater vehicle (AUV)* during a submarine, or sea-floor, survey for the detection of (a) *magnetic spherule* accumulations in deep waters and (b) *unexploded ordnance (UXO)*, pipelines, or power and communication cables prior to offshore and coastal constructions in shallow waters. **2.** *magnetic gradient* data acquired by a *magnetic gradiometer* composed of two or three synchronized magnetometers deployed as a fixed-*array* system and towed behind a survey ship or AUV. See also *submarine magnetics* and *marine gravity*.

marine magnetometer: a *scalar magnetometer*, commonly of the *Overhauser magnetometer* type, which is mounted inside a *fish* and towed behind a survey ship or *autonomous underwater vehicle (AUV)* during a marine or submarine *magnetic survey*. See also *marine survey* and *submarine survey*.

marine magnetometry: methods and instruments to record and process measurements of the *Earth magnetic field*, its *gradient*, and *gradient tensor* components on the surface of oceans and seas. Marine magnetic measurements are commonly made with the use of an *Overhauser magnetometer* or its adaptation for a *magnetic gradiometer*, such as a *transverse magnetic gradiometer*, that is towed behind a survey ship. Marine data are acquired along regularly spaced *acquisition* lines in the *marine survey* area under stable ocean/sea surface conditions. High-*precision* measurements with the use of *Global Positioning System (GPS)* are required to determine the ship coordinates and calculate all necessary corrections to obtain *marine magnetics*. See also *airborne magnetometry*, *satellite magnetometry*, and *submarine magnetometry*.

marine survey: measurements of the *Earth magnetic field* and relative changes in the *Earth gravity field*, or *relative gravity*, and their gradients on the surface of oceans and seas. Measurements are made along the regularly spaced *acquisition* lines in areas of *exploration* and scientific research interest. Sometimes referred to as *shipboard survey*. See also *airborne survey*, *ground survey*, *submarine survey*, and *add-on survey*.

marine terrain correction: *terrain correction* applied to *marine gravity* data to compensate for the *bathymetric effect* produced by variations of the bathymetric *relief* around observation points. Often, marine terrain corrections are calculated for Fourier-transformed line-oriented *observed gravity* data using the two-dimensional (2D) *Parker formula*. See also *Fourier transform*, *observed gravity*, *marine Bouguer correction*, and *airborne terrain correction*.

Markov chain: a sequence of random variables in a stochastic, or probabilistic, process that follows a chain of linked events, or state changes, where the probability of a future change of the process state depends on the present state only, i.e., regardless of the previous events. In *stochastic inversion*, the Markov chain is represented by a transition, or state changing, *matrix* composed of the probability distributions of each variable in the interconnected network of voxels in a three-dimensional (3D) *voxel model*. See also *Bayes theorem*.

Markov chain inversion: same as *stochastic inversion* with an emphasis on a probabilistic process based on a *Markov chain* of events, or changes in the *inversion* process state, where the probability outcomes satisfy the *Bayes theorem*.

Markov chain Monte Carlo (MCMC): the optimization algorithm implemented in *stochastic inversion*. It integrates the probabilistic, or stochastic, properties of the *Markov chain* and random *sampling* of the Monte Carlo methods with the deterministic relations between the individual *voxel* parameters in a *voxel model* to find iteratively their optimal combination for which the posterior probability distributions are maximized. The algorithm enables *inversion* of the *residual gravity* for both the rock property (*density contrast*) and *structure* (rock formation depth or isolated body shape) in the same inversion process, because the relationship between the property and structural probabilistic inversion solutions depends on the neighboring voxels only. Also referred to as *MCMC algorithm*.

mascon: 1. a planetary-scale area of rock mass concentration centered on a giant meteorite *impact structure* with a diameter up to several hundred kilometers and more. It has the *gravity signature* of a circular-shaped high-*amplitude* positive *anomaly* in *satellite gravity* data. Mascons have been discovered on the Moon and Mars, and their origin is assumed to be associated with a rebound, or recoil, of the impacted planetary *crust* that resulted in the rise of a pipe-like *igneous body*, or magmatic plug, from the underlying *mantle* material. The mass excess of the raised igneous body, composed of high-*density* rocks, is held against *isostasy* by the strength of the planetary lithosphere. **2.** the form of the gravity basis functions conceptually analogous to *spherical harmonics*. Each mascon cell represents the gravity signal of a particular area on the planetary surface, so that local cell-by-cell changes of *observed gravity* can be integrated into the regional- and planetary-scale studies of *temporal gravity variations* associated with the redistribution of flowing masses, such as groundwater storage changes in *aquifer* basins, upper *mantle* mass variations, and melting of ice sheets (Luthcke et al., 2013).

mass: a measure of the amount of a gaseous, liquid, or solid body material defined by the product of its *density* and volume. The larger the mass, the greater the *gravitational force* generated by this body. See also *mass deficiency*, *mass excess*, *mass differential*, and *gravitational mass*.

mass correction: same as *simple Bouguer correction* with an emphasis on a correction for the gravity effect of the rock *mass excess* above a *survey datum* (sea level or *reference ellipsoid*) onshore and seawater *mass deficiency* below a survey datum offshore.

mass correction factor: same as *Bouguer correction factor (BCF)* that represents the *Bouguer correction* parameter used for calculation of the gravity effect of a rock *mass* above a *survey datum*, i.e., sea level or *reference ellipsoid*.

mass deficiency: 1. a local- or regional-scale lateral decrease in a subsurface rock *mass* created either by a structural depression, such as *syncline*, *graben*, or *pull-apart basin* filled in with lower *density* rocks, or by a lateral change in the lithological composition of rock formations, such as that associated with (a) *allochthonous salt* intruded into the higher-density sedimentary formations and (b) the accumulation of reservoir-grade sands in a stratigraphic *trap*. **2.** an increase in the water mass below sea level that results in the *Bouguer effect* offshore. **3.** an increased thickness of the *crust* under the mountains and large topographic elevations. Because the crustal rock densities are lower than densities in the upper *mantle*, their increased thickness creates a lateral mass deficiency which, in accordance with the *Airy hypothesis* and *Vening Meinesz hypothesis*, maintains the isostatic balance, or *isostatic equilibrium*, between the crust and upper mantle. Rock mass deficiencies are sources of gravity anomalies at all scales: local, regional, and planetary. See also *gravity anomaly*, *isostatic antiroot*, *isostatic root*, *isostasy*, and *mass excess*.

mass differential: a parameter of *mass excess* or *mass deficiency* defined by product of the volume of the *source body* or rock formation and its *density contrast*. Mass differential and distance to observation points are two primary controls over the *amplitude* of a *gravity anomaly* generated by the rock body or rock formation creating the mass differential. According to the *Gauss law*, which relates the mass differential to a numerical integration of *anomaly energy*, the mass differential can be determined uniquely regardless of the source body shape, depth, and its internal distribution of *density*. Also referred to as *anomalous mass*.

mass excess: 1. a local- or regional-scale lateral increase in a subsurface rock *mass* created either by a structural elevation, such as *anticline*, *horst*, *uplift*, and *thrust*, composed of higher *density* rocks, or by a lateral change in the lithological composition of rock formations, such as that associated with (a) a high-density *igneous body* intruded into a sedimentary formation and (b) a *carbonate platform* or *reef* surrounded by younger lower-density sediments. **2.** rock mass above a *survey datum* that results in a *Bouguer effect* onshore. **3.** a decreased thickness of the *crust* under the oceans and seas and large topographic depressions, where the *mantle* is shallower than in the surrounding regions. Because the upper mantle densities are higher than the densities of crustal rocks, the decreased crust thickness creates a lateral mass excess which, in accordance with the *Airy hypothesis* and *Vening Meinesz hypothesis*, maintains the isostatic balance, or *isostatic equilibrium*, between the crust and upper mantle. Rock mass excesses are sources of gravity anomalies at all scales: local, regional, and planetary. See also *gravity anomaly*, *isostatic antiroot*, *isostatic root*, *isostasy*, and *mass deficiency*.

mass-spring accelerometer: an *accelerometer* the design of which is based on the use of *proof mass* suspended by a mechanical (quartz or metal) spring. Proof mass displacement is directly proportional to *acceleration* in the direction of the accelerometer sensing axis. The acceleration value is commonly measured by a capacitive readout system. The majority of accelerometers used in gravity and magnetic exploration applications are spring-mass accelerometers. See also *electromagnetic accelerometer* and *electrostatic accelerometer*.

mass-spring gravimeter: a *spring gravimeter* with an emphasis on its design based on the use of *proof mass* suspended by a quartz or metal spring.

matched filter: a spectral-domain *filter* designed to match the spectra of anomalies with specific spectral characteristics. Matched filter parameters are selected from the straight-line segments, or slopes, identified in the logarithmic plot of the *radial power spectrum* of a gridded magnetic field or, sometimes, *vertical derivative* of a gridded gravity field. Each slope corresponds to either (a) a group of magnetic sources occurring around a particular *grid*-averaged depth or (b) a group of magnetic or gravity anomalies within a particular *wavenumber* range. Anomalies generated by sources corresponding to a selected slope will be enhanced relative to other components of an *anomalous field* in a manner of a *band-pass filter*. Also known as *spectrum-matched band-pass filter* and *depth-slice filter*. See also *spectral domain* and *Spector-Grant method*.

matched-filtering method: a *data processing* technique based on the use of a *matched filter*. It is applied to enhance the anomalies generated by separate groups, or assemblages, of sources within a depth interval determined from the linear segments, or slopes, of the *radial power spectrum* of gridded *total field* magnetic data or, occasionally, *vertical derivative* of the anomalous gravity field. Anomalies generated by sources within the depth interval corresponding to a particular slope will be enhanced relative to other components of an *anomalous field*. Due to the spectral overlap, the filtering output includes the contributions from sources above and below the slope-selected depth interval and, thereby, decreases the *resolution* of this method. Also referred to as *depth-ensemble filtering*, *depth slicing*, *pseudodepth slicing*, and *spectrum-matched filtering*. See also *Spector-Grant method* and *spectral leakage*.

matrix: a rectangular assembly of rows and columns of data values representing the related mathematical entities, such as the second-order derivatives of the gravity or magnetic *potential* in a *gradient tensor* matrix. A square symmetric nine-element (3×3) matrix is the standard *format* for the presentation of *full tensor gradient (FTG)* data. See also *derivative*, *FTG matrix*, *Hessian matrix*, and *diagonalized tensor matrix*.

maximally flat filter: same as *Butterworth filter* with an emphasis on an *amplitude response* that is nearly constant, or flat, within its *passband*.

maximum curvature: a quantitative characteristic, or *attribute*, of the gravitational *equipotential surface* defined by a *curvature* in the direction along which the equipotential surface curves the most. Maximum curvature K_{max} is expressed as

$$K_{max} = K_m + (K_m^2 - K_G)^{1/2}$$
$$= -[g_{xx} + g_{yy} - 2(g_{xy}^2 + g_{UV}^2)^{1/2}]/2g_z$$
$$\approx (\lambda_1 + \lambda_2) + (\lambda_1^2 + \lambda_1\lambda_2 + \lambda_2^2)^{1/2},$$

where K_m is *mean curvature*; K_G is *Gaussian curvature*; $g_{xx}, g_{yy}, g_{xy}, g_{UV}$ are *differential curvature gradients* with $g_{UV} = (g_{xx} - g_{yy})/2$; g_z is *vertical gravity*; and λ_1 and λ_2 are *Hessian eigenvalues*. In integration with other *curvature attributes*, the maximum curvature is used for the high-*resolution* mapping of the *gradient* and *gradient tensor* anomalies generated by structural discontinuities, such as faults, fracture and shear zones, edges of igneous bodies, and other lithological contacts. The direction of the maximum curvature is orthogonal to the direction of *minimum curvature* (Li, 2015). See also *contact*, *fault*, *igneous body*, and *shear zone*.

maximum-depth rules: See *Smith depth rules*.

maximum-slope distance: same as *straight-slope length* that represents the horizontal distance corresponding to the maximum slope of an *anomaly* and its *gradient* value in the *profile* direction.

maximum-slope line: a straight line drawn through the *inflection point* on an anomaly *profile*. It approximates the maximum *slope* of an *anomaly* and its *gradient* value in the profile direction. Also known as *straight-slope line*. See also *Peters half-slope method* and *straight-slope method*.

maximum-slope method: same as *straight-slope method* in the context of a *graphical source depth estimation* based on measurements of *straight-slope length*, also known as *maximum-slope distance*.

maxwell: a unit of *magnetic flux* in the cgs system of measurements. The analogous unit in the SI system is *weber*, where 1 weber $= 10^8$ maxwell. See also *cgs units* and *SI units*.

MCMC algorithm: the *Markov chain Monte Carlo (MCMC)* optimization algorithm implemented in *stochastic inversion*. See also *Markov chain*.

mean curvature: a quantitative characteristic, or *attribute*, of a gravitational *equipotential surface* defined by the average value of *maximum curvature* K_{max} and *minimum curvature* K_{min}. Mean curvature K_m is expressed as

$$K_m = (K_{max} + K_{min})/2$$
$$= -(g_{xx} + g_{yy})/2g \cong g_{zz}/2g_z \approx \lambda_1 + \lambda_2,$$

where g_{xx}, g_{yy} are *curvature gradients*; g_z is *vertical gravity*; g_{zz} is *first vertical derivative (FVD)*; and λ_1 and λ_2 are *Hessian eigenvalues*. In integration with other *curvature attributes*, the mean curvature is used for mapping of the gravity anomalies generated by localized *mass* distributions, such as high-*density* carbonate reefs, *placer* deposits, low-density local *salt* pillows, and reservoir-grade turbidites. Negative values of the mean curvature are observed above sources with a positive *density contrast*, and the positive values are above sources with a negative density contrast (Li, 2015). See also *apparent curvature attributes* and *reef*.

mean filter: same as *averaging filter* with an emphasis on its operating principle based on the replacement of each *grid cell* value with the average, or arithmetic mean, value of all grid cells in a filter *kernel*.

mean-of-least-variance (MLV) filter: a *grid*-based space-domain *edge-sharpening filter* which is the improved version of the *Kuwahara filter*. For a *default value* of 5×5 grid cells in a *moving window*, Kuwahara and MLV filters have the same subwindow size of 3×3 grid cells, but the MLV filter uses nine overlapping 3×3 subwindows instead of the four subwindows in the Kuwahara filter. The filter output is the average, or arithmetic mean, value of all grid cells in the subwindow having the smallest *variance*. See also *grid cell* and *symmetric nearest neighbor (SNN) filter*.

mean sea level (MSL): the level of the sea and ocean surfaces averaged for all stages of the water tides and currents. Undisturbed mean sea level (no tides, no currents, no waves) represents the *geoid* and *equipotential surface* for the *gravitational field* of the Earth in the oceans and seas. The MSL has been a standard *vertical datum* for gravity and magnetic measurements for many years. Presently, the use of a *reference ellipsoid* for the vertical datum is much more common due to the widespread application of the satellite-based *Global Positioning System (GPS)*. High-accuracy measurements of the MSL variations by the microwave-range radar or laser *altimeter* installed on Earth-orbiting satellites are used for the calculation of *satellite-derived gravity*. Usually referred to simply as sea level. See also *nominal sea level*.

measured gradient: a *gradient* of the Earth's magnetic or gravity field obtained from measurements by a *magnetic gradiometer* or *gravity gradiometer*. In theory, the *accuracy* and resolving power, i.e., *lateral resolution* and *vertical resolution*, of gradient anomalies detected by measured gradients are higher than those of a *calculated gradient* due to: (a) the much larger *sampling interval* of the line-oriented *total field* measurements and, especially, *grid interval* of the gridded data in comparison with the *baseline* of a regular *gradiometer*; (b) spatial averaging of the *gridding* process; and (c) superposition of the regional components in the *total field* data. In practice, the effectiveness of using the measured gradients strongly depends on the *survey line spacing* and suppression of *acquisition noise*: the better the suppression of *noise* and the closer the line spacing, the higher the accuracy and *resolution* of the measured gradient anomalies. See also *gridding* and *full tensor gradient (FTG)*.

measured gravity: *raw gravity* in the context of the original gravity values measured at observation points during a *gravity survey*, i.e., gravity values before application of corrections.

measurement bandwidth: 1. a measurement parameter that describes the actual range of spatial frequencies or wavelengths in the measured data. It is estimated quantitatively by calculating the difference between the highest *spatial frequency* (shortest *wavelength*) and the lowest frequency (longest wavelength) components that can be identified in the data after standard *data processing* and *data reduction*. **2.** *sampling bandwidth*. See also *bandwidth*.

mechanical drift: *drift* that originates in the mechanical design and thermoelastic properties of the materials used in the measuring systems of a *relative gravimeter* and certain types of magnetometers.

median: the value in a given dataset that divides the data values into two halves: one half of the values exceed the median value and the other half of the values are below the median value. Also referred to as *median threshold*. See also *averaging filter*, *median filter*, and *SNN median filter*.

median filter: a *profile-* or *grid*-based space-domain *smoothing filter* for which the output is the *median* value calculated within a *moving window* of a specified length. It is used as a *spike*-removing and high-frequency *noise*-suppressing filter. See also *SNN median filter* and *differential median filtering*.

median threshold: See *median*.

Meissner effect: the physical phenomenon of the expulsion of an ambient magnetic field from a superconducting material, when it is cooled to temperatures close to absolute zero (–273.15°C) and loses its resistance to the flow of the induced electric currents. Once the material has become a superconductor, an external magnetic field cannot penetrate it, and the superconducting body will be suspended in the magnetic field generated by electric currents. The Meissner effect is utilized in the designs of the *superconducting gravimeter* and *superconducting gradiometer*, where the superconducting *proof mass* (usually a small sphere) is maintained in a suspended position, or levitates, as a result of the interaction between the magnetic fields generated by the applied electric currents in the external coils and induced electric currents inside the superconducting sphere. See also *ambient field* and *magnetic levitation*.

merging: See *grid merging*.

mesh: 1. a net-like layout of regularly spaced square or rectangular cells of the same size, such as *grid* cells comprising the continuous surface of an *anomalous field* or its attributes. **2.** a network of regularly spaced variable-width square cells around each observation point that approximates the *Digital Terrain Model (DTM)* in the calculation of *terrain correction*. The cell width increases with the distance from each *observation point* in order to adapt to the *inverse power law* decrease in the gravity effect of the distant topographic or bathymetric *relief* variations and to speed up computations. **3.** the arrangement of voxels in a three-dimensional (3D) *voxel model*, wherein a coarse mesh has large voxels and fine mesh has the small voxels. See also *attribute* and *voxel*.

mesh discretization: dividing the *Digital Terrain Model (DTM)* in a *gravity survey* area into square prisms whose width increases with the increasing distance from observation points and follows a user-determined rule of zone-based or linear changes in the *prism* widths. Prism height is defined by the difference between the average *elevation* within each prism and the elevation at the observation point location. Gravity effects of all prisms are calculated using the standard formulas and summed to obtain the *terrain correction* value at each observation point. See also *Hayford-Bowie zones*, *Hammer zones*, and *adaptive quadtree discretization*.

mesh points: See *grid nodes*.

metadata: supplementary digital and text files that describe a particular set of gravity or magnetic data. Metadata files contain all necessary information about the data to ensure their proper use, exchange, storage, and archiving. They include, but are not limited to, data *format*, parameters of the data *acquisition* and *data processing*, *precision* of the measuring instruments, mapping projection, origin of the data, point of contact for the data inquires, etc. See also *ASCII format* and *map projection*.

MHODE method: the multihomogeneous depth estimation (MHODE) method based on the concept of *multihomogeneity* and application of *multiridge analysis*. Method involves (a) forming a multiscale dataset by multiple upward continuations of the *anomalous field* or its derivatives with small increments of 100–200 m; (b) *multiridge Euler deconvolution* in integration with *reduced Euler deconvolution* to obtain local estimates of the depth $z_0(h)$ and local homogeneity degree $n(h)$, where h is the *upward continuation* level called the *altitude* of the calculations; and (c) analysis of the graphical plots $n(h)$ and $z_0(h)$ as functions of the altitude, wherein values of both variables in the asymptotic regions of a very close and distant (relative to the source size) proximity to the source represent the estimates of the source type and depth to its top or center. Once the multihomogeneous source model is established, then a system of local-homogeneity equations is formed and solved for the improved depth estimation and unknown source parameters (Fedi et al., 2015). See also *multihomogeneous field* and *multiscale analysis*.

microcgs (μcgs): the dimensionless unit of *magnetic susceptibility* in the cgs system, where susceptibility values are measured in units of 10^{-6} called microcgs units, or microcgs. Susceptibility in *SI units* is 4π times its value in *cgs units*, so the relationship between these units is expressed as

$$1000 \cdot 10^{-6} \text{ cgs} = 1000\ \mu\text{cgs} \cong 12.56 \cdot 10^{-3} \text{ SI} = 0.01256 \text{ SI}.$$

microfabricated atomic magnetometer (MFAM): the ultra-light-weight, high-performance *atomic magnetometer* with dimensions less than 1 cubic inch (<15 cm^3) for high-*accuracy* measurements of the *Earth magnetic field*, its *gradient*, *full tensor gradient (FTG)*, and *vector* components. Its reported *sensitivity* is up to $1\text{–}2\text{pT}\sqrt{Hz}$ with a low power consumption of 1.0–2.0 watts. The MFAM is used as a single-*sensor* instrument or assembled in a *multisensor gradient array* and attached to a small low-flying rotorcraft, or *drone*, for regular magnetic and micromagnetic measurements, *UXO detection*, locating abandoned wells, perimeter security, and engineering and environmental applications (Veloski et al., 2018). See also *picotesla (pT)* and *drone survey*.

microgal (µGal): a unit of *microgravity* and high-accuracy *ground gravity* measurements that is equivalent to one billionth part of the *nominal gravity* value $g \cong 9.8\ \mathrm{m \cdot s^{-2}}$, i.e., 1 µGal = 10^{-6} *Gal* = 10^{-3} *milligal (mGal)* = $10^{-8}\ \mathrm{m \cdot s^{-2}} = 10^{-2}\ \mu\mathrm{m \cdot s^{-2}} \cong 10^{-9} g$. Microgal is the current level of achievable *accuracy* of gravity measurements with the use of a high-precision *absolute gravimeter* or *relative gravimeter*. See also *precision*.

microgravimetry: methods and instruments to acquire, process, and interpret measurements of gravity anomalies within the range of 2–100 *microgal (µGal)*. Anomalies of this *magnitude* are measured by *ground gravity* surveys to detect the variations of the *Earth gravity field* associated with (a) depletion of heavy oil reservoirs under development; (b) changes in fluid volume and fluid distribution during oil or gas reservoir development with the injection of steam, water, CO_2, or other agents of enhanced recovery; (c) CO_2 sequestration; (d) underground water horizon or geothermal reservoir development; (e) small stratigraphic traps; (f) subsurface voids and cavities, i.e., *karst*; and (g) certain types of mineral deposits. A microgravimetric, or *microgravity*, survey is conducted with a supplementary theodolite geodetic survey for the most accurate measurements of *elevation* at each observation point, or *station*, required for adequate data corrections. For a planned *accuracy* of 1 µGal, the station elevations must be measured with an accuracy of 2–3 mm. See also *aquifer gravity*, *CO_2 sequestration gravity*, *geothermal gravity*, *volcano gravity*, and *time-lapse gravity*.

microgravity: 1. *gravity data* at the *microgal (µGal)* order of amplitudes. **2.** *absolute gravity* and *relative gravity* data obtained from high-*accuracy* ground measurements of the spatial variations of the *Earth gravity field* for the detection of small gravity anomalies within the range of 2–100 microgals. Among other applications, the microgravity data are used as *constraints* in the 3D *property inversion*-based *density model* built for monitoring oil-and-gas reservoirs under a waterflood development, steam-enhanced heavy oil recovery, depletion of groundwater aquifers and geothermal reservoirs, CO_2 sequestration, and growth of active volcanos. Estimates of changes in the observed anomalies and subsurface fluid distributions over months and years of monitoring are made by *forward modeling* with the incorporation of production data from wells and by *inverse modeling* of time-lapse *hybrid gravity* data. See also *anomaly*, *aquifer*, *time-lapse gravity*, and *4D gravity monitoring*.

microleveling: 1. *grid*-based minimization of the residual *leveling errors*, i.e., errors remaining after the standard *leveling* of line-oriented gravity and magnetic data acquired by marine (shipboard) and airborne surveys. **2.** elimination of mis-ties in aeromagnetic survey data caused by differences between the actual values of *diurnals* at the flight line intersections and their values recorded by a *base-station magnetometer*. **3.** *decorrugation*. Microleveling techniques are based on the use of low- and high-pass filters. The same as the standard leveling but to a lesser degree, microleveling changes the local field gradients and shapes of anomalies in the gridded data and these changes can influence the subsequent *inversion* results. **4.** high-*accuracy* geodetic measurements of changes in the *elevation* of observation points of a *time-lapse gravity* survey due to the depletion of oil-and-gas or geothermal reservoirs and underground water horizons. See also *gridding*, *herringbone effect*, *high-pass filter*, *low-pass filter*, and *mis-tie*.

microleveling corrections: data correction values obtained by *microleveling* of the gridded magnetic or gravity data. They are applied to the original line-oriented or gridded data prior to their *regridding*. See also *gridding* and *leveling corrections*.

micromagnetic anomalies: low-*amplitude* short-*wavelength* magnetic anomalies of the sub-*nanotesla (nT)* range from shallow sources. Their origin is associated primarily with (a) near-surface conversion of low-*susceptibility* sedimentary minerals into *magnetite* by hydrothermal alterations, *hydrocarbon seepage effect*, and other chemical changes in the mineral composition of rocks and (b) mineralization of faults and fracture zones by *juvenile water*. Micromagnetic anomalies are detected by a *high-resolution aeromagnetic survey* and visualized by application of *enhanced processing* techniques, such as *high-pass filter*, *Jacobsen filter*, and *wavelet filtering*. They are often displayed in the format of *stacked profiles* to highlight their location and wavelength content. Detection of micromagnetic anomalies clustered over and around oil-and-gas prospective structural highs and probable stratigraphic traps is used for prioritizing *exploration leads* identified by the *integrated interpretation* of gravity, magnetic, and seismic data. Referred to with the same meaning as *high-frequency magnetics*. See also *diagenetic magnetite*, *regolith*, *SRM anomaly*, *magnetic bright spot*, *structural high*, and *trap*.

micromagnetic aureoles: *micromagnetic anomalies* of circular and elliptical shapes with diameters up to several kilometers. They tend to cluster around oil-and-gas prospective structural highs and appear to coincide with locations of discovered oil fields. Their origin is assumed to be related to the *hydrocarbon seepage effect* and magnetic alterations at the margins of relatively shallow hydrocarbon accumulations (Stone et al., 2004). See also *magnetic bright spot* and *SRM anomaly*.

micromagnetic seep effect: See *hydrocarbon seepage effect*.

micropulsations: time-varying short-period fluctuations in the *external magnetic field* of the Earth from approximately 0.1 to 500–600 seconds caused by interactions between charged particles in the *magnetosphere* and *solar wind*. Continuous micropulsations are represented by sequences of the wave-like variations with periods from a few seconds to several minutes and rapid changes in *phase* and *amplitude* over distances comparable to the *magnetic survey* dimensions. Their amplitudes can increase with the increasing periods and range from a fraction of *nanotesla (nT)* at the mid-latitudes to several hundred nanoteslas in the auroral zones (±60°–80° latitude) and equatorial zone (±5° latitude). Irregular micropulsations are primarily night time, or nocturnal, disturbances with periods within 40–100 seconds and amplitudes up to several nanoteslas. Micropulsations are recorded by a *base-station magnetometer* and incorporated into *diurnal correction*. Also referred to as *magnetic pulsations* and *electromagnetic pulsations*. See also *bay*, *diurnals*, *auroral electrojets*, and *equatorial electrojets*.

microseismic noise: ground microvibrations caused by distant earthquakes, tree roots shifts under strong winds, railway and highway traffic, ocean storm waves, and other natural and industrial effects that result in errors in *ground gravity* measurements. See also *noise*.

milliEötvös (mE): a unit of high-*precision* measurements of the *full tensor gradient (FTG)* elements of the *Earth gravity field*, whereby 1 mE = 10^{-3} *Eötvös (E)*.

milligal (mGal): a standard unit of measurements in *gravity exploration*. It is equivalent to one millionth part of the average, or nominal, *Earth gravity field* value $g \cong 9.8\,\mathrm{m \cdot s^{-2}}$, i.e., 1 mGal = 10^{-3} *Gal* = $10^{-3}\,\mathrm{cm \cdot s^{-2}} = 10^{3}$ *microgal (μGal)* = $10^{-5}\,\mathrm{m \cdot s^{-2}} = 10\,\mathrm{\mu m \cdot s^{-2}} \cong 10^{-6} g$. See also *nominal gravity*.

milligal constant: same as *calibration factor* utilized as a constant for the conversion of *relative gravimeter* readings from the counter, or dial, units into the standard units of *milligal (mGal)*.

mineral nodule: a small, irregularly rounded solid mass, or concretionary lump, of a mineral aggregate which often contains manganese, cobalt, iron, nickel, and metals from the group of *rare earth elements (REE)*. Mineral nodules are found on the deep sea and ocean floors in proximity to the high-temperature thermal vents. Commercial accumulations of mineral nodules generate the detectable gravity and magnetic anomalies that can be measured and mapped by a high-resolution *submarine survey*. See also *anomaly*, *magnetic spherule*, *resolution*, *submarine gravity*, and *submarine magnetics*.

minimum curvature: 1. a quantitative characteristic, or *attribute*, of the gravitational *equipotential surface* defined by the *curvature* in the direction along which the equipotential surface curves the least. Minimum curvature K_{min} is expressed as

$$\begin{aligned} K_{min} &= K_m - (K_m^2 - K_G)^{1/2} \\ &= [g_{xx} + g_{yy} + 2(g_{xy}^2 + g_{UV}^2)^{1/2}]2g_z, \end{aligned}$$

where K_m is *mean curvature*; K_G is *Gaussian curvature*; g_z is *vertical gravity*; g_{xx}, g_{yy}, g_{xy}, g_{UV} are *differential curvature gradients*; and $g_{UV} = (g_{xx} - g_{yy})/2$. The direction of the minimum curvature is orthogonal to the direction of *maximum curvature*. The azimuthal direction of minimum curvature is defined as

$$\theta_{\min} = 0.5 \arctan[2g_{xy}/(g_{xx} - g_{yy})].$$

In integration with other *curvature attributes*, minimum curvature is used for the high-*resolution* mapping of the *gradient* anomalies generated by structural discontinuities, such as faults, fracture and shear zones, and lithological contacts (Li, 2015). **2.** a standard algorithm for *gridding* line-oriented gravity and magnetic data referred to as *minimum-curvature gridding*. See also *azimuth*, *contact*, *fault*, and *shear zone*.

minimum-curvature gridding: a method of *gridding* based on the iterative minimization of the square of the sum of the second-order horizontal derivatives which approximate the *curvature* of the continuous surface of a *grid*. The initial iteration is completed with the large grid cells, and all subsequent iterations use the progressively smaller grid cells. The sequence of iterations and final constructed surface of the grid are constrained by the original data values. See also *grid cell*, *horizontal derivative*, and *gradient-enhanced minimum-curvature*.

minimum-entropy method: a semi-automated method of estimating a gravity or magnetic source depth that is based on (a) *downward continuation* of the *anomalous field* to several levels above and below a probable source depth, (b) calculation of the normalized *analytic signal amplitude (ASA)* at each depth level within the square *window* centered at every anomaly of interest, and (c) calculation of the entropy at each level using the analytical expression of the entropy as a measure of randomness in the distribution of energy of the normalized ASA. The source depth corresponds to a minimum of the computed entropy-versus-depth curve (Florio and Fedi, 2018). See also *normalized derivatives*.

misclosure: *closure error* in the context of a measurement error accumulated in the gravity or magnetic observations along a single *loop* during a *ground survey*.

misfit: 1. a difference between the observed, or actual, values of an *anomalous field* or isolated *anomaly* and the values calculated from the constructed *model*, i.e., the difference between the observed and *inversion*-obtained values that is commonly referred to as *model misfit*. **2.** the difference between the *basement* depth interpreted from magnetic and gravity data and the actual basement depth penetrated by a borehole. See also *mis-tie*.

misleveling: residual *leveling errors*, i.e., errors remaining after *leveling* and *microleveling* of marine (shipboard) and airborne gravity and magnetic data. Misleveling is a source of low-*amplitude* artificial anomalies in *gridding* and errors in three-dimensional (3D) *inversion*, especially for deep exploration targets. See also *artificial anomaly* and *target*.

mis-tie: a difference between the measured values of the gravity or magnetic field at the same coordinate points, where *traverse lines* and *tie lines* intersect each other, i.e., difference between values measured at the crossover points. Flight height differences and *diurnals* are major sources of mis-ties in airborne surveys. Mis-ties are minimized by *leveling* and their remaining components, or residuals, are essentially eliminated by *microlevelling*. Referred to with the same meaning as *crossover error* and *crossover difference*. See also *crossover point* and *misfit*.

mks units: units in the meter-kilogram-second (mks) system of measurements, where distance is measured in meters, mass in kilograms, and time in seconds. These have been replaced by the nearly identical *SI units*. See also *cgs units*.

model: 1. an approximation of a geological *structure* or isolated *source body* that is geologically plausible, consistent with the available *constraints*, and generates the gravity or magnetic field matching the observed field within a specified range of errors called the *model tolerance*. **2.** a working hypothesis, or concept, used to

(a) describe an exploration *target* in terms of a set of parameters and (b) develop a better understanding of the observed gravity and magnetic fields by comparison with their concept-based simulations. **3.** the representation of an exploration target by a single body or multiple bodies of simple geometrical forms and their combinations. Local- and regional-scale gravity and magnetic models are built with the use of *two-dimensional (2D)* seismic depth sections, sedimentary horizons derived from *three-dimensional (3D)* seismic data, and available formation tops from boreholes. In the absence of seismic data, the geological structure is approximated by the individual source bodies of various shapes that constitute the model. Selection of the model parameters and components is based on: (a) available geological and geophysical information about the targeted geological structure and its elements and (b) the use of simple geometrical forms, such as *prism*, *sphere*, *horizontal cylinder*, *vertical cylinder*, *thick dike*, *thin dike*, *horizontal sheet*, and their combinations. There are four basic types of gravity and magnetic models: three *profile*-based models known as the *2D model*, *2.5D model*, and *2.75D model*, and one *grid*-based model referred to as the *3D model*. See also *full-Earth model* and *voxel model*.

model depth: the vertical extent of a *model* selected in accordance with the *modeling* objectives. Usually, it includes the *basement*. Where the *Moho boundary* is assumed to be nearly flat over the *model length*, the model depth is limited by the upper continental *crust* at 10–20 km below the estimated basement top. For *modeling* of regional geological structures and deep crustal anomaly sources over distances of several hundred kilometers, the model depth is extended to the Moho boundary and sub-Moho depths, and it is referred to as a *full-Earth model*.

model length: the lateral extent of a *model* selected in accordance with the structural objectives of *modeling*. As an empirical rule for gravity modeling, the minimization of edge effects requires the model length to be 6–8 times the estimated maximum depth of the exploration *target* for modeling with the *density contrast* of rocks and 10–12 times the maximum depth with the full *density* values of rocks. For modeling of the *Moho boundary*, the gravity model length is commonly several hundred kilometers. For *joint modeling*, the model length is defined by the lengths of the available seismic sections or selected in accordance with the general rules for the gravity models that require the model length to be, at least, 5–6 times the *model depth*.

model misfit: a difference between the calculated field of a constructed *model* and the observed (anomalous) gravity or magnetic field. Model misfit is iteratively reduced by the trial-and-error adjustments of the initial model parameters until all point-by-point misfit values fall within a specified range of errors called the *model tolerance*. Also referred to as *model mismatch* and *model residual*. See also *anomalous field*, *L1 norm*, *L2 norm*, and *observed field*.

model mismatch: See *model misfit*.

model residual: *model misfit* in the context of the remaining, or residual, difference between the *observed field* and the field generated by the constructed *model*.

model response: the gravity or magnetic field generated by a *model* which is constructed with the assumed geometry (depth and shape) and physical properties (*density* or *susceptibility*) of the targeted geological structure or isolated *source body*. Model response is calculated after every adjustment of the initial model parameters and compared to the *observed field* in the process of *modeling* in order to reduce the *model misfit* to the limits of *model tolerance*. Also referred to as *forward response*.

model strike: the orientation of a *profile*-based gravity or magnetic *model* with respect to the geographic north. See also *2D model*, *2.5D model*, and *2.75D model*.

model tolerance: a user-specified range, or margin, of errors within which the field generated by the constructed *model*, or *model response*, matches the observed gravity or magnetic field. See also *model misfit* and *stopping criterion*.

modeling: **1.** a quantitative *interpretation* technique based on the construction of a *model* with an assumed distribution of *density* and *susceptibility* of rocks, where the entire geological structure or its isolated elements are approximated by assemblages of layered and block-faulted sedimentary and *basement* rock formations or by *ideal source* geometries in accordance with the available geological and geophysical information, such as seismic and borehole data. **2.** the construction of a structural model which is consistent with the available *constraints*, is geologically plausible, and generates the gravity or magnetic field matching the *observed field* values within a user-specified margin of error called the *model tolerance*. The degree of matching between the calculated and observed horizontal derivatives along the model *profile* is used as the additional acceptance criterion for the constructed model. See also *Agerlin principle*, *horizontal derivative*, *forward modeling*, and *inverse modeling*.

modified-phase eigenvalue: an invariant-derived quantitative characteristic, or *attribute*, of an anomalous gravity or magnetic field, defined as

$$\lambda_{\text{ShI}} = \text{ShI} \cdot 2\sqrt{Q},$$

where ShI is *shape index (ShI)*, $2\sqrt{Q}$ is *eigenvalue magnitude*, $Q = -I_1/3$, and I_1 is one of the *principal invariants* of the *FTG matrix* constrained by the condition $I_1 \leq 0$. A modified-phase eigenvalue *grid* highlights the anomalies generated by sources with low-*density* and low-*susceptibility* contrasts and improves the *sharpness* of anomalies above the *source body* edges (Cevallos, 2017). See also *anomalous field*, *contrast*, *edge*, *eigenvalues*, *eigenvalue phase*, *invariants*, and *eigenvalue wheel*.

modified second vertical derivative: the spectral-domain *separation filtering* algorithm which combines a *Gaussian filter* and the *second vertical derivative (2VD)* of a anomalous gravity or magnetic field in the process of *cascaded filtering*. The Gaussian filter attenuates the highest wavenumbers near the *Nyquist frequency* to minimize *edge effects* associated with *Fourier transform*, and the 2VD separates the shallow part of the data *spectrum* from its intermediate and deep parts in order to enhance the short-*wavelength* anomalous components. See also *anomalous field*, *spectral domain*, and *wavenumber*.

modulus: **1.** the absolute value of the *amplitude* of a positive or negative *anomaly* in an observed gravity or magnetic field, also referred to as *total field modulus*. **2.** the *magnitude* |M| of a *complex variable*, defined by the square root of the sum of its squared real and imaginary parts as

$$|M| = (\mathrm{ReP}^2 + \mathrm{ImP}^2)^{1/2},$$

where ReP and ImP are real and imaginary parts of the complex variable. In *vector* space, it represents the length of a complex variable vector. The modulus of the *analytic signal (AS)* is known as the *analytic signal amplitude (ASA)*. As a measure of the amplitude content, or energy, of the complex variable, the modulus is equivalent to *local energy*. See also *observed field*.

Moho: See *Mohorovičić discontinuity*.

Moho boundary: **1.** *Mohorovičić discontinuity*. **2.** the boundary between the lowest part of the *crust* and uppermost *mantle* separating rock formations in different metamorphic states with different seismic and *density* characteristics. It is a basic component in the *Airy-Heiskanen model* and the *Vening Meinesz model* of *isostasy*.

Mohorovičić discontinuity: the seismic discontinuity between the Earth's *crust* and upper *mantle* that corresponds to the abrupt changes in the compressional P-wave velocity from 6.5–7.2 to 7.8–8.5 km · s^{-1} and shear S-wave velocity from 3.7–3.8 to 4.8 km · s^{-1}. Its globally approximated depth is predicted by the theory of *isostasy* with variations from 5–8 km below the ocean floors to 25–40 km under the low-*elevation* areas on the continents and 60–75 km below the high mountain ranges. Based on the predicted *density* of 3.30–3.45 g · cm^{-3} for the uppermost mantle and 2.95–3.05 g · cm^{-3} for the lowermost crust, the *density contrast* across this discontinuity is 0.25–0.50 g · cm^{-3}. Its regional-scale depth estimates are based on the seismic reflection and refraction measurements and calculations of the *isostatic root depth* and *isostatic antiroot depth* in accordance with the *Airy-Heiskanen model* and the *Vening Meinesz model* of isostasy, except for the mid-ocean ridge areas, where the *Pratt-Hayford model* is preferred. Commonly referred to as the *Moho boundary* or simply *Moho*.

moment: See *magnetic moment*, *magnetic dipole moment*, *vector magnetic moment*, and *integral magnetic moments*.

monitoring gravity survey: a *time-lapse gravity survey* in the context of monitoring the local changes in the *Earth gravity field* caused by the development of oil-and-gas and geothermal reservoirs, depletion of underground water horizons, and growth of active volcanos.

monogenic instantaneous amplitude: See *local amplitude* of *monogenic signal (MS)*.

monogenic instantaneous orientation: See *local orientation* of *monogenic signal (MS)*.

monogenic instantaneous phase: See *local phase* of *monogenic signal (MS)*.

monogenic signal (MS): **1.** representation of a *scalar* anomalous gravity or magnetic field by means of a three-component *vector* derived from the *Riesz transform*, which augments the *anomalous field* by the addition of two mutually orthogonal field components obtained by two 90° phase shifts at each *grid* point. **2.** a *complex variable* composed of three parts: one real part and two imaginary parts obtained by Riesz transforms of the real part. The monogenic signal at grid point (x, y) is expressed as

$$\mathrm{MS}(x,y) = f(x,y) + [i\mathrm{R}_{1f}(x,y) + j\mathrm{R}_2 f(x,y)],$$

where $f(x, y)$ is the real part defined by *RTP anomaly* or *Bouguer anomaly* value, $i = j = \sqrt{-1}$ are complex unit numbers indicating the orthogonality, and $R_1 f(x, y)$ and $R_2 f(x, y)$ are imaginary, or 90° *phase*-shifted, parts obtained by the Riesz transform of $f(x, y)$ in the mutually orthogonal complex planes (Felsberg and Sommer, 2001). Because the Riesz transform operators R_1 and R_2 are equivalent to H_1 and H_2 operators of the *3D Hilbert transform*, the monogenic signal is equivalent to the *direct analytic signal (DAS)*. At each grid point, the monogenic signal is decomposed into three vector components, or primary attributes, representing its *amplitude*, orientation, and *phase* and referred to as *local amplitude*, *local orientation*, *local phase*, respectively, and two scalar components, or secondary attributes, known as *Riesz transform attribute* and *directional Hilbert attribute*. These five calculated attributes are included into the integrated analysis of the anomalous field and *structural interpretation*. For magnetic data, *reduction-to-pole (RTP)* is required prior to calculation of the monogenic signal attributes. Also referred to as *complex monogenic signal*. See also *attribute*, *monogenic signal decomposition*, *scale-space monogenic signal*, and *zero-order ASA (ASA_0)*.

monogenic signal amplitude: a quantitative characteristic, or *attribute*, of a *monogenic signal (MS)* commonly referred to as the *local amplitude*.

monogenic signal attributes: quantitative characteristics, or attributes, of the anomalous gravity or magnetic field obtained by *monogenic signal decomposition*. They include but are not limited to the *local amplitude*, *local orientation*, *local phase*, *Riesz transform attribute*, and *directional Hilbert attribute*. See also *anomalous field*, *attribute*, and *complex attribute*.

monogenic signal decomposition: calculation of the primary and secondary complex attributes of a *monogenic signal (MS)* from its real and imaginary parts. The three primary complex attributes are the *local amplitude*, *local orientation*, and *local phase*. The secondary complex attributes include but are not limited to the *Riesz transform attribute* and *directional Hilbert attribute*. Decomposition of a monogenic signal into its complex attributes is performed with the following formulas:

1. local amplitude: $\mathrm{A}(x,y) = [|f(x,y)|^2 + |R_1 f(x,y)|^2 + |R_2 f(x,y)|^2]^{1/2}$;
2. local phase: $\Phi(x,y) = \arctan[(|R_1 f(x,y)|^2 + |R_2 f(x,y)|^2)^{1/2} / f(x,y)]$;
3. local orientation: $\theta = \arctan[R_2 f(x,y) / R_1 f(x,y)]$;
4. Riesz transform attribute: $Q_R = (|R_1 f(x,y)|^2 + |R_2 f(x,y)|^2)^{1/2}$;
5. directional Hilbert attribute: $\mathrm{H}_\theta = R_1 f(x,y)\cos\theta + R_2 f(x,y)\sin\theta$;

where $f(x,y)$ is the real part defined by the *RTP anomaly* or *Bouguer anomaly* value at *grid* point (x,y) and

$R_1f(x,y)$ and $R_2f(x,y)$ are imaginary, or 90° *phase*-shifted, parts obtained by *Riesz transform* of $f(x,y)$ in the mutually orthogonal complex planes (Felsberg and Sommer, 2001). Grids of the calculated complex attributes are used for the high-*resolution* mapping of structural discontinuities, such as faults, fracture and shear zones, lithological boundaries, and *igneous body* edges (Hassan and Yalamanchili, 2013). See also *complex attribute*, *complex plane*, *fault*, *grid*, *lateral resolution*, and *shear zone*.

monogenic signal orientation: a quantitative characteristic, or *attribute*, of the *monogenic signal (MS)* commonly referred to as the *local orientation*.

monogenic signal phase: a quantitative characteristic, or *attribute*, of the *monogenic signal (MS)* commonly referred to as the *local phase*.

monopolar anomaly: a bell-shaped symmetrical *gravity anomaly* generated by a localized gravity *source body* which is approximated by a *point source* or *sphere*. Its *amplitude* depends on the lateral *density contrast* created by the source body and its depth: the higher the contrast, the higher the amplitude, and the deeper the source, the lower the amplitude. A positive or negative sign of the anomaly is defined by the corresponding density contrast. Attenuation, or *fall-off*, of a monopolar anomaly is one power less than that of a *dipolar anomaly* generated by a *magnetic dipole*: $1/r^2$ versus $1/r^3$, where r is distance from the source. All gravity anomalies are *source geometry*-related versions of a monopolar anomaly. See also *anomaly shape (grav.)*.

monopole: a mathematical approximation of the dipolar magnetic source by a vertical column of magnetic dipoles that has *infinite depth*. Its magnetic field satisfies the *Euler homogeneity equation* with *Euler structural index* SI = 2. Referred to with the same meaning as *vertical cylinder*. See also *monopolar anomaly*, *dipolar anomaly*, *dipolar source*, and *magnetic dipole*.

montage: a color-coded computer display or printed copy showing the side-by-side *grid* images obtained by the application of different *enhanced processing* techniques, such as *filtering*, calculation of gradients, derivatives, and other attributes of anomalous gravity and magnetic fields, over the same area in order to compare and demonstrate their effectiveness in the detection of anomalies of the exploration interest. See also *anomalous field*, *attribute*, *derivative*, *gradient*, and *color-coded grid*.

mother wavelet: a basic, or prototype, *wavelet* whose stretched and compressed versions are elementary components in the *wavelet transform* of gravity and magnetic data. Unlike sinusoids in a *Fourier transform*, mother wavelets can be of many different shapes, and the choice of a particular mother wavelet depends, primarily, on its mathematical properties. Also referred to as *wavelet function*.

moving-base gravity gradiometer: See *rotating-disk gravity gradiometer*.

moving-platform corrections: corrections applied to the measured values to compensate for the non-geological effects interfering with the *relative gravity* and magnetic measurements on a moving *observation platform*, such as a *survey* ship, submarine, or aircraft. They include (a) *cross-coupling corrections*, (b) *Eötvös correction*, (c) *Harrison correction* or *Browne correction*, (d) *level arm correction*, (e) *vertical acceleration correction* for gravity data, and (f) *heading correction* and *lag correction* for aeromagnetic data. See also *aeromagnetic compensation*.

moving window: a line-based or *grid*-based calculation *window* of a specified length or size that is moved with a pre-selected distance increment (commonly multiples of the *sample interval* or *grid interval*) per each cycle of calculations along a survey line or *profile* and over a grid. Moving windows are used in various *data processing* and *inversion* techniques, such as *convolution filter*, *smoothing filter*, *Euler deconvolution*, *Werner deconvolution* and others. Also referred to as *sliding window* and *running window*.

multicomponent gravity: same as *FTG gravity* with an emphasis on its components representing the spatial derivatives of the *Earth gravity field* along the three mutually orthogonal *x*-, *y*-, *z*-directions in *Cartesian coordinates*.

multihomogeneity: **1.** a theoretical concept of the spatially variable homogeneity of the gravity or magnetic field generated by *source body*, where the homogeneity degree *n* is a function of the distance to the source and source distribution type (*point source*, 2D, 3D), so that *n* values can be either integer or fractional. For example, the magnetic field of a finite circular disk at a *profile* crossing its center will be characterized by a homogeneity degree value close to $n = -1$ (infinite *thin sheet*) at the shortest vertical distances relative to the disk size, $n = -2$ (extended *horizontal cylinder*) at the distances comparable to the disk size, the asymptotic value of $n = -3$ (*point source*) at relatively large vertical distances from the disk, and fractional *n* values in-between these three integer values. The distance-dependent, or multiscale, behavior of *n* and, therefore, *Euler structural index (SI)* and related source depth estimates, are studied and utilized by the *MHODE method* for improved estimations of source parameters (Fedi et al., 2015). **2.** a fundamental principle, or postulate, in the application of *Euler deconvolution* methods. It states that the non-homogeneous, or inhomogeneous, *anomalous field* in a *survey* area can be approximated within a *moving window* by the *homogeneous function* of a specific homogeneity degree *n* at each calculation position along a *profile* or over a *grid* and, therefore, it can be considered as a *field* of area-variable homogeneity, or *multihomogeneous field*, for the purposes of source parameter calculations. See also *Euler homogeneity equation*, *homogeneous field*, *multiscale analysis*, and *infinite length*.

multihomogeneous field: **1.** the approximation of a non-homogeneous *anomalous field* in application of *Euler deconvolution* for the calculation of source parameters: spatial coordinates x_0, y_0, z_0 and *Euler structural index (SI)*, where $SI = -n$ and *n* is the field homogeneity degree. It is based on the concept of *multihomogeneity* postulating that a *homogeneous field* can approximate locally a set of data points within a *moving window* at each calculation position along a *profile* or over a *grid*, wherein the best approximating homogeneity degree *n* value (integer or fractional) depends on the source depth z_0 and *source body* type: *3D source* (finite in all three dimensions), *2D source* (finite in two of three dimensions), or *1D source* (finite in one *dimension*). **2.** a local approximation of the anomalous field in the *MHODE method* which assumes that a degree of the field homogeneity n varies not only at each moving window position but also at different levels of *upward continuation* of the field, wherein the n values are generally fractional between the integer values corresponding to the asymptotic regions of a very close and distant (relative to the source size) proximity to the source. See also *homogeneous function*, *fractional structural index (fSI)*, and *infinite length*.

multimodal inversion: *inversion* for the *structure* and properties of an exploration *target* with the use of several datasets which represent measurements of the complementary physical properties of rocks, such as *density* and *magnetization*, or obtained by different *acquisition* techniques, such as measurements of the *total field*, *gradient*, *full tensor gradient (FTG)*, and *vector* components of the planetary gravity and magnetic fields. Often referred to with the same meaning as *joint inversion*. See also *integrated imaging* and *5C inversion*.

multimodel method: a *profile*-based automatic method of estimating the depth to magnetic sources approximated by five *two-dimensional (2D)* source geometries: dipping *magnetic contact*, dipping *thin sheet*, dipping *thick sheet*, sloping *finite step*, and *horizontal cylinder*. Source depth is obtained by least-squares matching the curves of the *multimodel wavenumber* and the characteristic bell-shaped function defined as

$$f(x) = \{H/[H^2 + (x - X)^2]\} + B,$$

where $H=h$ is depth to the top of a dipping contact, thin sheet, thick sheet, or center of a horizontal cylinder, and $X=x_0$ is the source location. For a sloping finite step,

$$H = h + t/2 \quad \text{and} \quad X = x_0 + t \cdot \operatorname{cotan}\delta/2,$$

where t is thickness of the sloping step and δ is its *dip* angle with the vertical; B is a base-level shift of the multimodel wavenumber curves that accounts for the interference with the distant sources, and it is zero for isolated anomalies. This method involves two main procedures: (1) calculation of the multimodel wavenumbers k_a and k_b and construction of their respective curves and (2) curve matching in the least-squares sense. Once both or either of the multimodel wavenumbers are calculated for each data point, then the fitting algorithm selects the most appropriate *source geometry* and values of H and x_0 for the best matching of the characteristic bell-shaped function to a particular multimodel wavenumber curve. Curve matching is performed with the use of a pre-selected calculation window centered at each data point. The window length defines the *lateral resolution* of the multimodel method, and it is selected to be more than one half of the approximate source depth obtained from the *Spector-Grant method* (Thurston et al., 2002). See also *least-squares fit* and *wavenumber*.

multimodel wavenumber: a *complex attribute* derived from the *analytic signal (AS)* and incorporated into algorithms of the automatic estimates of the magnetic source location and depth in the *grid*-based *iSPI™ method* and *profile*-based curve-matching *multimodel method*. There are two multimodel wavenumbers: k_a and k_b. Wavenumber k_a is defined as

$$k_a = k_2 - k_1,$$

where k_1 is the *first-order local wavenumber*, k_2 is the *second-order local wavenumber*, and its values represent a symmetric bell-shaped function for the magnetic sources approximated by three *two-dimensional (2D)* geometries of the dipping *magnetic contact*, dipping *thin sheet*, and *horizontal cylinder* with the maximum values positioned immediately above these sources. In the iSPI™ method, the depth to the top of a dipping contact, thin sheet, and center of a horizontal cylinder is obtained from the reciprocal, or inverse, of k_a at the location of its maximum value. Wavenumber k_b is defined as

$$k_b = 2k_1 - k_2,$$

and its values represent the symmetric bell-shaped function for five 2D source geometries: dipping *contact*, dipping *thin dike*, dipping *thick dike*, sloping *finite step*, and horizontal cylinder. See also *wavenumber*.

multiple-point source: a *source body* whose position is defined by two or more reference locations (x, y, or z) in *Cartesian coordinates*. For example, *finite contact* forms the two-point source body whose anomaly is effectively the difference between the anomalies generated by two infinite contacts. See also *infinite contact* and *single-point source*.

multiple-source Euler deconvolution: generalization of the *grid*-based *3D Euler deconvolution* (which assumes a single source in the calculation *window*) for the case of multiple sources with the same *Euler structural index (SI)*. User-specified variables are the *moving window* length, number of sources within each window, and their SI. First-order approximation of the *source body* boundaries and, hence, the required window length is defined by the maxima of one of the calculated and gridded *edge detectors* (Hansen and Suciu, 2002).

multiple-source Werner deconvolution: an extension of the *Werner deconvolution* algorithm to compute simultaneously the contributions of multiple source bodies, including those whose anomalies interfere with each other, that are approximated by *magnetic contact* source geometries. It is based on the use of the *analytic signal (AS)* and the representation of the *analytic signal amplitude (ASA)* of a two-dimensional (2D) polygonal *source body* of an arbitrary shape by the sum of the analytic signal amplitudes of the n magnetic contacts. For a single magnetic contact, two complex constants a_0 and b_0 can be found to satisfy the following linear equation for the Werner deconvolution of the ASA:

$$a_0 + b_0 A_1(x) + x A_1(x) = 0,$$

where $A_1(x)$ is ASA of the magnetic contact whose *vertex* is at the spatial position (x_1,z_1) with *strike* orthogonal to the ASA *profile*, and its value is expressed as

$$A_1(x) = \alpha_1/(x - x_1 - iz_1),$$

where $i = \sqrt{-1}$ is a complex unit number indicating the orthogonality and α_1 is a coefficient depending on the *susceptibility contrast*, direction and magnitude of the *main geomagnetic field*, and contact *dip*. The algorithm finds the least-squares solution for the overdetermined system of linear equations representing the multiple magnetic contact contributions required to fit a given ASA anomaly, wherein the obtained coefficients determine a complex polynomial whose roots define the locations and depths of the source body contacts. Calculations are run within a user-specified *moving window* along the profiles extracted from an ASA *grid* (Hansen and Simmonds, 1993). See also *contact*, *least-squares fit*, *source geometry*, and *3D Werner deconvolution*.

multiresolution analysis: *multiscale analysis* or *wavelet analysis* in the context of the analysis of anomalous gravity and magnetic fields at different levels, or scales, of *lateral resolution* of their *wavelength* content.

multiresolution decomposition: *discrete wavelet transform (DWT)* in the context of a decomposition of anomalous gravity and magnetic fields by *half-band filtering* into the successive levels, or scales, of *wavelength resolution* and using the *convolution* of the iteratively dilated and translated *mother wavelet* with the scaled data to determine the spatial locations of the individual *wavelength* components.

multiridge analysis: a method of the analysis of anomalous gravity and magnetic fields that is based on the use of *Euler deconvolution* and additional subsets of the data along the so-called ridges, which are built by joining the extreme points of the analyzed field recomputed with the small increments Δh at different levels of *upward continuation* that are referred to as altitudes. For estimates of the *homogeneous source* parameters, three types of ridges are built by joining the zeros of the first-order *horizontal derivative* and *vertical derivative* (ridge types I and II, respectively) and zeros of the *anomalous field* (type III). A multiridge subset is presented in the *format* of a *two-dimensional (2D)* multi-level vertical section, where the number of ridges at each level is defined either by three different types of ridges built for the field and its first-order derivatives or by the order of the calculated derivatives for a particular *ridge* type. An overdetermined system of *reduced Euler deconvolution* equations for every ridge type is solved within a one-dimensional (1D) *moving window* for (a) source depth z_0 and *Euler structural index (SI)* for the type I ridges, (b) horizontal position x_0 and SI for type II, and (c) x_0 and z_0 for type III (Fedi et al., 2009). See also *multiridge analytic method*, *multiridge geometric method*, *multiridge Euler deconvolution*, and *MHODE method*.

multiridge analytic method: a method of estimating the depth and *Euler structural index (SI)* of the gravity and magnetic sources approximated by the *ideal source* geometries of a *point source*, line of point sources, *thin sheet*, and *infinite contact*. It is implemented in *multiridge analysis* and based on solving an overdetermined system of the *reduced Euler deconvolution* equations at each position of a one-dimensional (1D) *moving window* along the corresponding *ridge* type. One subset of the reduced Euler deconvolution equations provides the solutions for three source parameters: source depth z_0 and SI for the type I ridges, horizontal position x_0 and SI for type II, and x_0 and z_0 for type III (Fedi et al., 2009). See also *multiridge geometric method* and *MHODE method*.

multiridge Euler deconvolution: an upgrade of the single-*ridge* calculations of two *homogeneous source* parameters in *multiridge analysis* to the simultaneous calculations of all three parameters (depth, *Euler structural index*, horizontal position) by using the data from several ridges of different types which correspond to the same source. Data along all selected ridges are sorted by their *altitude*, and calculations are repeated for the different altitudes and different *moving window* lengths (each defined by the amount of ridge points) to obtain solutions for each source parameter. The method can be applied to *analytic signal amplitude (ASA)* and *gradient tensor* or *Hilbert transform* components (Florio and Fedi, 2013). See also *reduced Euler deconvolution*.

multiridge geometric method: a graphical method of estimating the depth to a *source body* approximated by the *ideal source* geometries of a *point source*, line of point sources, *thin sheet*, and *infinite contact*. It is implemented in *multiridge analysis* and based on the use of the so-called ridges built by joining the zeros of the horizontal and vertical derivatives of the *anomalous field* and the field itself, wherein the anomalous field and its derivatives are calculated with the small increments Δh at different levels of *upward continuation* referred to as altitudes. For a *homogeneous field*, all ridges are straight lines which intersect, or converge, at the source depth position regardless of its *Euler structural index (SI)* and the *magnetization* direction of the magnetic sources. For the interfering anomalies, especially at high altitudes, the computed ridges are curved and cannot be interpreted (Fedi et al., 2009). See also *horizontal derivative*, *vertical derivative*, *multiridge analytic method*, and *MHODE method*.

multiscale analysis: **1.** *grid*-based analysis of the anomalous components of the gravity and magnetic fields at different scales of *lateral resolution* in order to accommodate the multiscale characteristics of the data. The larger scales emphasize the anomalies from deeper sources with a lower lateral resolution, whereas the smaller scales emphasize the anomalies from shallower sources with a higher lateral resolution. Analysis involves calculation of the horizontal and vertical derivatives of different orders at several levels of *upward continuation*, as the recomputing of the data at a higher *elevation* than their original level is analogous to increasing the *scale* at which the field is analyzed. The low-pass character of the upward continuation and high-pass properties of the Nth-order derivatives can be combined in a manner of a *wavelength filter*. **2.** *wavelet analysis* of line-oriented and gridded gravity or magnetic data. Also referred to as *multiresolution analysis* and *multiscale derivative analysis*. See also *anomaly*, *derivative*, *gridding*, *high-pass filter*, *low-pass filter*, and *smoothing-enhancing filter*.

multiscale derivative analysis: same as *multiscale analysis* with an emphasis on the calculation of the horizontal and vertical derivatives of an *anomalous field* at different levels of *upward continuation* that represent different scales of *lateral resolution*.

multiscale edge detection: the *edge detection* technique which is based on the correlation and mapping of the maxima of the *horizontal gradient amplitude (HGA)*, *tilt derivative (TDR)*, and other gridded edge-detection attributes at different levels of *upward continuation* of the anomalous gravity or magnetic field. Small-scale *edge*-generated anomalies highlight the shallow structural elements and isolated sources that can be further enhanced by vertical stacking, or summing, of several low levels, from tens up to hundreds of meters, of the upward-continued *anomalous field*. Large-scale edge-generated anomalies highlight the elements of a deep subsurface *structure* that can be enhanced by vertical stacking of several high levels, up to several kilometers, of the upward-continued anomalous field (Cameron and Goussev, 2010; Lahti and Karinen, 2010). See also *attribute*, *edge detectors*, *gridding*, and *multiscale methods*.

multiscale methods: methods of *interpretation* based on the analysis of anomalous gravity and magnetic fields and their derivatives recomputed with the small increments Δh at different levels of *upward continuation*. In this way, a *three-dimensional (3D)* data volume is built from the original *grid* data and *two-dimensional (2D)* vertical sections are composed from the *profile* data, so that both datasets can be used in *multiridge analysis* and *multiscale analysis* (Cella et al., 2009).

multisensor gradient array: 1. a *magnetic gradiometer* configuration where several synchronized magnetometers are mounted on the same reference frame to measure the horizontal and vertical gradients of the *Earth magnetic field*. **2.** *SQUID magnetic tensor gradiometer*. It is used primarily in high-*resolution* mineral exploration for the mapping of (a) mineral alteration zones within discovered ore bodies and alkaline igneous intrusions of the *carbonatite* type favorable for the accumulation of *rare earth elements (REE)*, (b) hydrothermal alterations along detected faults and their intersections, as well as in environmental, engineering, and military applications. **3.** a *gravity gradiometer* configuration where the paired and synchronized accelerometers are mounted on the same reference frame to measure the *full tensor gravity gradient (FTGG)* components of the *Earth gravity field*. Multisensor gradient array of accelerometers is implemented in the *GOCE gravity gradiometer* and *rotating-disk gravity gradiometer*. See also *accelerometer*, *gradient*, *igneous intrusion*, *ore body*, and *sensor*.

N

Nafe-Drake curve: a graphical plot of the relationship between the velocities of a seismic compressional wave, or P-wave, and rock densities within the Earth's *crust* for the velocity range of 1.5–8.5 $\mathrm{km \cdot s^{-1}}$. The graphical Nafe-Drake curve is approximated by the *Brocher formula*. Often referred to as *Nafe-Drake relation*. See also *Adams-Williamson equation* and *Gardner equation*.

Nafe-Drake relation: See *Nafe-Drake curve*.

nanogal (nGal): a unit of *sensitivity* of advanced relative gravimeters, such as a *superconducting gravimeter*, defined as $1\,\mathrm{nGal} = 10^{-3}$ *microgal* (μGal) $= 10^{-6}$ *milligal (mGal)* $= 10^{-9}$ *Gal* $= 10^{-11}\ \mathrm{m \cdot s^{-2}}$. See also *relative gravimeter*.

nanotesla (nT): a standard unit of measurements of the *Earth magnetic field* in exploration and research applications: $1\,\mathrm{nT} = 10^{-9}$ *tesla (T)* $= 10^{-5}$ *gauss* $= 1$ *gamma* $= 10^{3}$ *picotesla (pT)* $= 10^{6}$ *femtotesla (fT)*. In theory, it represents a unit of *magnetic flux density* that quantifies the *magnitude* of a magnetic field *vector* by the number of *magnetic flux* lines per unit area perpendicular to the vector direction.

natural density: the mass per unit volume of a fluid-saturated rock, including the solid rock mass and its void space filled with a fluid, defined as

$$\rho_N = \rho_B + V_{VF} \cdot \rho_F / 100,$$

where ρ_B is *bulk density* and V_{VF} is the percentage of the void space volume filled by fluid with the density ρ_F. See also *apparent density* and *true density*.

natural remanent magnetization (NRM): *remanent magnetization* acquired by rocks at the geological time and place of their natural occurrence. The NRM includes five main types: *chemical remanent magnetization (CRM), detrital remanent magnetization (DRM), isothermal remanent magnetization (IRM), thermoremanent magnetization (TRM)*, and *viscous remanent magnetization (VRM)*. Also referred to as *normal remanent magnetization*.

Naudy filter: a space-domain *smoothing filter* which replaces the values of anomalies with wavelengths shorter than the user-defined length of a *moving window* by values interpolated from the adjacent data. It is used for the elimination of high-frequency *noise* from gridded data and spikes from line-oriented data. See also *gridding, space domain*, and *wavelength*.

Naudy method: a *profile*-based automatic curve-matching method of estimating the magnetic source depth. It assumes (a) two-dimensional (2D) *thin dike* and *thick dike* source geometries, (b) *induced magnetization* of sources, and (c) a known source width determined from the anomaly parameters, such as the horizontal distance between the maxima of the *horizontal gradient*. At each observation point, the *total magnetic intensity (TMI)* anomalies are split into the symmetric and asymmetric components about the center of a user-defined calculation window, and the asymmetric component is reduced to the pole. Then, both the TMI symmetric and reduced-to-pole (RTP) profiles are cross-correlated with the theoretical profiles generated by dikes with different depth-to-width ratios in order to determine their best fit and corresponding source depth. The method can be upgraded to the use of derivatives of the TMI and RTP fields (Naudy, 1971). See also *derivative, depth rules, reduction-to-pole (RTP)*, and *source geometry*.

near-zone terrane correction: *terrain correction* which combines *innermost terrain correction* and *inner-zone terrain correction* in compensation for the *terrain effect* within either (a) six *Hayford-Bowie zones* A through F to the distance of 2290 m or (b) eight *Hammer zones* A through H to the distance of 2610 m around each *station* in high-accuracy *ground gravity* and *microgravity* surveys. See also *outer-zone terrain correction*.

NED reference frame: *Cartesian coordinates* where the x-axis is directed to the geographic north, the y-axis is to the geographic east, and the z-axis is vertical positive down. See also *right-sided Cartesian system*.

Neel point: See *Neel temperature*.

Neel temperature: the temperature at which the antiferromagnetic rock-forming minerals become essentially non-magnetic, i.e., *paramagnetic*. Also known as *Neel point*. See also *antiferromagnetism* and *Curie point*.

negative pole: 1. one of two magnetic poles in a *magnetic dipole* toward which the magnetic field lines of the magnetic dipole are oriented and concentrated. Field lines outside the magnetic dipole are directed from the positive, or north, pole to the negative, or south, pole of a magnetic dipole, whereas inside the dipole the field lines are directed from the negative to positive pole to close the loops formed by all field lines of the magnetic dipole. **2.** south-seeking *magnetic pole*. See also *positive pole*.

negative structural index: See *extended negative structural index*.

negative susceptibility: *diamagnetic susceptibility* in the two-dimensional (2D) and three-dimensional (3D) *modeling* of the diamagnetic effects of *salt* and *remanent magnetization* of source bodies.

Nelson method: a method of *gradient-enhanced gridding* of aeromagnetic data. It is based on the reconstruction of the total magnetic field in the *Fourier domain* from the measured and gridded *transverse gradient* and *longitudinal gradient* data. The obtained *grid* is used in lieu of a high-pass version of the *total magnetic intensity (TMI)* that does not require the standard tie-line *leveling*, because the leveling errors are assumed to be caused by *diurnals* which are suppressed by *gradient* measurements. If necessary, the regional long-*wavelength* component is added to the reconstructed *total field* using the Fourier-domain *vertical integral* of the low-pass filtered line-oriented gradient data. See also *high-pass filter*, *low-pass filter*, *tie lines*, and *Hardwick method*.

Nettleton method: a *profile*-based method of selecting the value of *Bouguer density* above the gravity *survey datum* (sea level or *reference ellipsoid*) for calculation of *simple Bouguer correction* and *terrain correction*. It is based on the construction of several *Bouguer gravity* profiles over a prominent topographic elevation which is not related to the known geological *structure*. Profiles represent the Bouguer gravity values obtained by application of the Bouguer correction calculated with different density values ranging from 1.6 to 2.6 $g \cdot cm^{-3}$ with the increment 0.1 $g \cdot cm^{-3}$. The Bouguer gravity profile showing the lowest correlation with the profile of the *topography* corresponds to the "true" Bouguer density value, which is the most appropriate for calculation of the Bouguer and terrain corrections and their application over the entire survey area. Bouguer gravity profiles obtained with densities lower than the "true" density will have a direct correlation with the topography due to an insufficient Bouguer correction, or undercorrection, whereas the Bouguer gravity profiles obtained with higher densities will have a mirror-like, or inverse, correlation with the topography due to an excessive Bouguer correction, or overcorrection. Sometimes referred to as *density-profile method*, *Nettleton profiling method*, and *Nettleton test*. See also *Siegert method*.

Nettleton profiling method: See *Nettleton method*.

Nettleton test: *Nettleton method* in the context of testing several *density* values for the selection of the one, i.e., *Bouguer density* value, that minimizes the correlation between the profiles of the *Bouguer gravity* and the *topography*.

network: **1.** a system of interconnected computer algorithms, such as the one implemented in an *artificial neural network (ANN)*. **2.** a set of interlocking closed loops and control points in a ground gravity or magnetic *survey* area, such as a *gravity loop network*. **3.** a set of the crossover, or tie, points in marine (shipboard) and airborne measurements that is used in *leveling* of the data acquired along the survey lines. See also *crossover point*, *loop*, *repeat station*, and *tie point*.

network adjustment: **1.** *leveling* in the context of adjustment, or minimization, of mis-ties between the measured values along *traverse lines* and *tie lines*. It is based on the use of the *network* of the intersection, or tie, points in marine (shipboard) and airborne gravity and magnetic surveys. **2.** the process of fitting together, or matching, the observation loops within a *gravity loop network*, where the control points represent the *datum* level values in a *ground gravity* survey. See also *loop*, *tie point*, and *mis-tie*.

network node: a ground observation point, such as *repeat station*, *tie station*, or *loop-base station*, where the repeated measurements are adjusted to a single gravity or magnetic value in the process of *network adjustment*. The choice and amount of the network nodes define how the adjustments are spread over the *network* of the control points in a ground gravity or magnetic *survey*.

neural network: See *artificial neural network (ANN)*.

neutral balance: a dynamically stable balance between the *gravitational force* acting on a *proof mass* and the restoring force of the metal or quartz spring in the *zero-length spring* suspension design of a *spring gravimeter*. The small displacement of the proof mass from one balanced point to another does not activate the restoring force and, thereby, makes the neutrally balanced proof mass suspension highly sensitive (in theory, infinitely sensitive) to very small changes in the gravitational force. See also *astatic balance* and *sensitivity*.

Newton law of gravitation: See *Newton law of gravitational attraction*.

Newton law of gravitational attraction: the fundamental physical law stating that the *magnitude* of the *gravitational force* between two masses is directly proportional to each *mass* and inversely proportional to the square of the distance between them. In *Cartesian coordinates*, the gravitational force F between mass m_1 centered at the point with coordinates (x_1, y_1, z_1) and mass m_2 with coordinates (x_2, y_2, z_2) is expressed as

$$\mathrm{F} = \mathrm{G}\, m_1 m_2 / \mathrm{r}^2,$$

where G is the *universal gravitational constant* and r is the distance between the two masses defined as

$$\mathrm{r} = \left[(x_2 - x_1)^2 + (y_2 - y_1)^2 + (z_2 - z_1)^2\right]^{1/2}.$$

Gravitational force is a *vector* quantity denoted by **F**, and its magnitude $|\mathbf{F}| = \mathrm{F}$ is defined by the preceding formula. For a *source body* with mass M acting upon *proof mass* m_0 in a *gravimeter*, its gravitational force **F** is expressed as

$$\mathbf{F} = -\mathbf{r}\mathrm{GM}m_0/\mathrm{r}^3 = -\check{\mathbf{r}}\mathrm{GM}m_0 / \mathrm{r}^2;$$

where a negative sign follows the convention that vector **r** (whose length, or magnitude, $|\mathbf{r}| = \mathrm{r}$) is directed from the source body to the gravimeter location, whereas the gravitational force **F** is directed toward the source body, i.e., in the opposite direction; and $\check{\mathbf{r}}$ is unit vector defined as

$$\check{\mathbf{r}} = (1/\mathrm{r})\left[(x_2 - x_1)\check{\mathbf{i}} + (y_2 - y_1)\check{\mathbf{j}} + (z_2 - z_1)\check{\mathbf{k}}\right];$$

where $\check{\mathbf{i}}$, $\check{\mathbf{j}}$, $\check{\mathbf{k}}$ are unit vectors in the mutually orthogonal *x*-, *y*-, *z*-directions, respectively. Also referred to as *Newton law of gravitation*, *Newton law of universal gravitation*, *Newton law of universal attraction* and other similar derivations. See also *Newton law of motion* and *Coulomb law*.

Newton law of motion: the physical law that relates the *magnitude* of the unbalanced force F to the *scalar* value of *acceleration a* of the moving mass *m* as

$$\mathrm{F} = m \cdot a.$$

In application to the *vector* field of *gravitational force*, this law describes *gravitational acceleration* **g** of a body in its free fall toward the attracting mass M by the ratio of the gravitational vector force **F** to the body mass *m*:

$$\mathbf{g} = \mathbf{F}/m = -\check{\mathbf{r}}\mathrm{GM}/\mathrm{r}^2,$$

where $\check{\mathbf{r}}$ is a unit vector directed from the attracting mass M to the mass *m*, a negative sign indicates that force **F** is directed toward the attracting mass M, G is a *universal gravitational constant*, and r is the distance between the body *m* and the center of the attracting mass M. According to this law, the gravitational acceleration ***g*** of a free-falling body in the absence of other forces is independent of its mass and directly proportional to the mass of the attracting body. See also *Newton law of gravitational attraction.*

Newton law of universal gravitation: See *Newton law of gravitational attraction.*

Newtonian gravitational acceleration: *gravitational acceleration* in the context of the *Newton law of motion* and *Newton law of gravitational attraction.*

Newtonian gravitational attraction: See *gravitational attraction.*

Newtonian gravitational constant: See *universal gravitational constant.*

Newtonian potential: See *gravitational potential.*

NGDC720 magnetic model: the global-scale model of the *lithospheric magnetic field* released by the U.S. National Geophysical Data Center (NGDC). It is parameterized by the *spherical harmonic expansion* of the scalar *magnetic potential* to the degree and order 720 with the *wavelength resolution* of magnetic anomalies up to 56 km and referenced to the *WGS84 datum.* The NGDC720 model is compiled from satellite, marine (shipboard), airborne, and ground magnetic data with incorporation of the vector measurements by Earth-orbiting satellites. The magnetic field *vector,* its direction, and other magnetic field parameters can be determined for any location on and any height above the Earth's surface. Also referred to as *Enhanced World Magnetic Model.* See also *geomagnetic field parameters, spherical harmonics, World Magnetic Model (WMM),* and *World Digital Magnetic Anomaly Map (WDMAM).*

nil zone: the vertical depth interval over which the densities of sedimentary rocks are within the range of *salt* densities. Without a *density contrast,* a salt body in this zone is undetectable, or transparent, for the gravity measurements, because the gravity effect of salt is either below the level of *noise* or zero, i.e., nil. A salt body that is shallower than the nil zone generates a positive *gravity anomaly,* whereas a salt body below the nil zone is a source of a negative gravity anomaly. Also known as *crossover depth* and *salt nil zone.*

nodule: See *mineral nodule.*

noise: any unwanted components in gravity and magnetic data, including the unpredictable measurement and data transformation errors and any gravity or magnetic anomalies other than the *target anomaly.* There are nine major types of noise: (1) *random noise,* such as *microseismic noise* in *ground gravity* measurements and high-frequency noise generated by surface and near-surface unconsolidated rocks with a heterogeneous distribution of *density* and *magnetization*; (2) *coherent noise,* such as *gridding* errors, *corrugations,* and *Gibbs ringing*; (3) residual, or unsuppressed, *dynamic noise* caused by the motion of an *observation platform* along marine and airborne *survey* lines; (4) *instrument noise* due to uncompensated *residual drift*; (5) errors in the *positioning* of observation points; (6) unidentified *cultural noise* generated by metal man-made objects and electric currents; (7) errors in approximation of the actual ground densities by a single constant value in *Bouguer correction*; (8) *survey design aliasing* caused by a large *sample interval* in a ground survey and large spacing between *traverse lines* in ground, marine, and airborne surveys; and (9) local and regional geological structures generating the anomalies which interfere with and/or superpose over the target anomalies. See also *aliasing, attitude noise, background noise, Gaussian noise, rms noise, white noise,* and *interference of anomalies.*

noise envelope: a quantitative characteristic of the *random noise* level in acquired data that sets a limit of detectability of the low-*amplitude* anomalies in the process of measurements on a moving *observation platform.* It is determined by calculating the difference between an individual, or test, *reading* and the average of the adjacent readings symmetrically positioned before and after the test reading. The number of the averaged readings is selected empirically to be not too large to interfere with the real anomalies and not too small to eliminate the statistical variation (Reeves, 2005). See also *fourth difference.*

noise grid: gridded *corrugations* and other line-oriented artificial anomalies removed by *directional filtering* in the process of *microleveling.* Noise grid is subtracted from the original gravity or magnetic *grid* to obtain the final microlevelled grid. See also *gridding.*

nominal free-air gradient: the *free-air gradient* value $0.3086\ \mathrm{mGal \cdot m^{-1}} = 0.3086\ \mu\mathrm{Gal \cdot mm^{-1}} = 3086\ \mathrm{E}$ at $\pm 45°$ latitude on a *reference ellipsoid.* Referred to with the same meaning as *average vertical gravity gradient* and *nominal vertical gravity gradient.* See also *milligal (mGal), microgal (μGal),* and *Eötvös (E).*

nominal gravity: **1.** a globally averaged value of *gravitational acceleration g* due to the *gravitational attraction* of a homogeneous rotating *reference ellipsoid* which has the same mass, center of mass, and *angular velocity* of rotation as those of the Earth. **2.** average *theoretical gravity.* Nominal gravity is defined as $g = 9.8\ \mathrm{m \cdot s^{-2}} = 980\ Gal = 980000\ milligal\ (mGal).$ Also referred to as *nominal gravitational acceleration* and *average gravity.*

nominal gravitational acceleration: same as *nominal gravity* with an emphasis on *acceleration* as the physical dimension, or quantitative measure, of a *gravitational field.*

nominal sea level: *geoid* in the context of the *equipotential surface* for the *gravitational field* of the Earth that closely approximates *mean sea level (MSL)* in the oceans and seas.

nominal vertical gradient: the vertical rate of change of the *theoretical gravity* or the *International Geomagnetic Reference Field (IGRF)* determined with the use of the first-order analytical expressions as $0.3086\ \mathrm{mGal \cdot m^{-1}}$ and $0.025\ \mathrm{nT \cdot m^{-1}}$, respectively. Also known as *normal vertical gradient.* See also *vertical gradient, milligal (mGal),* and *nanotesla (nT).*

nominal vertical gravity gradient: **1.** *average vertical gravity gradient.* **2.** *nominal free-air gradient.*

nominal vertical magnetic gradient: same as *average vertical magnetic gradient* with an emphasis on its difference from the actual *vertical magnetic gradient* measured at observation points or calculated from *observed magnetics*.

nondipole field: a part of the *main geomagnetic field* that is not representable by the global *magnetic dipole* approximation and, therefore, described by the quadripole and other magnetic source models. The nondipole field comprises approximately 10% of the main geomagnetic field. Its magnitudes are estimated at the order of 10^4 *nanotesla* (nT) with variable rates of change up to 100 nT per year. The nondipole field has both long-term stationary components and components which drift systematically westward at variable rates averaging approximately 0.18 degree per year. Spatial dimensions of the nondipole field anomalies are measured in thousands of kilometers and comparable to the size of continents but without apparent correlation with them. The nondipole field is included into the *IGRF* and *DGRF* models. See also *IGRF model* and *eccentric dipole field*.

nonhomogeneous field: a field of a real geological source in terms of the *Euler deconvolution* theory and concept of *multihomogeneity*. Its homogeneity degree n is spatially variable and generally fractional, while the actual value depends on the source depth and source type (*point source*, 2D, 3D). In asymptotic regions of the nearest and most distant (relative to the source size) proximity to the source, n varies slowly and tends to the integer values of the *homogeneous field* generated by an *ideal source*, which approximates most closely the real source at a particular *altitude*, or *scale*, of observations. The *Euler structural index (SI)* is a negative opposite of n, i.e., $\mathrm{SI} = -n$, and the nonhomogeneous field is considered locally homogeneous within a *moving window* at every scale of observations where the calculated SI estimate is an integer value. Altitude-dependent, or multiscale, spatial variations of the local homogeneity degree of a nonhomogeneous field are utilized by the *MHODE method*. Also referred to as *inhomogeneous field* (Fedi et al., 2015). See also *multihomogeneous field* and *multiscale analysis*.

nonhomogeneous source: an approximation of real sources in the theory of *Euler deconvolution* and concept of *multihomogeneity*. Differing from a *homogeneous source*, its *Euler structural index (SI)* is not a constant integer but a spatially variable fractional, and this fractional value is a function of the source-observation distance, i.e., *altitude* above the source, and source type (*point source*, 2D, 3D). See also *fractional structural index (fSI)* and *nonhomogeneous field*.

nonuniqueness problem: *inverse problem* in the context of its fundamental principle stating that a solution of the inverse problem is not unique in the absence of *constraints*.

normal correction: See *normal gravity correction* and *normal magnetic correction*.

normal distribution: See *Gaussian distribution*.

normal ellipsoid: a uniformly dense, or homogeneous, *ellipsoid* with the same mass, center of mass, and *angular velocity* of rotation as those of the Earth. Its *gravitational field* values are referred to as *normal gravity* and *theoretical gravity*. Also known as *equipotential ellipsoid* and *reference ellipsoid*.

normal gravitational potential: **1.** the *total gravitational potential* of a uniformly dense *reference ellipsoid* which has the same mass, center of mass, and *angular velocity* of rotation as those of the Earth. Its *self-gravitational potential* and *centrifugal potential* comprise a constant value at any point on the reference ellipsoid which is also referred to as *normal ellipsoid*. **2.** a *scalar* quantity U defined as

$$\mathrm{U} = V_{\mathrm{E}} + \Phi_{\mathrm{E}},$$

where V_{E} is self-gravitational potential of the stationary, or non-rotating, normal ellipsoid, and Φ_{E} is centrifugal potential of the rotating normal ellipsoid. It is used primarily in geodetic applications. Also known as *reference gravitational potential*. See also *anomalous gravitational potential*, *Earth gravity potential*, and *gravitational potential*.

normal gravity: same as *theoretical gravity* that represents the *gravitational field* of a rotating homogeneous *reference ellipsoid*, also known as *normal ellipsoid*. It accounts for (a) *gravitational acceleration* due to the attraction of the non-rotating reference ellipsoid mass and (b) *centrifugal acceleration* due to the repulsive action of *centrifugal force* resulting from rotation of the reference ellipsoid about its axis. Normal gravity is used as a standard, or norm, to which the measured *Earth gravity field* values are referenced in order to detect the gravity anomalies generated by local and regional geological structures. The normal gravity value at the latitude of each observation point is subtracted from the *observed gravity* to obtain the *free-air anomaly* and *Bouguer anomaly* values. Also known as *standard gravity*.

normal gravity components: *vector* components that define *normal gravity* at any point on the Earth's surface by the vector sum of *gravitational acceleration* due to attraction of the stationary, or non-rotating, Earth mass and *centrifugal acceleration* due to the repulsive action of *centrifugal force* resulting from the rotation of the Earth about its axis.

normal gravity correction: same as *theoretical gravity correction* that represents the correction for the *gravitational field* of the homogeneous *reference ellipsoid* used as a standard, or norm, relative to which the regional and local gravity anomalies are determined.

normal gravity field: See *normal gravity* and *theoretical gravity*.

normal gravity formula: *International Gravity Formula (IGF)* in the context of *theoretical gravity*, or *normal gravity*, representing the *gravity field* of the homogeneous *reference ellipsoid* to which the measured *Earth gravity field* values are referenced in order to determine the gravity anomalies generated by sources within the heterogeneous Earth *crust* and upper *mantle*.

normal gravity gradient: **1.** the horizontal rate of change of *theoretical gravity* per unit of the equator-to-pole distance. It describes the increase in the *gravitational acceleration* on the Earth's surface from the equator to the poles caused by (a) a decrease in a distance from the Earth's surface to the center of the Earth's mass and (b) a decrease in *centrifugal acceleration* due to a decrease of *gyration radius*. The gravity effect of changes in the normal gravity gradient values is eliminated by the application of *theoretical gravity correction*. For relatively small *ground gravity* surveys, where only the *base station* location is referenced to the global latitude, the normal gravity gradient is approximated by a linear function of the north–south distance x_{NS} from the base station, and it is determined in units of *milligal (mGal)* per kilometer (mGal·km^{-1}) by differentiating the *International Gravity Formula 1980* as

$$\partial g/\partial x_{NS} \cong 0.8144 \sin 2\theta,$$

where θ is the base station latitude. A calculated normal gravity gradient is multiplied by the distance between each observation point and base station, and the obtained value is referred to as *latitude correction* or *normal gravity gradient correction*. The normal gravity gradient has zero values at the equator and poles and a maximum value of 0.8144 mGal·km^{-1} at ±45° latitude (Hinze et al., 2013). **2.** the average vertical rate of decrease of *theoretical gravity* per unit of the vertical distance, or *nominal free-air gradient*, due to an increase in the distance between an *observation point* and the center of mass of the homogeneous *reference ellipsoid*, also known as *normal ellipsoid*. It is equal to 0.3086 mGal·m^{-1} = 0.3086 μGal·mm^{-1} = 3086 E, and it represents the theoretical *vertical gravity gradient* value at ±45° latitude on the reference ellipsoid. Referred to with the same meaning as *average vertical gravity gradient*. See also *equatorial gravity* and *polar gravity*.

normal gravity gradient correction: *latitude correction* in the context of its application to the *ground gravity* data acquired over relatively small areas, where the *normal gravity gradient* is approximated by a linear function of the north–south distance from the *base station*.

normal gravity vector: a *vector* for which the direction and *magnitude* are defined by the *gradient* of the scalar *normal gravitational potential*. A normal gravity vector is orthogonal to the mathematical surface of a *reference ellipsoid* which is also referred to as *normal ellipsoid* and *equipotential ellipsoid*. See also *scalar* and *Earth gravity vector*.

normal magnetic correction: same as *IGRF correction* applied to measured *Earth magnetic field* values to compensate for the *normal magnetic field* which is commonly referred to as the *International Geomagnetic Reference Field (IGRF)*.

normal magnetic field: 1. same as *International Geomagnetic Reference Field (IGRF)* that is considered the standard, or norm, to which the measured *Earth magnetic field* values are referenced in order to determine the magnetic anomalies generated by sources within the Earth's *crust*. **2.** a smooth component of the observed Earth magnetic field without anomalies of the *exploration* interest. It represents the *main geomagnetic field* and slowly varies over time and with position on the Earth's surface. See also *temporal magnetic variations* and *observed magnetics*.

normal polarity: 1. the *Earth magnetic field* configuration with the negative *magnetic pole* near the geographic south pole and positive magnetic pole near the geographic north pole. **2.** a *magnetization vector* orientation that is nearly parallel to the direction of the *main geomagnetic field* at the time of measurements. See also *induced magnetization* and *magnetic polarity*.

normal reference field: 1. the *theoretical gravity field* of the uniformly dense, or homogeneous, rotating *reference ellipsoid*. **2.** *International Geomagnetic Reference Field (IGRF)*.

normal remanent magnetization: See *natural remanent magnetization (NRM)*.

normal vertical gradient: See *nominal vertical gradient*.

normalization: 1. calculation of a dimensionless ratio of the measured or calculated quantities to some reference value used as a standard, or norm, relative to which their anomalous variations can be estimated. **2.** dividing, or scaling, a set of variables by a single value or combination of values that is specific for this dataset in such a way that the adjusted values fall within a particular range. For example, three *FTG anisotropic indices* are calculated from the magnitudes of the *FTG eigenvalues* and their linear combinations that are normalized, or scaled, by the largest eigenvalue λ_1 to make the dimensionless indices range between 0 and 1, so their sum equals 1 at every point on a *grid*, and variations of three *FTG anisotropy* parameters over the study area can be visualized by a color-coded *barycentric image*. See also *magnitude* and *numerical regularization*.

normalized derivatives: dimensionless quantities defined by the ratio of the orthogonal derivatives of an *anomalous field* and their combinations, such as the *vertical derivative* normalized by the *total horizontal derivative (THDR)* in the formula of the *tilt angle (TA)* and the THDR normalized by the *analytic signal amplitude (ASA)* in the *theta angle*. They retain no information about amplitudes of the original anomalies and respond to the different source depths and wide range of anomalous amplitudes in the manner of *amplitude balancing* analogous to *automatic gain control (AGC)*. Normalized derivatives are used to transform an anomalous gravity or magnetic field into the *field* of the angular, or phase, quantities, defined either by the arctangent of the ratio of the orthogonal derivatives, such as in calculation of the tilt angle and its modifications, or by the arccosine of the ratio of the THDR to the ASA in calculation of the theta angle and its modifications. The field of the calculated *phase* anomalies is more balanced, and it can be more resolving laterally and vertically than the field of the original anomalies, because the ratio of derivatives is much less dependent on the source depth and *amplitude* range of anomalies than the derivatives themselves. See also *derivative*, *directional tilt angle*, *horizontal tilt angle (TDX)*, and *resolution*.

normalized magnetic dipole moment: *point dipole* parameter μ defined by the *magnitude* of the *normalized magnetic moment vector* **μ**. It also can be expressed in terms of *eigenvalues* calculated from the *magnetic gradient tensor* elements to represent one of its *rotational invariants* as

$$\mu = |\boldsymbol{\mu}| = 3C_m|\mathbf{m}|/r^4 = (-\lambda_2^2 - \lambda_1\lambda_3)^{1/2},$$

where C_m is a magnetic constant of proportionality, $|\mathbf{m}|$ is a magnitude of the *magnetic moment* of the point dipole, r is the distance between the point dipole and *observation point*, and $\lambda_1 \geq \lambda_2 \geq \lambda_3$ are *eigenvalues* of the magnetic gradient tensor. The term on the right side of the equation, $(-\lambda_2^2 - \lambda_1\lambda_3)$, is always a non-negative quantity regardless of the source type (Beiki et al., 2012). Unlike the gradient tensor invariants I_1 and I_2 that depend on the magnetic moment direction, this particular combination of eigenvalues is independent from the *magnetization direction* of the point dipole source. It is directly proportional to the magnitude $|\mathbf{m}|$ and its peaks above the point dipole locations. Amplitude attenuation, or *fall-off*, of the normalized magnetic dipole moment is inversely proportional to the fourth power of the source-sensor distance r, i.e., $1/r^4$. Normalized dipole moment values are defined in units of *nanotesla (nT)* per meter ($nT \cdot m^{-1}$). Generalization of the normalized magnetic dipole moment for the *horizontal cylinder, thin sheet*, and *contact* source geometries is known as *normalized source strength (NSS)*. Also referred to as *scaled magnetic dipole moment*.

normalized magnetic moment: See *normalized magnetic dipole moment*.

normalized magnetic moment vector: an accessory *vector* parameter $\boldsymbol{\mu}$ of the magnetic field generated by a *point dipole*. It is defined as

$$\boldsymbol{\mu} = 3C_m\mathbf{m}/r^4 = \mu \cdot \bar{\mathbf{m}},$$

where $\mathbf{m}$ is *magnetic moment* of the point dipole; r is a distance between the point dipole and *observation point*; $\mu = |\boldsymbol{\mu}|$ is a vector length, also referred to as *magnitude*; $\bar{\mathbf{m}}$ is the unit vector parallel to the *magnetization vector*; and C_m is the magnetic constant of proportionality whose value depends on the system of measurements: $C_m = 1$ in the cgs system and $C_m = \mu_0/4\pi = 10^{-7}$ *henry*$\cdot$meter^{-1} in the SI system, where μ_0 is the *magnetic permeability* of a source-free space. Magnitude $|\boldsymbol{\mu}|$ of this vector parameter is called the *normalized magnetic dipole moment*. See also *vector magnetic moment*.

normalized source strength (NSS): a magnetic field *attribute* representing the generalization of *normalized magnetic dipole moment* $\boldsymbol{\mu}$ for the two-dimensional (2D) and three-dimensional (3D) magnetic sources approximated by the *point dipole*, *horizontal cylinder* (line of dipoles), vertical or dipping *thin sheet*, and *contact* source geometries. The NSS is computed by a specific combination of *eigenvalues* of the *magnetic gradient tensor* that compose one of its *rotational invariants*, expressed as

$$\mathrm{NSS} = \left(-\lambda_2^2 - \lambda_1\lambda_3\right)^{1/2},$$

where $\lambda_1 \geq \lambda_2 \geq \lambda_3$ are *eigenvalues* calculated at every observation point. For the 2D sources, $\lambda_2 = 0$, $\lambda_3 = -\lambda_1$, and $\mathrm{NSS} = |\lambda_1| = |\lambda_3|$. The NSS is independent of the *magnetization direction* of the source, so its *remanent magnetization* component, if any, has no effect on the NSS *amplitude* which is proportional to the *magnitude* of the source *magnetization* and reaches its maximum value immediately above the source location. The NSS is a *homogeneous function* which satisfies the *Euler homogeneity equation* and, therefore, NSS derivatives can be used as inputs to *Euler deconvolution* for the automatic estimation of the spatial position (horizontal position and depth) of 2D and 3D magnetic sources. Effective suppression of the *background magnetic field*, high *lateral resolution* of anomalies from closely spaced sources, and independence from remanent magnetization effects make the NSS attribute very efficient for mineral exploration, detection of *unexploded ordnance (UXO)*, and other high-*resolution* applications, especially in the low *magnetic latitude* and *magnetic equator* areas (Beiki et al., 2012; Clark, 2012). Also referred to as *scaled source strength*. See also *NSS method*.

North American Datum (NAD): the *regional geodetic datum* for the *positioning* of observation points in North America. There are two main NAD versions: (1) NAD27, which is based on the non-geocentric Clarke ellipsoid of 1866, and (2) NAD83, which is based on the geocentric *GRS80 ellipsoid* of 1980. The NAD83 supersedes the NAD27 as a reference frame for mapping applications.

north-seeking pole: the positive pole of a *magnetic dipole* or magnet that points in the direction of the *magnetic north pole*. See also *south-seeking pole*.

northing: **1.** the Y-coordinate of a particular *grid* or map point that represents its distance in meters along the *y*-axis from the grid or map reference point called *false northing*. **2.** the latitude difference between *survey* lines. See also *easting* and *XY coordinates*.

notch filter: a spectral-domain *filter* that rejects a narrow range, or *band*, of the spatial frequencies in the acquired data, such as those comprising the *spike*-like magnetic anomalies caused by the *high-line effect*. See also *acquisition*, *spatial frequency*, and *spectral domain*.

NSS method: **1.** a *grid*-based method of automatically estimating the horizontal position and depth of magnetic sources approximated by the *point dipole*, *horizontal cylinder* (line of dipoles), vertical or dipping *thin dike (thin sheet)*, and *contact* source geometries. It is based on the use of a rotational invariant called the *normalized source strength (NSS)*, which is calculated from the *eigenvalues* of *magnetic gradient tensor* data. The NSS derivatives are inputs to the *3D Euler deconvolution* algorithm which uses the data points enclosed by a square window centered at the NSS maxima for simultaneous estimates of the spatial positions and structural indices of sources. **2.** a *profile*-based graphical method of manually estimating a magnetic source depth. For a two-dimensional (2D) vertical or dipping thin sheet (thin dike), its depth d is defined as $d = \mathrm{HWHM}$; for a point dipole or *sphere*, $d = 1.55$ HWHM, where HWHM is the *half-width half-maximum* distance in meters on the NSS profile (Beiki et al., 2012). See also *rotational invariants* and *Euler structural index (SI)*.

nuclear-precession magnetometer: a *magnetometer* from the group of *quantum magnetometers*, such as a *proton magnetometer*, with an operating principle based on measuring the angular frequency of *precession* of the hydrogen nuclei.

nuclear-resonance magnetometer: a *magnetometer* from the group of *quantum magnetometers*, such as a *proton* *magnetometer*, with an operating principle based on the quantum physics phenomenon of *magnetic resonance* involving the spinning (rotating about axes of rotation) subatomic particles, such as the positively charged hydrogen nuclei (protons) and negatively charged electrons, and certain types of externally applied magnetic field.

null count: a *database* parameter that represents the amount of null values per *data channel* in a database. See also *null value*.

null point: a reference position of *proof mass* relative to which changes in the *Earth gravity field* values are determined by the measurement system of a *relative gravimeter*.

null value: a *database* parameter used to identify the data points without measured values. Commonly, the null value is shown as 99999 or 999999.

null zone: See *nil zone*.

numerical modeling: **1.** *forward modeling* in the context of numerical techniques used for calculation of the gravity or magnetic *field* generated by an assumed set of *model* parameters. **2.** *inverse modeling* in the context of using iterative numerical methods for the estimation of the structural and property (*density* or *magnetization* contrast) parameters of a particular *target* from the residual gravity and magnetic fields. See also *contrast* and *residual field*.

numerical regularization: application of numerical techniques, such as *least-squares optimization*, for minimization of the model *objective function* in solving the gravity or magnetic *inverse problem* to prevent *overfitting* and amplification of *noise*. It is based on finding the optimal numerical term, or norm, of differences between the observed and modeled values in the process of *inversion*. See also *L1 norm*, *L2 norm*, *model misfit*, and *stochastic inversion*.

Nyquist bandwidth: *bandwidth* in the context of the *sampling rate* of airborne or marine (shipboard) *survey* measurements, also known as the *sampling bandwidth*, that defines the highest *waveform* frequency, or the shortest *wavelength*, component which can be recovered from the acquired data in accordance with the *Nyquist theorem*.

Nyquist fractions: decimal fractions of the *Nyquist frequency* that are used instead of wavenumbers as the units of reference in the graphical plots of the *power spectrum* and *radial power spectrum* of the gravity or magnetic data. See also *wavenumber*.

Nyquist frequency: **1.** a *sampling* parameter that defines the highest *frequency*, or the shortest *wavelength*, that can be recovered from a regularly sampled continuous *field* in accordance with the *Nyquist theorem*. The Nyquist frequency is equal to one-half of the *sampling frequency*, e.g., for a sampling frequency 2 Hz (two measurements, or samples, per second), the highest recoverable *waveform* frequency is 1 Hz. **2.** *gridding* parameter f_N defined by the reciprocal, or inverse, of the *Nyquist wavelength* λ_N and expressed as

$$f_N = 1/\lambda_N = 1/2\Delta x,$$

where Δx is the space-domain *grid interval*. Any waveform with a *spatial frequency* higher than the Nyquist frequency by the amount Δf will be aliased as the lower artificial frequency $(f_N - \Delta f)$ from which the actual higher frequency $(f_N + \Delta f)$ is indistinguishable, because they have the same *sample* values. Data contamination with the aliased frequencies is prevented by the application of an *alias filter*. Sometimes referred to as *folding frequency*. See also *aliasing*, *grid aliasing*, and *space domain*.

Nyquist rule: *Nyquist theorem* in the context of a rule of reconstruction, or recovery, of a continuous *field* from its discrete measurements at regular time or distance intervals.

Nyquist sample rate: the *sample rate* based on the *Nyquist theorem* that defines the distance or time interval for the correct *sampling* of a continuous *field*.

Nyquist-Shannon sampling theorem: same as *Nyquist theorem* that defines the rate of *sampling*, or *sampling rate*, for the reconstruction of a continuous *field* from its discrete measurements at regular time or distance intervals.

Nyquist theorem: a theorem postulating the rule of reconstruction, or recovery, of a continuous *field* from its discrete measurements at regular time or distance intervals. It states that any field approximated by a *band-limited function*, such as the gravity or magnetic field, can be fully recovered if the *sampling frequency*, or *sample rate*, provides two or more discrete measurements, or samples, per cycle of the highest *frequency* component in the measured field. Also referred to as *Nyquist-Shannon sampling theorem*, *sampling theorem*, and *Nyquist rule*.

Nyquist wavelength: the shortest *wavelength* that can be recovered from the regularly sampled continuous *field* in accordance with the *Nyquist theorem*. Nyquist wavelength λ_N is twice the *sample interval* and expressed as

$$\lambda_N = 2\Delta x,$$

where Δx is the space-domain sample interval in the line-oriented data. For gridded data, the Nyquist wavelength value is twice a selected *grid interval*, and it is referred to as the *grid Nyquist wavelength*. The Nyquist wavelength is a major *sampling* parameter which defines the *lateral resolution* of anomalies in acquired gravity or magnetic data. See also *acquisition*, *gridding*, *space domain*, and *Nyquist frequency*.

Nyquist wavenumber: **1.** *Fourier domain* parameter k_N, which is the reciprocal, or inverse, of the *Nyquist wavelength* and expressed as

$$K_N = 1/\lambda_N = 1/2\Delta x,$$

where Δx is the space-domain *sample interval*. In *gridding*, Δx represents the *grid interval*. **2.** a *wavenumber* corresponding to the *wavelength* value of twice the sample interval and, therefore, the shortest wavelength that can be represented adequately by *Fourier transform*. Nyquist wavenumber k_N is equal to one-half of the *sampling wavenumber*, defined as

$$K_N = (2\pi/\Delta x)/2 = \pi/\Delta x.$$

It also can be expressed as

$$K_N = 2\pi/\lambda_N,$$

where λ_N is the Nyquist wavelength, i.e., the Nyquist wavenumber is the Fourier-domain analogue of the space-domain Nyquist wavelength. Also known as *folding wavenumber*. See also *space domain*.

O

objective function: a mathematical function of the data and parameters that is maximized or minimized by selection of specific parameters. The objective function for a *voxel model* is designed to incorporate information about the structural characteristics of an exploration or research *target* and available *constraints* in order to obtain an *inverse problem solution*. The inverse problem is solved by finding the *model* parameters that minimize the model objective function and ensure a *model misfit* within a specified *tolerance* (Oldenburg and Li, 2005). See also *inversion* and *model misfit*.

observation density: the amount of observation points, or stations, per square kilometer in a *ground gravity* survey. Also referred to as *sampling density*. See also *station*.

observation elevation: **1.** the height of an *observation point* above or below sea level in ground surveys and above the *reference ellipsoid* in airborne surveys. **2.** the water depth at each observation point in sea-bottom and underwater measurements. See also *elevation*.

observation platform: a ship, submarine, aircraft, *drone*, or Earth-orbiting satellite used as a carrier of the instruments for measuring the Earth's gravity and magnetic fields. See also *stabilized platform*, *Schuler-tuned inertial platform*, and *moving-platform corrections*.

observation point: a coordinate point on the ground, sea surface, or in the air at which a gravity or magnetic measurement is made. In marine (shipboard) and airborne surveys, the gravity or magnetic value recorded at a particular observation point corresponds to the average value of several discrete measurements made in accordance with the *sample rate* of a *gravimeter* or *magnetometer*. The ground observation point is called a *station*.

observation spacing: *sample interval* in the context of the distance between observation points in a ground, marine (shipboard), or airborne gravity or magnetic *survey*. See also *anisotropic spacing*.

observed anomaly: an *anomaly* obtained by application of all relevant corrections for the measurement effects and predictable planetary and non-geological gravitational and magnetic effects contributing to the measured gravity and magnetic values. Observed gravity anomalies are represented by the *free-air anomaly*, *Bouguer anomaly*, and *isostatic anomaly*, and the observed magnetic anomalies are represented by the *total magnetic anomaly (TMA)* and *RTP anomaly*. See also *observed field*, *observed gravity*, *observed magnetics*, *observed gravity components*, and *observed magnetic components*.

observed Bouguer anomaly: a *field* of the gravity anomalies obtained by application of *theoretical gravity correction*, *free-air correction* or *height correction*, *simple Bouguer correction*, *terrain correction*, and other *survey*-relevant corrections to *observed gravity* values. Commonly referred to as *Bouguer anomaly* or *Bouguer gravity*. See also *gravity anomaly*, *raw gravity corrections*, and *observed gravity corrections*.

observed field: **1.** a gravity or magnetic *field* obtained by application of all relevant corrections to the measured data. Referred to with the same meaning as *anomalous field*. **2.** a gravity or magnetic field in the two-dimensional (2D) or three-dimensional (3D) *modeling*. See also *raw gravity corrections*, *raw magnetic corrections*, *observed gravity corrections*, and *observed magnetic corrections*.

observed free-air anomaly: a *field* of the gravity anomalies obtained by application of *theoretical gravity correction* and *free-air correction* or *height correction* to the *observed gravity* values. Commonly referred to as *free-air gravity* or *free-air anomaly*.

observed gravity: **1.** *relative gravity* values after the application of *raw gravity corrections* for the instrument and *acquisition* effects, including: (a) conversion of the *gravimeter* counter, or dial, units into the standard gravity units of *milligal (mGal)* or *microgal (μGal)*; (b) compensation for *gravimeter drift*; (c) compensation for *cross-coupling errors* in measurements on a moving *observation platform*; and (d) *scale correction*. In *ground gravity* surveys, it is also referred to as *station gravity*. **2.** *absolute gravity* values after the application of corrections for the instrument effects in the stationary measurements by the *free-fall method*, including *diffraction correction*, *self-attraction correction*, *speed-of-light correction*, and *tilt correction*. Observed gravity is also referred to as *measured gravity* to highlight its incomplete *reduction* status, as opposed to *anomalous gravity* which is the final result of the *gravity reduction* process. **3.** a *modeling* term for the anomalous gravity field data, such as *Bouguer gravity* or *free-air gravity*, used in a two-dimensional (2D) or three-dimensional (3D) gravity *model*, where the gravity field of a constructed model, or *calculated gravity*, is compared with the anomalous gravity field via a sequence of iterations until the *model misfit* requirement is satisfied. Sometimes, the observed gravity is referred to with the same meaning as *raw gravity*.

observed gravity components: planetary, geological, and non-geological gravitational effects contributing to the *observed gravity* values. All but one should be compensated in order to obtain the gravity anomalies generated by the local and regional geological structures in a study area. The following list contains the 12 main components with the names of their 11 corresponding corrections shown in parentheses: (1) attraction of the homogeneous mass of a rotating *reference ellipsoid* (*theoretical gravity correction*); (2) effect of the *observation point* height above sea level (*free-air correction*) or reference ellipsoid (*height correction*); (3) effect of the submarine or sea floor observation point depth below sea level (*free-water correction*); (4) effect of the rock mass above sea level or reference ellipsoid (*simple Bouguer correction*); (5) effect of variations of the topographic or bathymetric relief around observation points (*terrain correction*); (6) time-dependent global and local variations (*tidal correction* and *atmospheric pressure correction*); (7) attraction of the Earth's atmosphere (*atmospheric mass correction*); (8) *Eötvös effect* of a moving observation platform (*Eötvös correction*); (9) regional-scale effect of the *curvature* of the Earth's surface (*Bullard B correction*); (10) regional-scale effect of *isostatic compensation* (*isostatic correction*); (11) regional-scale *indirect effect* (*indirect effect correction*); and (12) gravity effect of the local and regional geological structures and variations of *density* within the *crust* and upper *mantle*. The objective of the *gravity reduction* process is to obtain the gravity anomalies generated by the last component. See also *observed magnetic components*.

observed gravity corrections: corrections applied to the *observed gravity* to compensate for all predictable planetary and non-geological gravitational effects contributing to the measured *relative gravity* and *absolute gravity* values and obtain the *field* of gravity anomalies generated by the local and regional geological structures in a *gravity survey* area. The choice of corrections for the relative gravity data depends on the *survey* type (ground, marine, submarine, sea floor, airborne), survey scale (local or regional), survey objectives, and measurement *accuracy* requirements. Main relative gravity corrections include: (1) *theoretical gravity correction* for *gravitational attraction* of the homogeneous mass of the rotating *reference ellipsoid*; (2) *free-air correction* for the observation point height above sea level or *height correction* for the *GPS*-derived height above the reference ellipsoid; (3) *free-water correction* for the submarine and sea-floor observation point depth below sea level; (4) *simple Bouguer correction* for the rock mass above the sea level or reference ellipsoid; (5) *terrain correction* for variations of the topographic or bathymetric relief around observation points; (6) *Bullard B correction* for the regional-scale gravity effect of the *curvature* of the Earth's surface; (7) *tidal correction* for the time-dependent gravity effects related to changes in sea level and deformations of the Earth's surface caused by the cumulative effect of the gravitational attraction of the Moon and Sun; (8) *atmospheric mass correction* and *atmospheric pressure correction* for variations of the atmospheric mass and pressure during a high-accuracy *ground survey* covering a broad range of topographic heights; (9) *isostatic correction* for the regional-scale long-*wavelength* gravity effect of *isostatic compensation*; and (10) *indirect effect correction* for differences in the anomalous amplitudes obtained from measurements with different vertical datums: sea level or reference ellipsoid. *Eötvös correction* is commonly applied as a part of *raw gravity corrections*. See also *absolute gravity corrections*.

observed magnetic components: planetary, geological, and non-geological magnetic effects contributing to the *observed magnetics*. There are three main components: (1) the slow-varying *main geomagnetic field* generated inside the Earth's core and representing the major contribution to the measured *Earth magnetic field*, (2) the fast-varying *external magnetic field* generated outside the Earth in the *magnetosphere* and representing the source of *diurnals* and *magnetic storm*, and (3) the *lithospheric magnetic field* which is essentially constant over the time and location, or static, and represents the combined magnetic effects of *induced magnetization* and *remanent magnetization* of the igneous, metamorphic, and sedimentary rocks comprising the local- and regional-scale geological structures within the *crust* and upper *mantle*. In exploration applications, the lithospheric magnetic field is described by the *total magnetic intensity (TMI)*, also known as *total magnetic anomaly (TMA)*, and the *RTP magnetic field* that represent the final results of the *magnetic reduction* process and basic datasets for *enhanced processing* and *interpretation*. See also *observed gravity components*.

observed magnetic corrections: corrections applied to the *observed magnetics* in order to (a) compensate for all predictable planetary and non-geological magnetic effects contributing to the measured *Earth magnetic field* values, (b) compensate for *magnetic inclination* and *magnetic declination* of the Earth's magnetic field, and (c) obtain the *field* of the magnetic anomalies generated by the local and regional geological structures in a *magnetic survey* area. Observed magnetic corrections include: (1) *IGRF correction*; (2) *diurnal correction*; (3) *cultural editing*; (4) *magnetic terrain correction*, if necessary; and (5) *reduction-to-pole (RTP)* or *reduction-to-equator (RTE)*. The RTE is commonly replaced by *low-latitude reduction-to-pole (RTP-LL)* or by calculation of the *analytic signal amplitude (ASA)* and the ASA-derived attributes, such as *theta angle*, *second-order analytic signal amplitude* (ASA_2), and *tilt ASA*. See also *raw magnetic corrections* and *attribute*.

observed magnetics: 1. measured *Earth magnetic field* values after the application of *raw magnetic corrections* for the instrument and *acquisition* effects. Depending on the type of *magnetic survey* (ground, shipboard, submarine, airborne), these corrections include *cable correction*, *heading correction*, *lag correction*, and *scale correction*. **2.** a *modeling* term describing the anomalous magnetic field data, such as *total magnetic intensity (TMI)* or *RTP magnetic field*, used in a two-dimensional (2D) or three-dimensional (3D) magnetic *model*, where the magnetic field of a constructed model, or *calculated magnetics*, is compared with the TMA or RTP field via a sequence of iterations until the *model misfit* requirement is satisfied. See also *aeromagnetic compensation*, *raw magnetics*, and *observed magnetic components*.

observer: a person in charge of the monitoring and recording of gravity or magnetic measurements during ground, marine (shipboard), submarine, and airborne surveys. See also *interpreter*.

Occam razor: a general principle, or dictum, of the scientific reasoning that the simplest explanation of observable phenomena is the most probable: "It is vain to do with more what can be done with fewer." Named after English philosopher William of Ockham (1285–1347).

ocean-bottom gravimeter: See *underwater gravimeter*.

ocean floor striping: See *geomagnetic striping.*

ocean loading correction: the time- and location-variable correction applied to high-accuracy *ground gravity* measurements in the *microgravity* range, such as *time-lapse gravity* and *absolute gravity*, to compensate for the *ocean loading effect* due to changes in the ocean water mass distribution and related elastic deformation of the Earth's surface caused by *ocean tides.* Also referred to as *ocean tide loading correction* and *OTL correction.* See also *accuracy.*

ocean loading effect: the combined gravity effect caused by (a) elastic deformation of the Earth's surface under the load of the additional ocean/sea water mass, especially in coastal areas, and (b) direct *gravitational attraction* of the additional water mass in the *ocean tides.* Ocean loading may be out-of-phase with the *Earth tides*, and its highly variable gravity effect is commonly within the range of 3–5 to 15–20 microgals, but in some coastal areas it can be much greater. It is compensated by application of *ocean loading correction.* See also *microgal (μGal).*

ocean tide loading correction: See *ocean loading correction.*

ocean tides: changes in the water mass distribution in the oceans and seas caused by the different response of the ocean water mass and solid Earth to the combined *gravitation attraction* of the Sun and Moon. Highly variable gravity effects of the ocean tides are the strongest along the coastlines. Correction for the gravitational attraction of the additional water mass and elastic deformation of the Earth's surface related to the ocean tides depends on the time of day and observation point location, and it is known as *ocean loading correction.* See also *Earth tides*, *Earth tide correction*, and *ocean loading effect.*

oceanic magnetic lineations: See *geomagnetic striping.*

odd function: a function $f(x)$ for which its basic property is expressed as

$$f(-x) = -f(x),$$

i.e., it changes its sign when the sign of the variable x is changed. Any function can be split into even (symmetric) and odd (asymmetric) parts. Even and odd parts of a function represent its two independent averages. A *Fourier transform* of the odd part of a function is a *sine transform.* Also known as the asymmetric function. See also *even function* and *Naudy method.*

oersted (Oe): a unit of *magnetizing force (H)* in the cgs system. It is defined by the theoretical *magnetic field* which exerts the force of 1 dyne on a unit *magnetic pole.* Oersted is related to the equivalent unit of *ampere* per meter ($A \cdot m^{-1}$) in the SI system as: $1\,Oe = 10^3/4\pi\ A \cdot m^{-1} \approx 79.577\ A \cdot m^{-1}$. In a source-free space, $1\,Oe = 1$ *gauss* $= 10^5$ *nanotesla (nT).* See also *cgs units* and *magnetic permeability.*

off-level correction: *Harrison correction* or *Browne correction* in the context of a correction applied to *dynamic gravity* data to compensate for *off-level errors* caused by *platform tilt.*

off-level errors: errors in marine (shipboard) and airborne gravity measurements resulting from the deviation of the sensing axis of a *relative gravimeter* from its vertical orientation due to *platform tilt* caused by strong inertial accelerations of the moving *observation platform.* Off-level errors are compensated by application of *Harrison correction* or *Browne correction.* See also *inertial acceleration* and *stabilized platform.*

offset: **1.** the distance between a cable-towed *marine magnetometer* and *GPS* antenna of a *marine survey* ship. Also known as *antenna offset.* **2.** the lateral displacement of gravity or magnetic anomalies along an *interruption trend* associated with a structural discontinuity, such as a *fault*, regional wrench (strike-slip) fault system, and *shear zone.* **3.** the horizontal or vertical displacement of rocks measured along the *strike* of the identified fault or along its vertical plane, respectively. See also *strike-slip fault.*

offset correction: **1.** same as *cable correction* used as a correction for the actual distance, or *offset*, between a cable-towed *marine magnetometer* and the *GPS* antenna of the survey ship. **2.** a component of *AUV vertical acceleration correction* that accounts for the spatial separation of the *AUV gravimeter* and its depth sensor and time delay between receiving the depth sensor data and gravimeter data. See also *lever arm correction.*

offset trend: *interruption trend* in the context of a lateral *offset*, or displacement, of the gravity or magnetic anomalies along this trend on a *grid.*

offshore Airy depth: See *isostatic antiroot depth.*

omnidirectional sensor: a magnetometer *sensor* without a *dead zone.* Measurements made with such sensors are independent of the sensor orientation relative to direction of the *Earth magnetic field.* Also known as *isotropic sensor.* See also *magnetic sensor.*

one-dimensional (1D): a characteristic of the mathematical transformation method, *model*, or *source body* approximation that highlights its basic assumption that there is no change in the parameters and measured properties along two of three standard dimensions, such as the mutually orthogonal x-, y-, z-directions in *Cartesian coordinates.* See also *dimension*, *two-dimensional (2D)*, and *three-dimensional (3D).*

1D FFT: a one-dimensional (1D) *Fast Fourier Transform (FFT)* which converts line-oriented data and *grid*-extracted profiles from their original *space domain* into the equivalent *spectral domain.* See also *dimension* and *profile.*

1D FFT filters: *FFT filters* modified for application to line-oriented, or one-dimensional (1D), data and profiles extracted from gridded data. See also *dimension*, *gridding*, and *profile.*

1D Fourier transform: a one-dimensional (1D) *Fourier transform* of *grid*-extracted *profile* or line-oriented magnetic or gravity data. See also *2D Fourier transform.*

1D inversion: one-dimensional (1D) *inversion*, or *inverse modeling*, of gravity or magnetic data in terms of a *1D model*, which assumes variations of the modeled property in one *dimension* only, usually along the vertical z-axis in *Cartesian coordinates* that represents a depth below the measurement level.

1D model: a one-dimensional (1D) *model* which is based on the assumption that variations of the modeled property occur in one *dimension* only, such as along one of three mutually orthogonal x-, y-, z-axes in *Cartesian* *coordinates*. In 1D model calculations, the length of each modeled *source body* is assumed to approximate an *infinite strike length* along two axes orthogonal to the axis of the property variations. See also *1.5D model*, *2D model*, and *3D model*.

1D source: a *model* approximation of a gravity or magnetic *source body* that assumes variations of the source-modeled parameter (depth, thickness, *density* or *magnetization* contrast) in one *dimension* only, usually along the vertical z-axis in *Cartesian coordinates*. See also *contrast*, *2D source*, and *3D source*.

1D spectrum: a one-dimensional (1D) *spectrum* calculated for a *grid*-extracted *profile* or line-oriented gravity or magnetic data. See also *2D spectrum*.

1.5D model: a modification of a *1D model* with the assumption that the length of a modeled *source body* or series of rock layers along two axes orthogonal to the axis of property variations has a symmetric and finite lateral extent relative to the property variation axis. For example, the 1.5D *layered-Earth model* assumes (a) variations of *density* or *susceptibility* with depth along the vertical z-axis in *Cartesian coordinates* and (b) symmetric and limited lateral extent of the modeled layers along the x- and y-axes without changes in the modeled properties. See also *2.5D model*.

onshore Airy depth: See *isostatic root depth*.

open-end loop: a *loop* in which the first and last gravity measurements are made at different base stations, such as those in a marine (shipboard) *gravity survey* starting at one harbor and ending at another harbor. See also *base station*.

operating range: a characteristic of a *gravimeter* or *magnetometer* that defines the range of values, from the smallest value to the largest value, that can be measured by a particular instrument without resetting. See also *interval factor*.

optical-absorption magnetometer: same as *optically pumped magnetometer* with an operating principle utilizing the physical phenomena of *magnetic resonance* and absorption of the monochromatic laser light by protons and electrons.

optical free-fall gravimeter: same as *falling-mirror gravimeter* in which the optical fringes in a laser interferometer are used to track the free fall of the corner-cube reflector called "mirror."

optically pumped magnetometer: a *scalar magnetometer* for measuring the total intensity, or strength, of the *Earth magnetic field*. Its operational concept is based on two natural phenomena: (1) *magnetic resonance* and (2) absorption of the monochromatic waves by precessing protons and electrons in the vapor, or gaseous, alkali metal cells irradiated by the monochromatic laser light and light-polarizing transverse radio-frequency (RF) electromagnetic field. Absorption is at the maximum level when the RF field is tuned to the *precession* frequency, or *Larmor frequency*, which is proportional to the intensity of the Earth's magnetic field. Magnetometers of this type have extra-high *sensitivity* up to 0.001–0.0005 *nanotesla (nT)*, *sampling rate* 10–15 *hertz (Hz)*, and *absolute accuracy* 1–3 nT. They include the *cesium magnetometer*, *helium magnetometer*, *potassium magnetometer*, and *rubidium magnetometer*. Also referred to as *laser-pumped magnetometer*, *optical-absorption magnetometer*, *alkali-vapor magnetometer*, and *atomic magnetometer*. See also *proton-precession magnetometer* and *vector magnetometer*.

optimum filter: a spectral-domain *filter*, such as the *Wiener filter*, designed to minimize the sum of the squares of differences between the expected result of *filtering* and the actual filter output. Its *transfer function* H(k) is expressed as

$$H(k) = |P_s(k)|^2 \; W_{S/N}(k) \,/\, |P_F(k)|^2,$$

where k is the *wavenumber*, $P_S(k)$ is the *power spectrum* of the expected signal, $W_{S/N}(k)$ is a *wavenumber*-dependent weighting coefficient proportional to *signal-to-noise ratio (S/N)*, and $P_F(k)$ is the power spectrum of the anomalous gravity or magnetic field. Also known as *least-squares filter*. See *anomalous field*, *least-squares fit*, and *spectral domain*.

ore body: a continuous rock mass containing an accumulation of minerals of economic value in quantities sufficient for commercial extraction at a reasonable profit. It is often referred to with the name of its valuable constituent, such as iron ore. Usually, the ore body creates a high *density contrast* and *magnetization contrast* with the surrounding rocks and generates a detectable *anomaly* which can be additionally enhanced by calculation of the *residual field*. Igneous ore bodies are often remanently magnetized with a highly variable *magnetization direction*, so their correct locations are determined and mapped with the use of the calculated *analytic signal amplitude (ASA)*, magnitudes of *magnetic anomaly vector*, and other *scalar* and *vector* attributes of the gridded *anomalous field*. See also *attribute*, *gridding*, *igneous intrusion*, *magnitude*, *remanent magnetization*, and *carbonatite*.

oriented filter: a space-domain or spectral-domain *filter* for which the output is a function of the azimuthal direction or orientation relative to the reference frame. For example, the *directional filter* and *steerable filter* are oriented filters. See also *azimuth*.

oriented sensor measurements: measurements of the *vector* components of the Earth's gravity and magnetic fields using a single *sensor* oriented along the vertical or horizontal direction or several sensors oriented in the mutually orthogonal x-, y-, z-directions. Oriented single-sensor measurements are made, primarily, by ground-based surveys for archaeological and engineering surveys. In mineral exploration, multi-sensor airborne *gradiometer* systems are commonly used. In satellite missions, the sensor orientation is controlled by the star-imaging cameras and high-*accuracy* navigation systems. See also *geomagnetic observatory*.

orthogonal: **1.** perpendicular, rectangular, normal, i.e., involving right angles. **2.** A characteristic of functions that are linearly independent: they have independent dimensions and cannot be expressed as combinations of each other. See also *dimension*, *principal component analysis (PCA)*, and *quadrature pair*.

orthometric height: 1. the vertical distance from a *geoid* to an *observation point* on the Earth's surface that is used as the equivalent of the height above *mean sea level (MSL)*. Orthometric height H is calculated by (a) adding the *geoid undulation* N to the *GPS*-determined *ellipsoidal height h* over the oceans and (b) subtracting the geoid undulation value from the ellipsoidal height on the continents, i.e., $H = h \pm \text{N}$. **2.** the distance along a *plumb line*, or true vertical, from the surface of a geoid to an observation point. Sometimes referred to with the same meaning as a height, or *elevation*, above sea level.

orthonormal vectors: mutually *orthogonal* vectors.

OTL correction: See *ocean tide loading correction*.

outcrop: layers of a rock formation or geological *structure* exposed on the Earth's surface. Rock samples from outcrops provide the lower-limit estimates of the rock densities and susceptibilities for *modeling*. Outcrop locations are also used for tuning of *filtering* techniques in the high-*resolution* geological mapping of rock formations covered by surficial deposits via correlation of the *filter*-enhanced gravity and magnetic anomalies. The outcrop analogue for the subsurface unconformity truncation is called *subcrop*. See also *anomaly*, *density*, *susceptibility*, *lateral resolution*, and *clinker anomaly*.

outer-zone terrain correction: *terrain correction* applied to high-*accuracy* ground gravity and *microgravity* measurements to compensate for the topographic *relief* variations outside the range of an *inner-zone terrain correction* to the distance of 166.7 km, i.e., within the *Hayford-Bowie zones* G through O. Outer-zone terrain correction is complementary to *Bullard B correction* for the Earth's *curvature* at distances beyond 20 km around the observation points. Sometimes referred to as *far-zone terrain correction*. See also *near-zone terrain correction*.

overdetermined problem: a characteristic of the *inverse problem* when the number of assumed parameters of a *source body* generating an observed *anomaly* is less than the number of observations, so the *model* parameters approximating the particular source body can be determined uniquely. Estimates of source depth and other source parameters in *Euler deconvolution*, *Werner deconvolution*, and other *inversion* methods are based on solving an overdetermined system of equations relating the source body parameters to a pre-selected number of observations. See also *underdetermined problem*.

overfitting: a characteristic of a *modeling* result which indicates that (a) the *model misfit* is substantially less than the estimated level of *noise* and (b) the constructed *model* has excessive structural details of unreliable complexity. See also *underfitting*.

Overhauser effect: a quantum physics effect of the transfer of energy of the spinning orbital electrons to protons of the hydrogen atoms in the specially engineered hydrogen-rich fluid containing the atoms with free unpaired electrons. High-efficiency *polarization* of the hydrogen nuclei, i.e., alignment of protons, is achieved via the electron-proton coupling, wherein the unpaired electrons, instead of releasing their spinning energy by the emitted radiation, transfer this energy, boosted by the polarizing radio-frequency electromagnetic field, to the protons they are orbiting. The Overhauser effect provides much stronger *precession* signals and eliminates the need for polarization of protons by the high-energy external magnetic field required in a regular *proton magnetometer*.

Overhauser magnetometer: the advanced version of a *proton magnetometer* that implements the *Overhauser effect*. High efficiency polarization of protons stimulates them to precess continuously at the *Larmor frequency*, which is directly proportional to the intensity of the *Earth magnetic field*. The achievable *absolute accuracy* of measurements is 0.01 nT with *sensitivity* up to 0.005 nT. Its *omnidirectional sensor* and, hence, the absence of *heading error* make this magnetometer a preferred instrument for the continuous measurements in marine magnetic surveys and airborne *magnetic gradient* surveys with towed *gradiometer* systems. Sometimes referred to as *marine magnetometer*. See also *nanotesla (nT)* and *slung sensor*.

overhead point: a *sunshading* parameter that defines either the center of a sunshaded image in the display window or the current position of the artificial source of light. See also *sky map*.

P

padding: **1.** addition of extra values at each boundary of a gridded or line-oriented dataset to satisfy the *fast Fourier transform (FFT)* requirement of 2^N data points prior to the FFT application in order to minimize *Gibbs errors* in the transformation of data from one *domain* to another, such as from the *space domain* to the *spectral domain* and vice versa. Added values are commonly zeros or the original dataset values mirrored at the dataset boundaries. To ensure a smooth transition from the original data to the added data, a *cosine bell filter* and other *edge-smoothing filters* are applied. **2.** *tapering* datasets with zero values to make their length equal to the integer multiple of the *fundamental period* in order to minimize *wavenumber leakage*. **3.** outward expansion of the *three-dimensional (3D)* gravity or magnetic *model* to minimize *edge effects* in forward and inverse model calculations. For gridded data, padding is referred to with the same meaning as *grid expansion*. See also *symmetric boundary replication*.

paleochannel: a buried elongated or meandering-and-branching erosional structure formed by an ancient fluvial system and filled in with younger low-*density* sediments. A paleochannel filled in with reservoir-grade sands can be part of a stratigraphic *trap*, and it often has the distinct *gravity signature* of a residual *gravity low*. Shallow paleochannels can have the *magnetic signature* of clustered low-*amplitude* positive short-*wavelength* anomalies generated by localized accumulations of *detrital magnetite*. See also *clustering*, *residual gravity*, and *incised valley*.

paleomagnetism: a study of the ancient *Earth magnetic field* based on measurements of *natural remanent magnetization (NRM)* of rocks of known geological age. Measurements of the remanent magnetization *vector* are used to determine the direction and *magnitude* of the Earth's magnetic field at the geological time when the analyzed rock sample was magnetized. Magnetic pole positions, calculated in paleomagnetic studies, define the locations of the paleogeographic axis whose differences from the present axis location are interpreted in plate tectonic terms. See also *magnetochronology*, *magnetostratigraphy*, and *remanent magnetization*.

parallax correction: See *lag correction*.

parallel magnetization: the uniform direction of *magnetization* in a finite region that is used as an approximation of the actual distribution of magnetization within the localized, or compact, magnetic *source body* for evaluation of its volume, shape, and magnetic properties by *inversion* techniques.

paramagnetic: a characteristic of weakly magnetic rock-forming minerals having a small positive *magnetic susceptibility* and low *magnetic permeability*. Paramagnetic minerals cannot acquire *remanent magnetization*. See also *paramagnetic susceptibility*, *paramagnetism*, and *ferromagnetism*.

paramagnetic susceptibility: a small positive *magnetic susceptibility* of the rock-forming minerals, such as amphibolite, biotite, muscovite, olivine, pyroxene, tourmaline, and siderite, where the direction of the atomic magnetic moments is not random but partially aligned in parallel to the applied magnetic field and, thereby, produces a weak *magnetization* in the direction of the externally applied magnetic field, such as the *Earth magnetic field*. A typical value of the paramagnetic susceptibility is 10^{-3}–10^{-4} in *SI units*. According to the *Curie law*, the paramagnetic susceptibility is inversely proportional to the rock temperature. See also *magnetic dipole moment*, *paramagnetism*, and *diamagnetic susceptibility*.

paramagnetism: a weak positive *induced magnetization* of rocks that is parallel to the external magnetic field, such as the *Earth magnetic field*. Paramagnetism contributes only a few nanoteslas to the net, or total, observed magnetic field. Typical paramagnetic rock-forming minerals have a small positive *magnetic susceptibility* within the range of 10^{-3}–10^{-4} in *SI units*. Paramagnetic magnetization vanishes in the absence of the externally applied magnetic field, i.e., there is no *remanence* in paramagnetism. The great majority of rock-forming minerals are paramagnetic or diamagnetic. See also *diamagnetism*, *ferrimagnetism*, *ferromagnetism*, and *nanotesla (nT)*.

Parasnis method: a graphical method of estimating the *density* of near surface rocks. The average density value for an entire study area is derived from the formula of the terrain-corrected *Bouguer anomaly* and defined by the slope of the regression line obtained by plotting the *free-air anomaly* values (y-axis) against the difference between the *simple Bouguer correction* and *terrain correction* values (x-axis) calculated with a unit density, such as 1 g/cm^3, for all stations (Parasnis, 1952). An extension of this method for the case of a linear regional *gravity gradient* is presented by Rao and Murty (1973).

Parker algorithm: a computer algorithm based on the *Parker formula* which is used for the calculation of *marine terrain correction* for marine (shipboard) gravity data or *airborne terrain correction* for regional-scale *airborne gravity* surveys. Sometimes referred to as *Parker forward algorithm*.

Parker formula: a formula for the calculation of *gravitational attraction*, or gravity effect, of an undulating *two-dimensional (2D)* surface with a constant *density contrast*. Defining the *Fourier transform* of this gravity effect via the summation of the Fourier transforms of n powers of the undulating surface of the *topography* or *bathymetry*, it is expressed as

$$\mathcal{F}\{g(x)\} = 2\pi G \cdot \Delta\rho \cdot e^{-|k|Z_0} \sum_{n=1}^{\infty} |k|^{n-1}\ \mathcal{F}\{h^n(x)\}/n!,$$

where G is *universal gravitational constant*; $\Delta\rho$ is density contrast across the surface; k is the Fourier-domain *wavenumber* defined as $k=2\pi/\lambda$, where λ is space-domain *wavelength*; $h(x)$ is the undulating surface depth below each observation point with a positive down; Z_0 is a mean depth of the horizontal reference surface defined by an average of the minimum and maximum values of $h(x)$; $\mathcal{F}\{h^n(x)\}$ is the Fourier transform of $h^n(x)$; and $n!$ is factorial, i.e., $n!=1\cdot2\cdot3\ldots n$. The infinite series of this formula is truncated at the term for which the *rms value* is less than some convergence *tolerance*, usually 0.01 mGal (Parker, 1973). The Parker formula is implemented in the calculation of *terrain correction* for line-oriented marine (shipboard) and regional-scale *airborne gravity* data. See also *Fourier domain*, *space domain*, *marine Bouguer correction*, and *milligal (mGal)*.

Parker forward algorithm: a *Parker formula*-based algorithm for *forward calculation* of the gravity effect of an undulating *two-dimensional (2D)* surface with a constant *density contrast*.

partial derivative: the rate of change of a continuous function of several variables per unit of one selected variable. It is calculated with the assumption that there are no changes in the function relative to other variables. In a *1D model*, the partial derivative is represented by a *derivative* of the gravity or magnetic field along one of the mutually orthogonal axes in *Cartesian coordinates*, usually the vertical *z*-axis. See also *directional derivative*.

partial tensor gravity gradiometer: a *rotating-disk gravity gradiometer* system mounted on an inertial *stabilized platform* for measurements of the *differential curvature* components known as *differential curvature gradients* and denoted by g_{NE} and g_{UV}. In the basic design, it has a single horizontally rotating disk with two pairs of equally spaced and synchronized accelerometers, whose sensitive axes are tangential to the disk plane and orthogonal to the disk rotation axis. Each pair of accelerometers forms a *gradiometer arm* along one of the two mutually orthogonal axes of measurements. The disk rotates at a constant speed about the vertical axis, and accelerometer signals are modulated at twice the rotation frequency for the enhanced *common mode rejection* of inertial accelerations and filtering out the mis-match and mis-alignment errors. Outputs of all individual accelerometers are summed, and their combined output is expressed as

$$\begin{aligned}(a_1+a_2)-(a_3+a_4)\\ &= d(g_{xx}-g_{yy})\sin 2\Omega t + 2dg_{xy}\cos 2\Omega t\\ &= 2dg_{UV}\sin 2\Omega t + 2dg_{NE}\cos 2\Omega t,\end{aligned}$$

where (a_1+a_2) and (a_3+a_4) are output signals from pairs of the opposed accelerometers; d is their separation distance or *baseline*; $g_{NE}=g_{xy}$; $g_{UV}=(g_{xx}-g_{yy})/2$; g_{xx}, g_{yy}, g_{xy} are *curvature gradients*; Ω is rotation rate in radians per second ($\mathrm{rad\cdot s^{-1}}$); and t is measurement time. Measured g_{NE} and g_{UV} values are transformed into (a) the *first vertical derivative* g_{zz} often denoted by g_{DD} or *gDD* and (b) the residual component of *vertical gravity* g_z often denoted by g_D or *gD*. Obtained residual gravity g_D differs from *relative gravity* Δg_z measured with the use of a regular *relative gravimeter* by the amount of a regional long-*wavelength* component in the *gravity survey* area. The best reported *half-wavelength resolution* in airborne applications is 0.3–1.0 km with the operational *accuracy* of 3–7 *Eötvös (E)*. Partial tensor gravity gradiometer design is implemented in the *airborne gravity gradiometer (AGG)* systems deployable on fixed-wing (airplane) and helicopter platforms. Sometimes referred to as *horizontal tensor gradiometer* and *two-component tensor gradiometer*. See also *accelerometer*, *inertial acceleration*, *full tensor gravity gradiometer*, and *GOCE gravity gradiometer*.

pass filter: See *band-pass filter*, *high-pass filter*, and *low-pass filter*.

passband: a *filter* parameter that defines the range, or *band*, of spectral-domain wavenumbers or space-domain wavelengths in the input data that is retained, i.e., passed through without suppression, in the output of this filter. The suppressed portion of the input data is referred to as the *stopband*. See also *space domain*, *spectral domain*, *wavelength*, and *wavenumber*.

pattern: 1. a systematic distribution of line-oriented or gridded data values representing the characteristic parameters of *anomaly*, such as *amplitude*, *shape*, and *spatial frequency*, that are known to be associated with a particular *source body* or geological *structure* and, hence, can be used as guides for the identification and *interpretation* of the detected anomalies. **2.** a specific anomalous *signature* used as a criterion for the identification and discrimination of anomalies in automated *pattern recognition* techniques. **3.** a specific distribution of calculated *source depth solutions* associated with the elements of a subsurface geological *structure*, such as *magnetic basement*. See also *gridding*.

pattern recognition: 1. a concept of the automatic detection of structural elements and specific types of sources that is based on the recognition, or identification, of the characteristic patterns on a *grid*, *profile*, or other digital images of anomalous gravity and magnetic fields. **2.** an automated technique for analysis of the spatial distribution of anomalies with the objective of identifying their specific patterns associated with either a particular *source body*, such as *kimberlite intrusion*, *ore body*, and *unexploded ordnance (UXO)*, or elements of the geological *structure* favorable for the accumulation of hydrocarbons, such as a structural or stratigraphic *trap*. **3.** an automated method of *cultural editing* based on the use of anomalous signatures of ferrous man-made objects and electric current conductors. See also *pattern*, *signature*, *Keating method*, and *machine learning (ML)*.

PCA-based separation: a *regional-residual separation* method wherein the *regional field* is approximated by the first *principal component* determined by *principal component analysis (PCA)* of the anomalous gravity or magnetic field (Zhang et al., 2009). See also *anomalous field*.

peak: 1. the maximum positive deviation of a particular variable from its zero level. **2.** the maximum value of a positive *anomaly*. Peak is opposite of *trough* which is the minimum (often negative) value of a gravity or magnetic anomaly, especially in the *residual field*.

pearls: continuous *micropulsations* with periods from 0.1–0.2 to 5 seconds and amplitudes from 0.05 to 0.10 *nanotesla (nT).* Displayed records of pearls appear as *amplitude*-modulated sinusoidal waveforms that resemble a pearl necklace. See also *waveform.*

pendulum gravimeter: an *absolute gravimeter* that utilizes the concept of *reversible pendulum* to determine the swing period and pendulum length that are equal to those of the "ideal" pendulum. The relationship between the period T and length L of the ideal pendulum and *absolute gravity* g is defined as

$$T = 2\pi(L/g)^{1/2}.$$

The best reported *accuracy* of measurements is 0.1–0.2 *milligal (mGal).* Previously, the pendulum gravimeter was the standard instrument for absolute gravity measurements; currently, those measurements are made by an *absolute free-fall gravimeter.*

period: 1. the interval of time T for one cycle of the wave function or between reoccurrences of the *periodic function* and any periodic event in the *time domain.* The period is inversely related to *frequency* f as

$$T = 1/f = 2\pi/\omega,$$

where ω is *angular frequency.* **2.** the interval of time between the reoccurrence of the same *phase*, such as the *peak* or *trough* of a wave-like periodic function. In the *space domain*, the equivalent *waveform* parameter is *wavelength*, so the period of a periodic waveform is either its *fundamental wavelength* or the reciprocal of its *fundamental frequency.* **3.** a range of wavenumbers Δk in the *wavenumber domain* over which a *discrete Fourier transform (DFT)* repeats itself, defined as

$$\Delta k = 2\pi/\Delta x,$$

where Δx is the *sample interval* in the *space domain.* See also *dominant period, fundamental period, sampling wavenumber,* and *wavenumber.*

periodic function: a function that repeats itself at equal intervals of time or space. Sometimes referred to with the same meaning as *harmonic function.*

periodic noise: See *high-line effect.*

permanent dipole: a characteristic of *remanent magnetization* induced by a strong external *magnetic field* and remaining in the *ferrimagnetic* and *ferromagnetic* rock-forming minerals after the external magnetic field has been removed. Also referred to as *permanent magnetization.* See also *induced magnetization* and *Gauss law.*

permanent magnetization: *remanent magnetization* with an emphasis on it being *magnetization* that remains in a rock permanently after the removal of the external, or inducing, magnetic field.

permanent tidal deformation: See *permanent tide.*

permanent tide: the average value of the combined *gravitational attraction* of the Moon and Sun resulting in the elastic deformations of the solid Earth that are called *Earth tides.* Its effect on the equatorial bulge is called *permanent tidal deformation.* The latitude-dependent gravity effect of the permanent tide is compensated by *Honkasalo correction.*

permeability: See *magnetic permeability.*

permeability-susceptibility relation: the relationship between *magnetic permeability* μ and *magnetic susceptibility* k. It is expressed in the SI system as

$$\mu = \mu_0(1 + k)$$

and in the cgs system as

$$\mu = 1 + 4\pi k,$$

where μ_0 is magnetic permeability of a source-free space defined as $\mu_0 = 4\pi \cdot 10^{-7}$ *henry* per meter in *SI units* and $\mu_0 = 1$ *gauss* per *oersted* in *cgs units.*

Peters half-slope method: a graphical method of magnetic source depth estimation from the *principal profile* of a reduced-to-pole (RTP) *magnetic anomaly.* Three lines are drawn at any side of the symmetric anomaly or at the steeper side of the asymmetric anomaly: the *maximum-slope line* through the *inflection point* and two parallel half-maximum slope lines through two points of tangency before and after the maximum slope part. The horizontal distance between two tangent points of the half-maximum slope lines is called the *Peters length.* Its more accurate estimate can be obtained using the principal profile of the *horizontal derivative* of the same anomaly, where the *half-amplitude width* parameter equals the Peters length. Source depth is determined as the Peters length divided by the *Peters index.* A contour map or *grid* can be used for "quick-and-dirty" depth estimates without extracting the profiles, because the distance over which map contours are roughly equally spaced is proportional to the source depth. Here, two points are selected on both sides of the maximum slope area, where the contour line spacing becomes larger than the spacing of the maximum slope contours, so the distance between these two points approximates the Peters length which is divided by the Peters index *default value* 1.6 to obtain the magnetic source depth. See also *half-width rules* and *reduction-to-pole (RTP).*

Peters index: a denominator, or *dividing factor*, in the *Peters half-slope method.* Its value is based on the assumed approximation of a real *source body* by one of the *ideal source* geometries: 0.8–1.0 for a vertical *thin sheet*; 1.0 for a horizontal thin sheet; 1.6 for a vertical *thick sheet*; 1.8 for a pipe-like body; and 1.8–2.0 for a wide source body or *contact.* The Peter index *default value* is 1.6.

Peters length: 1. the horizontal distance between *half-slope points*: two points on the *principal profile* of the reduced-to-pole (RTP) *magnetic anomaly*, where two parallel half-maximum slope lines drawn before and after the inflection point are tangent to the principal profile. **2.** the *half-amplitude width* on the principal profile of the *horizontal derivative* of a magnetic anomaly. The Peters length is independent of the *amplitude* of an anomaly or *magnetization contrast* of its source, and it is directly proportional to the source depth: the deeper the source, the longer the Peters length. Also referred to as *half-slope length.* See also *Peters half-slope method* and *reduction-to-pole (RTP).*

Peters method: See *Peters half-slope method.*

Peters rule: *Peters half-slope method* in the general context of source depth estimation methods based on the use of *depth rules.*

phase: 1. the *waveform* parameter that represents the angular measure of difference, or angle of the relative direction, between a reference plane, such as the zero level, and a specific point on the waveform of an anomalous gravity or magnetic *field.* The phase of an *anomaly amplitude* is ±90° for positive and negative anomalies, respectively. **2.** a spatial orientation parameter, or spatial angle, of three-dimensional (3D) *geomagnetic vector* **B** defined by *magnetic inclination,* whereas *magnetic declination* defines the phase, or *azimuth,* of a 2D geomagnetic vector formed by its horizontal components $\boldsymbol{B}_x$ and $\boldsymbol{B}_y$. **3.** the parameter of 3D *gravity vector* $\boldsymbol{g}$ or *anomalous gravity vector* $\Delta\boldsymbol{g}$ that describes its orientation in the *vector* space. Phase is often referred to with the same meaning as *phase angle, local phase,* and *instantaneous phase.* See also *coherence, analytic signal phase, eigenvalue phase,* and *phase spectrum.*

phase angle: 1. same as *phase* with an emphasis on it being an angular quantity. **2.** a *waveform* characteristic defined by the arctangent of the ratio of the *quadrature,* or 90° out-of-phase, component to the in-phase component of the waveform. **3.** the angle of a *complex variable* vector indicating its direction relative to the real axis in a *complex plane,* defined as

$$\theta = \arctan(\mathrm{ImP} / \mathrm{ReP}),$$

where ImP is the imaginary part and ReP is the real part of a complex variable. Conceptual analogues of the phase angle are *analytic signal phase* and *tilt angle (TA).*

phase-based attributes: attributes of an *anomalous field* that are quantified by the ratio of the orthogonal derivatives of the gravity or magnetic *anomaly.* They define the spatial angle, or *local phase,* of the three-dimensional (3D) *analytic signal (AS), monogenic signal (MS),* and anomalous gravity or magnetic *vector.* The majority of these attributes are represented by the *tilt angle (TA)* and its numerous modifications, such as *ASTA filter, horizontal tilt angle (TDX), improved tilt angle (ITA), second-order tilt angle, tilt angle of horizontal gradient (TAHG), tilt ASA,* and *total horizontal derivative of tilt.* See also *attribute, normalized derivatives, anomalous gravity vector,* and *anomalous magnetic vector.*

phase congruency: a condition at which the *harmonics* of a *Fourier transform* have the same *phase,* i.e., Fourier harmonics are in-phase, and it is observed at a break of the mathematical step function. For gravity and magnetic data, the phase congruency algorithm finds the locations in a gridded gravity or magnetic field where the harmonics of the Fourier-transformed data are maximally in-phase and it maps these locations as *edge lineaments* representing the *edge*-like structural discontinuities, such as high-*offset* faults, high- and low-*contrast* lithological contacts, and *source body* edges. See also *contact, fault,* and *gridding.*

phase filter: 1. a spectral- or space-domain *phase*-changing operator, such as the first-order *horizontal derivative* operator of the $\partial/\partial x$ and $\partial/\partial y$ types or second-order *derivative* operator of the $\partial^2/\partial x\partial z$ and $\partial^2/\partial y\partial z$ types. It shifts the location of an anomalous *peak* and splits the original positive or negative *anomaly* into a positive-negative pair, or peak-to-*trough* anomalous doublet, along the calculation direction. **2.** a space-domain operator for calculation of the *tilt angle (TA).* See also *horizontal gradient amplitude (HGA), space domain, spectral domain,* and *zero-phase filter.*

phase response: 1. a characteristic of phases of anomalies in an observed gravity or magnetic *field.* **2.** a characteristic of the *transfer function* which defines the *phase* of an output signal in the *space domain* as a function of the *wavenumber* in the *spectral domain.* See also *amplitude response.*

phase reversal: a *phase* shift of 180° in a gravity or magnetic field *waveform* that indicates a lateral change in *density* and *magnetization* or structural characteristics. It can form the *peak*-to-*trough* anomalous doublet, such as that of a *shelf-edge gravity anomaly.*

phase shift: 1. a change in locations of anomalous peaks in line-oriented or gridded data after the application of a *phase filter* or *reduction-to-pole (RTP).* **2.** the relative difference in the spatial orientation of two vectors that is measured in degrees or in radians, e.g., the opposite vector orientation is 180° or π radians. See also *gridding* and *vector.*

phase spectrum: 1. *phase*-versus-*wavenumber* characteristic relation which describes the phase components of Fourier-transformed gravity or magnetic data as functions of their respective wavenumbers. **2.** wavenumber-dependent function $\theta(k)$ in the *complex spectrum* of Fourier-transformed data, defined as

$$\theta(k) = \arctan(\mathrm{ImP}/\mathrm{ReP}),$$

where ReP is the real part representing the *cosine transform* components of *Fourier transform* and ImP is the imaginary part representing the *sine transform* components. Also known as *Fourier phase spectrum.* See also *spectrum, amplitude spectrum,* and *power spectrum.*

picotesla (pT): a unit of *sensitivity* of advanced magnetometers, such as the *helium magnetometer* and *SQUID magnetometer,* defined as 1 pT = 10^{-12} *Tesla* = 10^{-3} *nanotesla (nT).* See also *femtotesla (fT)* and *magnetometer.*

pie-slice filter: a *directional filter* in the context of a characteristic *spectral domain* image of its selected parameters, i.e., azimuthal direction with the plus and minus deviations in radians, that is similar in appearance to a slice of pie or pizza. See also *radian.*

piezoelectric effect: a natural phenomenon observed in certain types of rock-forming minerals which generate temporal electric currents and, hence, localized temporal magnetic fields, in response to externally applied pressure.

piezomagnetism: a natural phenomenon of the dependence of rock *magnetization* parameters (*susceptibility, remanent magnetization,* and *magnetic permeability*) in certain types of rock-forming minerals on the amount of externally applied pressure. Also referred to as *inverse magnetostriction.* See also *magnetostriction* and *piezoelectric effect.*

piezoremanent magnetization (PRM): *pressure remanent magnetization* with an emphasis on the *piezoelectric effect* as its major source.

pill-box effect: See *bull's-eye effect.*

pipe: a *source body* approximation in the gravity or magnetic *modeling* of a long narrow *igneous intrusion* with a vertical or near-vertical occurrence, such as *kimberlite pipe*. Referred to with the same meaning as *point pole.* See also *vertical cylinder.*

pitch: the rising-and-falling rotational motion of a *survey* ship or aircraft about the horizontal axis caused by ocean waves or air turbulence during gravity and magnetic measurements along survey lines, i.e., it is rotation of a moving *observation platform* about the Y-axis in the XZ plane. Pitch is a source of magnetic interference effects in an aircraft frame and inertial accelerations of a shipboard (marine) or airborne observation platform. See also *roll*, *yaw*, *aeromagnetic compensation*, *inertial acceleration*, and *lever arm effect.*

pixel: a discrete graphical element represented by a colored dot in a digital image, such as a *raster image*. Pixel is a graphical analogue of *grid cell*, and it corresponds to a certain location on a gravity, magnetic, or topographic digital image or map. The number of pixels per inch, or PPI, defines *spatial resolution*: the smallest area that is distinguishable on a digital image. Pixel-based imaging of the detected linear and curvilinear alignments of anomalies is implemented in the visualization of the *skeletonization* and *steerable filter* results.

placer: a surface or near-surface deposit of high-*density* minerals, such as gold, beforsite (niobium ore), cassiterite (tin ore), rutile (titanium ore), and heavy *rare earth elements (REE)*, in fluvial systems and beach sands. It is formed by (a) separation and deposition of minerals in accordance with their densities and (b) local accumulations of the heavy components of weathered debris transported from nearby alkaline and other *igneous intrusion* types. Alluvial and beach placers can have the *signature* of the low-*amplitude* clustered residual magnetic anomalies generated by the accumulations of *detrital magnetite*. Placers also can be detected by high-accuracy *ground gravity* measurements within the depositional *fairway* areas mapped by a *high-resolution aeromagnetic survey*. See also *clustering*, *ore body*, and *residual magnetics*.

planar anisotropic index: a quantitative measure of the degree of *FTG planar anisotropy*, defined as

$$\mathrm{PAI} = (|\lambda_2| - |\lambda_3|)/|\lambda_1|,$$

where $|\lambda_1|$, $|\lambda_2|$, and $|\lambda_3|$ are magnitudes of *FTG eigenvalues*. The index is normalized by $|\lambda_1|$ to range between 0 and 1, so the sum of the linear, planar, and spherical indices is equal to 1 at every *grid* point. The relationship between three calculated anisotropic indices is visualized by a monochromatic or color-coded *barycentric image* to show the variations of the *FTG anisotropy* parameters associated with the elements of subsurface geological structures (Hassan and Goussev, 2020). See also *magnitude*, *linear anisotropic index*, *spherical anisotropic index*, and *normalization*.

planar anisotropy: See *FTG planar anisotropy*.

plate: a *source body* approximation in the gravity or magnetic *modeling* of a horizontal sheet-like rock body or formation with a finite thickness ranging from 0.1 to 1.0 times the depth to its top. See also *Bouguer plate* and *finite contact*.

platform: 1. *observation platform*. **2.** *stabilized platform*.

platform error correction: *Harrison correction* or *Browne correction* with an emphasis on it being the correction for *platform tilt* caused by a strong horizontal *inertial acceleration*.

platform tilt: the tilt, or deviation from the horizontal level, of a *stabilized platform* subjected to strong horizontal inertial accelerations. Platform tilt results in the artificial increase of the measured *relative gravity* values. Errors in *dynamic gravity* measurements due to platform tilt are referred to as *off-level errors*, and they are compensated by application of *Harrison correction* or *Browne correction*. See also *inertial acceleration* and *sensor tilt*.

platform tilt correction: 1. *Harrison correction*. **2.** *Browne correction*.

plumb line: a string with a small weight at the end that shows a direction of the actual, or true, vertical on the Earth's surface. It is perpendicular to a *geoid*, whereas the theoretical vertical is perpendicular to a *reference ellipsoid*. A plumb line also shows the orientation of the *Earth gravity vector* at a ground observation point. A theoretical plumb line is slightly curved, but its deviation from a straight vertical line is commonly ignored, including the geodetic calculations of *free-air gravity disturbance (FAD)* and *free-air gravity anomaly (FAA)* based on the use of *ellipsoidal height* and *orthometric height*, respectively. See also *deflection of the vertical* and *gravity field line curvature*.

plumb line curvature: See *gravity field line curvature*.

pluton: a large intrasedimentary or intrabasement *igneous intrusion*. Plutons often create high lateral *magnetization* and *density* contrasts with the surrounding sedimentary rocks and, thereby, generate high-*amplitude* positive magnetic and gravity anomalies. Granite plutons can create a negative lateral *density contrast* with the high-density *basement* rocks and, hence, have the *gravity signature* of a relative *gravity low* in the *residual field*. Upper sedimentary plutons, such as *batholith*, are prolific sources of thermal energy. See also *anomaly* and *magnetization contrast*.

plutonic body: See *igneous body*.

plutonic water: *juvenile water* with an emphasis on its origin from the process of cooling and solidification of *magma* within the upper *crust* and the conversion of the viscous magmatic body into a solid igneous, or plutonic, rock body.

Poincare-Prey reduction: a combined correction of the *observed gravity* that accounts for the *Bouguer effect* and *free-air gradient* and extrapolates the observed gravity values into the subsurface. Extrapolated gravity value as a function of depth D below the Earth's surface is expressed in units of *milligal (mGal)* as

$$g(D) \approx g_P - (\partial g/\partial h + 4\pi G\rho)D,$$

where g_P is observed gravity value at observation point P on the Earth's surface, $\partial g/\partial h$ is a negative free-air gradient, G is *universal gravitational constant* in *cgs units*, ρ is *density* of the *Bouguer plate*, and D is depth in meters. For the *nominal free-air gradient* of – 0.3086 mGal·m^{-1}, ocean/sea water *density* 1.027 g·cm^{-3}, and averaged upper *crust* density 2.67 g·cm^{-3}, the extrapolated gravity below the ocean/sea surface is defined as

$$g(D) \approx g_P + 0.2225D$$

and below the ground surface as

$$g(D) \approx g_P + 0.0848D.$$

Gravity field extrapolation can be improved by taking into account the gravity effect due to variations of the *topography* or *bathymetry* around observation points, i.e., *terrain effect*.

point dipole: **1.** a magnetic *point source*. **2.** a *magnetic dipole* with the infinitesimally small distance between its north and south poles that is used as a mathematical approximation for the calculation of magnetic fields and their effects. The amplitude of the *magnetic anomaly* generated by a point dipole falls off as the inverse cube of distance to the *observation point*, and its *Euler structural index* is SI = 3. The same as the *point mass* approximation for gravity sources, the point dipole is an elementary building block for magnetic sources of arbitrary shapes and a fundamental approximation applicable to all sufficiently compact magnetized bodies, both natural and man-made. See also *anomaly*, *fall-off*, and *fall-off rate*.

point-like anomalies: *spike*-like magnetic anomalies generated by man-made ferrous objects, such as abandoned wells, production wells, bridges, water tanks, sunken ships, etc., and by *unexploded ordnance (UXO)* and small-diameter *pipe*-like igneous intrusions. See also *igneous intrusion* and *magnetic anomaly*.

point mass: a gravity source approximation by the rock body or fluid *mass* concentrated at a point. The gravity effect, or *gravitational attraction*, of a uniform *sphere* is approximated by the mass concentrated at its center. The amplitude of the *gravity anomaly* generated by a point mass falls off as the inverse square of distance to *observation point*, and its *Euler structural index* is SI = 2. Any non-spherical local mass can be approximated by the point mass, when its distance from the measuring instrument is approximately 8–10 times the largest *dimension* of a non-spherical body. See also *anomaly*, *fall-off*, and *fall-off rate*.

point pole: a *source body* approximation in the magnetic *modeling* of long narrow axially magnetized *pipe*-like rock bodies of the vertical or near-vertical occurrence, such as *kimberlite pipe*, and presumably infinite depth extent. See also *infinite bottom*.

point source: **1.** a *source body* for which the actual shape and size can be disregarded as pertains to its observed gravity and magnetic effects, because they are closely approximated by those of a small *sphere*. **2.** the mathematical approximation of a magnetic or gravity source body positioned at a distance of approximately 8–10 times its largest vertical or horizontal *dimension* from an *observation point*. Attenuation, or *fall-off*, of the gravity field generated by a point source is a function of the inverse square of the distance r to its source, or 1/r^2, whereas for the magnetic field it is a function of the inverse cube of this distance, or 1/r^3. Any local-scale gravity or magnetic source body can be approximated by a point source, when its field is measured sufficiently far away from the source position. See also *point dipole*, *point mass*, *point pole*, and *single-point source*.

point source anomaly: a gravity or magnetic *anomaly* generated by a *source body* approximated by a *point source*. It is represented by a symmetric bell-shaped *monopolar anomaly* in the anomalous gravity field and an asymmetric *dipolar anomaly* in the *total magnetic intensity (TMI)* field. Asymmetry of the dipolar magnetic anomaly generated by a source with a small *remanent magnetization* component is eliminated by *reduction-to-pole (RTP)*. See also *anomalous field*.

Poisson equation: the differential equation that describes the relationship between the second-order derivatives of *gravitational potential* and continuous distribution of *density* for all points within a three-dimensional (3D) mass-filled space as

$$\nabla^2 U = \partial^2 U/\partial^2 x + \partial^2 U/\partial^2 y + \partial^2 U/\partial^2 z = -4\pi\rho G,$$

where ∇^2 is *Laplacian*, U is gravitational potential, ρ is density, G is *universal gravitational constant*, and the minus sign follows the theoretical convention. In the mass-free regions of space, i.e., outside of the mass distribution where $\rho = 0$, this equation is referred to as *Laplace equation* for the gravity field.

Poisson jump: a change in *apparent density*, measured by a *borehole gravity meter (BHGM)*, across a rock formation boundary penetrated by a borehole at 90° angle. Gravity effects of anomalous masses around a borehole are the same for measurements above and below the horizontal boundary, so the *magnitude* of their change across this boundary is preserved and, therefore, the apparent density change, or Poisson jump, is equal to a real *density* change, i.e., *density contrast* across the boundary. See also *BHGM density* and *Poisson equation*.

Poisson relation: the relation of a *scalar* proportionality between the anomalous magnetic and gravity fields postulated by the *Poisson theorem*: if the boundaries of gravity and magnetic sources are the same and their *magnetization* and *density* are uniform, then the *magnetic potential* of that source is directly proportional to the *derivative* of *gravitational potential* in the direction of the source magnetization, wherein this derivative defines a component of the gravity field *vector* in this direction. The Poisson relation implies that, for the *induced magnetization* distribution in the exact proportion to the *density* distribution, the *RTP magnetic field* and *vertical derivative* of *Bouguer gravity* will have anomalies with almost the same appearance but different units of measurement. See also *anomalous field*, *directional derivative*, and *magnetization direction*.

Poisson theorem: the theorem postulating that, for the same *source body* with a uniform *density* and *magnetization*, the *magnetic potential* is directly proportional to (a) the first-order *derivative* of the *gravitational* *potential* in the source magnetization direction and (b) the *magnitude* of magnetization *vector* **M**, whereby it is inversely proportional to the product of source density ρ and *universal gravitational constant* G. Its formula is expressed as

$$\mathcal{V} = -C_m(\partial U/\partial r)|\mathbf{M}|/\rho G = -g_r C_m M/\rho G,$$

where $\mathcal{V}$ is magnetic potential; U is gravitational potential; $|\mathbf{M}| = M$ is magnitude of the source magnetization vector **M**; the minus sign follows the theoretical convention; C_m is magnetic constant of proportionality: $C_m = 1$ in *cgs units* and $C_m = 10^{-7}$ *henry* per meter in *SI units*; r is direction of magnetization; and $\partial U/\partial r$ is *directional derivative* of the gravitational potential U that defines the component g_r of the *gravity vector* $\boldsymbol{g}$ in the *magnetization direction*, i.e., $\partial U/\partial r = g_r$. The theorem can be extended to relate: (a) the *RTP magnetic field* to the *vertical derivative* of an anomalous gravity field; (b) the *vertical integral* of the RTP magnetic field to an anomalous gravity field, i.e., the *field* of vertical component g_z of the gravity vector $\boldsymbol{g}$; and (c) the *N*th-order vertical derivative of the RTP magnetic field to the corresponding $(N+1)$-order vertical derivative of an anomalous gravity field: the *first vertical derivative* of the RTP magnetic field is directly proportional to the *second vertical derivative* of an anomalous gravity field. The theorem also holds for the derivatives and *tensor* components of a common source, so the vertical derivative of the magnetic potential, i.e., vertical component of the magnetic field B_z, is related to *gravity gradient tensor (GGT)* component g_{zz} as

$$B_z = -\partial \mathcal{V}/\partial z = -g_{zz} M/\rho G.$$

The Poisson theorem shows that the *fall-off rate* of the magnetic anomaly *amplitude* is one power higher than that of the gravity anomaly, which means that both the *lateral resolution* of the magnetic anomalies and their *sensitivity* to variations of the source depth are higher than those of the gravity anomalies. The Poisson theorem provides a theoretical basis for *pseudogravity transformation* and *pseudomagnetic transformation* and comparative analysis of the magnetic and gravity anomalies generated by the same source. See also *anomalous field.*

Poisson wavelet: *mother wavelet* obtained from the *horizontal derivative* of the *gravity anomaly* generated by a *point source*, wherein *wavelet scaling* is controlled either by changing the source depth or by *upward continuation*. See also *wavelet.*

polar flattening: the parameter of a *reference ellipsoid* defined as

$$f = (a - b)/a,$$

where a is the equatorial radius, or semi-major axis, and b is the polar radius, or semi-minor axis, of the reference ellipsoid. The approximate value is 1/298.25 or 0.0033. Polar flattening is a result of the equilibrium, or balance, between the *gravitational force* of attraction attempting to make the Earth's shape spherical and the rotational *centrifugal force* of repulsion trying to flatten the shape. Also known as *ellipsoidal flattening* and *geometrical flattening*. See also *gravitational flattening.*

polar form of complex number: *complex number* $z = x + iy$ expressed as $z = Ae^{i\theta}$, where A is the *modulus* of the complex number defined as $A = (x^2 + y^2)^{1/2}$, $i = \sqrt{-1}$ is a complex unit number indicating the orthogonality, and θ is the *phase* of the complex number representing the angle of the modulus direction relative to the real axis, defined as $\theta = \arctan(y/x)$. See also *complex spectrum.*

polar gravity: the *theoretical gravity* value at the northern and southern geographic poles making the axis of rotation of a *reference ellipsoid.* For the *GRS80 ellipsoid*, it is equal to 9.8321863685 $\text{m} \cdot \text{s}^{-2}$ or 983218.63685 *milligal (mGal).*

polar motion: a change in the position of the axis, or pole, of the Earth's rotation relative to the surface of a *geoid* or *reference ellipsoid.* Polar motions have two dominant components: (1) secular, or slow change, motion assumed to be caused by tectonic movements and convection in the *mantle* with the period of 435 days, referred to as the *Chandler wobble*; and (2) periodic fluctuations and side-to-side movements, or wobbling, of the rotation pole with the period of one year representing the annual seasonal motions. Averaged pole positions are determined at regular time intervals and issued daily by the International Earth Rotation Service (IERS) Bulletin. Changes in *absolute gravity* measurements due to the polar motions are compensated by application of *polar motion correction.* See also *secular variation.*

polar motion correction: a correction applied to the *absolute gravity* measurements by the *free-fall method* to compensate for changes in *centrifugal acceleration* due to a motion of the rotational axis of the Earth, or *polar motion*, resulting in errors of the measured absolute gravity values, i.e., values of *gravitational acceleration.* The pole position determined to be the nearest to the observation time at each *station* is used for calculation of this correction. The maximum correction value is 8–10 *microgal (μGal).* See also *atmospheric pressure correction, diffraction correction, ocean loading correction, self-attraction correction*, and *speed-of-light correction.*

polar motion effect: a planetary-scale gravity effect caused by the periodic and non-periodic movements of the rotational axis of the Earth, or *polar motion*, resulting in changes of *centrifugal acceleration* and, therefore, errors in the high-*accuracy* measurements of *absolute gravity* at the Earth's surface. Polar motion effect is compensated by the application of *polar motion correction.*

polar wandering: **1.** a regional-scale motion of the rotational axis of the Earth over the past and present times. Up-to-date estimates are used for the evaluation of *polar motion effect* and calculation of *polar motion correction.* **2.** slow and irregular rotation of the *geomagnetic poles* about the rotational axis of the Earth with the period of approximately 7000 years. Apparent pole locations in past geological times are determined from the paleomagnetic data acquired at different lithospheric plates and from their corresponding polar wandering curves. Estimates are based on measurements of the *remanent magnetization* direction in rock samples of known geological age and used in plate tectonic analysis. See also *lithosphere*, *paleomagnetism*, and *geomagnetic excursions*.

polarity: **1.** the positive or negative sign of a gravity or magnetic *anomaly*. **2.** a *skeletonization* parameter that defines the orientation of the anomalous *peak*-to-*trough* pair, i.e., peak-to-trough or trough-to-peak sequence. **3.** *magnetic polarity*.

polarity chrons: time intervals of the stable and nearly constant orientation, or *geomagnetic polarity*, of the *main geomagnetic field* in the geological past. Their typical duration is estimated at 10^5–10^6 years. Also known as *magnetic chrons*. See also *geomagnetic polarity reversals* and *geomagnetic excursions*.

polarity excursions: See *geomagnetic excursions*.

polarity reversals: See *geomagnetic polarity reversals*.

polarization: **1.** *magnetization* in the context of its property to line up, or polarize, the originally random-oriented magnetic dipoles within rock-forming minerals. **2.** the alignment of protons by an orthogonal oscillating electromagnetic field or by the monochromatic pulses of the laser light in *proton magnetometer* and other *quantum magnetometers*. See also *magnetic dipole*.

polarization vector: **1.** a *vector* of the rock *magnetization* **M** defined by the vector sum of *remanent magnetization* vector $\mathbf{M}_{RM}$ and *induced magnetization* vector $\mathbf{M}_{IND}$ as

$$\mathbf{M} = \mathbf{M}_{RM} + \mathbf{M}_{IND}.$$

2. a vector that shows the *magnetization direction* of a rock body at the present time, i.e., induced magnetization vector, or at the geological time when the rock body was formed and magnetized, i.e., remanent magnetization vector.

pole: See *magnetic pole* and *point pole*.

pole strength: See *magnetic pole strength*.

polyhedron: a *source body* approximation by a closed surface bounded by the flat triangular, square, or polygonal facets in the gravity and magnetic *modeling* of three-dimensional (3D) source bodies of arbitrary shapes. For homogeneous polyhedra, the gravity or magnetic field can be described by numerical integration of the facet contributions corresponding to the individual vertices of a particular polyhedron. See also *vertex*.

polynomial filtering: *polynomial fitting* in the context of its application for removing, or filtering out, the middle- and long-*wavelength* field components by subtracting the calculated polynomial-approximated fields from the original anomalous gravity or magnetic field in order to highlight the anomalies with shorter wavelengths and smaller amplitudes. See also *anomalous field*.

polynomial fitting: a method of approximating the *regional field* component of an anomalous gravity or magnetic field by the polynomial surfaces of different orders, wherein the first-order polynomial surface is a dipping flat plane and the theoretical *N*th-order surface represents a complete fit to the original *anomalous field*. The basic concept of the polynomial fitting is that successively higher order approximating surfaces fit the anomalous field components with successively shorter wavelengths. In practice, the highest order polynomial surface is commonly the seventh or ninth. Departures of the anomalous gravity or magnetic field values from a particular polynomial surface represent the *residual field* components of the polynomial-corresponding order that are known as polynomial residuals, such as the first-order polynomial residual, second-order polynomial residual, etc. Sometimes referred to as *polynomial filtering*. Also known as *surface fitting*. See also *amplitude*, *anomaly*, and *wavelength*.

polynomial residual: the *residual field* obtained by calculating the difference between an anomalous gravity or magnetic field and its approximation by the polynomial surface of a pre-selected order, such as the first-order residual, fifth-order residual, or any other order residual field chosen by an *interpreter*. Polynomial residuals highlight the short- and mid-*wavelength* low-*amplitude* sedimentary gravity and magnetic anomalies of an *exploration* interest, including those generated by local-scale geological structures, such as horsts, grabens, rollover anticlines, carbonate buildups (reefs), incised valleys, etc., and anomalies associated with depositional fairways, *hydrocarbon seepage effect*, *placer* deposits, structural and stratigraphic traps, subcrops, etc. See also *anticline*, *graben*, *horst*, *reef*, *incised valley*, *fairway*, *subcrop*, *trap*, *sedimentary gravity anomaly*, and *sedimentary magnetic anomaly*.

positioning: determining the geographic coordinates and *elevation* of stationary observation points on the ground or moving *observation platform*, such as a *survey* ship or aircraft, anywhere on or above the Earth's surface. Stationary *3D coordinates* (latitude, longitude, and elevation) relative to a ground geodetic reference point, or beacon, are determined with the use of a *reference ellipsoid* adopted by the national or regional *Geodetic Reference System*. The 3D coordinates of a moving observation platform are determined by *Global Positioning System (GPS)* and *differential GPS (DGPS)*. Positioning errors depend on the positioning system type and vary from 50–100 m to less than 1 m.

positive pole: **1.** one of two magnetic poles in a *magnetic dipole* toward which the magnetic field lines of the magnetic dipole are oriented and concentrated. Outside the dipole, the field lines are directed from the positive, or north, pole to the negative, or south, pole of the magnetic dipole, whereas inside the dipole the field lines are directed from the negative to the positive pole to close the loops formed by field lines of the magnetic dipole. **2.** north-seeking *magnetic pole*. See also *negative pole*.

potassium magnetometer: a *scalar magnetometer* of the *optically pumped magnetometer* type utilizing the gaseous, or vaporized, potassium isotopes. Its high-performance characteristics are among the best of all optically pumped magnetometers: *sensitivity* up to 0.0005 *nanotesla (nT)*, *precision* up to 0.0001 nT, and *absolute accuracy* of measurements 0.05 nT. Small *heading error*, ultra-high sensitivity, high absolute accuracy, and high *sampling rate* make this magnetometer applicable in both stationary and mobile measurements, including the *unmanned aerial vehicle (UAV)* and high-precision *magnetic gradiometer* systems. Also referred to as *potassium-vapor magnetometer.* See also *cesium magnetometer* and *UAV magnetometer.*

potassium-vapor magnetometer: a *potassium magnetometer* with an emphasis on its *magnetic sensor* utilizing the gaseous, or vaporized, potassium isotopes.

potential: 1. a continuous mathematical function of the physical *field* of force, such as the gravity or magnetic field, which represents a *scalar* quantity described in terms of the energy, or work, required to move a particle to a given position in the presence of the force acting upon this particle. The measurable part of the potential is presented in the relative values of the *potential difference* defined by the amount of work needed to move a particle from one point to another within this field. **2.** a theoretical physical quantity representing the amount of work, or energy, required to move a unit *magnetic dipole* or unit *mass* from infinity to a given position. As a scalar quantity, the potential is directionally independent, unlike the gravity or magnetic field which is a *vector field* the direction and *magnitude*, or strength, of which are defined by the *gradient vector* of the potential. The vector component of the gravity or magnetic field in any particular direction is equal to the *derivative* of the potential in that direction. The surface of the equal values of the potential is called the *equipotential surface.* See also *gravitational potential, normal gravitational potential, total gravitational potential, self-gravitational potential, centrifugal potential, disturbing potential, magnetic potential,* and *anomalous potential.*

potential coefficients: spherical harmonic coefficients in the global *geopotential model,* such as *Earth Gravitational Model 2008 (EGM2008).* Spherical harmonic expansion completed to the degree and order N involves $(N+1)^2$ spherical harmonic coefficients that quantify the relative importance, or weight, of each of the *spherical harmonics,* wherein the degree N describes a *wavelength resolution* of the geopotential model. See also *spherical harmonic expansion.*

potential difference: a measurable part of *potential* defined by the amount of work, or energy, required for the physical field of force to move a unit *magnetic dipole* or unit *mass* from one point to another within this field.

potential field: a *vector field* of the physical force, such as the gravity or magnetic field, whose direction and strength are uniquely described at every point within this field by the *gradient* of a scalar *potential,* wherein the *vector* component of the potential field in any particular direction in space is defined by the *derivative* of the scalar potential in that direction. In a source-free space, the potential field obeys the differential equation known as the *Laplace equation,* which formulates the relationship between the spatial derivatives of the potential field in terms of the second-order derivatives of the potential. Magnetic exploration is based on measurements of the numerical value, or *magnitude,* of the *Earth magnetic field* and its gradients that represent the second-order derivatives of *magnetic potential.* Gravity exploration is dominated by the *relative gravity* measurements of the vertical component of the *Earth gravity field,* or *vertical gravity,* and its vertical and horizontal gradients representing the second-order derivatives of *gravitational potential.* See also *scalar field* and *full tensor gradient (FTG).*

potential field method: a geophysical method, such as the *gravity method* or *magnetic method,* based on measurements of the quantities described by various-order spatial derivatives of *gravitational potential* or *magnetic potential,* respectively. See also *derivative.*

potential field transformation: a process of converting the anomalous gravity or magnetic *field* into a new and physically meaningful form, or *format,* in order to reveal new anomalies and facilitate the *interpretation* of all detected anomalies. See also *anomalous field* and *attribute.*

potential function: a continuous mathematical function representing the *scalar* quantity the *gradient vector* of which defines the direction and strength, or *magnitude,* of the *potential field* at any point within this field. Directional components of the potential field *vector* are quantified by derivatives (gradients) of the potential function in these directions, i.e., the *field* of the gradient vector of the potential function represents the gravity or magnetic *vector field.* Referred to with the same meaning as *potential.* See also *derivative* and *gradient.*

Potsdam Datum: the *absolute gravity* value of 981274 ±3 *milligal (mGal)* determined from high-*precision* measurements by a *pendulum gravimeter* in Potsdam, Germany, in 1906. Adopted as the international standard for the gravity measurements in 1930s, the Potsdam Datum has been proven to be higher by approximately 14 mGal than the current international standard adopted in 1971 and known as the *International Gravity Standardization Network 1971* or *IGSN71 datum.* Also referred to as *Potsdam gravity.* See also *milligal (mGal).*

Potsdam gravity: the *absolute gravity* value referred to as the *Potsdam Datum* which was the *gravity standard* before the introduction of the *IGSN71 datum.*

power-density spectrum: 1. *power spectrum* in the context of quantitative estimates of the distribution of the *spectrum* components of Fourier-transformed data represented by their squared amplitudes, or power, as functions of their respective wavenumbers. **2.** *power spectral density.* See also *wavenumber.*

power rosette: same as *map power spectrum* with an emphasis on its characteristic shape of the highest *power spectrum* values at its center and their gradual decrease along the orientation of the traverse and tie *survey* lines. See also *tie lines* and *traverse lines.*

power spectral density: 1. *power-density spectrum.* **2.** representation of the *power spectrum* components as functions of the degree N of *spherical harmonic expansion* in the global model of the *gravitational potential* of the Earth, such as the *Earth Gravitational Model 2008 (EGM2008).* See also *spherical harmonics.*

power spectrum: 1. an *amplitude* characteristic of the *anomalous field* as a function of *wavenumber.* It is displayed in the *format* of a two-dimensional (2D) logarithmic plot of the total energy, or power, quantified by the sum of the squared amplitudes of all individual wavenumber components of a Fourier-transformed field. Decimal fractions of the *Nyquist wavenumber* and reciprocals of wavenumbers are also used in power spectrum plots instead of wavenumbers. The lowest wavenumbers correspond to the longest wavelengths, which have the highest amplitudes and, hence, the greatest power, so the power spectrum has the descending shape of a decline curve. This curve is approximated by 3–5 linear segments, or slopes, for magnetic source depth estimates by the *Spector-Grant method* and the selection of parameters for *pseudodepth slicing*, *matched filter*, and other spectral-domain filters. The size, shape, and depth extent of the source bodies strongly influence their power spectra. Power spectrum is commonly computed for gridded data and, sometimes, for the separate *survey* lines and profiles extracted from a *grid*. There are two types of the power spectrum for gridded data: *radial power spectrum* and *map power spectrum*. **2.** a wavenumber-dependent function in the *complex spectrum* of Fourier-transformed data that is described by the squared *amplitude spectrum* and expressed as

$$|A(k)|^2 = (\mathrm{ReP})^2 + (\mathrm{ImP})^2,$$

where k is wavenumber, ReP is the real part representing the *cosine transform* components of *Fourier transform*, and ImP is the imaginary part representing the *sine transform* components. Also known as *energy-density spectrum*, *Fourier power spectrum*, *log-energy spectrum*, *log-power spectrum*, *power-density spectrum*, and *spectral density*.

power spectrum map: See *map power spectrum*.

Pratt gravity anomaly: See *Pratt isostatic anomaly*.

Pratt gravity effect: *isostatic compensation effect* calculated with parameters of the *Pratt-Hayford model*. Its *magnitude* comprises the amount of *isostatic correction* known as the *Pratt-Hayford gravity correction*.

Pratt-Hayford gravity correction: an *isostatic correction* calculated with parameters of the *Pratt-Hayford model* of regional-scale *isostatic compensation*.

Pratt-Hayford model: a mathematical model of the regional-scale *isostatic compensation* based on the *Pratt hypothesis* of *isostasy* which assumes that (a) crustal and upper *mantle* densities vary laterally in proportion to the topographic *elevation* or bathymetric depth and (b) the *isostatic compensation level* is within the lower part of the *lithosphere* at the constant depth of 100 km below sea level called the *depth of compensation*. See also *Airy-Heskanen model* and *Vening Meinesz model*.

Pratt hypothesis: one of three basic theories of *isostasy*. It assumes that crustal and upper *mantle* densities vary laterally with *topography* and *bathymetry* to maintain the isostatic balance: topographic highs have rock densities beneath them that are reduced in direct proportion to the topographic heights, and large topographic depressions and oceans have rock densities beneath them that are increased in direct proportion to the bathymetric depths. Accordingly, the lowest rock densities are under the highest surface elevations, and the highest rock densities are under the deepest parts of the oceans. Rock mass excesses above sea level and rock mass deficiencies below sea level are compensated within the lower *lithosphere* at the constant depth of approximately 100 km called the *depth of compensation*, i.e., below the *Moho boundary*. The Pratt hypothesis is preferred for calculation of the *isostatic compensation effect* over the mid-ocean ridges and other large areas of high heat flow and thermal expansion that correlate with the lateral decrease in rock densities within the lithosphere. See also *Airy hypothesis*, *Vening Meinesz hypothesis*, *mass excess*, *mass deficiency*, and *thermal gravity anomaly*.

Pratt isostatic anomaly: an *isostatic anomaly* obtained by application of *isostatic correction* calculated with parameters of the *Pratt-Hayford model* of *isostatic compensation*. Sometimes referred to as *Pratt gravity anomaly*. See also *Airy isostatic anomaly* and *flexural isostatic anomaly*.

precession: rotation of the axis of rotation, or rotation of the spin axis, of charged subatomic particles (protons and electrons) about the direction of and in the plane orthogonal to the stable ambient magnetic field, such as the *Earth magnetic field*, after their *polarization* by the oscillating electromagnetic field or by monochromatic pulses of laser light. The resonant *angular frequency* of precession of the spinning protons and electrons is directly proportional to the intensity, or strength, of the Earth's magnetic field, and it is called the *Larmor frequency*. The operational principle of the *proton-precession magnetometer*, *Overhauser magnetometer*, *potassium magnetometer*, and other *quantum magnetometers* is based on measurements of the Larmor frequency. See also *magnetic resonance*.

precision: an instrument characteristic which describes the range of deviations of the measured values from each other, i.e., repeatability or scattering of measurements: the lower the scattering, the higher the precision of the measured values. Precision is estimated by a mean (average) deviation of measurements from their average value. It is commonly calculated as the root-mean-square value, or *rms value*, averaged over a *measurement bandwidth*. For *random noise* approximations, precision is scaled inversely by the square root of the time for each measurement, so the higher precision is obtained for the longer duration of measurements. For magnetic measurements, the precision estimates are made for both the instrument and survey performance evaluation and presented in units of *nanotesla (nT)* per square root of *hertz* ($\mathrm{nT}/\sqrt{Hz}$). For airborne gravity and gravity gradient measurements, precision is presented in units of $\mathrm{m \cdot s^{-2}}/\sqrt{Hz}$, $g/\sqrt{Hz}$, $\mathrm{mGal}/\sqrt{Hz}$, and $\mathrm{E}/\sqrt{Hz}$, where g is *nominal gravity* value of 9.8 $\mathrm{m \cdot s^{-2}}$. Precision is different from *accuracy*, but sometimes it is used as its estimation substitute. See also *Eötvös (E)*, *sensitivity*, and *survey precision*.

preconditioning: See *data conditioning*.

predicted gravity: a gravity value which accounts for all known planetary and non-geological gravity effects contributing to the *observed gravity* values. Predicted gravity Δg_{PG} represents the total of the main *observed gravity corrections* and can be expressed in units of *milligal (mGal)* as

$$\Delta g_{PG} = \Delta g_{TG} - \Delta g_{atm} - \Delta g_{FA} + \Delta g_{sB} \pm \Delta g_{BB} \pm \Delta g_{Ter} \pm \Delta g_{Eo} + \Delta g_{Is} \pm \Delta g_{IE},$$

where Δg_{TG} is *theoretical gravity rrection*; Δg_{atm} is *atmospheric correction*; Δg_{FA} is *free-air correction* for elevations measured above sea level or *height correction* for elevations above a *reference ellipsoid*; Δg_{sB} is *simple Bouguer correction*; Δg_{BB} is *Bullard B correction*; Δg_{Ter} is *terrain correction*; Δg_{Eo} is *Eötvös correction*; Δg_{Is} is *isostatic correction*; Δg_{IE} is *indirect effect correction*; and $\pm$ indicates the variable sign of a particular correction that depends on (a) the *elevation* of the observation point for the Bullard B correction, (b) the measurement position relative to the surrounding bathymetry or topography for *marine terrain correction* and *airborne terrain correction*, (c) the azimuthal direction of a moving *observation platform* for the Eötvös correction, and (d) measurements on the continents or in the seas and oceans for the indirect effect correction. Predicted gravity is subtracted from observed gravity on a point-by-point basis in order to obtain the *gravity anomaly* value at each *observation point*. See also *azimuth* and *gravity data corrections*.

predictive IGRF model: See *IGRF model*.

preferential continuation: a spectral-domain *enhanced processing* technique based on the use of the *Green's equivalent layer* and *optimum filter* concepts for *upward continuation* and *downward continuation* of the anomalous gravity and magnetic fields. Unlike the regular upward and downward continuations, the preferential continuation (PC) selectively enhances a user-specified range of spatial frequencies in the *amplitude spectrum* of Fourier-transformed data, so downward PC can enhance the deep-sourced anomalies without amplifying the anomalies from shallow sources, and upward PC can suppress the high-frequency *noise* and shallow-sourced anomalies with a minimal attenuation of the medium- and long-*wavelength* components of the observed fields (Pawlowski, 1995). See also *anomalous field*, *Fourier transform*, *spatial frequency*, and *spectral domain*.

prefilter transformation: *data conditioning* in the context of the supplementary procedures applied to line-oriented and gridded data in order to improve the results of a subsequent *filtering*.

preprocessing: same as *data conditioning* that represents a sequence of preliminary *data processing* operations applied to line-oriented data to prepare them for subsequent *data reduction* and *gridding*.

pressure-remanent magnetization (PRM): permanent, or remanent, rock *magnetization* acquired under externally applied pressure, or stress. Its effect is much stronger when the temporal elastic deformation evolves into a permanent plastic deformation of rocks. Also referred to as *piezoremanent magnetization (PRM)*. See also *remanent magnetization* and *piezomagnetism*.

pre-survey flight tests: three standard instrument tests conducted prior to commencing an airborne magnetic survey: (1) *figure-of-merit (FOM)*, which evaluates the performance of *aeromagnetic compensation* for the magnetic interference effects generated by the aircraft frame, electronics, and engine; (2) *heading error test*, which estimates the total amount of *heading error* associated with the alternating directions of the aircraft flight relative to the direction of the *Earth magnetic field*; and (3) *lag test*, which is performed to determine the lag, or time delay, in the response of the electronic and *positioning* systems that results in displacement of the measured data values from their true location.

primary gravity base: See *absolute base station*.

primary lines: See *traverse lines*.

principal axis: 1. the *strike*, or long axis, of a gravity or magnetic *anomaly*. **2.** the direction, or orientation, of the unit length eigenvector paired with the largest eigenvalue. It is determined in the process of *principal component analysis (PCA)* and represents a local trend along which the similarity, or *coherence*, between the data parameters is maximized. Also referred to as *principal direction*. See also *eigenvalues* and *eigenvectors*.

principal component: 1. the largest eigenvalue paired with a unit length eigenvector and computed from a *covariance matrix* to quantify the degree of data correlation, or *covariance*, along the *principal axis* in a given dataset. **2.** a data *pattern* described by the eigenvector-eigenvalue pair and determined in the process of *principal component analysis (PCA)* which is applied to define the local trend in a given dataset along which the similarity, or *coherence*, between the data parameters is maximized. Eigenvector-eigenvalue pairs are computed for each data *dimension*, so the *N*-dimensional dataset will have *N* principal components. All principal components are arranged in the descending order of their covariance quantified by the corresponding eigenvalues, from the highest to the lowest, so the first principal component is represented by the eigenvector with the largest eigenvalue denoted by λ_1. **3.** the vertical component of the *Earth gravity field* denoted by g_z and referred to as *vertical gravity*. See also *eigenvalues* and *eigenvectors*.

principal component analysis (PCA): 1. a sequence of the data transformation procedures applied to a multidimensional dataset with the objective of identifying the data patterns, or principal components, which represent the orthogonal linear combinations of the original data that can be used in the *interpretation* of the geological structures. The PCA determines a new set of the orthogonal axes along which the covariant similarity, or *coherence*, between the data parameters, such as *amplitude* or *phase*, is maximized. The PCA includes four basic procedures: (1) assembling the *covariance matrix* by cross-correlating the data in each *dimension* with itself and with the data from all other dimensions; (2) decomposition of the computed covariance matrix into the eigenvector-eigenvalue pairs, where the unit length *eigenvectors* represent the new orthogonal axes for every dimension in the dataset and *eigenvalues* quantify the data *covariance* relative to the original *x*-, *y*-, *z*-axes; (3) projecting the data from every dimension onto the eigenvector-derived axis to obtain an *eigenimage* showing how much of the data coherence is achieved with each eigenvector; and (4) selection of the most representative eigenimages for combining them into the composite images, such as *RGB composite image*, for visualization of the coherent data trends. The most coherent eigenimages are associated with the largest eigenvalues representing the highest covariance, whereas the least coherent and incoherent, or random, eigenimages are associated with the smallest eigenvalues representing the lowest covariance, including the *random noise*. Suppression of *noise* and increase of *signal-to-noise ratio (S/N)* is achieved by rejecting the low-coherent and incoherent eigenimages from the final visualization of the PCA results. **2.** decomposition of a multidimensional dataset into the data components obtained by projecting the original data into the new mutually orthogonal coordinate axes along which the data coherence is maximized. See also *dimension*, *pattern*, *principal component*, and *PCA-based separation*.

principal curvatures: 1. curvatures of a continuous *three-dimensional (3D)* surface in two mutually orthogonal directions along which the surface curves, or bends, the most and the least. **2.** the *maximum curvature* and *minimum curvature* of the gravitational or magnetic *equipotential surface*. See also *curvature*, *curvature attributes*, and *rotated gradient tensor*.

principal direction: 1. *principal axis*. **2.** the direction along which the *curvature* of a *three-dimensional (3D)* continuous surface, such as *equipotential surface*, has its maximum or minimum value at a given observation point.

principal invariants: three basic *invariants* of a gravity or magnetic *FTG matrix* denoted by I_0, I_1, I_2 and defined as

$$I_0 = T_{xx} + T_{yy} + T_{zz} = 0,$$

$$I_1 = T_{xx}T_{yy} + T_{yy}T_{zz} + T_{xx}T_{zz} - T_{xy}^2 - T_{yz}^2 - T_{xz}^2,$$

$$I_2 = T_{xx}(T_{yy}T_{zz} - T_{yz}^2) + T_{xy}(T_{yz}T_{xz} - T_{xy}T_{zz}) + T_{xz}(T_{xy}T_{yz} - T_{xz}T_{yy});$$

where T_{ij} are elements of the FTG matrix, wherein each element represents the *derivative* of the *i*th component of the gravity or magnetic field *vector* in the *j*th direction; $i, j = (x,y,z)$. See also *FTG invariants*.

principal profile: a *profile* across the central part of a gravity or magnetic *anomaly* and orthogonal to the *strike*, or long axis, of this anomaly. Measurements of the *amplitude* attenuation, or *fall-off*, of a particular anomaly with distance from its source and source depth estimates are made along the principal profiles. See also *central amplitude* and *depth rules*.

prism: a semi-infinite vertical parallelepiped for which the vertical *dimension* is significantly larger than the lateral dimensions. As an *ideal source* approximation, the prism is used in various *model*-based applications, including source depth estimates, calculation of gravity *terrain correction* and *gradient terrain correction*, and as the elementary fill-in block in the approximation of arbitrary-shaped sources. Its *two-dimensional (2D)* version is referred to with the same meaning as a vertical or dipping *dike*. See also *dip*, *infinite bottom*, and *infinite length*.

processing: See *data processing*.

profile: 1. the graphical plot, or graph, of a measured quantity against the horizontal distance. **2.** two-dimensional (2D) representation of gravity or magnetic *field* values in the *format* of a graphical plot, where one *dimension* is the field value and another dimension is the distance along a particular direction. Profile-based source depth estimates by *depth rules* assume (a) the profile direction is orthogonal to the *strike* of the analyzed *anomaly* and (b) the profile is extracted across the central part of this anomaly. See also *principal profile*, *profile length*, and *profile line*.

profile curvature: a *curvature attribute* representing the rate of change of the *slope* of a gridded gravity or magnetic field in the direction of its steepest *gradient*. Profile curvature K_p is calculated at each *grid* point and expressed as

$$K_p = (f_{xx}f_x^2 + 2f_{xy}f_xf_y + f_{yy}f_y^2)/(f_x^2 + f_y^2) \times (f_x^2 + f_y^2 + 1)^{3/2},$$

where $f_x = \partial F/\partial x$, $f_y = \partial F/\partial y$, $f_{xx} = \partial^2 F/\partial x^2$, $f_{yy} = \partial^2 F/\partial y^2$, $f_{xy} = \partial^2 F/\partial x \partial y$, and $f_x^2 = (\partial F/\partial x)^2$, $f_y^2 = (\partial F/\partial y)^2$ are horizontal derivatives of the anomalous gravity or magnetic field F (Roberts, 2001). Zero contours of the profile curvature represent the laterally contrasting *density* and *magnetization* contacts associated with high-*offset* faults, *igneous body* edges, lithological boundaries, and other *edge*-like structural discontinuities. See also *anomalous field*, *balanced profile curvature*, *balanced ASA (ASAB)*, *contact*, and *horizontal derivative*.

profile length: a parameter defined by the requirements of the *profile* extraction objectives, such as *modeling*, source depth estimate, illustration, etc. For the *two-dimensional (2D)* modeling of anomalies generated by isolated sources, the minimum profile length is 6–10 times the expected source depth to represent the effective surface range of a particular gravity or magnetic *anomaly*. See also *gravity anomaly* and *magnetic anomaly*.

profile line: 1. the line on a gravity or magnetic *grid* along which the *profile* is extracted for *modeling* or illustration purposes. Profile lines for two-dimensional (2D) *modeling* are assumed to be orthogonal to the *strike* of the targeted geological *structure* in order to satisfy the 2D approximation requirements. **2.** the *acquisition* line along which gravity or magnetic measurements are made. See also *target*.

projection: 1. a concept of transferring the observation points from one surface to the corresponding positions on another surface. **2.** a method of displaying the gravity and magnetic *survey* data acquired in three-dimensional (3D) space on the two-dimensional (2D) planar surface of a *grid* or map. See also *acquisition* and *map projection*.

proof mass: a small body of *mass* suspended by a metal or quartz spring or levitating in the electromagnetic or electrostatic field in a *relative gravimeter*, *gravity gradiometer*, and *accelerometer*. The weight of the suspended mass in different types of instruments varies from 3–5 milligrams to 5–7 grams. Deviations of the proof mass from the *null point* in a relative gravimeter are directly proportional to changes in *gravitational attraction*. Also referred to as *test mass* and *reference mass*. See also *spring gravimeter*, *superconducting* *gravimeter*, *electrostatic gradiometer*, and *electromagnetic accelerometer*.

property inversion: *inversion* of the anomalous gravity or magnetic field for a physical property (*density* or *susceptibility* contrast) of the modeled exploration *target* whose structural parameters, such as depth and shape, are defined by *constraints* or approximated by the analogy-based assumptions. Property inversion is complementary to *structural inversion*, where the presumably known parameter of an inverted *model* is the density or susceptibility *contrast* of a targeted geological object. See also *anomalous field* and *stochastic inversion*.

prospect: 1. the probable location of a mineral deposit or hydrocarbon accumulation. **2.** a local area of the oil-and-gas or mineral exploration with the anomalous gravity or magnetic signatures indicating the presence of a geological *structure* favorable for the oil-and-gas accumulation or mineral deposit occurrence and recommended for the additional studies. Referred to with the same meaning as *exploration lead*. **3.** a local area selected for detailed oil-and-gas or mineral exploration. See also *detailed survey* and *signature*.

prospecting: *exploration* in the context of acquiring gravity and magnetic data with the objective of locating an oil-and-gas or mineral *prospect*. See also *acquisition*.

protocol: a set of computer codes which regulate the communication between the data transfer devices and systems in a computer network. For successful communication, all devices must recognize the same protocols.

proton magnetometer: a *scalar magnetometer* with an operating principle based on *magnetic resonance* and measurements of *precession* of spinning hydrogen nuclei (protons) in a hydrogen-rich fluid, such as kerosene or alcohol. In order to align the protons, the fluid is polarized by short-time exposure to a strong magnetic field generated by the alternating electric current in a wire coil around the fluid container. Polarization is oriented roughly orthogonal to the measured *Earth magnetic field*. When the polarizing field is removed, the spinning protons begin to precess about direction of the Earth's magnetic field at the *Larmor frequency* which is related to the intensity, or strength, of the Earth's magnetic field in a direct proportion defined by the *gyromagnetic ratio* of protons. The frequency of the electric current, induced by precessing protons into the measuring coil around the fluid container, is measured by a frequency counter to determine the intensity of the Earth's magnetic field. The achievable *absolute accuracy* of measurements is 0.05–0.10 *nanotesla (nT)*. Also referred to as *proton-precession magnetometer*, *proton-resonance magnetometer*, *proton-spin magnetometer*, and *nuclear-precession magnetometer*.

proton-precession magnetometer: *proton magnetometer* with an emphasis on its operational principle based on measuring the resonant *angular frequency* of the *precession* of hydrogen nuclei (protons).

proton-resonance magnetometer: *proton magnetometer* with an emphasis on *magnetic resonance* as the basic principle of its operation.

proton-spin magnetometer: *proton magnetometer* with an emphasis on its operating principle based on measuring the resonant *angular frequency* of the *precession* of the spinning (rotating about the axes of rotation) protons.

pseudoanomaly: the magnetic equivalent of a *gravity anomaly* calculated from a *magnetic anomaly* of the same source or vice versa, i.e., the gravity equivalent of a magnetic anomaly calculated from a gravity anomaly of the same source, in accordance with the *Poisson theorem*. See also *artificial anomaly*, *pseudogravity anomaly*, and *pseudomagnetic anomaly*.

pseudocolor: a multicolored display of gravity and magnetic fields as well as their anomalies and attributes, wherein each color shade corresponds to a particular range of values, from the lowest to the highest. By convention, the "hot" colors, such as red and pink, are assigned to the high values, whereas the "cold" colors, such as dark blue and magenta (reddish purple), are assigned to the low values. The grid in the pseudocolor *format* is referred to as a *color-coded grid* or *color-scaled grid*. Also known as *false color*. See also *anomaly* and *attribute*.

pseudodensity: the magnetic analogue of *density* used in lieu of the actual *density contrast* in the three-dimensional (3D) *inversion* of the *pseudogravity* anomalies generated by isolated source bodies. It is defined as

$$\rho_{psd} = k\mathrm{H}/\mathrm{G},$$

where k is *magnetic susceptibility*; H is total magnetic field intensity, i.e., *TMI field* or *RTP magnetic field* value; and G is *universal gravitational constant*. The relationship assumes *induced magnetization* of rocks with a negligible amount of *remanent magnetization* (Pratt and Shi, 2004). Also referred to as *pseudodensity contrast*. See also *pseudosusceptibility* and *Poisson theorem*.

pseudodensity contrast: *pseudodensity* with an emphasis on the *inversion*-based estimate of the magnetic analogue of a *density contrast* from a *pseudogravity anomaly*.

pseudodepth slicing: a *grid*-based spectral-domain *filtering* technique applied to anomalous gravity and magnetic fields to obtain several grids which highlight the different ranges of wavenumbers corresponding to different *wavelength* components generated by the grid-averaged assemblages of sources within 3–5 depth intervals determined by the *Spector-Grant method*. Wavelengths associated with the depth interval "sliced" by the selection of a particular linear segment, or slope, of a *radial power spectrum* will be enhanced relative to other wavelength components of the gridded data. Spectral overlap, or *spectral leakage*, between the different pseudodepth slices and grid averaging of different source depths in the assemblages make the depth estimates by this technique very approximate. Referred to with the same meaning as *matched-filtering method*. Also known as *depth slicing* and *spectral depth slicing*. See also *gridding*, *spectral domain*, and *wavenumber*.

pseudogravity: 1. the magnetic equivalent of *anomalous gravity* obtained by *pseudogravity transformation* of an anomalous *magnetic field* in accordance with the *Poisson theorem*. **2.** a field of pseudoanomalies obtained by the vertical integration of the reduced-to-pole (RTP) magnetic field, i.e., pseudogravity represents the result of transforming the RTP magnetic anomalies into the equivalent of the *vertical gravity* anomalies in accordance with the Poisson theorem assumptions. Pseudogravity calculations suppress the high-frequency *noise* and enhance the anomalies from deep magnetic sources. It is used in *structural inversion* for the *magnetic basement* depth, comparative analysis of the gravity and magnetic anomalies generated by the same source, and analysis of the regional-scale anomalies in the pseudogravity display *format* as an alternative to conventional magnetic maps. See also *pseudoanomaly, reduction-to-pole (RTP), differential pseudogravity*, and *vertical integral.*

pseudogravity anomaly: the magnetic equivalent of the *vertical gravity* anomaly calculated from a reduced-to-pole (RTP) *magnetic anomaly* under the following assumptions: (a) the actual *magnetization* distribution can be replaced by an equivalent *density* distribution in the same *source body*, (b) the density-to-magnetization ratio is constant throughout the source, and (c) the *density contrast* is proportional to the *magnetization contrast.* Pseudogravity anomalies are obtained by *pseudogravity transformation* in accordance with the *Poisson theorem* which postulates that *magnetic potential* is directly proportional to the *derivative* of *gravitational potential* in the direction of magnetization. A pseudogravity anomaly is calculated by *first vertical integral (FVI)* of the RTP magnetic field, as opposed to the *vertical derivative* of the anomalous gravity field in the case of a *pseudomagnetic anomaly.* The unit of evaluation of pseudogravity anomalies is sometimes referred to as *pseudomilligal* (*p-mGal*) instead of the original magnetic unit of *nanotesla* (*nT*). See also *anomalous field* and *reduction-to-pole (RTP).*

pseudogravity curvature: the magnetic equivalent of *gravity field curvature.* Its *structure*-related attributes are calculated from the *pseudogravity gradient tensor* components. See also *attribute, curvature, curvature attributes*, and *pseudogravity field.*

pseudogravity effect: *pseudogravity field* in the context of the *pseudogravity* anomalies obtained by *pseudogravity transformation* of a reduced-to-pole (RTP) magnetic field.

pseudogravity field: **1.** the magnetic equivalent of the anomalous gravity field obtained by *pseudogravity transformation* of a reduced-to-pole (RTP) magnetic field in accordance with the *Poisson theorem.* **2.** the *first vertical integral (FVI)* of an RTP magnetic field. According to the Poisson theorem, it is equivalent to the *anomalous field* of the vertical component g_z of the *Earth gravity field.* **3.** a *grid* or map of *pseudogravity* anomalies. Also referred to as *pseudogravity effect.* See also *reduction-to-pole (RTP)* and *pseudomagnetic field.*

pseudogravity filter: **1.** a *wavenumber*-domain data processing operator that transforms the reduced-to-pole (RTP) magnetic field into a *pseudogravity field.* **2.** *pseudogravity transformation* with an emphasis on its property to suppress the short-*wavelength* (high-frequency) components of the *RTP magnetic field* in a manner of a *low-pass filter.* See also *pseudomagnetic filter* and *reduction-to-pole (RTP).*

pseudogravity gradient: **1.** the *gradient* of a *pseudogravity field.* **2.** the magnetic equivalent of the *gravity gradient* calculated from a reduced-to-pole (RTP) magnetic field using the scaling factor Gρ/M, where G is *universal gravitational constant* in *cgs units*, ρ is *density* (often average crustal density of 2.67 $g \cdot cm^{-3}$), and M is anomalous magnetic *magnitude* |**M**| equal to the *RTP magnetic field* value. See also *reduction-to-pole (RTP), pseudogravity transformation*, and *pseudogravity gradient tensor.*

pseudogravity gradient tensor: a *gradient tensor* the elements of which are directional derivatives of the *pseudogravity field* obtained by *pseudogravity transformation* of a reduced-to-pole (RTP) magnetic field with the assumption that the *magnetization direction* of sources is vertical. Vertical pseudogravity component pg_z is calculated from the *RTP magnetic field* in accordance with the *Poisson theorem.* Horizontal components pg_x and pg_y are calculated as the orthogonal horizontal derivatives of *pseudogravity potential* obtained by *field-to-potential transformation* of the pseudogravity field in the *Fourier domain.* Calculated derivatives of the vertical and horizontal components of the pseudogravity field comprise the *full gradient tensor (FTG)* at each observation point for subsequent transformations, analysis, and *interpretation* with the use of *eigenvalues, eigenvectors*, and *invariants* (Beiki et al., 2011). See also *derivative, directional derivative, FTG attributes*, and *reduction-to-pole (RTP).*

pseudogravity potential: the magnetic equivalent of *gravitational potential* calculated from *pseudogravity* data by a *field-to-potential transformation* in the *Fourier domain.* Derivatives of the pseudogravity potential in the mutually orthogonal *x*-, *y*-, *z*-directions represent *vector* components of the pseudogravity field, or *pseudogravity vector* components, that are used in the calculation of elements of a *pseudogravity gradient tensor.* See also *derivative* and *directional derivative.*

pseudogravity transformation: transformation of reduced-to-pole (RTP) magnetic anomalies into the magnetic equivalent of the *vertical gravity* anomalies in accordance with the *Poisson theorem.* Pseudogravity transformation converts an anomalous magnetic field into the anomalous gravity field that would be observed when the *magnetization* distribution is replaced by the identical *density* distribution. Its application is based on: (a) the assumed *induced magnetization* of rocks in the direction of the *main geomagnetic field*, (b) the assumed constant density-to-magnetization ratio of rocks, (c) the vertical integration of the *RTP magnetic field* by *first vertical integral (FVI)*, and (d) the multiplication of the vertically integrated RTP values by the scaling factor Gρ/M, where G is *universal gravitational constant* in *cgs units*, ρ is density (often, average crustal density of 2.67 $g \cdot cm^{-3}$), and M is anomalous magnetic *magnitude* |**M**| representing the RTP magnetic field value. Integration amplifies the long-*wavelength* components generated by deep magnetic sources, whereas the short-wavelength components are suppressed. Due to its integrating properties, the pseudogravity transformation can be considered as a form of conversion of an anomalous

RTP magnetic field into a *format* equivalent to *magnetic potential*. Prior to the pseudogravity transformation, *detrending* is applied to remove a regional trend in the RTP magnetic field. Sometimes referred to as *pseudogravity filter*. See also *anomalous field*, *reduction-to-pole (RTP)*, and *pseudomagnetic transformation*.

pseudogravity vector: a three-component *vector* for which the direction and *magnitude* at every point in space are defined by the *gradient* of the scalar *pseudogravity potential* pU. It is expressed as

$$pg = \nabla p\mathrm{U} = (pg_x, pg_y, pg_z),$$

where ∇ is the gradient operator *del*. See also *gravity vector* and *pseudogravity gradient tensor*.

pseudo-inclination method: a method of *reduction-to-pole (RTP)* of magnetic data acquired at low magnetic latitudes and *magnetic equator* areas, where the standard RTP operator is unstable. In addition to the actual value of *magnetic inclination* I in the *magnetic survey* area, the second inclination I_P is used in the denominator of the RTP formula to control the RTP amount. By setting $|I_P| > |I|$, where I_P is the smallest empirically determined value producing the low-*noise* and stable RTP result, the artificially large correction values are avoided (Grant and Dodds, 1972; MacLeod et al., 1993). See also *low-latitude reduction-to-pole (RTP-LL)* and *magnetic latitude*.

pseudo-lines method: same as *Hardwick method* with an emphasis on the use of imaginary pseudo-lines between the actual marine (shipboard) or airborne *survey* lines.

pseudolithology model: a three-dimensional (3D) *voxel model* of a mapped *ore body* obtained by *joint inversion* of the gravity and magnetic *total field* or *gradient* data with *constraints* by the petrophysical data from available wells. Interpretation of the model-obtained data is based on selecting the ranges of the *inversion*-obtained physical properties (*density* and *magnetization* contrasts) and sorting them into the specific lithological units associated with the mineralization zones, mineral alteration types, and localized concentrations of the high-density and high-magnetization rock formations within the ore body (Li et al., 2019). See also *contrast*.

pseudomagnetic anomaly: the gravity equivalent of the reduced-to-pole (RTP) *magnetic anomaly* calculated from a *gravity anomaly* under the following assumptions: (a) the actual *density* distribution can be replaced by an identical distribution of *magnetization* in the same *source body*, (b) the density-to-magnetization ratio is constant throughout the source, and (c) the *magnetic contrast* is directly proportional to the *density contrast*. Pseudomagnetic anomalies are obtained by *pseudomagnetic transformation* in accordance with the *Poisson theorem*, which postulates that *magnetic potential* is directly proportional to the *derivative* of *gravitational potential* in the magnetization direction. Because the magnetization of a source of pseudomagnetic anomaly is assumed to be vertical, the transformation is based on calculation of the first-order *vertical derivative* of the gravity anomaly, as opposed to the *vertical integral* of the RTP magnetic field in the case of a *pseudogravity anomaly*. Due to the highly variable magnetization of rocks (up to 100–1000 times its lowest value), the calculation of *pseudogravity* anomalies is much more common. See also *reduction-to-pole (RTP)*.

pseudomagnetic effect: *pseudomagnetic field* in the context of the pseudomagnetic anomalies obtained by *pseudomagnetic transformation* of an anomalous gravity field, such as *Bouguer anomaly*, *free-air anomaly*, and *isostatic anomaly*.

pseudomagnetic field: a *field* of pseudomagnetic anomalies obtained by *pseudomagnetic transformation* of an anomalous gravity field, such as *Bouguer anomaly*, *free-air anomaly*, and *isostatic anomaly*, in accordance with the *Poisson theorem*. Sometimes referred to as *pseudomagnetic effect*. See also *pseudomagnetic anomaly* and *pseudogravity field*.

pseudomagnetic filter: **1.** a *wavenumber*-domain data processing operator that transforms an anomalous gravity field, such as *Bouguer anomaly*, *free-air anomaly*, and *isostatic anomaly*, into a *pseudomagnetic field*. **2.** *pseudomagnetic transformation* with an emphasis on suppressing the long-*wavelength* components of an anomalous gravity field in a manner analogous to a *high-pass filter*. See also *anomalous field*.

pseudomagnetic potential: the gravity equivalent of *magnetic potential* calculated from a *pseudomagnetic field* by the application of *field-to-potential transformation* in the *Fourier domain*.

pseudomagnetic transformation: transformation of *vertical gravity* anomalies into the pseudomagnetic anomalies in accordance with the *Poisson theorem*. Pseudomagnetic transformation involves (a) scaling of the anomalous gravity field by the assumed *density-to-magnetization* ratio for conversion of the gravity values into the pseudomagnetic values and (b) calculation of the *vertical derivative* of the scaled gravity field. Calculation of the *vertical derivative* amplifies the anomalous gravity field components in inverse proportion to their wavelengths: the shorter the *wavelength*, the higher the amplification. Inevitable amplification of the short-wavelength (high-*frequency*) *noise* can lead to data contamination by artificial high-frequency components. Sometimes referred to as *pseudomagnetic filter*. See also *reduction-to-pole (RTP)*, *pseudomagnetic anomaly*, and *pseudogravity transformation*.

pseudomagnetic vector: a three-component *vector* for which the direction and *magnitude* at every point in space are defined by the *gradient* of the scalar *pseudomagnetic potential* $p\mathcal{V}$. It is expressed as

$$\boldsymbol{pB} = \nabla p\mathcal{V} = (\boldsymbol{pB}_x, \boldsymbol{pB}_y, \boldsymbol{pB}_z),$$

where ∇ is the gradient operator *del*. See also *magnetic vector* and *pseudogravity vector*.

pseudomagnetics: **1.** the gravity equivalent of the anomalous magnetic field, or *anomalous magnetics*, obtained by *pseudomagnetic transformation* in accordance with the *Poisson theorem*. **2.** the *field* of the pseudomagnetic anomalies obtained by calculation of the *first vertical derivative (FVD)* of the scaled gravity field, i.e., pseudomagnetics represents the result of transformation of *vertical gravity* anomalies into the equivalent of the reduced-to-pole (RTP) magnetic anomalies in accordance with the Poisson theorem assumptions. **3.** a map or *grid* of pseudomagnetic anomalies. Due to the highly variable *magnetization* of rocks (up to 100–1000 times its lowest value), the calculation of *pseudogravity* is much more common. See also *anomalous field*, *pseudomagnetic anomaly*, and *reduction-to-pole (RTP)*.

pseudomilligal (p-mGal): an alternative unit of *pseudogravity* anomalies obtained by *pseudogravity transformation* of a reduced-to-pole (RTP) magnetic field in accordance with the *Poisson theorem*. See also *reduction-to-pole (RTP)* and *pseudodensity*.

pseudosusceptibility: a rock parameter defined by the ratio of the *magnitude* of the estimated *total magnetization* of a particular rock unit to the strength, or intensity, of the present *Earth magnetic field*. It is calculated primarily in magnetic data processing for mining applications. See also *magnetization*, *magnetization vector*, and *pseudodensity*.

pull-apart basin: a structural depression, or *structural low*, created by the regional-scale tectonic extension or local divergent-and-convergent movement of rocks along a *strike-slip fault*. Typically, the pull-apart basin has the *gravity signature* of a residual *gravity low* within the area of the branching-and-reconnecting, or *anastomosing*, alignment of the *gravity gradient* and *magnetic gradient* anomalies generated by normal and strike-slip faults bounding this structural depression. Sometimes, it can have the *magnetic signature* of a localized *magnetic high* or *magnetic low* depending on the dominant type of *magnetization* of the extension-driven igneous rock bodies intruded into the pull-apart basin: *induced magnetization* or *remanent magnetization*, respectively. See also *igneous intrusion*, *residual gravity*, and *rift basin*.

pulse: a *waveform* whose length is short relative to the wavelengths of the exploration or research interest and whose initial and final values are approximately the same and, usually, close or equal to zero. See also *spike*, *wavelength*, and *wavelet*.

Q

Q-factor: See *Koenigsberger ratio.*

quadrature: 90° out-of-*phase* relationship. Two functions are said to be in quadrature if they have the same *amplitude spectrum* but differ in phases by 90°. Often, the *vertical derivative* and *horizontal derivative* are assumed to be in quadrature. The imaginary and real parts of a *complex variable*, such as the *analytic signal (AS)* and *monogenic signal (MS)*, are in quadrature. See also *Hilbert transform* and *Riesz transform.*

quadrature filters: a pair of filters which have the same *frequency response* but differ in *phase* by 90°. See also *filter* and *steerable filter.*

quadrature pair: *Hilbert transform pair* in the context of a pair of functions which are in *quadrature* with each other.

quadrature spectrum: the *sine transform* components in the *Fourier transform* of gravity and magnetic data. See also *spectrum* and *cosine transform.*

qualitative interpretation: the initial part of the *interpretation* process that involves: (a) application of filters and calculation of attributes, derivatives, gradients, and *invariants* as well as the use of other *enhanced processing* techniques to detect as many anomalies of the exploration interest as necessary to minimize subjectivity in selection of alternative interpretation scenarios; (b) analysis of detected gravity and magnetic anomalies in order to find convincing evidence in support of the interpretation conclusions; and (c) identification of the specific elements of the geological structures that have generated the detected anomalies for the subsequent mapping, *forward modeling*, and *inversion* in the process of *quantitative interpretation.* See also *anomaly, attribute, derivative, filter, gradient*, and *structure.*

quality control (QC): a general term for the pre-*survey*, in-survey, and post-survey procedures applied to reveal the instrument malfunctions, *instrument drift*, deviations from *survey specifications*, and errors in data measurements, *data processing*, and *enhanced processing.* See also *airborne magnetometer calibration, airborne gravimeter calibration, ground gravimeter calibration, figure-of-merit (FOM)*, and *static test.*

quantitative interpretation: the final part of the *interpretation* process that presents all structural and physical property information that could be obtained from analysis of the detected gravity and magnetic anomalies, including: (a) mapped faults, fracture and shear zones, lithological boundaries, and other structural discontinuities; (b) interpreted *basement* depth structure, intrasedimentary igneous intrusions, depositional fairways, local sedimentary and basement structures favorable for accumulation of hydrocarbons, mineral deposits, and other objects of oil-and-gas or mineral exploration; (c) two-dimensional (2D), 2.5D, or 2.75D *forward modeling* along the regional transects, seismic sections, and profiles of anomalies of the exploration or research interest; and (d) 3D *structural inversion* and *property inversion* or 3D *objective function*-based inversion and *stochastic inversion* for the depth, shape, distribution of *density* and *magnetization*, and other parameters of the targeted geological objects. See also *fairway, fault, igneous intrusion, lineaments, shear zone, target*, and *integrated interpretation.*

quantum gravimeter: same as *absolute atomic gravimeter* with an emphasis on its operating principle based on the quantum wave-like nature of atoms and their use as identical test masses in a vacuum free fall to measure *absolute gravity.*

quantum gravity gradiometer: same as *atomic gravity gradiometer* with an emphasis on its operating principle based on the use of two or three synchronized *absolute atomic gravimeter* systems utilizing the quantum wave-like nature of atoms for (a) simultaneous measurements of *absolute gravity* at two or three points within the same reference frame and (b) calculation of the *absolute gravity gradient.*

quantum magnetometers: instruments of the *scalar magnetometer* type for measuring the total intensity, or strength, of the *Earth magnetic field.* Their operational principle is based on the quantum physics phenomenon of *magnetic resonance* and measurements of *angular frequency* of *precession* of the subatomic particles (protons or electrons) that is known as the *Larmor frequency*, whose value is directly proportional to the intensity of the Earth's magnetic field. Quantum magnetometers are dominant in *exploration* applications, and they include the *cesium magnetometer, helium magnetometer, Overhauser magnetometer, potassium magnetometer, proton magnetometer*, and *rubidium magnetometer.* Magnetometers of this type are also referred to as *resonance magnetometer, nuclear-resonance magnetometer, optically pumped magnetometer*, and *atomic magnetometer.* See also *vector magnetometer.*

quantum vector magnetometer: See *SQUID magnetometer.*

quasi-analytic signal (QAS): *analytical signal amplitude (ASA)* computed from one measured and two calculated gradients of the *Earth magnetic* field using the data acquired by a marine or airborne *magnetic survey* with a *transverse magnetic gradiometer*, where two synchronized magnetometers are deployed as a towed fixed-*array* system. The orthogonal *horizontal gradient* component, or *transverse gradient*, is determined from the difference between measurements of two magnetometers divided by their separation distance, or *baseline*, whereas the *longitudinal gradient* is the absolute value of the *horizontal derivative* along the *survey* line, and the *vertical derivative* is obtained by the *Hilbert transform* of one of the horizontal gradient components or via the *Laplace equation.* See also *magnetometer* and *magnetic gradient.*

R

radar altimeter: an instrument for measurements of the height above the ground, or *altitude*, of a *survey* aircraft. Its operational principle is based on the use of the reflected microwave-range electromagnetic pulses transmitted by an onboard radar. Measurement *accuracy* depends on the slope of the Earth's surface: the higher the slope, the lower the accuracy. The best reported accuracy estimates are 10–30 cm. See also *barometric altimeter* and *LIDAR*.

radial derivative: a *derivative* in the direction which is normal, or orthogonal, to a spherical surface along the Earth's radius or *reference ellipsoid*. In the spherical harmonic analysis of the *main geomagnetic field*, the radial derivative of *magnetic potential* $\mathcal{V}$ defines the vertical component B_z of the *geomagnetic field vector* **B** as

$$B_z = \partial \mathcal{V} / \partial r.$$

See also *spherical harmonics* and *gravity disturbance*.

radial log-power spectrum: See *radial power spectrum*.

radial power spectrum: a direction-independent *power spectrum* of the gridded and Fourier-transformed *anomalous field*. It is computed by averaging the power spectrum calculations along all 360° azimuthal directions over a *grid* and presented in the *format* of a *two-dimensional (2D)* graphical plot of the total spectral energy, or power, quantified by the squared amplitudes of each *wavenumber* component plotted in the logarithmic scale. The graphical plot has a characteristic descending shape with two dominant parts: (1) the high-power part of the steeply decreasing values over small wavenumbers corresponding to the deep and broad sources, and (2) the low-power part of the gently decreasing values over large wavenumbers corresponding to the shallow and localized sources, including the near-surface *noise*. Segments of the graphical plot approximated by straight lines, or slopes, are used in the *Spector-Grant method* for depth estimates of source ensembles, where each ensemble represents the grid-averaged depth of a group of sources. The parametric expression for *least-squares fit* and setting the minimum number of ensembles is defined as

$$P(k) = A_1 e^{-2kh_1} + A_2 k^2 e^{-2kh_2} + A_3 k^2 e^{-2kh_3} + P_n,$$

where A_i is the radially averaged *i*th ensemble power; h_i is the *i*th ensemble depth; *k* is *radial wavenumber* $k = (k_x^2 + k_y^2)^{1/2}$; k_x and k_y are wavenumbers along the mutually orthogonal *x*- and *y*- directions related to their respective wavelengths as $k_x = 2\pi/\lambda_x$ and $k_y = 2\pi/\lambda_y$; and P_n is the noise power. Depth estimates from the radial power spectrum calculated over the area covering more than one structural-tectonic *terrane* can produce artificial depth averages which are non-existent in these terranes. Also referred to as *radial log-power spectrum*, *radially averaged power spectrum*, and *total spectral power*. See also *map power spectrum*.

radial sunshading: a *grid*-based *sunshading* technique applied to highlight low-*amplitude* linear, curvilinear, and circular anomalies. Its algorithm is based on the use of a variable *azimuth* of the artificial sun, or *sun declination*, for different grid parts, where the reference circle center location and sun elevation, or *sun inclination*, can be selected interactively. The azimuth at an arbitrary grid point (*i*, *j*) is defined as

$$\theta_{ij} = \arctan(\Delta y_{ij} / \Delta x_{ij}),$$

where Δy and Δx are north–south and east–west distances between the selected reference circle center location and the *grid node* in the gridded data. Orthogonal azimuth value $\theta_{ij}^{OA} = \theta_{ij} - \pi/2$ is used to enhance the linear trends running through the center point of the reference circle regardless of their azimuth. Results of the application of regular and orthogonal azimuth options are combined in a composite grid to highlight the curvilinear, circular-shaped anomalies and linear anomalous trends (Cooper, 2003). See also *gridding* and *steerable filter*.

radial wavenumber: a *wavenumber* in the *Fourier transform* of gridded gravity and magnetic fields. It is expressed as

$$w_r = k_r = (k_x^2 + k_y^2)^{1/2} = (u^2 + v^2)^{1/2},$$

where *u* and *v* are wavenumbers in the mutually orthogonal *x*- and *y*-directions related to their respective wavelengths as $u = k_x = 2\pi/\lambda_x$ and $v = k_y = 2\pi/\lambda_y$. See also *gridding* and *wavelength*.

radially averaged power spectrum: *radial power spectrum* in the context of averaging the individual power spectra calculated along all 360° azimuthal directions over a *grid*.

radian: a unit of angular measurements defined by the angle between two radii of a circle that bound, or subtend, the circle perimeter arc whose length (subtended arc length) is equal to the circle radius, with 1 radian (rad) = 180/π = 57.2958 degrees, and 1 degree = 0.017453 radian. See also *steradian*.

random gridding: *gridding* of irregularly spaced gravity and magnetic data, i.e., creating the smoothest continuous surface composed of the regular-spaced cells each representing the interpolated or original data value assigned to a cell center. See also *grid* and *grid cell*.

random noise: uncorrelatable, predominantly short-*wavelength*, and highly variable *amplitude* components of a measured gravity or magnetic field. In ground and low-*altitude* airborne magnetic surveys, the random noise is generated primarily by the surface and near-surface accumulations of the unconsolidated magnetic rock material known as *regolith*. In airborne and marine (shipboard) magnetic surveys, the random noise is related mostly to the uncompensated components of *diurnals*, especially when the magnetic *base station* or nearest *geomagnetic observatory* is far from the *marine survey* area. In marine and airborne gravity surveys, the random noise is related primarily to the unsuppressed horizontal and vertical *inertial acceleration* components produced by the motion of the survey ship or aircraft. See also *noise*, *background noise*, *coherent noise*, *dynamic noise*, *white noise*, and *stacked sampling*.

rare earth elements (REE): a group of 17 metallic elements, mostly from lanthanum (atomic #57) through lutetium (atomic #71) in the Periodic Table of Elements, whose minable concentrations are rare. The REE-rich minerals are found in alkaline igneous bodies, primarily *carbonatite* ones, and associated pegmatites created by rock metamorphism along the *igneous body* margins. Heavy REE are also found in *placer* deposits, especially in beach sands, while the light REE are found in clay formations, where they are adsorbed onto the clay particles. The REE accumulations are known to be associated with the *mineral nodule* deposits on the deep sea or ocean floor in proximity to the high-temperature thermal vents. High-*resolution* airborne magnetic and gravity *total field* and *gradient* measurements are capable of providing REE *exploration leads* by: (a) mapping the boundaries of shallow igneous bodies and intersections of deep-rooted faults as the particular zones of high metamorphism and mineral alterations, (b) mapping the depositional fairways around known alkaline igneous bodies by correlation of the residual magnetic anomalies generated by accumulations of *detrital magnetite*, and (c) *inversion*-based mapping of the increased *density* locations in the internal structure of the igneous body and zones of high-density mineral alterations by the pegmatite metamorphism along the igneous body margins with the use of *gravity gradient tensor (GGT)* data. Alluvial and beach REE-bearing placers can be detected by high-accuracy *ground gravity* measurements in the depositional *fairway* areas mapped by a *high-resolution aeromagnetic survey*. In integration with temperature measurements, a high-resolution *submarine survey* with the use of a *sea-floor drone* is capable of detecting the gravity and magnetic anomalies generated by the commercial-grade REE accumulations on the deep sea and ocean floors. See also *anomaly*, *full tensor gradient (FTG)*, *residual gravity*, *residual magnetics*, *submarine gravity*, and *submarine magnetics*.

raster image: a digital image composed of discrete elements, or pixels, arranged in a rectangular set of multiple horizontal rows representing the regular-spaced samples of a given image or map. The arrangement, or array, of pixels in a raster image is analogous to the rectangular array of data cells in a *grid*. Raster images are stored as image files in various formats, including BMP, GIF, JPEG, and TIFF. A raster image file is also referred to as a *bitmap*. See also *pixel*.

raw gravity: digital records of the *Earth gravity field* or *gravity gradient* values measured during a ground, marine (shipboard), submarine, sea floor, or airborne *gravity survey*. Usually, this term refers to *relative gravity* values before the application of *raw gravity corrections*. Also known as *measured gravity*. See also *observed gravity*.

raw gravity corrections: corrections applied to *raw gravity* to compensate for the measurement effects accumulated in the process of *acquisition* and obtain *observed gravity*. Their application depends on the *gravity survey* type (ground, marine, submarine, sea floor, airborne) and measurement *accuracy* requirements. The main corrections for measured *relative gravity* values include: (1) *drift correction* for *gravimeter drift*; (2) *temperature correction* for *temperature drift*; (3) *tilt correction* for *sensor tilt* in *spring gravimeter* systems; (4) *reading correction* to convert *gravimeter* readings from the counter, or dial, units into the standard units of *milligal (mGal)* or *microgal (μGal)*; (5) *cross-coupling corrections* applied to gravity data acquired on a moving *observation platform*, such as a survey ship or aircraft; (6) *Harrison correction* or *Browne correction* for *platform tilt*; (7) *Eötvös correction* for the observation platform movement and its direction over the curved Earth's surface within the coordinate system fixed on the rotating Earth; (8) *lever arm correction* for the motion-related *lever arm effect* in *marine gravity* and *airborne gravity* surveys; and (9) *scale correction* to convert time-based marine and airborne digital records into the corresponding space-domain coordinates. In *ground gravity* surveys, the drift correction often incorporates *Earth tide correction*, and the obtained gravity values are tied to an *absolute base station*. For *absolute gravity* values obtained from stationary measurements by the *free-fall method*, these corrections include: (1) *diffraction correction*, (2) *self-attraction correction*, (3) *speed-of-light correction*, and (4) *tilt correction*. See also *automatic drift correction* and *observed gravity corrections*.

raw magnetic corrections: corrections applied to *raw magnetics* to compensate for the measurement effects accumulated in the process of *acquisition* and obtain *observed magnetics*. Depending on the type of *magnetic survey* (ground, marine, airborne), they include *cable correction*, *heading correction*, *lag correction*, *scale correction*, and, if necessary, correction for *magnetometer drift*. See also *aeromagnetic compensation* and *observed magnetic corrections*.

raw magnetics: digital records of the *Earth magnetic field* or *magnetic gradient* values measured during a ground, marine (shipboard), submarine, or airborne *magnetic survey*. Raw magnetics represent the absolute numerical values of the Earth's magnetic field or its gradients before application of *raw magnetic corrections*. Also known as *measured magnetics*. See also *observed magnetics*.

reading: a digital record of a numerical value of the *Earth gravity field* or *Earth magnetic field* and their *gradient* components measured at observation points by an *acquisition* instrument, such as a *gravimeter*, *magnetometer*, and *gradiometer*, during a gravity or magnetic *survey*.

reading correction: conversion of recorded *gravimeter* measurements, or readings, from the original counter, or dial, units into the standard units of *milligal* or *microgal* using the multiplication constant called the *calibration factor*. Conversion is unique for each gravimeter, and it is based on the calibration tables provided by the manufacturer of a particular gravimeter.

reading interval: same as *sample interval* with an emphasis on the time or space interval between the successive instrument readings during a gravity or magnetic *survey*.

reading resolution: 1. the smallest value that can be displayed or downloaded by a measuring instrument. **2.** the minimum increment, or step, in the reading device of *quantum magnetometers* for measurements of the *precession* frequency and its conversion into a magnetic field value. **3.** the minimum increment of a *gravimeter* counter in measurements of *relative gravity*.

real time: a characteristic of (a) processing of acquired data at the same time and the same rate as those at which they are recorded or (b) measurements with the same time reference as the actual time of day. See also *acquisition* and *real-time mode*.

real-time compensation: 1. an adaptive *aeromagnetic compensation* system based on the three-dimensional (3D) mathematical model of the *magnetic signature* of a *survey* aircraft. **2.** minimization of the motion-related *sensor tilt* of an *autonomous underwater vehicle (AUV)* by a *stabilized platform*. See also *real-time*.

real-time DGPS: a *positioning* system which computes the latitude, longitude, and *elevation* of a moving *observation platform* at the same rate as the *differential GPS (DGPS)* data are received.

real-time mode: 1. a mode of operation of data recording, data processing, navigation, and *positioning* that is characterized by the capability of the corresponding computer systems to respond instantly, i.e., at the same time and the same rate of change, to variations of the input data these systems are designed to monitor, control, and process. **2.** the processing of acquired data at the time of their measurement and application of the processed results to control other processes occurring simultaneously, such as a real-time adaptive *aeromagnetic compensation* system.

real-time sunshading: the interactive *sunshading* which instantly updates a sunshaded image with every change in position of an artificial source of light by the movement of a cursor, so the sunshading effects are observed in *real-time mode*.

real-time video: a color video recording of the ground surface below a *survey* aircraft synchronized with *airborne survey* measurements. Real-time video records are used for the identification of sources of *cultural magnetic anomalies* in the process of *cultural editing* and control over *positioning* errors.

reciprocity principle: a physical law postulating that the *gravity potential* or *magnetic potential* at observation point "A" due to the *source body* at point "B" remains the same when the points of observation and source are reversed. See also *equivalence principle* and *superposition principle*.

reconnaissance survey: a gravity or magnetic *airborne survey* conducted at a relatively high flight *altitude* (500–1000 m) along widely spaced (2–3 km or more) survey lines with the objective of obtaining regional-scale information about the major tectonic and geologic characteristics of an *exploration* area, such as the dominant *fault* trends, sedimentary basins, strikes of the dominant anomalies, magnetic terranes, etc. See also *detailed survey*, *regional survey*, *magnetic terrane*, *strike*, and *survey design aliasing*.

rectangular coordinates: *Cartesian coordinates* representing a three-dimensional (3D) system of linear coordinates, where the angles between three coordinate *x*-, *y*-, *z*-axes are right angles, such as those of a rectangle. See also *spherical coordinates*.

reduced Euler deconvolution: a method of estimating source parameters in the *multiridge analysis* of anomalous fields. It is based on reducing the three-dimensional (3D) *Euler deconvolution equation* to the two-dimensional (2D) subsets of data which are called ridges and obtained by joining the zero points of the horizontal and vertical derivatives of an *anomalous field* and field itself calculated at the different levels of *upward continuation* called altitudes. With the *y* coordinate fixed at the *grid* point (y–y_0) and assuming that the unknown background B is null or substantially suppressed by differentiation (i.e., calculation of derivatives), 3D Euler deconvolution equation is reduced to

$$(z - z_0)\partial F_Q/\partial z = -N_Q F_Q,$$
$$(x - x_0)\partial F_Q/\partial x = -N_Q F_Q,$$
$$(x - x_0)\partial F_Q/\partial x = -(z - z_0)\partial F_Q/\partial Z,$$

for ridges of the type I (zeros of *horizontal derivative*), type II (zeros of *vertical derivative*), and type III (zeros of the anomalous field), respectively. Here, Q is the order of differentiation of the anomalous field and N_Q is the *Euler structural index (SI)* defined as

$$N_Q = N_0 + Q,$$

where N_0 = SI = *N* in the 3D Euler deconvolution equation. Each set of the reduced Euler equations for a specific *ridge* type is solved within a *moving window* for two source parameters: depth z_0 and SI for the type I ridges, horizontal position x_0 and SI for the type II ridges, and x_0 and z_0 for the type III ridges (Fedi et al., 2009). See also *altitude*.

reduction: 1. a correction procedure applied to measured gravity and magnetic values in order to compensate for the *acquisition*, planetary, and non-geological effects and predictable sources of errors. **2.** a multi-step process of converting the measured values of the *Earth gravity field* or *Earth magnetic field* into the *field* of gravity or magnetic anomalies, or *anomalous field*, generated by elements of subsurface geological structures. Reduction involves (a) numerical removing of the planetary-scale components, such as the *theoretical gravity* and *International Geomagnetic Reference Field*, from the measured gravity or magnetic values, and (b) elimination of, or compensation for, the known and predictable effects, such as *diurnals*, *drift*, *Bouguer effect*, *curvature effect*, *Eötvös effect*, *free-air effect*, *isostatic effect*, *terrane effect*, and *tidal effect*. See also *gravity reduction* and *magnetic reduction*.

reduction density: same as *correction density* with an emphasis on it being a numerical component of the *gravity reduction* procedure for calculation of *simple Bouguer correction* and *terrain correction.*

reduction filter: a *data processing* operator applied to *total magnetic intensity (TMI)* data in order to perform *reduction-to-pole (RTP)* or *reduction-to-equator (RTE).*

reduction-to-equator (RTE): the mathematical procedure which transforms the obtained *total magnetic intensity (TMI)* anomalies into the anomalies that would be observed at the *magnetic equator,* where both the *induced magnetization* of sources and *ambient field* are directed horizontally. The RTE algorithm recalculates the TMI anomalies from their original *magnetic inclination* and *magnetic declination* to those of the magnetic equator, where the *main geomagnetic field* is horizontal. The standard RTE algorithm tends to center the TMI anomalies over their respective sources, but its application results in (a) amplified stretching of anomalies in the east–west direction relative to the horizontal dimensions of their sources, (b) the addition of artificial side lobes to the anomalous shapes, and (c) negative *residual field* anomalies over sources with a positive *magnetization contrast.* A high-pass alternative to the RTE for enhancement of anomalies at the magnetic equator and low *magnetic latitude* areas is calculation of the *analytic signal amplitude (ASA)* and ASA-based attributes, such as *theta angle, second-order analytic signal amplitude* (ASA_2), and *tilt ASA.* See also *low-latitude reduction-to-pole (RTP-LL)* and *pseudo-inclination method.*

reduction-to-pole (RTP): the mathematical procedure which transforms the obtained *total magnetic intensity (TMI)* anomalies into the anomalies that would be observed at the *Earth magnetic poles,* where both the *induced magnetization* of sources and *ambient field* are directed vertically. The RTP is a *phase*-shifting procedure changing the positions and amplitudes of the TMI anomalies to center them above their respective sources and, thereby, eliminate the asymmetry of TMI anomalies caused by the *vector* nature of *magnetization* and non-vertical direction of the *main geomagnetic field* in the *survey* area. The RTP algorithm recalculates TMI anomalies from their original *magnetic inclination* I and *magnetic declination* D to those of the *magnetic north pole* or *magnetic south pole,* where the main geomagnetic field is vertical, i.e., $I = \pm 90°$ and $D = 0°$. The amount of the RTP shift is proportional to the anomaly *wavelength* (which is a function of the source depth): the longer the wavelength, the larger the RTP shift. The RTP is based on two assumptions: (1) the induced magnetization of rocks is dominant and the effects of *remanent magnetization* can be ignored and (2) the rock magnetization direction is collinear, or aligned, with the direction of the main geomagnetic field. When both assumptions are satisfied, the asymmetric peak-to-trough TMI anomalies are transformed into the symmetric bell-shaped RTP anomalies centered above their sources. At low geomagnetic latitudes within ±15°–20° from the *magnetic equator,* the RTP transformation becomes unstable due to the horizontal or near-horizontal direction of the main geomagnetic field and, therefore, it requires the application of a specific RTP algorithm, such as *equivalent-source reduction-to-pole,* or substitution for the alternative *data enhancement* methods, such as calculation of the *analytic signal amplitude (ASA)* and its versions. The RTP does not affect the *power spectrum* of the data. The standard RTP algorithm utilizes the average value of the magnetic inclination and declination for the entire *survey* area, so the RTP results of the northern and southern *survey* area margins are less accurate than in the central area. The range of inclination and declination values in the *survey* area is determined from the *IGRF* or *DGRF* model appropriate for the year, months, and days of the *magnetic survey.* For regional-scale surveys extending over several degrees of latitude, *differential reduction-to-pole (DRTP)* is applied instead of the standard RTP. See also *low-latitude reduction-to-pole (RTP-LL)* and *pseudo-inclination method.*

reef: a ridge- or mound-like geological *structure,* or buildup, composed of the highly porous calcareous remains of reef-building organisms, often intermingled with carbonate sands cemented by calcium carbonates and enclosed in the sediments of a different lithology and laterally lower *density.* Buried pinnacle and other reef types are targets of oil-and-gas *exploration* due to their high reservoir qualities. High *density contrast* of reefs with surrounding sedimentary formations often generates a distinct positive *gravity anomaly,* especially in the *residual gravity* field. Commonly, reefs have no *magnetic signature* due to the low *magnetization* of carbonates and other sedimentary rocks. Also referred to as *carbonate buildup.* See also *target.*

reference crustal density: the *density* value of 2.67 g · cm^{-3} relative to which density contrasts of modeled rock formations are often determined for the two-dimensional (2D) or three-dimensional (3D) *gravity inversion* and *gravity stripping* of *profile*-oriented or gridded anomalous gravity data. It is used as the *default value* for calculation of *Bouguer correction* in regional-scale gravity surveys. See also *anomalous field, density contrast, gridding,* and *global crustal correction.*

reference crustal thickness: a parameter in the *Airy-Heiskanen model* of *isostasy* that defines the thickness of the *crust,* or depth to the *mantle,* at the zero *elevation* of a shoreline. Its value is assumed to be 30–40 km for calculation of the *isostatic root depth* or *isostatic anti-root depth* representing the predicted depth to the crust-mantle boundary known as the *Moho boundary.*

reference elevation: same as *vertical datum* with an emphasis on its use as the reference surface, or reference level, relative to which an observation point height, or *elevation,* is determined.

reference ellipsoid: 1. a homogeneous *ellipsoid* which is the best-fitting mathematical approximation of the rotating Earth and its heterogeneous internal composition. It has the same mass, center of mass, and *angular velocity* of rotation as those of the Earth. **2.** an ellipsoid for which the surface is the *equipotential surface* for the gravity field generated by the uniformly dense, or homogeneous, Earth. The reference ellipsoid is used as a *geodetic datum* for three-dimensional (3D) geographic coordinates (latitude, longitude, and *elevation*) determined by the satellite-based *Global Positioning System (GPS)* and for calculation of *theoretical gravity* and the *theoretical geomagnetic field*. If the Earth was uniformly dense, the reference ellipsoid and *geoid* would have the same shape. The theoretical gravity field on the reference ellipsoid is a function of the geographic latitude only, and it is defined by the *International Gravity Formula (IGF)*. The reference ellipsoid has four basic parameters: (1) semi-major axis equal to the equatorial radius, (2) semi-minor axis equal to the polar radius, (3) *angular velocity* of rotation, and (4) *geocentric gravitational constant*. Currently, the most accurate reference ellipsoids are the *GRS80 ellipsoid* and *WGS84 ellipsoid*. Also known as *Earth ellipsoid, equipotential ellipsoid, geocentric reference ellipsoid, geocentric equipotential ellipsoid*, and *terrestrial ellipsoid*.

reference gravitational potential: same as *normal gravitational potential* which is the *gravitational potential* of a uniformly dense, or homogeneous, *reference ellipsoid*.

reference gravity field: *theoretical gravity* used as a standard, or norm, to which measured gravity field values are referenced in order to determine the anomalies of an exploration or research interest.

reference gravity value: the value of the Earth's gravity field at a *base station* relative to which the *relative gravity* measurements are made over a *gravity survey* area. This value is also used for (a) calculation of *drift correction*, (b) ties of a local gravity dataset to the regional gravity network, and (c) merging with the gravity datasets acquired in previous years.

reference magnetic field: See *International Geomagnetic Reference Field (IGRF)*.

reference mass: *proof mass* in the *relative gravity* measurements by a *spring gravimeter* and *superconducting gravimeter*.

reference plane: See *datum*.

reference station: a *base station* in the context of a reference point on the Earth's surface for *relative gravity* and magnetic measurements and *GPS* receiver corrections. See also *Differential GPS (DGPS)* and *GPS base station*.

regional anomaly: 1. any gravity or magnetic *anomaly* which has a longer *wavelength* and, usually, a higher *amplitude* than the wavelengths and amplitudes of the anomalies of the exploration interest, i.e., this term does not necessarily refer to the depth of a regional anomaly source but rather to its size relative to the assumed size of the exploration *target*. In order to define properly the regional contribution to the *target anomaly*, the gravity or magnetic *survey* is planned to cover an area which is several times larger than the predicted wavelength of the target anomaly. **2.** an anomalous component of a *regional field*. See also *residual anomaly* and *feasibility modeling*.

regional field: 1. a part of the anomalous gravity or magnetic field that has longer wavelengths and, usually, higher amplitudes than those of a *target anomaly*. Quantitative estimate of the *wavelength* content of the regional fields depends on the *scale* of the exploration project: the smaller the scale, the shorter the regional field wavelengths. Generally, the regional field components are assumed to be associated with any geological *structure* or isolated *source body* which is much larger than the lateral extent of a targeted subsurface object or too deep for the present exploration interest. **2.** a *field* of the gravity or magnetic anomalies generated by elements of the regional geology. **3.** components of an *anomalous field* remaining after the application of a *low-pass filter* and other *regional-residual separation* techniques to suppress relatively short wavelengths. See also *residual field*.

regional filter: a version of the spectral-domain *Jacobsen filter* applied for the separation of the *regional field* generated by all sources that are deeper than the user-specified depth h_0. Its *transfer function* is expressed as

$$F(k) = e^{-2|k|h_0},$$

where k is the *wavenumber* with $k = |k| = (k_x^2 + k_y^2)^{1/2}$, and k_x and k_y are wavenumbers in the mutually orthogonal x- and y-directions related to their respective wavelengths as $k_x = 2\pi/\lambda_x$ and $k_y = 2\pi/\lambda_y$. See also *spectral domain, regional-residual separation*, and *residual filter*.

regional geodetic datum: a reference surface for *positioning* observation points that represents a specific *reference ellipsoid*, its applicable regional extent, location relative to the Earth's center, and orientation of its coordinate axes. Regional and national geodetic reference systems in various global areas implement different reference ellipsoids, most of which closely approximate, or conform to, the ground surfaces of those particular regions rather than the whole Earth surface. See also *Geodetic Reference System*.

regional gradient: 1. the dominant *trend* of the long-*wavelength* components in an *anomalous field*. **2.** the horizontal or vertical rate of change, or *gradient*, of a *regional field* obtained by application of a *low-pass filter* or other *regional-residual separation* techniques.

regional gravity: long-*wavelength* components of *free-air gravity* or *Bouguer gravity* generated by depth and *density* variations in the regional-scale geological structures, such as *basement, rift*, thrust-and-fold deformations in the sedimentary basins, deep crustal lithological complexes, and *Moho boundary*. Regional gravity is assumed to be associated with the sources that are deeper or larger than the sources of anomalies of the exploration interest, and it is commonly suppressed in order to enhance the *target*-generated anomalies. See also *enhanced processing, isostatic regional anomaly*, and *residual gravity*.

regional grid: the *grid* of a gravity or magnetic *regional field*, where the short-*wavelength* components of the original *anomalous field* are suppressed by application of a *low-pass filter* or other method of *regional-residual separation*. See also *residual grid*.

regional isostasy: *flexural isostasy* in the context of a regional bending, or flexing, of the Earth's *crust* in response to the rock mass loading and unloading due to the topographic and bathymetric relief variations as opposed to *local isostasy* assuming the local-scale breaking of the crust into the rigid *fault*-bounded blocks in the *Airy hypothesis* and the *Pratt hypothesis*.

regional magnetics: long-*wavelength* components of the *total magnetic intensity (TMI)* or *RTP magnetic field* generated by regional-scale geological structures, such as *basement*, extensive basalt lava flows, *rift*, and deep intrasedimentary plutons. Deep regional magnetic effects are often less important than those of *regional gravity* because of their lower amplitudes due to a higher *attenuation*, or *fall-off*, of the magnetic anomalies. See also *magnetic anomaly*, *pluton*, and *residual magnetics*.

regional map: a printed or digital map of an anomalous gravity or magnetic *regional field*. See also *anomalous field*, *regional-residual separation*, and *residual map*.

regional-residual separation: separation of the *anomalous field* into two parts: *residual field* of the relatively short-*wavelength* anomalies and *regional field* of the longer wavelengths and, commonly, the higher *amplitude* anomalies. Quantitative estimates of the regional and residual components are based on the exploration project objectives, *survey resolution*, and evaluation of the *target anomaly* and its source size, i.e., "regional" and "residual" are relative terms. Local-*scale*, or residual, anomalies are obtained by suppressing the gravity and magnetic effects associated with the structures which are larger than the project *target* or too deep to be of the exploration interest. Methods of regional-residual separation include but are not limited to: (1) *polynomial fitting* which approximates the regional components of the gridded field by polynomial surfaces of different orders: the higher the polynomial order, the more localized the residual components; (2) space-domain or spectral-domain *filtering*, where the regional or residual components of the anomalous field are suppressed by application of high-pass or low-pass filters, respectively; (3) *upward continuation*, which attenuates anomalies from the shallow and small sources and enhances anomalies generated by the deeper and larger sources; (4) *separation filtering*, which divides the original field into the *wavelength* components generated by sources above and below a user-specified depth level in order to approximate the residual and regional fields; (5) *inversion-based separation*; (6) *gravity stripping*, where the calculated field of the two-dimensional (2D) or three-dimensional (3D) gravity *model*, which is built with parameters constrained by the independent geological-geophysical data (seismic, magnetic, electromagnetic, deep wells), is subtracted from *Bouguer anomaly* or *free-air anomaly*; and (7) *PCA-based separation*. Also referred to as *residualizing*. See also *constraints*, *high-pass filter*, and *low-pass filter*.

regional survey: an airborne or marine (shipboard) gravity or magnetic *survey* designed with the objective of obtaining structural information about a particular regional-scale sedimentary *basin* (e.g., trends of the regional faults, thickness of accumulated sediments, igneous intrusions, volcanic terranes, etc.) for oil-and-gas exploration and mapping the areas of a shallow *basement* and its outcrops for mineral exploration. The orientation of *traverse lines* is commonly orthogonal to the *strike* of the regional geological *trend* determined by a *reconnaissance survey*. The flight *altitude* and *survey line spacing* of the regional airborne survey are selected in accordance with the *regional field* properties known from the reconnaissance survey and satellite data. Flight line spacing is constant to preserve a consistency in the gravity and magnetic signatures, or patterns, over a similar geology. Regional marine gravity and magnetic surveys are often complementary to *two-dimensional (2D)* seismic exploration, and their line spacing is defined by the seismic acquisition program. See also *detailed survey*, *signature*, *terrane*, and *survey design aliasing*.

regional trend: a relatively smooth component of an *anomalous field* that represents the dominant increase or decrease in its values that is associated with a regional-scale geological *structure*. A regional trend surface is either a dipping plane (first-order trend) or a smoothly curving surface approximated, among other methods, by the low-order polynomials. See also *polynomial fitting* and *trend removal*.

regional wavelength filter: same as *low-pass filter* (*high-cut filter*) with an emphasis on its property to retain, or pass, the long-*wavelength* components of an *anomalous field* and suppress, or reject, all wavelengths which are shorter than its *cutoff wavelength*. See also *residual wavelength filter*.

regolith: surface and near-surface accumulations of the unconsolidated rock material, both residual and transported, that overlays a non-weathered bedrock. Regolith can contain the magnetic minerals originating from the weathered part of the bedrock and from chemical changes during the weathering process. Magnetic effects of the regolith are observed as random spikes and relative magnetic highs generated by local regolith assemblages, and they can obscure the short-*wavelength* low-*amplitude* anomalies from sources within the sedimentary section, including those associated with *hydrocarbon seepage effect*. See also *micromagnetic anomalies* and *random noise*.

regression: a statistical method of determining the relationship between values of two similar functions. Scaled values of the *apparent density* distribution defined by *terracing* algorithms can be converted into geologically meaningful densities using the multiplicative factor determined by least-squares regression between the observed gravity *profile* and the profile calculated from the terraced *model*. Regression is also implemented in calculation of *diurnal correction* for marine (shipboard) magnetic data with the use of stationary measurements by a *base station* or *geomagnetic observatory* which is located far from the *marine survey* area. See also *least-squares approximation*.

regridding: an auxiliary post-*gridding* procedure which changes the *grid interval* and, if necessary, *XY coordinates* of the *grid* origin. Depending on the objective of regridding, a new grid interval can be larger or smaller than the original grid interval. Also referred to as *grid resampling*.

regular noise: *coherent noise* in the context of the correlated *noise* components that exhibit a systematic, or regular, *phase* relation between the waveforms of gridded data or between the profiles of data along adjacent *survey* lines. See also *grid*, *profile*, and *waveform*.

relative accuracy: a performance characteristic of certain types of *quantum magnetometers* that is used as a substitute for *absolute accuracy* to describe the repeatability of measurements in terms of the percentage of a measured value. For example, the relative accuracy of the ultra-sensitive *helium magnetometer* is approximately 0.1% of the measured values in the *vector* mode of operation and 0.001% in the *scalar* mode. See also *accuracy*, *precision*, and *sensitivity*.

relative base station: one or several reference stations where the *relative gravity* value is known and tied to an *absolute base station*. It is established in the *ground gravity* survey area to control *gravimeter drift*. Gravity measurements are repeated at the relative base stations with regular time intervals. Referred to with the same meaning as *relative gravity base*. See also *base station*.

relative Bouguer anomaly: *Bouguer anomaly* calculated from the *observed gravity* data presented in the *relative gravity* values. See also *absolute Bouguer anomaly*.

relative gravimeter: a *gravimeter* which measures the amount of change in the *Earth gravity field*, or *relative gravity*, in relation to some reference level, such as a gravimeter *null point*. Differing from an *absolute gravimeter* which measures the full value of *gravitational acceleration g*, or *absolute gravity*, a relative gravimeter measures the change in the downward *gravitational force* acting on a *proof mass* suspended in a *spring gravimeter* by a metal or quartz spring, i.e., measurements are based on the same concept as the one implemented in the weighing devices. Change in the Earth's gravity field is determined by change in the spring length that is proportional to and calibrated in the units of gravitational acceleration g. Alternatives to using the mechanical spring include: (1) *virtual spring* design, where the proof mass is a superconducting sphere levitating in an electromagnetic or electrostatic field, and its deviations from the *null point* are analogous to changes in the spring length; (2) *electromagnetic accelerometer*; (3) *electrostatic accelerometer*; and (4) *vibrating-string gravimeter*. All gravimeters currently used in airborne, shipboard (marine), sea floor, borehole, and ground gravity exploration applications are relative gravimeters. The best reported *accuracy* of measurements is 1 *microgal (μGal)*, which is one billionth part of the *nominal gravity* value $g = 9.8\ \mathrm{m \cdot s^{-2}}$. See also *superconducting gravimeter*, *zero-length spring gravimeter*, *scale factor*, and *relative magnetometer*.

relative gravity: a gravity value measured relative to some reference level, i.e., it is a relative value Δg which is different from the full, or absolute, value of the *Earth gravity field* called *absolute gravity* and denoted by g. Every relative gravity value represents the amount of change in the Earth's gravity field at the observation points that is measured by a *relative gravimeter*. Depending on the gravimeter type, the relative gravity value is defined by the difference Δg between (a) the observed value and *null point* of the gravimeter at each observation point, (b) the gravimeter readings at two successive observation points, and (c) the *reading* at each observation point and the gravity value established at the location of the *base station*. All gravity surveys in *exploration* applications acquire relative gravity data, and their values are commonly measured relative to the gravimeter null point. See also *absolute gravimeter* and *acquisition*.

relative gravity base: a temporary gravity *base station* with the known *relative gravity* value which is used as a reference value, or *datum*, for measurements of changes in the *Earth gravity field* over a *survey* area by a regular, or conventional, exploration *gravimeter* referred to as a *relative gravimeter*. Also known as *arbitrary gravity base*, *secondary gravity base*, and *relative base station*. See also *absolute gravity base* and *absolute gravimeter*.

relative gravity corrections: *observed gravity corrections* in the context of their application to the *observed gravity* presented in *relative gravity* values, i.e., these corrections are applied to relative gravity data. See also *absolute gravity corrections*.

relative gravity datum: a numerical value of *relative gravity* used as a reference value, or *datum*, for the gravity measurements during ground, marine (shipboard), submarine, and airborne gravity surveys. See also *gravity datum* and *static test*.

relative magnetometer: same as *vector magnetometer* with an emphasis on its use for measurements of the *vector* components of the *Earth magnetic field* relative to (a) the spatial orientation of its sensors and (b) *baseline* value which is calibrated periodically by a *scalar magnetometer*. See also *magnetic sensor*.

relative permeability: a proportionality constant represented by the dimensionless quantity μ_R which is defined by the ratio of *magnetic permeability* μ of a particular rock material to the magnetic permeability μ_0 of a source-free space as

$$\mu_R = \mu/\mu_0.$$

In a medium characterized by relative permeability μ_R, the relationship between the *magnetic induction* **B** and inducing magnetic field strength, or *magnetizing force*, **H** is expressed as

$$\mathbf{B} = \mu_R \mathbf{H}.$$

relativistic correction: same as *speed-of-light correction* with an emphasis on compensation for the measurement effect of a relative *phase* shift in the optical fringes of a laser interferometer that is produced by the accelerated motion of a corner-cube reflector in a *free-fall gravimeter* and known as the Doppler shift.

relief: a difference between the highest and lowest topographic heights or the deepest and shallowest parts of the ocean or sea in a *survey* area. High relief indicates an area of highly variable *topography* or *bathymetry*, whereas low relief indicates an area of smooth topography or bathymetry. Sometimes, it is used in expressions such as "gravity relief" to highlight *amplitude* variations in observed gravity anomalies. See also *anomaly*.

remanent magnetism: a natural phenomenon of the permanent rock *magnetization* which remains in rock-forming minerals after the removal of an externally applied, or inducing, magnetic field.

remanence: See *remanent magnetization*.

remanence dating: See *archaeomagnetism*.

remanent magnetization: the permanent *magnetization* remaining in rock-forming minerals after the removal of an externally applied, or inducing, magnetic field. Differing from *induced magnetization*, which falls to zero in the absence of the external magnetic field, remanent magnetization retains the strength and orientation obtained in the geological past, so its *vector* direction is commonly different from the vector direction of the induced magnetization defined by the present *Earth magnetic field*, i.e., the remanent magnetization vector shows the direction of the Earth's magnetic field at the geological time when this rock was magnetized. Remanent magnetization intensity depends on: (a) the *magnetic susceptibility* of rocks; (b) the volume percentage of the *ferrimagnetic* minerals, such as *magnetite*, in the rock composition; (c) the strength and direction of the Earth's magnetic field at the time of rock magnetization; and (d) the tectonic and thermal history of rocks. There are several types of remanent magnetization, including *chemical remanent magnetization (CRM)*, *detrital remanent magnetization (DRM)*, *isothermal remanent magnetization (IRM)*, *normal remanent magnetization (NRM)*, *pressure remanent magnetization (PRM)*, *thermoremanent magnetization (TRM)*, and *viscous remanent magnetization (VRM)*. Generally, the remanent magnetization is approximately 0.1–1.0 of the induced magnetization value in the same rock, but it can be much larger in kimberlite intrusions, basalts, young volcanic (extrusive) rocks, mafic and ultramafic igneous intrusions, some iron and steel objects, and various special rock types, such as the ilmenite-bearing anorthositic (monomineralic) igneous rocks. Because *reduction-to-pole (RTP)* assumes a small amount of remanent magnetization and results in a nearly symmetric RTP anomaly, the presence of a strong remanent magnetization is indicated by the distinct asymmetric negative troughs on the opposite sides of an RTP anomaly over a compact three-dimensional (3D) *source body* or pipe-like *igneous intrusion*, wherein the *magnitude* of one of two troughs is often the largest in this trough-peak-trough anomalous triplet. A small horizontal shift between the maximum values of the *total gradient (TG)* and *vertical gradient (VG)* of the RTP anomaly over a 3D source body indicates that its remanent magnetization is negligible. Also referred to as *remanence*, *residual magnetization*, and *reverse magnetization*. See also *kimberlite pipe*, *Koenigsberger ratio*, and *cross-correlation remanence*.

remote sensing: planetary, regional, and local-*scale* measurements of the specific physical properties and monitoring of the spatial characteristics of a study object, such as the Earth and other planets, without a direct physical contact with its surface. Airborne and spaceborne (satellite) measurements of the Earth's gravity and magnetic fields and *satellite altimetry* are examples of the remote sensing in gravity and magnetic applications (Berger, 2012). See also *airborne gravimetry*, *drone survey*, *high-resolution aeromagnetic survey*, *satellite gradiometry*, *satellite gravimetry*, and *satellite magnetometry*.

reoccupations: **1.** repeated gravity measurements at a primary or secondary *base station* or at selected observation points during a *ground gravity* survey in order to determine the *drift* of a *spring gravimeter.* Reoccupation periods in a ground survey are minimized to approximate *gravimeter drift* by a linear time-dependent function and incorporate the *Earth tide effect* into the *drift correction*. **2.** repeated magnetic measurements at selected observation points during a ground *magnetic survey* to determine the *magnetometer drift* in some types of magnetometers. **3.** repeated gravity measurements at a fixed airfield location on the ground to determine the value of *static correction* for *accelerometer*-based *airborne gravity* measurements. Also referred to as *gravity repeats*. See also *base-check method*, *cross-check method*, and *static test*.

repeat station: an *observation point* on the ground where the gravity or magnetic measurements are made two or more times within a *loop*, but not as consecutive measurements and not in any other loop. Repeat stations are used for monitoring *drift* and other *gravimeter* or *magnetometer* performance characteristics. See also *tie station*.

repeatability: a quantitative estimate of consistency in the performance characteristics of a *spring gravimeter* or *scalar magnetometer* during a ground gravity or magnetic *survey*. Differences between the linear *drift*-corrected randomly distributed *reoccupations* are used for calculation of *rms error*, which is also referred to as *repeatability error* and expressed as

$$\mathrm{RE} = \{\Sigma(x_0 - x_i)^2 / n\}^{1/2},$$

where x_0 is the first measurement value at a *repeat station*, x_i is repeated measurement, $1 < i < n$, and n is total number of repeated measurements in the *ground survey* area. See also *gravity survey precision*, *static test*, and *standard deviation (SD)*.

repeatability error: the *rms error* between random *reoccupations*.

replacement density: a *density* value for calculation of *simple Bouguer correction* for the marine (shipboard) gravity measurements by an onboard *gravimeter* and for the stationary measurements by an *underwater gravimeter.* Replacement density is defined by calculating the difference between the assumed *density* of rocks below the sea floor and mean seawater density 1.03 g · cm^{-3}. For *global bathymetric correction*, the replacement density value is defined by the difference between the reference crustal density 2.67 g · cm^{-3} and mean seawater density. See also *Nettleton method*.

resampling: a *data conditioning* operation applied to change the original *sampling* of line-oriented or gridded data. There are two types of resampling: (1) *decimation*, or *downsampling*, which reduces the amount of data samples by dropping every second, third, fourth, etc. sample, and (2) *upsampling*, which increases the amount of samples by adding the samples obtained by interpolation between the original data samples. See also *gridding*.

reservoir monitoring: application of *time-lapse gravity* and *time-lapse gravity gradiometry* for monitoring the natural gas production, steam-enhanced heavy oil recovery, waterflood in enhanced oil recovery, geothermal reservoir development, long-term underground CO_2 storage, and groundwater *aquifer* depletion. Monitoring methodology is based on (a) construction of a three-dimensional (3D) reservoir *model* from the seismic and well data and (b) evaluation of the required *accuracy* of measurements, *lateral resolution* and *vertical resolution* of the *absolute gravity* and *residual gravity* anomalies that can be generated by changes in fluid content and their distribution within a reservoir (Krahenbuhl et al., 2011). See also *aquifer gravity*, *CO_2 sequestration gravity*, *geothermal gravity*, *feasibility study*, and *4D gravity monitoring*.

residual anomaly: 1. any *anomaly* remaining in an anomalous gravity or magnetic *field* after the subtraction of a *regional field* obtained by one of the *regional-residual separation* techniques suppressing the relatively long-*wavelength* field components. **2.** an anomaly generated by a *source body* of the exploration or research interest. **3.** an anomalous component of a *residual field*. See also *anomalous field* and *regional anomaly*.

residual Bouguer anomaly: the remaining part of a *Bouguer anomaly* after the subtraction of its regional long-*wavelength* component. Also referred to as *local gravity*. See also *regional-residual separation*.

residual drift: short-term (~10–20 μGal per hour) and long-term (~10–30 μGal per day) mostly non-linear components of a *gravimeter drift* that remain after removing the *drift* component approximated by a linear function of time. See also *automatic drift correction*, *gravimeter*, and *microgal (μGal)*.

residual disturbance: small-*amplitude* fluctuations of the *external magnetic field* of the Earth observed during the relatively quiet days after a *magnetic storm*. See also *micropulsations*.

residual field: 1. short- and mid-*wavelength* components of the observed, or anomalous, field remaining after the *grid*-based subtraction of the longer-wavelength *regional field*. The wavelength content of a residual field is defined by the applied approximation of the regional field, and it depends on the exploration project objectives, *survey resolution*, evaluation of *target anomaly*, and its *source geometry*. **2.** *anomalous field* components remaining after the application of a *high-pass filter* or other *regional-residual separation* technique suppressing the relatively long-wavelength field components. Calculation of the residual field highlights the anomalous gravity and magnetic signatures of the local intrasedimentary structures targeted in the mineral or oil-and-gas *exploration*. See also *residual anomaly*, *polynomial residual*, *gravity signature*, *magnetic signature*, *target*, and *feasibility modeling*.

residual filter: a version of the spectral-domain *Jacobsen filter* applied for the separation of the *residual field* generated by sources that are shallower than the user-specified depth h_0. Its *transfer function* is expressed as

$$F(k) = 1 - e^{-2|k|h_0},$$

where k is the *wavenumber* with $k = |k| = (k_x^2 + k_y^2)^{1/2}$ and k_x and k_y are wavenumbers in the mutually orthogonal x- and y-directions related to their respective wavelengths as $k_x = 2\pi/\lambda_x$ and $k_y = 2\pi/\lambda_y$. See also *regional-residual separation*, *spectral domain*, *wavelength*, and *regional filter*.

residual gravity: 1. the *field* of the gravity anomalies remaining after the subtraction of *regional gravity* from *free-air gravity* or *Bouguer gravity*. **2.** *stripped gravity*. **3.** the field of the gravity anomalies obtained by space-domain and spectral-domain *filtering* or by *regional-residual separation*. Residual gravity is calculated in order to enhance the local-scale anomalies of an exploration or research interest and obtain the most accurate *gravity inversion* results. Differing from *residual magnetics*, the residual gravity field is dominated by anomalies of structural rather than lithological origin that are generated by local-scale sedimentary geological structures, such as anticlines, synclines, horsts, grabens, and reefs. Anomalies generated by the low-*density* and high-density source bodies, such as reservoir sands accumulated in incised valleys and shallow *ore body*, will be enhanced too. Also referred to as *local gravity* and *residual gravity anomaly*. See also *anticline*, *syncline*, *horst*, *graben*, *incised valley*, and *reef*.

residual gravity anomaly: 1. the *field* of local-scale gravity anomalies obtained by the subtraction of the *regional field* component from an original *anomalous field*, such as *free-air anomaly* or *Bouguer anomaly*. **2.** *residual gravity*. **3.** a *gravity anomaly* generated by the *source body* of an exploration or research interest. See also *regional-residual separation*.

residual grid: a *grid* of the gravity or magnetic *residual field* where the long-*wavelength* components of the original *anomalous field* are suppressed by application of a *high-pass filter* or other method of *regional-residual separation*. See also *regional grid*.

residual magnetic anomaly: 1. *residual magnetics*. **2.** the *field* of local-scale magnetic anomalies obtained by the subtraction of *regional magnetics* from an original *anomalous field*, such as *TMI field* or *RTP magnetic field*. **3.** a *magnetic anomaly* generated by the *source body* of an exploration or research interest. See *regional-residual separation*.

residual magnetic vector: a *magnetic vector* calculated by subtracting the constant *vector*, which represents the *main geomagnetic field* or average magnetic vector for the entire *survey* area, from the observed magnetic vector at each *observation point*. As the numerical, or *scalar*, measures of differences between the local vectors and their regional approximation, the residual *magnetic vector attributes* are very sensitive to local variations of the magnetic vector components. Grids of the calculated residual vector attributes highlight the magnetic vector anomalies associated with heterogeneously magnetized rock formations (Christensen and Dransfield, 2002). See also *anomaly*, *anomalous magnetic vector*, *grid*, *magnetization*, *magnetization vector*, *regional-residual separation*, *sensitivity*, and *airborne vector magnetometry*.

residual magnetics: 1. the *field* of magnetic anomalies remaining after the subtraction of *regional magnetics* from a *TMI field* or *RTP magnetic field*. **2.** the field of TMI or RTP magnetic anomalies obtained by space-domain or spectral-domain *filtering* or by *regional-residual separation*. Differing from *residual gravity*, the residual magnetic field is dominated by the anomalies of lithological rather than structural origin that are generated by igneous intrusions, shallow ore bodies, near-surface *magnetite*-bearing rock formations, localized high-*magnetization* mineral deposits, and igneous and metamorphic rocks in the upper part of the *basement*. The residual magnetic field is calculated in order to enhance the local-scale anomalies of an exploration or research interest and obtain the most accurate results of *magnetic inversion* and *UXO detection*. Also referred to as *residual magnetic anomaly*. See also *total magnetic intensity (TMI)*, *space domain*, *spectral domain*, *igneous intrusion*, and *ore body*.

residual magnetization: same as *remanent magnetization* that remains in rocks after an externally applied, or inducing, magnetic field is removed.

residual map: a printed or digital map of an anomalous gravity or magnetic *residual field*. See also *anomalous field*, *regional-residual separation*, and *regional map*.

residual potential: See *anomalous potential* and *disturbing potential*.

residual wavelength filter: a *high-pass filter* (*low-cut filter*) that will retain, or pass, the short- and mid-*wavelength* components of an *anomalous field* and suppress, or reject, all wavelengths which are longer than the *cutoff wavelength* of this filter. See also *regional wavelength filter*.

residualizing: *regional-residual separation* in the context of its objective to obtain the *residual field* of the short- and mid-*wavelength* anomalies of an exploration or research interest that are not visible due to superposition of the regional anomalies which have longer wavelengths and higher amplitudes.

resolution: 1. a characteristic of an *acquisition* system that describes the shortest *wavelength* and the smallest anomalous *amplitude* that can be detected, or resolved, with the use of a particular instrument or acquisition method. **2.** the minimal *anomaly width* that can be identified on a *grid*, map, or line-oriented data. **3.** the ability of an acquisition system to distinguish between two closely spaced source bodies of an exploration or research interest. The resolution of source bodies depends on the *sensor*-source distance: when two source bodies are closer than this distance, their anomalies will merge into a single *anomaly*, and when they are apart from each other by a much larger distance, their anomalies will be resolved separately. For measurements on a moving *observation platform*, such as a ship or aircraft, the resolution is defined primarily by the following *survey* parameters: (1) *sensitivity* of the measuring instruments, (2) *precision* of measurements, (3) *sampling interval* along the survey lines, (4) *survey line spacing*, (5) length of the *low-pass filter* or other *filter* applied to suppress *dynamic noise*, and (6) flight elevation, or *altitude*, of the airborne survey. There are *vertical resolution* and *lateral resolution* that describe the *spatial resolution* of acquired gravity and magnetic data. Often but not always, the vertical resolution is estimated at two or three times the averaged precision limit, and the lateral resolution is defined by the *filter length* of the *noise*-suppressing filter. For ground measurements with irregular *station* intervals, the lateral resolution is approximated by an average sampling interval over the whole survey area. See also *anomaly resolution*, *full-wavelength resolution*, *half-wavelength resolution*, *grid resolution*, *reading resolution*, *source resolution*, *survey resolution*, and *interference of anomalies*.

resonance magnetometer: See *nuclear-resonance magnetometer*.

reverse magnetization: *remanent magnetization* in the context of rock *magnetization* the *vector* direction of which can be the opposite, or reverse, of the direction of the present *main geomagnetic field* in the *magnetic survey* area.

reverse polarity: a strong *remanent magnetization* wherein its *vector* is directed in the opposite, but not necessarily diametrically opposite, direction of the present *main geomagnetic field*. See also *geomagnetic polarity reversals*.

reverse spectral transform: a data processing operation which converts Fourier-transformed data from the *spectral domain* back to their original *space domain*. Commonly referred to as *inverse Fourier transform*.

reversible pendulum: a pendulum composed of a metal bar and adjustable weight with two pivots from which the pendulum can be suspended interchangeably for measuring its swing periods. Weight is moved along the bar to adjust the period of swinging until two periods of the opposite, or reversed, pivot suspensions become equal. The obtained swing period value is equal to the swing period T of an "ideal pendulum," and the distance between the adjusted pivot points is equal to the length L of the ideal pendulum. The concept of the reversible pendulum is implemented in the design of a *pendulum gravimeter* for measurements of *absolute gravity*.

RGB composite image: a red-green-blue (RGB) multicomponent, or composite, image for visualization of the gridded data variations in multidimensional datasets, where shades of each color represent a user-defined range of data parameters, such as *amplitude*, *coherence*, and *phase*, in any three selected dimensions. See also *dimension* and *gridding*.

ribbon: a *source body* approximation in gravity and magnetic *modeling*. Often referred to with the same meaning as a low-dipping *thin dike* or *thin sheet*. See also *dip*.

ridge: 1. an alignment of the *peak* values of a linear or curvilinear *anomaly* on a *grid*. Automatic correlation of ridges is implemented in *skeletonization* and *steerable filter* for the detection and mapping of *structure*-related *lineaments*. **2.** a line joining the maximum or zero points of an *anomalous field* and its derivatives calculated at different levels of *upward continuation*, or altitudes. Ridges obtained by joining the zero points of the horizontal and vertical derivatives of an anomalous field and field itself are used in *multiridge analysis*. Ridges of the *homogeneous field* generated by an *ideal source* are straight lines intersecting at the source position, when they are extrapolated into the depth section of a *two-dimensional (2D)* multiridge dataset (Fedi et al., 2009). **3.** a steep-sided *elevation* on the Earth's surface or a long and narrow *uplift*, such as *horst*, in the subsurface. See also *altitude*, *Blakely-Simpson method*, *horizontal derivative*, *vertical derivative*, and *reduced Euler deconvolution*.

Riesz transform: generalization of the one-dimensional (1D) *Hilbert transform* for dimensions D $\geq$ 2. A two-dimensional (2D) Riesz transform adds to an *anomalous field* two *quadrature*, or mutually orthogonal, field components obtained by two 90° *phase* shifts at each observation point and, thereby, creates a new dataset, where the anomalous field values are represented by the three-component *vector* which can be defined at every point in space by the mutually orthogonal field attributes, such as the *local orientation*, *local phase*, and *local amplitude* of the three-component *monogenic signal (MS)*. For the function $f(x, y)$, such as the gridded gravity or magnetic field, the Riesz transform R can be expressed as

$$Rf(x,y) = [iR_1 + jR_2]f(x,y),$$

where $i = j = \sqrt{-1}$ are complex unit numbers indicating the orthogonality, and R_1 and R_2 are the first-order Riesz transform operators equivalent to the mutually orthogonal H_x and H_y operators of the *3D Hilbert transform* (Langley and Anderson, 2010). In the high-*resolution* processing of airborne gravity and magnetic data, the Riesz transform is applied to convert the gridded *RTP anomaly* or *Bouguer anomaly* values into the monogenic signal (MS), which is then decomposed into three primary and two secondary complex attributes used in the structural interpretation: *local amplitude*, *local orientation*, *local phase*, *directional Hilbert attribute*, and *Riesz transform attribute* (Hassan and Yalamanchili, 2013). See also *attribute*, *complex attribute*, *dimension*, and *gridding*.

Riesz transform attribute: the secondary *complex attribute* of a *monogenic signal (MS)*. It is a measure of the *amplitude* content, or energy, of the imaginary components in a complex monogenic signal. Riesz transform attribute Q_R can be expressed as

$$Q_R = \{[R_1 f(x,y)]^2 + [R_2 f(x,y)]^2\}^{1/2},$$

where $f(x, y)$ is the real part representing the *RTP anomaly* or *Bouguer anomaly* value at *grid* point (x, y) and $R_1 f(x, y)$ and $R_2 f(x, y)$ are two imaginary, or 90° *phase*-shifted, components of the monogenic signal obtained by the first-order *Riesz transform* operators R_1 and R_2 in the mutually orthogonal complex planes. Because R_1 and R_2 are equivalent to the orthogonal operators H_x and H_y of the *3D Hilbert transform*, the Riesz transform attribute is equivalent to the *horizontal DAS (HDAS)* and analogous to the *horizontal gradient magnitude (HGM)* of the *vertical integral* of the magnetic or gravity *anomaly*, i.e., it is analogous to the HGM of *anomalous potential*. Also referred to as *Riesz transform magnitude*. See also *complex plane*, *gridding*, and *monogenic signal decomposition*.

Riesz transform magnitude: See *Riesz transform attribute*.

rift: 1. a long narrow structural depression, or trough, bounded by the normal faults, i.e., *graben* of a regional extent. **2.** an extensional sedimentary or upper *basement* structure associated with a pull-apart zone created by a regional strike-slip, or wrench, *fault*. **3.** a long-extension *strike-slip fault* with a negative *flower structure*. Rifts are identified by the linear gravity and magnetic *gradient* anomalies bounding the areas of (a) reduced or negative *residual Bouguer anomaly* values due to the thick accumulated sediments whose densities are lower than those of the surrounding older rock formations and (b) the positive or negative residual *RTP magnetic anomaly* generated by *volcanics* and other high-*magnetization* igneous rocks intruded into the rift under the extensional tectonic regime. See also *pull-apart basin*, *igneous intrusion*, and *residual magnetic anomaly*.

rift basin: a regional-scale structural depression bounded by the normal or strike-slip faults that mark either the region where the *crust* is split apart, such as in the mid-Atlantic *rift*, or zone of crustal stretching and thinning, such as those in the continental margins undergoing thermal subsidence. Its *gravity signature* is defined by the superposition and interference of the broad long-*wavelength* positive anomalies due to the regionally thinned crust and relatively narrower positive and negative anomalies generated by thick accumulations of sediments interbedded with the localized intrusions of high-*density* igneous rocks. See also *igneous intrusion* and *interference of anomalies*.

right-sided Cartesian system: *Cartesian coordinates* with the x-axis directed to the geographic east and the y-axis to the geographic north. See also *NED reference frame*.

rim syncline: a synclinal *structure* which partially or completely surrounds a *salt dome* and is formed by the withdrawal of *salt* in the process of its movement into the dome. A shallow salt dome can have the *gravity signature* of a local *gravity high* surrounded by the lower values of *residual Bouguer anomaly* generated by accumulations of low-*density* sediments in the rim *syncline*. See also *nil zone*.

ringing: *Gibbs phenomenon* in the context of the high-frequency oscillations at the *grid* margins and around the *cutoff wavelength* value of the spectral-domain *FFT filters* (especially narrow *band-pass filter)* that represent the accumulated errors of the forward and inverse parts of the *Fourier transform*.

rms accuracy: the root-mean-square (rms) *accuracy* of measured gravity and magnetic values. It is calculated as (a) the averaged *rms error* at mis-ties after *leveling* and *microleveling* in estimates of airborne, marine (shipboard), and submarine *survey accuracy*, and (b) the square root of the average of the sum of the squares of differences between the individual measured values and the corresponding set of reference, or true, values, such as those of *ground truth* for *airborne gravity* measurements. See also *precision*, *noise*, *standard deviation (SD)*, and *mis-tie*.

rms crossover difference: *rms accuracy* in the context of the airborne, marine (shipboard), and submarine *survey accuracy* estimated by calculation of the *crossover error* after *leveling* and *microleveling* of the acquired gravity and magnetic data.

rms error: **1.** a root-mean-square error used as the equivalent of *standard deviation (SD)* for the statistical estimates of (a) instrument *precision* and (b) accumulated gravity and magnetic survey errors due to *positioning* errors and data contamination by *dynamic noise* and unaccountable *diurnals*, respectively. **2.** a statistical estimate of the *repeatability error* based on the calculated differences between the linear *drift*-corrected randomly distributed repeated measurements, or *reoccupations*, in a ground gravity or magnetic *survey* area.

rms noise: *rms error* referred to with the same meaning as *standard deviation (SD)* in the statistical estimates of the level of *noise* in measured data.

rms value: the square root of the average, or arithmetic mean, of the squares of values in a dataset, i.e., root-mean-square value. It is used as (a) the equivalent of *standard deviation (SD)* for the statistical estimates of *sensitivity* of a particular instrument and the *precision* of its measurements, (b) the *tolerance* value for truncation of an infinite series, (c) the estimate of the accumulated survey error due to *positioning* errors and data contamination by *dynamic noise* and unaccountable *diurnals*, and (d) the *target rms* for the automatic termination of an *inversion* process.

rock magnetism: a general term referring to the natural phenomena associated with *magnetization* and magnetic properties of sedimentary, igneous, and metamorphic rocks and rock-forming minerals.

roll: the side-to-side rotational motion of a *survey* ship or aircraft about the horizontal axis caused by the ocean/sea waves or air turbulence during the gravity and magnetic measurements along the survey lines, i.e., it is rotation of a moving *observation platform* about the X-axis in the YZ plane. Roll is a source of magnetic interference effects in an aircraft frame and inertial accelerations of a shipboard (marine) or airborne observation platform. See also *pitch*, *yaw*, *aeromagnetic compensation*, and *inertial acceleration*.

roll-off ramp: See *roll-off window*.

roll-off window: **1.** *taper* in the context of a *window* within which the data values gradually decrease, or roll-off, to zero at the *grid* edges and *survey* line ends. **2.** the transition zone between the pass and reject modes of a *band-pass filter* operation, where the data values gradually decrease to the level of suppression but not necessarily to zero. Also referred to as *roll-off ramp*. See also *cosine roll-off filter*.

root: See *isostatic root*.

root gravity effect: *Airy gravity effect* on the continents, i.e., in the areas of *isostatic root*.

root isostatic effect: *isostatic compensation effect* on the continents in the context of the *Airy-Heiskanen model* of *isostasy*.

rotated gradient tensor: **1.** the *gravity gradient tensor (GGT)* for which the fixed Cartesian coordinate frame of the original measurements has been rotated at each observation or *grid* point to a *local reference frame*, and the measured values of the tensor components have been converted into the values corresponding to the new orientation of the Cartesian axes for calculation of the *true curvature attributes* and other applications. For the point, spherical, and infinite *horizontal cylinder* source geometries, the rotated tensor components $T_{xz} = T_{zx} = T_{yz} = T_{zy} = 0$. When the *x*- and *y*- axes of the local reference frame are aligned with the directions of *maximum curvature* and *minimum curvature*, $T_{xy} = 0$ and calculations of the *curvature attributes* for all source geometries involve only the diagonal elements T_{xx}, T_{yy}, and T_{zz}, which is then transformed by vertical integration into the *vertical gravity* g_z and used in lieu of the local gravity field strength (Li, 2018). **2.** the *full tensor gradient (FTG)* whose components are recomputed in a reference frame aligning the primary FTG axes with the dominant *trend* of a geological structure or *strike* of a targeted *source body* (Dickinson et al., 2009). See also *apparent curvature attributes*, *vertical integral*, and *window inversion*.

rotating accelerometer: a *gravity gradiometer* design concept based on the use of pairs of synchronized electromagnetic or electrostatic accelerometers symmetrically mounted on a rotating disk in order to separate the *gravity gradient* signals from *dynamic noise*. This concept is implemented in the *rotating-disk gravity gradiometer* for measurements of the *gravity gradient tensor (GGT)* components on a moving *observation platform*, such as a ship, submarine, and aircraft. See also *electromagnetic accelerometer* and *electrostatic accelerometer*.

rotating-disk gravity gradiometer: a *gravity gradiometer* that utilizes the assemblages of paired accelerometers mounted on one or several rotating disks, where the disk rotation provides the *angular frequency*-based separation of the *gravity gradient* signal from *dynamic noise* during the measurements of the *gradient* and *gradient tensor* components of the *Earth gravity field* on a moving *observation platform*, such as a ship, submarine, and aircraft. Gradients are determined from a numerical difference between the *acceleration* values measured by the opposing pairs of synchronized accelerometers which form the *gradiometer arm* along the mutually orthogonal axes of the measurements. Two design concepts are currently implemented in exploration applications: (1) *partial tensor gravity gradiometer* (PTGG), which measures two *differential curvature* components of the *gravity gradient tensor (GGT)* known as *differential curvature gradients*, and (2) *full tensor gravity gradiometer* (FTGG), which measures six gradient tensor components in three separate coordinate frames with conversion of the measured components into all independent elements of the *full tensor gradient (FTG)*. The basic PTGG has a single disk with four equally spaced electromagnetic accelerometers, whose sensitive axes are orthogonal to the axis of the disk rotation and tangential to the disk plane. The disk rotates at a constant speed about the vertical axis, wherein the accelerometer signals are modulated at twice the rotation frequency to ensure the enhanced *common mode rejection* and filtering-out of the mismatch and misalignment errors. The FTGG has three rotating disks with a total of 12 electrostatic accelerometers in a symmetrical arrangement similar to the one used in PTGG. Three FTGG disk assemblages are tilted toward each other, and their axes of rotation are mutually orthogonal. An assembly of three rotating disks is rotated on the mounting sub-platform (carousel) with a constant angular frequency about the vertical axis. In total, six gradient tensor components are measured in three separate coordinate frames to obtain six independent FTG components. Sometimes referred to as *moving-base gravity gradiometer.* See also *electromagnetic accelerometer* and *electrostatic accelerometer.*

rotating gradiometer: See *rotating-disk gravity gradiometer* and *rotating-sensor magnetic gradiometer.*

rotating-sensor magnetic gradiometer: *SQUID magnetic tensor gradiometer.* Its design is based on the use of three rotating single-axis SQUID sensors in an umbrella configuration. See also *axial gradiometer* and *SQUID sensor.*

rotational invariants: 1. *invariants.* **2.** invariants I_1 and I_2 denoted by R_1 and R_2 to highlight their property to retain the shape, sign, and orientation of the detected *gradient* anomalies regardless of the direction of their rotation about the *z*-axis in *Cartesian coordinates.* **3.** invariants comprising the *normalized source strength (NSS).* See also *FTG invariants* and *horizontal invariants.*

rotational potential: same as *centrifugal potential* with an emphasis on it being the *potential* of the rotational *centrifugal force.*

RTE: See *reduction-to-equator (RTE).*

RTP: See *reduction-to-pole (RTP).*

RTP magnetic anomaly: the *field* of the magnetic anomalies obtained by the application of *reduction-to-pole (RTP)* to *total magnetic intensity (TMI)* data. Usually, it is displayed in the format of a *grid* and referred to with the same meaning as *RTP magnetic field.* Also known as *RTP magnetics.* See also *anomalous field* and *format.*

RTP-FFT transform: *reduction-to-pole (RTP)* of *total magnetic intensity (TMI)* data performed with the use of *fast Fourier transform (FFT).* See also *FFT filters.*

RTP filter: a spectral-domain operator for recalculating *total magnetic intensity (TMI)* data from their original *inclination I* and *declination D* to those of either the *north magnetic pole* or *south magnetic pole*, where the Earth's magnetic field is vertical, i.e., $I = \pm 90°$ and $D = 0°$. Its application is analogous to the *phase* shifting which (a) changes positions and amplitudes of the TMI anomalies to center them above their respective sources, (b) converts the asymmetric *peak*-to-*trough* TMI anomalies to the nearly symmetric RTP anomalies in accordance with the positive or negative lateral *magnetization contrast* of their sources, and (c) increases the *lateral resolution* of magnetic anomalies. See also *reduction-to-pole (RTP)* and *spectral domain.*

RTP grid: reduced-to-pole (RTP) magnetic data in the format of a *grid.* The RTP grid represents anomalous components of the *Earth magnetic field* after *leveling, microleveling, cultural editing*, and *gridding* of the *total magnetic intensity (TMI)* data and application of *reduction-to-pole (RTP).* It is a final result of the magnetic *data reduction* process which provides a basic dataset for the subsequent *enhanced processing*, analysis, and *interpretation* of the detected magnetic anomalies. See also *anomalous field* and *format.*

RTP magnetic field: the anomalous magnetic field obtained by the application of *reduction-to-pole (RTP)* to *total magnetic intensity (TMI)* data. In the absence or negligible amount of *remanent magnetization*, whose direction is different from the direction of the present *Earth magnetic field*, the maximum values of the nearly symmetric anomalies in an RTP magnetic field are located immediately above their sources. The *TMI field* and RTP magnetic field are the basic datasets for the *enhanced processing* and *interpretation* of magnetic data in *exploration* applications. Usually, it is displayed in the format of a *grid* which is referred to as an *RTP grid.* See also *anomalous field, format*, and *magnetization direction.*

RTP magnetics: See *RTP magnetic field.*

rubidium magnetometer: a *scalar magnetometer* of the *optically pumped magnetometer* type, where the gaseous isotopes of rubidium are used as the basic components of the *magnetic sensor.* The high-*sensitivity* of a rubidium magnetometer enables the detection of very small variations of the *Earth magnetic field* as low as 0.001 *nanotesla (nT).* Also referred to as *rubidium-vapor magnetometer.* See also *absolute magnetometer* and *quantum magnetometers.*

rubidium-vapor magnetometer: same as *rubidium magnetometer* with an emphasis on its *magnetic sensor* utilizing the gaseous, or vaporized, rubidium isotopes.

rule of thumb: a pragmatic generalization in approximate quantitative estimates which is based on experience rather than theoretical justification, such as the *wavelength-depth relation* based on observations relating the *anomaly wavelength* to 3–4 source depth values for shallow sources and 2–3 source depth values for deep sources.

running window: See *moving window.*

S

salt: the sedimentary rock, most commonly halite (NaCl), deposited as a result of the evaporation of saline water in ancient oceans and seas. In exploration applications, the term "salt" refers to rocks composed primarily of halite, as distinct from evaporites which include all salts precipitated from sea water. Salt *density* is essentially independent from the depth of occurrence, and it depends, primarily, on the volume of impurities other than salt. Salt density varies from 2.10–2.12 $g \cdot cm^{-3}$ for pure halite to 2.24–2.28 $g \cdot cm^{-3}$ for *allochthonous salt*, which is a "dirty" salt moved by a viscous flow from its place of deposition to shallower depths under the pressure of accumulated overlaying sediments. Average salt density is 2.14–2.18 $g \cdot cm^{-3}$. Salt domes and salt diapirs rising above a *salt nil zone* generate the positive *residual gravity* anomalies similar to those produced by *caprock effect*. In all other cases, both shallow and deep salt bodies generate the negative gravity anomalies whose amplitudes depend on the salt body volume and *density contrast* with the surrounding rocks. Salt is *diamagnetic* with a near-zero negative *magnetic susceptibility*, so large salt bodies can generate negative magnetic anomalies up to several *nanotesla (nT)* and greater. The boundaries of salt bodies are mapped by *edge-detection methods*. The base, or bottom, of allochthonous salt is a common *target* of *gravity inversion*, especially in offshore areas. See also *diapir*.

salt anticline: See *salt dome*.

salt dome: a dome-shaped concordant upwelling or discordant piercement, or diapiric, geological *structure* with a columnar *salt* plug at its core, anhydrite or gypsum *cap rock* in the shallow uppermost part, and upturned folded and faulted sediments adjacent to the salt plug. A salt dome often generates a large negative *gravity anomaly*, whereas sufficient amounts of the high-density anhydrite and/or gypsum in the cap rock can generate a small positive anomaly within the large negative anomaly. Depending on its volume, a salt dome produces the local minima of various amplitudes in the *RTP magnetic field* due to a near-zero negative *magnetic susceptibility*. Boundaries of a salt dome are mapped by *edge-detection methods*. Sometimes referred to as *salt anticline*. See also *caprock effect*, *edge-detector attribute*, and *shale dome*.

salt nil zone: the depth interval, or zone, where the *density contrast* between a *salt* body and the surrounding sediments is zero or negligibly small to generate a detectable *gravity anomaly*. The presence of salt shallower than the salt nil zone results in a positive *residual gravity* anomaly, whereas salt bodies below the nil zone generate negative gravity anomalies. Transition from the positive to negative residual gravity values marks the salt nil zone location. Also referred to as *crossover depth* and *nil zone*.

salt residual: *stripped gravity* obtained by subtracting the *model*-calculated gravity effect of a *salt dome* and other known salt bodies from *Bouguer gravity* in order to highlight the local-scale *residual gravity* anomalies. See also *sediment flooding*.

sample: **1.** the measured value of a gravity or magnetic *field* along the regular-spaced *survey* lines in airborne, marine (shipboard), and submarine surveys or at the irregularly spaced observation points on the ground or shallow sea floor. **2.** the value of a gravity or magnetic field at a particular *grid nod*. See also *gridding*.

sample interval: **1.** the distance or time interval between the discrete measurements of the gravity or magnetic field along the regular-spaced *survey* lines in airborne, marine (shipboard), and submarine surveys or at the irregularly spaced ground or sea floor observation points. **2.** the distance between the equally spaced data points, or *grid interval*, which is selected for the conversion of discrete measured values into the continuous surface of a *grid*. For *airborne gravity* and airborne *gravity gradient* surveys, the nominal sample interval is determined from the aircraft speed in meters per second divided by the *bandwidth* of the airborne measurements in *hertz (Hz)*. Actual along-the-line sample interval is defined by the length of the *low-pass filter* applied to suppress *acquisition noise*, whereas the across-the-line sample interval is defined by the *survey line spacing*, unless the low-pass filter length exceeds the line spacing. Sample interval and survey line spacing are the basic parameters that define the *lateral resolution* of anomalies in gravity and magnetic surveys. Also referred to as *sampling*, *sample period*, and *reading interval*. See also *sample rate* and *sampling bandwidth*.

sample period: **1.** *sample interval* in the context of a regular time or distance interval between discrete measurements. **2.** the time interval between discrete records of measured values. As a part of *data conditioning*, the continuous measurements during airborne or marine (shipboard) surveys are averaged over a pre-selected sample period in order to attenuate *acquisition noise*.

sample rate: **1.** an *acquisition* parameter representing the number of discrete measurements, or samples, per unit distance or per regular time interval. In time-based airborne and marine (shipboard) *survey* measurements, the sample rate is used with the same meaning as *sampling frequency* and described in the *frequency* units of *hertz (Hz)*, e.g., sample rates of 5 Hz and 0.2 Hz correspond to five measurements per second and one measurement every 5 seconds, respectively. In distance-based ground surveys, the sample rate is used with the same meaning as *sample interval* along the *acquisition line*. **2.** *sampling speed*. **3.** *sampling bandwidth* in airborne gravity and *gravity gradient* surveys. Sample rate defines the highest frequency, or the shortest *wavelength*, of the gravity or magnetic field that can be recovered from the recorded data. Also referred to as *sampling rate* and *Nyquist sample rate*. See also *Nyquist theorem*.

sample spacing: *sample interval* in the context of (a) the distance, or space interval, between the measurement points in a *ground survey*, or (b) the along-the-line distance between the time-based discrete measurements during airborne, marine, and submarine surveys.

sampling: 1. the process of obtaining the discrete measurements of a continuous gravity or magnetic *field* at (a) regular distance or time intervals along the ground, marine, submarine, or airborne *survey* lines, or (b) irregular distance intervals over a *ground survey* area. **2.** *grid interval* in the context of the conversion of the line-oriented or irregularly spaced acquisition data into equally spaced data points on a *grid*. According to the *Nyquist theorem*, the sampling should provide two or more measured values per cycle of the highest *frequency*, or the shortest *wavelength*, present in the measured field in order to recover its anomalies without distortions. **3.** iterative retrieving of *model* parameters in the process of *inversion*. See also *downsampling*, *resampling*, *undersampling*, *upsampling*, and *stacked sampling*.

sampling bandwidth: *bandwidth* in the context of *airborne gravity* and airborne *gravity gradient* measurements, wherein the length of the *noise*-suppressing *low-pass filter* defines the *sample interval* along the flight lines and, hence, the shortest *wavelength* (the highest *spatial frequency*) recoverable from the acquired airborne data.

sampling density: 1. a gravity or magnetic *ground survey* parameter defined by the number of the irregularly spaced measurement points per square kilometer in the *survey* area. **2.** *sample interval*. Also known as *observation density*.

sampling frequency: *sample rate* in the context of (a) using *frequency* units of *hertz (Hz)* to describe the number of the discrete measurements per regular time interval along airborne and marine *survey* lines, e.g., sampling frequencies of 2 Hz and 0.2 Hz correspond to two measurements per second and one measurement every 5 seconds, respectively, or (b) using *angular frequency* units in radians per second, when the discrete samples of a continuous function are described by its separate *phase* values. See also *radian* and *Nyquist theorem*.

sampling rate: *sample rate* in the context of the discrete measurements, or *sampling*, of a continuous gravity or magnetic *field*.

sampling speed: an instrument performance characteristic that describes the fastest speed of discrete measurements and their recording. Referred to with the same meaning as *sample rate*.

sampling theorem: the theorem that establishes a rule of the reconstruction, or recovery, of a continuous *band-limited function* from its discrete measurements at regular time or distance intervals. Sampling theorem postulates that the *sampling frequency*, or *sample rate*, should provide two or more samples per cycle for the highest *frequency* component of the measured function in order to ensure its full recovery. In application to gravity and magnetic fields, the sampling theorem is commonly referred to as the *Nyquist theorem* and *Nyquist-Shannon sampling theorem*. See also *aliasing* and *sample*.

sampling wavenumber: a *discrete Fourier transform (DFT)* parameter defined by its *period*, i.e., range of wavenumbers over which the DFT repeats itself. The sampling *wavenumber* is expressed as

$$k_S = 2\pi/\Delta x,$$

where Δx is the *sample interval* for the line-oriented data or *grid interval* for gridded data. Half of the sampling wavenumber is called the *Nyquist wavenumber*. See also *gridding*.

satellite-altimeter gravity: 1. *free-air gravity* over oceans and seas computed from *satellite altimetry*, i.e., high-*precision* measurements of the ocean/sea surface undulations by the radar or laser altimeters installed on Earth-orbiting satellites. Satellite altimeter measurements over several years are combined to obtain the high data density per square kilometer and minimize temporal ocean/sea surface disturbances (waves, currents, tides, etc.) by averaging and other data transformations. Statistically averaged height variations of the ocean/sea surface closely approximate *mean sea level (MSL)* and represent undulations of the *marine geoid*, which is the *equipotential surface* for the *Earth gravity field*, wherein the ocean/sea surface slope of 1 microradian (μrad) corresponds to a change in the free-air gravity value of 1 *milligal (mGal)*. These data form the basis for calculation of marine free-air gravity anomalies by the *geoid-to-gravity method* and the *slope-to-gravity method*. The achievable *lateral resolution* of anomalies is 8–15 km with *accuracy* of 3–5 mGal and shelf-water waveform tracking to 10–15 km off the coastline. Also known as *marine free-air gravity*. **2.** offshore *Bouguer gravity* calculated with the use of the free-air gravity data obtained from satellite altimetry. Sometimes referred to as *altimetry-derived gravity*. See also *radian* and *satellite gravity*.

satellite altimetry: instruments and methodology for the high-*precision* measurements of the Earth's surface *topography* and ocean/sea surface undulations by the radar and laser altimeters installed on the space satellites orbiting the Earth. The precision of a satellite-mounted *radar altimeter* varies from 40–60 cm to 1.5–4.0 m under all atmospheric conditions. The precision of a satellite *laser altimeter* varies from 2–10 to 20–30 cm, but a laser altimeter cannot make measurements through thick clouds. Satellite altimetry data over oceans and seas are used for the calculation of marine *free-air gravity* anomalies. See also *marine geoid*.

satellite-derived gravity: same as *satellite gravity* with an emphasis on its origin from both direct (*satellite-to-satellite tracking*) and indirect (*satellite altimetry*) measurements of the *Earth gravity field* with the use of Earth-orbiting space satellites.

satellite gradiometry: instruments and methodology for measurements of the gradients of the *Earth gravity field* and the *Earth magnetic field* by satellites orbiting the Earth at altitudes of 250–270 km (*GOCE mission*) and 400–550 km (*CHAMP mission* and *GRACE mission*) above the Earth's surface. At the *altitude* of the GOCE mission, the amplitudes of the measured gradients are approximately ±0.5 *Eötvös (E)*. Data acquired by a

GOCE gravity gradiometer has proven its capability to measure the *full tensor gradient (FTG)* elements with *precision* of several units of mE/$\sqrt{\text{Hz}}$, where mE is *milliEötvös* and Hz is *hertz*. The GOCE *gravity gradient* data have been transformed into *vertical gravity* g_z and integrated with the long-*wavelength* components of the Earth's gravity field measured by *satellite-to-satellite tracking*. The combined dataset enables the mapping of the Earth's gravity field with *lateral resolution* of anomalies up to 80–100 km at the Earth's surface and *accuracy* 1–2 *milligal (mGal)* in the long-*wavelength* range. See also *derivative* and *gradient*.

satellite gravimetry: instruments and methodology for measuring the long-*wavelength* components of the *gravitational field* of the Earth and other planets with the use of data acquired by satellite missions. Because the orbiting satellites are effectively in a free fall, their onboard accelerometers do not sense the direct *acceleration* due to the *gravitational attraction* of the planetary mass, but they sense changes in the *gradient* components of its gravitational field, so the gravity measurements by the orbiting satellites are limited fundamentally to *satellite gradiometry*. In general, satellite gravimetry is based on four measurement concepts: (1) conversion of satellite radar or laser *altimeter* measurements of the oceanic and sea surface undulations to *free-air gravity*; (2) high-*precision* tracking of a satellite orbit relative to the Earth's surface using the ground-based laser ranging stations or high-orbiting constellation of the *Global Positioning System (GPS)* satellites with the subsequent calculation of deviations of the satellite orbit from the theoretical orbit model for the homogeneous Earth, wherein the amount of deviation is directly proportional to a change in the long-*wavelength* components of the *Earth gravity field*; (3) *satellite-to-satellite tracking*; and (4) *gradient* and *full tensor gradient (FTG)* measurements by an onboard *gravity gradiometer* system. See also *accelerometer*, *gravitational acceleration*, and *satellite magnetometry*.

satellite gravity: gravity data obtained from measurements by the satellites orbiting the Earth and other planets. There are four basic types of satellite gravity and *gravity gradient* data: (1) *satellite-altimeter gravity* computed from the high-*precision* measurements of the oceanic and sea surface undulations by satellite-mounted radar or laser *altimeter*; (2) long-*wavelength* components of the *Earth gravity field* calculated from data obtained by tracking satellite orbital variations relative to ground reference stations; (3) *satellite-to-satellite tracking*; and (4) *gradient* and *full tensor gradient (FTG)* data acquired by an onboard *gravity gradiometer* system, such as the *GOCE gravity gradiometer*. Satellite orbit tracking provides *full-wavelength resolution* of anomalies at the order of the orbital *altitude* (~250–550 km), wherein the *accuracy* of such measurements varies from 0.01–0.1 *milligal (mGal)* for the longest wavelengths to 2–3 milligals for the shortest of the long wavelengths. In integration with the long-*wavelength* components of the Earth's gravity field measured by *satellite-to-satellite tracking*, the achievable *lateral resolution* of anomalies obtained from *satellite gradiometry* measurements is 80–100 km with the highest achievable *accuracy* of 1–2 mGal. The best reported *lateral resolution* of the satellite-derived gravity field over oceans and seas is within the *wavelength* range of 8–15 km with accuracy of 3–5 mGal. Satellite gravity is complementary to airborne and marine (shipboard) gravity data in the analysis of global-scale geological structures, such as the lithospheric plate boundaries, tectonic provinces, transform faults, and regional sedimentary basins. At the scale of a local *exploration* project, the satellite gravity is used to determine (a) the regional component in *regional-residual separation* for the *profile*-based and *grid*-based *inversion* of the *residual gravity* anomalies and (b) the regional component for *vertical gravity* g_z obtained by the transformation of airborne *gravity gradient* measurements. Also referred to as *spaceborne gravity* and *satellite-derived gravity*. See also *GOCE gravity*.

satellite magnetics: magnetic data obtained from measurements of the gradients and long-*wavelength* components of the *main geomagnetic field* and short-wavelength variations of the *external magnetic field* by the satellites orbiting the Earth and other planets. The *IGRF*-corrected magnetic anomalies computed from the data acquired by Earth-orbiting satellites represent the cumulative magnetic effects produced by the structural and *magnetization* heterogeneity within the regional-scale tectonic provinces in the *crust* and *lithosphere* and more localized effects of the large near-surface iron ore deposits, such as the Kursk iron-bearing formation in Russia. See also *altitude*, *gradient*, *satellite magnetometry*, *CHAMP mission*, and *Swarm mission*.

satellite magnetometer: a *magnetometer* installed on space satellites orbiting the Earth or other planets for measurements of the magnetic field, its gradients, and *vector* components. It is attached to a rigid 4–5 m long pole, or boom, to minimize the interference effects of the onboard magnetic fields generated by the electronic equipment and magnetized components of the satellite. There are three types of satellite magnetometers: *scalar magnetometer*, *vector magnetometer*, and *absolute scalar magnetometer (ASM)* operating in both the *scalar* and *vector* modes. The required high *accuracy* in the orientation of the vector magnetometer sensors is provided by multiple-head star-imaging cameras.

satellite magnetometry: instruments and methodology for measuring the long-*wavelength* components of the magnetic field of the Earth and other planets, its gradients, and *vector* components by orbiting satellites. Earth-orbiting satellites with magnetometers are commonly positioned at altitudes from 350–400 km to 550–600 km. Typical onboard instrumentation includes: (a) a *vector magnetometer* or *absolute scalar magnetometer (ASB)*, which operates in both the *scalar* and vector modes, to measure the *vector* components; (b) a *scalar magnetometer* of the high-*sensitivity* type, such as a *helium magnetometer*, to measure the *magnitude* of the planetary magnetic field and calibrate the *baseline* value in vector magnetometers of the fluxgate type; and (c) multiple-head star-imaging cameras to orient the sensing axes of the vector magnetometer. At the altitudes of the Earth-orbiting satellites, the *main geomagnetic field* represents approximately 98% of the measured magnetic field values, wherein the *external magnetic field* and *lithospheric magnetic field* comprise 1.6% and 0.3%, respectively. Vector measurements are made along three mutually orthogonal directions, and these measurements are essential for separating the magnetic field components from above and below the satellite observation plane. The achievable *accuracy* of measurements is 0.5–1.0 *nanotesla (nT)*. The highest *full-wavelength resolution* is estimated at the order of the orbital *altitude* of a particular satellite and, thereby, provides the data representing the long-wavelength (>350–600 km) components of the continental-scale magnetic anomalies. See also *CHAMP mission*, *Swarm mission*, and *fluxgate magnetometer*.

satellite-to-satellite tracking: 1. high-*accuracy* microwave-range measurements of changes in the distance between two satellites on the same orbit with the use of laser interferometry and *precision* up to 0.002 mm. When the leading satellite passes over the areas of increased *gravitational attraction* of the Earth or other planetary mass, it speeds up and increases its distance from the trailing satellite, whereas over areas of decreased gravitational attraction the leading satellite slows down and, thereby, decreases the distance between two satellites. The amount of relative displacement of two satellites is directly proportional to changes in the long-*wavelength* component of the *gravitational field* of the Earth or other planet. **2.** three-dimensional (3D) continuous tracking of the satellite nearly free-fall trajectory with the use of the ground-based laser ranging stations and/or high-orbiting constellation of the *Global Positioning System (GPS)* satellites. The amount of deviation of the satellite orbit from the theoretical orbit calculated for the homogeneous Earth is directly proportional to changes in the long-wavelength component of the *Earth gravity field*. See also *GOCE mission*, *GRACE mission*, and *CHAMP mission*.

scalar: a quantity which has only a numerical value without association with any direction as in the case of a *vector*. For example, *gravitational potential* is represented by a *field* of scalar quantities, or *scalar field*, whereas a *gravitational field* (whose directional components are defined by derivatives of the scalar gravitational potential) is the field of *vector* quantities, or *vector field*. See also *derivative*.

scalar field: 1. a field of *scalar* quantities whose variations are described by functions of the spatial coordinates and time. **2.** the *field* of a single component of a three-component *vector* quantity. The scalar field has only one numerical value at every coordinate point within the space at a given time. Numerical values of the vertical component of the *Earth gravity field* are measured by a *gravimeter*, and these values constitute the scalar field, whereas the Earth gravity field in general is a field of *vector* quantities, or *vector field*, whose direction and *magnitude* at every point in space are defined by the *gradient* of the scalar *gravitational potential*. See also *gravity vector* and *magnetic vector*.

scalar gravimetry: methods and instruments for measurements of (a) the full, or absolute, numerical value of *gravitational acceleration* that is known as *absolute gravity* and measured by a *free-fall absolute gravimeter* and (b) the relative changes of the vertical component g_z of the three-component *Earth gravity vector* $\mathbf{g}$ that is referred to as *vertical gravity* or *relative gravity*, whose changes Δg relative to the *null point* are measured by a conventional, or regular, *gravimeter* known as a *relative gravimeter*. See also *scalar magnetometry*.

scalar magnetometer: a *magnetometer* which measures the absolute numerical value, or *magnitude*, of the *Earth magnetic field* regardless of its *vector* direction with the assumption that the vector of the anomalous component of the measured field and vector of the *main geomagnetic field* have nearly the same direction at an *observation point*. Approximation errors due to the actual differences in the vector directions are mostly negligible, because the typical crustal magnetic anomalies are much less than the main geomagnetic field values. The assumption is not valid for magnetic anomalies with amplitudes ≥10,000 *nanotesla (nT)* observed over large near-surface iron ore deposits. Scalar magnetometers are either omnidirectional or require approximate orientation of their *magnetic sensor* axes. The great majority of the exploration magnetometers for measurements on moving observation platforms are scalar magnetometers. Also referred to as *absolute magnetometer* and *total field magnetometer*. See also *absolute scalar magnetometer (ASM)* and *vector magnetometer*.

scalar magnetometry: methods and instruments for measuring the absolute numerical value, or *magnitude*, of the *Earth magnetic field* regardless of its *vector* orientation. See also *scalar*, *scalar magnetometer*, and *vector magnetometer*.

scalar product: a *scalar* quantity which represents the result of multiplying, or product, of the magnitudes of two vectors **A** and **B** scaled by the cosine of the angle θ between the vectors, expressed as

$$\mathbf{A} \cdot \mathbf{B} = |\mathbf{A}| \cdot |\mathbf{B}| \cos\theta.$$

The scalar product is the largest when vectors are parallel, and it is zero when vectors are orthogonal. The scalar product of *vector* **A** and unit vector **r** is equal to the component of **A** in the direction of the unit vector. In *Cartesian coordinates*, the scalar product of two vectors is expressed as

$$\mathbf{A} \cdot \mathbf{B} = A_x B_x + A_y B_y + A_z B_z,$$

where A_x, A_y, A_z, B_x, B_y, B_z are components of vectors **A** and **B** defined by projections of each vector on the mutually orthogonal *x*-, *y*-, *z*-axes. Due to the mathematical symbol •, the scalar product is also referred to as the *dot product*. See also *vector product* and *directional derivative*.

scalar-to-vector transformation: 1. transformation of *scalar* gravity or magnetic anomalies into the three *vector* components of the Earth's gravity or magnetic field along the mutually orthogonal *x*-, *y*-, *z*-directions in *Cartesian coordinates*. It is a three-step procedure involving: (1) *forward Fourier transform* of the space-domain *anomalous field*; (2) application of a *component filter*, the algorithm of which is based on *field-to-potential transformation* and calculation of the corresponding directional derivatives of the gravity or magnetic *potential* that represent the vector components of the anomalous field; and (3) *inverse Fourier transform* of the filtered data back to the *space domain*. **2.** calculation of the directional derivatives of the *anomalous potential* of the *Bouguer anomaly* or *RTP anomaly* in the mutually orthogonal *x*-, *y*-, *z*-directions. **3.** reconstruction of the gravity vectors of the local gravity fields from the measured *gravity gradient tensor (GGT)* components in the *wavenumber domain* (Blakely, 1996; Li, 2018). See also *directional derivative*, *gravity anomaly vector*, and *magnetic anomaly vector*.

scale: 1. the ratio between a linear distance on a map, *grid*, or image and its corresponding distance on the Earth's surface. Scale is displayed as a bar-graph or double-headed arrow with space-dimension units and as a numerical ratio, e.g., 1:50,000. The map scale of 1:50,000 shows that one unit on the map represents 50,000 identical units on the ground. **2.** a *wavelet transform (WT)* parameter which defines a specific range of spatial frequencies, or wavelengths, in transformed data. The sequence of WT scales represents the original data at the successive ranges of *wavelength resolution*: the smaller the scale, the shorter the resolvable wavelengths, and the larger the scale, the longer the resolvable wavelengths, i.e., wavelet transform scaling is analogous to the different levels of *upward continuation* of the original data. See also *resolution*, *scaling function*, and *spatial frequency*.

scale correction: conversion of continuous time-based marine (shipboard) and airborne gravity and magnetic measurement records into the corresponding space-domain coordinates, or space *scale*, of the standard *data processing* in accordance with the instantaneously measured speed of a *survey* ship or aircraft. Sometimes referred to as *format correction*. See also *space domain* and *time domain*.

scale factor: 1. a proportionality constant, or coefficient, for the conversion of *relative gravimeter* readings ΔR relative to the *null point* into the corresponding gravity change Δg in units of *milligal (mGal)* or *microgal (μGal)* as

$$\Delta g = k \cdot \Delta \mathrm{R},$$

where k is a scale factor. Every gravimeter has its own scale factor tabulated by the manufacturer of a particular gravimeter. Scale factors tend to vary with the gravimeter aging and might be changed due to level misalignments, vacuum leak, or thermostat malfunction, i.e., *gravimeter drift* can be considered as a temporary change of the scale factor. The scale factor is monitored by a *ground gravimeter calibration* procedure. **2.** a *grid* or map *projection* parameter which defines the coefficient of deviation of the actual Earth surface from the *grid* or map plane relative to the nearest central meridian of the *Universal Transverse Mercator (UTM)* coordinate system. See also *reading*.

scale-space monogenic signal: the *monogenic signal (MS)* of a gridded *anomalous field* which is filtered by an *upward continuation*-based *Jacobsen filter* prior to the MS calculation in order to suppress both the high-frequency *noise* and *regional field* components (Hidalgo-Gato and Barbosa, 2015). See also *gridding* and *edge detection*.

scale-space representation: representation of a gridded *anomalous field* at different levels, or scales, of *anomaly resolution* by changing the dimensions of anomalies without changing their shapes. It is based on the use of *upward continuation*, where the continuation height h controls the *lateral resolution* of the anomalies: the larger the h value, the larger the *scale* and the lower the lateral resolution. See also *dimension*, *multiscale analysis*, and *wavelet analysis*.

scaled magnetic dipole moment: See *normalized magnetic dipole moment*.

scaled source strength: See *normalized source strength (NSS)*.

scaling function: a *wavelet function*, also known as a *mother wavelet*, two scaling parameters of which define a specific range of *wavelength resolution* in the process of the *wavelet decomposition* of anomalous fields with the use of the *discrete wavelet transform (DWT)* and *continuous wavelet transform (CWT)*: (1) a dilation parameter for stretching or compressing the mother wavelet horizontally and, if necessary, compressing it vertically; and (2) a translation parameter for shifting the dilated mother wavelet along the data points at the successive ranges, or scales, of the wavelength resolution. See also *anomalous field*, *wavelet dilation*, and *wavelet translation*.

scatter point data: ground gravity or magnetic data obtained by measurements at observation points distributed irregularly over a *survey* area.

Schiaparelli body: a hypothetical "gravitationally transparent" heterogeneous body of *mass*, where *density* is distributed in such a way that the body does not generate a detectable *gravity anomaly* on the observation plane above or below this body. See also *forward problem* and *self-demagnetization*.

Schuler frequency: the reciprocal of *Schuler period*.

Schuler period: the *period* of a mathematical pendulum that defines the rate of precession, or period of rotation about the axis of rotation, for gyroscopes in the servo-system of a *stabilized platform*. It is equal to the period of a hypothetical satellite orbiting the Earth along its smoothed surface and expressed as

$$T_{Sch} = 2\pi(\mathrm{R}/g)^{1/2} \cong 84.4 \text{ minutes},$$

where R is the average Earth radius and g is *nominal gravitational acceleration* of 9.8 m · s^{-2}.

Schuler-tuned inertial platform: the airborne gyroscope- and *accelerometer*-controlled *stabilized platform* tuned to the *Schuler period* of 84.4 minutes, so the platform does not tilt when it is subjected to low-turbulence horizontal inertial accelerations during a flight. The sensing axis of an *airborne gravimeter* is held within ±10 arc-seconds (±0.0028 degrees) off-level to ensure its near-vertical orientation independently of low-turbulence aircraft maneuvers. See also *arc-second, inertial acceleration*, and *off-level errors*.

sea-bottom gravimetry: methodology and instruments for measurements of the *Earth gravity field* with the use of submergible automated *gravimeter* systems lowered onto the sea floor. See also *AUV gravimeter, towed deep ocean gravimeter (TOWDOG)*, and *underwater gravimeter*.

sea-floor drone: See *autonomous underwater vehicle (AUV)*.

sea-floor spreading: a continuous process in which the convective rising, or upwelling, of *magma* along the axes of the mid-oceanic ridges (where the Earth oceanic *crust* is split apart) creates a new oceanic crust, so the old lithospheric plates are moved away from the mid-oceanic ridges at rates from 0.4–0.5 cm to 9–10 cm per year. See also *lithosphere, asthenosphere*, and *thermal gravity anomaly*.

sea level: See *mean sea level (MSL)*.

sea-tide correction: a correction applied to underwater gravity measurements to compensate for the increase or decrease of measured gravity values caused by changes in the water column above an *underwater gravimeter*. Usually, it is integrated into *double Bouguer correction*. See also *AUV gravimeter* and *towed deep ocean gravimeter (TOWDOG)*.

seamount: a steep-sided, often carrot-shaped, *igneous body* within the ocean or sea water with its top below the sea level and height up to 1–3 km above the ocean/sea floor. Seamounts are either isolated and arranged in linear or random groupings, or they are connected at their bases and aligned along the sea-floor ridge. They generate high-*amplitude*, short-*wavelength*, circular- and oval-shaped positive gravity anomalies and variable-sign magnetic anomalies. The amplitude and sign (positive or negative) of a *magnetic anomaly* depend on the direction of *remanent magnetization* in a particular seamount. Flat-topped seamount is referred to as guyot, and a peaked seamount is called seapeak.

second-derivative map: a printed or digital map of the *second vertical derivative (SVD)* of an anomalous gravity or magnetic field. The SVD calculation increases the *lateral resolution* of anomalies and enhances the short- and mid-*wavelength* anomalies generated by local-scale geological structures and shallow sources while suppressing the regional long-wavelength components of an *anomalous field*. Residual gravity anomalies associated with the oil-and-gas structural traps of the *anticline, horst*, and *thrust* types and stratigraphic traps of the *reef* and thinning-out types can be highlighted on SVD maps of *Bouguer gravity*. See also *enhanced processing, residual gravity, trap*, and *trap-door structure*.

second horizontal derivative (SHD): the second-order *horizontal derivative*, or horizontal derivative of the *first horizontal derivative (FHD)*, of anomalous gravity or magnetic field F denoted by

$$\partial^2 F/\partial x^2 \quad \text{and} \quad \partial^2 F/\partial y^2,$$

where x and y are mutually orthogonal horizontal directions in *Cartesian coordinates*. Asymmetry of the *horizontal gradient* anomaly above a dipping *source body* provides the basis for using SHD grids in estimates of the *dip* of lithological contacts, high-*offset* faults, dip and vertical extent of shallow finite-bottom ore bodies. Also abbreviated as 2HD. Sometimes referred to as *second horizontal difference*. See also *contact, fault, infinite bottom, ore body*, and *total horizontal derivative of tilt*.

second horizontal difference: See *second horizontal derivative (SHD)*.

second-order analytic signal (AS$_2$): the *analytic signal (AS)* of the first-order *vertical derivative* of a gravity or magnetic field. Its components are represented by the second-order derivatives of the *anomalous field*, and its *amplitude* |AS$_2$| is expressed as

$$|AS_2| = \{[\partial(\partial M/\partial z)/\partial x]^2 + [\partial(\partial M/\partial z)/dy]^2 + [\partial(\partial M/\partial z)/\partial z]^2\}^{1/2},$$

where M is *magnitude* of the anomalous gravity or magnetic field. The amplitude of the two-dimensional AS$_2$ is used in the calculation of *second-order local wavenumber k_2* and magnetic source parameters by the *improved SPI^{TM} method*. The AS$_2$ is equivalent to the *first-order enhanced AS (eAS$_1$)*. See also *local wavenumber (LW)* and *AS-LW relationship*.

second-order ASA (ASA$_2$): 1. the *analytic signal amplitude (ASA)* of the second-order *vertical derivative* of the *TMI field* in the *AN-EUL method*. **2.** the *amplitude* of the three-dimensional (3D) *second-order analytic signal (AS$_2$)*, which is equivalent to the *first-order enhanced ASA (eASA$_1$)*. **3.** an *attribute* of a gravity or magnetic field calculated by the square root of the sum of the squared derivatives of the 3D ASA along the mutually orthogonal x-, y-, z-directions in *Cartesian coordinates*, expressed as

$$ASA_2 = [(\partial ASA/\partial x)^2 + (\partial ASA/\partial y)^2 + (\partial ASA/\partial z)^2]^{1/2},$$

where $\partial ASA/\partial x$, $\partial ASA/\partial y$, $\partial ASA/\partial z$ are first-order derivatives of the ASA that represent the second-order derivatives of the *anomalous field*. See also *zero-order ASA (ASA$_0$), first-order ASA (ASA$_1$)*, and *derivative*.

second-order enhanced ASA (eASA$_2$): the *analytic signal amplitude (ASA)* of the second-order *vertical derivative* of an anomalous gravity or magnetic field expressed as

$$eASA_2 = \{\partial[(\partial^2 M/\partial z^2)/\partial x]^2 + \{\partial[(\partial^2 M/\partial z^2)/\partial y]^2 + \{\partial[(\partial^2 M/\partial z^2)/\partial z]^2\}^{1/2},$$

where M is *magnitude* of the anomalous gravity or magnetic field. The field of eASA$_2$ values is represented by the third-order derivatives of the *anomalous field* that can be efficient in resolving the interfering anomalies from closely spaced sources and in highlighting the low-*amplitude* anomalies generated by local geological structures. Prior to computations, the anomalous magnetic field (a) can be, but not necessarily, reduced to the pole, and (b) should be recomputed at the lowest empirically determined level of *upward continuation* to suppress the high-frequency *noise* components (Hsu et al., 1996). See also *first-order enhanced AS (eAS$_1$), reduction-to-pole (RTP)*, and *resolution*.

second-order free-air correction: the most accurate value of a *free-air correction* calculated with the use of the *Heiskanen-Moritz formula* and *orthometric height*. It is used primarily in geodetic studies of the *geoid*. See also *first-order free-air correction* and *height correction*.

second-order local wavenumber: a space-domain *attribute* defined by the *horizontal derivative* of the *local phase* of the first-order *vertical derivative* of an anomalous gravity or magnetic field. The second-order local wavenumber is denoted by k_{xx} or k_2 and expressed as

$$k_{xx} = k_2 = \partial\{\arctan[(\partial^2 M/\partial z^2)/(\partial^2 M/\partial z\partial x)]\}/\partial x$$
$$= |AS_2|^{-2}[(\partial^2 M/\partial z\partial x)\cdot(\partial^3 M/\partial z^2\partial x) - (\partial^2 M/\partial z^2)\cdot(\partial^3 M/\partial x^2\partial z)],$$

where $|AS_2| = [(\partial^2 M/\partial z\partial x)^2 + (\partial^2 M/\partial z^2)^2]^{1/2}$ is the *amplitude* of the two-dimensional (2D) *second-order analytic signal (AS_2)* which is equivalent to the 2D *first-order enhanced ASA ($eASA_1$)*, and M is the *magnitude* of the *anomalous field*. Based on the equivalence between the local phase and *analytic signal phase*, the second-order local wavenumber is also expressed as

$$k_{xx} = k_2 = \left(\partial|AS_2|/\partial z\right)/|AS_2|.$$

The second-order local wavenumber is represented by the normalized third-order derivatives, and it is used in the *improved SPI™ method* and the *multimodel method* for estimating the depth and location of magnetic sources approximated by the 2D source geometries of *magnetic contact*, *thin dike*, and *horizontal cylinder* (Smith et al., 1998; Pilkington and Keating, 2006). See also *first-order local wavenumber* and *normalization*.

second-order theta map: a modified *theta map* technique, where the first-order *vertical derivative* of an anomalous gravity or magnetic field is used instead of the field value in the calculation of cosine of *theta angle* θ_2 as

$$\cos\theta_2 = [(\partial f_z/\partial x)^2 + (\partial f_z/\partial y)^2]^{1/2}/[(\partial f_z/\partial x)^2 + (\partial f_z/\partial y)^2 + (\partial f_z/\partial z)^2]^{1/2},$$

where $f_z = \partial f/\partial z$ is the first-order vertical derivative, or *first vertical derivative (FVD)*, of the anomalous field f, and $\partial f_z/\partial z = \partial^2 f/\partial z^2$ is the second-order vertical derivative, or *second vertical derivative (SVD)*. Modification increases both the *lateral resolution* and high-frequency *noise*. The amount of noise is minimized by *upward continuation* and calculation of the SVD in the *space domain* with the use of the *Laplace equation* instead of the standard spectral-domain SVD calculation (Ma, 2013). See also *Gibbs phenomenon*, *spectral domain*, and *improved theta map*.

second-order tilt angle: an *attribute* of the anomalous gravity or magnetic field defined as *tilt angle (TA)* of the original, or first-order, TA. Because the TA theory is based on the use of the *vertical derivative* of the *potential field*, whereas the TA is not a derivative, the vertical derivative substitute is obtained from the difference between the first-order TA values calculated at two different elevations, or upward continuations, of the *anomalous field*. Sensitivity to *noise* is reduced, among other ways, by *smoothing* of the original tilt values prior to calculations of the second-order tilt angle (Cooper and Cowan, 2006). See also *elevation* and *upward continuation*.

second vertical derivative (SVD): the second-order *vertical derivative*, or vertical derivative of *first vertical derivative (FVD)*, of the anomalous gravity or magnetic field F denoted by $\partial^2 F/\partial z^2$. The SVD zeroes at the *inflection point* of an anomaly *profile* and represents a measure of the field *curvature*. It enhances the short- and mid-*wavelength* components of an *anomalous field* more effectively than FVD, and it can resolve the interfering anomalies generated by closely spaced sources. The SVD is often cascaded with a *low-pass filter* or *upward continuation* to suppress the high-frequency *noise* components amplified by its calculation. The lowest noise contamination in the SVD calculation is obtained with the use of the space-domain *Laplace equation*. A spectral-domain calculation alternative is a three-step transformation which is faster but contaminated by *Gibbs ringing*: (1) *Fourier transform* of the anomalous field; (2) multiplying by $|k|^2$, where k is *wavenumber*; and (3) *inverse Fourier transform* of the obtained result. Zero contours on an SVD *grid* or map closely approximate the structural margins and edges of isolated source bodies. Also abbreviated as 2VD. See also *cascading*, *contour map*, *second-derivative map*, and *variable order vertical derivative*.

secondary gravity base: same as *relative gravity base* with an emphasis on it being different from a *primary gravity base* which is also known as *absolute gravity base*.

sectoral harmonics: *spherical harmonics* which depend on the longitude and do not depend on the latitude. See also *tesseral harmonics* and *zonal harmonics*.

secular accelerations: same as *geomagnetic jerks* with an emphasis on them being the fastest components of *secular variation* with variable rates of changes identified by the second-order time derivatives of the observed *Earth magnetic field* at the time scale of several months.

secular change: *secular variation* in the context of a slow continuous change of the dipole and non-dipole components of the *main geomagnetic field* in *magnitude* and direction over the Earth's surface.

secular variation: a slow continuous change in the *magnitude* and direction of the dipole and non-dipole components of the *main geomagnetic field* at a time scale ranging from months and years to millennia. The rate of the year-range change is a function of time of the year and position on the Earth's surface, and it is estimated at 80–90 *nanotesla (nT)* per year with the overall westerly drift at the rate of 0.20°–0.35° per year. Secular variation components with variable rates of their relatively fast changes are referred to as *geomagnetic jerks* and *secular accelerations*. The origin of secular variation is thought to be associated with fluid convection in the liquid outer core of the Earth. Measurements by the worldwide network of geomagnetic observatories are summarized in *isoporic maps* and *IGRF* and *DGRF* models. Predictive annual components are incorporated into an *IGRF model* for the following *epoch*. Secular variation is removed from the measured *Earth magnetic field* values by application of *IGRF correction*. The term is derived from the Latin "seculum" referring to the time interval of a century. Also known as *geomagnetic secular variation* and *secular change*. See also *eccentric dipole field*, *nondipole field*, and *geomagnetic observatory*.

sediment correction: **1.** adjustment to *isostatic correction* and *flexural isostatic correction* in offshore areas that incorporates the estimated load of accumulated sediments for a more accurate calculation of the *Moho boundary* depth within rifted continental margins. **2.** correction for the gravitational effect of the sediments accumulated in rifted continental margins, where the sediment thickness is determined from the seismic data or interpreted *magnetic basement* depth structure smoothed out to the regional-scale approximation. See also *smoothing*.

sediment flooding: salt-free sedimentary fill-in, or flooding, of the three-dimensional (3D) *model* space occupied by *salt*. It is a model transformation procedure applied to obtain the 3D sedimentary model without the gravitational effects of all salt bodies. The calculated gravity field of this model is used as a reference in addition to the available *constraints* in the 3D *gravity inversion* for the base of *allochthonous salt*. See also *salt residual*.

sedimentary gravity anomaly: a *gravity anomaly* generated by intrasedimentary gravity sources associated with *density*-related lateral changes in rock composition, including (a) elements of the subsurface geological *structure*, such as *anticline*, *carbonate platform*, *graben*, *horst*, *reef*, *rift*, *salt dome*, *syncline*, *thrust*, *uplift*, and (b) localized *source body*, such as *igneous intrusion*, *salt dome*, and *salt* or shale *diapir*. The amplitude of the sedimentary gravity anomaly depends on the volume, lateral *density contrast*, and depth of a particular source body: the larger the volume, the higher the density contrast, and the shallower the source, the higher the amplitude, whereas its *polarity* (positive or negative sign) is defined by the sign of the lateral density contrast created by this source. Low-amplitude gravity anomalies associated with the depositional fairways, buried fluvial systems, and accumulations of reservoir-grade sands in the stratigraphic traps of the *incised valley* and *paleochannel* types can be detected by calculation of *residual gravity* and its derivatives. See also *derivative*, *fairway*, *placer*, and *trap*.

sedimentary magnetic anomaly: a *magnetic anomaly* generated by intrasedimentary magnetic sources associated with *magnetization*-related lateral changes in rock composition. Sources of the dominant sedimentary magnetic anomalies include: (a) *igneous intrusion*, *ore body*, shallow *volcanics*, *magnetite*-rich *subcrop*, and *banded iron formation (BIF)* generating the high-*amplitude* positive anomalies in both the total and residual magnetic fields; (b) salt domes and salt or shale diapirs generating the low-amplitude negative anomalies; and (c) localized accumulations of *detrital magnetite*, *diagenetic magnetite*, and other *diagenetic magnetic minerals* generating the low-amplitude clustered positive anomalies. The amplitude of a sedimentary magnetic anomaly depends on the volume, lateral *magnetization contrast*, and depth of a particular *source body*: the larger the volume, the higher the magnetization contrast, and the shallower the source, the higher the amplitude; whereas its *polarity* (positive or negative sign) is defined by a sign of the lateral magnetization contrast created by this source. In comparison with gravity gradients, the magnetic gradients of detected anomalies are much more sensitive to the presence of intrasedimentary structural discontinuities, such as a normal or strike-slip *fault*, salt dissolution or fracture zone, *shear zone*, and lithological *contact* formed by *salt* or *igneous body*. See also *clinker anomaly*, *diagenetic magnetic anomaly*, *syngenetic magnetic anomaly*, *SRM anomaly*, *strike-slip fault*, *gradient*, and *sensitivity*.

seismic basement: same as *acoustic basement* with an emphasis on it being the deepest relatively continuous seismic reflection horizon which often, but not always, represents the top of a thick carbonate formation or extensive *lava* flow rather than the *geological basement*.

self-attraction correction: a correction applied to measured *absolute gravity* values to compensate for the *self-attraction effect*. Its calculation is based on the gravity effect, or *signature*, of a particular *free-fall absolute gravimeter* obtained by the three-dimensional (3D) *forward modeling* of the absolute gravimeter mass and bulk approximation of its mounting platform components. The estimated range of correction values is 1.0–1.8 *microgal (μGal)*. The self-attraction correction is subtracted from the measured absolute gravity values. See also *atmospheric pressure correction*, *diffraction correction*, *ocean loading correction*, *self-gradient correction*, *speed-of-light correction*, and *tilt correction*.

self-attraction effect: **1.** the gravity effect resulting from the *gravitational attraction* of a free-falling *proof mass* by the mass of *free-fall absolute gravimeter* and the nearest components of its mounting platform. It is a source of systematic errors in the high-*accuracy* measurements of *absolute gravity* that cause a measured absolute gravity value to be higher than its actual value at an *observation point*. The self-attraction effect is compensated by application of *self-attraction correction*. **2.** the *self-gradient effect* in measurements by a *rotating-disk gravity gradiometer*.

self-calibration: three-dimensional (3D) *forward modeling* of the time-variable *self-gradient effect* which is a source of systematic errors in the high-*accuracy* airborne and marine (shipboard) *gravity gradient* measurements with the use of a *rotating-disk gravity gradiometer*. Self-calibration is usually combined with the calculation of *centripetal acceleration correction* for another type of systematic errors associated with the rotation of the assemblage of three tilted rotating disks with sensing accelerometers on its mounting subplatform (carousel). See also *accelerometer* and *gravity sensor*.

self-demagnetization: a *shape*-dependent property of magnetized rock bodies that controls the *magnitude* and direction of the *magnetization vector* within a rock body. It is a result of the magnetic effect produced by pairs of positive and negative magnetic charges, or magnetic poles, emerging on the opposite boundaries of a rock body whenever the body is placed into an external magnetic field, such as the *Earth magnetic field*. The magnetic poles produce an internal *demagnetizing field* which aligns with the direction of the rock body long axis, so the superposition of the external and internal fields can decrease the magnitude and change the direction of *total magnetization* in the rock volume. These changes result in a decrease of *amplitude* and change of shape of the measured magnetic anomaly relative to its amplitude and shape computed without accounting for self-demagnetization in the process of *forward modeling* or *inversion*. The self-demagnetization effect is the strongest when the internal demagnetizing field of an iron-rich elongated rock body is comparable to the ambient magnetic field and aligned nearly opposite to it. The magnitude of the self-demagnetization effect depends on the shape of a rock body, its magnetic properties, and the orientation of its long axis relative to the direction of the inducing magnetic field. Self-demagnetization is assumed to be negligible, when *bulk susceptibility* is less than approximately 0.1 SI units or 800 *microcgs* (*μcgs*). It is commonly ignored in magnetic *modeling*, except for highly magnetic sources, such as *banded iron formation (BIF)*, ultramafic igneous rocks, volcanic rocks with high *remanent magnetization* components, and some types of ore bodies. Also referred to as *shape demagnetization*, *shape effect*, and *internal demagnetization*. See also *demagnetization factor* and *ore body*.

self-gradient effect: the *self-attraction effect* taken into account in high-*accuracy* airborne and marine (shipboard) *gravity gradient* measurements with the use of a *rotating-disk gravity gradiometer*. The time-variable self-gradient effect is caused, primarily, by the relative movements between the permanently oriented *stabilized platform* and *attitude* motions of a survey aircraft or ship, such as *pitch*, *roll*, and *yaw*. It also includes the gradient effects due to the decreasing amount of fuel in the aircraft tank and water balance and fuel tanks of the survey ship. Self-gradient effects are determined by *self-calibration* with the use of the forward-modeled gradient *signature* of the survey ship or aircraft and moving components of the stabilized platform. It is compensated by application of *self-gradient correction*. See also *forward modeling*, *gradient*, and *lever arm effect*.

self-gradient correction: a correction applied to the *gravity gradient* values obtained by high-*accuracy* airborne and marine (shipboard) measurements with the use of a *rotating-disk gravity gradiometer* to compensate for systematic errors due to the *self-gradient effect*. Its calculation is based on a three-dimensional (3D) *model* of the bulk-approximated mass of the survey ship or aircraft, including the movable components of the *stabilized platform* and amount of liquids in the fuel and water tanks. Correction values are adjusted to account for changes in the modeled variables during the survey and subtracted from the measured gradients on a point-by-point basis. See also *self-calibration* and *gradient*.

self-gravitational potential: *gravitational potential* of the stationary, or non-rotating, Earth whose shape, size, and mass are approximated by the *normal ellipsoid* also known as the *reference ellipsoid*. Self-gravitational potential and *centrifugal potential* comprise the *total gravitational potential* which is referred to as the *Earth gravity potential* and used, primarily, in geodetic applications.

sensel correction: same as a *flight elevation correction* which accounts for the *elevation* of a gravity or magnetic *sensor*, i.e., ***sens***or ***el***evation correction.

sensitivity: 1. an instrument characteristic that defines the smallest change in a measured quantity, such as the gravity or magnetic field and its gradients, that can be detected by a *gravimeter*, *magnetometer*, or *gradiometer*. Sensitivity depends on the *signal-to-noise ratio (S/N)*, which controls the instrument response to variations of the measured values: the higher the S/N, the higher the sensitivity. Statistical estimates of the sensitivity are averaged over the measurement *bandwidth* and used for the instrument performance evaluation. For magnetic measurements, the sensitivity is quoted in units of *nanotesla (nT)* per square root of *hertz (Hz)*, i.e., $\mathrm{nT}/\sqrt{\mathrm{Hz}}$. For gravity measurements, it is quoted in units of $\mathrm{m \cdot s^{-2}}/\sqrt{\mathrm{Hz}}$ and $g/\sqrt{\mathrm{Hz}}$ for the gravity field and in units of $\mathrm{E}/\sqrt{\mathrm{Hz}}$ for gravity gradients, where g is *nominal gravity* value $g = 9.8\ \mathrm{m \cdot s^{-2}}$. A measuring instrument can have a high sensitivity and yet a low *accuracy*. **2.** a characteristic of a calculated *attribute* that describes its *amplitude response* or *phase response* to the smallest changes in a geological *structure* or physical properties, such as a source depth or its *density* or magnetization *contrast*. See also *precision*, *Eötvös (E)*, *density contrast*, and *magnetization contrast*.

sensor: a device with mechanical, electromagnetic, or natural components (e.g., *proof mass* suspended by a metal or quartz spring or levitating in the electromagnetic field in a *gravimeter* or hydrogen-rich fluid or vaporized chemical element in a *magnetometer*) that can detect, or sense, very small changes in a gravity or magnetic field. A sensor generates an electronic signal which is measured and recorded digitally. See also *gravity sensor* and *magnetic sensor*.

sensor compensation: elimination of the static and dynamic magnetic interference effects on a *magnetic sensor* that are generated by the aircraft components (frame, engine, etc.) during an aeromagnetic *survey*. See also *aeromagnetic compensation* and *figure-of-merit (FOM)*.

sensor tilt: a deviation from the vertical orientation, i.e., incorrect vertical alignment, of the *gravity sensor* axis in a *spring gravimeter*. Sensor tilt results in an artificial increase of the measured *relative gravity* values. Within the range of ±200 arc-seconds, it is compensated by *automatic tilt correction* which is a part of *automatic drift correction* incorporated into the *ground gravity* measurements by automated *relative gravimeter* systems. See also *arc-second* and *platform tilt*.

sensor tilt correction: See *tilt correction*.

separation filter: 1. a *data processing* operator that separates the regional and residual components of an anomalous gravity or magnetic field, often by *upward continuation*. **2.** the *Jacobsen filter* in the context of its property to separate via a relative *enhancement* the anomalous components of a gravity or magnetic field generated by assemblages of sources at a user-specified depth level. See also *anomalous field* and *regional-residual separation*.

separation filtering: a group of the *enhanced processing* methods based on the use of *Jacobsen filter, matched filter, modified second vertical derivative*, and other *filtering* techniques that attempt to separate the anomalies generated by the layer-approximated assemblages of sources occurring at sufficiently different depth levels. The obtained results are used in *regional-residual separation, cascaded filtering*, and other techniques for the *enhancement* of anomalies of the exploration interest. Sometimes referred to as *layer filtering*. See also *anomaly*.

shaded relief: *sunshading* in the context of representing the gridded gravity or magnetic field in a way of variations in the topographic *relief* composed of hills and valleys illuminated by an artificial source of light from different directions.

shadowgraph: a *grid* image obtained by *sunshading*.

shadowing: See *sunshading*.

shale dome: a dome-shaped concordant or discordant piercement, or diapiric, geological *structure* with a columnar shale plug at its core and upturned complex faulted and folded sediments adjacent to the shale plug. Commonly, these structures occur in deltaic settings. Density of the shale dome can be almost as low as the salt density, if the original water content is preserved. Often, shale diapirs dewater after *intrusion*, and the resulting range of density contrasts produces a wide range of gravity anomalies. See also *salt dome*.

shape: a characteristic of a gravity or magnetic *anomaly* described by its spatial or geometric form on a *grid* or by a continuity between profiles in adjacent *survey* lines. A source *dip* shifts the axis of an anomaly in the down-dip direction and, thereby, results in a steeper anomaly *slope*, or a higher *gradient*, in the up-dip direction of the source. This asymmetry in the anomaly shape and its gradients decreases with a decrease in the source depth extent and with an increase of its dip. The slope of an anomaly is directly related to the depth of its source: the shallower the source, the steeper the slope. Unlike *amplitude*, slope is independent from the *density contrast* or *magnetization contrast*, and this property is used in *graphical source depth estimation* methods. At distances relatively close to observation points, the source shape is reflected in the shape of its anomaly, so compact sources generate sharp and narrow anomalies. A broad anomaly has a narrower *amplitude spectrum* than that of a narrow anomaly. For the purposes of initial *modeling* and *inversion*, the approximate *source geometry* can be assessed from the shapes and gradients of observed anomalies. See also *anomaly shape (grav.)* and *anomaly shape (mag.)*.

shape demagnetization: same as *self-demagnetization* with an emphasis on it being a *shape*-dependent property of all magnetized rock bodies.

shape effect: *self-demagnetization* in the context of a *shape*-related internal demagnetizing effect that can reduce *amplitude* and change the direction of the *magnetization vector* within a rock body.

shape factor: See *fall-off*.

shape index (ShI): a dimensionless *attribute* of the *curvature* of the gravitational *equipotential surface*. The shape index describes the *shape* of the equipotential surface for the gravity field generated by individual sources and their local assemblages, and it is used for approximation of the dominant structural shapes in a *gravity survey* area. It is expressed as

$$\begin{aligned}\mathrm{ShI} &= (2/\pi)\arctan[(K_{max} + K_{min})/(K_{max} - K_{min})] \\ &= (2/\pi)\arctan(2K_m / K_d) \\ &= (2/\pi)\arctan[(-g_{xx} - g_{yy}) / 2(g_{xy}^2 + g_{UV}^2)^{1/2}],\end{aligned}$$

where K_{max} is *maximum curvature*, K_{min} is *minimum curvature*, K_m is *mean curvature*, K_d is *differential curvature*, and g_{xx}, g_{yy}, g_{xy}, g_{UV} are *differential curvature gradients* with $g_{UV} = (g_{xx} - g_{yy})/2$. The shape index varies from −1 to +1, and its color-scaled *grid* is commonly displayed with five color ranges to describe the different shape morphologies: bowl (−1), valley (−0.5), flat (0.0), ridge (+0.5), and dome (+1). Superposition of the regional-scale anomalies from deep sources can produce false shape indications and, hence, the more accurate ShI estimates are obtained with the use of *residual gravity*. The shape index has positive values above sources with a positive *density contrast* and negative values above sources with a negative density contrast (Li, 2015). An improved version of the *terracing* algorithm is based on the use the shape index, whose zero values mark the boundaries between the pseudo-lithological units (Cooper, 2020). See also *curvature attributes* and *apparent curvature attributes*.

sharpening filter: **1.** a space-domain *filter*, such as the *Laplace filter* or *Kuwahara filter*, which increases the *sharpness* of anomalies and enhances edges of the structural elements obtained by *inversion*. **2.** a version of a *grid*-based *averaging filter*, where the sharpening operation is performed by subtracting the grid smoothed by the averaging filter from the original grid. A sharpening filter can be applied for the *terracing* of anomalous gravity and magnetic fields. See also *anomalous field* and *edge-sharpening filters*.

sharpness: a qualitative characteristic of the gravity or magnetic *anomaly* described as a measure of its horizontal rate of change. Sharp anomalies have the high values of their horizontal gradients, and they are easily distinguished on grids and maps. Sharpness is a function of the depth, *shape*, orientation, and *geomagnetic latitude* of an anomaly source. It increases with an increasing similarity of the source shape to a *sphere*, and it decreases with an increasing source depth. Sharpness attenuates one power faster than the *amplitude* of an anomaly and, unlike the amplitude, it is independent from the *mass differential* or *magnetization contrast* of a source. High-pass filtering and calculation of high-order derivatives, such as the *second vertical derivative (SVD)*, increase the sharpness of anomalies. See also *attenuation, high-pass filter, horizontal gradient, grid*, and *sharpening filter*.

shear zone: a tabular, parallel-sided zone of intense fracturing and crushing of rocks by the shearing deformations that cause rock bodies to break into fragments and slide relative to each other. It is often mineralized by ore-forming fluids and repeated hydrothermal flows with the precipitation of magnetic minerals. Shear zones are approximated by the vertical or dipping *thin sheet* source geometries, and they often generate linear and curvilinear variable-*amplitude* short-*wavelength* positive magnetic anomalies. Fracturing and fragmentation of rocks within the shear zone reduce their original densities and create a negative *density contrast* resulting in the local *gravity low* which is coincident with the alignment of positive magnetic anomalies. Shear zones are common boundaries of the tectonic terranes. See also *dip*, *magnetized fault*, *strike-slip fault*, and *terrane*.

shelf-edge gravity anomaly: a *gravity anomaly* caused by the *free-air edge effect*. In its simplest form, it has the characteristic *profile* signature of a *peak* over the shelf edge and a *trough* over the continental slope, i.e., peak-to-trough anomalous doublet, and it can form the laterally extended paired anomaly belts on *free-air anomaly* maps within the rifted continental margins. Amplitudes of the anomalous doublet can reach tens of milligals. Variations of its shape and amplitude depend on the bathymetric *relief*, thickness of the rift-accumulated sediments, and *Moho* depth structure. Also referred to as *edge-effect gravity anomaly*. See also *milligal (mGal)* and *signature*.

shipboard gravimeter: a *relative gravimeter* and supporting system for measurements of the *Earth gravity field* variations, or *relative gravity*, on board a ship during marine (shipboard) gravity surveys. A gravimeter *sensor* is heavily damped against the high-frequency vibrations to which a moving ship is subjected. The supporting system includes: (a) *stabilized platform* to minimize the inertial accelerations; (b) several monitor channels to record *acceleration* and velocity data for the calculation of corrections for the motion-related effects, such as *cross-coupling effect* and *Eötvös effect*; and (c) digital recording of the measured data in *real-time mode*. Also known as *marine gravimeter*. See also *damping*, *inertial acceleration*, and *moving-platform corrections*.

shipboard gravity: *marine gravity* in the context of the *relative gravity* measurements by a *shipboard gravimeter* installed on board a ship as opposed to such measurements by a *relative gravimeter* on board a submarine or by an *underwater gravimeter* on the sea floor.

shipboard survey: See *marine survey*.

shock demagnetization: **1.** *demagnetization* of rocks on the cratered planetary surfaces caused by the hypervelocity impact of meteorites and other cosmic bodies. High pressure and heating effects of the impact-generated shock wave introduce the crystallographic defects in the rock-forming minerals that result in the irreversible changes to their original *magnetization*. In a weak ambient magnetic field, the magnetization of rocks in the impact area can be destroyed or randomized significantly. The combined effects of *thermal remanent magnetization* from the impact-related heating, *chemical remanent magnetization* from the post-impact hydrothermal mineralization of the fractured rocks, and shock demagnetization produce the highly variable anomalous magnetic signatures observed in and around impact depressions on the Martian and lunar surfaces. **2.** the local demagnetization of rocks by explosion or impact of *unexploded ordnance (UXO)* that can result in uniaxial *magnetic anisotropy* effects (Paoletti et al., 2019). See also *ambient field*, *astrobleme*, and *signature*.

SI units: units of measurement in the International System of Units (Systeme International d'Unites) accepted worldwide by governmental agencies and professional societies. Basic SI units are the same as *mks units*, where distance is measured in meters, mass in kilograms, and time in seconds. In gravity exploration, the SI unit of *gravitational acceleration* is meter per second squared ($m \cdot s^{-2}$), which is related to the cgs unit of *Gal* as $1\ m \cdot s^{-2} = 100$ Gal. In magnetic exploration, *magnetic susceptibility* is measured in dimensionless SI units related to the corresponding cgs unit of *microcgs (μcgs)* as $1000 \cdot 10^{-6}$ cgs $= 1000$ μcgs $= 4\pi \cdot 10^{-3}$ SI $\cong 0.01256$ SI.

Siegert method: a method for calculation of the *Bouguer correction factor (BCF)* and estimation of its probable error for each *traverse* in a *ground gravity* survey. It assumes that the surface *elevation* does not correlate with a geological *structure*, and it is based on the use of triplets of data, i.e., sets of three gravity readings and three elevation values for each three consecutive stations along the traverse. The BCF value k and its error ε are defined, respectively, as

$$k = -\left(\sum_{i=2}^{N-1} g_i h_i\right) / \sum_{i=2}^{N-1} h_i^2,$$

$$\varepsilon = 0.67\left[\left(\sum_{i=2}^{N-1} g_i^2 / \sum_{i=2}^{N-1} h_i^2 - k^2\right) / n\right]^{1/2};$$

where g_i is the difference between the gravity *reading* at the *station* in the middle of each triplet and the weighted average of readings at stations on either side of the middle station, wherein the difference is positive when the reading is higher than its interpolation by the weighted average, and weights are inversely proportional to the distance between stations in each triplet; h_i is the difference between the station elevation in the middle of each triplet and the weighted average of station elevations on either side of the middle station; N is the number of stations along the traverse; and $n = N - 2$. The addition of the *free-air factor* to the calculated BCF value k will comprise the value of the *elevation correction factor* (Siegert, 1942). See also *Nettleton method*.

signal-to-noise ratio (S/N): a quantitative characteristic of line-oriented and gridded gravity or magnetic data acquired on a moving *observation platform*. It is defined by the ratio of the power (squared amplitudes) of the measured anomalies ("signal") to either the total power of the data, including *noise*, or to the estimated average power of noise. The S/N is expressed in units of *decibel (dB)* as

$$\mathrm{S/N} = 10\,log_{10}\left(P_S / P_{S+N}\right) \quad \text{or}$$

$$\mathrm{S/N} = 10\,log_{10}\left(P_S / P_N\right),$$

where P_S, P_N, P_{S+N} are average powers of the signal, noise, and total power, respectively. The S/N ratio is one of the major factors in the achievable *vertical resolution* of anomalies and the *sensitivity* of measuring instruments. See also *acquisition*, *anomaly*, *gridding*, and *stacked sampling*.

signature: 1. a characteristic of the *total field* or *gradient* anomaly described by its quantitative and qualitative parameters (*amplitude*, *shape*, *sharpness*, and *wavelength*) that makes it distinctly different from other anomalies. Signatures that are known to be associated with a particular geological body, geological *structure*, structural boundary, or group of structural elements can be used as complementary guides for the identification and *interpretation* of anomalies. For example, the magnetic and gravity signatures of *kimberlite intrusion* are often expressed as either "bright spot" or "dim spot," i.e., localized circular-shaped positive or negative anomaly whose sign is defined by the sign of the *magnetization contrast* and *density contrast* created by this intrusion. **2.** the magnetic field of a three-dimensional (3D) *model* of an aircraft components (frame, engine, etc.) calculated by *forward modeling* and applied in *aeromagnetic compensation*. **3.** the gravity field of the 3D model of a *free-fall absolute gravimeter* computed by forward modeling and used for the calculation of *self-attraction correction*. **4.** the gravity field of a 3D model of the bulk-approximated mass of a *survey* ship or aircraft, including the movable components of a *stabilized platform* and the amount of liquids in the water and fuel tanks, that is computed by forward modeling and used for the calculation of *self-gradient correction*. See also *gravity signature*, *magnetic signature*, and *self-calibration*.

sill: a tabular, or sheet-like, *igneous intrusion* which is much longer in the horizontal *dimension* than in its vertical dimension. Differing from a *dike*, it parallels the sedimentary layers and, usually, it has a horizontal occurrence. The same as for a dike, the high *density* and high *magnetization* of the sill make it one of the main intrasedimentary sources of gravity and magnetic anomalies. In *modeling* and source depth estimates, the sill is approximated by a horizontal *thin sheet*. See also *stock*.

simple Bouguer anomaly: *Bouguer anomaly* which is compensated for the *Bouguer effect* and, if necessary, for the *indirect effect*, but it is not compensated for (a) the local gravity effect of the variations of *topography* and *bathymetry* around observation points known as the *terrain effect* and (b) the regional gravity effect of the Earth's curvature known as the *curvature effect*, i.e., the simple Bouguer anomaly represents the Bouguer gravity field before the application of *terrain correction* and *Bullard B correction*. See also *complete Bouguer anomaly*.

simple Bouguer correction: a type of *Bouguer correction* which compensates for the *Bouguer effect* only, as opposed to the *complete Bouguer correction* which compensates for the Bouguer effect, *terrain effect*, and, in regional gravity surveys, *curvature effect*. It is calculated by the *slab formula*, where the assumed *density contrast* of the slab is the *Bouguer density* value. Sometimes referred to as *incomplete Bouguer correction*.

simple Bouguer plate value: a *scalar* quantity representing the gravity effect, or *gravitational attraction*, of the infinite horizontal *Bouguer plate*. It is expressed in units of *milligal (mGal)* of the equivalent *gravitational acceleration* Δg_{BP} as

$$\Delta g_{BP} = 2\pi G\rho h \cong 0.04193\rho h,$$

where G is *universal gravitational constant* in *cgs units*, ρ is assumed *Bouguer density* in $g \cdot cm^{-3}$, and h is Bouguer plate thickness in meters defined by the vertical distance between a *station* on the ground and the *survey datum* of sea level or *reference ellipsoid*. The simple Bouguer plate value comprises the amount of *simple Bouguer correction*. See also *Bouguer correction factor (BCF)* and *slab formula*.

sine transform: 1. *Fourier transform* of the odd, or asymmetric, part of *potential field* from its *waveform* format in the *space domain* into a series of sine waves in the *wavenumber domain*. The mathematical expression of the sine transform is given by the sine integral in the Fourier transform formula. Sometimes, its result is referred to as the *quadrature spectrum*. **2.** the imaginary part of a *complex spectrum*. See also *cosine transform*.

single-point source: a *source body* whose position is described by a single spatial location, such as the center of a *sphere*, a two-dimensional (2D) *thin sheet* edge, the axis of an infinite *horizontal cylinder*, the top of an *infinite contact*, etc. Its shape parameters (thickness, width, depth extent) are assumed to be either negligible or infinite, similar to those of an infinite contact or plate edge. The amplitude attenuation, or *fall-off*, of its gravity or magnetic field is described by the *inverse power law* for ideal sources with the integer *Euler structural index (SI)* in *Euler deconvolution equation* (Reid and Thurston, 2014). See also *ideal source*, *infinite length*, *point source*, and *multiple-point source*.

skeletonization: a *grid*-based method of mapping the anomalous *peak*-and-*trough* pairs, or *waveform* cycles, by linear and curvilinear alignments that are one pixel in width. Automatic correlation of these pairs is based on the characteristic parameters of the adjoining cycles that include (a) peak *amplitude*, (b) peak-to-trough width and *polarity*, (c) cycle length, (d) *strike* direction, and (e) linearity. Skeletonization transforms the original grid values into a binary *format*, where the black or colored pixels are assigned to the alignment location and its strike direction, and the white pixels are assigned to the background parameter defined by a pre-selected threshold value. Parameters are stored in a *database*, or catalogue, which can be used for the parameter-based analysis and *filtering* of magnetic and gravity data (Eaton and Vasudevan, 2004). Method is the most efficient in its application to high-*resolution* aeromagnetic data for the detection and mapping of the local- and regional-scale *structural grain*, high-fracturing areas in sedimentary rocks, and boundaries of mineral alteration zones in ore bodies. See also *steerable filter*.

skewness: a horizontal displacement of the maximum value of a dipolar, or peak-and-trough, *total magnetic intensity (TMI)* anomaly relative to the location of its *source body*. Peak-and-trough asymmetry of an TMI anomaly shape and the *offset* of its peak value are caused by the non-vertical direction of the inducing *geomagnetic field*. When the *remanent magnetization* component in the source body is small, the anomaly maximum is displaced toward the *magnetic equator* in both the northern and southern hemispheres. Skewness is minimized by the calculation of the *analytic signal amplitude (ASA)* and its derivatives or eliminated by the application of *reduction-to-pole* (*RTP*). See also *dipolar anomaly* and *induced magnetization*.

sky map: a circular display window showing the current position of the artificial source of light, or "sun," in the application of *sunshading*. The center of the sky map corresponds to the *overhead point*. The optimal sunshading of a *grid* is determined by interactive testing of different "sun" positions on the sky map.

slab correction: *simple Bouguer correction* in the context of its approximation of rocks above a *survey datum* by the infinite horizontal plate, or slab, called the *Bouguer plate* or *Bouguer slab*.

slab formula: a formula for calculation of the *gravity anomaly* generated by the infinite horizontal plate, or slab. It is expressed in units of *milligal (mGal)* as

$$\Delta g_S = 2\pi G \Delta\rho h \cong 0.04193\Delta\rho h,$$

where G is *universal gravitational constant* in *cgs units*, $\Delta\rho$ is assumed *density contrast* of the plate in $g \cdot cm^{-3}$, and h is plate thickness in meters. Among other applications, this formula is used for the calculation of *simple Bouguer correction*, the amount of which is also referred to as the *simple Bouguer plate value*. See also *Bott method*.

sliding window: See *moving window*.

slope: **1.** a *profile* parameter that defines the rate of change of a function per unit horizontal distance. Differing from *amplitude*, the slope of a gravity or magnetic *anomaly* is independent of the *density contrast* or *magnetization contrast*, but it depends on the source shape and direction of *magnetization*, and it is proportional to the source depth: the shallower the source, the steeper the slope. This depth-related property of the slope is used in source depth estimates by the graphical *slope methods*, such as *straight-slope method*, *Sokolov method*, *Peters half-slope method*, and *demi-pentes method*. **2.** the first-order horizontal *derivative* of a gravity or magnetic *field* in the direction of calculation. **3.** a *horizontal gradient* which is represented graphically by the slope of a gravity or magnetic anomaly.

slope-distance rules: *depth rules* relating the source depth to the projection of the *slope* of an anomaly *profile* on a horizontal plane.

slope methods: graphical, or manual, source depth estimation methods based on the use of *slope*-related characteristic points on a *profile* or *contour map* of an isolated (non-interfering) gravity or reduced-to-pole (RTP) magnetic *anomaly*. They include *demi-pentes method*, *Peters half-slope method*, *Sokolov method*, *Smith depth rules*, and *straight-slope method*.

slope-to-gravity method: a method of calculating the marine gravity anomalies, known as the *marine free-air gravity*, from *satellite altimetry* data over oceans and seas. Because *mean sea level (MSL)* variations correspond to changes in the height of the *geoid*, which is the *equipotential surface* for the *Earth gravity field*, the method involves the following basic procedures: (a) calculation of the first- and second-order horizontal derivatives of the altimetry-derived ocean/sea surface that represent the slopes and curvatures of the geoid, respectively, in the mutually orthogonal x and y directions in *Cartesian coordinates* using the satellite altimetry track data; (b) calculation of the second-order *vertical derivative* via the *Laplace equation*; (c) calculation of the corresponding *vertical gravity gradient*; (d) conversion of the calculated vertical gravity gradients into *free-air gravity* at the ocean/sea surface; (e) *gridding* of the obtained free-air gravity values; and (f) adding the regional long-*wavelength* free-air gravity component from the global reference models. See also *geoid-to-gravity method*.

sloping-wedge method: a method for calculation of *inner-zone terrain correction* within the first four *Hammer zones* (2.0–170.1 m) around each *station*, where the surrounding *topography* is approximated by wedge-shaped sectors with the uniform slope θ. Correction for the gravity effect of the uniform slope wedge above or below a station within a given radius R, which is rotated about the vertical axis through the station, is expressed in units of *milligal (mGal)* as

$$\Delta g = 2\pi\rho R(1 - \cos\theta),$$

where ρ is rock *density* assumed to be equal to *Bouguer density*. See also *Hammer method*.

slung sensor: **1.** a *multisensor gradient array* suspended below a *survey* aircraft on a cable with electrical connections to power and operate an airborne *magnetic gradiometer*. Cable suspension at 30–40 m below a helicopter enables magnetic measurements to be taken without the *aeromagnetic compensation* system, which might suppress the small magnetic anomalies of an exploration interest during high-*resolution* mineral exploration surveys or anomalies generated by buried anthropogenic materials, such as abandoned wells, *unexploded ordnance (UXO)*, pipelines, etc. **2.** a lightweight *magnetometer* suspended below a *drone* in a low-*altitude* high-resolution airborne *magnetic survey*. See also *drone survey* and *magnetic sensor*.

Smith depth rules: three rules for source depth estimates with the use of the horizontal derivatives of gravity anomalies generated by *two-dimensional (2D)* and *three-dimensional (3D)* sources. Rule 1: depth h to the source of an isolated (non-interfering) *gravity anomaly* is estimated as

$$h \leq K(A_{max}/|\partial A/\partial x|_{\max}),$$

where K is the *index factor* with $K = 0.65$ for a 2D source and $K = 0.86$ for a 3D source; A_{max} is maximum *amplitude* of the anomaly; $|\partial A/\partial x|_{\max} = \Delta A/\Delta x$ is absolute maximum of the *horizontal derivative* or maximum *slope* of the anomaly; and $\Delta A/\Delta x$ is change of amplitude A over the distance Δx on the anomaly *profile*. Rule 2: depth h to the source of a partially (one side) isolated anomaly is estimated as

$$h \leq K(A_{IP}/(\Delta A/\Delta x),$$

where $K = 1.0$ for a 2D source and $K = 1.5$ for a 3D source and A_{IP} is anomaly profile value at the *inflection point* over which the slope $\Delta A/\Delta x$ is determined. Rule 3: when the absolute maximum *density contrast* $\Delta\rho_{max}$ of the source and absolute maximum of the *second horizontal derivative* of its anomaly are known, the depth h to the top of this source is estimated as

$$h \leq 5.4G(\Delta\rho_{max}/|\partial A^2/\partial x^2|_{max}),$$

where G is *universal gravitational constant* in *cgs units*. Also referred to as *Bott-Smith rules* and *maximum-depth rules*.

smoothing: a line-based or *grid*-based statistical averaging (mean, median, etc.) of adjacent data values that is applied to suppress high-frequency *noise* generated by irregularly distributed shallow sources of natural or artificial origin. The degree of smoothing is defined, primarily, by the length of a smoothing operator: the longer the length, the smoother the obtained field and the longer its retained *wavelength* components. Smoothing is performed with the use of a *moving window* the size and shape of which are defined by the *kernel* of a particular *smoothing filter*. See also *averaging filter* and *grid smoothing*.

smoothing-enhancing filter: a spectral-domain composite, or combined, *data processing* operator which involves *smoothing* by *upward continuation* and enhancing anomalies by the vertical or horizontal differentiation of an *anomalous field*. Its *wavenumber* response is described by the bell-shaped *transfer function* of a band-pass type, whose central wavenumber at a given *altitude* of upward continuation depends on the order of the applied vertical or horizontal *derivative* (Fedi et al., 2009). See also *band-pass filter* and *spectral domain*.

smoothing filter: a line-based or *grid*-based operator that eliminates the high-frequency components in an *observed field* by replacing each original data value with a new value calculated by the statistical averaging (mean, median, etc.) of a given value and its neighbors. Smoothing filters are applied for *grid smoothing*, where the length and shape of a filter operator are defined by the total number of the grid cells surrounding each particular cell at the center of a *moving window*, e.g., nine grid cells in a 3 × 3 square window. Grid-based smoothing filters can be used in two modes of operation: *smoothing* and *residualization*. In the latter mode, the smoothed *grid* is subtracted from the original grid in order to enhance the remaining, or residual, short-wavelength components of an anomalous gravity or magnetic field. See also *anomalous field*, *Griffin method*, and *edge-sharpening filter*.

SNN mean filter: an option in the application of the *grid*-based *symmetric nearest neighbor (SNN) filter*. It is based on calculation of the arithmetic mean, or average, of the *grid cell* values nearest to a value at the central grid cell in a *moving window*. The calculated value is assigned to the central grid cell as the filter output. It is used as a *spike*-removing and high-frequency *noise*-suppressing filter.

SNN median filter: an option in the application of the *grid*-based *symmetric nearest neighbor (SNN) filter*. It is based on calculation of the *median threshold* of the *grid cell* values nearest to a value at the central grid cell in a *moving window*. The calculated value is assigned to the central grid cell as the filter output. It is used as a *spike*-removing and high-frequency *noise*-suppressing filter.

Sobel filter: a space-domain *data processing* operator for calculation of the *horizontal gradient* in gridded data. Its algorithm is implemented in *edge-detection methods* and other *enhanced processing* algorithms. See also *filter*, *gridding*, and *space domain*.

Sokolov length: the horizontal distance between two points of intersection of a *maximum-slope line* with (a) the asymptotic *baseline* for both ends of an anomaly *profile* or zero level of the profile and (b) the line which is parallel to the asymptotic baseline or zero level and tangent to the maximum profile value. It is a main parameter for estimating a source depth by the *Sokolov method*.

Sokolov method: a graphical method of estimating a source depth from the *slope* of an isolated (non-interfering) reduced-to-pole *magnetic anomaly*. Source depth is calculated as *SL*/*df*, where *SL* is *Sokolov length* and *df* is a *dividing factor* whose value is based on the *source body* approximation: 1.54 for *thin dike* and 2.0 for *thick dike* or *prism*. The dividing factor *default value* is 2.0. Also referred to as the *Sokolov rule*. See also *reduction-to-pole (RTP)*.

Sokolov rule: *Sokolov method* in the context of a specific rule that relates magnetic source depth to the anomaly *profile* parameter called the *Sokolov length*.

solar daily variation: same as *diurnals* with an emphasis on the time-variable interaction between the *solar wind*, rotation of the Earth within its *magnetosphere*, and transient electric currents in the *ionosphere*.

solar wind: a constant supersonic flow of Sun-radiated charged particles, such as the negatively and positively charged ions of hydrogen and helium. The solar wind is largely deflected from the Earth by the *main geomagnetic field*, but it is a major factor in the time-variable irregular changes, or fluctuations, of the *external magnetic field* of the Earth that define the *magnitude* and *wavelength* content of *diurnals*. See also *magnetic storm*, *magnetosphere*, and *van Allen belts*.

solenoidal field: a divergenceless *vector field*, i.e., vector field with zero *divergence* of its field lines, or *flux* lines. According to the *Gauss law* for magnetic fields, once the divergence is zero, then (a) the number of flux lines entering the volume inside a closed surface is equal to the number of the flux lines leaving this volume, and (b) the normal, or orthogonal, component of the vector field integrates to zero over any closed surface within the magnetic field region. The gravity field is solenoidal, or divergenceless, in the regions outside the *mass* material, whereas the magnetic field is solenoidal both outside and within the magnetic material. See also *conservative field*, *irrotational field*, and *potential field*.

solid angle: the cone-shaped *three-dimensional (3D)* analogue of a planar *two-dimensional (2D)* angle. The solid angle is described by the surface area S on a sphere, where the solid angle cone expanding from its top at the center of the sphere intersects its surface. The numerical value is defined as

$$\Omega = S/r^2,$$

where r is radius of the sphere. The solid angle is measured in units of *steradian*, where one steradian corresponds to the surface area of r^2 on the sphere. For the whole sphere surface of $4\pi r^2$, its full solid angle is 4π steradians. See also *radian*.

solid Earth tides: *Earth tides* in the context of the time-varying elastic deformations of the solid Earth caused by changes in the cumulative effect of the *gravitational attraction* of the Moon and Sun.

Somigliana equation: See *Somigliana-Pizzetti formula.*

Somigliana formula: See *Somigliana-Pizzetti formula.*

Somigliana-Pizzetti equation: See *Somigliana-Pizzetti formula.*

Somigliana-Pizzetti formula: **1.** *International Gravity Formula (IGF).* **2.** a mathematical formula that defines *theoretical gravity*, also known as *normal gravity*, at any point on the surface of a *reference ellipsoid* as a function of the geographic latitude

$$\begin{aligned} g_{TG} &= g_o \\ &= (ag_E \cos^2\theta + bg_P \sin^2\theta)/(a^2\cos^2\theta + b^2\sin^2\theta)^{1/2} \\ &= g_E(1 + k\sin^2\theta)/(1 - e^2\sin^2\theta)^{1/2}; \end{aligned}$$

where a and b are equatorial and polar radii, or semi-axes, of the reference ellipsoid; g_E is theoretical gravity value at the equator; g_P is theoretical gravity at the geographic poles; θ is latitude; k is constant defined as $k = (bg_P/ag_E) - 1$; and e is the first numerical eccentricity defined as $e = (a^2 - b^2)^{1/2}/a$. Based on this formula and successive updates of the reference ellipsoid parameters, five standard versions of the International Gravity Formula (IGF) have been adopted in 1930, 1967, 1980, 1984, and 1999: *IGF1930, IGF1967, IGF1980, IGF1984,* and *IGF1999.* The last three versions are based on parameters of the *GRS80 ellipsoid* and *WGS84 ellipsoid*, and their differences are negligible for exploration applications. Also referred to as *Somigliana-Pizzetti equation, Somigliana-Pizzetti gravity, Somigliana equation,* and *Somigliana formula.*

Somigliana-Pizzetti gravity: See *Somigliana-Pizzetti formula.*

sonic log: a well log of the transit time per unit distance for ultra-high *frequency* acoustic waves. It is measured in microseconds per foot or meter and represents a reciprocal of the seismic compressional wave, or P-wave, velocity. Sonic log measurements with frequencies from 1 to 50 kHz are averaged to obtain interval seismic velocities, which are converted into the apparent densities of rocks using one of the empirical relationships between the seismic P-wave velocity and *density* of rocks: *Gardner equation, Brocher formula,* and *Nafe-Drake curve.* Also known as *acoustic log.* See also *apparent density.*

source body: **1.** a geological body of the laterally contrasting *density* or *magnetization* that is the source of a gravity or magnetic *anomaly.* **2.** any *structure* and structural or lithological discontinuity in the subsurface that create a lateral *density contrast* or *magnetization contrast* and, thereby, generating the gravity or magnetic *total field* and *gradient* anomalies. For the purposes of *forward modeling* and *inversion* of the anomalous fields, individual source bodies are approximated by the simplified geometrical shapes of *contact, dike, prism, horizontal cylinder, sphere,* and their combinations. A source body is considered wide or thick, like a *thick dike*, when its width is the same or larger than its depth, and it is considered narrow or thin, like a *thin dike*, when its width is much less than its depth. Also referred to as *causative body.* See also *anomalous field, homogeneous source,* and *ideal source.*

source depth analysis: the analysis of *source depth solutions* obtained by line-, *profile*-, and *grid*-based *automated source depth estimation* techniques and profile-based *graphical source depth estimation* methods in order to determine (a) depths of the isolated source bodies, (b) *magnetic basement* depth structure, and (c) locations of faults, shear and fracture zones, and lithological contacts identified by alignments of the source depth solutions on a *grid.* See also *depth estimation methods, shear zone,* and *Magnetic Fault Interpretation Cube (MaFIC).*

source depth solutions: estimates of depths to the sources of magnetic and gravity anomalies obtained by *automated source depth estimation* techniques that are based on solving the overdetermined system of equations relating the depth of a particular *ideal source* to the parameters of an observed *anomaly.* Depth solutions can be obtained on the condition that there is a lateral discontinuity in the physical property (*density* or *magnetization*) or structural discontinuity created by the intrasedimentary and *basement* faults. Different patterns in distributions of the calculated source depth solutions above and below a *basement* surface are used in *integrated interpretation* of its depth structure (Goussev and Peirce, 2010). See also *depth estimation methods, fault,* and *pattern.*

source geometry: the geometrical shape of a *source body.* For the purposes of *forward modeling* and *inversion*, sources of anomalies are approximated by simplified geometrical shapes, such as *contact, thin dike* (*thin sheet*), *thick dike, prism, sphere, vertical cylinder,* and *horizontal cylinder.* Any arbitrary shape of a three-dimensional (3D) *source* can be represented by an assemblage of simple shapes which fill-in its volume, wherein their corresponding formulas for the individually generated fields are used to calculate the total gravity or magnetic effect of the whole body. When an *anomaly* is measured sufficiently far from its source, this source is considered to be a *point source* regardless of its actual shape. The initial estimate of a source shape for approximation by a simple source geometry can be made from the *shape* of an anomaly and its gradients. For example, an equidimensional anomaly is most likely approximated by a *sphere* or *vertical cylinder*, whereby a vertical cylinder anomaly will have lower horizontal and vertical gradients than those of a sphere anomaly. See also *gradient, general source,* and *ideal source.*

source magnetic moment: *vector magnetic moment* representing the parameter of a localized, or compact, magnetic *source body* quantified by the volume integral of its *magnetization vector* and related via the first-order *Helbig integrals* to the *vector* components of the magnetic field generated by this body.

source mass: a *source body* parameter defined by the product of the source's *density* and volume. See also *mass deficiency*, *mass excess*, and *mass differential*.

source parameter imaging (SPI™) method: a *grid*-based method of the automatic estimation and imaging of three magnetic source parameters: depth, *dip*, and *susceptibility contrast*. The method assumes: (a) two-dimensional (2D) *source geometry* of the sloping *magnetic contact* or dipping *thin sheet* (*thin dike*); (b) non-interfering magnetic anomalies generated by isolated sources; and (c) negligible *remanent magnetization* in the calculation of source dip and susceptibility contrast. Using the *total magnetic intensity (TMI)* for the thin-sheet source model and *horizontal gradient* for the magnetic contact, three attributes are calculated at *grid nodes* in the gridded data: *analytic signal amplitude (ASA)*, *local phase*, and *horizontal local wavenumber*. From these attributes, three source parameters are determined: (1) depth h to the top of the magnetic contact or thin sheet:

$$h = 1/k_{max},$$

where k_{max} is maximum value of *horizontal local wavenumber* k_x; (2) source dip:

$$D = \theta + 2\beta - 90^\circ;$$

where θ is local phase and $\tan\beta = \tan I/\cos\delta$, where I is *magnetic inclination* and δ is angle between the positive x-axis and *magnetic north*; and (3) susceptibility contrast:

$$K = \left|\text{ASA}\right|/2k_x M \cdot C \cdot \sin D,$$

where M is *magnitude* of the *Earth magnetic field*, $C = 1 - \cos^2 I \cdot \sin^2\delta$, and D is dip angle measured from the x-axis. The calculated parameters for each identified source in a *magnetic survey* area are presented in *pseudocolor* on four grids: (1) source location from the maximum values of the horizontal local wavenumbers; (2) source depth; (3) source dip; and (4) susceptibility contrast at the source location. Estimates of a 2D source location and depth are independent from magnetic inclination, *declination*, dip, *strike*, and any remanent magnetization. Unlike 2D *Werner deconvolution*, 2D and 3D *Euler deconvolution* and other line- or grid-based methods, the results of this method are independent from the orientation of the magnetic survey lines and length of the calculation operator (Thurston and Smith, 1997). See also *grid*, *advanced source-parameter imaging method*, *improved SPI™ method*, and *enhanced local wavenumber method*.

source resolution: the minimum horizontal distance between sources that allows the identification of individual anomalies generated by these sources. Source *resolution* is a function of the depth, *shape*, and *density contrast* or *magnetization contrast* of each source. Generally, the anomalies from similar sources at the same depth can be resolved when their sources are separated laterally by a horizontal distance of 1–2 times their depth.

south-seeking pole: the negative pole of a *magnetic dipole* or magnet that points in the direction of the *magnetic south pole*. See also *north-seeking pole*.

space domain: 1. a *domain* where the measured quantity is expressed as a function of the spatial coordinates or distance from some reference point. **2.** a mathematical set of values that defines a physical location in space as the independent variable and corresponding values of a particular function as the dependent variables. For example, a gravity or magnetic *field* is defined by a function which varies depending on three independent space variables x, y, z in *Cartesian coordinates*. See also *frequency domain* and *time domain*.

space-frequency localization: See *wavelet analysis*.

space-varying filtering: *filtering* with the use of specific filters having the *frequency response* adaptable to local variations of the spectral properties of an *anomalous field*. Ridsdill-Smith (2000) describes the application of the space-varying filtering concept to the design of *wavelet*-based filters for the stabilized variable-level *downward continuation* of aeromagnetic data into the subsurface down to the seismic-derived *acoustic basement* depth varying from 170 to 3150 m below sea level. See also *spectrum* and *wavelet transform*.

spaceborne gravity: See *satellite gravity*.

spatial aliasing: 1. distortions of the measured gravity or magnetic field components by *undersampling*, whereby the original high-frequency (short-*wavelength*) components are represented by lower frequency (longer wavelength) substitutes on a *grid*. **2.** *gridding* artifacts that appear in gridded data, when the *line spacing* is larger than a width of the anomaly striking across several *survey* lines. Also referred to as *cross-track aliasing*. See also *anomaly width*, *aliasing*, *boudinage artifact*, *strike*, and *grid aliasing*.

spatial derivative: 1. a three-dimensional (3D) *vector* quantity which defines the orientation and rate of change of an *anomalous field* along the x-, y-, z-axes in *Cartesian coordinates*. Three mutually orthogonal axes of measurements are required to obtain the entire vector *magnitude*. **2.** a *derivative* in 3D space. See also *directional derivative*.

spatial filter: a *filter* operating in the *space domain*. Targeted components of a gravity or magnetic *field* are enhanced by the suppression of all components other than those passed without suppression by this filter. See also *enhanced processing*, *target*, and *spectral filter*.

spatial frequency: 1. a space-domain *waveform* parameter defined by the amount of waveform cycles per unit distance in a given direction or along the coordinate axis in *Cartesian coordinates*. Spatial frequency f is a reciprocal, or inverse, of *wavelength* λ, i.e., $f = 1/\lambda$, so the long- and short-wavelength components of the gravity or magnetic field are also referred to as the low- and high-frequency components, respectively. **2.** the space-domain equivalent of the time-domain *frequency*. **3.** a spectral-domain parameter called the *wavenumber* and expressed as $f = k = 2\pi/\lambda$ in *Fourier transform* applications. See also *space domain, spectral domain*, and *time domain.*

spatial resolution: 1. the shortest distance over which the differences between gravity or magnetic anomalies can be distinguished, or resolved, in a *three-dimensional (3D)* frame of *Cartesian coordinates.* Spatial resolution includes *lateral resolution* and *vertical resolution.* Lateral resolution depends on the *wavelength* content of a measured field: the shorter the measured wavelengths, the higher the lateral resolution of anomalies and, hence, *resolution* of the geological *structure* elements. Vertical resolution depends on the *precision* of the measuring instruments and suppression of *noise*: the higher the precision and the lower the level of noise, the smaller the amplitudes of the detected anomalies. **2.** *grid spacing* in the global-scale grids of gravity and magnetic anomalies, such as the *Earth Magnetic Anomaly Grid-2 (EMAG2),* wherein the lateral resolution is defined in units of *arc-minute.* **3.** the smallest area which is distinguishable on a digital image, such as a *raster image*, and defined by the number of pixels per inch (PPI). In reference to the *wavelength resolution* of measurements, the terms spatial resolution and lateral resolution are often used with the same meaning. See also *grid resolution* and *pixel.*

spatial sampling: measurements at different locations with regular intervals along airborne, marine (shipboard), or submarine *survey* lines or at irregular locations over a *ground survey* area. Measurements, especially those with irregular intervals of *sampling*, can involve *spatial aliasing* problems in the subsequent *gridding* of the acquired data. See also *acquisition.*

Spector-Grant method: the magnetic source depth estimation method based on the use of a *radial power spectrum* which is calculated for the *total field* magnetic data and presented in the logarithmic scale as a function of *wavenumber* or decimal fraction of the *Nyquist frequency.* A *two-dimensional (2D)* radial power spectrum graphical plot shows the amount of power (squared amplitudes) at each wavenumber, and it has the characteristic shape of the descending curve, or decline of the power spectrum, approximated by 3–5 straight-line segments called slopes. Each *slope* represents the squared amplitudes of anomalies associated with a particular wavenumber range, wherein the slope angle relative to the wavenumber axis is directly proportional to the depth of a corresponding assemblage of magnetic sources. The slope-obtained depth value is interpreted as a single approximation of the multiple source depths in an assemblage that are averaged over the total area of the radial power spectrum calculation. For gravity sources, the analogous depth estimates are obtained from the radial power spectrum calculated for the *vertical derivative* of the anomalous gravity field. Spectral shapes that do not exhibit the distinct straight-line slopes are interpreted only qualitatively. Slope intercept points are used to select the parameters of a *matched filter,* also known as a *spectrum-matched band-pass filter.* Wavelengths associated with the depth which is "sliced" by the selection of a particular slope will be enhanced relative to other *wavelength* components, and this property is used in *pseudodepth slicing.* The method is commonly applied for gridded magnetic data. Also known as *spectral depth method* and *spectral slope method.* See also *anomalous field, gridding*, and *spectral leakage.*

spectral analysis: 1. *Fourier analysis.* **2.** spectral-domain analysis of the *spatial frequency* content of an *anomalous field* and components of its *spectrum* computed from Fourier-transformed data. **3.** source discrimination analysis based on the *Spector-Grant method.* Spectral analysis involves two principal data transformation procedures: (1) *forward Fourier transform,* which decomposes the original data into a series of sinusoids of different spatial frequencies, or wavenumbers, wherein each sinusoid has its own *amplitude, phase*, and *wavenumber*; and (2) *inverse Fourier transform* to convert the decomposed data back to their original waveforms in the *space domain.* Decomposed, or Fourier-transformed, data are presented as an *amplitude spectrum, phase spectrum, power spectrum*, and their modifications. They are analyzed for the differential and integral properties of gravity and magnetic fields with various objectives, including: (a) identification of the spectral components associated with the dominant wavelengths at different depth levels; (b) selection of parameters for *filtering*; (c) identification of the dominant spatial trends of anomalies; and (d) evaluation of the specific *data processing* errors and overall *noise* contamination of the data. For example, the graphical plot of a *radial power spectrum* of gridded magnetic data is analyzed for the identification of the linear segments associated with assemblages of magnetic sources at different depths, whereas a *map power spectrum* is analyzed for the evaluation of *leveling* errors, *culturing editing* effects, and other noise-related characteristics of the data. See also *gridding, waveform*, and *spectral domain.*

spectral correlation filtering: *filtering* which passes without suppression the positive-, negative-, or null-correlated *anomalous field* components or attributes which can be identified between two or more datasets based on their correlation characteristics in the *spectral domain.* See also *attribute* and *spectrum.*

spectral decomposition: 1. separation of the space-localized components of a *spectrum* of gridded or line-oriented data by application of *window Fourier transform (WFT).* **2.** *eigenvalue decomposition* of the *FTG matrix* or *Hessian matrix.* **3.** *wavelet analysis.* See also *gridding* and *wavelet transform.*

spectral density: 1. squared *amplitude spectrum.* **2.** *energy-density spectrum.* **3.** *power spectrum.*

spectral depth estimation: two main types of source depth estimates obtained by the *Spector-Grant method* and its versions: (1) the averaged depth of the assemblage of sources determined from a *radial power spectrum* of gridded *total field* magnetic data or the *vertical derivative* of a *Bouguer anomaly* grid; and (2) the depth of a single source of an isolated magnetic anomaly from its *power spectrum* calculated in a square *window* whose length is approximately 4–6 times the expected source depth. See also *gridding*, *spectral leakage*, and *pseudodepth slicing*.

spectral depth method: *Spector-Grant method* which is an approximate source depth estimation method based on the use of the *radial power spectrum*.

spectral depth slicing: same as *pseudodepth slicing* with an emphasis on the obtained *grid*-averaged depth estimates that correspond to the certain ranges of wavenumbers associated with linear slopes in a *radial power spectrum*. See also *wavenumber* and *Spector-Grant method*.

spectral domain: *Fourier domain* or *wavenumber domain* in the context of a mathematical *domain* in which the *spectrum* of Fourier-transformed gravity and magnetic data is computed. See also *Fourier transform*.

spectral-domain method: 1. a method of data transformation which is based on the use of the *Fourier transform*, e.g., *spectral filtering*. **2.** a method of gravity and magnetic data analysis, where parameters of an exploration *target* are estimated with the use of characteristics of its *amplitude spectrum* and *phase spectrum* computed from the Fourier-transformed data in the *spectral domain*. See also *FFT filters*, *Parker formula*, and *Spector-Grant method*.

spectral filter: a *filter* operating in the *spectral domain*, where the *enhancement* of the targeted *wavenumber* components of a gravity or magnetic field *spectrum* is achieved by the suppression of all wavenumber components other than those passed by this filter. See also *FFT filters*, *spatial filter*, and *target*.

spectral filtering: *filtering* of the Fourier-transformed gravity or magnetic data in the *spectral domain*. Spectral filtering of space-domain line-oriented or gridded gravity and magnetic data involves three data transformation procedures: (1) data conversion into the equivalent *wavenumber* components by *forward Fourier transform*; (2) filtering with the use of a particular algorithm which involves the multiplication of each wavenumber amplitude in the Fourier-transformed data by coefficients of a selected *filter* in its pass, reject, or correlation mode; and (3) conversion of the filtered data back to their original *space domain* by *inverse Fourier transform*. Spectral filtering is the most commonly used method of *enhancement* and visualization of the gravity and magnetic anomalies of an exploration or research interest. See also *anomaly*, *FFT filters*, *Gibbs phenomenon*, *grid*, and *spectral correlation filtering*.

spectral leakage: the overlap of the low-frequency spectral components of anomalies generated by sources of different sizes occurring at sufficiently different depth levels. High-frequency components are always associated with shallow sources, whereas the low-frequencies are not necessarily related to the deep sources only, because the laterally extensive shallow structures can generate the low-frequency components contributing to the *shape* and *spectrum* of the observed anomalies. Spectral leakage is an inherent property of gravity and magnetic fields, and it is a source of errors in the spectral depth estimates obtained by the *Spector-Grant method* and its versions. Sometimes referred to as *energy leakage* and *wavenumber leakage*. See also *spatial frequency* and *spectral depth estimation*.

spectral modeling: 1. *modeling* of the sources of the gravity and magnetic anomalies with the use of *Fourier transform*. **2.** area-averaged depth estimates for assemblages of the magnetic sources by *Spector-Grant method*. **3.** FFT-based calculation of the gravity field generated by *model* approximating the real geological *structure*, such as the one used in calculation of *gravitational admittance* from the gridded *bathymetry* or *topography* and calculation of *3D Bouguer correction* for the marine (shipboard) gravity data with *Parker formula*. See also *Fast Fourier Transform (FFT)*, *spectral domain*, and *gridding*.

spectral sensitivity: a quantitative characteristic of data that describes their *amplitude* response, or *sensitivity*, to the components of an *anomalous field* within a particular *wavenumber* range of its *spectrum*, such as the range of *spherical harmonics* in the spherical harmonic expansion of the *gravitational potential* of the Earth in a global *geopotential model*. See also *Earth Gravitational Model 2008 (EGM2008)*.

spectral signature: the spectral-domain *amplitude*, *phase*, or *wavenumber* characteristic by which a gravity or magnetic *anomaly* can be identified and interpreted. See also *Fourier transform*, *spectral domain*, *spectrum*, and *signature*.

spectral slope method: *Spector-Grant method* in the context of using the straight-line segments, or slopes, in the logarithmic plot of the *radial power spectrum* for estimating the *grid*-, and occasionally, line-averaged depths of the assemblages of magnetic sources.

spectral transform: *Fourier transform* that transforms, or converts, gravity and magnetic data from their original *space domain* to the equivalent *spectral domain* and subsequently calculates the *spectrum* of the Fourier-transformed data.

spectrum: 1. *amplitude*, *phase*, and power (squared amplitudes) characteristics of Fourier-transformed gravity and magnetic data as functions of *wavenumber*. **2.** a *two-dimensional (2D)* graphical plot of the amplitude, phase, or power characteristics of gravity or magnetic field components presented in the order of their corresponding wavenumbers in the *Fourier domain*. **3.** an amplitude-versus-wavenumber graphical plot of a *filter response* showing the range of wavenumbers separated by this filter in the pass or reject mode of its operation. The spectrum is commonly computed for gridded data and, sometimes, for the individual profiles or line-oriented data. There are three basic types of the spectrum: (1) *amplitude spectrum*, (2) *phase spectrum*, and (3) *power spectrum*. Also known as *Fourier spectrum*. See also *complex spectrum*, *Fourier transform*, *gridding*, *1D spectrum*, *2D spectrum*, and *spherical harmonic spectrum*.

spectrum-matched band-pass filter: same as *matched filter* that operates in the manner of a *band-pass filter* for which the parameters are selected from the *radial power spectrum* of the gridded data.

spectrum-matched filtering: *matched-filtering method* in the context of selecting its parameters from the slope intercept points in a logarithmic plot of a *radial power spectrum*.

speed-of-light correction: a correction applied to the *absolute gravity* measurements by the *free-fall method* with the use of a laser interferometer and free-falling corner-cube reflector. Correction compensates for the time delay effect of the Doppler *phase* shift, which is produced by the accelerated motion of the corner-cube reflector, on the interference of the reflecting and reference laser beams in a laser interferometer. This effect is a source of systematic errors in the high-*accuracy* measurements that make the measured absolute gravity value to be up to 10 *microgal (μGal)* lower than its actual value at an *observation point*. Sometimes referred to as *relativistic correction*. See also *atmospheric pressure correction*, *diffraction correction*, *polar motion correction*, *self-attraction correction*, and *tilt correction*.

sphere: the *three-dimensional (3D)* approximation of a compact *source body*, such as a small *ore body* and other subsurface source bodies whose vertical and horizontal dimensions are much less than their depth. Source approximation by a sphere is used when (a) an observed anomaly has an equidimensional *shape*, i.e., the anomaly attenuates approximately the same way in all horizontal directions, and (b) the source-to-observation distance is sufficiently long and, thereby, allows the source body of any shape to be considered a *point source*. See also *dimension*, *horizontal cylinder*, *vertical cylinder*, *thick dike*, *thin sheet*, and *prism*.

spherical anisotropic index: a quantitative measure of the degree of *FTG spherical anisotropy* defined as

$$\mathrm{SAI} = |\lambda_3| / |\lambda_1|,$$

where $|\lambda_1|$ and $|\lambda_3|$ are magnitudes of the largest and smallest *FTG eigenvalues*, respectively. Index ranges between 0 and 1, where the values close to 1 indicate the very small lateral changes in the rock *density* and *magnetization* contrasts that are below the *resolution* of the FTG measurements. Due to *normalization* by the same magnitude $|\lambda_1|$, the sum of the linear, planar, and spherical anisotropic indices is equal to 1 at every *grid* point. The relationship between three calculated indices is visualized by a monochromatic or color-coded *barycentric image* to show the variations of the *FTG anisotropy* parameters associated with elements of a subsurface geological *structure* (Hassan and Goussev, 2020). See also *contrast*, *magnitude*, *linear anisotropic index*, and *planar anisotropic index*.

spherical anisotropy: See *FTG spherical anisotropy*.

spherical cap effect: the *gravitational attraction*, or gravity effect, of a *Bouguer spherical cap*. The difference between the gravity effects of a *Bouguer plate* and a Bouguer spherical cap is compensated by application of *Bullard B correction* calculated with the use of the *Cogbill formula* or *LaFehr formula*.

spherical coordinates: a three-dimensional (3D) *coordinate system* where the position of any observation point on the surface of a spherical body or planet, such as the Earth, is determined by its radius r and two angles θ and ϕ equivalent to those of the geographic *colatitude* and longitude, respectively, with $0 \le \vartheta \le \pi$ and $0 \le \phi \le 2\pi$. The relationship between the spherical coordinates and planar *Cartesian coordinates x, y, z* is defined as

$$x = r\sin\theta \cdot \cos\phi, \quad y = r\sin\theta \cdot \sin\phi, \quad z = r\cos\theta.$$

spherical harmonic decomposition: See *spherical harmonic expansion*.

spherical harmonic expansion: representation of the *harmonic function* defined on a spherical surface, such as the global-scale *potential* of the Earth's gravity or magnetic *field*, by a linear combination of weighted *spherical harmonics*, wherein each weighting coefficient quantifies the amount of variation of the function described by its respective spherical harmonic. Integration of discrete gravity and magnetic measurements into the mathematical *format* of a spherical harmonic expansion makes possible the separate description and analysis of the specific components contributing to the observed gravity and magnetic fields over the entire Earth's surface, such as the dipole and nondipole components of the *main geomagnetic field* or stationary and time-varying parts of the *Earth gravity field*, using selective combinations of the weighted spherical harmonics of various degrees and orders. Sometimes referred to as *spherical harmonic decomposition* and *spherical harmonic transform*. See also *temporal gravity variations*.

spherical harmonic spectrum: weighting coefficients indicating the relative significance of their respective *spherical harmonics* in a *spherical harmonic expansion*. By analogy with the *Fourier spectrum*, a set of weighting coefficients is considered as the *spectrum* of a *harmonic function*, such as the gravity or magnetic *potential*. Weighting coefficients can be determined uniquely, if the *gravitational potential* or *magnetic potential* is known on a spherical surface bounding the gravity or magnetic sources. See also *Gauss coefficients* and *Stokes coefficients*.

spherical harmonic transform: See *spherical harmonic expansion*.

spherical harmonics: 1. a set of *orthogonal* basis functions composed of the associated Legendre polynomials and defined in *spherical coordinates*. **2.** the spherical analogue of a planar *Fourier series*. Spherical harmonics are identified by their order m and degree n: if $m = 0$, spherical harmonics depend on the latitude only and are called *zonal harmonics*; if $n = m$, spherical harmonics depend on the longitude only and are called *sectoral harmonics*; if $m > 0$ and $n - m > 0$, spherical harmonics depend on both the latitude and longitude and are called *tesseral harmonics* (Blakely, 1996). Linear combination of the spherical harmonics of the various degrees and orders, wherein each harmonic is weighted by the appropriate coefficient, is used to represent the gravity or magnetic *potential* in global mathematical models, such as the *Earth Gravitational Model 2008 (EGM2008)* and *World Magnetic Model (WMM)*. There is a direct correspondence between the spherical harmonic degree n and spatial *wavelength*: the higher the degree, the higher the *wavelength resolution* of the spherical harmonic representation. The relationship between the spherical harmonic degree *n* and the shortest resolvable wavelength is expressed as

$$\lambda = 2\pi R/n \cong 360° \cdot 111/n \approx 40000/n,$$

where λ is *wavelength* in kilometers (km), and R = 6371 km is the average Earth radius. Also referred to as *surface spherical harmonics.*

spherule: See *magnetic spherule.*

SPI™ method: See *source parameter imaging (SPI™) method.*

SPI™ structural index: the *source geometry* parameter N_{SPI} implemented in the *improved SPI™ method*, defined as

$$N_{SPI} = k_1/(k_2 - k_1) - 1,$$

where k_1 is *first-order local wavenumber* and k_2 is *second-order local wavenumber.* Both local wavenumbers are independent of *inclination*, *declination*, intensity of *main geomagnetic field*, *susceptibility contrast*, and *dip* of two-dimensional (2D) sources. Here, $N_{SPI} = 0$ for a *magnetic contact*, $N_{SPI} = 1$ for a *thin dike*, and $N_{SPI} = 2$ for a *horizontal cylinder* (Smith et al., 1998).

spike: a short-*wavelength* single-point or several-points *anomaly* whose high positive or negative *amplitude* deviates significantly from the local trend and amplitude level of an anomalous gravity or magnetic *field.* Spike anomalies include measurement errors, *positioning* errors, short-duration *diurnals*, and *cultural magnetic anomalies* generated by metal bridges, isolated buildings, oil rigs, well heads, sunken ships, etc. See also *high-line effect.*

spike filtering: elimination of *spike*-like anomalies in measured data by (a) replacing their abnormally high or low values with the average or median value of the surrounding measurement points and (b) applying a *wavelet denoising* algorithm. Also referred to as *despiking.*

spinner magnetometer: a *magnetometer* for measurements of *normal remanent magnetization (NRM)* in rock samples. An equidimensional sample is rotated about a fixed axis inside a ring-shaped magnetometer. The amplitude and *phase* of the magnetic field in a rock sample are used to determine the intensity and direction of the NRM component orthogonal to the spin axis. Total *remanent magnetization* and its direction are determined by spinning (rotating) a rock sample in six orthogonal orientations. See also *SQUID magnetometer.*

spin-precession magnetometer: a *magnetometer* from the group of *quantum magnetometers*, such as *Overhauser magnetometer* and *proton-precession magnetometer*, whose operational principle is based on measurements of the *angular frequency* of the *precession* of spinning hydrogen nuclei (protons), i.e., measurements of the spin-precession angular frequency of protons.

spline: a polynomial function of the order *m* that approximates the whole dataset, satisfies all specified point values, and maintains the continuity of their derivatives up to the order *m*–1, e.g., the quadratic spline has a continuous first-order derivative and the cubic spline has the continuous first- and second-order derivatives. Splines are used in (a) curve-fitting methods, (b) predicting the *drift* value between measurement points, and (c) interpolation between discrete data values in *gridding* algorithms. Approximations of gravity and magnetic fields by polynomials of different orders are used to determine the regional components of anomalous fields in the process of *regional-residual separation.* See also *anomalous field*, *polynomial fitting*, and *polynomial residual.*

spring balance: the design concept of a *spring gravimeter* wherein the weight of a hinged beam with a *proof mass* on its free end is balanced by the tension of a mechanical spring, so the relative change in the measured gravity field results in a change of the spring length. See also *zero-length spring gravimeter.*

spring-balanced gravimeter: same as *spring gravimeter* with an emphasis on the *spring balance* concept implemented in its design. See also *zero-length spring gravimeter.*

spring gravimeter: a *relative gravimeter* which measures the amount of change in the equilibrium position of a *proof mass*, suspended by a metal or quartz spring, that is caused by change of the *Earth gravity field* between the measurement points. In its simplest design, the proof mass is hanging at the end of a vertical spring, such as in a *straight-line gravimeter.* More sophisticated designs are based on the use of a proof mass at the end of a lever arm with an inclined spring and *astatic balance* for the higher *sensitivity*, such as the one implemented in a *zero-length spring gravimeter.* The stretch of the metal or quartz spring under the varying *gravitational force* acting on the proof mass is used to determine the *relative gravity* value. Depending on the gravimeter model, this value is obtained by measuring (a) the change in the physical length of the spring, i.e., spring tension; (b) the deflection of the proof mass from its reference position, or *null point*, using the light beam reflecting off a mirror connected to or mounted on the proof mass; and (c) the restoring force in a capacitive or electromagnetic feedback system, such as the one implemented in an *accelerometer*-based design, that is required to bring the proof mass back to its null point. The spring gravimeter on a moving *observation platform* measures the combined effect of both the *gravitational acceleration* and *inertial acceleration* components that are parallel to its sensing axis. Inertial accelerations are determined by independent accelerometer systems or calculated from high-accuracy *differential GPS (DGPS)* data and their values are subtracted from values measured by the spring gravimeter. Achievable *accuracy* of the modern zero-length spring gravimeters is up to 1 *microgal (μGal).* Sometimes referred to as *spring-balanced gravimeter* and *mass-spring gravimeter.* See also *gravity sensor*, *virtual spring*, and *superconducting gravimeter.*

spring hysteresis: the spring relaxation effect which occurs when spring tension is changed, such as in the process of measurements, and then returned to the original tension at *null point.* A quartz or metal spring tends to maintain its last position, so the spring length is not restored instantly but only after some time, thereby resulting in a delayed response of the *spring gravimeter.* This effect also can be caused by multiple resets and clamping of the *proof mass* for the transportation between measurements. Also known as *elastic hysteresis*. See also *hysteresis*.

SQUID magnetic tensor gradiometer: a triaxial *magnetic gradiometer* for measurements of the *full tensor gradient (FTG)* and *vector* components of the *Earth magnetic field*. Its design is based on the use of three rotating single-axis SQUID sensors in an umbrella configuration: a central sensor rotates about its vertically oriented axis, and two side sensors rotate about their axes tilted at 45° to the vertical. The different high-frequency rotation of sensors minimizes the cross-interference *noise* and reduces the effects of the low-frequency mechanical vibrations of a moving *observation platform*. Fourier-domain decomposition and other mathematical transformations of the acquired data are used to separate the vector components from the FTG components, so that all three vector components and nine FTG components of the Earth's magnetic field can be recovered from the measurements by three single-axis SQUID sensors. See also *axial magnetometer, Fourier domain, Fourier transform*, and *SQUID sensor*.

SQUID magnetometer: an *axial magnetometer* whose *sensor* is a superconducting quantum interference device, or SQUID, which has the super-high *sensitivity* of 10^{-5} *nanotesla (nT)* to very small changes in the *Earth magnetic field*. The SQUID magnetometer is equipped with mutually orthogonal sensing modules and can be used as a *vector magnetometer* and *magnetic gradiometer*. There are versions designed to measure the *remanent magnetization* of rock samples. Sometimes referred to as *cryogenic magnetometer*, because the superconducting loop in the sensing module is maintained at temperatures below −180°C. Also known as *superconducting magnetometer* and *quantum vector magnetometer*. See also *SQUID sensor* and *SQUID magnetic tensor gradiometer*.

SQUID sensor: a *magnetic sensor* capable of detecting very small changes in the strength and direction of a planetary *magnetic field* by a superconducting quantum interference device, or SQUID, which has its *sensitivity* within the *femtotesla (fT)* range and dimensions of a few cubic centimeters. A SQUID sensor includes a superconducting loop with one or two non-conductive barriers, called Josephson junctions, that are too thin to prevent completely the flow of electrons. Measurements are based on the property of the through-barrier-flow of electrons to oscillate in response to the applied voltage with an oscillation *frequency* proportional to the ambient magnetic field strength, wherein the *phase* difference between the oscillations at two barriers on opposite sides of a superconducting loop is proportional to the change of the ambient magnetic field in the direction of the barriers' separation. The loop axis can be oriented to measure the *vector* component of the field orthogonal to the superconducting loop. Superconductive properties of a loop are maintained at temperatures below −180°C. The sensor is used in the measuring modules of a *SQUID magnetometer* and *SQUID magnetic tensor gradiometer*. See also *Josephson junction* and *ambient field*.

SRM anomaly: a low-*amplitude* short-*wavelength* sedimentary residual magnetic (SRM) *anomaly* remaining in a *total magnetic intensity (TMI)* or *RTP magnetic field* after *cultural editing* and suppression of the high-amplitude magnetic field components generated by the intrasedimentary igneous intrusions and heterogeneous *basement* rocks. Line-to-line correlated clusters of SRM anomalies may be indicators of *a magnetic bright spot* associated with the *hydrocarbon seepage effect* or accumulations of *detrital magnetite* in *placer* deposits. See also *chimney, clustering, igneous intrusion*, and *residual anomaly*.

stabilized downward continuation: 1. *downward continuation* applied jointly with some preceding *smoothing* procedure in the manner of *cascaded filtering*. The stability of the process is secured by one of the low-pass *filtering* operations, such as the high-cut option of a *Butterworth filter* or *match filter*, that eliminates the high-frequency (short-*wavelength*) components causing instability in the downward continuation. **2.** downward continuation between uneven surfaces, such as between an *airborne survey* flight *altitude* and ground surface, with the use of the *equivalent-source method*. **3.** variable-level downward continuation into the subsurface with the use of *wavelet*-based *space-varying filtering*. See also *high-cut filter*.

stabilized platform: a mechanical self-leveling platform on which the *gravimeter* or *gravity gradiometer* is mounted during marine (shipboard) and airborne gravity surveys. Its design is based on the concept of a damped, or slow-swinging, long-*period* pendulum. The platform is supported with gimbals and controlled by gyroscopes coupled with *accelerometer*-activated servomotors on each gimbal axis to counteract the rotational motions of a *survey* ship or aircraft, such as *roll*, *pitch*, and *yaw*, and to keep the platform horizontal, or levelled, within the range of a few arc-seconds. Also referred to as *inertial platform* and *attitude compensation platform*. See also *arc-second, damping*, and *Schuler-tuned inertial platform*.

stacked profiles: a method of the visualization and line-to-line correlation of low-*amplitude* short-*wavelength* anomalies, such as *SRM anomaly, micromagnetic anomalies*, diagenetic and detrital magnetic anomalies associated with ancient fluvial systems, including the anomalies generated by the localized accumulations of reservoir-grade sands, beach and alluvial placers, etc. Anomalies are displayed by a series of profiles of the high-pass-filtered magnetic or gravity field along the selected *survey* lines, wherein the location of each selected line represents the *baseline* for the displayed low-amplitude anomaly. The screen image or printed map of the stacked profiles mimics a *3D view* and highlights the very small anomalies that might be difficult to identify on a regular *grid*. See also *high-pass filter, detrital magnetite, diagenetic magnetite*, and *placer*.

stacked sampling: summing up, or *stacking*, data samples recorded with a short *sample interval* along an airborne or marine (shipboard) *survey* line. It is based on the use of a *moving window* of a pre-selected length and applied to suppress *random noise* and increase the *signal-to-noise ratio (S/N)*. In the stacking of *n* samples, the coherent in-*phase* signal components are amplified by $\sqrt{n}$, whereas the out-of-phase uncorrelated *noise* components tend to cancel out each other. See also *coherence*, *line averaging*, and *smoothing*.

stacking: a *data processing* technique based on the summing of (a) several data grids obtained by *upward continuation* or *differential continuation* within a specified continuation range and (b) data samples of line-oriented data, such as those obtained by the high-speed *sampling* of airborne or shipboard (marine) measurements, in order to suppress *random noise* and increase the *signal-to-noise ratio (S/N)* of the acquired data. See also *acquisition*, *grid*, and *stacked sampling*.

standard deviation (SD): the square root of the average of the sum of the squared differences between the individual values in a given dataset and their arithmetic mean, or average, value, expressed as

$$SD = \{\Sigma(x_i - m)^2/n\}^{1/2},$$

where m is the average value, x_i is a measured value with $1 < i < n$, and n is total number of measured values. Standard deviation is a measure of *precision*, not *accuracy*, of measurements, and it is often used in estimates of the upper bound of *noise* in acquired data with the assumption that noise is random within a *bandwidth* of the measurements. The square of the standard deviation is known as the *variance*. Often referred to with the same meaning as *standard error*, *RMS error*, *RMS value*, and *RMS noise*. See also *acquisition* and *random noise*.

standard error: See *standard deviation (SD)*.

standard gravity: a numerical value of *gravitational acceleration* due to the *gravitational attraction* of a *reference ellipsoid* approximating the Earth's mass. It is accepted by the International Bureau of Weights and Measures (BIPM) as the standard reference value of the theoretical *Earth gravity field*. The standard gravity value is equal to $9.80665\ m \cdot s^{-2} = 980.665\ cm \cdot s^{-2} = 980.665$ *Gal* = 980665 *milligal (mGal)* = 980665000 *microgal (μGal)*. Also known as *normal gravity*.

standard gravity formula: the *International Gravity Formula (IGF)* for calculation of the theoretical *gravity field* generated by a homogeneous rotating *reference ellipsoid* approximating the heterogeneous Earth. This field is referred to as *theoretical gravity* and is used as a standard to which the measured gravity values are referenced in order to detect the anomalies of an exploration or research interest. See also *anomaly*.

standard vertical gravity gradient: a *free-air gradient* that represents the vertical rate of change of the theoretical, or standard, gravity field of a rotating homogeneous *reference ellipsoid* due to a change in the distance between an observation point and an ellipsoidal center of mass, regardless of the local or regional geological *structure*.

static correction: 1. a correction applied to airborne gravity measurements to compensate for (a) the *drift* of an *accelerometer*-based *airborne gravimeter* and (b) systematic errors due to irregular changes in the airborne system control parameters. **2.** *drift correction* for airborne gravity data. The static correction value is determined from the results of a daily *static test*. See also *gravity-derived static corrections*.

static gravimeter: a *relative gravimeter* which measures changes in the *Earth gravity field* by balancing the force of gravity, or *gravitational force*, against the known and adjustable static force, such as the extended spring and electrostatic (capacitive) or electromagnetic force, utilized in the design of a *spring gravimeter*, *superconducting gravimeter*, and *accelerometer*-based airborne gravimeters. Except for the *vibrating-string gravimeter*, all currently used gravimeters for measurements of *relative gravity* are static gravimeters. See also *dynamic gravimeter*.

static reading: See *still reading*.

static test: daily pre-flight and post-flight digital records, or readings, of the *accelerometer*-based *airborne gravimeter* system at a fixed airfield location with a known local *relative gravity* value which is tied to the national gravity standardization network. A static test is conducted in order to (a) determine the amount of *drift* in the airborne gravity measurements and (b) identify the systematic errors, or data offsets, due to irregular changes in the airborne system control parameters. To meet the survey *acceptance criteria*, the deviations of the pre-flight and post-flight static readings from a local gravity value should not exceed 1 *milligal (mGal)*. A static test is a part of the daily *airborne gravimeter calibration* procedures, and its results are used for calculation of *static correction*. See also *International Gravity Standardization Network 1971 (IGSN71)* and *still reading*.

station: a pre-planned position on the ground or sea bottom with known geographic coordinates, where a *gravimeter* or *magnetometer* is set up to measure the Earth's gravity or magnetic field.

station density: the ground gravity or magnetic *survey* parameter that defines the average number of observation points, or stations, per square kilometer in a *ground survey* area.

station gravity: 1. *raw gravity*. **2.** *observed gravity* data obtained from the measured gravity values, or raw gravity, at each *station* in a ground *gravity survey* area.

station spacing: the distance between measurement points in a ground gravity or magnetic *survey*. Differing from marine (shipboard) and airborne surveys, the station spacing can be highly irregular, and it depends on the accessible pathways and assumed depths and sizes of the geological structures generating the gravity and magnetic anomalies. To avoid *aliasing*, the average station spacing is selected to be less than one-half of the expected depth to the shallowest *target*. See also *anomaly* and *structure*.

Steenland-Vacquier rule: See *straight-slope method*.

steerable filter: a *grid*-based *oriented filter* for which the output is synthesized by a linear combination of the basis filters designed to detect the edges (abrupt changes in data values) and ridges (*peak* values) in gridded data using the first-, third-, and fifth-order derivatives of the *Gaussian function*. A steerable filter highlights the linear and curvilinear alignments of gravity and magnetic anomalies by correlating their edges and ridges along all 360° azimuthal directions relative to each of the *grid nodes*. Orientation-dependent filtering results are visualized in the *pixel*-based binary format of *skeletonization*. Steerable filters can be designed in a *quadrature pair* to make the filter output a function of *azimuth* and *phase* (Hassan and Peirce, 2007). See also *derivative*, *edge-detector attribute*, *gridding*, and *radial sunshading*.

steered function: a function which is presented by a linear combination, or sum, of the rotated versions of itself. The concept of a steered function is implemented in the design of a *steerable filter* for application to gridded gravity and magnetic data. See also *gridding*.

steradian: the dimensionless unit of a *solid angle*. One steradian corresponds to the surface area of r^2 on a sphere with radius r. For the whole sphere surface of $4\pi r^2$, its full solid angle is equal to 4π steradians. See also *radian*.

still reading: 1. the stationary digital record, or *reading*, of a marine (shipboard) or airborne *gravimeter* obtained under a stable environment, such as a pier in a sea port or fixed location at an airfield. Several still readings are obtained during ground, borehole, marine, and airborne gravity surveys to assess and compensate *gravimeter drift* and facilitate a tie of the measured *relative gravity* values to *absolute gravity* in a regional or national gravity network. Also referred to as *static reading* and *still value*. **2.** the reference relative gravity value at a *relative base station* from which the *null point* value of the *relative gravimeter* is determined. See also *base constant* and *International Gravity Standardization Network 1971 (IGSN71)*.

still value: See *still reading*.

stinger: a rigid long pole, or beam, extending from the tail of a fixed-wing airplane or attached to the landing skids of a helicopter. A stinger holds either the single-*sensor* total field *magnetometer* or multiple-sensor *magnetic gradiometer* to keep the magnetic sensors away from the electromagnetic environment of the survey aircraft. See also *magnetic sensor*, *slung sensor*, and *towed bird system*.

stochastic inversion: *grid*-based three-dimensional (3D) *inversion* of the residual *free-air gravity* or *Bouguer gravity*, wherein all variables, including the *residual gravity* values and unknown 3D *voxel model* parameters, are considered as the mathematical sequences of the random variables described by means of the multivariate probability distributions which satisfy the *Bayes theorem*. A subsurface model volume is divided into elementary unit volumes, or voxels, and each *voxel* is characterized by two random variables: *density* and "label," which is a parameter of the depth structure. Available geological and geophysical data are presented in terms of the initial, or prior, probability distributions at all *grid nodes* for every model layer. Solution of an *inverse problem* is obtained with the use of the *Markov chain Monte Carlo (MCMC)* optimization algorithm which samples randomly the density-label pairs from their *conditional probability* distributions to find the combinations of the density and structural (label) parameters for which the posterior probability distributions are maximized and the estimated *model misfit* is within a pre-selected *tolerance*. Unlike a traditional *property inversion* or *structural inversion*, the stochastic inversion iteratively inverts the residual gravity for both the rock formation properties (*density contrast*) and geological *structure* (rock formation depth) in the same inversion process, because the relationship between the property and structural probabilistic inversion solutions depends on the neighboring voxels only (Rossi, 2017). Also referred to as *Bayesian stochastic inversion*, *Markov chain inversion*, and *stochastic modeling*. See also *Markov chain*.

stochastic modeling: same as *stochastic inversion* which is a method of *quantitative interpretation* of gravity data that is based on the use of a three-dimensional (3D) *voxel model*, where the *model* parameters are considered as the mathematical sequences of random variables and described by means of the multivariate probability distributions that satisfy the *Bayes theorem*.

stock: a vertical or high-angle *igneous body* with a highly variable lateral extent (sometimes, up to 60–80 km^2) and unknown bottom. Due to the strong *density* and *magnetization* contrasts with the surrounding sedimentary or metamorphic rocks, the intrasedimentary and upper *basement* stocks commonly generate high-*amplitude* gravity and magnetic anomalies. See also *anomaly*, *contrast*, *infinite bottom*, and *sill*.

Stokes coefficients: weighting coefficients in the *spherical harmonic expansion* of *gravitational potential* that quantify the relative importance, or weight, of each of the *spherical harmonics* representing the components of the gravitational potential. They can be determined uniquely, if the gravitational potential values are known on a spherical surface, such as the Earth's surface, bounding the gravity sources. See also *Gauss coefficients*.

stopband: a *filter* parameter which defines a range, or *band*, of the space-domain wavelengths or spectral-domain wavenumbers suppressed or attenuated substantially by the application of a filter. The part of the input data that is retained, i.e., passed through without *attenuation*, is called the *passband*. See also *space domain*, *spectral domain*, *wavelength*, and *wavenumber*.

stopping criterion: a *model misfit* value, such as the one equal to the estimated *accuracy* of measurements, at which the process of *inversion* is terminated, i.e., stopped. It is used with the same meaning as *tolerance* in the forward and inverse *modeling* calculations. See also *forward modeling* and *inverse modeling*.

straight-line gravimeter: a type of *spring gravimeter* in which the *proof mass* is hanging at the end of a vertical spring, and the displacement of the proof mass, i.e., change of the mechanical spring length, is proportional to a change in the measured *Earth gravity field*. See also *zero-length spring gravimeter*.

straight-slope distance method: same as *straight-slope method* with an emphasis on measurements of the horizontal distance over which the *slope* of an anomaly *profile* before and after an *inflection point* is approximately a straight line.

straight-slope length: the horizontal distance over which the *profile* of a magnetic anomaly before and after the *inflection point* is approximated by a straight line corresponding to the maximum *slope* of the *anomaly* and, hence, a constant *gradient* value in the profile direction. It is one of two basic parameters in estimating the source depth by the graphical *straight-slope method.* Also referred to as *maximum-slope distance.*

straight-slope line: a straight line drawn through the *inflection point* of a *principal profile* that approximates the maximum *slope* of a *magnetic anomaly* and, therefore, a constant value of its *gradient* in the profile direction.

straight-slope method: a *profile*-based graphical method of estimating the magnetic source depth from the measured *straight-slope length* divided by the source index value. Source indices are 1.9 for the horizontal or vertical *thin sheet*, 1.7 for the vertical *thick sheet*, 1.4 for the plug-like body, 1.2 for the top of *contact*, and 1.5 is the *default value.* The straight-slope method can be used for the "quick-and-dirty" estimate of the prismatic source depth, defined as

$$D \approx X_{SSD}/0.8,$$

where X_{SSD} is the straight-slope distance measured perpendicular to the contour lines at the flank of an elongated anomaly on a *contour map*, where the contour lines are relatively equally spaced. Also referred to as *Steenland-Vacquier rule*, *Vacquier rule*, *maximum-slope method*, and *straight-slope distance method.* See also *Peters half-slope method*, *Sokolov method*, *demi-pentes method*, and *prism*.

stratigraphic trap: See *trap.*

strike: 1. the direction of the long axis of a gravity or magnetic *anomaly* on a map or *grid.* **2.** the azimuthal direction of a structural surface, such as a formation top, *fault* plane, or intrusive body, where it intersects the horizontal plane. Strike is perpendicular to *dip.* **3.** the orientation of a two-dimensional (2D) *model* with respect to the geographic north. See also *azimuth*, *strike correction*, and *principal profile.*

strike-angle invariant: a combination of the *full tensor gradient (FTG)* components for estimating the *strike* direction of two-dimensional (2D) geological structures and enhancing the dominant trends of anomalies in the manner of *directional filtering.* It is defined as

$$\tan 2\theta_S = 2[T_{xy}(T_{xx}+T_{yy})+T_{xz}T_{yz}]/\left(T_{xx}^2-T_{yy}^2+T_{xz}^2-T_{yz}^2\right),$$

where θ_s is strike angle, or *azimuth*, and T_{ij} are the FTG components. Linear and curvilinear alignments of anomalies enhanced by calculation of the strike-angle invariant at each *grid* point are referred to as *strike lineaments* (Dickinson et al., 2009). See also *enhanced processing*, *invariants*, and *lineaments.*

strike correction: 1. a correction applied to the calculated source depth to compensate for the non-orthogonal *strike* of a source body relative to the direction of the *profile* of a gravity or magnetic *anomaly* used in *graphical source depth estimation.* The strike correction value ΔD_{SC} is defined as

$$\Delta D_{SC} = D\sin\varphi,$$

where D is initial depth estimate and φ is angle between the profile direction and source body strike, i.e., direction of the long axis of a *source body.* When the angle β between the actual profile direction and *principal profile* is considered, the strike correction is referred to as *cosine depth correction* and defined as

$$\Delta D_{CDC} = D\cos\beta.$$

2. rotation of the original coordinate axes of a *full tensor gradient (FTG)* in order to compensate the measured gradient tensor components for the non-orthogonality of the angle between the *two-dimensional (2D)* source body strike and direction of the *survey* lines, so the corrected FTG data can be used for more accurate estimates of the source body parameters (Dickinson et al., 2009). See also *rotated gradient tensor* and *FTG rotation.*

strike filter: same as *directional filter* with an emphasis on its property to pass or reject components of the gravity or magnetic *field* along a particular azimuthal direction, such as *strike.* See also *azimuth.*

strike lineaments: *lineaments* in a gridded gravity or magnetic field that are obtained by calculation of the *strike-angle invariant* at each *grid* point. See also *gridding*, *edge lineaments*, and *line lineaments.*

strike-slip fault: a *fault* across which the movement, or shifting, of rock formations relative to each other is parallel to the *strike* of the fault and predominantly horizontal. Convergent movement of rock formations along a strike-slip fault creates the positive *flower structure*, whereas their divergent movement creates the negative flower structure and *pull-apart basin.* A regional-scale strike-slip fault is often referred to as a wrench fault. Strike-slip faults have their most distinct signatures in the fields of gravity and magnetic *gradient* anomalies. See also *anomaly*, *gravity gradient*, *magnetic gradient*, and *signature.*

striping: See *geomagnetic striping.*

stripped gravity: the remaining part of *Bouguer gravity* or *free-air gravity* after removing, or stripping-off, the model-calculated gravity effects of the sedimentary layers or isolated geological structures whose depths and densities are known from seismic and borehole data. In two-dimensional (2D) and three-dimensional (3D) *inversion*, the stripped gravity is calculated in order to (a) improve the *lateral resolution* of the inversion results and (b) enhance the anomalies from sources modeled with incorrect *density* values or not included into *model.* For example, the stripping-enhanced local-scale positive *anomaly* on a *difference gravity* profile or *grid* can be generated by a previously unknown *structural high* or buried *reef*, whereas an enhanced local negative anomaly can be related to an undetected *structural low* or accumulation of low-density reservoir sands in an *incised valley* or *paleochannel.* See also *enhanced processing*, *gravity stripping*, *stripping boundary*, and *salt residual.*

stripping: See *gravity stripping* and *inversion-based stripping*.

stripping boundary: **1.** the seismic horizon or sedimentary formation top above which the depth *structure* and *density* distribution of the overlaying sediments are known or approximated, so their gravity effect, or *gravitational attraction*, can be calculated and then eliminated by *gravity stripping* in order to highlight the anomalies from sources below this boundary. **2.** the bottom of the local-source volume in an *inversion-based separation*.

stripping correction: same as *geological correction* for removing, or stripping-off, the *model*-calculated gravity effects of all known or approximated elements of the subsurface geological *structure* in a *gravity survey* area.

stripping filtering: a *filtering* technique based on the use of upward continuations which remove, or strip-off, the local gravity and magnetic anomalies of shallow origin in order to enhance the regional-scale anomalies and anomalies generated by deep sources. See also *enhanced processing*, *Jacobsen filter*, and *upward continuation*.

structural grain: *fabric* or *grain* in the context of the area-specific characteristics of anomalies generated by the geological structures and structural discontinuities within a sedimentary section and upper *basement*.

structural high: a part of the Earth's *crust* that is shallower, i.e., structurally higher, than surrounding rock formations of the same geological age. It can be associated with *anticline*, *carbonate platform*, positive *flower structure*, *horst*, *thrust*, *uplift*, and other local- or regional-scale structurally elevated rock formations. Gravity and magnetic anomalies associated with the intrasedimentary and *basement* structural highs are enhanced by *regional-residual separation* and other *enhanced processing* techniques. See also *anomaly*, *gravity high*, *magnetic high*, and *structural low*.

structural index (SI): See *Euler structural index (SI)* and *fractional structural index (fSI)*.

structural interpretation: *interpretation* of gravity and magnetic anomalies while focusing on the elements of the subsurface geological *structure* of an exploration or research interest, such as faults, fracture and shear zones, lithological boundaries, *basement* depth, *structural high*, *structural low*, deep and shallow igneous intrusions, and prospective mineral ore bodies. Structural interpretation results provide a basis for the selection of *exploration leads* in accordance with the *survey* objectives for oil-and-gas, mineral, and hydrothermal exploration. See also *fault*, *igneous intrusion*, *ore body*, *shear zone*, and *structural inversion*.

structural inversion: a method of *quantitative interpretation* that is based on the *inversion* of the residual gravity or magnetic field for a *structure* of an exploration *target*, such as the *shape* of a shallow *ore body*, base of *allochthonous salt*, and depth of *basement*, for which a physical property (*density* or *susceptibility* contrast) is known from the available information or approximated by analogy-based assumptions. Structural inversion is complementary to *property inversion*, in which the assumably known parameter of an inverted *model* is the target structure. Sometimes referred to as *interface inversion*. See also *anomalous field*, *contrast*, and *stochastic inversion*.

structural low: a part of the Earth's *crust* that is deeper, i.e., structurally lower, than surrounding rock formations of the same geological age. It can be associated with *syncline*, *graben*, *pull-apart basin*, negative *flower structure*, and other local- or regional-scale structural depressions. Gravity and magnetic anomalies associated with intrasedimentary and *basement* structural lows are highlighted by *regional-residual separation* and other *enhanced processing* techniques. See also *anomaly*, *gravity low*, *magnetic low*, and *structural high*.

structural model: a *profile*-based two-dimensional (2D) *model* or *grid*-based three-dimensional (3D) model of a subsurface geological *structure* showing the assumed distribution of *density* and *susceptibility* of rocks within the modeled rock layers and *fault*-bounded blocks. A structural model is developed from gravity, magnetic, seismic, and well (borehole) data or approximated by analogy with the known geological structures in a *survey* region. See also *forward model*, *inverse model*, *full-Earth model*, *2.5D model*, and *2.75D model*.

structural trap: See *trap*.

structure: **1.** the subsurface composition, or configuration, of rocks formed in the process of sedimentation and subsequent tectonic deformations, such as faulting, folding, thrusting, *salt* or shale diapirism, intrusions of igneous bodies, etc. **2.** the regional or local subsurface arrangement of rocks (e.g., *fault*, fold, *thrust*, *anticline*, salt or shale *diapir*, *igneous intrusion*, *ore body*, *reef*, *subcrop*, basement *uplift*, etc.) that can be associated with oil-and-gas accumulation or mineralization conditions favorable for mineral deposits. **3.** part of the Earth's *crust* characterized by a particular rock configuration formed as a result of tectonic deformations. **4.** the geological object of an exploration or research interest. **5.** variations of a physical property (*density* or *magnetization*) in the spatial directions. See also *block structure*, *trap-door structure*, *target*, and *prospect*.

subcrop: the subsurface erosional truncation of a geological *structure*, such as *anticline* or *syncline*, or dipping layers of sedimentary rock formations that is covered by younger overlaying sediments. See also *dip*, *unconformity*, and *outcrop*.

subcrop anomaly: a gravity or magnetic *anomaly* associated with a *subcrop* of the sedimentary formations or geological *structure*. Upper- and mid-sedimentary subcrops of *magnetite*-rich layers generate high-*amplitude* magnetic anomalies, which are correlated for mapping lithological boundaries and truncated anticlinal structures. Possible stratigraphic traps associated with the reservoir-grade sands of a truncated *anticline* can be identified by the correlation of low-amplitude negative *residual gravity* anomalies between the lithological contacts mapped by a *high-resolution aeromagnetic survey*. See also *contact* and *trap*.

submarine gravimetry: instruments and methodology for measuring the relative changes in the *Earth gravity field*, or *relative gravity*, with the use of an *autonomous underwater vehicle (AUV)*, also known as a *sea-floor* *drone*. Submarine gravity data are acquired along the regularly spaced survey lines, or traverses, in a *submarine survey* area, where the AUV (a) follows the software-controlled pre-programmed traverses in shallow waters or (b) operates under remote control by an acoustic communication system in deep water areas. The achievable *vertical resolution* and *lateral resolution* of the gravity anomalies are much higher than those of the shipboard measurements on the ocean/sea surface, especially in the deep water areas, because the AUV-based measurements are made at low heights (~50 m) above the ocean/sea floor under the reduced *dynamic noise* environment of the deep waters and slow-moving (~3.7 km · h^{-1}) AUV (Ishihara et al., 2018). See also *acquisition* and *gravity anomaly*.

submarine gravity: *relative gravity* data obtained from measurements by an *autonomous underwater vehicle (AUV)* along the regularly spaced *acquisition* lines in a *submarine survey* area at low heights (~50 m) above the ocean/sea floor under the reduced *dynamic noise* environment of the deep waters. The achievable *amplitude resolution* is 1–2 mGal with *rms crossover difference* of 0.1 mGal after *leveling*, wherein the *wavelength resolution* is limited by the deep water *positioning* accuracy and estimated at approximately 100 m (Ishihara et al., 2018). See also *airborne gravity*, *ground gravity*, *marine gravity*, and *satellite gravity*.

submarine magnetics: *Earth magnetic field* data obtained from measurements by an *Overhauser magnetometer* mounted inside a *fish* and towed behind an *autonomous underwater vehicle (AUV)* during a submarine magnetic survey for the discovery and mapping of (a) accumulations of magnetic spherules and mineral nodules on the deep sea and ocean floors or (b) *unexploded ordnance (UXO)*, pipelines, power and communication cables in shallow waters prior to offshore or coastal construction. See also *magnetic spherule*, *mineral nodule*, and *submarine survey*.

submarine magnetometry: instruments and methodology for measuring the *Earth magnetic field* and its gradients with the use of an *autonomous underwater vehicle (AUV)*, also known as a *sea-floor drone*. Submarine magnetic data are acquired along the pre-planned traverses in a *submarine survey* area. The achievable *vertical resolution* and *lateral resolution* of magnetic anomalies are much higher than those of the shipboard data, especially in the deep sea and ocean areas, because the AUV-based measurements are made at low heights (~50 m) above the ocean/sea floor and under reduced *dynamic noise* conditions. See also *acquisition*, *magnetic anomaly*, and *traverse*.

submarine survey: measurements of the *Earth magnetic field* or relative changes in the *Earth gravity field* and its gradients along the *acquisition* lines in the area of an exploration or research interest with the use of an *autonomous underwater vehicle (AUV)*, which can navigate following the software-controlled pre-programmed traverses in shallow waters or under remote control by an acoustic communication system in deep water areas. See also *AUV gravimeter*, *AUV magnetometer*, *gradient*, *traverse*, *submarine gravity*, *submarine magnetics*, *airborne survey*, *ground survey*, and *marine survey*.

subsampling: See *downsampling*.

substorm: See *magnetic substorm*.

sun angle: a *sunshading* parameter that specifies the *sun declination* and *sun inclination*. Changing the sun angle creates "shadows" on a sunshaded *grid* that emphasize different anomalous trends in the gridded gravity or magnetic field. See also *anomalous field*, *gridding*, and *trend*.

sun declination: a *sunshading* parameter that defines the *azimuth* of a straight line from an *overhead point* to the position of an artificial source of illumination, or "sun." Sun declination of 0° means that the "sun" is positioned to the north from an overhead point, and 90° means that the "sun" is positioned to the east from an overhead point.

sun elevation: *sun inclination* in the context of the height of an artificial source of illumination, or "sun," above a horizon.

sun inclination: a *sunshading* parameter that defines the vertical angle between the position of an artificial source of illumination, or "sun," and the horizon. Sun inclination of 0° corresponds to a "sun" position on the horizon, and 90° is for the "sun" at the *overhead point* directly above a reference site. Also referred to as *sun elevation*.

sunshading: the image processing technique which involves the calculation of two mutually orthogonal gradients of data and creates the three-dimensional (3D) effect of the illumination of a gravity or magnetic *grid* by an artificial source of light, or "sun," from different directions, as if the grid represents a topographic surface of "hills" (high grid values) and "valleys" (low grid values). Each direction of illumination is defined by (a) *azimuth*, or *sun declination*, and (b) height of the "sun" above the horizon, or *sun inclination*. The interactive sunshading process updates in *real-time mode* every change of the "sun" position, thereby allowing the instant illumination and visual enhancement of the anomalous trends with the different orientations. Sunshading highlights the high-*gradient* areas and low-*amplitude* residual anomalies, whereas the regional anomalies are relatively attenuated. Sunshading is also useful in the visualization of *data processing* errors, such as *leveling errors*, *grid merging* effects, *acquisition footprint*, and other line-oriented and *gridding* artifacts. Also known as *artificial sun illumination*, *false-sun illumination*, *hillshading*, *shaded relief*, *shadowgraph*, and *shadowing*. See also *artifact*, *regional field*, *residual field*, *trend*, and *radial sunshading*.

sunshading filter: an image processing operator of the *directional filter* type implemented in *sunshading*. A sunshading filter highlights, primarily, the linear trends of anomalies oriented at 90° to the user-selected *sun declination* and attenuates the anomalous trends parallel to the sun declination. See also *trend*.

superconducting gradiometer: **1.** a high-sensitivity *gravity gradiometer* the design of which is based on the use of a superconducting *proof mass* levitating in an artificial magnetic or electrostatic field maintained in each *gradiometer arm* module. **2.** a high-sensitivity *magnetic gradiometer* utilizing a superconducting quantum interference device, or *SQUID*, in its gradiometer arms. See also *magnetic levitation*, *Meissner effect*, and *sensitivity*.

superconducting gravimeter: a high-precision and high-sensitivity *relative gravimeter*. In its measurement system, *magnetic levitation* and cryogenic superconductivity are used in lieu of a *virtual spring*, or ideal spring. It has a cryogenic chamber with temperatures close to absolute zero (−273.15°C), where the superconducting niobium sphere, acting as a *proof mass*, is suspended, or levitates, by the balanced interaction between the magnetic fields generated by the electric currents in the external coils and the induced electric currents inside the superconducting sphere. The electric current voltage required to generate a sphere-suspending magnetic field is directly proportional to the change in the *Earth gravity field* between observation points. Very small gravity changes are determined by measuring: (a) the calibrated feedback voltage required to maintain the superconducting sphere in a fixed suspended position or (b) the vertical displacement of the sphere from its *null point* using either a high-*accuracy* optical system, such as in a *cryogenic optical gravimeter*, or an electrostatic capacitance device. Superconducting gravimeters can measure *relative gravity* with the very high *precision* of 1 *nanogal (nGal)*, *accuracy* up to 0.1 *microgal (μGal)*, *sensitivity* up to $10^{-12}g/\sqrt{\text{Hz}} = 10^{-3}\ \mu\text{Gal}/\sqrt{\text{Hz}} = 1\ \text{nGal}\sqrt{\text{Hz}}$, and very low *drift* rates less than 0.3–0.5 μGal per month. Also known as *cryogenic gravimeter*, *virtual-spring gravimeter*, *ideal-spring gravimeter*, and *magnetic accelerometer*. See also *Meissner effect*.

superconducting magnetometer: *SQUID magnetometer*.

superposition principle: the physical rule postulating that a result of two or more simultaneous effects, or causes, can be obtained by summing the results of the individual causes. Accordingly, (a) a composite nature of the measured fields and observed anomalies is defined by the numerical or *vector* sum of the individual gravity or magnetic effects produced by all subsurface sources, each having its own depth, size, *density contrast* or *magnetization contrast*, and vector orientation, and (b) *gravitational potential* of all gravity sources and *magnetic potential* of all magnetic sources are defined by the *scalar* sum of the individual gravitational and magnetic potentials of each of the gravity and magnetic sources. See also *equivalence principle*.

suprabasement anomalies: gravity and magnetic anomalies generated by the geological structures rising above the average *basement* relief, such as *horst*, *uplift*, thick-skin *thrust*, high-*offset* normal *fault* or *strike-slip fault*, and igneous intrusions of the *stock* or *sill* types at or immediately above the basement surface. In spite of the high-*density* and high-*magnetization* properties of basement rocks, their lateral density and magnetization contrasts are often low, thereby generating the relatively low-*amplitude* suprabasement anomalies which can be highlighted by *enhanced processing*. In *modeling*, the sources of suprabasement anomalies are approximated by *ideal source* geometries bounded by interpreted basement faults. See also *block structure*, *fault*, *density contrast*, *magnetic contrast*, *igneous intrusion*, *intrabasement anomalies*, and *intrasedimentary anomalies*.

surface anomaly: a gravity or magnetic *anomaly* generated by lateral changes in the *density* or *magnetization* of rock formations at or near the Earth's surface. See also *subcrop anomaly*.

surface density: a hypothetical quantity defined by the rock *mass* per unit area. A surface *density* distribution generating the same *gravity field* as the mass volume contained by this surface is known as the *Green's equivalent layer*. Surface density is used in data transformation by the *equivalent-source method* and for calculation of the *Bouguer correction* in lieu of *reduction density*. See also *magnetostatic charge*.

surface fitting: *polynomial fitting* with an emphasis on it being a *grid*-based method of approximating the regional component of an anomalous gravity or magnetic *field*, where the polynomial surfaces of the different orders (usually, from first to seventh through ninth) represent different approximations of a *regional field*, and deviations of the original gridded field from a particular polynomial surface represent the *residual field* components. See also *anomalous field* and *gridding*.

surface spherical harmonics: See *spherical harmonics*.

survey: measurements of the *Earth magnetic field* or changes in the *Earth gravity field*, its gradients, and *gradient tensor* components over the area of an *exploration* or research interest. Airborne and marine (shipboard) measurements are made along the regularly spaced, usually parallel, *acquisition* lines with the *survey line spacing* selected in accordance with the survey objectives. Depending on the surface relief and vegetation, the ground measurements are often made at irregular locations and along the accessible roads. Except for *microgravity* surveys, a *ground survey* is commonly carried out when the exploration or research *target* requires a much higher *accuracy* and *lateral resolution* of anomalies than those achievable and economical by an *airborne survey*. Virtual data spacing in airborne and marine (shipboard) surveys is determined by averaging the actual time-based measurements over some interval of time that is necessary to minimize *dynamic noise*. A marine gravity or magnetic survey can be carried out as a low-cost addition to a seismic acquisition program, and this survey is referred to as *add-on survey*. There are three basic types of gravity and magnetic surveys: *reconnaissance survey*, *regional survey*, and *detailed survey*. Airborne *survey height*, or flight *altitude*, and *survey line spacing* define the *lateral resolution* of anomalies: the lower the altitude and the closer the line spacing, the higher the resolution of anomalies. Detection of the long-*wavelength* anomalies generated by regional-scale geological structures is limited by the area extent of each survey. See also *gradient*, *tie lines*, and *time filter*.

survey accuracy: a quantitative characteristic of the ability of a gravity or magnetic *survey* to obtain the true values of a measured field and its components. Survey *accuracy* is a function of several key factors, including: (a) performance of the measuring instruments, e.g., *bandwidth*, *drift*, *precision*, *sampling rate*, and *sensitivity*; (b) *sample interval* of measurements; (c) errors in the *GPS* navigation; (d) *survey layout*; (e) *acquisition noise*;

(f) *data reduction*, especially *terrain correction* for gravity data; (g) *survey design aliasing*; and (h) effectiveness of applied algorithms in *noise* suppression, *leveling*, and other *data processing* operations. For ground gravity surveys, the survey accuracy is often estimated in terms of *survey precision*. For marine (shipboard), submarine, and airborne surveys, the survey accuracy estimate is based on the *intersection statistics* and the averaged *rms error* at mis-ties after *leveling* and *micro-leveling*. Airborne gravity and *gravity gradient* data are compared to the available *ground gravity* data in the survey area via a *ground truth test*. See also *mis-tie*, *static test*, *even-odd grid comparison*, *gravity survey accuracy*, and *magnetic survey accuracy*.

survey datum: a reference surface, such as sea level, *reference ellipsoid*, or other regional geodetic *datum*, that is used for (a) calculation of gravity corrections in lieu of *gravity datum* and (b) determination of the latitude, longitude, and *elevation* of observation points in the gravity or magnetic *survey*. See also *Geodetic Reference System*.

survey design: selection of *survey specifications* in accordance with *survey* objectives. There are four basic survey design parameters: *survey line spacing*, line length, line orientation, and flight height, or *altitude*, for an *airborne survey*. The number of survey lines and their lengths control the survey cost, as its estimate is based on the total length of all survey lines. Usually, all *traverse lines* in airborne surveys have the same orientation selected to be overly orthogonal to the dominant *strike* of the geological structures known in the survey area. Line spacing defines the *lateral resolution* of anomalies in the gridded data: the closer the line spacing, the higher the lateral resolution. Flight height is selected in accordance with the survey type: *reconnaissance survey*, *regional survey*, or *detailed survey*. The lower the flight height, the higher the amplitudes of the targeted anomalies and their *resolution* along the flight lines. However, because the *anomaly wavelength* decreases with a decrease in the *sensor*-source separation, low flight heights with wide line spacing can lead to *grid aliasing* in the direction orthogonal to the orientation of the survey lines. See also *gridding*, *target*, and *prospect*.

survey design aliasing: **1.** *aliasing* of the short-*wavelength* components of a measured field caused by an inadequately selected ratio of the *survey height* to the *survey line spacing*. For the *detailed survey*, this ratio is commonly 1:1 or 1:2, but not less. For aeromagnetic *total field* surveys, the minimum ratio is 0.5, i.e., line spacing should not exceed the doubled survey height value; whereas for the aeromagnetic *gradient* surveys, the optimal ratio is 1:1 (Reid, 1980). **2.** *gridding* errors in the restoration of the amplitudes and shapes of anomalies over a *survey* area from the gravity or magnetic field values recorded along the survey lines. Errors are caused, primarily, by a large survey line spacing: large intervals between *traverse lines* result in *undersampling* in the direction orthogonal to the orientation of the survey lines, thereby distorting the actual amplitudes, shapes, and *spectrum* of anomalies between the survey lines. Sometimes referred to as *survey design noise*. See also *aliasing*, *grid aliasing*, and *grid*.

survey design noise: See *survey design aliasing*.

survey height: an *airborne survey* parameter that specifies the flight *altitude*, i.e., height of the aircraft flight above the ground. Survey height and *survey line spacing* define the *lateral resolution* of anomalies and control the level of interference between anomalies generated by the adjacent shallow and mid-depth sources: the lower the height and the closer the line spacing, the higher the lateral resolution and the less the *interference of anomalies*. Selection of the survey height depends on: (a) the type of the survey: *reconnaissance survey*, *regional survey*, or *detailed survey*; (b) the topographic relief; (c) flight characteristics of the survey aircraft; and (d) flight safety regulations. See also *anomaly*, *survey specifications*, and *survey design aliasing*.

survey layout: *survey* parameters which specify: (1) coordinates of the survey location; (2) the orientation and length of the *acquisition* lines; (3) *survey line spacing*; and (4) the number of *tie lines* for a ground, marine (shipboard), and airborne magnetic or gravity survey. See also *traverse lines* and *survey specifications*.

survey line spacing: a *survey* parameter that specifies the distance between the *traverse lines* of a gravity or magnetic survey. Selection of the survey line spacing is guided by the survey type and its objectives. For mineral exploration, it is based on the estimated lateral extent of a prospective *target*, such as an *ore body* or *kimberlite intrusion*. For petroleum exploration, survey line spacing is based on estimates of the shallowest depth interval from which the gravity and magnetic anomalies should be resolved. In all types of surveys, the survey line spacing is a major factor which defines the *lateral resolution* of anomalies: the closer the line spacing, the higher the lateral resolution of anomalies. For aeromagnetic surveys, the survey line spacing should not exceed the doubled *survey height* to avoid *aliasing* of the short-*wavelength* components (Reid, 1980). Sometimes referred to as *track spacing*. See also *reconnaissance survey*, *regional survey*, *detailed survey*, and *survey design aliasing*.

survey precision: See *gravity survey precision*.

survey resolution: *lateral resolution* of anomalies in the data acquired by a gravity or magnetic *survey*. For ground surveys, the survey resolution depends on the distance between observation points, and its estimate is based on the *station spacing* and *station density*. For marine (shipboard) and airborne surveys carried out along regularly spaced *acquisition* lines, the resolution is a function of several factors, including: (a) *survey line spacing*; (b) number of *tie lines*; (c) airborne *survey height*; (d) *cutoff* parameter of the *low-pass filter* applied to suppress a *dynamic noise*; (e) type of survey: gravity or magnetic; (f) type of measurements: *total field* or *gradient*; and (g) effectiveness of the applied *data processing* algorithms, such as *leveling*, *filtering*, and combination of the *full-tensor gradient (FTG)* components, *invariants*, and other calculated attributes. See also *attribute*, *full-wavelength resolution*, and *half-wavelength resolution*.

survey specifications: specific parameters of a gravity or magnetic *survey*, including: (1) schedule: time of the year and duration; (2) survey area: location, coordinates, and size; (3) instrument type: *gravimeter*, *magnetometer*, or *gradiometer*; (4) *survey layout*; (5) *survey line spacing*; (6) *sampling interval* along *acquisition* lines; (7) *survey height* for the airborne survey; (8) preprogrammed *drape flight* or constant *elevation* flight for the *airborne gravity* survey; (9) *positioning* system: *GPS* or *differential GPS (DGPS)*; (10) *survey datum*; and (11) maximum maneuver *noise* level for airborne and marine surveys. Key factors in selecting the survey specifications are: (a) terrain elevations or water depths; (b) estimated *target* characteristics: depth, lateral extent, and orientation; and (c) geological environment. Survey specifications provide a basis for estimating the projected *accuracy* of measurements and *lateral resolution* of anomalies. See also *anomaly* and *drone survey*.

surveying: 1. measurements of the *elevation* of each *station* in a ground *gravity survey* area with the use of high-*precision* geodetic instruments to obtain elevation data with a higher *accuracy* than those provided by the *Global Position System (GPS)* and *differential GPS (DGPS)*. **2.** *survey* in the context of carrying out gravity and magnetic measurements.

susceptibility: See *magnetic susceptibility*.

susceptibility contrast: 1. the difference in *magnetic susceptibility* values between two laterally adjacent rock units or formations. The degree of the lateral susceptibility contrast is one of the primary controls over the *amplitude* of a *magnetic anomaly*: the higher the contrast, the higher the anomaly amplitude. The susceptibility contrast can be as high as thousands of *microcgs* (*μcgs*), such as the contrast between igneous intrusions and sedimentary rocks within a sedimentary section and between igneous and metamorphic rocks within the *basement*. In the absence of, or assumably negligible amount of, *remanent magnetization* in a *source body*, the susceptibility contrast is referred to with the same meaning as the *magnetization contrast* and *anomalous magnetization*. **2.** the difference in the susceptibility values across a lithological or structural non-horizontal boundary between two layers of rocks: the younger overlying rock formations and the older underlying rocks. See also *igneous intrusion* and *density contrast*.

susceptibility filter: a *grid*-based operator applied for the calculation of the *apparent susceptibility* of targeted rock formations under specific assumptions and with the use of a residual *RTP magnetic field*. See also *residual magnetics* and *target*.

susceptibility meter: a hand-held high-*sensitivity* and high-*accuracy* instrument for measurements of the *magnetic susceptibility* in rock outcrops and core samples recovered from drilled wells. Measured values are stored in the digital memory and can be transferred to a *workstation* for subsequent processing and analysis. Also known as *susceptometer*. See also *outcrop*.

susceptometer: See *susceptibility meter*.

Swarm mission: the European Space Agency (ESA) satellite mission launched in November 2013 with the objective of obtaining high-*accuracy* measurements of the long-*wavelength* and *vector* components of the *Earth magnetic field*, its temporal variations, and electric currents in the *magnetosphere* and *ionosphere* by three identical satellites in near-polar orbits. Two satellites orbit the Earth at the initial *altitude* of 470 km, and a third satellite orbits at approximately 520 km. All satellites are equipped with an *absolute scalar magnetometer (ASM)* and a triaxial *vector magnetometer*. See also *CHAMP mission*, *GOCE mission*, *GRACE mission*, *temporal magnetic variations*, and *triaxial magnetometer*.

symmetric boundary replication: a *data conditioning* procedure based on the replicating, or mirroring, of the near-boundary values at each end of line-oriented data or *grid* edge in order to minimize the data distortions by the *Gibbs phenomenon* in the transformation of the data from one *domain* to another, such as from the *space domain* to the *Fourier domain* and from the Fourier domain back to the space domain.

symmetric nearest neighbor (SNN) filter: a *grid*-based space-domain *edge-sharpening filter* that operates by comparing the values of the *grid cell* pairs, or neighbors, symmetric to the center of a *moving window*. The default window of 3 × 3 grid cells has eight neighbors which form four pairs. The filter algorithm inspects all cells in the symmetric pairs around the central grid cell in the moving window and retains the grid cell for which the value is the nearest to that of the central cell. There are two SNN filter options: (1) the *SNN mean filter*, which calculates the arithmetic mean, or average, of all retained grid cell values and assigns it to the central grid cell, and (2) the *SNN median filter*, which assigns a *median threshold* of the retained grid cell values to the central grid cell as a filter output. See also *default* and *mean-of-least-variance (MLV) filter*.

syncline: a downward-curved, or concave, intrasedimentary fold the core of which contains stratigraphically younger and, usually, lower *density* rock formations. A synclinal fold creates a negative lateral *density contrast*, the *magnitude* of which depends on the size, age, and composition of the sedimentary rocks in the syncline. This density contrast generates a local- or regional-scale *gravity low* in the *free-air anomaly* and *Bouguer anomaly* fields. Unless it is related to a structural depression in the *basement*, a mid-depth or deep syncline usually has no *magnetic signature* of its own due to the low *magnetization contrast* between the sedimentary rocks. When synclinal rock formations include *magnetite*-rich layers, a shallow syncline can be identified in *residual magnetics* by the concentric curvilinear anomalous alignments, continuous or segmented, of the high-*frequency* (short-*wavelength*) content. See also *residual field* and *anticline*.

syngenetic magnetic anomaly: a low-*amplitude* short-*wavelength* magnetic anomaly in the residual *RTP magnetic field* generated by high concentrations of *syngenetic magnetite*. Syngenetic anomalies can be used as indicators of structural and stratigraphic trends in the upper sedimentary section for: (a) mapping of regional and local geological structures, such as faults, subcrops, eroded/truncated anticlines, sand channels and lenses, shale diapirs, etc., and (b) mapping of drilling hazards, such as buried drainage channels. When integrated with the seismic horizon and interval velocity data, syngenetic magnetic anomalies can be used to map the upper sedimentary reservoir sands in the stratigraphic traps of the *incised valley* type. See also *anticline*, *diapir*, *fault*, *residual field*, *SRM anomaly*, *subcrop*, and *trap*.

syngenetic magnetite: *magnetite* deposited simultaneously with its sedimentary host formation. Certain types of sandstones, limestones, conglomerates, glacial till, sands, and erosional debris in paleochannels and other areas of the incised topography can accumulate high concentrations of syngenetic magnetite sufficient for generating a detectable *syngenetic magnetic anomaly*. Referred to with the same meaning as *detrital magnetite*. See also *paleochannel*, *incised valley*, and *diagenetic magnetite*.

synthetic anomaly: a gravity or magnetic field, or anomalous effect, of an isolated *ideal source* or a combination of ideal sources that is calculated in the process of *synthetic modeling*.

synthetic data: gravity and magnetic data representing the calculated anomalies and their *field* generated by sources of simple geometries, such as *sphere*, *prism*, *thin sheet*, *vertical cylinder*, and *horizontal cylinder*, that are referred to as ideal sources and implemented in *synthetic modeling*. See also *ideal source*.

synthetic modeling: *forward modeling* and *inverse modeling* of the gravity and magnetic fields generated by sources of simple geometries, such as *sphere*, *prism*, *horizontal sheet*, *vertical cylinder*, etc., whose gravity and magnetic effects are described by closed-form analytical expressions (formulas). It is used to study the properties and demonstrate the effectiveness of mathematical transformations of the data and new algorithms for *enhanced processing* and *inversion*. Also referred to as *synthetic test*. See also *ideal source*.

synthetic test: *synthetic modeling* in the context of its use for testing new algorithms for *enhanced processing* and *inversion*.

T

tabular body: **1.** a *source body* in which two dimensions are much longer than the third *dimension*. Often, the implication is that its lateral extent, or length, is more than 8–10 times its thickness. **2.** a source body of finite thickness with a horizontal top, parallel sides, and an *infinite strike length*. **3.** a source body with a horizontal top and bottom, parallel dipping or vertical sides, and a rectangular or rhomboidal cross-section. It is used as (a) elementary building block in the *modeling* of arbitrary-shaped geological bodies and (b) *model* approximation of isolated source bodies and lithological units, such as *dike*, *sill*, *volcanics*, and *banded iron formation (BIF)*.

TAHG attribute: See *tilt angle of horizontal gradient (TAHG)*.

Talwani algorithm: **1.** an algorithm for calculation of the gravity or magnetic field, or anomalous effect, of a two-dimensional (2D) arbitrary-shaped *source body* the cross-section of which is approximated by an *N*-sided polygon. Its gravity and magnetic effects are calculated by the line integral of *N*-contributions, each associated with a side of the polygon. All calculated effects are summed at every observation point placed at the origin of the coordinate system, wherein all geometric elements of the line integral are analytically expressed in terms of the *x*- and *z*-coordinates of the end points, or vertices, of the polygon-bounding line segments (Talwani et al., 1959). Also known as *2D Talwani algorithm*. **2.** an algorithm for calculation of the anomalous effect of three-dimensional (3D) arbitrary-shaped source bodies. Each body is represented by a pack, or stack, of vertical or horizontal polygonal plates, or laminas, whose shapes are approximated by the *N*-sided polygons bounded by the straight line segments. The thickness of laminas is usually one-tenth of their depth or less. The gravity or magnetic effect of each *lamina* is calculated separately by the line integral of each polygon, wherein the line integral components are analytically expressed in terms of the *x*-, *y*-, *z*-coordinates of the polygon vertices. The gravity or magnetic field of the whole 3D source body is obtained by numerical integration of the individual gravity or magnetic effects of all laminas in the source body (Talwani and Ewing, 1960). Also known as *3D Talwani algorithm*.

Talwani method: **1.** a method of the two-dimensional (2D) *forward modeling*. Model bodies are represented by uniformly dense or uniformly magnetized polygons bounded by straight-line segments. Calculation of the gravity or magnetic field of the constructed model is based on the use of the *Talwani algorithm*. The difference between the model-calculated and observed fields, or *model misfit*, is minimized by multiple iterative adjustments of (a) structural positions of the polygon vertices in the separate geological bodies, continuous sedimentary formations, and *basement* surface, and (b) *density* or *susceptibility* values within the individual polygons. Also known as the *2D Talwani method*. **2.** a method of the three-dimensional (3D) forward modeling and *3D inversion* based on the use of the Talwani algorithm for calculation of the gravity or magnetic field of 3D arbitrary-shaped source bodies, wherein each body is represented by a pack, or stack, of horizontal or vertical laminas. Also known as *3D Talwani method*. See also *lamina*, *vertex*, and *Hubbert method*.

Talwani modeling: See *Talwani method*.

taper: **1.** rows and columns of artificial data added to *grid* margins to provide a smooth transition of the grid values to zero and prevent the abrupt discontinuity of data at the grid edges in order to minimize *Gibbs ringing*. **2.** artificial data added to the ends of *survey* lines to ensure a gradual decrease of the data values, or roll-off, to zero outside the survey area. Also referred to as *roll-off window*.

taper filter: a *filter* from the group of *edge-smoothing filters* applied to minimize *Gibbs ringing* in the process of *Fourier transform*. See also *cosine taper filter* and *tapering*.

tapering: addition of *taper* to gridded or line-oriented datasets in order to minimize *edge effects* and *wavenumber leakage*. Referred to with the same meaning as *grid extension* and *padding*. Also known as *window carpentry*.

tare: a step-like change, or jump, in a sequence of *spring gravimeter* readings that is caused by a change in *drift rate* or reset in the reference value relative to which measurements are made. Tare can be temporary, when the gravimeter readings move up or down and then return to the *baseline* trend, or it can be permanent, when the readings show a permanent baseline shift. Large tares up to 0.05 *milligal (mGal)* can be caused by a slippage at the measuring spring hooks. Most tares are less than 10–15 *microgal (μGal)*, and they are often related to the mechanical or thermal shock and temporal variations of the *proof mass* caused by contamination of the *sensor* compartment. See also *gravimeter drift* and *reading*.

target: **1.** the geological object of an *exploration* or research interest. A ground, marine (shipboard), submarine or airborne *survey* is designed to study this object. **2.** a specific geological *structure* or isolated *ore body* selected for the gravity or magnetic *forward modeling* or *inversion*. **3.** *prospect*. See also *survey design*.

target anomaly: **1.** a gravity or magnetic *residual anomaly* associated with a particular geological *structure* or isolated *source body* of an *exploration* or research interest. **2.** a gravity or magnetic anomaly calculated by *forward modeling* of the targeted structure whose parameters were estimated from the available geological and geophysical information. See also *tuning filter* and *feasibility modeling*.

target rms: *rms accuracy* selected as a *tolerance* value in the gravity or magnetic *inversion* for properties of an exploration *target*, i.e., it is a value of the difference between the observed and model-generated fields below which the inversion process is automatically terminated.

Taylor series: a mathematical representation of a function in terms of the value of this function and its derivatives at the point x for extrapolation to nearby points. For the 1D function $f(x)$, its extrapolation to the location $(x+\Delta x)$ is expressed as

$$f(x+\Delta x) = f(x) + [\partial f(x)/\partial x]\Delta x/1! + [\partial^2 f(x)/\partial x^2](\Delta x)^2/2! + [\partial^3 f(x)/\partial x^3](\Delta x)^3/3! + \ldots,$$

where ! denotes the factorial: $2! = 2 \cdot 1 = 2, 3! = 3 \cdot 2 \cdot 1 = 6$, etc. High-order derivative terms are dropped when Δx is considered sufficiently small for the purposes of a particular extrapolation. It is used in various applications, including the interpolation in *gridding* by the *gradient-enhanced minimum-curvature* method and the optimization criteria in an *inverse problem solution*. See also *Fourier series*.

TDX attribute: See *horizontal tilt angle (TDX)*.

TDX derivative: same as *horizontal tilt angle (TDX)* and *TDX attribute* with an emphasis on its calculation with the use of the *total horizontal derivative* normalized by the *vertical derivative*.

temperature coefficient: a *spring gravimeter* characteristic that describes the rate of change of its *temperature drift* in units of *microgal (μGal)* per °C or °K. The temperature coefficient is negative, because the quartz or metal spring becomes weaker with increasing temperature, thereby producing an artificial increase in the measured values. It is unique for each spring gravimeter and determined during the manufacturing process.

temperature correction: a correction applied to the *relative gravity* measurements by a *spring gravimeter* to compensate for *temperature drift*. In the automated *gravimeter* systems used in *ground gravity* surveys, it is integrated into the *automatic drift correction* and referred to as *automatic temperature correction*.

temperature drift: a gradual change in the readings of a *spring gravimeter* and certain types of magnetometers that is caused, primarily, by an increase in the ambient air or water temperature. In spring gravimeters, the quartz or metal spring becomes weaker with the increasing temperatures, and the *reading*-quantifying deviation of a spring-suspended *proof mass* from its *null point* becomes artificially increased. It is compensated by application of a *temperature correction*. See also *drift* and *temperature coefficient*.

temperature range: an instrument characteristic that specifies the lowest and highest ambient temperatures acceptable for making correct measurements by the *gravimeter* or *magnetometer* during a gravity or magnetic *survey*. See also *temperature coefficient*.

temporal gravity variations: 1. time-dependent changes in the observed *Earth gravity field*. They are caused, primarily, by the cumulative effect of the *gravitational attraction* of the Moon and Sun and, to a much lesser degree, by other planetary and regional gravity effects, including: (a) surface *elevation* changes due to shallow seasonal and deep tectonic processes; (b) changes in the groundwater levels associated with groundwater consumption; (c) *ocean loading effect*; (d) *glacial isostatic adjustment*; (e) *polar motion effect*; (f) volcanic mass changes and *lava* flows on the Earth's surface; (g) local variations associated with the *atmospheric effect*; and (h) local-scale *density* changes related to depletion of geothermal reservoirs and development of oil-and-gas accumulations. Except for *microgravity* surveys and measurements of *absolute gravity* by the *free-fall method*, temporal gravity variations are essentially compensated by *tidal correction*. **2.** time-dependent changes in the planetary gravity field observed in the *real-time mode*. See also *aquifer*, *atmospheric mass effect*, *atmospheric pressure effect*, *injection effect*, and *volcanic mass effect*.

temporal magnetic variations: time-dependent changes in the observed *Earth magnetic field*. By their source, *magnitude*, and rate of change, they can be divided into five main categories: (1) regular daily variations of the *external magnetic field* of the Earth with magnitudes from several *nanotesla (nT)* to tens and hundreds of nanoteslas, known as *diurnal variation* or *diurnals*; (2) rapid irregular variations of the external magnetic field with magnitudes up to several thousands of nanoteslas that are caused by effects of the active Sun radiation, known as *magnetic storm*; (3) highly variable in magnitude and duration changes of the external magnetic field within the space- and time-varying zones in the *van Allen belts*; (4) time- and magnitude-varying magnetic effect of the electric currents in the *ionosphere*, known as *tidal effect*; and (5) slow long-term variations of the *main geomagnetic field* presumably associated with a convection in the liquid outer core of the Earth, known as *secular variation*. Diurnals and all other irregular variations (which are magnitude-permitting for conducting the magnetic surveys) are recorded by a *base-station magnetometer* and integrated into *diurnal correction*. Secular variation is incorporated into the *IGRF model* and compensated by the *IGRF correction*. See also *bay*, *micropulsations*, *auroral electrojets*, *equatorial electrojets*, and *K-index*.

tensor: the mathematical entity defined by a rectangular assemble of jointly related variables, such as the second-order derivatives of gravity or magnetic *potential* along the mutually orthogonal x-, y-, z-directions in *Cartesian coordinates*, that are functions of coordinates at any point in space. A gravity or magnetic *gradient tensor* is presented in the *format* of a 3×3 symmetric *matrix* of the measured or calculated *gradient* values, referred to as a *tensor matrix*. See also *derivative* and *full tensor gradient (FTG)*.

tensor Bouguer anomaly: *Bouguer anomaly* obtained by application of (a) *free-air correction* or more accurate *height correction* and (b) *simple Bouguer correction* to the *tensor free-air gravity* data.

tensor deconvolution: a method of processing the *full tensor gradient (FTG)* gravity data for estimating the spatial position (location and depth) of sources of *residual* *gravity* anomalies. It is based on the theory of *Euler deconvolution* and calculation of *eigenvalues, invariants, invariant ratio*, and *Euler structural index (SI)* at each observation point without the use of a *moving window* (Mikhailov et al., 2007). See also *tensor Euler deconvolution.*

tensor-derived gravity: three directional components of the *relative gravity,*

$$\Delta g_x = \Delta T_x, \quad \Delta g_y = \Delta T_y, \quad \Delta g_z = \Delta T_z$$

obtained from the integration-based transformation of the *full tensor gradient (FTG)* components T_{xx}, T_{xy}, T_{xz}, T_{yx}, T_{yy}, T_{yz}, T_{zx}, T_{zy}, T_{zz} that are measured by a *full tensor gravity gradiometer* and represent the second-order derivatives of *gravitational potential* along the mutually orthogonal x-, y-, z-directions in *Cartesian coordinates.* The most often used vertical component, $\Delta T_z = \Delta g_z$, is referred to as *tensor-derived vertical gravity*, and it also can be obtained from the *partial tensor gravity gradiometer* data, where the measured gradients g_{NE} and g_{UV} are transformed into the *vertical derivative* g_{zz} and then integrated into the *vertical gravity* Δg_z. See also *derivative, gradient*, and *FTG gravity.*

tensor-derived HGM: *horizontal gradient magnitude (HGM)* calculated from the *full tensor gradient (FTG)* or partial *gradient tensor* measurements. See also *full tensor gravity gradiometer* and *partial tensor gravity gradiometer.*

tensor-derived vertical gravity: the vertical component Δg_z of the *relative gravity* obtained from measurements of the *full tensor gradient (FTG)* elements by a *full tensor gravity gradiometer* or by a *partial tensor gravity gradiometer.* For the FTG data, it is calculated by the integration-based transformation of: (1) one measured component T_{zz}, which represents the *vertical derivative* of the *vertical gravity* $\Delta g_z = \Delta T_z$; (2) the two measured components T_{xz} and T_{yz}, which represent the horizontal derivatives of ΔT_z; and (3) the three measured components T_{zz}, T_{xz}, T_{yz}, which represent the vertical and horizontal derivatives of ΔT_z. In the *noise*-free theory, the third option is the most resolving, both vertically and laterally. For partial tensor gradient data, the two measured *differential curvature gradients* g_{NE} and g_{UV} are transformed into the vertical derivative g_{zz} and then integrated into the vertical gravity Δg_z. The field of the tensor-derived *vertical gravity* anomalies differs from the vertical gravity field measured with the use of a conventional, or regular, *relative gravimeter* by the amount of the regional long-*wavelength* component of the relative gravity values in the *survey* area, because this component is suppressed by the *gradient* measurements. See also *derivative, lateral resolution*, and *vertical resolution.*

tensor Euler deconvolution: a modification of the *Euler deconvolution*, where the measured *full tensor gradient (FTG)* components are used instead of the calculated first-order derivatives of the gravity or magnetic field F to form three Euler deconvolution equations that are solved for the spatial position (location and depth) of structural boundaries and isolated source bodies. In addition to the standard *Euler deconvolution equation*, where field F represents the *vertical derivative* of the gravity or magnetic *potential* P, i.e., $F = \partial P/\partial z = P_z$, two more equations are added for its horizontal x- and y-components. Three tensor Euler deconvolution equations are expressed as

$$(x - x_0)P_{zx} + (y - y_0)P_{zy} + (z - z_0)P_{zz} = N(B_z - P_z),$$

$$(x - x_0)P_{xx} + (y - y_0)P_{xy} + (z - z_0)P_{xz} = N(B_x - P_x),$$

$$(x - x_0)P_{yx} + (y - y_0)P_{yy} + (z - z_0)P_{yz} = N(B_y - P_y),$$

where (x_0, y_0, z_0) is the spatial position of the source whose magnetic or gravity field F is measured at the observation point with coordinates (x, y, z) and B_x, B_y, and B_z are constant background values for the orthogonal x-, y-, z-components, respectively, of the field F. With the assumed value for *Euler structural index N*, six unknown parameters x_0, y_0, z_0, B_x, B_y, B_z are obtained by solving the overdetermined system of equations for each position of a *moving window* along the survey line, *grid*, or selected *profile* in the same manner as done in the application of standard Euler deconvolution (Zhang et al., 2000). See also *extended Euler deconvolution* and *tensor deconvolution.*

tensor free-air gravity: *free-air gravity* calculated from *tensor-derived gravity* by a two-step procedure: (1) transformation of the *gravity gradient tensor (GGT)* components, obtained from measurements by a *full tensor gravity gradiometer* or a *partial tensor gravity gradiometer*, into *vertical gravity*; and (2) addition of the *regional gravity* values obtained from regional *airborne gravity* or *ground gravity* survey data or from measurements by Earth-orbiting satellites. See also *regional survey, satellite gravity*, and *tensor Bouguer anomaly.*

tensor gradient: See *gradient tensor.*

tensor gradiometry: instruments and methodology for measuring the *gradient tensor* components of the Earth's gravity or magnetic field. Tensor gradiometry measurements suppress the long-*wavelength* low-*gradient* anomalies generated by deep sources and highlight the short-wavelength high-gradient anomalies generated by shallow and mid-depth sources. See also *gradiometer* and *full tensor gradient (FTG).*

tensor gravity: See *tensor-derived gravity.*

tensor invariants: *invariants* calculated by mathematical combinations of measured *full tensor gradient (FTG)* components or calculated *FTG eigenvalues.*

tensor matrix: *gradient tensor* data presented in the *format* of a 3 × 3 symmetric *matrix*, i.e., a square assembly of three rows and three columns of either the measured or calculated gravity or magnetic *gradient* values. Referred to with the same meaning as *Hessian matrix* and *FTG matrix.* See also *diagonalized tensor matrix.*

ternary image: 1. *barycentric image.* **2.** *RGB composite image.*

terracing: a *profile*- or *grid*-based technique which sharpens the edges of anomalies by increasing their slopes and simultaneously flattening the anomaly values between slopes into stepped, or terraced, sectors with abrupt edges approximating the steeply dipping lithological boundaries and faults. Terracing algorithms, also known as terracing filters, are based on the use of the following options: (1) algorithm which increases or decreases the *anomalous field* values at the center of a *moving window*, wherein the upward or downward adjustment over each sector is guided by the positive or negative sign of a local field *curvature* determined within the moving window; (2) *Walsh transform*; (3) *edge-sharpening filters*; (4) statistical *clustering*; (5) *shape index (ShI)*. Terraced profile values can be converted into scaled *density* units using the infinite *Bouguer plate* approximation and scaling factor $1/2\pi hG$, where h is plate thickness and G is *universal gravitational constant*. Scaled density values are adapted to geologically meaningful *apparent density* values by the multiplicative factor determined by a least-squares *regression* between the observed profile and the profile calculated from an obtained terraced *model*. Prior to magnetic data terracing, the *RTP magnetic field* is converted into *pseudogravity* to minimize the dipolar magnetic effects and directional components of *magnetization* (Cordell and McCafferty, 1989; Li, 2016; Florio and Le Re, 2018; Cooper, 2020). See also *fault*, *filter*, and *least-squares approximation*.

terrain: the area over which a particular *topography*, or surface *relief*, is dominant. High relief terrain is the area of a highly variable topography and low relief terrain is the area of a smooth topography. See also *terrane*.

terrain chart: same as a *Hammer chart* used for the calculation of *terrain correction* by *Hammer method*.

terrain clearance: the aircraft *altitude* during an airborne gravity or magnetic *survey*. Terrain clearance acts as a *low-pass filter* on the *wavelength* content of the measured data: the larger the clearance, the longer the measured wavelengths. The minimum terrain clearance of an *airborne survey* is determined based on three main factors: (1) terrain *relief*, (2) aircraft flight characteristics, and (3) flight safety regulations. Also referred to as *ground clearance*.

terrain correction: **1.** a correction applied to *observed gravity* values to account for the *terrain effect* which is not compensated by *simple Bouguer correction*. It is calculated with the use of the *Digital Terrain Model (DTM)* and the common assumption of a constant *density* of rocks above a *survey datum*. The DTM is divided into concentric ring-shaped zones around each *station*, wherein the number of zones and their lateral extent are referenced to *Hammer zones* or *Hayford-Bowie zones*. Zones are subdivided into flat-top square vertical prisms whose widths increase with the zone distance from the station in order to adapt to the *inverse power law* decrease in the gravity effect of prisms and increase the speed of computations. The height of each *prism* is defined by the difference (regardless of its sign) between each station *elevation* and prism elevation obtained by averaging the DTM values within the prism extent. Gravity effects of all individual prisms around every station are calculated with the standard formula for *gravitational attraction* of a vertical prism and summed to obtain the terrain correction value at the station location. For ground and sea-bottom stations, the terrain correction is always positive, i.e., it is added to the observed gravity values. For marine (shipboard), submarine, and onshore airborne measurements, the sign of each correction is defined by the sign of a lateral change in the bathymetric or topographic *relief* relative to the observation point location: the terrain correction is positive for deepening relief, because the horizontal *Bouguer plate* approximation does not account for the decreased gravitational attraction due to the lateral *mass deficiency*; and it is negative for rising relief, because the Bouguer plate does not account for the increased gravitational attraction due to the lateral *mass excess*. Correction values and the *accuracy* of their calculation are strongly dependent on (a) distance from an *observation point*, (b) *terrain* deviation from the horizontal Bouguer plate, and (c) *lateral resolution* of the available DTM. For *ground gravity* surveys, terrain correction is partitioned into the *innermost terrain correction*, the *inner-zone terrain correction*, and the *outer-zone terrain correction*. For *airborne gravity* and airborne *gravity gradient* surveys, the DTM discretization and range of the terrain correction calculations are functions of *acquisition filtering*, *terrain clearance*, and estimated airborne survey *noise* (Nowell, 1999; Li et al., 2010a). Often referred to as *bathymetric correction* for offshore gravity data and *topographic correction* for onshore gravity data. **2.** *magnetic terrain correction*. See also *mesh discretization*, *airborne terrain correction*, *marine terrain correction*, and *gradient terrain correction*.

terrain correction calculation: an automated procedure utilizing an algorithm based on the concept implemented in the *Hammer method*. The algorithm parameters are selected in accordance with the *gravity survey* type (ground, shipboard, airborne), survey objectives, measurement *accuracy* requirements, estimated *acquisition noise* level, and available *Digital Terrain Model (DTM)*. For high-*accuracy* ground observations, the calculations of the *innermost terrain correction* and *inner-zone terrain correction* require high-*precision* measurements of the surface *elevation* with the use of geodetic instruments or a *drone*-attached *altitude*-measuring device. Standard terrain corrections are supplemented with *Bullard B correction* for the Earth's *curvature* at distances beyond 20 km around each *station*. For *airborne gravity* and airborne *gravity gradient* surveys, the calculations are based on the gridded DTM data obtained from high-resolution *LIDAR* or satellite measurements. For *marine gravity* surveys and regional-scale onshore airborne gravity surveys, terrain corrections are calculated for line-oriented bathymetric and topographic data using the two-dimensional (2D) *Parker formula*. For satellite-based regional and planetary gravity studies, terrain corrections are calculated with the use of 15-*arc-second*, 1-*arc-minute*, and 2-minute global DTM grids. See also *grid*, *gridding*, *terrain surveying*, and *Agerlin principle*.

terrain correction density: a constant *density* value used in the calculation of *terrain correction* in a *gravity survey* area. Its selection is based on the concept implemented in the *Nettleton method*, where the terrain correction is computed and gridded with several *density* values, and the value producing the minimal overall correlation between the gridded *topography* or *bathymetry* and a particular terrain correction *grid* is selected as the terrain correction density. In airborne gravity and *gravity gradient* surveys, the gridded *free-air anomaly*, obtained from *vertical gravity* g_z or *vertical gradient* g_{zz}, can be used instead of the topography in selection of the optimal "minimum correlation" density value. In marine (shipboard) surveys, the same approach is followed in selection of the *density contrast* for calculation of *marine terrain correction* with the use of *Parker formula*. See also *gridding*.

terrain correction range: an extent of the *terrain correction* variations defined by two main factors: (1) distance from an *observation point* and (2) amount of deviation of the topographic or bathymetric *relief* from the horizontal *Bouguer plate* around the observation points. The distance-dependent decrease in the gravity effect of the relief variations is described by the *inverse power law*. For relief deviations within tens of meters, the correction values vary from approximately ±0.1 to ±1.0 *milligal (mGal)* within the distance of several hundred meters. For the relief deviations up to several hundred meters, correction values are on the order of ±2.0 to ±10.0 mGal. In the mountains and offshore areas with sea bottom variations in excess of 1000 m, terrain corrections can reach several tens of milligals within 1–2 km.

terrain effect: 1. a three-dimensional (3D) gravity effect of deviations of the topographic or bathymetric *relief* around observation points from the horizontal *Bouguer plate* on which the calculation of a *simple Bouguer correction* is based, i.e., it is a non-geological gravity effect that is not compensated by the simple Bouguer correction. **2.** an integral of the gravity effects of *mass excess* above and *mass deficiency* below a reference horizontal plane, i.e., *survey datum*. For ground stations, this effect is always negative, and the measured gravity value is reduced due to the combined effect of the increased upward *gravitational attraction* of a *gravity sensor* by the adjacent rock mass excesses above the *station* elevation and decreased downward attraction of a gravity sensor due to the lack of adjacent rock masses below the station *elevation*. For marine (shipboard), submarine, and onshore airborne gravity measurements, the terrain effect depends on the sign of the bathymetric or topographic deviations relative to the observation point location: deepening relief decreases the measured value (terrain correction is positive), and rising relief increases the measured value (terrain correction is negative). The terrain effect is strongest for *airborne gravity* and airborne *gravity gradient* measurements due to the high *density contrast* between the surface rocks and air. For local-scale surveys, the near terrain effects are the most important, because the longer *wavelength* contributions to measured values by the distant *terrain* variations can be minimized or eliminated by *regional-residual separation*. The calculated terrain effect comprises the amount of *terrain correction* for each observation point in a *gravity survey* area.

terrain surveying: high-*accuracy* measurements of a topographic *relief* to the distance of 70–90 m around each gravity *station* with the use of geodetic instruments or a *drone*-attached *altitude*-measuring device and other airborne digital imaging systems (Kaub et al., 2018). Terrain surveying data are used for calculation of the *innermost terrain correction* and *inner-zone terrain correction* in proximity to the observation points, where variations of *topography* have the strongest effect on the gravity values measured in high-accuracy *ground gravity* and *microgravity* surveys. See also *terrain correction* and *terrain effect*.

terrane: a *fault*-bounded area of a regional extent for which the structural-tectonic history differs from that of the surrounding or adjacent areas. Often, both the *magnetization contrast* and *density contrast* across terrane boundaries are sufficiently strong to map these boundaries by correlation of the maximum values of the *total horizontal gradient* of the gridded and low-pass-filtered anomalous gravity or magnetic field over a study area. See also *anomalous field*, *gridding*, *low-pass filter*, *magnetic terrane*, and *terrain*.

terrestrial ellipsoid: same as *reference ellipsoid* with an emphasis on it being the best-fitting mathematical approximation of the rotating Earth.

terrestrial gravity: 1. the *gravity field* generated by the *gravitational attraction* of the Earth mass, commonly referred to as the *Earth gravity field*. **2.** the gravity field of a rocky planet, i.e., planetary body with extensive land masses on its surface ("terra" is Latin for "earth").

terrestrial magnetism: *Earth magnetic field* in the context of its properties and their studies.

Tesla (T): a unit of *magnetic induction* which is commonly referred to with the same meaning as *magnetic field*. It is defined as $1\text{T} = 10^4$ *gauss* $= 1$ *weber* $\cdot\text{m}^{-2}$. In magnetic exploration, the standard unit of the *Earth magnetic field* measurements is *nanotesla (nT)*: $1\text{ nT} = 10^{-9}\text{ T} = 10^{-5}$ gauss $= 1$ *gamma*. It is named after Nikola Tesla (1856–1943), a Serbian-American physicist and inventor. See also *picotesla (pT)* and *femtotesla (fT)*.

tesseral harmonics: *spherical harmonics* that depend on the latitude and longitude. See also *sectoral harmonics* and *zonal harmonics*.

test line: 1. a line on the ground with known coordinates over which *airborne gravity* data are acquired at the beginning and end of each *acquisition* cycle in order to (a) estimate the amount of accumulated errors in the airborne measurements and *GPS* navigation and (b) find indicators of the measurement system performance that are known to be damaging for the data quality. **2.** the *ground truth* line. Test line data are acquired in the process of *quality control (QC)* and evaluation of *survey accuracy* for airborne gravity measurements. See also *even-odd grid comparison* and *static test*.

test mass: See *proof mass*.

texture: 1. a characteristic of the spatial distribution of anomalies that describes their specific alignments, amplitudes, shapes, and *wavelength* content. Referred to with the same meaning as *fabric* and *grain*. **2.** a measure of the data variations within a regional- or local-scale area. Texture differences between the adjacent and surrounding areas provide a basis for the identification of regional terranes and local elements of the geological structures. See also *anomaly*, *amplitude*, *shape*, and *terrane*.

texture analysis: an *enhanced processing* technique for the detection of textural features on grids and digital images of an *anomalous field*. It is based on the use of statistical tools, such as the gray-level co-occurrence *matrix*, to obtain the qualitative (coarseness, smoothness, homogeneity, etc.) and quantitative (contrast, entropy, etc.) statistical measures of *texture* for the identification and classification of *structure*-related textural features (Zhang et al., 2009). See also *grid* and *gridding*.

Thalen method: a graphical method of estimating the depth to the three-dimensional (3D) *source* of an isolated *magnetic anomaly*. Source depth is determined at 0.7 times the horizontal distance between the maximum and minimum amplitudes at the *principal profile* of a TMI anomaly, wherein the minimum amplitude is a *trough* value adjacent to the *peak* value. Also referred to as the *Thalen rule*. See also *graphical source depth estimation* and *total magnetic intensity (TMI)*.

Thalen rule: *Thalen method* in the context of a *graphical source depth estimation* method based on the use of *depth rules*.

theoretical Earth gravity field: *theoretical gravity* that represents the gravity field of the theoretical Earth whose actual size, shape, and heterogeneous internal composition are approximated by the best-fitting uniformly dense, or homogeneous, *ellipsoid* of revolution known as the *reference ellipsoid*.

theoretical geomagnetic field: 1. *International Geomagnetic Reference Field (IGRF)*. **2.** a mathematical *model* of the *main geomagnetic field* generated by the *geomagnetic dynamo* processes within the Earth's core.

theoretical gravity: 1. the theoretical, or predicted, *gravitational acceleration* at any point on the Earth's surface. It represents a numerical value of the *gravity field* of the homogeneous rotating *reference ellipsoid* which is the best-fitting mathematical approximation of the actual size, shape, and heterogeneous internal composition of the rotating Earth. **2.** the theoretical gravity field generated by an *equipotential ellipsoid* of revolution which has the same mass, center of mass, and *angular velocity* of rotation as those of the Earth. Theoretical gravity accounts for: (a) the total mass and size of the Earth; (b) the shape of the Earth flattened at the poles, so its surface points are not equally distant from the center of the Earth's mass; and (c) the rotation of the Earth, as it rotates with the different angular velocities at different latitudes resulting in different values of *centrifugal acceleration* and, therefore, different measured values of gravitational acceleration representing the *Earth gravity field* values. The theoretical *gravity vector*, or *normal gravity vector*, is perpendicular to a *geoid*, and its *magnitude* is defined by the *vector* sum of the gravitational acceleration due to attraction of the Earth's mass and centrifugal acceleration due to the Earth's rotation about its polar axis. Theoretical gravity is described by the *Somigliana-Pizzetti formula* and calculated at any point on the Earth's surface using its periodically updated versions within the *International Gravity Formula (IGF)*. The gravity effect of the Earth's atmospheric mass, i.e., *atmospheric mass effect*, is included in the *International Gravity Formula 1980 (IGF1980)*, *International Gravity Formula 1987 (IGF1987)*, and *International Gravity Formula 1999 (IGF1999)* with the assumption that the atmospheric mass is within the reference ellipsoid. The theoretical gravity value increases by approximately 5186 mGal from the equator to the poles due to (a) the decrease of the Earth's radius by roughly 21 km and (b) the decrease of the centrifugal acceleration from its maximum value at the equator to zero at the poles. The calculated theoretical gravity value comprises the amount of *theoretical gravity correction* at each observation point. Also referred to as *theoretical Earth gravity field*, *ellipsoidal theoretical gravity*, *ellipsoidal gravity*, *reference gravity field*, *normal gravity*, and *standard gravity*. See also *milligal (mGal)*.

theoretical gravity correction: a correction applied to *observed gravity* values to compensate for *theoretical gravity* representing the *gravitational field* of a homogeneous rotating *reference ellipsoid* which is the best-fitting mathematical approximation of the actual size, shape, and heterogeneous internal composition of the rotating Earth. Currently used correction values are defined by the *International Gravity Formula 1980 (IGF1980)*, *International Gravity Formula 1987 (IGF1987)*, or *International Gravity Formula 1999 (IGF1999)*. Theoretical gravity correction removes the gravity effect of the rotating reference ellipsoid, including the *atmospheric mass effect*, and highlights both the regional-scale and local gravity anomalies generated by the heterogeneous components of the Earth's geological structure and regional-scale *density* variations within the Earth's *crust* and upper *mantle*. As a function of the geographic latitude, it is also referred to as *latitude correction*. Sometimes referred to as *normal gravity correction*.

theoretical gravity formula: *International Gravity Formula (IGF)* in the context of a mathematical formula for the calculation of *theoretical gravity* representing the *gravity field* of a homogeneous rotating *reference ellipsoid* which is the best-fitting mathematical approximation of the actual size, shape, and heterogeneous internal composition of the rotating Earth.

theoretical magnetic correction: See *IGRF correction*.

thermal density: *density* reduced by the high-temperature thermal expansion of rocks. It can produce a local *gravity low* over the pipe-like conduit (main vent) within a volcano crater and generate a high-*amplitude* regional-scale negative *thermal gravity anomaly* over the mid-oceanic ridges and transition zones between the thick continental and thinned oceanic parts of the *lithosphere*. Thermal density ρ_T is expressed as

$$\rho_T = \rho_0(1 - \mathrm{k}\Delta T),$$

where ρ_0 is regular density of a particular rock, k is its thermal expansion coefficient, and ΔT is temperature increment. See also *thermal isostasy* and *gravity high*.

thermal gravity anomaly: a regional-scale high-*amplitude* (up to −320 mGal) negative *gravity anomaly* generated by the lateral decrease of densities within the *lithosphere* as a result of the thermal expansion of rocks in the areas of the elevated lithospheric *isotherm*, such as over the mid-oceanic ridges and transition zones between the thick continental and thinned oceanic parts of the lithosphere. Also referred to as *lithospheric thermal gravity anomaly* (Chappell and Kusznir, 2008). See also *mantle residual anomaly*, *milligal (mGal)*, and *thermal density*.

thermal gravity correction: a correction applied to satellite-derived *marine free-air gravity* to compensate for the gravity effect of a *thermal gravity anomaly* prior to three-dimensional (3D) *gravity inversion* for the *Moho* depth structure. The correction method is based on the use of a lithospheric isothermal *model* from which the corresponding *thermal density* and thermal gravity anomaly are calculated (Chappell and Kusznir, 2008). See also *isotherm*, *mantle residual anomaly*, and *satellite gravity*.

thermal isostasy: a theory of *isostasy* based on the assumption that regional-scale differences in the *topography* or *bathymetry* depend on differences between the subsurface distribution of temperatures which control the *density* variations due to the thermal expansion of rocks within the isostatically balanced rock columns, so the "hot" regions have lower densities and higher elevations than the "cold" regions. The estimated *magnitude* of the *elevation* differences between the "hot" and "cold" continental regions can be up to 3000 m (Chapman and Hasterok, 2011). See also *thermal density* and *thermal gravity anomaly*.

thermal remanent magnetization: See *thermoremanent magnetization (TRM)*.

thermoremanence: See *thermoremanent magnetization (TRM)*.

thermoremanent magnetization (TRM): 1. *remanent magnetization* acquired by rocks in the process of their cooling to temperatures below the *Curie point* under exposure to an external magnetic field, such as the *Earth magnetic field*. Magnetization of this type is very stable, and it contributes significantly to the *total magnetization* of igneous and metamorphic rocks. **2.** the permanent *magnetization* of organic-rich and iron-bearing sedimentary formations that is acquired as a result of *combustion magnetism*. Also known as *thermal remanent magnetization*. See also *isothermal remanent magnetization (IRM)*.

theta angle: 1. the *edge*-enhancing *attribute* of an *anomalous field* defined by the angle between the horizontal plane and the direction of the *3D analytic signal (3DAS)* vector. **2.** a *phase* of the 3DAS. Theta angle and *tilt angle (TA)* represent the same angle of a hypothetical right-angled triangle whose hypotenuse is the *analytic signal amplitude (ASA)*, but they are expressed differently: theta angle θ is the arccosine of the ratio of the *total horizontal derivative (THDR)* to the ASA, i.e., $\theta = \arccos(\mathrm{THDR/ASA})$, whereas the TA is the arctangent of the ratio of the *first vertical derivative (FVD)* to the total horizontal derivative, i.e., $\mathrm{TA} = \arctan(\mathrm{FVD/THDR})$. The theta angle varies from 0° to 90° with the zero value above the source edges and other lithological contacts, and it responds with a maximum bracketed by two symmetric minima to form the 0°–90°–0° triplet above a vertical *thin dike* (Wijns et al., 2005). See also *contact*, *theta map*, and *3D gradient vector*.

theta map: a *grid*-based *enhanced processing* algorithm that calculates the cosine of the *theta angle* by the ratio of the *total horizontal derivative (THDR)* to the *analytic signal amplitude (ASA)*:

$$\cos\theta = \mathrm{THDR/ASA} = [(\partial F/\partial x)^2 + (\partial F/\partial y)^2]^{1/2} / [(\partial F/\partial x)^2 + (\partial F/\partial y)^2 + (\partial F/\partial z)^2]^{1/2},$$

where θ is theta angle, F is *total magnetic intensity (TMI)*, and $\partial F/\partial x$, $\partial F/\partial y$, $\partial F/\partial z$ are first-order derivatives of the TMI along the mutually orthogonal x-, y-, z-directions in *Cartesian coordinates*. Theta map values vary between 0 and 1. Above a vertical *contact*, the theta map shows a maximum bracketed by two symmetric minima (0–1–0 triplet), whereas above a vertical *thin dike* it shows a minimum bracketed by two maxima (1–0–1 triplet). Dipping contacts and dikes move these triplets in the down-dip direction and make the bracketing minima and maxima asymmetric, wherein this asymmetry indicates the direction of *dip*. A theta map equalizes the high- and low-amplitude anomalies in the manner of *amplitude balancing*, and its application is efficient in *edge detection*, especially in low-latitude and *magnetic equator* areas, where the direction of the *main geomagnetic field* is nearly or completely horizontal (Wijns et al., 2005). See also *improved theta map* and *second-order theta map*.

thick contact: a *model* approximation of a structural *edge*, such as the one associated with a deep-rooted high-*offset* fault, by a *two-dimensional (2D)* horizontal plate, or slab, that has a thickness sufficiently larger than its depth. See also *thick step* and *extended negative structural index*.

thick dike: a two-dimensional (2D) *model* approximation of a *tabular body* for which a width, or thickness, is approximately the same as the depth to its top. The difference between the thick-dike and thin-dike approximations depends on the source depth and distance from an *observation point*: the larger the depth and the distance, the lesser the difference between these approximations, so a shallow and nearby *thick dike* will become a deep and distant *thin dike*. Also referred to as *thick sheet*. See also *horizontal cylinder*, *sphere*, and *thick contact*.

thick plate: *finite contact* or *thick step* in the context of *source body* approximation by an infinite horizontal plate, or slab, with a finite thickness.

thick sheet: See *thick dike*.

thick step: same as *finite contact*, *finite step*, and *thick plate*.

thin dike: 1. a two-dimensional (2D) *model* approximation of the *igneous body* of a tabular shape with an *infinite bottom* and *infinite strike length* whose width, or thickness, is 4–6 times less than the depth to its top. **2.** a 2D approximation of a deep-rooted *magnetized fault*. Both intrasedimentary and upper *basement* igneous thin dikes generate high-*amplitude* and relatively short-*wavelength* positive anomalies in observed gravity and magnetic fields. The principal profile of the *horizontal gradient (HG)* over a thin-dike magnetic or gravity *anomaly* has a characteristic trough-between-peaks shape, wherein the *trough* is positioned immediately above the vertical or high-angle *dike*. Referred to with the same meaning as *vertical sheet*. See also *thick dike*, *observed gravity*, *observed magnetics*, and *principal profile*.

thin sheet: a two-dimensional (2D) *model* approximation of (a) structural discontinuity, such as dipping or vertical *fault* or *salt* dissolution zone, and (b) tabular *source body*, such as a horizontal *sill*-like or vertical *dike*-like *igneous intrusion*, with *infinite strike length* and thickness which is 4–6 times less than the depth to its top. Referred to with the same meaning as *thin dike*, *thin step*, and *vertical sheet*. See also *dip*, *horizontal cylinder*, *sphere*, and *tabular body*.

thin slab: *horizontal sheet* in the context of any flat-lying intra-sedimentary *source body* of a finite thickness which is much less than its depth and lateral extent.

thin step: a *fault*-truncated low-angle dipping or horizontal *thin sheet*. See also *dip*.

3C magnetic survey: *magnetic survey* with the use of a *3C magnetometer* for direct measurements of the three mutually orthogonal *vector* components of the *Earth magnetic field*.

3C magnetics: magnetic data obtained from the three-component (3C) measurements of the *geomagnetic vector* by ground, airborne, and satellite *magnetic vector* systems. See also *vector magnetometer*.

3C magnetometer: a three-component (3C) *magnetometer*, such as a triaxial *fluxgate magnetometer* or *SQUID magnetometer*, which measures the three mutually orthogonal *vector* components of *Earth magnetic field* in the direction of orientation of three mutually orthogonal axes of its magnetic sensors. Also referred to as *triaxial magnetometer* and *vector magnetometer*. See also *magnetic sensor*, *absolute scalar magnetometer (ASM)*, and *airborne vector magnetometry*.

three-dimensional (3D): 1. a characteristic of the mathematical transformation method, *grid*-based gravity or magnetic *model*, grid-based *inversion* of a residual gravity or magnetic field, and *source body* approximation that highlights their property to account for the spatial variations of parameters along three standard dimensions, such as the mutually orthogonal x-, y-, z-directions in *Cartesian coordinates*. **2.** a characteristic of an *acquisition* method for measurements of the 3D *gravity vector*, *magnetic vector*, and *full tensor gradient (FTG)* components. See also *dimension*, *one-dimensional (1D)*, *two-dimensional (2D)*, and *residual field*.

3D accelerometer: See *vector accelerometer*.

3D analytic signal (3DAS): 1. the *analytic signal (AS)* of gridded *anomalous field* F, where two horizontal derivatives $\partial F/\partial x$ and $\partial F/\partial y$ represent the real part and the *vertical derivative* $\partial F/\partial z$ is the imaginary part of the AS. At the *grid* point (x, y), a three-dimensional (3D) analytic signal can be expressed as

$$\begin{aligned}3DAS &= \mathbf{AS}(x,y)\\ &= (\partial F/\partial x)\boldsymbol{e}_x + (\partial F/\partial y)\boldsymbol{e}_y + i(\partial F/\partial z)\boldsymbol{e}_z,\end{aligned}$$

where $i=\sqrt{-1}$ is a complex unit number indicating the orthogonality and $\boldsymbol{e}_x$, $\boldsymbol{e}_y$, $\boldsymbol{e}_z$ are unit vectors parallel to the mutually orthogonal x-, y-, z-directions, respectively, in *Cartesian coordinates*. **2.** the analytic signal of a *3D source*. In *enhanced processing* and *interpretation* of gravity and magnetic data, 3DAS is often referred to with the same meaning as *3D analytic signal amplitude (3DASA)* and *total gradient (TG)*. See also *gridding*, *2D analytic signal (2DAS)*, and *direct analytic signal (DAS)*.

3D analytic signal amplitude (3DASA): the absolute value, or *magnitude*, of a *3D analytic signal (3DAS)*, expressed as

$$\begin{aligned}3DASA &= |3DAS|\\ &= [(\partial F/\partial x)^2 + (\partial F/\partial y)^2 + (\partial F/\partial z)^2]^{1/2},\end{aligned}$$

where $\partial F/\partial x$, $\partial F/\partial y$, $\partial F/\partial z$ are mutually orthogonal derivatives of *anomalous field* F in *Cartesian coordinates*. The 3DASA components of the *vertical integral* of a magnetic or gravity *anomaly*, i.e., 3DASA components of *anomalous potential*, represent the orthogonal *vector* components F_x, F_y, F_z of anomalous field F. The 3DASA of a gravity or magnetic anomaly is equivalent to the magnitude of a *3D gradient vector*. The 3DASA of a 3D source is dependent on everything its *anomaly* might depend: source *contrast* and depth, source depth extent, source *dip* angle, source *magnetization direction*, and spatial orientation of the *Earth magnetic field*, i.e., *inclination* and *declination*. The shape and value of a 3DASA also depend on the actual shape of an observation surface, because both are affected by an undulating surface and *drape flight*. The rate of attenuation, or *fall-off rate*, of a gravity or magnetic 3DASA is proportional to $1/r^3$ or $1/r^4$, respectively, where r is distance from a *point source*. References to a calculated 3DASA, 3D analytic signal (3DAS), and *total gradient (TG)* are often made interchangeably with the same meaning. See also *direct analytic signal amplitude (DASA)* and *2D analytic signal amplitude (2DASA)*.

3D Bouguer anomaly: the *field* of gridded gravity anomalies obtained by application of all relevant instrument and data corrections to ground, marine (shipboard), and airborne gravity data, including *terrain correction* for the three-dimensional (3D) *terrain effect*. It is the final result of the *gravity reduction* process and a basic dataset for *enhanced processing*, analysis, and *interpretation*. Referred to with the same meaning as *complete Bouguer anomaly*, especially in *marine gravity* surveys. Also referred to as *3D Bouguer gravity*. See also *gravity anomaly*, *gridding*, and *3D Bouguer correction*.

3D Bouguer correction: 1. a combined correction which represents two (for local-scale surveys) or three (for regional-scale surveys) gravity corrections applied to the *observed gravity* values obtained from ground and airborne gravity data: *simple Bouguer correction*, *terrain correction*, and *Earth curvature correction (Bullard B correction)*. **2.** *marine Bouguer correction* in the context of compensating for the three-dimensional (3D) gravity effects of the water *mass deficiency* and variations of *bathymetry* around each observation point.

See also *dimension, raw gravity corrections, observed gravity corrections*, and *Parker formula*.

3D Bouguer gravity: See *3D Bouguer anomaly*.

3D coordinates: the latitude, longitude, and *elevation* of observation points determined relative to a specific *reference ellipsoid* in the *Geodetic Reference System* implemented in a gravity or magnetic *survey*. See also *XY coordinates*.

3D Euler deconvolution: *Euler deconvolution* in application to gridded magnetic and gravity data. Calculations of the source location and depth are made for several *Euler structural index (SI)* values within a square window moving over a *grid* with a pre-selected increment. All points in a *moving window* are used to solve the overdetermined system of Euler deconvolution equations for the source position coordinates x_0, y_0, z_0. Calculations are run with different moving window sizes defined by the amount of the corresponding grid points. For every window position, the *standard deviation (SD)* of the obtained depth solutions is calculated, and solutions having deviations less than a user-specified SD value are accepted for the display. Calculated depth values are proportional to the window size: the larger the window, the deeper the calculated depth. The method is much more accurate in mapping the locations of structural discontinuities (faults, lithological contacts, etc.) than in estimating their depth. It is also dependent on the interpolated field values obtained by *gridding* and, hence, subjected to calculation errors due to *grid aliasing* (Reid et al., 1990; Reid et al., 2003). See also *fault, contact*, and *3D Werner deconvolution*.

3D Euler deconvolution equation: See *Euler deconvolution equation*.

3D gradient vector: a three-dimensional (3D) *vector* for which the components in the *x*-, *y*-, *z*-directions in *Cartesian coordinates* are mutually orthogonal horizontal and vertical derivatives of a gravity or magnetic *anomaly*. The length, or *magnitude*, of the 3D gradient vector at each observation point is defined by the *3D analytic signal amplitude (3DASA)* of an anomalous gravity or magnetic field. A 3D gradient vector is oriented in the direction of the maximum change of an *anomalous field*, and its projection on the XY plane is orthogonal to the *contour* lines of the equal field values on a *contour map*. Often referred to simply as *gradient vector*. See also *gradient, horizontal derivative, vertical derivative*, and *2D gradient vector*.

3D gravity: *FTG gravity* in the context of measuring *gravity gradient* anomalies in the three dimensions of *Cartesian coordinates*. See also *dimension* and *tensor-derived gravity*.

3D gravity survey: 1. *FTG survey* with the use of an *airborne gravity gradiometer (AGG)*. **2.** a high-*resolution* marine (shipboard) *gravity survey* with *survey line spacing* of 1 km or less and high-*accuracy* positioning provided by the *Global Positioning System (GPS)* or *differential GPS (DGPS)*. Often conducted as an *add-on survey* during a three-dimensional (3D) marine seismic program on the same ship. See also *marine gravimetry*.

3D gravity vector: See *gravity vector*.

3D Hilbert transform: 1. the spectral-domain equivalent of the two-dimensional (2D) *Riesz transform*. **2.** a generalization of the *Hilbert transform* for three dimensions in *Cartesian coordinates* in application to Fourier-transformed data, where it is composed of two parts: one part transforms the *x*-component and the other part transforms the *y*-component of the gridded data. 3D Hilbert transform components of the function $f(x, y) = f$ representing gravity or magnetic *grid* values can be expressed as

$$H_x(f) = \mathrm{F}^{-1}[H_1\mathcal{F}(f)] \quad \text{and} \quad H_y(f) = \mathrm{F}^{-1}[H_2\mathcal{F}(f)],$$

where F^{-1} is *inverse Fourier transform*; $\mathcal{F}(f)$ is *forward Fourier transform* of $f(x, y)$; and H_1 and H_2 are mutually orthogonal 3D Hilbert transform operators (Nabighian, 1984; LaFehr and Nabighian, 2012). One vertical and two horizontal derivatives of a gridded field f are related via the 3D Hilbert transform as

$$\mathcal{F}(\partial f/\partial z) = H_1\mathcal{F}(\partial f/\partial x) + H_2\mathcal{F}(\partial f/\partial y).$$

Also referred to as *generalized Hilbert transform*. See also *derivative, dimension, wavelength, wavenumber*, and *Cauchy-Riemann equations*.

3D inversion: 1. an *inversion* methodology applied to the constructed *3D model* of a local-scale geological *structure* or isolated *target*, such as an *ore body*. It is based on the use of the gridded gravity or magnetic *residual field* and available *constraints*, such as formation top depths from wells (boreholes), seismic depth horizons, rock densities and susceptibilities from core samples, and densities derived from seismic interval velocities. **2.** *voxel inversion*. See also *density, gridding, Gardner velocity-density relationship, joint inversion*, and *stochastic inversion*.

3D magnetic vector: See *magnetic vector*.

3D magnetics: See *full tensor magnetic gradiometry*.

3D model: 1. a *grid*-based three-dimensional (3D) gravity or magnetic *model* of a geological *structure* or isolated *target*, where one *dimension* is depth and the two other dimensions are horizontal distances along the mutually orthogonal *x*- and *y*-directions in *Cartesian coordinates*. Geological boundaries and isolated bodies are represented by their respective 3D surfaces and 3D shapes of contrasting densities and susceptibilities. A 3D model is constructed with the use of available seismic depth horizons, seismic interval velocities converted into apparent densities, formation top data from wells (boreholes), and other relevant geological information. The degree of detail in a constructed 3D model depends on the information content of the *constraints*. A 3D *forward model* shows the gravity or magnetic field generated by a constructed model. A 3D *inverse model* represents the targeted structure and distribution of *density* or *susceptibility* contrasts obtained by the *3D inversion* of a residual gravity or magnetic field. **2.** *voxel model*. See also *apparent density, density contrast, susceptibility contrast*, and *residual field*.

3D perspective plot: See *3D view*.

3D source: 1. a *source body* with a finite extent along each of three mutually orthogonal *x*-, *y*-, *z*-directions in *Cartesian coordinates*. **2.** a source body for which the horizontal and vertical dimensions are smaller than the flight *altitude* in an *airborne survey* or water depth in a *marine survey*. It is used as a *model* approximation of an *ore body*, shallow deposits of commercial-grade minerals, and local-scale geological structures, such as *anticline*, *graben*, *horst*, *reef*, and *uplift*, that are favorable for the accumulation of hydrocarbons. See also *dimension*, *infinite length*, *1D source*, and *2D source*.

3D Talwani algorithm: See *Talwani algorithm*.

3D Talwani method: See *Talwani method*.

3D view: an interactive three-dimensional (3D) visualization *format* which provides an out-of-scale perspective view of *interpretation* results, such as the *inversion*-obtained *structure* or interpreted *magnetic basement* depth, at a user-specified *azimuth* and angle above the horizontal level. A digital 3D view image selected for illustration purposes is also referred to as *isometric display*, *volume-rendered image*, and *3D perspective plot*.

3D Werner deconvolution: **1.** a generalization of *2D Werner deconvolution* for three dimensions in *Cartesian coordinates* applied to gridded magnetic and gravity data. It is based on the use of the *3D Hilbert transform* and data transformation concept implemented in *extended Euler deconvolution* (Nabighian and Hansen, 2001). **2.** a generalization of the two-dimensional (2D) *multiple-source Werner deconvolution* for three Cartesian dimensions and gridded magnetic and gravity data for calculation of the depth and location of 3D sources of arbitrary shapes. Its algorithm is based on the 3D Hilbert transform and approximation of the 3D *source body* by a homogeneous *polyhedron*, wherein the gravity or magnetic field of each homogeneous polyhedron is obtained in the *Fourier domain* by the sum of the contributions corresponding to each *vertex* of the polyhedron. Data are assumed to be acquired on a level surface or continued to it before calculations. Depth solutions are calculated within a *moving window*, the length of which is approximately twice the expected *target* depth (Hansen and Wang, 1988; Hansen, 2005). See also *gridding*, *3D source*, *downward continuation*, and *upward continuation*.

thrust: a geological *structure* created by the overriding movement of one sedimentary or *basement* rock unit over another along a low-angle (less than 45°) *fault* as a result of the regional-scale compressional tectonic deformations. Intrasedimentary and basement thrusts create positive lateral *density* and *magnetization* contrasts between rock formations and generate *total field* gravity and magnetic anomalies of various amplitudes. Thrust-generated anomalies are highlighted by calculation of the *residual field*. Intrasedimentary, or thin-skin, thrust can have a weak *magnetic signature* due to the low magnetization contrasts between the sedimentary rocks. Upper basement, or thick-skin, thrust commonly has a strong *gravity signature* and magnetic signature. See also *anomaly*, *density contrast*, and *magnetization contrast*.

Tiburg method: a graphical method of estimating the depth of isolated magnetic sources approximated by the *point dipole* geometry. The source depth is equal to 2/3 of the horizontal distance at one-half the maximum *amplitude* on the *principal profile* of a reduced-to-pole *magnetic anomaly*. Also referred to as *half-maximum method* and *Tiburg rule*. See also *depth rules* and *reduction-to-pole (RTP)*.

Tiburg rule: See *Tiburg method*.

tidal accelerations: vertical and horizontal components of *gravitational acceleration* due to the combined *gravitational attraction* of the Sun and Moon. They are calculated with the use of the *Longman formulas* for any time and any point on the Earth's surface regardless of the *reference ellipsoid* version. See also *Earth tide effect*.

tidal correction: **1.** *Earth tide correction*. **2.** a correction applied to *observed gravity* obtained from marine (shipboard) *gravity survey* data in order to compensate for the time-varying and latitude-dependent *Earth tide effect* caused by the combined *gravitational attraction* of the Moon and Sun. The correction value is calculated by the *Longman algorithm* which defines the *amplitude* and sign of the *tidal effect* at any time and any place on the Earth's surface. See also *ocean loading correction*.

tidal effect: **1.** the *Earth tide effect* on gravity measurements. **2.** a time-varying magnetic effect of the electric currents in the *ionosphere* that are generated by ionization of the moving atmospheric particles by the *solar wind* and by deformation of the ionospheric mass in response to the cumulative effects of the *gravitational attraction* of the Moon and Sun. See also *ocean loading effect*.

tidal gravity variations: periodic changes in the measured *Earth gravity field* values caused by the cumulative effects of the *gravitational attraction* of the Moon and Sun. They are compensated by application of the *Earth tide correction*. See also *temporal gravity variations*.

tidal tensor: a *gravity gradient tensor (GGT)* described by the conceptual analogy of a water droplet which in the absence of external gravity forces has the shape of a sphere, whereas in the presence of a *gravity field* the water droplet acquires the shape of an *ellipsoid* for which the axial parameters are related to the *gradient* components of the gravity field. In the case of *full tensor gradient (FTG)*, the stretching and compressing of the water droplet along the ellipsoidal axes is quantified by *FTG eigenvalues*, wherein the spatial orientation of the ellipsoidal axes is defined by the *FTG eigenvectors* at each observation point. See also *eigenvalue ellipsoid* and *FTG anisotropy*.

tie: **1.** the gravity or magnetic measurement repeated at a pre-selected observation point in a *ground survey* area in order to monitor the accumulated *instrument drift* and measurement errors. **2.** *tie point*. **3.** *gravity tie*.

tie-in: relating the gravity or magnetic values measured at new observation points, or stations, to the data acquired at previously established stations in a *ground survey* area. See also *acquisition* and *station*.

tie-line leveling: *leveling* with the use of *tie lines* required for the standard leveling of line-oriented shipboard (marine) and airborne gravity or magnetic data.

tie lines: *survey* lines which are orthogonal or, sometimes, oblique to *traverse lines*. The distance between tie lines, or tie-line spacing, is much larger than that of the traverse lines. The ratio of the tie-line spacing to the traverse-line spacing is often 4 or 5, but it can be more, depending on the survey objectives. The primary purpose of tie lines is to provide a set of intersections, or tie points, with the traverse lines for *leveling* of line-oriented shipboard (marine) and airborne gravity or magnetic data. The secondary purpose is minimizing *grid aliasing* and increasing *lateral resolution* of gridded data. In ground surveys, the tie line is the one which closes a *traverse loop*. Also referred to as *control lines*. See also *gridding*.

tie point: **1.** a coordinate point in the *survey* area where *traverse lines* and *tie lines* cross each other and their values should be tied, i.e., made equal to each other or differ within the *tolerance* limits. Also referred to as *crossover point*. **2.** the location in a two-dimensional (2D) or three-dimensional (3D) *model* where the values of the observed and calculated gravity or magnetic field are equal to each other. **3.** the *observation point*, commonly the ground *station*, where measurements are repeated in order to establish a relationship between two datasets acquired at different times. Also referred to as *repeat station*. See also *acquisition*, *gravity repeats*, and *static test*.

tie station: **1.** the *station* where gravity or magnetic measurements are made in two or more observation loops or in two or more ground gravity and magnetic surveys. **2.** *base station*. See also *loop* and *repeat station*.

tiling: a method of the interactive data visualization of several display windows showing the grids of derivatives, gradients, and other calculated attributes of an anomalous gravity or magnetic field or sub-windows of the same *grid* at different scales. See also *anomalous field*, *attribute*, *derivative*, *gradient*, and *montage*.

tilt: **1.** *tilt angle (TA)*. **2.** *platform tilt*. **3.** *sensor tilt*. **4.** incorrect vertical alignment of a *free-fall absolute gravimeter* during its setup.

tilt angle (TA): **1.** the *derivative*-based quantitative characteristic, or *attribute*, of an anomalous magnetic or gravity *field* defined by the arctangent of the ratio of the *vertical derivative* to the *total horizontal derivative (THDR)*. For *grid*-based calculations, its formula is expressed as

$$\mathrm{TA}\,(x,y) = \arctan\{(\partial F/\partial z)/[(\partial F/\partial x)^2 + (\partial F/\partial y)^2]^{1/2}\}$$
$$= \arctan\{(\partial F/\partial z)/\mathrm{THDR}\},$$

where $\partial F/\partial x$, $\partial F/\partial y$, $\partial F/\partial z$ are first-order derivatives of *anomalous field* F in the mutually orthogonal x-, y-, z-directions, respectively, in *Cartesian coordinates* (Miller and Singh, 1994). For line-oriented and *profile* data, the absolute value of the *horizontal derivative*, i.e., *horizontal gradient (HG)*, is often used:

$$\mathrm{TA}\,(x) = \arctan(\partial F/\partial z/|\partial F/\partial x|) = \arctan(\partial F/\partial z/\mathrm{HG}).$$

2. the angle between the adjacent side and hypotenuse of a hypothetical right-angled triangle whose adjacent sides are the first-order vertical and horizontal derivatives, respectively. **3.** the local, or instantaneous, *phase* of a *3D analytic signal (3DAS)*. The tilt angle retains no information about the amplitudes of anomalies, and degree of its *amplitude response* varies in accordance with the amplitudes of local gradients. The TA calculation has an *amplitude balancing* effect analogous to *automatic gain control (AGC)* that allows the detection, or resolution, of both shallow- and deep-sourced anomalies equally well. The TA values vary between −90° and +90° or $-\pi/2$ and $+\pi/2$ radians. For the inclination $I=90°$, i.e., for the reduced-to-pole (RTP) magnetic field, the TA has its positive values above the *source body*, zero values above the source edges, and negative values outside the source body. For the inclination $I=0°$, i.e., near the *magnetic equator*, TA values are negative above the source body and positive outside the source body. In the *Bouguer anomaly* field, TA is positive above the *mass excess* and negative above the *mass deficiency*. The TA is independent from the *density contrast* or *magnetization contrast* created by a source body. As a local characteristic of the *phase* of an *anomalous field* rather than its *amplitude*, the tilt angle is analogous to the *local phase*, but the local phase utilizes a sign of the horizontal derivative, whereas tilt angle is based on the use of its absolute value. Also referred to as *tilt*, *first-order tilt angle*, and, occasionally, *tilt derivative (TDR)*. See also *adaptive tilt angle (ATA)*, *ASTA filter*, *directional tilt angle*, *horizontal tilt angle (TDX)*, *improved tilt angle (ITA)*, *second-order tilt angle*, *tilt angle of horizontal gradient (TAHG)*, *tilt ASA*, *total horizontal derivative of tilt*, *tilt-depth method*, *tilt-derivative method*, and *zero-order tilt ASA*.

tilt angle of horizontal gradient (TAHG): the *edge*-enhancing *attribute* of an anomalous magnetic or gravity field used in the high-*resolution* mapping of structural edges. It is defined by the arctangent of the ratio of the *vertical derivative* of the *horizontal gradient (HG)* to the *total horizontal derivative (THDR)* of the horizontal gradient, expressed as

$$\mathrm{TAHG}$$
$$= \arctan\{(\partial \mathrm{HG}/\partial z)/[(\partial \mathrm{HG}/\partial x)^2 + (\partial \mathrm{HG}/\partial y)^2]^{1/2}\}$$
$$= \arctan\{(\partial \mathrm{THDR}/\partial z)/[(\partial \mathrm{THDR}/\partial x)^2 + (\partial \mathrm{THDR}/\partial y)^2]^{1/2}\},$$

where HG is the horizontal gradient, which has the same meaning as the total horizontal derivative THDR. Angular values of TAHG anomalies vary from $-\pi/2$ to $+\pi/2$ radians and reach their maxima above the *source body* edges. The TAHG calculation is much less influenced by the *dip* and *magnetization direction* effects, and it tends to equalize the *amplitude response* for shallow and deep sources in a manner similar to *automatic gain control (AGC)*. Because the obtained anomalies are represented by the second-order derivatives, *noise* in the data is also enhanced. Often referred to as *TAHG attribute* (Ferreira, F. et al., 2013). See also *edge*, *total horizontal derivative of tilt*, and *normalized derivatives*.

tilt ASA: the *edge*-enhancing *attribute* of anomalous gravity and magnetic fields defined by the *tilt angle (TA)* of the *analytic signal amplitude (ASA)*:

$$\mathrm{TA\_ASA} = \arctan\{-(\partial \mathrm{ASA}/\partial z)/[(\partial \mathrm{ASA}/\partial x)^2 + (\partial \mathrm{ASA}/\partial y)^2]^{1/2}\}.$$

For three-dimensional (3D) sources, its *amplitude response* is much less dependent on the *magnetization direction* of sources than the amplitude response of the standard ASA. For two-dimensional (2D) *contact* and *dike* source approximations, the tilt ASA is independent from the *magnetization vector* direction and *dip* of these sources. The TA calculation is not a linear transformation, and the tilt ASA is not the same as the ASA of the tilt angle (TA) calculated by an *ASTA filter*. Reduced dependence on the magnetization vector orientation improves the amplitude response of this attribute in high-resolution *edge detection* applications, especially in areas of low *magnetic inclination* and horizontal *induced magnetization* of sources at and near the *magnetic equator*. Because anomalies are represented by the second-order derivatives of the anomalous field, *noise* in the tilt ASA data is amplified (Cooper, 2014b). See also *anomalous field*, *derivative*, *lateral resolution*, and *zero-order tilt ASA*.

tilt correction: **1.** a correction applied to the *absolute gravity* measurements by the *free-fall method* to compensate for the *verticality error* caused by the incorrect vertical alignment of a *free-fall absolute gravimeter* during its setup at an *observation point*. **2.** a correction applied to the *relative gravity* measurements by a *spring gravimeter* to compensate for *sensor tilt*, i.e., for deviation of the sensing axis of the spring gravimeter from the vertical alignment. Also known as *sensor tilt correction*. **3.** *Harrison correction* or *Browne correction* applied to gravity values measured on a moving observation platform to compensate for *platform tilt*. Also known as *platform tilt correction*. Verticality error and tilt of the sensor axis or platform result in the artificial increase of the measured absolute and relative gravity values. See also *automatic tilt correction*.

tilt-depth method: a method of estimating the depth of isolated magnetic sources from the *contour* lines on *tilt angle (TA)* map or *grid*. Source depth is determined by one-half of the distance between +45° and −45° contours. The method assumes two-dimensional (2D) *source geometry* of a vertical or high-angle *magnetic contact* and its vertical *magnetization*, i.e., the magnetic field is reduced to the pole, with a negligible *remanent magnetization* component (Salem et al., 2007). Analytical expressions for the tilt-depth estimates and their *accuracy* in application for both magnetic and gravity sources approximated by the *contact* and *sphere* geometries are presented by Liu et al. (2017). See also *reduction-to-pole (RTP)*, *contact-depth method*, and *adaptive tilt-depth method*.

tilt derivative (TDR): **1.** the space-domain *attribute* of an anomalous gravity or magnetic field that represents the horizontal or vertical *derivative* of the *tilt angle (TA)* along one of the mutually orthogonal x-, y-, z-directions in *Cartesian coordinates*. As a qualitative estimate of the shape-related variations of the *local phase* of a gravity or magnetic *anomaly* and a quantitative characteristic describing its rate of change in the horizontal and vertical directions, the tilt derivative is equivalent to the *horizontal local wavenumber* and *vertical local wavenumber*, respectively. First-order tilt derivatives are used in estimating the depth, location, and *Euler structural index (SI)* of magnetic sources by the *tilt-derivative method*. **2.** *tilt angle (TA)*, often referred to as *tilt*, in the context of its origin, or derivation, from the concept of *normalized derivatives*. See also *derivative*, *space domain*, and *total horizontal derivative of tilt*.

tilt-derivative method: a *grid*-based automated method of estimating the depth, location, and *Euler structural index (SI)* of magnetic sources. The method utilizes the second-order derivatives of a reduced-to-pole (RTP) magnetic field and involves the calculation of three attributes: (1) the first-order derivatives of the *Euler deconvolution equation* along the mutually orthogonal x-, y-, z-directions in *Cartesian coordinates*; (2) the first-order derivatives of the *tilt angle (TA)* along the x-, y-, z-directions that are called local wavenumbers k_x, k_y, and k_z, respectively; and (3) the *total horizontal derivative of tilt* expressed as $k_h = (k_x + k_y)^{1/2}$, where h is an arbitrary horizontal direction. Rearranging the obtained derivatives and local wavenumber expressions yields a linear equation which does not require any prior SI information and relates the x-, y-, z-coordinates of the data points to the source depth z_0 and location (x_0, y_0) as

$$k_x x + k_y y + k_z z = k_x x_0 + k_y y_0 + k_z z_0.$$

The obtained equation is solved for the source position parameters (x_0, y_0, z_0) using the groups of consecutive observations, or moving data windows, centered at locations of the k_h maxima found by the *Blakely-Simpson method* or other similar technique. Window size selection depends on the data quality and degree of *interference of anomalies* from nearby sources, wherein the number of grid points in a *moving window* defines the overdetermined system of linear equations. Once the source position parameters have been determined, then the SI value can be calculated from the corresponding Euler deconvolution equation (Salem et al., 2008). See also *attribute*, *directional derivative*, and *local wavenumber (LW)*.

tilt filter: a *derivative*-based *filter* the algorithm of which calculates the *tilt angle (TA)* to transform the gridded magnetic or gravity field, such as the *Bouguer gravity* or *total magnetic intensity (TMI)*, into the field of anomalies generated by the *source body* edges and other structural discontinuities in the subsurface. The degree of *enhancement* of these anomalies varies in accordance with the ratio of amplitudes of the vertical and horizontal derivatives of the original *anomalous field*. See also *gridding*, *horizontal derivative*, and *vertical derivative*.

tilt THD: See *total horizontal derivative of tilt*.

time averaging: averaging of line-oriented gravity data acquired in the *time domain* in order to suppress the high-frequency components of *dynamic noise*. Data values are averaged over an empirically selected *period* of the gravity signal with the assumption that this period is longer than the periods of high-frequency vibrations of the *observation platform* during marine (shipboard) and airborne gravity surveys. See also *line averaging*.

time-differenced gravity: *time-lapse gravity* in the context of observing the differences between the gravity values measured at the same coordinate points with preselected time intervals.

time domain: **1.** a mathematical set of variables where time is an independent variable and the corresponding values of a particular function, such as the values of time-based measurements of the gravity or magnetic field along the airborne and marine (shipboard) *acquisition* lines, are the dependent variables, i.e., functions of the time of their measurement. **2.** a *domain* where variables are expressed as functions of time in contrast to the *Fourier domain* and the *space domain*, where variables are expressed as functions of *wavenumber* and distance from a reference point, respectively.

time filter: a time-based *low-pass filter* applied to time-domain *airborne gravity* and *gravity gradient* measurements along *airborne survey* lines in order to suppress the *dynamic noise* resulting, primarily, from the uncompensated inertial accelerations of the survey aircraft during a flight. The filter length is defined in seconds, so a time filter is referred to as 60-s filter, 100-s filter, etc. The spatial, or *wavelength*, equivalent of time filtering is determined by multiplying the aircraft speed and filter length. The obtained value defines the *full-wavelength resolution* of the airborne gravity or gravity gradient data: at an aircraft speed of 50 m · s^{-1} (180 km per hour) with a 80-s time filter, the full-wavelength resolution is 4.0 km. See also *half-wavelength resolution*, *inertial acceleration*, *space domain*, and *time domain*.

time-lapse gravity: high-accuracy *ground gravity* data acquired by repeated measurements at the same coordinate points on the ground with time intervals of several months or once per year to monitor the time-dependent changes in an anomalous gravity field caused by the development of an oil-and-gas or geothermal reservoir, groundwater-storage changes in natural (alluvial) aquifers, the growth of active volcanos, and CO_2 sequestration. Variations of the gravity field anomalies associated with changes in the fluid volume are generally less than 80–100 *microgal (μGal)*, and they can be reliably determined and mapped with the *precision* of gravity measurements better than 2–4 μGal and the *accuracy* of *elevation* measurements of 2–3 mm. Also referred to as *time-differenced gravity* and *4D gravity*. See also *acquisition*, *anomalous field*, *aquifer*, *microgravity*, *4D gravity monitoring*, and *hybrid time-lapse gravity*.

time-lapse gravity gradiometry: instruments and methods for measuring and processing of the *gradient* and *full tensor gradient (FTG)* data acquired at the same coordinate points on the ground with pre-defined time intervals. Time-dependent changes in the location and boundaries of the fluid-injector interfaces in relatively shallow oil-and-gas reservoirs, including those developed by steam-assisted gravity drainage (SAGD) technology to recover the bitumen from oil sands, can be determined with a higher *lateral resolution* of anomalies and higher *accuracy* of their estimation than those of *time-lapse gravity* due to a higher *sensitivity* of the gradient measurements. A shallow (~200 m) steam-injected heavy oil reservoir can have the gravity gradient change of 1–3 *Eötvös (E)* over a period of a few months, thereby requiring the sub-Eötvös *accuracy* of measurements (DiFrancesco and Talwani, 2002). Also known as *4D gravity gradiometry*.

time-lapse gravity survey: high-accuracy *ground gravity* or *microgravity* survey conducted with the objective of acquiring the *time-lapse gravity* data. At first, measurements are made to obtain the *baseline* gravity data, usually prior to the fluid injection into the oil or gas reservoir and initiation of the underground water horizon or geothermal reservoir development. Regular observations are repeated at the same coordinate points every 6–8 months or once per year. The survey requires high-*accuracy* measurements of *elevation* (~2–3 mm) at and around observation points and *precision* of gravity measurements better than 2–4 *microgal (μGal)*. Also referred to as *monitoring gravity survey* and *4D gravity survey*. See also *acquisition*, *observed gravity components*, and *4D gravity monitoring*.

time-lapse measurements: high-*accuracy* ground gravity and *gravity gradient* measurements at the same coordinate points with pre-defined time intervals in order to determine time-dependent changes in the *density* and volume of an exploration or research *target* by monitoring the related changes in the anomalous gravity field and its gradients. See also *anomalous field*, *gradient*, *time-lapse gravity*, *temporal gravity variations*, and *4D gravity monitoring*.

time-varying corrections: **1.** corrections applied to *observed gravity* to compensate for the time-varying gravity effects, or *temporal gravity variations*, which interfere with the *absolute gravity* and *relative gravity* measurements. They include *atmospheric mass correction*, *atmospheric pressure correction*, *Earth tide correction*, *ocean loading correction*, and *polar motion correction*. **2.** corrections applied to *observed magnetics* to compensate for the short-term variations of the *external magnetic field* and long-term variations of the *main geomagnetic field* that are known as *temporal magnetic variations* and compensated by application of *diurnal correction* and *IGRF correction*, respectively.

time-varying gravity: **1.** time-varying components of measured *relative gravity* that are known as *temporal gravity variations*. **2.** time-varying gravity effects which interfere with the *absolute gravity* measurements by the *free-fall method* and are compensated by application of *atmospheric pressure correction*, *ocean loading correction*, *polar motion correction*, and *Earth tide correction*. **3.** *time-lapse gravity*. See also *observed gravity components*.

time-varying magnetics: time-varying components of the measured *Earth magnetic field* that are known as *temporal magnetic variations*. See also *observed magnetic components*.

TMA field: same as *TMI field* with an emphasis on it being a *field* of all local- and regional-scale magnetic anomalies, or the *total magnetic anomaly (TMA)*, originating primarily in the upper *crust*.

TMI: See *total magnetic intensity (TMI)*.

TMI field: *total magnetic intensity (TMI)* values which represent the *field* of the magnetic anomalies obtained by (a) application of *diurnal correction* and *IGRF correction* to *observed magnetics* and (b) *cultural editing* of the corrected line-oriented or gridded data. Along with the *RTP magnetic field*, the TMI field provides a basis for subsequent *enhanced processing*, analysis, and *interpretation* of magnetic data. It is displayed in the *format* of either line-oriented data or a *grid*. Also referred to as *TMA field*.

TMI grid: *total magnetic intensity (TMI)* data in the format of a *grid* after *leveling*, *cultural editing* and, if necessary, *microleveling* of gridded magnetic data. TMI grid represents the anomalous components of the measured *Earth magnetic field* obtained by application of *observed magnetic corrections* without *reduction-to-pole (RTP)* or *reduction-to-equator (RTE)*. See also *gridding*.

TMI magnetics: *total magnetic intensity (TMI)* data obtained by application of *observed magnetic corrections* to *observed magnetics* which represent the measured magnetic values after application of *raw magnetic corrections* for the instrument and *acquisition* effects.

tolerance: 1. a range of permitted variability of observed variables, such as: (a) the maximum level of *diurnals* or *drift* acceptable for a magnetic or gravity *survey*; (b) the maximum deviations from the criteria established by the *figure-of-merit (FOM)* for the aeromagnetic measurements or *static test* for *airborne gravity* measurements; (c) the minimum number of source depth solutions for *clustering*; (d) the convergence parameter for *structural inversion*, *property inversion*, *stochastic inversion*, and *inversion* with the use of an *objective function*; and (e) truncation of an infinite series in analytical expressions. **2.** a *modeling* parameter that defines the *model misfit* value at which a difference between the observed and model-generated fields is considered acceptable. Also known as *stopping criterion*. Tolerance is one of the *acceptance criteria* in *quality control (QC)* over gravity and magnetic measurements.

topographic correction: same as *terrain correction* with an emphasis on its use as the gravity correction for variations of the *topography* around observation points in a *ground gravity* survey and below observation points in *airborne gravity* and airborne *gravity gradient* surveys.

topographic height: *elevation* in the context of a topographic, or surface, *relief* parameter at a ground or airborne observation point. See also *orthometric height*.

topographic magnetic anomalies: magnetic anomalies of various amplitudes and, mostly, short wavelengths generated by the magnetized rocks and unconsolidated rock material that comprise the topographic surface and fill-in its low-laying parts in a *magnetic survey* area. Topographic magnetic anomalies can be identified by a strong correlation of their shapes with elements of the *topography*. See also *regolith* and *wavelength*.

topography: a general configuration of the ground surface, or ground *relief*, representing the shapes and relative positions of the land parts on the Earth's surface. In exploration applications, the topography is presented in the format of the *Digital Terrain Model (DTM)* with surface elevations determined relative to sea level or, predominantly, a *reference ellipsoid*. Carrying out ground measurements of the topographic relief variations with the use of high-*precision* geodetic instruments is called *surveying*. See also *elevation*, *LIDAR*, and *bathymetry*.

torsion balance: 1. the design concept for measuring the horizontal gradients of the horizontal and vertical components of the *Earth gravity field*. Developed at the end of the 19th century by Lorand von Eötvös, the torsion balance measuring system is based on the use of a horizontal bar, or beam, suspended by a platinum-iridium or tungsten wire, and two equal proof masses at opposite ends of the beam: one is commonly attached to the beam and another is hanging at a lower height. In a non-uniform gravity field, the horizontal components of *gravitational attraction* acting on these masses are different and the beam rotates, or twists, about the vertical axis, until its rotation is counter-balanced by a torque, or twisting force, of the wire that is proportional to the *gravitational force*. Because of the different heights of two proof masses, four gradients can be determined in the reference frame of *Cartesian coordinates*: (1) the *horizontal gradient* of the horizontal component g_x of *gravity vector* **g** along the orthogonal *y*-axis, denoted in symbols of the *gradient tensor* by T_{xy}; (2) the difference between two horizontal gradients of the mutually orthogonal horizontal components g_x and g_y, denoted by $T_{xx} - T_{yy}$; (3) the horizontal gradient of the vertical component g_z along the *x*-axis, denoted by T_{zx}; and (4) the horizontal gradient of the vertical component g_z along the *y*-axis, denoted by T_{zy}. Quantities T_{xy} and $T_{xx} - T_{yy}$ are known as *differential curvature gradients*. When both proof masses are hanging at the same height, the instrument is called a *curvature variometer*. **2.** *gravity gradiometer* which measures the second-order horizontal derivatives of *gravitational potential* that represent the horizontal gradients of the horizontal and vertical components of the Earth's gravity field. Also known as *Eötvös torsion balance*. See also *derivative*, *differential curvature*, and *proof mass*.

torsion-balance gravity gradiometer: a *gravity gradiometer* the design of which is based on the *torsion balance* concept for the stationary measurements of the horizontal gradients of the horizontal and vertical components of the *Earth gravity field* on the ground. For a single beam instrument, five observations with different azimuths of the beam orientation are made at each *station* to determine (a) three horizontal gradients, or *tensor gradient* elements, denoted by T_{xy}, T_{zx}, T_{zy}, and (b) the difference between the two horizontal gradients of the horizontal components denoted by $T_{xx} - T_{yy}$. The direction θ_H and *magnitude* of the *horizontal gradient* |HG| of the vertical component g_z are calculated as

$$\theta_H = \arctan(T_{zy}/T_{zx}) \quad \text{and} \quad |\text{HG}| = (T_{zx} + T_{zy})^{1/2},$$

where θ_H is *azimuth* of the horizontal component of the *gradient vector*. The achievable *accuracy* of measurements is 1–3 *Eötvös (E)*. See also *rotating-disk gravity gradiometer*.

torsion magnetometer: See *variometer*.

total absolute magnetic intensity (TAMI): the strength, or intensity, of the *main geomagnetic field* measured at a particular time and fixed location of a *geomagnetic observatory*. It is a *scalar* quantity that represents the *magnitude* of *geomagnetic vector* **B**, expressed as

$$|\mathbf{B}| = (B_z^2 + B_x^2 + B_y^2)^{1/2},$$

where B_z, B_x, B_y are mutually orthogonal vertical and horizontal components of the *main geomagnetic field* in *Cartesian coordinates*. The TAMI is calculated from the geomagnetic vector measurements at geomagnetic observatories and onboard Earth-orbiting satellites by a *fluxgate magnetometer, helium magnetometer*, or *absolute scalar magnetometer (ASM)*. The TAMI is different from the *total magnetic intensity (TMI)* obtained from measurements by a regular, or conventional, *scalar magnetometer* which measures a numerical value of the Earth's magnetic field regardless of its *vector* direction. Referred to with the same meaning as *geomagnetic field strength* and *total geomagnetic intensity*. See also *absolute magnetic direction*.

total acceleration: a *vector* quantity $\boldsymbol{g}$ which defines the *gravitational acceleration* on the Earth's surface. Its *magnitude* $g = |\boldsymbol{g}|$ and spatial orientation are defined by a vector sum of the *Newtonian gravitational acceleration* $\boldsymbol{g}_N$ due to the attraction of the stationary, or non-rotating, Earth mass and *centrifugal acceleration* $\boldsymbol{g}_C$ due to the Earth's rotation about its axis, expressed as $\boldsymbol{g} = \boldsymbol{g}_N + \boldsymbol{g}_C$. Total acceleration on the Earth's surface is oriented in the direction of a *plumb line* at a ground *observation point*. See also *deflection of the vertical* and *gravity vector*.

total Bouguer reduction: same as *complete Bouguer correction* with an emphasis on it representing the sum, or total, of a *simple Bouguer correction, terrain correction*, and, if necessary, *Bullard B correction*.

total curvature: See *Gaussian curvature*.

total Eötvös acceleration: same as *Eötvös acceleration* with an emphasis on the motion-related *inertial acceleration* representing the sum, or total, of the vertical components of the direction-dependent *Coriolis acceleration* and direction-independent *centrifugal acceleration*.

total Eötvös correction: *Eötvös correction* in the context of its compensation for the combined, or total, gravity effect of the vertical components of *Coriolis acceleration* and *centrifugal acceleration* of an *observation platform* moving over the curved surface of the rotating Earth.

total field: a *scalar field* of gravity or magnetic anomalies obtained from measurements of the absolute or relative numerical values of the Earth's gravity or magnetic field as opposed to the anomalous fields obtained from the *gradient, gradient tensor*, and *vector* measurements. *Bouguer anomaly, free-air anomaly, TMI field*, and *RTP magnetic field* are examples of total fields. In *enhanced processing*, the total field is divided into the *residual field* and the *regional field* by the *regional-residual separation* methods for subsequent analysis and *interpretation*. See also *anomalous field, total field gradiometry*, and *total field modulus*.

total field anomaly: a gravity or magnetic *anomaly* obtained from measurements of the absolute or relative numerical values of the *Earth gravity field* or *Earth magnetic field* as opposed to anomalies obtained from the *gradient, gradient tensor*, and *vector* measurements. Total field anomalies are determined by application of all relevant corrections to the measured values and represented by the *total magnetic intensity (TMI)* and *RTP magnetic field* for magnetic data and *free-air anomaly, Bouguer anomaly*, and *isostatic anomaly* for gravity data.

total field gradient: 1. a horizontal or vertical *gradient* obtained from measurements by two regular, or conventional, gravimeters or magnetometers separated horizontally or vertically by a distance called the *baseline*. **2.** *calculated gradient*. See also *gravimeter, magnetometer, gradiometer*, and *measured gradient*.

total field gradiometry: a methodology for the *enhanced processing* and *interpretation* of the gravity or magnetic gradients calculated after application of all relevant corrections to and *gridding* of the values obtained from measurements by a regular, or conventional, *gravimeter* or *magnetometer*. The achievable *lateral resolution* of *calculated gradient* anomalies is less than that of *measured gradient* anomalies obtained from measurements by *gradiometer* systems due to (a) the much larger *grid interval* in comparison with the *baseline* of a gravity or magnetic gradiometer, (b) the spatial averaging of the *gridding* process, and (c) the superposition of the regional components in the *total field* data. See also *full tensor gradiometry, observed gravity corrections*, and *observed magnetic corrections*.

total field magnetometer: *scalar magnetometer* in the context of its measurements of the total numerical value of a magnetic field, such as the *Earth magnetic field*, regardless of its *vector* orientation at an *observation point*.

total field modulus: the absolute maximum value of a positive or negative *anomaly* in a gravity or magnetic *total field*. In the *noise*-free theory, it corresponds to the zero value of the *horizontal derivative* of an *anomalous field* at the same coordinate point. See also *background noise* and *modulus*.

total geomagnetic intensity: *total absolute magnetic intensity (TAMI)* or *geomagnetic field strength*.

total gradient (TG): a quantitative characteristic, or *attribute*, of the anomalous gravity or magnetic *field* F, defined by the square root of the sum of its squared derivatives along the mutually orthogonal *x*-, *y*-, *z*-directions in *Cartesian coordinates*, expressed as

$$\mathrm{TG} = [(\partial \mathrm{F}/\partial x)^2 + (\partial \mathrm{F}/\partial y)^2 + (\partial \mathrm{F}/\partial z)^2]^{1/2}.$$

TG is always positive and its maxima are centered above the vertical *thin dike* and *contact* source geometries. Over a *thick dike*, it has the *signature* of two maxima equidistant from the local minimum which is located directly above the thick dike center. In the case of a dipping dike or contact, the TG maximum is reached at the contact top, while the maximum of the *horizontal gradient (HG)* is shifted downdip, and this property is used by the *gradient-dip estimation* method. For two-dimensional (2D) *source body* approximation, the TG is independent from the directions of the *main geomagnetic field* and *remanent magnetization*, thereby making its calculation efficient for mapping source bodies in the low *magnetic latitude* and *magnetic equator* areas. For three-dimensional (3D) source bodies, the total gradient is dependent on the directions of both the source magnetization and main geomagnetic field. A decrease in the *amplitude*, or *fall-off*, of the gravity and magnetic TG with distance r from the source is proportional, respectively, to $1/r^2$ and $1/r^3$ for 2D sources and $1/r^3$ and $1/r^4$ for a *point source*. The total gradient (TG) and its amplitude (TGA) and the *3D analytic signal (3DAS)* and its amplitude (*3DASA*) are often referred to with the same meaning; however, the total gradient is a *scalar* variable, whereas the *analytic signal (AS)* is a *complex variable*, and in the case of measured gradients their amplitudes and *lateral resolution* of anomalies are different. Sometimes referred to as *full gradient*. See also *anomalous field*, *derivative*, *gradient*, *calculated gradient*, and *measured gradient*.

total gradient method: a line- or *grid*-based automated method of *enhanced processing* and *interpretation* of anomalous magnetic and gravity fields that is based on calculation of the *total gradient (TG)* and its various processing versions. In the case of calculated gradients, it is referred to with the same meaning as *analytic signal method*. See also *calculated gradient* and *measured gradient*.

total gravitational potential: *gravitational potential* of a rotating planetary body of *mass*, such as the Earth. It is a *scalar* quantity defined by the sum, or total, of the *self-gravitational potential* and *centrifugal potential* at any point on or above the Earth's surface within the coordinate frame rotating with the same *angular velocity* as the Earth. In geodetic applications, the total gravitational potential is referred to as the *Earth gravity potential* and is used for mapping a *geoid* and calculating a *geoid undulation*. The total gravitational potential of a *reference ellipsoid* (also known as *normal ellipsoid*) is called *normal gravitational potential*.

total horizontal component: 1. a *vector* quantity **H** defined by the sum of two horizontal components B_x and B_y of three-dimensional (3D) *geomagnetic vector* **B** or the horizontal components g_x and g_y of *3D gravity vector* $\mathbf{g}$. **2.** a *scalar* quantity denoted by |H|, or B_H and g_H, that represents the length, or *magnitude*, of the vector **H**, expressed as

$$|\mathrm{H}| = B_H = (B_x^2 + B_y^2)^{1/2}, \quad |\mathrm{H}| = g_H = (g_x^2 + g_y^2)^{1/2}.$$

By convention, B_x and g_x are the north-oriented components and B_y and g_y are the east-oriented components of geomagnetic vector **B** and gravity vector $\mathbf{g}$, respectively. The direction of the total horizontal component **H** defines the direction of the *magnetic meridian* at an observation point. Sometimes referred to as *horizontal intensity*. See also *geomagnetic field parameters* and *geomagnetic observatory*.

total horizontal derivative (THDR): a *scalar* quantity defined by the square root of the sum of the squared horizontal derivatives of a gridded anomalous gravity or magnetic field F that are calculated along the mutually orthogonal *x*- and *y*-directions in *Cartesian coordinates*:

$$\mathrm{THDR} = [(\partial \mathrm{F}/\partial x)^2 + (\partial \mathrm{F}/\partial y)^2]^{1/2}.$$

Referred to with the same meaning as *horizontal gradient (HG)*, *total horizontal gradient*, and *horizontal gradient magnitude (HGM)*. See also *anomalous field*, *derivative*, *gradient*, and *gridding*.

total horizontal derivative of tilt: the *attribute* of an anomalous magnetic or gravity field defined by the square root of the sum of the squared horizontal derivatives of the *tilt angle (TA)* along the mutually orthogonal *x*- and *y*-directions in *Cartesian coordinates*:

$$\mathrm{THDR\_TA} = [(\partial \mathrm{TA}/\partial x)^2 + (\partial \mathrm{TA}/\partial y)^2]^{1/2}.$$

Its *amplitude* is essentially independent from the amplitudes of the *total magnetic intensity (TMI)* anomalies. For *two-dimensional (2D)* source bodies, it is independent from *geomagnetic inclination*, source *dip*, and *magnetization direction*, but it tends to produce low-amplitude secondary edges. The same as *analytic signal amplitude (ASA)*, THDR_TA has its maximum values above the *source body* edges and *thin dike* source geometries, but its maxima are better defined and more resolving laterally than those of the ASA. Because the obtained anomalies are represented by second-order derivatives, *noise* components of the original TA are enhanced (Verduzco et al., 2004). Also referred to as *horizontal gradient of tilt* and *tilt THD*. See also *derivative*, *horizontal tilt angle (TDX)*, *tilt angle of horizontal gradient (TAHG)*, and *tilt ASA*.

total horizontal gradient: See *total horizontal derivative (THDR)*.

total horizontal wavenumber: 1. an *attribute* of the reduced-to-pole (RTP) magnetic field implemented in the *tilt-derivative method*, defined as

$$k_h = (k_x^2 + k_y^2)^{1/2},$$

where h is an arbitrary horizontal direction on the XY plane and k_x and k_y are horizontal local wavenumbers representing the first-order derivatives of the *tilt angle (TA)* in the mutually orthogonal *x*- and *y*-directions in *Cartesian coordinates*. **2.** *total horizontal derivative of tilt*. **3.** the length, or *magnitude*, of a *wavenumber vector*. See also *reduction-to-pole (RTP)* and *horizontal local wavenumber*.

total intensity: 1. a quantitative characteristic of the strength of a gravity or magnetic *field* defined by the *magnitude* of the three-dimensional (3D) *gravity vector* or *magnetic vector*, respectively. **2.** *total magnetic intensity (TMI)* obtained from single-component measurements of the *Earth magnetic field* by a *scalar magnetometer*. **3.** *total absolute magnetic intensity (TAMI)* obtained from three-component measurements of the *geomagnetic vector* by a *vector magnetometer* at a *geomagnetic observatory* or onboard an Earth-orbiting satellite. See also *vertical intensity* and *horizontal intensity*.

total magnetic anomaly (TMA): same as *total magnetic intensity (TMI)* with an emphasis on its components representing the sum, or total, of all regional and local magnetic anomalies generated, primarily, in the upper *crust*. Long-*wavelength* contributions from sources in the lower crust and upper *mantle* can be significant in terms of the total *amplitude*, but their horizontal rate of change is negligible for exploration applications.

total magnetic field (TMF): *total magnetic intensity (TMI)* in the context of the anomalous magnetic *field* representing the sum, or total, of all magnetic anomalies generated predominantly in the upper part of the *crust.*

total magnetic intensity (TMI): 1. a numerical, or *scalar,* value that represents the *magnitude* of the *vector* sum of the *main geomagnetic field* and local- and regional-scale magnetic fields at each *observation point.* **2.** the *field* of the magnetic anomalies obtained from *observed magnetics* by application of *IGRF correction, diurnal correction* and, if necessary, *culturing editing* of magnetic data, i.e., it is the anomalous magnetic field after compensation for the planetary and non-geological components of the measured *Earth magnetic field* values. The TMI represents the sum, or total, of the magnetic fields generated by all sources in the *lithosphere*, wherein the major contributors are the main geomagnetic field and regional and local fields generated by magnetized rocks in the upper *crust.* The TMI closely approximates the anomalous *amplitude* components in the direction of the present main geomagnetic field with the assumption that the *vector* of each *magnetic anomaly* and the vector of the main geomagnetic field have the same orientation at observation points. Approximation errors due to the actual differences in the vector directions are mostly negligible, because the typical crustal magnetic anomalies are much less than the main geomagnetic field values. For iron ore deposits and certain ultramafic igneous rocks generating anomalies in excess of 10,000 *nanotesla (nT)*, the amplitude errors of this approximation can be up to 10%. The TMI anomalies have asymmetric trough-to-peak *profile* signatures with peaks shifted to the south from troughs in the northern hemisphere and to the north in the southern hemisphere. At the *magnetic equator,* where the main geomagnetic field is horizontal, the TMI profile over a *sphere*-like source with *induced magnetization* and positive *magnetization contrast* has the shape of a localized *trough* between two small artificial *peak* lobes that corresponds to a negative anomaly in the *residual field.* Generally, the TMI field contains small contributions from sources within the lower crust and upper *mantle*, because the area extent of regular magnetic surveys is not sufficient for a proper *sampling* of the deep-sourced long-*wavelength* components of the *lithospheric magnetic field.* The TMI is often the final result of standard magnetic *data processing* and is presented in the *format* of line-oriented and gridded data. Grids of the TMI and reduced-to-pole TMI data are two basic datasets for *enhanced processing* and *interpretation.* Also referred to as *total magnetic anomaly (TMA)* and *total magnetic field (TMF).* See also *gridding, reduction-to-pole (RTP)*, and *total absolute magnetic intensity (TAMI).*

total magnetic vector: See *geomagnetic vector.*

total magnetization: 1. a rock *magnetization* characteristic that describes the *magnitude* and direction of rock magnetization *vector* $\mathbf{M}$. It is defined by the vector sum of the *induced magnetization* vector $\mathbf{M_I}$ and *remanent magnetization* vector $\mathbf{M_R}$ as

$$\mathbf{M} = \mathbf{M_I} + \mathbf{M_R}.$$

Due to heterogeneity of a remanent magnetization resulting from the repeated partial realignments of the atomic magnetic moments by all major changes of the *Earth magnetic field* during the past geological times, the induced magnetization component is often, but not always, dominant. **2.** a characteristic of rock magnetization that describes a dominant alignment of the elementary magnetic dipoles in the rock-forming minerals. Also known as *total polarization.* See also *magnetic dipole moment, diamagnetism, ferromagnetism,* and *paramagnetism.*

total polarization: *total magnetization* in the context of the rock *magnetization* property to line up, or polarize, originally random-distributed magnetic dipoles within rock-forming minerals.

total spectral power: *radial power spectrum* in the context of the total spectral energy, or power (squared amplitudes), of Fourier-transformed anomalies in a gridded *anomalous field.*

towed bird system: an airborne system for magnetic field measurements with the use of a single-sensor *magnetometer* or multi-sensor *magnetic gradiometer* mounted in a streamlined housing, or *bird*, which is towed at 40–50 m below the aircraft to keep the magnetic sensors away from the electromagnetic interference effects of a *survey* aircraft. See also *magnetic sensor, slung sensor,* and *stinger.*

towed deep ocean gravimeter (TOWDOG): the automated submarine *relative gravimeter* system mounted inside a watertight pressure housing designed for underwater towing stability. It is towed in close proximity to the sea floor, thereby providing a much higher lateral and vertical *resolution* of the *relative gravity* anomalies than the resolution achievable by shipboard gravity measurements on the sea surface. See also *free-water gradient, free-water anomaly,* and *AUV gravimeter.*

tower gradiometer: an instrument for measurements of the *vertical gradient (VG)* of the Earth's gravity or magnetic field by a single *gravimeter* or *magnetometer* which is placed at two vertically separated fixed observation positions on a measurement stand called "tower." The difference between the instrument readings divided by their separation distance, or *baseline*, defines a rate of change of the Earth's gravity or magnetic field along the vertical direction. Also referred to as *vertical gradiometer.*

trace: See *gradient tensor trace.*

track: an imaginary line on the Earth's surface along which marine (shipboard), airborne, and satellite (spaceborne) measurements of the *Earth gravity field, Earth magnetic field,* ocean/sea surface, and ground *topography* are made. See also *banded flight tracks* and *satellite altimetry.*

track line data: gravity or magnetic data acquired along imaginary lines on the Earth's surface referred to as tracks. Primarily, this term is used in reference to satellite and marine (shipboard) data. See also *acquisition.*

track plot: a map showing the coordinates and locations of marine (shipboard) and airborne gravity or magnetic *survey* lines and observation points. Also known as *index map.*

track spacing: *survey line spacing* in the context of marine (shipboard) or airborne *survey* and satellite (spaceborne) missions.

tracking speed: an airborne *magnetometer* characteristic that describes the maximum change of the *Earth magnetic field* in units of *nanotesla (nT)* per second ($nT \cdot s^{-1}$) that can be recorded coherently, i.e., without significant distortions of its true value, but not necessarily within the full specifications regarding the magnetometer *sensitivity* and *accuracy*. Similar to *gradient tolerance*, the tracking speed defines the operational limits for each particular type of *airborne magnetometer*.

transfer characteristic: See *transfer function*.

transfer function: **1.** a *filter* characteristic that describes its *amplitude response* and *phase response* to variations of the space-domain or spectral-domain parameters of the input data, such as *phase* and *wavelength* or *wavenumber*, respectively. Also referred to as *filter response*. **2.** a characteristic of a data processing algorithm that describes the transformation of an *anomalous field* with the use of (a) specific mathematical properties of the field-approximating harmonic functions, such as those implemented in *Fourier transform*, *Hilbert transform*, and *wavelet transform*, and (b) closed-form analytical expressions, or formulas, relating the observed anomalies to the specific *source geometry* and physical properties (*density* or *susceptibility* contrast) of the source bodies. Also known as *transfer characteristic*. See also *contrast*, *harmonic function*, *space domain*, *spectral domain*, and *convolution*.

transfer height correction: See *datum transfer correction*.

transform: a change, or conversion, of observed gravity and magnetic data from one mathematical representation to another. For example, the *Fourier transform* converts observed data from their original *space domain* to the equivalent *spectral domain*. See also *Hilbert transform*, *Riesz transform*, *Walsh transform*, and *wavelet transform*.

transient disturbances: short-duration, highly variable *amplitude* fluctuations of the *external magnetic field* of the Earth that are caused, primarily, by the interaction between the solar radiation (*solar wind*, solar flares, etc.) and deep-space radiation effects with the *magnetosphere*. Transient disturbances are recorded by a *base-station magnetometer*, and their contribution to the measured magnetic field values is compensated by application of *diurnal correction*. See also *magnetic disturbances*.

transients: See *transient disturbances*.

translational acceleration: *inertial acceleration* in the context of its interference with, or translation into, the measurements of *gravitational acceleration* and its components by the gravity and *gravity gradient* systems on a moving *observation platform*, such as a *survey* ship, aircraft, and satellite.

transpose: *matrix* denoted by A^T the rows of which are identical to the columns of matrix A.

transverse gradient: **1.** the *horizontal gradient* measured in the direction which is orthogonal, or transverse, to the flight lines in airborne magnetic surveys. For *magnetic gradient* surveys with two wingtip-mounted sensors and one *sensor* at the tail *stinger*, it is measured by calculating the difference between two wingtip measurements of the *total magnetic field* divided by the distance between the aircraft wingtips. **2.** the horizontal gradient measured by a *transverse magnetic gradiometer* during a seabed (submarine) *survey* for the detection of *UXO* objects, pipelines, power cables, etc. Transverse gradients can be used in *gradient-enhanced gridding*. Also referred to as *lateral gradient*. See also *cross-line gradient*, *longitudinal gradient*, and *Nelson method*.

transverse magnetic gradiometer: a fixed-*array* airborne, ship-towed, or towed by *autonomous underwater vehicle (AUV)* instrument system composed of two synchronized magnetometers for measurements of the *horizontal gradient* of the Earth's magnetic field in the direction which is orthogonal, or transverse, to the orientation of the survey lines that is known as the *transverse gradient*. See also *AUV magnetometer*, *gradient array*, and *quasi-analytic signal*.

transverse Mercator projection: *map projection* which transforms points on the curved Earth's surface onto the global-scale cylinder at the right angle, or transverse, to the Earth axis plane. Transverse Mercator projection is the basis for the *Universal Transverse Mercator (UTM)* planar, or projected, *coordinate system*.

trap: an intrasedimentary geological formation, or *structure*, where hydrocarbons can be accumulated. A trap has three main components: (1) porous rock capable of holding the hydrocarbon fluids, i.e., reservoir rock; (2) sufficiently thick overlaying impermeable rock, such as shale or *salt*, which forms a barrier for the upward movement of hydrocarbons, i.e., sealing rock; and (3) upward curved, or convex, shape or laterally impermeable structural-lithological boundary, such as a *fault*, that prevents the lateral migration of hydrocarbons. The structural trap is associated with a *structural high*, such as *anticline*, *carbonate platform*, *flower structure*, *horst*, *thrust*, and *uplift*, that generates a positive *gravity anomaly* in the *residual field*. Stratigraphic traps are associated with the local lithological variations within layered sediments, such as the sand lenses, sand-filled fluvial paleochannels and incised valleys, eroded granite intrusions at the top of the *basement* ("granite wash"), *subcrop*-sealed reservoir-grade sands, and porous reefs overlain with impermeable rock formations. Sand-filled stratigraphic traps have the *gravity signature* of a low-*amplitude* negative *residual gravity* anomaly, whereas carbonate buildup, or *reef*, generates a positive gravity anomaly in the residual field. Mid-depth and deep intrasedimentary traps usually have no detectable magnetic signatures due to a low *magnetization contrast* between the sedimentary rocks. A stratigraphic trap within the uppermost sedimentary section can have the clustered short-*wavelength* anomalous *magnetic signature* produced by localized accumulations of *detrital magnetite* and *diagenetic magnetite*. Sometimes, shallow traps can be identified by *micromagnetic anomalies*, such as *magnetic bright spot*, associated with the *hydrocarbon seepage effect*. See also *clustering*, *igneous intrusion*, *incised valley*, *paleochannel*, and *subcrop anomaly*.

trap-door structure: a *fault*-bounded *edge*-like geological *structure* representing the local *structural high* formed by a hinged sedimentary or *basement* block on the upthrown side of each of two intersecting faults. With sufficient lateral extent, it has the *gravity signature* of a low-*amplitude* positive *residual gravity* anomaly. On the downdip side, it is bounded by edge-generated *gravity gradient* anomalies. See also *dip*, *residual field*, and *trap*.

trapezoidal filter: same as *Butterworth filter* with an emphasis on the profile shape of its *band-pass filter* option.

traveling-base-station method: *cross-check method* in the context of establishing the secondary base stations over a *ground gravity* survey area in order to decrease the time intervals between the successive *reoccupations*. See also *base station*.

traverse: **1.** a gravity or magnetic *survey* line along which measurements of the *Earth gravity field* or *Earth magnetic field* and their gradients are made in accordance with *survey specifications*. Referred to with the same meaning as *acquisition line* in marine (shipboard), submarine, and airborne surveys. **2.** one cycle of *ground gravity* measurements that begins from a *base station*, continues at the pre-planned observation points in the *gravity survey* area, and ends at the base station. **3.** a pre-programmed acquisition line along which an *autonomous underwater vehicle (AUV)* navigates during a *submarine survey*.

traverse lines: *acquisition* lines along which the great majority of measurements of the Earth's gravity or magnetic field are made over a *survey* area. The orientation of the traverse lines is preferred to be mostly orthogonal to the dominant *strike* of the geological structures to provide the observation conditions for the highest achievable *lateral resolution* of anomalies. Generally, the orientation and distance, or spacing, between the traverse lines for a *detailed survey* are selected in accordance with the *survey* objectives and guided by the geological characteristics of a survey area obtained from a *regional survey*. Sometimes referred to as *primary lines*. See also *tie lines*.

traverse loop: *loop* in the context of one cycle of ground gravity or magnetic measurements that begins and ends at the same coordinate point.

trench: a long and narrow depression formed in the subduction zone by one lithospheric plate plunging under another plate. If the subduction zone involves the continental and oceanic plates, then the oceanic plate, being denser, subducts under the continental plate. In this case, the trench is oriented parallel to the continental margin, and it can be deeper than the surrounding ocean floor by two or more kilometers and thousands of kilometers long. A trench has the *gravity signature* of a distinct *gravity low* in the *total field* (free-air or Bouguer anomaly) and a negative anomaly in the *residual field*. Over the transition zone from an *accretionary prism* to the trench bottom, it can have the *signature* of a peak-to-trough *shelf-edge gravity anomaly* in the *free-air gravity* field. See also *lithosphere*.

trend: **1.** the orientation of a regional-*scale* alignment of gravity and magnetic anomalies on a *grid*. **2.** the general direction of the dominant faults and geological structures on a tectonic map. **3.** *regional trend*. **4.** the general direction of a local-scale deposition or mineralization. **5.** the *azimuth*, or direction of *strike*, of a two-dimensional (2D) *model* relative to the geographic north. See also *fairway* and *trend removal*.

trend analysis: **1.** a *regional-residual separation* methodology based on the use of *polynomial fitting* and decomposition of an anomalous gravity or magnetic field into (a) the relatively smooth surface of a low polynomial order approximating the *regional field* components and (b) the higher polynomial order surface representing the *residual field* components associated with the geological structures of an exploration interest. Sometimes referred to as *surface fitting*. **2.** *Fourier analysis* for the detection of the dominant alignments of gravity or magnetic anomalies and their *wavenumber* content. **3.** identification of trends in a gridded *anomalous field* for the selection of *directional filter* parameters and optimal orientation of the *principal profile* or two-dimensional (2D) *model*. **4.** *principal component analysis (PCA)*. See also *trend* and *interruption trend*.

trend removal: *detrending* in the context of removing the *regional trend* from data in order to highlight the low-*amplitude* anomalies of an exploration interest and minimize *Gibbs errors* in *Fourier transform*.

trend replacement: restoration of the *regional trend* in gridded gravity or magnetic data that has been removed by *detrending* prior to *Fourier transform*. See also *gridding*.

triangle plot: See *barycentric image*.

triangular diagram: See *barycentric image*.

triaxial magnetometer: a *vector magnetometer* which measures three orthogonal *vector* components of a planetary magnetic field, such as the *Earth magnetic field*, in the direction of orientation of the mutually orthogonal axes of its three magnetic sensors. Also referred to as *3C magnetometer*. See also *magnetic sensor*.

trimming: **1.** a *grid*-based procedure applied to remove, or cut away, *edge effects* in order to reduce the color-coded *dynamic range* of a grid and improve the lateral and vertical *resolution* of anomalies. Removal of the high-*amplitude* artificial edge anomalies from a *color-coded grid* is equivalent to subtraction of the *regional field* component from an *anomalous field*. **2.** selecting a rectangular window within the original grid and removing the rest of the grid in order to retain the window data for subsequent *enhanced processing* and detailed analysis. See also *clipping*, *DC shift*, *tapering*, and *windowing*.

triplet method: **1.** a method of *acquisition* of the airborne gravity and magnetic data in a *reconnaissance survey* flown along *banded flight tracks*, where three closely spaced flight lines are separated from the next set, or band, of three flight lines by a relatively large distance. **2.** *Siegert method* for calculation of the *Bouguer correction factor (BCF)*.

tripod correction: a correction applied to the gravity data acquired by a *ground gravity* survey to compensate for the height of the *gravimeter* tripod or other mounting platform. The correction value is expressed as

$$\Delta g_{tr} = h\,|\partial g/\partial h|,$$

where h is tripod height and $\partial g/\partial h$ is *free-air gradient*. Tripod correction is added to *observed gravity*, when the *observation point* is above sea level, and it is subtracted from observed gravity values when the observation point is below sea level. See also *mean sea level (MSL)*.

trough: **1.** a regional-scale structural depression filled-in with sedimentary rocks. **2.** the minimum (often, negative) value of an *anomaly*. Trough is the opposite of *peak*, which is the maximum value of a positive gravity or magnetic anomaly. See also *depocenter* and *graben*.

true bearing: same as *azimuth*, which represents the angle between the *true north* direction and a given direction.

true curvature attributes: *curvature attributes* calculated from a *rotated gradient tensor* for which the fixed Cartesian coordinate frame of measurements has been rotated to a *local reference frame*, and the measured values of the *gravity gradient tensor (GGT)* components have been converted into values corresponding to the new orientation of the coordinate axes at each observation or *grid* point. True curvature attributes provide the most accurate information about anomalous characteristics of the gravitational *equipotential surface* that are related to lateral changes in the *structure* and *density contrast* of rock formations and used in the analysis and *interpretation* of *gradient tensor* data (Li, 2018). See also *apparent curvature attributes*.

true density: the rock sample *density* defined by a rock mass per unit of the rock sample volume without porosity and other voids. See also *apparent density*, *bulk density*, *formation density*, and *natural density*.

true north: the direction to the geographic north pole. Referred to with the same meaning as the geographic meridian direction. See also *magnetic north pole* and *magnetic meridian*.

truncation error: **1.** error in the calculation of the total mass or *density* and *magnetization* of a *source body* that results from incomplete structural or property *inversion*, when it is not performed to the full extent of a gravity or magnetic *anomaly*. **2.** *clipping*.

truncation trend: same as *interruption trend*, which will truncate, or cut off, gravity or magnetic anomalies along a particular *trend* associated with a structural discontinuity, such as a *fault*, *strike-slip fault*, and *shear zone*.

tuning filter: a *grid*-based optimum-type *band-pass filter* for which the parameters are determined from, or tuned to, the *power spectrum* of a *target anomaly* calculated by the *forward modeling* of an exploration *target* (e.g., local anticlinal *structure*, *reef* or other prospective stratigraphic object, *kimberlite intrusion* or *ore body* confirmed by drilling) using the available geological and geophysical *constraints*. A tuning filter is applied to the *residual field* of gravity or magnetic anomalies to detect those anomalies in a *survey* area that are similar to the target anomaly. See also *optimum filter*, *Keating method*, and *residual anomaly*.

turbulence effect: a distortion of airborne gravity measurements by interference with the inertial accelerations of a *survey* aircraft produced by air turbulence during a flight. The turbulence effect can result in the vertical inertial accelerations of an *observation platform* up to tens of thousands of milligals. See also *inertial acceleration*, *lever arm effect*, *milligal (mGal)*, and *wave-motion effect*.

two-component tensor gradiometer: same as *partial tensor gravity gradiometer* with an emphasis on measurements of two *differential curvature* components of *gravitational potential* U: cross-line component U_{xy} and inline component $U_{xx} - U_{yy}$, which are referred to as *differential curvature gradients*.

two-dimensional (2D): a characteristic of the mathematical transformation method, *profile*-based *model*, and *source body* approximation that highlights the assumption that there is no variation, i.e., parameters are invariant, along: (a) one of the mutually orthogonal x-, y-, z-directions in *Cartesian coordinates*; (b) the direction perpendicular to the line of measurements, the *profile* of anomaly, or the model plane; (c) the plane perpendicular to the *strike* of the "infinitely" long source body. See also *dimension*, *infinite length*, *one-dimensional (1D)*, and *three-dimensional (3D)*.

2D analytic signal (2DAS): **1.** the *analytic signal (AS)* of a *2D source*. **2.** the AS of two-dimensional (2D) line- or *profile*-oriented magnetic or gravity data, where the *horizontal derivative* $\partial F/\partial x$ of *anomalous field* F is the real part and the *vertical derivative* $\partial F/\partial z$ is the imaginary part of the AS. The 2D analytic signal can be expressed as

$$2DAS = \partial F/\partial x + iH\{\partial F/\partial x\} = \partial F/\partial x + i\partial F/\partial z,$$

where $i = \sqrt{-1}$ is a complex unit number indicating the orthogonality and $H\{\partial F/\partial x\}$ is the *Hilbert transform* of $\partial F/\partial x$ applied for calculation of the vertical derivative $\partial F/\partial z$. See also *dimension*, *direct analytic signal (DAS)*, and *3D analytic signal (3DAS)*.

2D analytic signal amplitude (2DASA): the absolute value, or *magnitude*, of a *2D analytic signal (2DAS)*, expressed as

$$2DASA = |2DAS| = [(\partial F/\partial x)^2 + (\partial F/\partial z)^2]^{1/2},$$

where $\partial F/\partial x$ and $\partial F/\partial z$ are the horizontal and vertical derivatives of an *anomalous field* F in *Cartesian coordinates*. See also *horizontal derivative* and *vertical derivative*.

2D dike: *igneous intrusion* with a flat top, variable width, and nearly parallel sides extending to "infinity" along its *dip* and *strike*. It is used for approximation of the magnetic and gravity sources of the relatively simple geometric forms (faults, mineral-filled veins, igneous bodies, etc.) in source depth estimations, *forward modeling*, and *inversion*. Complex geometries of the deep-rooted sources and heterogeneous block-faulted composition of the *basement* can be approximated by assemblages of the multiple dike-like elementary forms filling-in the source shapes. See also *infinite bottom*, *infinite strike length*, *thin dike*, and *thick dike*.

2D Euler deconvolution: *Euler deconvolution* in application to the line-oriented and arbitrary-directed profiles of magnetic or gravity data, where the source parameter calculations are made in two dimensions: depth and horizontal distance along a *profile* line. See also *3D Euler deconvolution*.

2D FFT filters: *FFT filters* adapted for application to the gridded gravity and magnetic data. See also *gridding* and *1D FFT filters*.

2D Fourier transform: *Fourier transform* of the gridded gravity or magnetic data. See also *gridding* and *1D Fourier transform*.

2D gradient vector: a two-dimensional (2D) *vector* showing the orientation of the highest rate of change of a gridded gravity or magnetic field. At each *grid* point, its mutually orthogonal vector components are represented by two horizontal derivatives of an *anomalous field*. Its length, or *magnitude*, is quantified by the *horizontal gradient amplitude (HGA)*. Its *phase*, or direction on the grid, is defined by the *horizontal gradient azimuth*. On a *contour map*, the 2D gradient vector points to the center of a local positive *anomaly*, and it is everywhere orthogonal to the contour lines. Also referred to as *horizontal gradient vector*. See also *derivative*, *gradient*, and *3D gradient vector*.

2D model: a *profile*-based two-dimensional (2D) gravity or magnetic *model*, where one *dimension* is depth and another dimension is the horizontal distance along a modeling line. All 2D model bodies and structures are assumed to have *infinite strike length* and invariant properties in the direction perpendicular to the model plane, i.e., there is no structural and property changes in that direction. The degree of detail in a modeled geological *structure* depends on the availability and information content of the modeling *constraints*. See also *2.5D model*, *2.75D model*, and *3D model*.

2D source: a gravity or magnetic *source body* with *infinite strike length* along one of its horizontal axes. When the long axis of the source body is 8–10 times larger than its short axis, the gravity and magnetic effects of both long ends are assumed to be negligible for the purposes of two-dimensional (2D) *modeling* and source depth calculations along the *profile* perpendicular to the body *strike*. See also *1D source*, *2.5D source*, *2.75D source*, and *3D source*.

2D spectrum: *spectrum* calculated from gridded gravity or magnetic data. Referred to with the same meaning as *radial power spectrum*. See also *gridding* and *1D spectrum*.

2D Talwani algorithm: See *Talwani algorithm*.

2D Talwani method: See *Talwani method*.

2D Werner deconvolution: *Werner deconvolution* in the context of its application to profiles and line-oriented magnetic data, where calculations are made in two dimensions: depth and horizontal distance along the *survey* line or *profile* extracted from a *grid*. See also *3D Werner deconvolution*.

2D Werner inversion: same as *Werner deconvolution* which is a two-dimensional (2D) method of *inversion*, i.e., method of calculation of the *source body* parameters from its magnetic field represented by line-oriented data or profiles extracted from a *grid*.

2.5D model: a *profile*-based two-dimensional (2D) *model* with symmetric and finite *strike* lengths of the off-profile parts of each *source body*, i.e., length of each source body in the direction orthogonal to the model plane is limited by the same lateral extent. See also *infinite strike length*, *2D model*, *2.75D model*, and *3D model*.

2.5D source: a gravity or magnetic *source body* which has a finite and symmetric extent in the direction orthogonal to the *profile* line used in a *2.5D model*. See also *2D source*, *2.75D source*, and *3D source*.

2.75D model: a *profile*-based two-dimensional (2D) *model* with asymmetric and finite *strike* lengths of the off-profile parts of each *source body*, i.e., length of each source body in the direction orthogonal to the model plane is limited by a different lateral extent. This model is preferred for representing the off-profile source body parts with variable strike lengths and oblique strike angles. See also *infinite strike length*, *2D model*, *2.5D model*, and *3D model*.

2.75D source: a gravity or magnetic *source body* which has a finite and asymmetric extent in the direction orthogonal to the *profile* line used in a *2.75D model*. See also *2D source*, *2.5D source*, and *3D source*.

U

UAV magnetometer: a light-weight *magnetometer* towed on a cable or attached to an *unmanned aerial vehicle (UAV)*, also known as a *drone*, which can fly as low as 1–2 m above the ground. The UAV systems with a *fluxgate magnetometer* or *microfabricated atomic magnetometer (MFAM)* are used in *magnetic archaeology*, *forensic magnetism*, detection and mapping of *unexploded ordnance (UXO)*, and other high-*resolution* exploration, engineering, and environmental applications. Also referred to as *drone magnetometer*.

UAV survey: a low-*altitude* high-*resolution* aeromagnetic *survey* with the use of a light-weight *fluxgate magnetometer* or *quantum magnetometer*, such as a *microfabricated atomic magnetometer (MFAM)*, towed on a cable or attached to an *unmanned aerial vehicle (UAV)*. Survey measurements are made along the preprogrammed closely spaced flight lines using high-accuracy *differential GPS (DGPS)*. Among the various *exploration*, engineering, and environmental applications, UAV surveys are conducted for the detection and mapping of anthropogenic targets, such as underground metal pipe lines, abandoned oil and gas wells, and *unexploded ordnance (UXO)*.

unconformity: a structural relationship between rock formations, where the younger overlying formations do not "conform" to the *dip* and *strike* of the older underlying rocks having different tectonic history and lithological composition, such as the relationship between the older igneous and metamorphic rocks of the *geological basement* and the younger sedimentary rocks. See also *subcrop*.

underdetermined problem: a characteristic of an *inverse problem* which indicates that the number of assumed parameters of a *source body* generating an observed *anomaly* is larger than the number of observations, so the *model* parameters approximating the source body cannot be determined uniquely. See also *overdetermined problem*.

underfitting: a characteristic of a *modeling* result which indicates that (a) *model misfit* is substantially greater than the estimated level of *noise* in the observed data and (b) obtained *model* has a lower *resolution* of the structural elements than the observed data are capable to provide. See also *overfitting*.

underplating: **1.** accumulation, or ponding, of the high-*density* igneous (magmatic) intrusions within the lower *crust* and uppermost *mantle* around the *Moho boundary*, wherein a partial mixing of the original mafic (basaltic) *magma* with the remolten crustal rocks results in the geochemical alterations and formation of a felsic (granitic) magma (Thybo and Artemieva, 2013). **2.** addition of oceanic crust rocks to the base of an *accretionary prism* by scrapping-off the oceanic lithospheric plate descending into the subduction zone and producing the *thrust*-related structures within the accretionary prism. Underplating contributes to the fields of long-*wavelength* gravity and magnetic lithospheric anomalies. See also *igneous intrusion* and *lithosphere*.

undersampling: a characteristic of discrete measurements with the *sample interval* which is larger than the short-*wavelength* components of a measured field. Undersampling is detrimental for the *resolution* of anomalies, both vertical and lateral, because it distorts the actual amplitudes, shapes, and *spectrum* of anomalies and produces artificial, or aliased, anomalies in the process of their recovery and *gridding*. In the case of an *anomaly* striking at an angle oblique to adjacent survey lines, the regular algorithm of interpolation between measurement points in the direction of the *strike* of the short-wavelength anomaly will produce a sequence of isolated circular or elongated anomalies, which can resemble a "string of beads" or "bicycle chain," instead of a continuous alignment of the anomalous values on a *grid*. This property of undersampled gridding is known as the *bull's-eye effect*. Undersampling in the direction orthogonal to orientation of *survey* lines is common for airborne and marine (shipboard) measurements, and it is a main reason for a much lower *lateral resolution* of gridded data as compared to the lateral resolution of the original line-oriented data. See also *aliasing*, *anisotropic sampling*, and *Nyquist theorem*.

underwater gravimeter: a remote-controlled *relative gravimeter* system for measurements of changes in the *Earth gravity field* on an ocean/sea or lake bottom. Commonly, it includes: (a) a gravity *sensor* adapted to the deep-water operations; (b) underwater pressure housing; (c) electronic cable connections; (d) automatic or remote-controlled *leveling* and clamping-unclamping; (e) a display unit on the survey ship; and (f) a cable winch to submerge and lift the system. The achievable *accuracy* of the measurements is 0.01–0.3 *milligal (mGal)* depending on the ocean/sea state and *drift* control. Modifications of an underwater gravimeter operate remotely on the cable suspended from a helicopter during a *gravity survey* over rough and forested terrains without a need to land the helicopter. Sometimes referred to as *ocean-bottom gravimeter*. See also *towed deep ocean gravimeter (TOWDOG)* and *AUV gravimeter*.

underwater gravity: gravity data obtained from the *relative gravity* measurements by an *AUV gravimeter*, *underwater gravimeter*, and *towed deep ocean gravimeter (TOWDOG)*. See also *autonomous underwater vehicle (AUV)*, *AUV gravity*, and *submarine gravimetry*.

underwater magnetics: magnetic data obtained from *total field* or *gradient* measurements with the use of an *autonomous underwater vehicle (AUV)*. See also *AUV magnetometer* and *submarine magnetometry*.

unexploded ordnance (UXO): unexploded landmines and bombs, artillery and mortar projectiles, and other military explosives from wartime actions or training exercises that are lost or abandoned, thereby creating potential hazards. The *UXO detection* methodology is based on high-resolution *magnetic gradient* measurements with the use of high-sensitivity *magnetic gradiometer* systems. The UXO objects are mapped by the calculation of various *attributes*, such as *analytic signal amplitude (ASA)*, *tilt angle (TA)*, and their derivatives, wherein the UXO depth is often estimated from results of a two-dimensional (2D) or three-dimensional (3D) *Euler deconvolution* with the *Euler structural index* SI = 3. Airborne UXO detection requires closely spaced *acquisition* lines, high-accuracy *differential GPS (DGPS)* navigation, high *amplitude resolution* of the *magnetic gradient* measurements, and a low-flying helicopter or *drone*. See also *drone survey*, *resolution*, *sensitivity*, and *UAV magnetometer*.

unit calibration: conversion of *gravimeter* readings from its counter, or dial, units into the standard gravity units of *milligal (mGal)* or *microgal (μGal)*. See also *reading* and *calibration factor*.

universal gravitational constant: **1.** the numerical value of the constant of proportionality in the formula of the *Newton law of gravitation*. It is commonly denoted by the capital letter G and defined as $G = (6.674215 \pm 0.000092) \cdot 10^{-11}\ m^3/kg \cdot s^2$ in *SI units* and $G = (6.674215 \pm 0.000092) \cdot 10^{-8}\ cm^3/g \cdot s^2$ in *cgs units*. **2.** *magnitude* of the force of attraction, or *gravitational force*, between two unit masses separated by a unit distance anywhere in the Universe. Also known as *Newtonian gravitational constant*. Often referred to simply as *gravitational constant*. See also *geocentric gravitational constant (GM)*.

Universal Transverse Mercator (UTM): the global projected, or planar, *coordinate system* based on a rectangular map frame, wherein the unit of distance is the meter (m). The Earth's surface is divided into 60 meridional zones with the longitudinal width of 6°, and each zone has a central meridian which splits it into two parts that are 3° wide. Each meridional zone is projected separately onto the cylinder, which is tangential to the Earth's surface along the equator across the globe, to provide the projection of a curved surface of each meridional zone onto a flat map with orthogonal x- and y-coordinate axes. The UTM zones are numbered from the west (W) to the east (E) beginning at –180° latitude (called the International Date Line), so that Zone 1 is between –180°W and –174°W longitudes with its central meridian at –177°W longitude, and Zone 60 is between 174°E and 180°E longitudes with its central meridian at 177°E longitude. The Greenwich Meridian (GM) corresponds to 0° longitude. Numerical values of the X- and Y-coordinates of every point on the map are called *easting* and *northing*, respectively. Each central meridian is assigned with the *false easting* value of 500,000 m, whereas the equator is assigned with the *false northing* value of 0 m in the northern hemisphere and 10,000,000 m in the southern hemisphere. Within each zone, the easting values increase toward the east, so the point located 50 m west of a central meridian will have the X-coordinate value of 499,950 m and the point located 50 m east of a central meridian will have the X-coordinate value of 500,050 m; the northing values increase from the south to the north in both the southern and northern hemispheres, so the point located 75 m north of the equator will have the Y-coordinate value of 0,000,075 m and the point located 75 m south of the equator will have the Y-coordinate value of 9,999,925 m. Application of the UTM coordinate system is limited to the global area between 84° latitude in the northern hemisphere (84°N) and 80° latitude in the southern hemisphere (80°S). Beyond these areas, the Universal Polar Stereographic (UPS) coordinate system is used. The UTM is the standard coordinate system for a map or rectangular *grid*. See also *Geodetic Reference System*, *map projection*, and *XY coordinates*.

unmanned aerial vehicle (UAV): a pilotless, or unmanned, mini-aircraft equipped with a *differential GPS (DGPS)* receiver and onboard sensors. It is directed by remote control or can fly autonomously as low as 1–2 m above the ground following a software-controlled pre-programmed flight plan. Depending on the design and application, the UAV dimensions vary from several centimeters to several meters. Commonly referred to as a *drone*. See also *sensor*, *UAV magnetometer*, and *autonomous underwater vehicle (AUV)*.

uplift: a structurally elevated area, or *structural high*, in the upper part of the Earth's *crust* created by the upward-directed tectonic movements that raise or upthrust the rocks above the surrounding rock formations of the same age. Intrasedimentary and *basement* uplifts commonly generate the positive gravity and magnetic anomalies of various amplitudes in a *total field*, and they can be made better visible, or enhanced, by calculation of the *residual field*. Uplifts are often associated with structural and stratigraphic traps. See also *anomaly*, *horst*, *thrust*, and *trap*.

upsampling: **1.** increasing the number of samples in a dataset by adding new samples obtained by interpolation between the original data values. When the wavelengths of the high-*frequency* (short-*wavelength*) components of a gridded *TMI field* are nearly the same as a *grid cell* size, the upsampling for the decrease of the original *grid interval* by 2–4 times is required before the Fourier-domain *reduction-to-pole (RTP)* in order to minimize the *Gibbs ringing* in the obtained *RTP grid*. **2.** *padding* by the addition of extra values at each margin of gridded and line-oriented datasets or *tapering* of the datasets with zero values so that their length is equal to the integer multiple of the *fundamental period* in order to minimize errors associated with the *Gibbs phenomenon* in the transformations of gravity and magnetic data from the *space domain* to the *Fourier domain* and back to their original space domain. See also *resampling*, *undersampling*, and *downsampling*.

upward continuation: calculation of the anomalous gravity or magnetic field at an *elevation* which is higher than the elevation of the original measurements, i.e., further away from all sources. Its basic properties are generalized by the mathematical expression in *Fourier domain* as

$$\mathcal{F}(k,h) = \mathcal{F}(k,0) \cdot e^{-hk} = \mathcal{F}(k,0)/e^{hk},$$

where $\mathcal{F}(k,h)$ is the *Fourier transform* of the *anomalous field* at the elevation $z = h$, $\mathcal{F}(k,0)$ is the Fourier transform of the field at the elevation $z=0$, k is *wavenumber*, and h is a continuation distance. Upward continuation attenuates all components of an anomalous field. The degree of attenuation increases exponentially with increasing k and h, thereby emphasizing the longer *wavelength* components, because the shorter the wavelengths, the higher their wavenumbers, i.e., anomalies from shallower sources attenuate more rapidly than those from deeper sources. There are no *phase* changes in the upward-continued field, and shapes of anomalies are preserved in the level-to-level continuation. Upward continuation is analogous to a *low-pass filter* applied for the suppression of high-frequency *noise* and other near-surface effects. It is also used for *merging* of airborne surveys flown at different elevations and for *enhancement* of regional field components in the process of *regional-residual separation.* The spectral-domain *Jacobsen filter* is based on calculation of the difference between two upward continuations referred to as *differential upward continuation.* See also *downward continuation.*

upward continuation residual: a portion of the anomalous gravity or magnetic field remaining after subtraction of the upward-continued field from the original *anomalous field.* It is calculated with the objective of enhancing the high-*frequency*, or short-*wavelength*, components of the anomalous field. See also *anomaly, enhancement, regional-residual separation*, and *residual field.*

UTM: See *Universal Transverse Mercator (UTM).*

UTM coordinates: standard coordinates of a gravity or magnetic *grid* or map that are based on the *Universal Transverse Mercator (UTM)* planar *coordinate system.* Referred to with the same meaning as *XY coordinates.*

UTM projection: the *map projection* of the gravity and magnetic maps implemented in the *Universal Transverse Mercator (UTM)* planar, or projected, *coordinate system.* Referred to with the same meaning as *transverse Mercator projection.*

UXO: See *unexploded ordnance (UXO).*

UXO detection: a high-*resolution* methodology for the detection of *unexploded ordnance (UXO).* Its basic principles, such as the use of high-sensitivity *magnetic gradiometer* systems, close *survey line spacing*, low-*altitude* airborne *acquisition*, high-*accuracy* positioning by *differential GPS (DGPS)*, and *enhanced processing* of the *gradient* and *full tensor gradient (FTG)* data, are implemented in the mapping of buried anthropogenic objects, including the abandoned oil and gas wells, buried weapons, underground metal pipe lines, sunken ships, etc. This methodology is also used in the diamond, gold, REE, and other high-resolution mineral exploration applications due to its capability to detect very small changes in *magnetization* of rocks within faults, fracture zones, and *fault* intersections as well as along *igneous intrusion* margins and other mineral alteration zones. See also *lateral resolution, sensitivity, vertical resolution*, and *rare earth elements (REE).*

V

Vacquier rule: See *straight-slope method.*

van Allen belts: space- and time-varying zones within the inner part of the *magnetosphere* where the high-velocity charged subatomic particles (protons and electrons), captured by the *Earth magnetic field,* are circulating around the Earth. Most of the belt-forming particles originate from the *solar wind,* while others are cosmic rays from outside the solar system. There are two permanent belts: the inner belt dominated by the slower moving, lower energy protons at altitudes 700–11,000 km and the outer belt dominated by much faster moving electrons at altitudes 14,000–60,000 km. One or more temporal belts emerge from time to time in a gap between two permanent belts. The Earth's magnetic field appears to act as a planet-scale accelerator which can increase the speed of the trapped electrons to nearly speed-of-light values. With larger distances from the Earth, the belts increase their size and acquire the shape of a torus, whose cross-section diameter gradually increases from a few hundred to several thousand kilometers. Belts change their shapes, dimensions, and the *flux density* of charged particles in response to high-energy fluctuations of the solar and deep-space radiation. In accordance with the *Ampere law,* moving charged particles generate the magnetic fields, so the van Allen belts generate their own highly variable magnetic fields that contribute to *diurnals* and *magnetic storm.* See also *external magnetic field* and *ionosphere.*

variable order vertical derivative: a *vertical derivative* for which the order is dependent on the amount of *standard deviation (SD)* of the gridded *anomalous field* values within a square *moving window*: the higher the standard deviation, the lower the vertical derivative order. It is calculated to obtain an optimal balance between the less-resolving lower vertical derivative orders and the more-resolving higher orders of the vertical derivative with a minimized amplification of the high-frequency *noise* (Cooper and Cowan, 2004). See also *gridding, resolution,* and *fractional vertical derivative.*

variance: a quantitative estimate of the deviation of measured values from the average value for a given dataset, i.e., it is a measure of the dispersion of measurements. Variance is defined by the sum of the squared differences between the individual values and their arithmetic mean, or average, value divided by the total number of the measured values:

$$var(x) = \sigma^2 = \Sigma(x_i - m)^2/n,$$

where m is the average value, x_i is a measured value with $1 < i < n$, and n is total number of measured values. The square root of the variance is known as the *standard deviation (SD).* It is used as a statistical estimate of the *precision* of the measured values and the distribution of *random noise* within a *bandwidth* of measurements. See also *covariance.*

variogram: a measure of the spatial continuity and correlation of *anomalous field* values. It is described by a graphical curve of the difference between pairs of field values as a function of their separation distance, expressed as

$$\gamma(d) = (1/2k)\Sigma_{i=1}^{n}[Z(x_i) - Z(x_{i+d})]^2,$$

where k is the number of pairs of measured values separated by distance d and $Z(x_i)$ is the field value. There are three curve parameters: (1) "range" specifying the distance beyond which there is no correlation between two values and the curve levels off, or flattens; (2) "sill" which is the upper limit variogram value at a distance equal to the range; and (3) "nugget" which represents the total *random noise* remaining in the data after initial processing, including the *positioning* errors, measurement errors, and any field variability unresolved by *sampling.* The relationship between the variogram, *variance,* and *covariance* is expressed as

$$\gamma(d) = \sigma^2 - C(d),$$

where σ^2 is variance and $C(d)$ is covariance. Zero covariance indicates the absence of correlation between field values. Variogram is used in *kriging* and other interpolation and correlation techniques. See also *coherence, regression,* and *resolution.*

variometer: **1.** an instrument for measuring the relative changes, or variations, of the horizontal and vertical components of the *Earth magnetic field* by a magnet or other *magnetic sensor* suspended on a torsion fiber and capable of rotating horizontally or vertically relative to the plane of a *magnetic meridian.* Electronic versions of the variometer are based on the use of saturation coils and provide a measurement *accuracy* up to 1 *nanotesla (nT).* Variometers have been replaced almost entirely in exploration and research applications by *quantum magnetometers.* Also known as *torsion magnetometer.* **2.** a *vector magnetometer* in the context of measuring the variations of the *vector* components of the Earth's magnetic field relative to a *baseline* value. See also *curvature variometer, horizontal variometer,* and *torsion balance.*

vector: a mathematical quantity with a numerical value and direction, or orientation, as distinct from a *scalar* which has a numerical value only. In *Cartesian coordinates,* a vector is described in terms of its three orthogonal components along the mutually orthogonal *x-, y-, z-*axes. A vector is shown by a boldface character, such as **A**, whereas a scalar quantity is not emphasized in text. The length of a vector, or vector *magnitude,* is expressed as

$$|\mathbf{A}| = (A_x^2 + A_y^2 + A_z^2)^{1/2},$$

where A_x, A_y, A_z are mutually orthogonal vector components representing the projections of vector **A** onto the *x*-, *y*-, *z*-axes. See also *gravity vector*, *magnetic vector*, *gradient vector*, and *eigenvectors*.

vector accelerometer: an assemblage, or *array*, of inertial-grade accelerometers which have their sensing axes oriented along the mutually orthogonal *x*-, *y*-, *z*-directions within the reference frame of a *survey* aircraft or ship. Its measurements are integrated with high-accuracy *positioning* data from *differential GPS (DGPS)* and *inertial acceleration* data from onboard accelerometers to determine changes in the directional *vector* components of the *Earth gravity field*, which is a field of *gravitational acceleration* due to attraction of the Earth mass. Also referred to as *3D accelerometer* and *vector gravimeter*. See also *inertial-grade accelerometer* and *GOCE gravity gradiometer*.

vector field: a *field* of *vector* quantities for which the variations are described by functions of the spatial coordinates and time. A vector field has only one value at every coordinate point in the space at a given time, and that value has both *magnitude* and direction. Gravity and magnetic fields are vector fields described in *Cartesian coordinates* by the horizontal and vertical vector components along three mutually orthogonal *x*-, *y*-, *z*-directions. A regular, or conventional, *gravimeter* or *magnetometer* measures one component of a corresponding vector field: the vertical component of a three-dimensional (3D) *gravity vector* or magnitude, or *total intensity*, of a 3D *magnetic vector*, respectively. Values of a measured component of the vector field comprise the *scalar field* which is presented for *enhanced processing* and *interpretation* in the standard *format* of a *grid*. See also *total absolute magnetic intensity (TAMI)* and *total magnetic intensity (TMI)*.

vector gravimeter: *vector accelerometer* in the context of its use for measuring changes of three mutually orthogonal *vector* components of the *Earth gravity field* within the reference frame of a *survey* aircraft or ship.

vector inversion: See *vector magnetic inversion*.

vector magnetic inversion: a three-dimensional (3D) *magnetic inversion* of *magnetic vector* data, obtained from *vector* or transformed *scalar* measurements, to determine the orientation of a *magnetization vector* and *magnitude* of *total magnetization* for every *voxel* in a 3D magnetic *voxel model*. Knowledge of the distribution of the magnetization vector orientation and its magnitude in an *ore body* improves the *inversion*-based estimates of its physical properties (volume, shape, *susceptibility contrast*, high-mineralization zones, etc.), especially for the magnetic data strongly affected by *remanent magnetization* and *self-demagnetization* (Li et al., 2021; Sampson and Williams, 2021). See also *magnetic vector calculation*, *scalar-to-vector transformation*, and *scalar magnetometer*.

vector magnetic moment: the *vector* parameter of a localized, or compact, magnetic *source body* that is defined by the volume integral of its *magnetization vector* **M**. Mutually orthogonal components of a vector magnetic moment **m** in the reference frame of *Cartesian coordinates* are represented by volume integrals of the vector components M_x, M_y, M_z of the magnetization vector **M** as

$$m_x = \int_V M_x dv, \quad m_y = \int_V M_y dv, \quad m_z = \int_V M_z dv.$$

The magnitude $|\mathbf{m}|$ of a vector magnetic moment is defined as

$$|\mathbf{m}| = (m_x^2 + m_y^2 + m_z^2)^{1/2}.$$

Vector magnetic moment components are determined with the use of first-order *Helbig integrals* and interpreted for the *magnitude* $|\mathbf{m}|$ and *total magnetization* direction of a compact source body (Caratori Tontini and Pedersen, 2008). Also referred to as *source magnetic moment*. See also *magnetic moment*, *integral magnetic moments*, and *Helbig method*.

vector magnetometer: a *magnetometer* for measuring the *vector* components of the *Earth magnetic field* in the direction of orientation of its *magnetic sensor*. Depending on the design and application, a vector magnetometer can measure one vector component of the Earth's magnetic field by orienting its sensor in the desired direction or three mutually orthogonal vector components by three separate sensors mounted at right angles to each other. Except for an *absolute scalar magnetometer (ASM)*, which is capable of operating in both *vector* and *scalar* modes, the vector magnetometers for measurements on a moving *observation platform* are typically modifications of a *fluxgate magnetometer*, which measures the vector components relative to the spatial orientation of its sensors and reference, or *baseline*, value. A fluxgate magnetometer drifts with time, and its baseline value is calibrated by a separate *scalar magnetometer*. The operational *accuracy* of vector measurements is strongly dependent on the errors in orientation of their sensors in the process of *acquisition*. In order to ensure a subnanotesla *noise* level for the anomalous vector components in the Earth's magnetic field of 50000 nT, the sensor orientation should be determined with the accuracy 0.001°. The orientation sensitivity is minimized in a *SQUID magnetic tensor gradiometer*, which can measure both the vector components of the Earth's magnetic field and its *full tensor gradient (FTG)* elements. Three-component, or triaxial, vector magnetometers are mounted on Earth-orbiting satellites and oriented by the onboard star image cameras. Also referred to as *relative magnetometer*, *3C magnetometer*, and, occasionally, *variometer*. See also *attitude noise*, *satellite magnetometer*, *microfabricated atomic magnetometer (MFAM)*, and *airborne vector magnetometry*.

vector product: a *vector* quantity that represents the result of multiplying, or product, of two vectors. Vector product **C** of two vectors **A** and **B** is expressed as

$$\mathbf{C} = \mathbf{A} \times \mathbf{B} = |\mathbf{A}| \cdot |\mathbf{B}| \cdot \sin\theta,$$

where $|\mathbf{A}|$ and $|\mathbf{B}|$ are the magnitudes of two vectors, θ is the angle between the two vectors, $\mathbf{A} \times \mathbf{B} = -\mathbf{B} \times \mathbf{A}$, and **C** is a new vector whose direction is orthogonal to the original **A** and **B** vectors and pointed in the direction of advance of a right-turning screw rotating from **A** to **B**. New vector *magnitude* $|\mathbf{C}|$ is equal to the area of a parallelogram defined by the two original vectors. In *Cartesian coordinates*, the vector product can be expressed as

$$\mathbf{A}\times\mathbf{B}=\begin{vmatrix}\mathbf{i} & \mathbf{j} & \mathbf{k}\\ A_x & A_y & A_z\\ B_x & B_y & B_z\end{vmatrix}=(A_yB_z-A_zB_y)\mathbf{i}+(A_zB_x-A_xB_z)\mathbf{j}+(A_xB_y-A_yB_x)\mathbf{k}$$

where **i**, **j**, **k** are unit vectors parallel to the mutually orthogonal *x*-, *y*-, *z*-axes; and A_x, A_y, A_z, B_x, B_y, B_z are components of vectors **A** and **B** along *x*-, *y*-, *z*-axes defined by projections of each vector on these axes. Due to the use of the standard mathematical symbol ×, it is also referred to as the *cross product*. See also *scalar product*.

vector survey: measurements of the *vector* components of the *Earth magnetic field* for, primarily, mineral exploration applications using a *vector magnetometer*. Magnetic vector data provide additional quantitative *constraints* for the structural and property *inversion*, especially in areas where the shapes of anomalies are distorted by a strong *remanent magnetization* of sources and low *magnetic latitude* effects. The contributions of induced and remanent rock magnetizations to observed magnetic anomalies, estimates of the *Koenigsberger ratio*, and other parameters of a localized *source body* also can be constrained. See also *SQUID magnetic tensor gradiometer* and *airborne vector magnetometry*.

Vening Meinesz hypothesis: one of three basic theories of *isostasy*. It assumes a regional balance, or compensation, of the topographic and bathymetric loads rather than their local and, mostly, vertical compensation in the Airy and Pratt hypotheses. The Vening Meinesz isostatic balance is accommodated within a broader region with longer approximating wavelengths and smaller amplitudes of crustal thickness variations, and it is achieved by a flexural downwarping or upwarping of the *crust* and upper *mantle*, i.e., load of the topographic and bathymetric masses is supported by the flexural strength of the *lithosphere* and buoyancy forces of the underlying *asthenosphere* within the upper mantle. The Vening Meinesz hypothesis predicts shallower depths to the *Moho boundary* than the Moho depths predicted by the *Airy hypothesis* and the *Pratt hypothesis*. Also known as *crustal flexure hypothesis*.

Vening Meinesz model: the mathematical model of *isostatic compensation* based on the *Vening Meinesz hypothesis* of *isostasy*. Similar to the *Airy-Heskanen model*, it predicts the crustal thickening or thinning associated with the topographic or bathymetric mass load, respectively. The primary difference is that the isostatic compensation in the Airy-Heiskanen model is local, whereas the isostatic compensation in the Vening Meinesz model is distributed over a broad region. The principal parameter of this model is a flexural strength of the *lithosphere*, defined by its elastic thickness required to produce the predicted flexural deformations of the lithosphere under a given load of the topographic or bathymetric mass. The model parameters are used for calculation of the gravity effect of the regional, or flexural, isostatic compensation known as the *isostatic compensation effect*, which comprises the amount of *flexural isostatic correction*. See also *Pratt-Hayford model*.

vertex: the end point of a line segment bounding the polygonal *source body* in two-dimensional (2D) *model* or the prismatic *lamina* in a three-dimensional (3D) model. Multiple line segments bound the arbitrary-shaped source bodies and rock formations in both the 2D and 3D models, wherein the coordinates of their vertices, i.e., end-point coordinates, are used in calculation of the gravity or magnetic effects of the individual 2D polygons and 3D laminas in *forward modeling* and *inversion*. See also *Talwani modeling, 2D model, 3D model,* and *voxel*.

vertical acceleration: *inertial acceleration* resulting from the vertical motion of an *observation platform*, such as an aircraft, ship, or *autonomous underwater vehicle (AUV)*. It is compensated by *vertical acceleration correction*. See also *acceleration*.

vertical acceleration correction: 1. a correction applied to *raw gravity* data obtained from *airborne gravity* measurements to compensate for the vertical inertial accelerations of a *survey* aircraft resulting from air turbulence and referred to as *turbulence effect*. Its calculation is based on (a) high-*accuracy* spatial *positioning* data provided by *differential GPS (DGPS)*, (b) mathematical expression of *vertical acceleration* in the general motion equation, and (c) *inertial acceleration* data compensated for the relevant measurement effects. **2.** a correction applied to marine (shipboard) gravity data to compensate for the vertical inertial accelerations produced by *wave-motion effect*. Correction value is calculated from the direct vertical acceleration measurements by an onboard *accelerometer* system that are compensated for the relevant measurement effects. **3.** a correction applied to submarine gravity data to account for the vertical inertial accelerations resulting from the vertical motions of an *autonomous underwater vehicle (AUV)* within the deep water currents. Correction value is based on the conversion of water pressure *sensor* data into the depth changes and compensation of the obtained acceleration values for measurement effects. See also *AUV vertical acceleration correction*, *inertial acceleration*, and *horizontal acceleration correction*.

vertical accelerometer: a high-precision *inertial-grade accelerometer* used as a *gravity sensor* in an *airborne gravimeter (AG)* system. It measures the sum of the *gravitational acceleration* and *inertial acceleration* components along its vertical sensing axis. Relative gravity values, i.e., changes in gravitational acceleration due to attraction of the Earth mass, are obtained by subtracting the vertical inertial acceleration values, determined with the use of high-accuracy *differential GPS (DGPS)* and inertial *sensor* data, from the values measured by a vertical accelerometer at each observation point in the *airborne gravity* survey area. See also *electromagnetic accelerometer, electrostatic accelerometer, accuracy*, and *relative gravity*.

vertical attraction: *vertical gravity* in the context of the vertical component of *gravitational attraction* due to, or caused by, a body of *mass*, i.e., *causative body*.

vertical continuation: recalculation of an anomalous gravity or magnetic field, commonly in the format of a *grid*, from the *elevation* of its original measurement to a higher or lower level by *upward continuation* or *downward continuation*, respectively. Upward continuation is inherently stable, and it acts similar to a *smoothing filter*, whereas the downward continuation becomes unstable at depths approaching the source depth. Differential vertical continuations are used in *data enhancement*, because they can preferentially amplify either the long- or short-*wavelength* components of an *anomalous field* generated by sources at different depths. See also *differential upward continuation*, *differential downward continuation*, and *Jacobsen filter*.

vertical cylinder: 1. one of the *ideal source* geometries equivalent to a vertical line of point sources, or small spheres, that generates a *homogeneous field* which is used as approximation of the actual *anomalous field* in *Euler deconvolution*. Its *Euler structural index (SI)* is the integer value of 2. **2.** a *model* approximation of sources generating circular-like gravity and magnetic anomalies, such as those produced by volcanic and kimberlite pipes, intrusive igneous plugs, mine shafts, and other high-angle and vertical objects whose horizontal dimensions are much less than their vertical extent, i.e., the source bottom is assumed to be too deep to have a measurable gravity or magnetic effect. See also *infinite bottom*, *kimberlite pipe*, *point source*, and *horizontal cylinder*.

vertical datum: 1. a reference surface for establishing the vertical coordinate of observation points, which is called the *elevation* and defined by the height above sea level or *reference ellipsoid*. As a new standard for the vertical datum, the most recent models of the reference ellipsoid, the *GRS80 ellipsoid* and *WGS84 ellipsoid*, are recommended for gravity and magnetic surveys instead of the previously used *mean sea level (MSL)*. **2.** a reference surface for establishing the elevation and *theoretical gravity* values at observation points. Sea level remains a standard vertical datum for source depth estimates and *basement* depth interpretation by gravity and magnetic techniques and for *modeling* applications, where the depth values below sea level are negative and values above sea level are positive. Also known as *reference elevation* and *elevation datum*. See also *ellipsoidal height*, *orthometric height*, and *horizontal datum*.

vertical derivative: a calculated rate of change of the gravity or magnetic field F per unit distance in the vertical direction z (positive down) in *Cartesian coordinates*. It is a rate of decrease of this field, because the field strength decreases with increasing distance from a source. The vertical derivative highlights the local-scale short- and mid-*wavelength* anomalies generated by relatively shallow sources and suppresses anomalies from deep sources and regional-scale long-wavelength anomalies. The vertical derivative of the Nth-order is denoted by $\partial^N F/\partial z^N$, and it is calculated in the *Fourier domain* as

$$\mathcal{F}\,|\partial^N F/\partial z^N| = |k|^N \cdot \mathcal{F}\,(F),$$

where $\mathcal{F}$(F) is the *Fourier transform* of *anomalous field* F, k is the *wavenumber* for line-oriented data, $|k| = (k_x^2 + k_y^2)^{1/2}$ is the *radial wavenumber* for the gridded data, k_x and k_y are wavenumbers in the mutually orthogonal x- and y-directions, respectively. The vertical derivative order N can be either integer or fractional. Calculation of the vertical derivatives narrows the width of anomalies, improves their *lateral resolution*, and effectively suppresses or eliminates the regional components of anomalous fields. Often referred to with the same meaning as the first-order vertical derivative, also known as *first vertical derivative (FVD)*. See also *ASA vertical derivative*, *fractional vertical derivative*, *variable order vertical derivative*, and *vertical integral*.

vertical gradient (VG): a measured rate of change of a gravity or magnetic *field* per unit distance in the vertical direction. Its theoretical value for the Earth is derived from the analytical expression of the vertical rate of change of *theoretical gravity* or the *International Geomagnetic Reference Field (IGRF)*. A calculated vertical gradient is often referred to as the *first vertical derivative (FVD)* and is computed in the *Fourier domain* using the formula for the *vertical derivative*. It also can be calculated in the *space domain* from two second-order horizontal derivatives via the *Laplace equation* or by the *finite-difference method*. The actual value, or measured vertical gradient, is determined from the difference between the field values measured by two vertically separated sensors and divided by their separation distance called the *baseline*. The theoretical vertical gradient is used for calculation of (a) the *free-air correction* or *height correction* applied to *observed gravity* and (b) the optional *drape correction* for high-*resolution* aeromagnetic data. A decrease in the *amplitude*, or *fall-off*, of gravity and magnetic vertical gradients with the distance r from a *source body* is proportional to $1/r^2$ and $1/r^3$ for a two-dimensional source (*2D source*) and to $1/r^3$ and $1/r^4$ for a *point source*, respectively. Unlike the *horizontal gradient (HG)*, which is efficient in mapping the structural discontinuities and lithological boundaries between rock formations, the *vertical gravity gradient* highlights the anomalies associated with local sedimentary structures, such as *anticline*, *flower structure*, *horst*, *igneous intrusion*, *reef*, *thrust*, *uplift*, etc., whereas the *vertical magnetic gradient* highlights, primarily, the high-*magnetization* lithological components of the *basement* and intrasedimentary igneous intrusions and *volcanics*. See also *sensor*, *calculated gradient*, *measured gradient*, *average vertical gravity gradient*, and *average vertical magnetic gradient*.

vertical gradiometer: *tower gradiometer* in the context of measurements of the *vertical gradient (VG)* of an anomalous gravity or magnetic *field* using two vertically separated gravimeters or magnetometers. See also *anomalous field*.

vertical gravity: 1. a *scalar* quantity that is a measure of the *gravitational field* strength defined by a *magnitude*, or length, of the vertical component of the *Earth gravity vector* $\boldsymbol{g}$ and denoted by g_z. **2.** a *vector* quantity $\boldsymbol{g}_z$ representing the vertical component of the Earth gravity vector $\boldsymbol{g}$ in *Cartesian coordinates*. **3.** the vertical *gravitational attraction* of a body of *mass*, or *source body*, quantified by the *vertical derivative* of its scalar *gravitational potential*, expressed as

$$\Delta g_z = \partial U/\partial z,$$

where U is the anomalous gravitational potential of a source body. **4.** *relative gravity* Δg measured by a conventional, or regular, *gravimeter* known as a *relative gravimeter.* See also *tensor-derived vertical gravity* and *anomalous potential.*

vertical gravity gradient: the rate of change of the *Earth gravity field* per unit distance along the direction orthogonal to sea level or *reference ellipsoid.* The theoretical, or nominal, vertical *gravity gradient* is referred to as the *free-air gradient,* and it is averaged by the 45° latitude value equal to 0.3086 mGal · m^{-1} = 3.086 μGal · cm^{-1} = 0.3086 μGal · mm^{-1} = 3086 E, i.e., it is a theoretical average regardless of the local and regional geological structures. In high-accuracy *ground gravity* surveys, it is measured by a *gradiometer* with two vertically separated gravity sensors. Without direct measurements, its anomalies are closely approximated by calculation of the *first vertical derivative (FVD)* of gridded or line-oriented data. The relationship between the vertical derivatives of anomalous magnetic and gravity fields is described by the *Poisson theorem.* See also *accuracy, anomalous field, Eötvös (E), gravity sensor, gridding, microgal (μGal), milligal (mGal),* and *structure.*

vertical integral: 1. a data transformation operator for the numerical vertical integration of gravity and magnetic fields. It is performed in the *Fourier domain* and expressed as

$$\mathcal{F}(k) = \mathcal{F}(u,v) = (1/|k|^N)\mathcal{F}(\mathrm{F}),$$

where $\mathcal{F}$(F) is the *Fourier transform* of *anomalous field* F; k is the *wavenumber* for line-oriented data; $|k| = (k_x^2 + k_y^2)^{1/2}$ is the *radial wavenumber* for the gridded data; $u = k_x$ and $v = k_y$ are wavenumbers in the mutually orthogonal x- and y-directions, respectively; and N is the vertical integration order which can be either integer or fractional. For $N = 1$, the vertical integral is referred to as the *first vertical integral (FVI).* **2.** an algorithm of the *pseudogravity transformation* which converts *RTP magnetic field* values into the equivalent *pseudogravity* values in accordance with the *Poisson theorem.* **3.** a *data processing* operator applied to an anomalous gravity or magnetic field to suppress the effects of *noise* and local surface or near-surface anomalies in a manner analogous to *upward continuation.* The vertical integral of a gravity or magnetic *anomaly* quantifies its *potential,* which is called *anomalous potential.* See also *vertical derivative.*

vertical local wavenumber: a *complex attribute* derived from the *analytic signal amplitude (ASA)* and defined by the *vertical derivative* of the *local phase.* It describes the rate of change of the local phase θ along the vertical direction and represents the 90° *phase*-rotated version of the *horizontal local wavenumber.* The vertical local wavenumber can be expressed as

$$\begin{aligned} k_z &= \partial\theta/\partial z \\ &= \{\partial\{\arctan[(\partial M/\partial z)/(\partial M/\partial x)]/\partial z\} \\ &= -|ASA|^{-2}\,[\partial^2 \mathrm{M}/\partial x\partial z)(\partial \mathrm{M}/\partial z) \\ &\quad - (\partial^2 \mathrm{M}/\partial z^2)(\partial \mathrm{M}/\partial x)], \end{aligned}$$

where M is the *magnitude* of an anomalous gravity or magnetic field. Because the local phase is equivalent to the *analytic signal phase,* the vertical local wavenumber is also expressed as

$$k_z = -(\partial|ASA|/\partial x)/|\mathrm{ASA}|,$$

i.e., the vertical local wavenumber is equivalent to the normalized *horizontal derivative* of the ASA (Pilkington and Keating, 2006). Combinations of the horizontal local wavenumber and vertical local wavenumber are implemented in the *source parameter imaging (SPI™) method,* the *improved SPI™ method,* the *advanced source-parameter imaging method,* and the *enhanced local wavenumber method* for estimates of the depth, *Euler structural index (SI),* and other parameters of the two-dimensional (2D) magnetic sources approximated by the *magnetic contact, thin dike,* and *horizontal cylinder* source geometries (Smith and Salem, 2005; Salem et al., 2005). See also *local wavenumber, multimodel wavenumber,* and *normalized derivatives.*

vertical magnetic gradient: the rate of change of the *Earth magnetic field* per unit distance along the direction orthogonal to sea level or *reference ellipsoid.* Its theoretical value is derived from the first-order analytical expression of the vertical rate of change of the *International Geomagnetic Reference Field (IGRF),* and it equals 0.025 *nanotesla* per meter (nT · m^{-1}) regardless of the local and regional geological structures. In high-*accuracy* ground magnetic surveys, it is measured by a *gradiometer* with two vertically separated magnetic sensors. Without direct measurements, its anomalies are closely approximated by calculation of the *first vertical derivative (FVD)* of gridded or line-oriented magnetic data. The relationship between the vertical derivatives of anomalous magnetic and gravity fields is described by *Poisson theorem.* See also *anomalous field, gridding, magnetic sensor,* and *structure.*

vertical magnetic intensity: the vertical component of a *3D magnetic vector* measured by the three-component, or triaxial, *vector magnetometer* also known as a *3C magnetometer.* See also *horizontal magnetic intensity, total magnetic intensity (TMI), total absolute magnetic intensity (TAMI),* and *geomagnetic field parameters.*

vertical resolution: 1. the smallest *amplitude* of a gravity or magnetic *anomaly* that can be identified above the *noise* level in line-oriented or gridded data. **2.** the vertical separation, or minimal vertical distance, between similar sources at which they are resolved as individual sources. Sources occurring one above another can be resolved if their vertical separation is significantly larger than the depth to an upper source. **3.** a statistical estimate of the *precision* of measurements quantified by a value which is equal to two or three times the *standard deviation (SD)* value. Referred to with the same meaning as *amplitude resolution.* See also *gridding, signal-to-noise ratio (S/N),* and *lateral resolution.*

vertical sheet: a *model* approximation of the vertical or near-vertical *source body* with a tabular shape and *infinite strike length* wherein the width, or thickness, is much smaller (in practice, 6–8 times) than other spatial dimensions of the source body and depth to its top. It is used in *modeling* of magnetized faults, *salt* dissolution zones, dikes, and other structural discontinuities. Referred to with the same meaning as a vertical *thin* *dike*. See also *dike*, *horizontal sheet*, *magnetized fault*, *tabular body*, and *2D source*.

verticality correction: a correction applied to *absolute gravity* measurements by a *free-fall method* to compensate for the artificial increase in measured values caused by the incorrect vertical alignment of a *free-fall absolute gravimeter* during its setup. Correction value is defined by the estimated amount of *verticality error*. Commonly referred to as *tilt correction*.

verticality error: error in *absolute gravity* measurements by the *free-fall method* with the use of a corner-cube reflector ("mirror") and laser interferometer. It is caused by the incorrect vertical alignment, or *tilt*, of a *free-fall absolute gravimeter* during its setup. For small angles, the verticality error is approximated as

$$\Delta g/g \cong -\theta^2/2,$$

where θ is the angular vertical alignment error. A laser beam verticality error of 10 arc-seconds results in 1 *microgal (μGal)* error of the measured *absolute gravity* value. It is compensated by application of *verticality correction*. See also *arc-second* and *off-level errors*.

vesselness (VS): an *attribute* of an anomalous gravity or magnetic field calculated by the combination of *eigenvalues* λ_1, λ_2, λ_3 obtained from *eigenvalue decomposition* of a two-dimensional (2D) or three-dimensional (3D) *Hessian matrix*. Its algorithm incorporates (a) second-order derivatives of an *anomalous field* to enhance its short-*wavelength* components and (b) field parameters related to shapes of anomalies to increase their *lateral resolution*. For the 3D Hessian matrix, it is expressed as

$$\mathrm{VS} = \{[1 - \exp(-\mathrm{R}_\mathrm{A}^2/2\alpha^2)]\exp(-\mathrm{R}_\mathrm{B}^2/2\beta^2)\} \cdot [1 - \exp(-\mathrm{S}^2/2\gamma^2)],$$

where R_A, R_B, and S are shape-dependent field parameters defined as $\mathrm{R}_\mathrm{A} = |\lambda_2|/|\lambda_3|$, $\mathrm{R}_\mathrm{B} = |\lambda_1|/(|\lambda_2\lambda_3|)^{1/2}$, $S = (\lambda_1^2 + \lambda_2^2 + \lambda_3^2)^{1/2}$; and α, β, γ are user-selected *sensitivity* thresholds (Frangi et al., 1998). Both the suppression of high-frequency *noise* by the inverse exponential *transfer function* in the manner of *upward continuation* and *amplitude balancing* by the ratio of variables increase the *sensitivity* of this attribute to low-*amplitude* anomalies generated by low-*contrast* structural elements and improve the *coherence* of the detected alignments. Vesselness is calculated at the stage of *enhanced processing* for the high-*resolution* mapping of faults, fracture and *salt* dissolution zones, shear zones, and other structural and lithological *lineaments* (Hassan and Goussev, 2018). See also *fault*, *shear zone*, and *normalized derivatives*.

vesselness filter: **1.** same as *vesselness (VS)* when it is considered as a *filter* to convert an original *anomalous field* into a field of new anomalies unavailable before the calculation of this *attribute*. **2.** a data processing operator for calculating the VS attribute.

vibrating-string gravimeter: a *relative gravimeter* the operational concept of which is based on the use of the resonant frequency of a vertically suspended and mass-loaded conducting ribbon, or string. The string is kept vibrating transversely in the magnetic field of permanent magnets by an alternating electric current of a constant frequency passing through the string. The rate of vibration, or resonant frequency, of the string is directly proportional to the square root of *gravitational acceleration g*. The resonant frequency *f* of the vibrating-string gravimeter is expressed as

$$f = (mg/\rho)^{1/2}/2L,$$

where m is mass suspended at the string end, ρ is string *density* per unit length, and L is string length. The relationship between a relative change in the resonant frequency and gravity change is defined as

$$\partial f/f = 0.5\,\partial g/g.$$

A vertically mounted double-string system, where the second mass-loaded string is separated from the first string by an additional spring and both strings are constrained at their ends, has been used for the gravity measurements on the Moon by the Apollo 17 astronauts. In the double-string system, the difference between the resonant frequencies of the two strings is directly proportional to the *relative gravity*. Depending on the reading time of the observations, the achievable *accuracy* of measurements is within the range of 0.1–1.0 *milligal (mGal)*. It is adaptable for marine (shipboard), land, and borehole gravity measurements.

video reviewing: a *cultural editing* procedure in which flight video records of the ground are synchronized with time-scaled line-oriented data and reviewed in order to identify the man-made ferrous objects, such as power lines, power stations, farm buildings, drilling rigs and wellheads, oil pumping units, bridges, railroads, etc., that correlate with *spike*-like anomalies in the aeromagnetic data.

vintage data: geophysical data (gravity, magnetic, seismic, well logging, etc.) acquired many years prior, so they are not considered contemporaneous with recent or current surveys. Usually, they are assumed to have poor quality *positioning* data compared to the present *GPS* navigation. Also referred to as *legacy data*.

virtual geomagnetic dipole: *geomagnetic dipole* in the context of approximating the *main geomagnetic field* by the magnetic field of a slow-rotating *magnetic dipole* located at about 400 km from the Earth's geometrical center with its axis of rotation tilted by 11° relative to the rotational axis of the Earth.

virtual spring: a design concept implemented in several types of *relative gravimeter*. It is based on the use of *magnetic levitation* or *electrostatic levitation* as the simulated, or "virtual", spring instead of a mechanical spring. Changes in the gravity field values between observation points, or *relative gravity*, are determined by measuring the voltage required to maintain the *proof mass* in a suspended position at the *null point*. Also referred to as *ideal spring*.

virtual-spring gradiometer: a *gravity gradiometer* the measurement system of which implements gravity sensors with the *virtual spring* design, such as a *GOCE gravity gradiometer*. See also *gravity sensor*.

virtual-spring gravimeter: **1.** a *relative gravimeter* with the *gravity sensor* design based on the *virtual spring* concept. **2.** *superconducting gravimeter.*

viscous magnetization: See *viscous remanent magnetization (VRM).*

viscous remanent magnetization (VRM): *remanent magnetization* acquired by rocks over long periods of time in the presence of the ancient *Earth magnetic field.* Its amount varies widely depending on the mineral content and grain size of the rock-forming minerals. The VRM is found in all rocks, and it can reach magnitudes close to *induced magnetization.* Also referred to as *viscous magnetization.* See also *magnitude* and *magnetic viscosity.*

volcanic mass effect: a local-scale change in *observed gravity* values caused by the growth of a volcanic dome and the flow of *lava* onto the Earth's surface. Measured gravity changes due to the volcanic mass effect can reach tens and hundreds of microgals. See also *microgal (μGal).*

volcanics: strongly magnetic igneous rocks ejected explosively or extruded as *lava* flows onto the Earth's surface from a volcano connected to a deep *magma* chamber. Basalt lava flows can cover regional-scale areas, thereby creating an analogue of the *magnetic basement* for younger sedimentary rocks. Accumulations of shallow volcanics are identified by their relatively high-*amplitude*, high-*frequency* (short-*wavelength*) chaotic *grain* on residual magnetic grids. Mid-depth and deep volcanics generate the irregularly shaped, localized anomalies of various amplitudes and wavelengths. In *profile*-based *modeling*, volcanics are often approximated by *2.5D source* geometries, such as *horizontal sheet.* The subsurface distribution of volcanics is mapped by the three-dimensional (3D) *magnetic inversion* of gridded *residual magnetics* with the use of a *voxel model* and calculated *magnetic anomaly vector* whose length, or *magnitude*, is very weakly dependent on the *remanent magnetization* effects and changes in the *magnetization direction* of locally accumulated volcanics (Li et al., 2012). See also *grid, gridding, residual field, magnetic signature*, and *spatial frequency.*

volcano gravity: *absolute gravity* and *relative gravity* data from *time-lapse gravity* measurements of the *Earth gravity field* and its changes for monitoring the growth of active volcanos. High-*accuracy* time-lapse gravity data are referenced to the traditionally corrected *baseline* gravity data, integrated with high-*resolution* surface deformation data, and compensated for (a) the *Earth tide effect,* (b) free-air and Bouguer effects due to the vertical displacement of the ground surface, and (c) the *water-table effect* due to groundwater level variations. Obtained *microgravity* values represent the time-dependent *residual gravity* changes caused by changes in the subsurface *mass* and/or *density* inside and around an active volcano (Battaglia et al., 2008). See also *Bouguer effect, free-air effect, aquifer gravity, CO_2 sequestration gravity, geothermal gravity*, and *4D gravity monitoring.*

volume-rendered image: *3D view* in the context of a viewpoint image of the spatial relationship between elements of an *inversion*-derived or interpreted depth structure, such as an *ore body* shape or *basement* depth in a *survey* area.

volume susceptibility: *magnetic susceptibility* per unit rock volume. It is a magnetic analogue of the rock *density.* See also *apparent susceptibility, bulk susceptibility*, and *effective susceptibility.*

voxel: a volume element, usually an elementary cube, representing the unit volume of a constant physical property, such as *density* or *magnetic susceptibility*, in a three-dimensional (3D) *voxel model.* Voxels replace the constant density and susceptibility approximations in the layered models of geological structures by a more realistic 3D distribution which accounts for the vertical and horizontal variations of the density and *magnetization* in rock formations. Voxels are used primarily in 3D gravity modeling, where the volume density element is referred to as *density voxel.* The term is a combination of the words volume and *pixel.* Sometimes spelled as *voxet.* See also *gravity model* and *magnetic model.*

voxel inversion: a three-dimensional (3D) *inversion* for complex geological structures, such as the base of *salt* or isolated *ore body*, with the use of a 3D *voxel model* of the rock *density* or *magnetic susceptibility* distribution. Voxel inversion of *residual gravity* is implemented in *stochastic inversion* for both the density and depth *structure* of rock formations, and it is also performed to obtain *constraints* for a seismic velocity *model* used in the prestack depth migration (PSDM) of seismic reflection data. See also *gravity-constrained velocity model* and *vector magnetic inversion.*

voxel model: **1.** a *voxel*-based three-dimensional (3D) *model* of a particular subsurface *structure* or isolated *ore body* that accounts for the vertical and horizontal variations of densities or susceptibilities in the sedimentary rock formations and heterogeneous composition of an ore body. A density voxel model is built with the use of the available seismic horizons, *sonic log* or *density log* data, and densities from core sample measurements. Rock *density* distributions also can be obtained by conversion of a 3D seismic velocity model (velocity cube) using the empirical relationships between the seismic interval velocities and rock densities, such as the *Gardner equation, Brocher formula*, and *Nafe-Drake curve.* Voxel model *resolution* is selected in accordance with the *lateral resolution* of the gridded *residual gravity* or *residual magnetics* used in the process of *3D inversion* which in this case is referred to as *voxel inversion.* **2.** a voxel-based 3D *density model* built for the purposes of *4D gravity monitoring.* See also *gridding, hybrid density model*, and *pseudolithology model.*

voxet: See *voxel.*

W

Walsh transform: transformation of the continuous line-oriented gravity or magnetic data into a stepped, or terraced, function. Similar to the *Fourier transform* and its elementary sinusoids, the elementary components of the Walsh transform are rectangular functions which assume only the values of +1 or −1. Successively higher orders of Walsh functions, from zero-order to *N*th-order, are arranged in sequences defined by the number of zero crossings per unit interval of distance that are analogous to the sequence of the discrete positive and negative boxcar functions with ±1 peak values. Data *terracing* can be executed by a low-pass option of the Walsh transform, where the high-order sequences above a certain threshold are replaced with zeros, while all of the lower order sequences remain unchanged. A density distribution approximated by Walsh-transformed gravity data corresponds to a blocky structure with sharp density transitions that are consistent with the rock boundaries in a geological *structure*, wherein the block sizes are defined by selection of the *cutoff* sequence (Lanning and Johnson, 1983; Keating, 1992). See also *boxcar function*.

washboard effect: See *corrugations*.

water-table correction: a correction applied to *microgravity* data, obtained from a *time-lapse gravity survey* for monitoring active volcanos, geothermal reservoirs, and groundwater-storage changes in unconfined aquifers, to compensate for the *water-table effect*. Its estimate Δg_{WT} is based on the *slab formula* and expressed in units of *microgal (μGal)* as

$$\Delta g_{WT} \cong 2\pi G \rho_W \phi \Delta z,$$

where G is *universal gravitational constant* in *SI units*; ρ_W is groundwater *density*; ϕ is approximated, or *apparent*, porosity of the *aquifer*; and Δz is vertical change of the groundwater level (Battaglia et al., 2008; Pool, 2008). See also *aquifer gravity*, *geothermal gravity*, *volcano gravity*, and *4D gravity monitoring*.

water-table effect: a *microgravity* effect produced by changes of groundwater level in the hydrothermal system of an unconfined aquifer, a geothermal reservoir under development, and around active volcanos. It is compensated by the application of *water-table correction*. See also *volcano gravity* and *aquifer gravity*.

wave-motion effect: a distortion of the measured marine (shipboard) gravity values by inertial accelerations of a *survey* ship resulting from its motion on the ocean/sea waves. Wave heights of 2–3 m can produce vertical inertial accelerations up to hundreds of milligals. At short wave periods of 1–10 minutes, wave heights of a few centimeters can generate *noise* at the level of 1 *milligal (mGal)*. The analogous motion-related effect for *airborne gravity*, but with a much higher *magnitude*, is produced by air turbulence and referred to as the *turbulence effect*. See also *inertial acceleration*, *cross-coupling effect*, and *lever arm effect*.

waveform: a wave-shaped component of a measured *field* described by a mathematical function of distance or time. Spatial variations of anomalous gravity and magnetic fields are represented by waveforms of various amplitudes and shapes that can be decomposed by *forward Fourier transform* into a series of sinusoids (sine and cosine wave functions) each having its own *amplitude*, *phase*, and *wavenumber*. See also *spatial frequency*.

wavelength: the horizontal distance between the successive equivalent points, such as *peak* or *trough*, in a periodic wave function and between the principal adjacent peaks or troughs in a non-periodic *waveform*. Denoted by λ, the space-domain wavelength is inversely related to the spectral-domain *wavenumber* k as $\lambda = 2\pi / k$ and to the space-domain *spatial frequency* f as $\lambda = 1/f$. The wavelength of *anomaly* depends on the size, shape, and, most importantly, depth of a *source body*, and it is independent from the *density contrast* or *magnetization contrast* created by a source body. For the same source shape and size, the wavelength is proportional to the source depth: the deeper the source, the longer the wavelength. The slope, or *gradient*, of an anomaly is proportional to its wavelength: the shorter the wavelength, the steeper the slope and the higher the gradient. See also *anomaly slope*, *dominant wavelength*, *space domain*, *spectral domain*, and *wavelength-depth relation*.

wavelength attenuation: a decrease in the *amplitude* of a gravity or magnetic *anomaly* described by the exponential function of the anomalous *wavelength* λ and source depth. Attenuation is inversely proportional to the wavelength and directly proportional to the source depth, so the short-wavelength anomalies are attenuated faster than the long-wavelength anomalies. In *satellite gravity* studies of the planetary *gravitational field*, the wavelength attenuation factor (WAF) is approximated by a dimensionless exponential function as

$$WAF = e^{-2\pi h/\lambda},$$

where h is the height, or *altitude*, of a satellite above the planetary surface. When $h = \lambda$, the wavelength attenuation factor is approximately 535. The WAF is applicable to *gravitational potential* and its first- and second-order derivatives representing the gravity anomalies and gravity gradients, respectively. There is a critical wavelength (approximately, the altitude h), such that wavelengths longer than the critical wavelength can be resolved, whereas much shorter wavelengths cannot be resolved by the satellite gravity measurements (NRC, 1997). See also *derivative*, *gradient*, and *resolution*.

wavelength-depth relation: an empirical relation which roughly approximates the observed *anomaly wavelength* by 3–4 source depth values for shallow sources and 2–3 source depth values for deep sources. See also *Jacobsen filter*, *matched-filtering method*, and *wavenumber-depth relation*.

wavelength filter: a *filter* which passes or rejects components of an anomalous gravity or magnetic field based on their *wavelength* parameters. Low-pass, or high-cut, filters suppress the short-wavelength components associated with *noise* and small or shallow sources, whereas high-pass, or low-cut, filters suppress the longer wavelength components generated by large-volume and deep sources. Band-pass wavelength filters suppress noise and enhance the anomalies within a user-specified range of wavelengths and highlight the anomalies from targeted sources within their approximately estimated depth ranges. See also *anomalous field, band-pass filter, high-pass filter, low-pass filter,* and *target.*

wavelength resolution: *lateral resolution* in the context of the shortest *wavelength* component of an anomalous gravity or magnetic field that can be detected on the condition that the *amplitude* of this wavelength component exceeds the *accuracy* of measurements.

wavelet: a mathematical function of several wave cycles and finite length that oscillates about the horizontal axis. Its average value is zero and its *amplitude* decays rapidly with distance from the wavelet center. Wavelets are used in a *wavelet transform (WT)* to detect, or resolve, individual *wavelength* components and determine their locations in line-oriented or gridded gravity and magnetic data. Differing from sinusoids in the conceptually analogous *window Fourier transform (WFT)*, wavelets can be of many different forms, and the choice of a particular wavelet depends primarily on its mathematical properties. Wavelets are described by two functions: (1) a *mother wavelet*, which defines the wavelet shape and other properties, and (2) a *scaling function*, which defines the dilation and translation parameters of the mother wavelet. At each *scale* of a wavelet transform, the *wavelet dilation* parameter specifies the stretching or compression of the mother wavelet, thereby controlling the wavelet *bandwidth* and the *lateral resolution* of the bandwidth-matching wavelength components, while the *wavelet translation* parameter specifies the incremental shifting of the dilated wavelets along the data points and localization of the wavelet-resolved wavelengths. A mother wavelet can be obtained from the *horizontal derivative* of a gravity or magnetic *anomaly* generated by a *point source*, wherein the wavelet scaling is managed either by changing the source depth parameter or by *upward continuation* of the point-source anomaly. See also *Gaussian wavelet*, *Poisson wavelet*, and *wavelet domain*.

wavelet analysis: application of the *wavelet transform (WT)* for decomposition of an *anomalous field* into different ranges, or scales, of *wavelength resolution* with the use of variable length wavelets to determine the individual *wavelength* components of the field and their locations. Within the line or *grid* intervals where the operating *wavelet* and a particular wavelength component match each other, *wavelet coefficients* reach their maximum values, and thereby represent information about the space-localized wavelength (*spatial frequency*) content of the data at each WT scale. Sometimes referred to as *multi-resolution analysis* and *space-frequency localization*. See also *convolution, wavelet dilation, wavelet translation,* and *wavelet synthesis*.

wavelet coefficients: weighting coefficients in the *wavelet transform (WT)*. They are obtained from *convolution* of the iteratively dilated and translated *mother wavelet* with the line-oriented or gridded data at different ranges, or scales, of *wavelength resolution*. Wavelet coefficients reach their maximum values within the line or *grid* intervals where the bandwidths of a particular *wavelength* component and operating *wavelet* closely match each other. Coefficients are large in the areas where the data values change rapidly, and this property is used for *despiking* and elimination of other high-*amplitude* high-frequency *noise* components by *wavelet denoising*. Wavelet coefficients at large, or coarse, scales are called approximation coefficients, and coefficients at smaller, or finer, scales are called detail coefficients. At each *scale* of a wavelet transform, the wavelet coefficients describe the distribution of the local power, or energy, which reaches its maximum values at locations of the wavelet-resolved wavelengths and, therefore, they can be used to map the space-localized wavelength content of the data. Wavelet coefficients are (a) displayed on a space-scale plane for visualization of the scales contaminated by *random noise* and *coherent noise* and (b) arranged in the format of a *matrix*, where the rows are coefficients determined at one scale and the columns are coefficients at other scales, for the subsequent processing of the transformed data. The separation of coefficients by a *wavelet threshold* and their locations is used in wavelet denoising and other *wavelet filtering* applications. Also referred to as *wavelet transform coefficients*. See also *bandwidth, wavelet dilation,* and *wavelet translation*.

wavelet decomposition: *wavelet transform (WT)* in the context of decomposing the original data, such as an *anomalous field*, into different scales, or ranges, of *wavelength resolution* using the dilated wavelets of variable lengths (the larger the *wavelet* stretching, the longer the resolved *wavelength*) and, hence, bandwidths in order to determine the spatial locations of the individual wavelength components. See also *bandwidth, wavelet dilation, wavelet translation,* and *wavelet coefficients*.

wavelet denoising: elimination of a high-frequency *noise* in line-oriented gravity and magnetic data by application of the *discrete wavelet transform (DWT)*. At each *scale* of transformation, the high-pass part of the data is compared with a pre-selected *wavelet threshold*, and all *wavelet coefficients* below this threshold are replaced by zero values, so the corresponding noise-contaminated *wavelength* components can be eliminated. In the *despiking* option, the coefficients above a certain threshold value are zeroed to eliminate the spikes and other high-*amplitude*, high-*frequency* noise components. Differing from a regular, or conventional, space-domain or Fourier-domain *filtering*, the wavelet denoising provides the high-*resolution* localization, separation, and elimination of the noise components, thereby preserving the high-frequency anomalies of the geological origin in both the noise-contaminated and noise-free parts of the data. After wavelet denoising, which is performed in the *wavelet domain*, the line-oriented data are transformed by *wavelet synthesis* back to their original *space domain*, where they are gridded

(Leblanc and Morris, 2001). See also *denoising*, *Fourier domain*, *space domain*, *gridding*, and *wavelet*.

wavelet dilation: a *wavelet transform (WT)* parameter that specifies the stretching or compression of a *mother wavelet* and, therefore, controls the *wavelet* length, *bandwidth*, and its matching with the *wavelength* components of the data at different scales, or ranges, of *wavelength resolution*. Also referred to as *wavelet scaling*. See also *wavelet translation*.

wavelet domain: *domain* where the *x*- and *y*-coordinates of line-oriented or gridded data are independent variables and the scaled wavelets with the corresponding *wavelet coefficients* are dependent variables. See also *gridding*, *wavelet*, *wavelet dilation*, *wavelet scaling*, and *wavelet transform (WT)*.

wavelet Euler deconvolution: *Euler deconvolution* with the use of the horizontal and vertical derivatives calculated after application of a *wavelet transform (WT)* and *wavelet denoising*. Higher accuracy of the denoised derivatives provides a better *clustering* of the Euler depth solutions around the true tops of sources (Ridsdill-Smith, 2000). See also *horizontal derivative*, *vertical derivative*, and *source depth solutions*.

wavelet filtering: 1. *wavelet transform* in the context of its application for (a) selective filtering-out, or *denoising*, of the *noise*-contaminated scales and local parts of the data and (b) high-*resolution* detection, localization, and *enhancement* or suppression of the individual *wavelength* components of the data by scaling wavelets via the process of *convolution* in a manner of band-pass filters with the different bandwidths at different locations. Wavelet filtering is spatially variable, or local, because it allows the suppression of high-*frequency* (short-wavelength) components in some areas and their retention in others, whereas the traditional space- or Fourier-domain *filtering* is spatially constant, or global, because it has the same effect over the entire data area. **2.** *wavelet-based space-varying filtering*. See also *band-pass filter*, *bandwidth*, *convolution filter*, *Fourier domain*, *space domain*, *wavelet domain*, and *wavelet scaling*.

wavelet function: 1. same as *mother wavelet* with an emphasis on it representing a basis, or prototype, function which defines the mathematical properties of wavelets implemented in a *wavelet transform (WT)*. **2.** *scaling function*. See also *wavelet*.

wavelet scaling: *wavelet dilation* in the context of changing the *wavelet* length and, hence, its *bandwidth* to enable the matching of wavelets with the individual *wavelength* components at different ranges, or scales, of *wavelength resolution* and different locations. See also *scaling function*.

wavelet synthesis: a space-domain reconstruction of line-oriented gravity and magnetic data after *wavelet denoising* and other *wavelet filtering* operations performed in the *wavelet domain*. Reconstruction of the data decomposed by a *discrete wavelet transform (DWT)* involves *upsampling*. Also referred to as *inverse wavelet transform*. See also *space domain* and *wavelet analysis*.

wavelet threshold: a *wavelet denoising* parameter that defines the range of *amplitude* deviations from the zero value and the range-corresponding limiting value for the *wavelet coefficients* at the *wavelength resolution* scales with identified *random noise* and *coherent noise*. All wavelet coefficients below this value are replaced by zeros, whereas the larger coefficients are preserved, as they are assumed to be related to components of the *anomalous field*. In the *despiking* option, wavelet coefficients above the selected wavelet threshold are zeros.

wavelet transform (WT): a mathematical operation which uses the scaling wavelets to decompose line-oriented or gridded *anomalous field* data into different ranges, or scales, of *wavelength resolution* in order to determine the individual *wavelength* components of the field and their locations. Scaling wavelets are obtained by *wavelet dilation* which changes the *wavelet* length and its *bandwidth* by stretching or compressing the original *mother wavelet*, whereas *wavelet translation* defines the increment of shifting of a dilated mother wavelet along the data points. The wavelength content and localization of the wavelength components are determined via *convolution* of the iteratively dilated and translated wavelets with the data in a manner of a variable bandwidth *convolution filter*. Within the line or *grid* intervals where the dilated wavelets and individual wavelength components closely match each other, the convolution-obtained *wavelet coefficients* reach their maximum values and, therefore, they can be used to map the space-localized wavelength content of the data. There are two basic WT types: *continuous wavelet transform (CWT)* and *discrete wavelet transform (DWT)*. The main difference between them is in the wavelet dilation and translation (shifting): in the CTW, the dilation and shifting are arbitrary within a range of user-specified scales, whereas in the DWT they are restricted to powers of two, so the operating wavelet length is either doubled or halved by dilation, and the dilated wavelet is shifted twice or one-half the increment at a previous scale of the wavelength resolution (Hornby et al., 1999; Ridsdill-Smith and Dentith, 1999). Also known as *wavelet decomposition* and *multiresolution decomposition*.

wavelet transform coefficients: See *wavelet coefficients*.

wavelet translation: a *wavelet transform (WT)* parameter that specifies the incremental shifting of the dilated wavelets along data points at different scales of *wavelength resolution*. See also *wavelet dilation*.

wavenumber: 1. an independent variable in the *Fourier domain*. It is defined in units of inverse distance and used to describe the *amplitude* and *phase* characteristics of the Fourier-transformed gravity or magnetic data. For the one-dimensional (1D) function $f(x)$ representing line-oriented data, the Fourier-domain wavenumber is related to the space-domain *wavelength* λ as

$$k = 2\pi/\lambda.$$

For the 2D function $f(x,y)$ representing gridded data, it is called the *radial wavenumber* and expressed as

$$k = (k_x^2 + k_y^2)^{1/2} = (u^2 + v^2)^{1/2},$$

where u and v are wavenumbers in the mutually orthogonal x- and y-directions in *Cartesian coordinates* related to their respective wavelength components as $u=k_x=2\pi/\lambda_x$ and $v=k_y=2\pi/\lambda_y$. Logarithmic plots of the *amplitude spectrum*, *power spectrum*, and *radial power spectrum* are commonly displayed in wavenumbers. **2.** *spatial frequency.* **3.** a space-domain angular measure of the *phase* change per unit distance expressed in units of *radian* per meter or per kilometer. See also *Fourier transform*, *gridding*, *local wavenumber*, *multi-model wavenumber*, *total horizontal wavenumber*, and *wavenumber vector.*

wavenumber aliasing: errors in the calculation of the *power spectrum* of gridded data caused by *grid aliasing*. When the *grid interval* is too large to represent the short-*wavelength* components of line-oriented data, the artificial, or aliased, long-wavelength components are included into the *gridding* and subsequent *Fourier transform*, thereby producing false wavenumbers in the power spectrum. See also *aliasing*, *Nyquist frequency*, *wavenumber*, and *wavenumber leakage.*

wavenumber-depth relation: a general relationship describing the *amplitude* content of gravity and magnetic anomalies in terms of their Fourier-domain *wavenumber* components: the deeper the source, the lesser the contributions of the large wavenumbers corresponding to the short wavelengths generated by shallow sources. See also *Fourier domain*, *wavelength*, and *wavelength-depth relation.*

wavenumber domain: same as *Fourier domain* with an emphasis on the *wavenumber* being an independent variable which describes the *amplitude* and *phase* characteristics of Fourier-transformed data. See also *domain* and *Fourier transform.*

wavenumber filtering: the spectral-domain *filtering* which separates a user-defined *wavenumber* range from other wavenumbers in a Fourier-transformed *observed field.* The separation is based on the area-averaged spectral power at each wavenumber rather than on the space-point values in the space-domain filtering. See also *Fourier transform*, *power spectrum*, *space domain*, and *spectral domain.*

wavenumber leakage: 1. *Fourier transform* errors in the *wavenumber* components of an *amplitude spectrum* that occur when the length of a gridded or line-oriented dataset is different from the integer multiple of the *fundamental period.* It is minimized by *padding.* **2.** *spectral leakage.* See also *gridding*, *Gibbs phenomenon*, and *wavenumber aliasing.*

wavenumber vector: two-component spectral-domain *vector* **k**, which represents an independent variable in the *Fourier domain.* Its length, or *magnitude*, is referred to as the *radial wavenumber* and defined as

$$|\mathbf{k}|=(k_x^2+k_y^2)^{1/2},$$

where k_x and k_y are wavenumbers in the mutually orthogonal x- and y-directions in *Cartesian coordinates* related to their respective space-domain wavelength components as $k_x=2\pi/\lambda_x$ and $k_y=2\pi/\lambda_y$. See also *Fourier transform*, *gridding*, *spectral domain*, *wavelength*, *wavenumber*, and *total horizontal wavenumber.*

weber: a unit of *magnetic flux* in the SI system. The analogous unit in the cgs system is *maxwell*, where 1 maxwell $=10^{-8}$ weber.

weighted Euler deconvolution: *Euler deconvolution* of irregularly spaced and gridded *ground gravity* data with the capability of eliminating erroneous depth solutions resulting from *grid aliasing* and low-*accuracy* measurements. It is based on the weighting of the individual Euler equations by an error factor proportional to the measurement accuracy and distance between the actual data locations and equally spaced *grid nodes.* The larger the measurement error at each *station* and the larger the distance between a particular grid node and the nearest station, the smaller the weight assigned to the *Euler deconvolution equation* solved for the source location and depth at each grid node (Keating, 1998). See also *gridding*, *source depth solutions*, and *located Euler deconvolution.*

Weiss theory: a theory of rock *magnetization* according to which the *ferrimagnetic* and *ferromagnetic* rock-forming minerals are composed of microscopic-scale regions, or domains, magnetized to saturation against the tendency of thermal agitation to disorient the alignments of magnetic dipoles within the domains. An external, or ambient, magnetic field polarizes, or lines up, these domains in the field direction and, if sufficiently strong, it can align all domains irreversibly and create the permanent magnetization of rocks known as *remanent magnetization*, which often has its *vector* orientation different from the present *Earth magnetic field.* See also *ambient field*, *hysteresis loop*, *magnetic dipole*, *magnetic domain*, *magnetization vector*, and *Barkhausen effect.*

Werner deconvolution: an automated method of two-dimensional (2D) *inversion* of the line-oriented *total magnetic intensity (TMI)* or reduced-to-pole (RTP) magnetic data for the depth, location, and other parameters of magnetic sources. All sources are approximated by two source geometries: *thin dike* and *magnetic contact.* For thin-dike sources, the *total field* values, i.e., values of the *TMI field* or *RTP magnetic field*, are used. For magnetic contact sources, the values of the *horizontal derivative* of the TMI or RTP magnetic field are used. Calculations are made with different *moving window* lengths to obtain the inversion solutions for the different depths. Windows are moved with a pre-selected increment along the *survey* lines or profiles extracted from a *grid* for each set of calculations. The basic assumption is that values of the magnetic field or its horizontal derivative within the calculation window are related to a single *source body.* Each calculation window includes 5–7 data points, and the window operator solves the overdetermined system of equations at each moving window position for the source depth, horizontal location, *dip*, and *susceptibility*-times-thickness values for the thin-dike sources (Hartman et al., 1971; Ku and Sharp, 1983). Werner deconvolution is applied to the vertical derivatives of *Bouguer gravity* or *free-air gravity* in the manner of *pseudomagnetic field* inversion for the source parameter estimates. Because the relationship between an *anomaly* and its source can be described as *convolution*, wherein the Earth is the *transfer function*, the method is called deconvolution. Also referred to as *2D Werner deconvolution*, *2D Werner inversion*, and *2D Werner method.* See also *multiple-source Werner* *deconvolution*, *3D Werner deconvolution*, and *Euler deconvolution*.

Werner depth estimates: 1. magnetic *source depth solutions* obtained within a *moving window* by the application of two-dimensional (2D) *Werner deconvolution* or *multiple-source Werner deconvolution* to (a) line-oriented *total magnetic intensity (TMI)* data, (b) profiles extracted from gridded TMI or reduced-to-pole (RTP) magnetic data, and (c) line-oriented *horizontal derivative* data obtained from a TMI or RTP magnetic field. Profiles are extracted along user-specified directions with a preference for those that are orthogonal to the dominant structural trends in a *magnetic survey* area. Depth solutions calculated from the horizontal derivative data are approximately 20% shallower than depth solutions calculated from a TMI or RTP field. When the profile line is not perpendicular to the *strike* of a particular 2D source, *strike correction* is applied to the obtained source depth values. **2.** source depth estimates obtained by the application of three-dimensional *(3D) Werner deconvolution* to gridded magnetic or gravity *anomalous field* data. The large number of *noise*-generated solutions is reduced by the application of *clustering* and *tolerance* criteria. See also *gridding*, *profile*, *reduction-to-pole (RTP)*, and *Euler depth estimates*.

Werner inversion: same as *Werner deconvolution* with an emphasis on it being a two-dimensional (2D) or three-dimensional (3D) method of *inversion* of line-oriented or gridded magnetic and pseudomagnetic data for the depth, location, and other parameters of a *source body*. See also *pseudomagnetic field* and *Euler inversion*.

Werner method: same as *Werner deconvolution* with an emphasis on it being an automated method of estimating the depth, location, *dip*, and other parameters of a magnetic *source body*. The term is used in references to both *profile*-based two-dimensional (2D) and *grid*-based three-dimensional (3D) applications known as *2D Werner deconvolution* and *3D Werner deconvolution*, respectively.

WGS84: See *World Geodetic System 1984*.

WGS84 datum: same as *WGS84 ellipsoid* with an emphasis on it representing a *geodetic datum* for the calculation of the three-dimensional (3D) coordinates of gravity and magnetic observation points: latitude, longitude, and *ellipsoidal height*, which is also known as *elevation*.

WGS84 ellipsoid: *reference ellipsoid* for which the defining parameters (equatorial radius, *geometrical flattening*, angular velocity of rotation, and *geocentric gravitational constant*) have been adopted by the *World Geodetic System 1984 (WGS84)* as the most accurate mathematical approximation of the Earth's shape. It is a uniform reference surface, or *geodetic datum*, implemented in a *Global Positioning System (GPS)* for the three-dimensional (3D) coordinates of observation points: latitude, longitude, and *ellipsoidal height*, which is commonly referred to as *elevation*. The WGS84 ellipsoid and the previous standard reference ellipsoid version, known as *GRS80 ellipsoid*, have insignificant differences for exploration applications.

whiskers: spikes of *noise* in airborne and marine (shipboard) magnetic data, such as the short-duration *diurnals* and certain types of *cultural magnetic anomalies* generated by bridges, buildings, oil rigs, wellheads, sunken ships, and other relatively small man-made objects. See also *spike*.

white depth: the depth at which a *radially averaged power spectrum* of the downward-continued *total magnetic intensity (TMI)* or *RTP magnetic field* data becomes flattened. See also *downward continuation*.

white noise: *random noise* containing all *spatial frequency* components in equal proportions within a measurement *bandwidth* but with randomly distributed phases. See also *phase* and *noise*.

white spectrum: a *spectrum* which contains all spatial frequencies in equal proportions within a specified *band*. See also *spatial frequency*.

width: 1. *anomaly width*. **2.** *pulse* width defined by the width of a *boxcar window* which has the same *peak* amplitude and contains the same area as the pulse.

Wiener filter: an optimum-type zero-*phase* spectral-domain *filter* which takes into account the spectral characteristics of a signal and variations of *signal-to-noise ratio (S/N)* in the presence of *random noise* in order to optimize the *filter response* by *least-squares minimization*. Its *transfer function* is expressed as

$$F_W(k) = [S(k)]^2 / [S(k)]^2 + [N(k)]^2,$$

where $[S(k)]^2$ and $[N(k)]^2$ are the estimated power spectra of the signal and noise, respectively. When the S/N value is high, the filter response tends to 1; for low S/N values, it tends to 0. The Wiener filter can be adapted to the geologically constrained *filtering* mode by incorporating the *power spectrum* of a *target anomaly*, calculated from the results of *forward modeling*, into its transfer function for the subsequent application in the manner of a *tuning filter*. The low-pass filter version is known as the *Wiener noise filter*. Sometimes referred to as *least-squares filter*. See also *constraints*, *spectral domain*, *optimum filter*, and *zero-phase filter*.

Wiener noise filter: a low-pass version of the *Wiener filter* applied for the suppression of high-frequency *noise* and stabilization of *total magnetic intensity (TMI)* data before the application of *reduction-to-pole (RTP)*.

window: 1. a portion of a line-oriented dataset or *grid* selected as a basis for designing *data processing* and *inversion* operators. Window parameters include the size, shape, edge slopes, weighting of values within the window, type of processing (*smoothing*, averaging, *residualizing*, etc.), and distribution of data (mean, median, etc.) within the window. Grid-based calculation windows are typically square or rectangular. The effective width, or length, of a non-rectangular window is defined by the width of a *boxcar function* with the same peak *amplitude* and the same energy. **2.** a rectangular portion of a *grid* extracted in accordance with specified *XY coordinates*. **3.** a part of the computer screen which displays a portion of data, especially when several datasets are displayed simultaneously, such as *montage* grids, to assist in the visual comparative analysis and *interpretation*. Sometimes referred to as *gate*. See also *Gaussian window*, *moving window*, *windowing*, and *zoom*.

window carpentry: **1.** *tapering* in the context of adding the rows and columns of artificial data along the margins of a *grid* to minimize *edge effects*. **2.** designing the boundaries of grids to minimize *Gibbs ringing* in Fourier-transformed data with the use of *edge-smoothing filters* and other techniques to eliminate the abrupt discontinuities at data edges. See also *Fourier transform* and *window function*.

window Fourier transform (WFT): *Fourier transform* of time- or space-domain data with the use of a sliding *window function*, such as *Gaussian window*, so only the data within a time or space interval of this window function are Fourier-transformed. The WFT is applied to determine the time- or space-localized *spectrum* content of gridded and line-oriented data at a particular location of a window function. The window length controls the *resolution* of the data components: large windows provide a good *frequency* resolution but a poor *spatial resolution*, whereas small windows provide a good spatial resolution but a poor frequency resolution. See also *wavelet transform*.

window function: **1.** a mathematical function which defines the type of processing within a *window* and its size and shape that minimize *edge effects* and errors in the *spectrum* calculations during *Fourier transform* of a selected portion of the gridded or line-oriented data. **2.** *scaling function* implemented in a *wavelet transform*. See also *Gaussian window* and *window carpentry*.

window inversion: *inversion* of the line-oriented or gridded g_{xz} and g_{zz} components of a rotated *gravity gradient tensor (GGT)* for estimating the parameters of sources approximated by the two-dimensional (2D) geometries of *dike* and *contact*: horizontal position, depth to top, thickness, *dip*, and *density contrast*. The method combines the dimensionality and eigenvector analyses of the GGT data to transform the measured GGT components into a *strike*-oriented coordinate frame for the subsequent *joint inversion* of the transformed g_{xz} and g_{zz} components within a variable size square *moving window*, which is centered at the maxima of g_{zz}, with the use of a *least-squares optimization* algorithm and available *constraints* (Beiki and Pedersen, 2011). See also *dimensionality indicator*, *eigenvalues*, *eigenvectors*, *rotated gradient tensor*, and *located Euler deconvolution*.

windowing: **1.** selection of a square or rectangular area, or *window*, within a *grid* by (a) user-defined *XY coordinates* or (b) interactive *zooming* into a particular portion of the grid. **2.** using a *window function* to select a portion of the *profile* or line-oriented data prior to *Fourier transform* in order to determine the *spatial frequency* content of a particular area with minimized errors in its *spectrum* calculation. **3.** *tapering* in the context of an edge-smoothing procedure applied to gridded data prior to *Fourier transform* in order to minimize *edge effects*. See also *Gibbs phenomenon*, *edge-smoothing filters*, and *window Fourier transform (WFT)*.

wire-frame representation: a graphical representation of a *three-dimensional (3D)* object, such as an *ore body* or *igneous intrusion* in a 3D gravity or magnetic *model*, by line-connected depth points.

workstation: a stand-alone or network-connected interactive computer terminal for *data processing*, *enhanced processing*, *modeling*, analysis, *interpretation*, and visualization of gravity and magnetic data.

World Digital Magnetic Anomaly Map (WDMAM): a global compilation of the public-domain *magnetic anomaly* data acquired by airborne, marine (shipboard), and ground magnetic surveys and satellite missions over a period of more than 50 years. The WDMAM 1.0 version presents *total magnetic intensity (TMI)* data in the format of a *grid* with a spatial *grid interval* of 3 arc-minutes at 5 km *elevation* above *mean sea level (MSL)*. The WDMAM 2.0 version has the same *grid spacing* of 3 arc-minutes, but the TMI data are referenced to the *WGS84 datum* and presented at two elevations: marine (shipboard) data at 0 km and all other compilations are upward continued to 5 km above the *WGS84 ellipsoid*. See also *arc-minute*, *upward continuation*, *World Magnetic Model (WMM)*, and *Earth Magnetic Anomaly Grid-2 (EMAG2)*.

World Geodetic System 1984 (WGS84): the Earth-centered and Earth-fixed *Geodetic Reference System*, wherein the three-dimensional (3D) coordinates of any point on or above the Earth's surface (latitude, longitude, and *elevation*) are determined relative to the *WGS84 ellipsoid*, and the Earth's mass center is defined for the whole Earth, including the oceans and atmosphere. The WGS84 is the reference frame for the satellite-based *Global Positioning System (GPS)*, and it is also the standard global *geodetic datum* for the representation of geospatial information. The WGS84 is compatible with the *International Terrestrial Reference Frame (ITRF)*.

World Magnetic Model (WMM): a global-scale predictive model of the *main geomagnetic field* described by the *spherical harmonic expansion* of *magnetic potential* to the degree and order 12 with *wavelength resolution* of 3,000 km and referenced to the *WGS84 ellipsoid*. The WMM is equivalent to and based on the same principles as the predictive *IGRF model*, but it has been produced independently by the U.S. National Geophysical Data Center jointly with the British Geological Survey. The WMM2015 is the latest model which predicts the *geomagnetic field parameters* along with the estimates of *secular variation* for the years 2015–2020 at any location on and any height above the Earth's surface. See also *International Geomagnetic Reference Field (IGRF)*, *spherical harmonics*, and *NGDC720 magnetic model*.

wraparound effect: the edge effect caused by *circular convolution* in the process of spectral-domain *filtering* and other transformations of the gridded gravity and magnetic data. It is minimized by *detrending* prior to *Fourier transform*. See also *edge effects*, *gridding*, and *spectral domain*.

X

XY coordinates: coordinates of the gravity and magnetic data on a standard *grid* or map, where the X-coordinate indicates the distance in meters along the *x*-axis from a reference point called *false easting*, and the Y-coordinate indicates the distance in meters along the *y*-axis from a reference point called *false northing*. Referred to with the same meaning as *UTM coordinates*. See also *Universal Transverse Mercator (UTM)*.

XYZ format: a standard three-column *format* of data archiving that is compatible with all common *gridding* algorithms. The first two columns are *XY coordinates* of the observation points and the third column represents the corresponding observed gravity or magnetic values obtained by application of all relevant corrections to the measured values. See also *ASCII format*, *observed gravity*, and *observed magnetics*.

Y

yaw: the rotational motion of a *survey* ship or aircraft about the vertical axis that is caused by the ocean waves or air turbulence during gravity and magnetic measurements along survey lines, i.e., it is the rotation of a moving *observation platform* about the Z-axis in the XY plane. Yaw is a source of magnetic interference effects in the aircraft frame and inertial accelerations of a shipboard (marine) or airborne observation platform. See also *pitch*, *roll*, *aeromagnetic compensation*, and *inertial acceleration*.

Z

Z-gradiometer: an arrangement of two gravimeters or single-axis magnetometers aligned vertically and placed at a pre-selected distance from each other for measurements of the *vertical gradient* of the *Earth gravity field* or *Earth magnetic field*, respectively. In magnetic measurements, the Z-gradiometer suppresses both the *main geomagnetic field* components and fluctuations of the *external magnetic field* of the Earth, thereby highlighting the *gradient* anomalies generated by elements of a local geological *structure*. See also *diurnals*, *gravimeter*, *magnetometer*, and *gradiometer*.

Zeeman effect: a quantum physics effect of splitting the energy radiated by atoms into several sub-levels in the presence of a static magnetic field. When the magnetic field is sufficiently strong, the *spectrum* of the energy levels in the radiation emitted by atoms is split into two or more separate spectrum lines. The line separation distance represents the energy gap between the sub-levels, and it is directly proportional to the *magnitude* of the static magnetic field. The Zeeman effect is implemented in high-*sensitivity* versions of the *optically pumped magnetometer*, such as the *helium magnetometer*, to measure the magnetic field of the Earth and other planets by orbiting satellites. It is also used in space spectrometry for the detection of strong magnetic fields in a deep space. See also *satellite magnetometry*.

zero crossing: the point on a *profile* or contour line on a color-coded *residual grid* of an anomalous gravity or magnetic field where the value of a particular *anomaly* is zero. See also *anomalous field*, *color-coded grid*, and *gridding*.

zero-length spring: the spring suspension design concept implemented in the high-*sensitivity* types of *spring gravimeter*. Unlike the dynamically unstable *astatic balance* suspension, the zero-length spring suspension provides a stable *neutral balance*, or neutral dynamic equilibrium, between the *gravitational force* acting on a *proof mass* and the restoring force of the metal or quartz spring which suspends the horizontal beam with a proof mass at its end. In the state of neutral equilibrium, there is no restoring spring force, and the smallest change in the gravitational force will cause a relatively large displacement of the proof mass to a new equilibrium position. The term "zero length" refers to the mechanical characteristics of a specially manufactured pre-stressed spring for which the linear stress-length graph passes through its origin, i.e., zero-length and zero-stress point, rather than the actual length of spring in a gravimeter. See also *virtual spring*.

zero-length spring gravimeter: a type of a high-*sensitivity relative gravimeter* in which the *proof mass* is maintained in a stable *neutral balance* by the *zero-length spring* suspension with the use of a metal spring (LaCoster & Romberg gravimeters) or quartz spring (Worden gravimeters). Without a change in the *gravitational force*, the proof mass remains at the "null" position of the neutral balance. Constrained to move along the vertical axis, the proof mass deflects upward to a new balanced position in response to a decrease in the gravity force and downward from the "null" position in response to an increase in the gravitational force. A micrometer screw or electronic nulling system restores the original "null" position of the proof mass after the completion of each measurement. Proof mass deflections are calibrated in standard units of *milligal (mGal)* or *microgal (μGal)*. Zero-length spring gravimeters are very sensitive to variations of the *Earth gravity field*, and their versions for *microgravity* measurements have a *sensitivity* up to 0.1 μGal. The achievable *accuracy* of measurements is 1 μGal. See also *superconducting gravimeter*.

zero-order ASA (ASA_0): 1. a quantitative characteristic, or *attribute*, of an anomalous gravity or magnetic field, defined as

$$ASA_0 = \{f^2 + H_x^2(f) + H_y^2(f)\}^{1/2},$$

where f is *anomalous field* and $H_x(f)$ and $H_y(f)$ are x- and y-components of the *3D Hilbert transform* of an anomalous field in the mutually orthogonal directions (Cooper, 2014a). The ratio of ASA_0 to *first-order ASA* (ASA_1), i.e., $R_{ASA} = ASA_0/ASA_1$, is used to determine the depth to the top of a *thin dike* by the *ASA-ratio method*. The ASA_0 is equivalent to (a) *direct analytic signal amplitude (DASA)*, (b) *local amplitude* of *monogenic signal (MS)*, an (c) analytic signal amplitude (ASA) of the *vertical integral* of a magnetic or gravity *anomaly*, i.e., it is equivalent to the ASA of an *anomalous potential*. **2.** *analytic signal amplitude (ASA)* of the original *TMI field* with an emphasis on its difference from ASA_1, which is the ASA of the first-order *vertical derivative* of the TMI or RTP field. See also *total magnetic intensity (TMI)*, *RTP field*, *fractional-order ASA*, and *AN-EUL method*.

zero-order tilt ASA: an upgraded version of the *tilt ASA* edge-detection *attribute* based on the use of the first-order derivatives of the *zero-order ASA* (ASA_0) instead of the standard ASA:

$$\mathrm{TA\_ASA}_0 = \arctan\{-(\partial \mathrm{ASA}_0/\partial z)/[(\partial \mathrm{ASA}_0/\partial x)^2 + (\partial \mathrm{ASA}_0/\partial y)^2]^{1/2}\}.$$

For two-dimensional (2D) *contact* and *dike* source approximations, the zero-order tilt ASA is independent from the *magnetization vector* direction and *dip* of these sources. For three-dimensional (3D) sources, its *amplitude response* is less dependent on the *magnetization direction* of the source than the amplitude response of the standard ASA. Because anomalies are represented by the first-order derivatives of an anomalous gravity or magnetic field, the *sensitivity* of this attribute to *noise* is much less than the noise sensitivity of the tilt ASA

(Cooper, 2014a). See also *anomalous field*, *derivative*, and *edge detection*.

zero-phase filter: a spectral- or space-domain *filter* that operates without a change of *phase* of the input signal, i.e., its *phase shift* is zero for all spatial frequencies. This class of filters includes *upward continuation* and *differential continuation*, first- and second-order *vertical derivative* filters of the $\partial/\partial z$ and ∂^2/z^2 types, and *Wiener filter*. Differing from a *phase filter*, the zero-phase filter does not change the original location of anomalies. Similar to a phase filter, its spectral-domain *frequency response* is proportional to the *wavenumber* for the first-order *derivative* types and to the square of wavenumbers for the second-order derivative types. See also *anomaly*, *filter*, *space domain*, and *spectral domain*.

zonal harmonics: *spherical harmonics* which depend on the latitude and do not depend on the longitude. See also *sectoral harmonics* and *tesseral harmonics*.

zone chart: 1. *Hammer chart* for the calculation of *terrain correction*. **2.** a transparent template for the manual calculation of *elevation* differences between the average elevation within each terrain zone and *station* elevation that are used in the terrain correction and *isostatic correction*. It is laid over a topographic map with its center at the station being corrected, and the obtained elevation differences (regardless of their signs, because terrain and isostatic corrections are always positive for ground observation points) are referenced to their tabulated values to determine a corresponding correction amount for each zone. See also *Hammer method* and *Jachens-Roberts method*.

zoom: an interactive display function which magnifies a user-defined square or rectangular portion of gridded data in the zoom-in option and returns to the original display in the zoom-out, or unzoom, option. See also *gridding*.

References

Abaco, C. I., and D. C. Lawton, 2003, Magnetic anomalies in the Alberta foothills, Canada: 73rd Annual International Meeting, SEG, Expanded Abstracts, https://doi.org/10.1190/1.1818004.

Alexander, A. L., K. Hasan, G. Kindlmann, D. L. Parker, and J. S. Tsuruda, 2000, A geometric analysis of diffusion tensor measurements of the human brain: Magnetic Resonance in Medicine, **44**, no. 2, 283–291, https://doi.org/10.1002/1522-2594(200008)44:2<283::AID-MRM16>3.0.CO;2-V.

Allis, R. G., and T. M. Hunt, 1986, Analysis of exploitation-induced gravity changes at Wairakei Geothermal Field: Geophysics, **51**, no. 8, 1647–1660, https://doi.org/10.1190/1.1442214.

Alnes, H., O. Eiken, and T. Stenvold, 2008, Monitoring gas production and CO_2 injection at the Sleipner field using time-lapse gravimetry: Geophysics, **73**, no. 6, WA155–WA161, https://doi.org/10.1190/1.2991119.

Ander, M. E., T. Summers, and M. E. Gruchalla, 1999, LaCoste & Romberg gravity meter: System analysis and instrumental errors: Geophysics, **64**, no. 6, 1708–1719, https://doi.org/10.1190/1.1444675.

Ansari, A. H., and K. Alamdar, 2011, A new edge detection method based on the analytic signal of tilt angle (ASTA) for magnetic and gravity anomalies: Iranian Journal of Sci. & Tech., **A2**, 81–88.

Ardalan, A. A., and E. W. Grafarend, 2001, Somigliana-Pizzetti gravity: The international gravity formula accurate to the sub-nanoGal level: Journal of Geodesy, **75**, no. 7, 424–437, https://doi.org/10.1007/PL00004005.

Argast, D., M. Bacchin, and R. Tracey, 2009, An extension of the closed-form solution for the gravity curvature (Bullard B) correction in the marine and airborne cases: ASEG Technical Program Extended Abstracts, https://doi.org/10.1071/ASEG2009ab129.

Arisoy, M. Ö., and Ü. Dikmen, 2013, Edge detection of magnetic sources using enhanced total horizontal derivative of tilt angle: Bulletin of the Earth Sciences Application and Research Centre of Hacettepe University, **34**, 73–82.

Arkani-Hamed, J., 1988, Differential reduction-to-pole of regional magnetic anomalies: Geophysics, **53**, no. 12, 1592–1600, https://doi.org/10.1190/1.1442441.

Arora, K., 2011, Geodesy, Figure of the Earth, *in* K. Gupta, ed., Encyclopedia of Solid Earth Geophysics: Springer, 313–316, https://doi.org/10.1007/978-90-481-8702-7_83.

Bain, J. E., T. R. Horscraft, J. Weyand, A. H. Saad, and D. N. Bulling, 1993, Complex salt features resolved by integrating seismic, gravity and magnetics: 5th EAPG Conference and Technical Exhibition.

Bainbridge, G., C. Musselman, N. Whitehead, and M. McDonald, 2002, Euler 3D: Processing, analysis and visualization system for 3D inversion of potential field, Tutorial and User Guide: Geosoft Publication.

Baranov, V., 1957, A new method for interpretation of aeromagnetic maps: Pseudo-gravimetric anomalies: Geophysics, **22**, no. 2, 359–382, https://doi.org/10.1190/1.1438369.

Barbosa, V. C. F., J. B. C. Silva, and W. F. Medeiros, 1999, Stability analysis and improvement in structural index estimation in Euler deconvolution: Geophysics, **64**, no. 1, 48–60, https://doi.org/10.1190/1.1444529.

Barriot, J.-P., and L. Sichoix, 2011, Gravity Modeling, Theory and Computation, *in* K. Gupta, ed., Encyclopedia of Solid Earth Geophysics: Springer, 518–524, https://doi.org/10.1007/978-90-481-8702-7_233.

Barrows, L. J., and J. D. Fett, 1991, A sloping wedge technique for calculating gravity terrain corrections: Geophysics, **56**, no. 7, 1061–1063, https://doi.org/10.1190/1.1443116.

Battaglia, M., J. Gottsmann, D. Carbone, and J. Fernandez, 2008, 4D volcano gravimetry: Geophysics, **73**, no. 6, WA3–WA18, https://doi.org/10.1190/1.2977792.

Bean, R. J., 1966, A rapid graphical solution for the aeromagnetic anomaly of the two-dimensional tabular body: Geophysics, **31**, no. 5, 963–970, https://doi.org/10.1190/1.1439827.

Beiki, M., 2010, Analytic signals of the gravity gradient tensor and their application to estimate source location: Geophysics, **75**, no. 6, I59–I74, https://doi.org/10.1190/1.3493639.

Beiki, M., D. A. Clark, J. R. Austin, and C. A. Foss, 2012, Estimating source location using normalized magnetic source strength calculated from magnetic gradient tensor data: Geophysics, **77**, no. 6, J23–J37, https://doi.org/10.1190/geo2011-0437.1.

Beiki, M., and L. B. Pedersen, 2011, Window constrained inversion of the gravity gradient tensor data using dike and contact models: Geophysics, **76**, no. 6, I59–I72, https://doi.org/10.1190/geo2010-0368.1.

Beiki, M., and L. B. Pedersen, 2010, Eigenvector analysis of gravity gradient tensor to locate geologic bodies: Geophysics, **75**, no. 6, I37–I49, https://doi.org/10.1190/1.3484098.

Beiki, M., L. B. Pedersen, and H. Nazi, 2011, Interpretation of aeromagnetic data using eigenvector analysis of pseudogravity gradient tensor: Geophysics, **76**, no. 3, L1–L10, https://doi.org/10.1190/1.3555343.

Bellott, A., J. Corpel, and R. Millon, 1991, Contribution of magnetic modeling to the discovery of a hidden massive sulfide body at Hajar, Morocco: Geophysics, **56**, no. 7, 983–991, https://doi.org/10.1190/1.1443131.

Berger, Z., 2012, Satellite hydrocarbon exploration: Interpretation and integration techniques: Springer.

Berger, Z., Fortin, D., and Wang, X., 2004, High-resolution aeromagnetic (HRAM) surveys: Exploration applications from the Western Canada Sedimentary Basin – Exploration in highly deformed terrains using fixed-wing aircraft and helicopter mounted systems: AAPG Search and Discovery Article #40123.

Besse, J., V. Courtillot, and M. Greff, 2011, Paleomagnetism, polar wander, *in* K. Gupta, ed., Encyclopedia of Solid Earth Geophysics: Springer, 945–955, https://doi.org/10.1007/978-90-481-8702-7_125.

Best, M. E., H. J. Abercrombie, and J. W. Peirce, 1998, Interpreted faulting patterns in northeast Alberta using high resolution aeromagnetic data: Canadian Journal of Exploration Geophysics, **34**, 49–57.

Biegert, E., J. Ferguson, and X. Li, 2008, 4D gravity monitoring – Introduction: Geophysics, **73**, no. 6, WA1–WA2, https://doi.org/10.1190/1.3010377.

Blakely, R. J., 1996, Potential theory in Gravity and Magnetic Applications: Cambridge University Press.

Blakely, R. J., and V. J. S. Grauch, 1983, Magnetic models of crystalline terrane: Accounting for the effect of topography: Geophysics, **48**, no. 11, 1551–1557, https://doi.org/10.1190/1.1441437.

Blakely, R. J., and R. W. Simpson, 1986, Approximating edges of source bodies from magnetic or gravity anomalies: Geophysics, **51**, no. 7, 1494–1498, https://doi.org/10.1190/1.1442197.

Bosch, M., and J. McGaughey, 2001, Joint inversion of gravity and magnetic data under lithologic constraints: The Leading Edge, **20**, no. 8, 877–881, https://doi.org/10.1190/1.1487299.

Bott, M. H. P., 1960, The use of rapid digital computing methods for direct gravity interpretation of sedimentary basins: Geophysical Journal of the Royal Astronomical Society, **3**, no. 1, 63–67, https://doi.org/10.1111/j.1365-246X.1960.tb00065.x.

Bouman, J., and M. J. Fuchs, 2012, GOCE gravity gradients versus global gravity field models: Geophysical Journal International, **189**, no. 2, 846–850, https://doi.org/10.1111/j.1365-246X.2012.05428.x.

Braga, M. A., I. Endo, H. F. Galbiatti, and D. U. Carlos, 2014, 3D full tensor gradiometry and Falcon systems data analysis for iron ore exploration: Bau Mine, Quadrilatero Ferrifero, Minas Gerais, Brazil: Geophysics, **79**, no. 5, B213–B220, https://doi.org/10.1190/geo2014-0104.1.

Braitenberg, C., and J. Ebbing, 2009, The GRACE-satellite gravity and geoid fields in analyzing large-scale, cratonic or intracratonic basins: Geophysical Prospecting, **57**, no. 4, 559–571, https://doi.org/10.1111/j.1365-2478.2009.00793.x.

Brenner, A. C., J. P. DiMarzio, and H. J. Zwally, 2007, Precision and accuracy of satellite radar and laser altimeter data over the continental ice sheets: IEEE Transactions on Geoscience and Remote Sensing, **45**, no. 2, 321–331, https://doi.org/10.1109/TGRS.2006.887172.

Brewster, J., 2017, Gravity gradiometer design comparison by analysis of parametric inversion: 87th Annual International Meeting, SEG, Expanded Abstracts, https://doi.org/10.1190/segam2017-17647742.1.

Brocher, T. M., 2005, Empirical relations between elastic wavespeeds and density in the Earth's crust: Bulletin of the Seismological Society of America, **95**, no. 6, 2081–2092, https://doi.org/10.1785/0120050077.

Bronner, A., M. Munschy, D. Sauter, J. Carlut, R. Searle, and A. Maineult, 2013, Deep-tow 3C magnetic measurement: Solutions for calibration and interpretation: Geophysics, **78**, no. 3, J15–J23, https://doi.org/10.1190/geo2012-0214.1.

Brown, G., N. H. Platt, and A. McGrandle, 1994, The geophysical expression of Tertiary dikes in the southern North Sea: First Break, **12**, no. 3, 137–146, https://doi.org/10.3997/1365-2397.1994011.

Busse, F. H., 2011, Geomagnetic field, Theory, *in* K. Gupta, ed., Encyclopedia of Solid Earth Geophysics: Springer, 394–401, https://doi.org/10.1007/978-90-481-8702-7_103.

Butler, D. K. 2005, Near-surface Geophysics: Society of Exploration Geophysicists.

Butler, D. K., 1984, Microgravimetric and gravity gradient techniques for detection of subsurface cavities: Geophysics, **49**, no. 7, 1084–1096, https://doi.org/10.1190/1.1441723.

Cameron, B., and S. Goussev, 2010, 3D Goussev filter – a signal separation and edge detection filter applied to aeromagnetic data in the Great Australian Bight: The Leading Edge, **29**, no. 8, 902–905, https://doi.org/10.1190/1.3480000.

Capriotti, J., and Y. Li, 2015, Integrating gravity and gravity gradiometry data for joint inversion: A case study at the Kauring test site: 85th Annual International Meeting, SEG, Expanded Abstracts, https://doi.org/10.1190/segam2015-5925814.1.

Caratori Tontini, F., and L. B. Pedersen, 2008, Interpreting magnetic data by integral moments: Geophysical Journal International, **174**, no. 3, 815–824, https://doi.org/10.1111/j.1365-246X.2008.03872.x.

Cascone, L., C. Green, S. Campbell, A. Salem, and D. Fairhead, 2017, ACLAS – A method to define geologically significant lineaments from potential-field data: Geophysics, **82**, no. 4, G87–G100, https://doi.org/10.1190/geo2016-0337.1.

Caté, A., L. Perozzi, E. Gloaguen, and M. Blouin, 2017, Machine learning as a tool for geologists: The Leading Edge, **36**, no. 3, 215–219, https://doi.org/10.1190/tle36030215.1.

Cazenave, A., G. Ramillien, and R. Biancale, 2011, Gravity Field, temporal variations from space techniques, *in* K. Gupta, ed., Encyclopedia of Solid Earth Geophysics: Springer, 484–489, https://doi.org/10.1007/978-90-481-8702-7_96.

Cella, F., M. Fedi, and G. Florio, 2009, Toward a full multiscale approach to interpret potential fields: Geophysical Prospecting, **57**, no. 4, 543–557, https://doi.org/10.1111/j.1365-2478.2009.00808.x.

Cevallos, C., 2017, Mathematical properties and physical meaning of the gravity gradient tensor eigenvalues: Geophysics, **82**, no. 6, G115–G124, https://doi.org/10.1190/geo2016-0418.1.

Cevallos, C., 2016, Interpreting the direction of the gravity gradient tensor eigenvectors: Their relation to curvature parameters of the gravity field: Geophysics, **81**, no. 3, G49–G57, https://doi.org/10.1190/geo2015-0331.1.

Cevallos, C., M. Dransfield, J. Hope, and H. Carey, 2012, Interpretation of airborne gravity gradiometry and magnetic data using cross-gradients: ASEG Technical Program Extended Abstracts, https://doi.org/10.1071/ASEG2012ab273.

Cevallos, C., P. Kovac, and S. J. Lowe, 2013, Application of curvatures to airborne gravity gradient data in oil exploration: Geophysics, **78**, no. 4, G81–G88, https://doi.org/10.1190/geo2012-0315.1.

Chambodut, A., 2011, Geomagnetic field, IGRF, *in* K. Gupta, ed., Encyclopedia of Solid Earth Geophysics: Springer, 379–380, https://doi.org/10.1007/978-90-481-8702-7_111.

Chapin, D. A., 1998, Gravity instruments: past, present, future: The Leading Edge, **17**, no. 1, 100–112, https://doi.org/10.1190/1.1437806.

Chapin, D. A., 1997, Wavelet transforms: A new paradigm for interpreting gravity and magnetic data?: 67th Annual International Meeting, SEG, Expanded Abstracts, https://doi.org/10.1190/1.1885940.

Chapin, D. A., 1996a, A deterministic approach toward isostatic gravity residuals – a case study from South America: Geophysics, **61**, no. 4, 1022–1033, https://doi.org/10.1190/1.1444024.

Chapin, D. A., 1996b, The theory of the Bouguer gravity anomaly: A tutorial: The Leading Edge, **15**, no. 5, 361–363, https://doi.org/10.1190/1.1437341.

Chapman, D. S., and D. Hasterok, 2011, Isostasy, Thermal, *in* K. Gupta, ed., Encyclopedia of Solid Earth Geophysics: Springer, 662–668, https://doi.org/10.1007/978-90-481-8702-7_223.

Chappell, A. R., and N. J. Kusznir, 2008, Three-dimensional gravity inversion for the Moho depth at rifted continental margins incorporating a lithosphere thermal gravity anomaly correction: Geophysical Journal International, **174**, no. 1, 1–13, https://doi.org/10.1111/j.1365-246X.2008.03803.x.

Chen, A.-G., T.-F. Zhou, D.-J. Liu, and S. Zhang, 2017, Application of an enhanced theta-based filter for potential field edge detection: a case study of the Luzong ore district: Chinese Journal of Geophysics, **60**, no. 2, 203–218, https://doi.org/10.1002/cjg2.30039.

Chen, D.-X., 2004, High-field ac susceptometer using Helmholtz coils as a magnetizer: Measurement Science & Technology, **15**, no. 6, 1195–1202, https://doi.org/10.1088/0957-0233/15/6/021.

Chen, S., Q.-S. Wang, Y.-Q. Zhu, C.-S. Jiang, W.-X. Wang, D.-F. Lu, and F.-Y. Guo, 2011, Temporal and spatial features of isostasy anomaly using gravitational admittance model at eastern margin of Tibetan plateau: Chinese Journal of Geophysics, **54**, no. 1, 11–24, https://doi.org/10.1002/cjg2.1582.

Childers, V. A., R. E. Bell, and J. M. Brozena, 1999, Airborne gravimetry: An investigation of filtering: Geophysics, **64**, no. 1, 61–69, https://doi.org/10.1190/1.1444530.

Chowdhury, P. R., and C. Cevallos, 2013, Geometric shapes derived from airborne gravity gradiometry data: New tools for the explorationist: The Leading Edge, **32**, no. 12, 1468–1474, https://doi.org/10.1190/tle32121468.1.

Christensen, A. N., and M. H. Dransfield, 2002, Airborne vector magnetometry over banded iron formations: 72nd Annual International Meeting, SEG, Expanded Abstracts, https://doi.org/10.1190/1.1816921.

Clark, D. A., 2014, Methods for determining remanent and total magnetizations of magnetic sources – A review: Exploration Geophysics, **45**, no. 4, 271–304, https://doi.org/10.1071/EG14013.

Clark, D. A., 2012, New methods for interpretation of magnetic vector and gradient tensor data: Eigenvector analysis and the normalized source strength: Exploration Geophysics, **43**, no. 4, 267–282, https://doi.org/10.1071/EG12020.

Close, D. J., 2010, Isostasy and gravity modeling: integrating potential field data in interpretation workflows: CSEG Recorder, **35**, no. 6, 35–39.

Close, D. J., A. B. Watts, and H. M. J. Stagg, 2009, A marine geophysical study of the Wilkes Land rifted continental margin, Antarctica: Geophysical Journal International, **177**, no. 2, 430–450, https://doi.org/10.1111/j.1365-246X.2008.04066.x.

Cogbill, A. H., 1990, Gravity terrain corrections calculated using digital elevation models: Geophysics, **55**, no. 1, 102–106, https://doi.org/10.1190/1.1442762.

Cooper, G. R. J., 2020, An improved terracing algorithm for potential-field data: Geophysics, **85**, no. 5, G109–G113, https://doi.org/10.1190/geo2019-0129.1.

Cooper, G. R. J., 2015, Using the analytic signal amplitude to determine the location and depth of thin dikes from magnetic data: Geophysics, **80**, no. 1, J1–J6, https://doi.org/10.1190/geo2014-0061.1.

Cooper, G. R. J., 2014a, Reducing the dependence of the analytic signal amplitude of aeromagnetic data on the source vector direction: Geophysics, **79**, no. 4, J55–J60, https://doi.org/10.1190/geo2013-0319.1.

Cooper, G. R. J., 2014b, The automatic determination of the location, depth, and dip of contacts from aeromagnetic data: Geophysics, **79**, no. 3, J35–J41, https://doi.org/10.1190/geo2013-0181.1.

Cooper, G. R. J., 2009, Balancing images of potential-field data: Geophysics, **74**, no. 3, L17–L20, https://doi.org/10.1190/1.3096615.

Cooper, G. R. J., 2004, Euler deconvolution applied to potential field gradients: Exploration Geophysics, **35**, no. 3, 165–170, https://doi.org/10.1071/EG04165.

Cooper, G. R. J., 2003, Feature detection using sunshading: Computers & Geosciences, **29**, no. 8, 941–948, https://doi.org/10.1016/S0098-3004(03)00091-8.

Cooper, G. R. J., and D. R. Cowan, 2006, Enhancing potential field data using filters based on the local phase: Computers & Geosciences, **32**, no. 10, 1585–1591, https://doi.org/10.1016/j.cageo.2006.02.016.

Cooper, G. R. J., and D. R. Cowan, 2005, Differential reduction to the pole: Computers & Geosciences, **31**, no. 8, 989–999, https://doi.org/10.1016/j.cageo.2005.02.005.

Cooper, G. R. J., and D. R. Cowan, 2004, Filtering using variable order vertical derivatives: Computers & Geosciences, **30**, no. 5, 455–459, https://doi.org/10.1016/j.cageo.2004.03.001.

Cooper, G. R. J., and D. R. Cowan, 2003, Application of fractional calculus to the potential field data: Exploration Geophysics, **34**, no. 1–2, 51–56, https://doi.org/10.1071/EG03051.

Cooper, G. R. J., and R. C. Whitehead, 2016, Determining the distance to magnetic sources: Geophysics, **81**, no. 2, J25–J34, https://doi.org/10.1190/geo2015-0142.1.

Coraggio, F., P. Bernardelli, and G. Gabbriellini, 2012, Structural reconstruction using potential field data in hydrocarbon exploration: 82nd Annual International Meeting, SEG, Expanded Abstracts, https://doi.org/10.1190/segam2012-0983.1.

Cordell, L., 1985, Applications and problems of analytical continuation of New Mexico aeromagnetic data between arbitrary surfaces of very high relief: Proceedings of the International Meeting on Potential Fields in Rugged Topography, Bulletin, **7**, 96–101, Institut de Geophysique de Universite de Lausanne.

Cordell, L., and A. E. McCafferty, 1989, A terracing operator for physical property mapping with potential field data: Geophysics, **54**, no. 5, 621–634, https://doi.org/10.1190/1.1442689.

Cordell, L., Y. A. Zorin, and G. R. Keller, 1991, The decompensative gravity anomaly and deep structure of the region of the Rio Grande rift: Journal of Geophysical Research, **96**, B4, 6557–6568, https://doi.org/10.1029/91JB00008.

Cowan, D. R., and S. Cowan, 1993, Separation filtering applied to aeromagnetic data: Exploration Geophysics, **24**, no. 3-4, 429–436, https://doi.org/10.1071/EG993429.

Cracknell, M. J., and A. M. Reading, 2013, The upside of uncertainty: Identification of lithology contact zones from airborne geophysics and satellite data using random forests and support vector machines: Geophysics, **78**, no. 3, WB113–WB126, https://doi.org/10.1190/geo2012-0411.1.

Dampney, C. N. G., 1969, The equivalent source technique: Geophysics, **34**, no. 1, 39–53, https://doi.org/10.1190/1.1439996.

Dannemiller, N., and Y. Li, 2006, A new method for determination of magnetization direction: Geophysics, **71**, no. 6, L69–L73, https://doi.org/10.1190/1.2356116.

Davies, G. F., 2011, Mantle convection for geologists: Cambridge University Press, https://doi.org/10.1017/CBO9780511973413.

Davis, K., and Y. Li, 2007, A fast approach to magnetic equivalent source processing using the adaptive quadtree mesh discretization: ASEG Technical Program Extended Abstracts, https://doi.org/10.1071/ASEG2007ab030.

Davis, K., Y. Li, and M. Batzle, 2008, Time-lapse gravity monitoring: A systematic 4D approach with application to aquifer storage and recovery: Geophysics, **73**, no. 6, WA61–WA69, https://doi.org/10.1190/1.2987376.

Davis, K., Y. Li, and M. Nabighian, 2010, Automatic detection of UXO magnetic anomalies using extended Euler deconvolution: Geophysics, **75**, no. 3, G13–G20, https://doi.org/10.1190/1.3375235.

de Angelis, M., A. Bertoldi, L. Cacciapuoti, A. Giorgini, G. Lamporesi, M. Prevedelli, G. Saccorotti, F. Sorrentino, and G. M. Tino, 2009, Precision gravimetry with atomic sensors: Measurement Science & Technology, **20**, no. 2, 022001, https://doi.org/10.1088/0957-0233/20/2/022001.

Debeglia, N., and J. Corpel, 1997, Automatic 3-D interpretation of potential field data using analytic signal derivatives: Geophysics, **62**, no. 1, 87–96, https://doi.org/10.1190/1.1444149.

Dentith, M., 2011, Magnetic methods, Airborne, *in* K. Gupta, ed., Encyclopedia of Solid Earth Geophysics; Springer, 761–766, https://doi.org/10.1007/978-90-481-8702-7_119.

de Saint-Jean, Verdun J., H. Duquenne, J. P. Barriot, S. Melachroinos, and J. Cali, 2007, Fine analysis of lever arm effects in moving gravimetry, *in* P. Tregoning and C. Rizos, eds., Dynamic Planet. International Association of Geodesy Symposia, **130**, 809–816.

Dickinson, J. L., J. R. Brewster, J. W. Robinson, and C. A. Murphy, 2009, Imaging techniques for full tensor gravity gradiometry data: SAGA Bienalle Technical Program Expanded Abstracts.

DiFrancesco, D., and M. Talwani, 2002, Time-lapse gravity gradiometry for reservoir monitoring: 72nd Annual International Meeting, SEG, Expanded Abstracts, https://doi.org/10.1190/1.1817377.

Dimri, V. P., 1998, Fractal behavior and detectability limits of geophysical surveys: Geophysics, **63**, no. 6, 1943–1946, https://doi.org/10.1190/1.1444487.

Dolmaz, M. N., T. Ustaömer, Z. M. Hisarli, and N. Orbay, 2005, Curie point depth variations to infer thermal structure of the crust at the African-Eurasian convergence zone, SW Turkey: Earth, Planets, and Space, **57**, no. 5, 373–383, https://doi.org/10.1186/BF03351821.

Doo, W.-B., S.-K. Hsu, C.-H. Tsai, and Y.-S. Huang, 2009, Using analytic signal to determine magnetization/density ratio of geological structures: Geophysical Journal International, **179**, no. 1, 112–124, https://doi.org/10.1111/j.1365-246X.2009.04297.x.

Douch, K., B. Foulon, B. Christophe, M. Diament, G. Pajot-Metivier, and I. Panet, 2013, A new planar electrostatic gradiometer for airborne surveys: 83rd Annual International Meeting, SEG, Expanded Abstracts, https://doi.org/10.1190/segam2013-1122.1.

Dransfield, M. H., and T. Chen, 2019, Heli-borne gravity gradiometry in rugged terrain: Geophysical Prospecting, **67**, no. 6, 1626–1636, https://doi.org/10.1111/1365-2478.12736.

Dransfield, M. H., and A. N. Christensen, 2013, Performance of airborne gravity gradiometers: The Leading Edge, **32**, no. 8, 908–922, https://doi.org/10.1190/tle32080908.1.

Dransfield, M. H., and Y. Zeng, 2009, Airborne gravity gradiometry: Terrain corrections and elevation error: Geophysics, **74**, no. 5, I37–I42, https://doi.org/10.1190/1.3170688.

Drenth, B. 2014, Airborne gravity gradient expression of a buried niobium and rare earth element deposit: Elk Creek carbonatite, Nebraska: 84th Annual International Meeting, SEG, Expanded Abstracts, https://doi.org/10.1190/segam2014-0230.1.

Eaton, D., and K. Vasudevan, 2004, Skeletonization of aeromagnetic data: Geophysics, **69**, no. 2, 478–488, https://doi.org/10.1190/1.1707068.

Ebbing, J., J. Bouman, M. Fuchs, V. Lieb, R. Haagmans, J. A. C. Meekes, and R. A. Fattah, 2013, Advancements in satellite gravity gradient data for crustal studies: The Leading Edge, **32**, no. 8, 900–906, https://doi.org/10.1190/tle32080900.1.

Ebner, E., 1997, Effective quality control of HRAM surveys: CSEG HRAM Forum Expanded Abstracts.

Elieff, S., and S. Sander, 2004, AIRGrav airborne gravity survey in Timmins: ASEG-PESA Technical Program Extended Abstracts.

Erkan, K., C. Jekeli, and C. K. Shum, 2012, Fusion of gravity gradient and magnetic field data for discrimination of anomalies using deformation analysis: Geophysics,

77, no. 3, F13–F20, https://doi.org/10.1190/geo2010-0184.1.

Fairhead, J. D., 2015, Generating a high-resolution global gravity model for oil exploration: Part 2 – Marine satellite altimeter-derived gravity: The Leading Edge, **34**, no. 5, 566–571, https://doi.org/10.1190/tle34050566.1.

Fairhead, J. D., G. R. T. Cooper, and S. Sander, 2017, Advances in airborne gravity and magnetics, *in* Proceedings of Exploration 17: Sixth Decennial International Conference on Mineral Exploration, V. Tcshirhart and M. D. Thomas, eds., 113–127.

Fairhead, J. D., and S. E. Williams, 2006, Evaluating normalized magnetic derivatives for structural mapping: 76th Annual International Meeting, SEG, Expanded Abstracts, https://doi.org/10.1190/1.2370388.

Fedi, M., and G. Florio, 2001, Detection of potential fields source boundaries by enhanced horizontal derivative method: Geophysical Prospecting, **49**, no. 1, 40–58, https://doi.org/10.1046/j.1365-2478.2001.00235.x.

Fedi, M., G. Florio, and V. Paoletti, 2015, MHODE: A local-homogeneity theory for improved source-parameter estimation of potential fields: Geophysical Journal International, **202**, no. 2, 887–900, https://doi.org/10.1093/gji/ggv185.

Fedi, M., G. Florio, and T. A. M. Quarta, 2009, Multiridge analysis of potential fields: Geometric method and reduced Euler deconvolution: Geophysics, **74**, no. 4, L53–L65, https://doi.org/10.1190/1.3142722.

Fedi, M., and M. Pilkington, 2012, Understanding imaging methods for potential field data: Geophysics, **77**, no. 1, G13–G24, https://doi.org/10.1190/geo2011-0078.1.

Fedi, M., and T. Quarta, 1998, Wavelet analysis for the regional-residual and local separation of potential field anomalies: Geophysical Prospecting, **46**, no. 5, 507–525, https://doi.org/10.1046/j.1365-2478.1998.00105.x.

Fehringer, M., G. Andre, D. Lamarre, and D. Maeusli, 2008, GOCE and its gravity measurement systems: ESA Bulletin, **133**, 15–23.

Felsberg, M., and G. Sommer, 2001, The monogenic signal: IEEE Transactions on Signal Processing, **49**, no. 12, 3136–3144, https://doi.org/10.1109/78.969520.

Ferré, E. C., S. A. Friedman, F. Martin-Hernandez, J. M. Feinberg, J. L. Till, D. A. Ionov, and J. A. Conder, 2014, Eight good reasons why the uppermost mantle could be magnetic: Tectonophysics, **624–625**, 3–14, https://doi.org/10.1016/j.tecto.2014.01.004.

Ferreira, F. J. F., J. de Souza, A. de B. e S. Bongiolo, and L. G. de Castro, 2013, Enhancement of the total horizontal gradient of magnetic anomalies using the tilt angle: Geophysics, **78**, no. 3, J33–J41, https://doi.org/10.1190/geo2011-0441.1.

Ferreira, W., and W. Shukowsky, 2013, Estimation of the total magnetization direction using Helbig's method: A comparative study using the reduction-to-pole operator: 13th International Congress of Brazilian Geophysical Society and EXPOGEF, 492–497.

Filina, I., E. K. Biegert, L. Sander, V. Tschirhart, N. Bundalo, and C. Schiek-Stewart, 2019, Integrated imaging: A powerful but undervalued tool: The Leading Edge, **38**, no. 9, 720–724, https://doi.org/10.1190/tle38090720.1.

Finlay, C. C., S. Maus, C. D. Beggan, M. Hamoudi, F. J. Lowes, N. Olsen, and E. Thébault, 2010, Evaluation of candidate geomagnetic field models for IGRF-11: Earth, Planets, and Space, **62**, no. 10, 787–804, https://doi.org/10.5047/eps.2010.11.005.

FitzGerald, D. J., and H. Holstein, 2006, Innovative data processing methods for gradient airborne geophysical data sets: The Leading Edge, **25**, no. 1, 87–94, https://doi.org/10.1190/1.2164762.

Flis, M. F., and D. R. Cowan, 2000, Aeromagnetic drape corrections applied to the Turner Syncline, Hamersley Basin: Exploration Geophysics, **31**, no. 1–2, 84–88, https://doi.org/10.1071/EG00084.

Floberghagen, R., M. Drinkwater, R. Haagmans, and M. Kern, 2008, GOCE measurements of the gravity field and beyond: ESA Bulletin, **133**, 25–31.

Florio, G. 2020, The estimation of depth to basement under sedimentary basins from gravity data: Review of approaches and the ITRESC method, with an application to the Yucca Flat basin (Nevada): Surveys in Geophysics, Springer online publication.

Florio, G., 2018, Mapping the depth to basement by iterative rescaling of gravity or magnetic data: Journal of Geophysical Research, Solid Earth, **123**, no. 10, 9101–9120, https://doi.org/10.1029/2018JB015667.

Florio, G., and M. Fedi, 2018, Depth estimation from downward continuation: An entropy-based approach to normalized full gradient: Geophysics, **83**, no. 3, J33–J42, https://doi.org/10.1190/geo2016-0681.1.

Florio, G., and M. Fedi, 2013, Multiridge Euler deconvolution: Geophysical Prospecting, **62**, no. 2, 333–351, https://doi.org/10.1111/1365-2478.12078.

Florio, G., M. Fedi, and R. Pašteka, 2014, On the estimation of the structural index from low-pass filtered magnetic data: Geophysics, **79**, no. 6, J67–J80, https://doi.org/10.1190/geo2013-0421.1.

Florio, G., M. Fedi, and R. Pašteka, 2006, On application of Euler deconvolution to the analytic signal: Geophysics, **71**, no. 6, L87–L93, https://doi.org/10.1190/1.2360204.

Florio, G., and D. Lo Re, 2018, Terracing of potential fields by clustering methods: Geophysics, **83**, no. 4, G47–G58, https://doi.org/10.1190/geo2017-0140.1.

Foote, R. S., 1996, Relationship of near-surface magnetic anomalies to oil-and-gas producing areas, *in* Hydrocarbon Migration and Its Near-Surface Expression, D. Schumacher and A. Abrams, eds., AAPG Memoir **66**, 111–126.

Foss, C., 2011, Magnetic data enhancements and depth estimation, *in* K. Gupta, ed., Encyclopedia of Solid Earth Geophysics: Springer, 736–746, https://doi.org/10.1007/978-90-481-8702-7_104.

Frangi, A. F., W. J. Niessen, K. L. Vincken, and M. A. Viergever, 1998, Multiscale vessel enhancement filtering, *in* Medical Image Computing and Computer-Assisted Intervention-MICCAI98, W. M. Wells, A. C. Colchester, and S. L. Delp, eds., Lecture Notes in Computer Science, vol. **1496**, 130–137, Springer-Verlag.

Fratter, I., J.-M. Leger, F. Bertrand, T. Jager, G. Hulot, L. Brocco, and P. Vigneron, 2016, Swarm absolute scalar magnetometers first in-orbit results: Acta Astronautica, **121**, 76–87, https://doi.org/10.1016/j.actaastro.2015.12.025.

Freeman, W. T., and E. H. Adelson, 1991, The design and use of steerable filters: IEEE Transactions on Pattern Analysis and Machine Intelligence, **13**, no. 9, 891–906, https://doi.org/10.1109/34.93808.

Fregoso, E., and L. A. Gallardo, 2009, Cross-gradients joint 3D inversion with applications to gravity and magnetic data: Geophysics, **74**, no. 4, L31–L42, https://doi.org/10.1190/1.3119263.

Freymueller, J. F., 2011, GPS, Tectonic Geodesy, *in* K. Gupta, ed., Encyclopedia of Solid Earth Geophysics: Springer, 431–449, https://doi.org/10.1007/978-90-481-8702-7_77.

Fullagar, P. K., and G. A. Pears, 2010, From gravity gradients to density gradients: ASEG-PESA Airborne Gravity 2010 Workshop Abstracts, Geoscience Australia Record, no. 23.

Fullagar, P. K., G. A. Pears, and B. McMonnies, 2008, Constrained inversion of geological surfaces – pushing the boundaries: The Leading Edge, **27**, no. 1, 98–105, https://doi.org/10.1190/1.2831686.

Gallardo, L. A., and M. A. Meju, 2004, Joint two-dimensional DC resistivity and seismic travel time inversion with cross-gradients constraints: Journal of Geophysical Research, **109**, B3, B03311, https://doi.org/10.1029/2003JB002716.

Gardner, G., L. Gardner, and A. Gregory, 1974, Formation velocity and density – the diagnostic basics for stratigraphic traps: Geophysics, **39**, no. 6, 770–780, https://doi.org/10.1190/1.1440465.

Gasperikova, E., and G. M. Hoversten, 2008, Gravity monitoring of CO_2 movement during sequestration: Model studies: Geophysics, **73**, no. 6, WA105–WA112, https://doi.org/10.1190/1.2985823.

Gavazzi, B., P. Le Maire, M. Munschy, and A. Dechamp, 2016, Fluxgate vector magnetometer: A multisensor device for ground, UAV, and airborne magnetic surveys: The Leading Edge, **35**, no. 9, 795–797, https://doi.org/10.1190/tle35090795.1.

Genda, H., 2016, Origin of Earth's oceans: An assessment of the total amount, history and supply of water: Geochemical Journal, **50**, no. 1, 27–42, https://doi.org/10.2343/geochemj.2.0398.

Gerovska, D., and P. Stavrev, 2006, Magnetic data analysis at low latitudes using magnitude transforms: Geophysical Prospecting, **54**, no. 1, 89–98, https://doi.org/10.1111/j.1365-2478.2006.00518.x.

Gerovska, D., P. Stavrev, and M. J. Araúzo-Bravo, 2005, Finite-difference Euler deconvolution algorithm applied to the interpretation of magnetic data from northeJanern Bulgaria: Pure and Applied Geophysics, **162**, 591–608, https://doi.org/10.1007/s00024-004-2623-1.

Gilbert, L. A., and H. P. Johnson, 1999, Direct measurements of oceanic crustal density at the northern Juan de Fuca Ridge: Geophysical Research Letters, **26**, no. 24, 3633–3636, https://doi.org/10.1029/1999GL008391.

Götze, H.-J., 2011a, Gravity Method, Principles, *in* K. Gupta, ed., Encyclopedia of Solid Earth Geophysics: Springer, 500–504, https://doi.org/10.1007/978-90-481-8702-7_93.

Götze, H.-J., 2011b, International Gravity Formula, *in* K. Gupta, ed., Encyclopedia of Solid Earth Geophysics: Springer, 611–612, https://doi.org/10.1007/978-90-481-8702-7_102.

Glenn, W. E., J. A. Rhodes, M. N. Benedict, R. A. Charters, S. A. Goussev, and J. W. Peirce, 2002, Aeromagnetic depth solutions indicate numerous magnetic sources within the sedimentary section in the Western Canadian Sedimentary Basin: CSEG Technical Program Expanded Abstracts.

Goussev, S. A., R. A. Charters, J. W. Peirce, and W. E. Glenn, 2003, Jackpine Creek magnetic anomaly: Identification of a buried meteorite impact structure: The Leading Edge, **22**, no. 8, 740–741, https://doi.org/10.1190/1.1605073.

Goussev, S. A., and J. W. Peirce, 2010, Magnetic basement: Gravity-guided magnetic source depth analysis and interpretation: Geophysical Prospecting, **58**, no. 2, 321–334, https://doi.org/10.1111/j.1365-2478.2009.00817.x.

Goussev, S. A., M. Rauch-Davies, P. A. Smith, and J. Weber, 2014, Prediction of deep stratigraphic targets from integrated interpretation of the gravity, magnetic, seismic, and well data: Concept, methodology and application: The Leading Edge, **33**, no. 1, 40–46, https://doi.org/10.1190/tle33010040.1.

Goussev, S. A., G. Shields, J. Rowe, and A. H. Saad, 2008, Origin and enhancement of the gravity and magnetic signatures of the continental-oceanic boundary (COB): Examples from the West Africa passive margin: 78th Annual International Meeting, SEG, Expanded Abstracts, https://doi.org/10.1190/1.3063759.

Grant, F., and J. Dodds, 1972, MAGMAP FFT processing system development notes: Grant and Watson, Ltd.

Grauch, V. J. S., 1987, A new variable-magnetization terrain correction method for aeromagnetic data: Geophysics, **52**, no. 1, 94–107, https://doi.org/10.1190/1.1442244.

Grauch, V. J. S., and L. Cordell, 1987, Limitations of determining density or magnetic boundaries from the horizontal gradient of gravity or pseudogravity data: Geophysics, **52**, no. 1, 118–121, https://doi.org/10.1190/1.1442236.

Grauch, V. J. S., and M. R. Hudson, 2007, Guides to understanding the aeromagnetic expression of faults in sedimentary basins: Lessons learned from the central Rio Grande rift, New Mexico: Geosphere, **3**, no. 6, 596–623, https://doi.org/10.1130/GES00128.1.

Gravrand, O., A. Khokhlov, J. L. LeMouel, and J. M. Leger, 2001, On calibration of a vectorial ^{4}He pumped magnetometer: Earth, Planets, and Space, **53**, no. 10, 949–958, https://doi.org/10.1186/BF03351692.

Grieve, R. A. F., and G. R. Osinski, 2011, Impact Craters on Earth, *in* K. Gupta, ed., Encyclopedia of Solid Earth Geophysics: Springer, 593–599, https://doi.org/10.1007/978-90-481-8702-7_174.

Griffin, W. R., 1949, Residual gravity in theory and practice: Geophysics, **14**, no. 1, 39–56, https://doi.org/10.1190/1.1437506.

Griffiths, D. J., 1987, Introduction to Elementary Particles: John Wiley & Sons, Inc., https://doi.org/10.1002/9783527618460.

Gunn, P. J., and R. Almond, 1997, A method for calculating equivalent layers corresponding to large aeromagnetic and radiometric grids: Exploration Geophysics, **28**, no. 1–2, 72–79, https://doi.org/10.1071/EG997072.

Gunn, P. J., D. Fitzgerald, N. Yassi, and P. Dart, 1997, New algorithms for visually enhancing airborne geophysical data: Exploration Geophysics, **28**, no. 1–2, 220–224, https://doi.org/10.1071/EG997220.

Guo, W., M. C. Dentith, R. T. Bird, and D. A. Clark, 2001, Systematic error analysis of demagnetization and implications for magnetic interpretation: Geophysics, **66**, no. 2, 562–570, https://doi.org/10.1190/1.1444947.

Guo, H., K. Marfurt, and J. Liu, 2009, Principal component spectral analysis: Geophysics, **74**, no. 4, P35–P43, https://doi.org/10.1190/1.3119264.

Guspí, F., and I. Novara, 2012, Generalized Hilbert transforms of the effect of single magnetic sources: Geophysics, **77**, no. 3, J7–J14, https://doi.org/10.1190/geo2011-0188.1.

Gyulai, A., M. K. Baracza, and E. E. Tolnai, 2013, The application of joint inversion in geophysical exploration: International Journal of Geosciences, **4**, no. 02, 283–289, https://doi.org/10.4236/ijg.2013.42026.

Hackney, R., 2011, Gravity, Data to Anomalies, *in* K. Gupta, ed., Encyclopedia of Solid Earth Geophysics: Springer, 524–533, https://doi.org/10.1007/978-90-481-8702-7_78.

Hammer, S., 1963, Deep gravity interpretation by stripping: Geophysics, **28**, no. 3, 369–378, https://doi.org/10.1190/1.1439186.

Hammer, S., 1939, Terrain corrections for gravimeter stations: Geophysics, **4**, no. 3, 184–194, https://doi.org/10.1190/1.1440495.

Hammer, S., and R. Anzoleaga, 1975, Exploring for stratigraphic traps with gravity gradients: Geophysics, **40**, no. 2, 256–268, https://doi.org/10.1190/1.1440523.

Hansen, R. O., 2005, 3D multiple-source Werner deconvolution for magnetic data: Geophysics, **70**, no. 5, L45–L51, https://doi.org/10.1190/1.2073883.

Hansen, R. O., and E. de Ridder, 2006, Linear feature analysis for aeromagnetic data: Geophysics, **71**, no. 6, L61–L67, https://doi.org/10.1190/1.2357831.

Hansen, R. O., L. Racic, and V. J. S. Grauch, 2005, Magnetic methods in near-surface geophysics *in* Near-Surface Geophysics: D. K. Butler, ed., 151–175, Society of Exploration Geophysicists.

Hansen, R. O., and M. Simmonds, 1993, Multiple-source Werner deconvolution: Geophysics, **58**, no. 12, 1792–1800, https://doi.org/10.1190/1.1443394.

Hansen, R. O., and L. Suciu, 2002, Multiple-source Euler deconvolution: Geophysics, **67**, no. 2, 525–535, https://doi.org/10.1190/1.1468613.

Hansen, R. O., and X. Wang, 1988, Simplified frequency-domain expressions for potential fields of arbitrary three-dimensional bodies: Geophysics, **53**, no. 3, 365–374, https://doi.org/10.1190/1.1442470.

Hardwick, C. D., 2004, Two methods of gradient-enhanced gridding – a discussion: SEG Magnetic Gradiometry Workshop Expanded Abstracts.

Hardwick, C. D., 1999, Gradient-enhanced total field gridding: 69th Annual International Meeting, SEG, Expanded Abstracts, https://doi.org/10.1190/1.1821029.

Harlan, R. B., 1968, Eotvos corrections for airborne gravimetry: J. Geophysical Research, **73**, no. 14, 4675–4679, https://doi.org/10.1029/JB073i014p04675.

Hartman, R. R., K. J. Teskey, and J. L. Friedberg, 1971, A system for rapid digital aeromagnetic interpretation: Geophysics, **36**, no. 5, 891–918, https://doi.org/10.1190/1.1440223.

Hassan, H. H., and S. A. Goussev, 2020, A new approach to visualize and interpret eigenvalues of airborne gravity gradiometry (AGG) tensor data: GeoConvention 2020 Technical Program Poster Presentation.

Hassan, H. H., and S. A. Goussev, 2019a, Deep learning approach to automatic detection of faults and fractures from magnetic data using the convolutional neural network (CNN): GeoConvention 2019 Technical Program Poster Presentation.

Hassan, H. H., and S. A. Goussev, 2019b, Fusion of multiphysics data with machine learning independent component analysis (ICA) improves detection of geological features: GeoConvention 2019 Technical Program Expanded Abstracts.

Hassan, H. H., and S. A. Goussev, 2018, Delineation of curvilinear structures on magnetic images using eigenvalues of the Hessian-based vesselness filter: GeoConvention 2018 Technical Program Expanded Abstracts.

Hassan, H. H., and J. W. Peirce, 2008, Fusion of airborne gravity and magnetic images for improved detection of structural control: GeoConvention 2008 Technical Program Expanded Abstracts.

Hassan, H. H., and J. W. Peirce, 2007, Mapping geological features in HRAM data using 2D steerable filters and its comparison to 2D wavelet transform: 77th Annual International Meeting, SEG, Expanded Abstracts, https://doi.org/10.1190/1.2792531.

Hassan, H. H., and S. V. Yalamanchili, 2013, Monogenic signal decomposition: A new approach to enhance magnetic data: 83rd Annual Meeting, SEG, Expanded Abstracts, https://doi.org/10.1190/segam2013-1014.1.

Hatch, D., 2004, Evaluation of a full tensor gravity gradiometer for kimberlite exploration: ASEG-PESA Airborne Gravity 2004 Workshop Abstracts, Geoscience Australia: Record, 2004, no. 18.

Hayford, J. F., and W. Bowie, 1912, The effect of topography and isostatic compensation upon the intensity of gravity: U.S. Coast and Geodetic Survey Special Publication No. 10.

Heiskanen, W. A., and F. A. Vening Meinesz, 1958, The Earth and its Gravity Field: McGraw-Hill.

Helbig, K., 2001, Surfaces of equal potential and physics behind the torsion balance and gradiometers: The Leading Edge, **20**, no. 8, 840–847, https://doi.org/10.1190/1.1487293.

Helbig, K., 1963, Some integrals of magnetic anomalies and their relation to the parameters of the disturbing body: Zeitschrift für Geophysik, **29**, 83–96.

Herceg, M., I. M. Artemieva, and M. Thybo, 2016, Sensitivity analysis of crustal correction for calculation of lithospheric mantle density from gravity data: Geophysical Journal International, **204**, no. 2, 687–696, https://doi.org/10.1093/gji/ggv431.

Herring, A. T., 1985, Increased resolution of shipborne measurements using GPS: 55th Annual International Meeting, SEG, Expanded Abstracts, 218–220, https://doi.org/10.1190/1.1892710.

Hidalgo-Gato, M. C., and V. C. F. Barbosa, 2015, Edge detection of the potential-field sources using scale-space monogenic signal: Fundamental principles: Geophysics, **80**, no. 5, J27–J36, https://doi.org/10.1190/geo2015-0025.1.

Hinze, W. J., C. Aiken, J. Brozena, B. Coakley, D. Dater, G. Flanagan, R. Forsberg, T. Hildenbrand, G. R. Keller, J. Kellogg, R. Kucks, X. Li, A. Mainville, R. Morin, M. Pilkington, D. Plouff, D. Ravat, D. Roman, J. Urrutia-Fucugauchi, M. Veronneau, M. Webring, and D. Winester, 2005, New standards for reducing gravity data: The North American gravity database: Geophysics, **70**, no. 4, J25–J32, https://doi.org/10.1190/1.1988183.

Hinze, W. J., R. R. B. von Frese, and A. H. Saad, 2013, Gravity and Magnetic Exploration: Cambridge University Press, https://doi.org/10.1017/CBO9780511843129.

Holme, R., M. A. James, and H. Luhr, 2005, Magnetic field modeling from scalar-only data: Resolving the Backus effect with the equatorial electrojet: Earth, Planets, and Space, **57**, no. 12, 1203–1209, https://doi.org/10.1186/BF03351905.

Hornby, P., F. Boschetti, and F. Horowitz, 1999, Analysis of potential field data in the wavelet domain: Geophysical Journal International, **137**, no. 1, 175–196, https://doi.org/10.1046/j.1365-246x.1999.00788.x.

Hovde, D. C., M. D. Prouty, I. Hrvoic, and R. E. Slocum, 2013, Commercial magnetometers and their application, *in* F. Kimball, ed., Optical Magnetometry: Cambridge University Press, 387–405, https://doi.org/10.1017/CBO9780511846380.021.

Hrvoic, D., 2007, SeaSpy Overhauser magnetometer technical application guide: Marine Magnetics Corp. Website.

Hrvoic, I., 2011, Magnetometers, *in* K. Gupta, ed., Encyclopedia of Solid Earth Geophysics: Springer, 810–816, https://doi.org/10.1007/978-90-481-8702-7_122.

Hrvoic, I., and G. M. Hollyer, 2013, Brief review of quantum magnetometers: GEM Systems Website.

Hsu, S.-K., J.-C. Sibuet, and C.-T. Shyu, 1996, High-resolution detection of geologic boundaries from potential field anomalies: An enhanced analytic signal technique: Geophysics, **61**, no. 2, 373–386, https://doi.org/10.1190/1.1443966.

Hubbert, M. K., 1947, A line-integral method of computing gravimetric effects of two-dimensional masses: Geophysics, **12**, 215–225, https://doi.org/10.1190/1.1437395.

Hugill, A., 2011, Gravimeters, *in* K. Gupta, ed., Encyclopedia of Solid Earth Geophysics: Springer, 449–456, https://doi.org/10.1007/978-90-481-8702-7_87.

Ikizoglu, S., and Y. Kamer, 2015, Acceleration data correction of an inertial navigation unit using turntable test bed: Proceedings of the EECSS World Congress (EECSS2015).

Ioane, D., C. Calota, and D. Ion, 2005, Deep geological features as revealed by 3D gravity stripping: Western part of the Moesian Platform, Romania: Journal of Balkan Geophysical Society, **8**, 129–138.

Ishihara, T., M. Shinohara, H. Fujimoto, T. Kanazawa, A. Araya, T. Yamada, K. Iizasa, S. Tsukioka, S. Omika, T. Yoshiume, M. Mochizuki, and K. Uehira, 2018, High-resolution gravity measurements aboard an autonomous underwater vehicle: Geophysics, **83**, no. 6, G119–G135, https://doi.org/10.1190/geo2018-0090.1.

Isles, D. J., and L. R. Rankin, 2013, Geological Interpretation of Aeromagnetic Data: Australian Society of Exploration Geophysicists.

Jachens, R. C., and C. W. Roberts, 1981, Documentation of a FORTRAN program "isocomp" for computing isostatic residual gravity: USGS Open-File Report, https://doi.org/10.3133/ofr81574.

Jacobsen, B. H., 1988, Reply by the authors to "A case for upward continuation as a standard separation filter for potential-field maps" by B. H. Jacobsen: Geophysics, **53**, 723–724, https://doi.org/10.1190/1.1486683.

Jacobsen, B. H., 1987, A case for upward continuation as a standard separation filter for potential-field maps: Geophysics, **52**, no. 8, 1138–1148, https://doi.org/10.1190/1.1442378.

James, D. E., 2011, Lithosphere, Continental, *in* K. Gupta, ed., Encyclopedia of Solid Earth Geophysics: Springer, 675–681, https://doi.org/10.1007/978-90-481-8702-7_32.

Jain, S., 1976, An automated method of direct interpretation of magnetic profiles: Geophysics, **41**, no. 3, 531–541, https://doi.org/10.1190/1.1440631.

Jekeli, C., 2011a, Gravity, Gradiometry, *in* K. Gupta, ed., Encyclopedia of Solid Earth Geophysics: Springer, 547–561, https://doi.org/10.1007/978-90-481-8702-7_80.

Jekeli, C., 2011b, Gravity Field of the Earth, *in* K. Gupta, ed., Encyclopedia of Solid Earth Geophysics: Springer, 471–484, https://doi.org/10.1007/978-90-481-8702-7_91.

Jekeli, C., 2007, Potential theory and static gravity field of the Earth, *in* Treatise on Geophysics: Elsevier, G. Schubert, ed., **3**, 11–42.

Jekeli, C., 2000, Inertial Navigation Systems with Geodetic Application: Walter de Gruyter.

Johnson, E. A. E., 1998, Use higher resolution gravity and magnetic data as your resource evaluation progresses: The Leading Edge, **17**, no. 1, 99, https://doi.org/10.1190/1.1437846.

Jordan, T. A., and A. B. Watts, 2005, Gravity anomalies, flexure and the elastic thickness structure of the India-Eurasia collisional system: Earth and Planetary Science Letters, **236**, no. 3–4, 732–750, https://doi.org/10.1016/j.epsl.2005.05.036.

Jorgensen, C., 2012, Project scale exploration using GRACE satellite data: Two case studies: 82nd Annual International Meeting, SEG, Expanded Abstracts, https://doi.org/10.1190/segam2012-1326.1.

Kaban, M. K., 2011, Gravity Anomalies, Interpretation, *in* K. Gupta, ed., Encyclopedia of Solid Earth Geophysics: Springer, 456–461, https://doi.org/10.1007/978-90-481-8702-7_88.

Kaub, L., C. Seruge, S. D. Chopra, J. M. G. Glen, and M. Teodorescu, 2018, Developing an autonomous unmanned aerial system to estimate field terrain corrections for gravity measurements: The Leading Edge, **37**, no. 8, 584–591, https://doi.org/10.1190/tle37080584.1.

Keating, P., 1998, Weighted Euler deconvolution of gravity data: Geophysics, **63**, no. 5, 1595–1603, https://doi.org/10.1190/1.1444456.

Keating, P., 1995, A simple technique to identify magnetic anomalies due to kimberlite pipes: Exploration and Mining Geology Journal, **4**, 121–125.

Keating, P., 1992, Density mapping from gravity data using the Walsh transform: Geophysics, **57**, no. 4, 637–642, https://doi.org/10.1190/1.1443276.

Keating, P., and M. Pilkington, 2013, Analysis and reprocessing of airborne gravity gradiometer data over the Strange Lake rare Earth deposit, Quebec-Labrador: The Leading Edge, **32**, no. 8, 940–947, https://doi.org/10.1190/tle32080940.1.

Keating, P., and M. Pilkington, 2004, Euler deconvolution of the analytic signal and its application to magnetic interpretation: Geophysical Prospecting, **52**, no. 3, 165–182, https://doi.org/10.1111/j.1365-2478.2004.00408.x.

Keating, P., and L. Zerbo, 1996, An improved technique for reduction to the pole at low latitudes: Geophysics, **61**, no. 1, 131–137, https://doi.org/10.1190/1.1443933.

Khokhlov, A., G. Hulot, and J.-L. Le Mouel, 1999, On the Backus Effect – II: Geophysical Journal International, **137**, no. 3, 816–820, https://doi.org/10.1046/j.1365-246x.1999.00843.x.

Khokhlov, A., G. Hulot, and J.-L. Le Mouel, 1997, On the Backus Effect – I: Geophysical Journal International, **130**, no. 3, 701–703, https://doi.org/10.1111/j.1365-246X.1997.tb01864.x.

Kilty, K. T., 1983, Werner deconvolution of profile potential field data: Geophysics, **48**, no. 2, 234–237, https://doi.org/10.1190/1.1441463.

Kim, Y., and N. Nakata, 2018, Geophysical inversion versus machine learning in inverse problems: The Leading Edge, **37**, no. 12, 894–901, https://doi.org/10.1190/tle37120894.1.

Kindlmann, G., 2007, Tensor invariants and their gradients *in* Visualization and Processing of Tensor Fields, J. Weicket and H. Hagen, eds.:Springer.

Korte, M., 2011, Geomagnetic Field, Secular Variation *in* K. Gupta, ed., Encyclopedia of Solid Earth Geophysics: Springer.

Krahenbuhl, R. A., Y. Li, and T. Davis, 2011, Understanding the applications and limitations of time-lapse gravity for reservoir monitoring: The Leading Edge, **30**, no. 9, 1060–1068, https://doi.org/10.1190/1.3640530.

Ku, C. C., and J. A. Sharp, 1983, Werner deconvolution for automated magnetic interpretation and its refinement using Marquart's inverse modeling: Geophysics, **48**, no. 6, 754–774, https://doi.org/10.1190/1.1441505.

LaCoste, L., 1988, The zero-length spring gravity meter: The Leading Edge, **7**, no. 7, 20–21, https://doi.org/10.1190/1.1439525.

LaFehr, T. R., 1991, An exact solution for the gravity curvature (Bullard B) (correction): Geophysics, **56**, no. 8, 1179–1184, https://doi.org/10.1190/1.1443138.

LaFehr, T. R., 1983, Rock density from borehole gravity surveys: Geophysics, **48**, no. 3, 341–356, https://doi.org/10.1190/1.1441472.

LaFehr, T. R., and M. N. Nabighian, 2012, Fundamentals of Gravity Exploration: Society of Exploration Geophysicists.

Lahti, I., and T. Karinen, 2010, Tilt derivative multiscale edges of magnetic data: The Leading Edge, **29**, no. 1, 24–29, https://doi.org/10.1190/1.3284049.

Lakhina, G. S., and B. T. Tsurutani, 2011, Magnetic Storms and Electromagnetic Pulsations, *in* K. Gupta, ed., Encyclopedia of Solid Earth Geophysics: Springer, 792–796, https://doi.org/10.1007/978-90-481-8702-7_156.

Langley, K., and S. J. Anderson, 2010, The Riesz transform and simultaneous representations of phase, energy and orientation in spatial vision: Vision Research, **50**, no. 17, 1748–1765, https://doi.org/10.1016/j.visres.2010.05.031.

Lanning, E. N., and D. M. Johnson, 1983, Automated identification of rock boundaries: An application of the Walsh transform to geophysical well-log analysis: Geophysics, **48**, no. 2, 197–205, https://doi.org/10.1190/1.1441458.

Laske, G., G. Masters, Z. Ma, and M. Pasyanos, 2013, Update on CRUST 1.0: A 1-degree global model of Earth's crust: Geophysical Research Abstracts, **15**, EGU2013–EGU2658.

Leão-Santos, M., Y. Li, and R. Moraes, 2015, Application of 3D magnetic amplitude inversion to iron oxide-copper-gold deposits at low magnetic latitudes: A case study from Carajas Mineral Province, Brazil: Geophysics, **80**, no. 2, B13–B22, https://doi.org/10.1190/geo2014-0082.1.

Leblanc, G. E., and W. A. Morris, 2001, Denoising of aeromagnetic data via the wavelet transform: Geophysics, **66**, no. 6, 1793–1804, https://doi.org/10.1190/1.1487121.

LeCun, Y., Y. Bengio, and G. Hinton, 2015, Deep learning: Nature, **521**, no. 7553, 436–444, https://doi.org/10.1038/nature14539.

Lee, J. B., 2001, FALCON gravity gradiometer technology: Exploration Geophysics, **32**, no. 3–4, 247–250, https://doi.org/10.1071/EG01247.

Lee, M., B. Morris, G. Leblanc, and J. Harris, 2012, Curvature analysis for geological mapping and mineral resource assessment: 82nd Annual International Meeting, SEG, Expanded Abstracts, https://doi.org/10.1190/segam2012-1358.1.

Lee, M., B. Morris, and H. Ugalde, 2010, Effect of signal amplitude on magnetic depth estimations: The Leading Edge, **29**, no. 6, 672–677, https://doi.org/10.1190/1.3447778.

Leger, J.-M., F. Bertrand, T. Jager, M. Le Prado, I. Fratter, and J.-C. Lalaurie, 2009, Swarm absolute and vector magnetometer based on helium-4 optical pumping: Procedia Chemistry, **1**, no. 1, 634–637, https://doi.org/10.1016/j.proche.2009.07.158.

Leger, J.-M., T. Jager, F. Bertrand, G. Hulot, L. Brocco, P. Vigneron, X. Lalanne, A. Chulliat, and I. Fratter, 2015, In-flight performance of the Absolute Scalar Magnetometer vector mode on board the Swarm satellites: Earth, Planets, and Space, **67**, no. 1, 57, https://doi.org/10.1186/s40623-015-0231-1.

Lelièvre, P. G., and D. W. Oldenburg, 2009, A 3D total magnetization inversion applicable when significant, complicated remanence is present: Geophysics, **74**, no. 3, L21–L30, https://doi.org/10.1190/1.3103249.

Lemoine, F. G., S. Goossens, T. J. Sabaka, J. B. Nicholas, E. Mazarico, D. D. Rowlands, B. D. Loomis, D. S. Chinn, G. A. Neumann, D. E. Smith, and M. T. Zuber, 2014, GRGM900C: A degree 900 lunar gravity model from GRAIL primary and extended mission data: Geophysical Research Letters, **41**, no. 10, 3382–3389, https://doi.org/10.1002/2014GL060027.

Lesur, V., M. Hamoudi, Y. Choi, J. Dyment, and E. Thébault, 2016, Building the second version of the World Digital Magnetic Anomaly Map (WDMAM): Earth, Planets, and Space, **68**, no. 1, 1–13, https://doi.org/10.1186/s40623-016-0404-6.

Li, X., 2016, Terracing gravity and magnetic data using edge-preserving smoothing filters: Geophysics, **81**, no. 2, G41–G47, https://doi.org/10.1190/geo2015-0409.1.

Li, X., 2015, Curvature of a geometric surface and curvature of gravity and magnetic anomalies: Geophysics, **80**, no. 1, G15–G26, https://doi.org/10.1190/geo2014-0108.1.

Li, X., 2008, Magnetic reduction-to-pole at low latitudes: Observations and considerations: The Leading Edge, **27**, no. 8, 990–1002, https://doi.org/10.1190/1.2967550.

Li, X., 2006, Understanding 3D analytic signal amplitude: Geophysics, **71**, no. 2, L13–L16, https://doi.org/10.1190/1.2184367.

Li, X., 2003, On the use of different methods for estimating magnetic depth: The Leading Edge, **22**, no. 11, 1090–1099, https://doi.org/10.1190/1.1634912.

Li, X., 2001, Vertical resolution: Gravity versus vertical gravity gradient: The Leading Edge, **20**, no. 8, 901–904, https://doi.org/10.1190/1.1487304.

Li, X., and H.-J. Götze, 2001, Ellipsoid, geoid, gravity, geodesy, and geophysics: Geophysics, **66**, no. 6, 1660–1668, https://doi.org/10.1190/1.1487109.

Li, X., and M. Pilkington, 2016, Attributes of the magnetic field, analytic signal, and monogenic signal for the gravity and magnetic interpretation: Geophysics, **81**, no. 6, J79–J86, https://doi.org/10.1190/geo2015-0697.1.

Li, Y., 2018, Understanding curvatures of the equipotential surface in gravity gradiometry: Geophysics, **83**, no. 4, G35–G45, https://doi.org/10.1190/geo2017-0612.1.

Li, Y., 2017, From Susceptibility to Magnetization: Advances in the 3D inversion of magnetic data in the presence of significant remanent magnetization *in* Proceedings of Exploration 17: Sixth Decennial International Conference on Mineral Exploration, V. Tschirhart and M. D. Thomas, eds., 239–260.

Li, Y., Z. He, and Y. Liu, 2012, Application of magnetic amplitude inversion in exploration for volcanic units in a basin environment: Geophysics, **77**, no. 5, B219–B225, https://doi.org/10.1190/geo2012-0008.1.

Li, Y., M. A. Kass, K. Davis, M. Braga, and C. Martinez, 2010a, Terrain correction and its effect on 3D inversion of airborne gravity gradiometry data: Abstracts from the ASEG-PESA Airborne Gravity 2010 Workshop, Geoscience Australia Record, 2010/23.

Li, Y., A. Melo, C. Martinez, and J. Sun, 2019, Geology differentiation: A new frontier in quantitative geophysical interpretation in mineral exploration: The Leading Edge, **38**, no. 1, 60–66, https://doi.org/10.1190/tle38010060.1.

Li, Y., S. E. Shearer, M. M. Haney, and M. Dannemiller, 2010b, Comprehensive approaches to 3D inversion of magnetic data affected by remanent magnetization: Geophysics, **75**, no. 1, L1–L11, https://doi.org/10.1190/1.3294766.

Li, Y., J. Sun, S.-L. Li, and M. Leao-Santos, 2021, A paradigm shift in magnetic data interpretation: Increased value through magnetization inversions: The Leading Edge, **40**, no. 2, 89–98, https://doi.org/10.1190/tle40020089.1.

Li, Y., and D. W. Oldenburg, 2001, Stable reduction to the pole at the magnetic equator: Geophysics, **66**, no. 2, 571–578, https://doi.org/10.1190/1.1444948.

Li, Y., and D. W. Oldenburg, 1998a, 3-D inversion of gravity data: Geophysics, **63**, no. 1, 109–119, https://doi.org/10.1190/1.1444302.

Li, Y., and D. W. Oldenburg, 1998b, Separation of regional and residual magnetic field data: Geophysics, **63**, no. 2, 431–439, https://doi.org/10.1190/1.1444343.

Li, Y., and D. W. Oldenburg, 1996, 3-D inversion of magnetic data: Geophysics, **61**, no. 2, 394–408, https://doi.org/10.1190/1.1443968.

Liu, P., T. Liu, P. Zhu, Y. Yang, Q. Zhou, H. Zhang, and G. Chen, 2017, Depth estimation for magnetic/gravity anomaly using model corrections: Pure and Applied Geophysics, **174**, no. 4, 1729–1742, https://doi.org/10.1007/s00024-017-1509-y.

Lockwood, A. M., 2004, Western Australia: Isostatic residual gravity anomaly and depth to basement model: Geological Survey of Western Australia, Record 2004, no. 14.

Long, L. T., and R. D. Kaufmann, 2013, Acquisition and Analysis of Terrestrial Gravity Data: Cambridge University Press, https://doi.org/10.1017/CBO9781139162289.

Long, K. P., B. S. Van Gosen, N. K. Foley, and D. Cordier, 2010, The principal rare earth element deposits of the United States: USGS Scientific Investigation Report 2010–5220.

Longman, I. M., 1959, Formulas for computing the tidal acceleration due to the Moon and Sun: Journal of Geophysical Research, **64**, no. 12, 2351–2355, https://doi.org/10.1029/JZ064i012p02351.

Love, J. J., 2008, Magnetic monitoring of Earth and space: Physics Today, **61**, no. 2, 31–37, https://doi.org/10.1063/1.2883907.

Lourenco, J. S., and H. F. Morrison, 1973, Vector magnetic anomalies derived from measurements of a single component of the field: Geophysics, **38**, no. 2, 359–368, https://doi.org/10.1190/1.1440346.

Lowes, F. J., 2000, An estimate of the errors in the IGRF/DGRF field 1945–2000: Earth, Planets, and Space, **52**, no. 12, 1207–1211, https://doi.org/10.1186/BF03352353.

Lowrie, W., 2011, Paleomagnetism, Principles, *in* K. Gupta, ed., Encyclopedia of Solid Earth Geophysics: Springer, 955–964, https://doi.org/10.1007/978-90-481-8702-7_107.

Luo, Y., M. Wang, F. Luo, and S. Tian, 2011, Direct analytic signal interpretation of potential field data using 2D Hilbert transform: Chinese Journal of Geophysics, **54**, no. 4, 551–559, https://doi.org/10.1002/cjg2.1637.

Luthcke, S. B., T. J. Sabaka, B. D. Loomis, A. A. Arendt, J. J. McCarthy, and J. Camp, 2013, Antarctica, Greenland and Gulf of Alaska land-ice evolution from an iterated GRACE global mascon solution: Journal of Glaciology, **59**, no. 216, 613–631, https://doi.org/10.3189/2013JoG12J147.

Lyrio, J. C. S. de O., L. Tenorio, and Y. Li, 2004, Efficient automatic denoising of the gravity gradiometry data: Geophysics, **69**, no. 3, 772–782, https://doi.org/10.1190/1.1759463.

Ma, G., 2013, Edge detection of potential field data using improved local phase filter: Exploration Geophysics, **44**, no. 1, 36–41, https://doi.org/10.1071/EG12022.

Ma, G., and X. Du, 2012, An improved analytic signal technique for the depth and structural index from 2D magnetic anomaly data: Pure and Applied Geophysics, **169**, no. 12, 2193–2200, https://doi.org/10.1007/s00024-012-0484-6.

Machel, H. G., and E. A. Burton, 1991, Chemical and microbial processes causing anomalous magnetization in environments affected by hydrocarbon seepage:

Geophysics, **56**, no. 5, 598–605, https://doi.org/10.1190/1.1443076.

MacLeod, I. N., K. Jones, and T. F. Dai, 1993, 3-D analytic signal in the interpretation of total field data at low magnetic latitudes: Exploration Geophysics, **24**, no. 3–4, 679–687, https://doi.org/10.1071/EG993679.

Macmillan, S., 2011, Geomagnetic Field, Global Pattern, *in* K. Gupta, ed., Encyclopedia of Solid Earth Geophysics: Springer, 373–379, https://doi.org/10.1007/978-90-481-8702-7_115.

Mallick, K., A. Vasanthi, and K. K. Sharma, 2011, Gravity Data, Regional-Residual Separation, *in* K. Gupta, ed., Encyclopedia of Solid Earth Geophysics: Springer, 466–471, https://doi.org/10.1007/978-90-481-8702-7_90.

Mandal, A., W. K. Mohanty, S. P. Sharma, A. Biswas, J. Sen, and A. K. Bhatt, 2015, Geophysical signatures of uranium mineralization and its subsurface validation at Beldih, Purulia District, West Bengal, India: A case study: Geophysical Prospecting, **63**, no. 3, 713–726, https://doi.org/10.1111/1365-2478.12205.

Mandea, M., C. Gaina, and V. Lesur, 2011, Magnetic Modeling, Theory and Computation, *in* K. Gupta, ed., Encyclopedia of Solid Earth Geophysics: Springer, 781–792, https://doi.org/10.1007/978-90-481-8702-7_232.

Mandea, M., and A. Isac, 2011, Geomagnetic field, Measurement Techniques, *in* K. Gupta, ed., Encyclopedia of Solid Earth Geophysics: Springer, 381–386, https://doi.org/10.1007/978-90-481-8702-7_117.

Martinez, C., Y. Li, R. Krahenbuhl, and M. A. Braga, 2013, 3D inversion of airborne gravity gradiometry data in mineral exploration: A case study in Quadrilatero Ferrifero, Brazil: Geophysics, **78**, no. 1, B1–B11, https://doi.org/10.1190/geo2012-0106.1.

Mataragio, J., and J. Kieley, 2009, Application of full tensor gradient invariants in detection of intrusion-hosted sulphide mineralization: Implications for deposition mechanisms: First Break, **27**, no. 7, 95–98, https://doi.org/10.3997/1365-2397.27.1301.29032.

Maunde, A., F. A. Rufa'l, A. S. Raji, and M. B. Saleh, 2017, Determination of subsurface bulk density distribution for geotechnical investigation using gravity technique: Journal of Earth Sciences and Geotechnical Engineering, **7**, 63–69.

Mauring, E., and O. Kihle, 2006, Leveling aerogeophysical data using a moving differential median filter: Geophysics, **71**, no. 1, L5–L11, https://doi.org/10.1190/1.2163912.

Maus, S., 2010, An ellipsoidal harmonic representation of Earth's lithospheric magnetic field to degree and order 720: Geochemistry Geophysics Geosystems, **11**, no. 6, Q06015, https://doi.org/10.1029/2010GC003026.

S. Maus, U. Barckhausen, H. Berkenbosch, N. Bournas, J. Brozena, V. Childers, F. Dostaler, J. D. Fairhead, C. Finn, R. R. B. von Frese, C. Gaina, S. Golynsky, R. Kucks, H. Lühr, P. Milligan, S. Mogren, R. D. Müller, O. Olesen, M. Pilkington, R. Saltus, B. Schreckenberger, E. Thébault, and F. Caratori Tontini, 2009, EMAG2: A 2arc-min resolution Earth Magnetic Anomaly Grid compiled from satellite, airborne, and marine magnetic measurements: Geochemistry Geophysics Geosystems, **10**, no. 8, Q08005, https://doi.org/10.1029/2009GC002471.

Maus, S., and V. P. Dimri, 1994, Scaling properties of potential fields due to scaling sources: Geophysical Research Letters, **21**, no. 10, 891–894, https://doi.org/10.1029/94GL00771.

McGuirk, J. M., G. T. Foster, J. B. Fixler, M. J. Snadden, and M. A. Kasevich, 2002, Sensitive absolute-gravity gradiometry using atom interferometry: Physical Review A. **65**, no. 3, 033608, https://doi.org/10.1103/PhysRevA.65.033608.

McLean, M., and G. Reitmayr, 2005, An airborne gravity survey south of the Prince Charles Mountains, East Antarctica: Terra Antarctica, **12**, 99–108.

McMillan, M., E. Haber, B. Peters, and J. Fohring, 2021, Mineral prospectivity mapping using a VNet convolutional neural network: The Leading Edge, **40**, no. 2, 99–105, https://doi.org/10.1190/tle40020099.1.

Medeiros, W. E., and J. B. C. Silva, 1995, Gravity source moment inversion: A versatile approach to characterize position and 3-D orientation of anomalous bodies: Geophysics, **60**, no. 5, 1342–1353, https://doi.org/10.1190/1.1443870.

Mendonça, C. A., 2004, Inversion of gravity-field inclination to map the basement relief of sedimentary basins: Geophysics, **69**, no. 5, 1240–1251, https://doi.org/10.1190/1.1801940.

Merriam, J. B., 1992, Atmospheric pressure and gravity: Geophysical Journal International, **109**, no. 3, 488–500, https://doi.org/10.1111/j.1365-246X.1992.tb00112.x.

Meyer, U., 2011, Gravity Method, Airborne, *in* K. Gupta, ed., Encyclopedia of Solid Earth Geophysics: Springer, 497–500, https://doi.org/10.1007/978-90-481-8702-7_179.

Mickus, K. I., and J. H. Hinojosa, 2001, The complete gravity gradient tensor derived from the vertical component of gravity: A Fourier transform technique: Journal of Applied Geophysics, **46**, no. 3, 159–174, https://doi.org/10.1016/S0926-9851(01)00031-3.

Mikhailov, V., G. Pajot, M. Diament, and A. Price, 2007, Tensor deconvolution: A method to locate equivalent sources from full tensor gravity data: Geophysics, **72**, no. 5, I61–I69, https://doi.org/10.1190/1.2749317.

Miller, H. G., and V. Singh, 1994, Potential field tilt: A new concept for location of potential field sources: Journal of Applied Geophysics, **32**, no. 2–3, 213–217, https://doi.org/10.1016/0926-9851(94)90022-1.

Minty, B. R. S., 1991, Simple micro-levelling for aeromagnetic data: Exploration Geophysics, **22**, no. 4, 591–592, https://doi.org/10.1071/EG991591.

Mishra, D. C., and V. M. Tiwari, 2011, Gravity Method, Surface, *in* K. Gupta, ed., Encyclopedia of Solid Earth Geophysics: Springer, 513–518, https://doi.org/10.1007/978-90-481-8702-7_95.

Mitášová, H., and J. Hofierka, 1993, Interpolation by regularized spline with tension: II. Application to terrain modeling and surface geometry analysis: Mathematical Geology, **25**, no. 6, 657–669, https://doi.org/10.1007/BF00893172.

Mjelde, R., T. Raum, Y. Murai, and T. Takanami, 2007, Continent-ocean-transitions: Review, and a new tectono-magmatic model of the Voring Plateau, NE Atlantic: Journal of Geodynamics, **43**, no. 3, 374–392, https://doi.org/10.1016/j.jog.2006.09.013.

Mogren, S., A. M. Al-Amri, K. Al-Damegh, D. Fairhead, A. Jassim, and A. Algamdi, 2008, Sub-surface geometry of Ar Rika and Ruwah faults from gravity and magnetic surveys: Arabian Journal of Geosciences, **1**, no. 1, 33–47, https://doi.org/10.1007/s12517-008-0003-3.

Murphy, C. A., 2007, Interpreting FTG gravity data using horizontal tensor components: EGM Capri 2007 International Workshop Expanded Abstracts.

Murphy, C. A. 2004, The Air-FTG™ airborne gravity gradiometer system: ASEG-PESA Airborne Gravity 2004 Workshop Extended Abstracts, Geoscience Australia Record, 2004/17.

Murphy, C. A., and J. L. Dickinson, 2010, Geological mapping and targeting using invariant tensor analysis on full tensor gravity data: EGM Capri 2010 International Workshop Expanded Abstracts.

Murphy, C., J. Dickinson, and A. Salem, 2012, Depth estimating full tensor gravity data with the adaptive tilt angle method: ASEG 2012 Technical Program Extended Abstracts.

Murray, A. S., and R. M. Tracey, 2001, Best Practice in Gravity Surveying: Australian Geological Survey Organization.

Mushayandebvu, M. F., V. Lesur, A. B. Reid, and J. D. Fairhead, 2004, Grid Euler deconvolution with constraints for 2D structures: Geophysics, **69**, no. 2, 489–496, https://doi.org/10.1190/1.1707069.

Mushayandebvu, M. F., P. van Driel, A. B. Reid, and J. D. Fairhead, 2001, Magnetic source parameters of two-dimensional structures using extended Euler deconvolution: Geophysics, **66**, no. 3, 814–823, https://doi.org/10.1190/1.1444971.

Mushayandebvu, M. F., P. van Driel, A. B. Reid, and J. D. Fairhead, 1999, Magnetic imaging using extended Euler deconvolution: 69th Annual International Meeting, SEG, Expanded Abstracts, https://doi.org/10.1190/1.1821035.

Nabighian, M. N., 1984, Toward a three-dimensional automatic interpretation of potential field data via generalized Hilbert transforms: Fundamental relations: Geophysics, **49**, no. 6, 780–786, https://doi.org/10.1190/1.1441706.

Nabighian, M. N., 1972, The analytic signal of two-dimensional magnetic bodies with polygonal cross-section: its properties and use for automated anomaly interpretation: Geophysics, **37**, no. 3, 507–517, https://doi.org/10.1190/1.1440276.

Nabighian, M. N., M. E. Ander, V. J. S. Grauch, R. O. Hansen, T. R. LaFehr, Y. Li, W. C. Pearson, J. W. Peirce, J. D. Phillips, and M. E. Ruder, 2005, 75th Anniversary: Historical development of the gravity method in exploration: Geophysics, **70**, 63ND–83ND, https://doi.org/10.1190/1.2133785.

Nabighian, M. N., V. J. S. Grauch, R. O. Hansen, T. R. LaFehr, Y. Li, J. W. Peirce, J. D. Phillips, and M. E. Ruder, 2005, 75th Anniversary: Historical development of the magnetic method in exploration: Geophysics, **70**, no. 6, 33ND–61ND, https://doi.org/10.1190/1.2133784.

Nabighian, M. N., and R. O. Hansen, 2001, Unification of Euler and Werner deconvolution in three dimensions via the generalized Hilbert transform: Geophysics, **66**, no. 6, 1805–1810, https://doi.org/10.1190/1.1487122.

Nagarajan, N., 2011, Magnetic Methods, Surface, *in* K. Gupta, ed., Encyclopedia of Solid Earth Geophysics: Springer, 774–781, https://doi.org/10.1007/978-90-481-8702-7_121.

Naudy, H., 1971, Automatic determination of depth on aeromagnetic profiles: Geophysics, **36**, no. 4, 717–722, https://doi.org/10.1190/1.1440207.

Nelson, J. B., 1994, Levelling total-field aeromagnetic data with measured horizontal gradients: Geophysics, **59**, no. 7, 1166–1170, https://doi.org/10.1190/1.1443673.

Nelson, J. B., 1988, Calculation of the magnetic gradient tensor from total field gradient measurements and its application to geophysical interpretation: Geophysics, **53**, no. 7, 957–966, https://doi.org/10.1190/1.1442532.

Nettleton, L. L., 1971, Elementary Gravity and Magnetics for Geologists and Seismologists: Society of Exploration Geophysicists.

Neuendorf, K. K. E., J. P. Melh, and J. A. Jackson, eds., 2005, Glossary of Geology, 5th ed.: American Geological Institute.

Niebauer, T. M., R. Billson, A. Schiel, D. van Westrum, and F. Klopping, 2013, The self-attraction correction for the FG5X absolute gravity meter: Metrologia, **50**, no. 1, 1–8, https://doi.org/10.1088/0026-1394/50/1/1.

Niebauer, T. M., G. S. Sasagawa, J. E. Faller, R. Hilt, and F. Klopping, 1995, A new generation of absolute gravimeters: Metrologia, **32**, no. 3, 159–180, https://doi.org/10.1088/0026-1394/32/3/004.

Nikulin, A., and T. S. de Smet, 2019, A UAV-based magnetic survey method to detect and identify orphaned oil and gas wells: The Leading Edge, **38**, no. 6, 447–452, https://doi.org/10.1190/tle38060447.1.

Noriega, G., 2011, Performance measures in aeromagnetic compensation: The Leading Edge, **30**, no. 10, 1122–1127, https://doi.org/10.1190/1.3657070.

Nowell, D. A. G., 1999, Gravity terrain corrections – an overview: Journal of Applied Geophysics, **42**, no. 2, 117–134, https://doi.org/10.1016/S0926-9851(99)00028-2.

O'Connell, M. D., R. S. Smith, and M. A. Vallee, 2005, Gridding aeromagnetic data using longitudinal and transverse horizontal gradients with the minimum curvature operator: The Leading Edge, **24**, no. 2, 142–145, https://doi.org/10.1190/1.1876036.

NRC (National Research Council), 1997, Satellite Gravity and the Geosphere: Contributions to the Study of the Solid Earth and Its Fluid Envelope: National Academy Press, Washington, D.C.

Oldenburg, D. W., and Y. Li, 2005, Inversion for Applied Geophysics: A Tutorial *in* D. K. Butler, ed., Near-Surface Geophysics: Society of Exploration Geophysicists, 89–150.

Oruç, B., İ. Sertçelik, O. Kafadar, and H. H. Selim, 2013, Structural interpretation of the Erzurum Basin, eastern Turkey, using curvature gravity gradient tensor and gravity inversion of basement relief: Journal of Applied Geophysics, **88**, 105–113, https://doi.org/10.1016/j.jappgeo.2012.10.006.

Panepinto, S., L. DeLuca, M. Mantovani, M. Sfolciaghi, and B. Garcea, 2014, Using the pseudo-gravity functional transform to enhance deep-magnetic sources and enrich regional gravity data: 84th Annual Meeting, SEG, Expanded Abstracts, https://doi.org/10.1190/segam2014-1323.1.

Paoletti, V., A. Buggi, and R. Pašteka, 2019, UXO detection by multiscale potential field methods: Pure and Applied Geophysics, **176**, no. 10, 4363–4381, https://doi.org/10.1007/s00024-019-02202-7.

Paoletti, V., M. Fedi, F. Italiano, G. Florio, and S. Ialongo, 2016, Inversion of gravity gradient tensor data: Does it provide better resolution?: Geophysical Journal International, **205**, no. 1, 192–202, https://doi.org/10.1093/gji/ggw003.

Parasnis, D. S., 1952, A study of rock densities in the English Midlands: Geophysical Supplements to the Monthly Notices of the Royal Astronomical Society, **6**, 252–271, https://doi.org/10.1111/j.1365-246X.1952.tb03013.x.

Parker, R. L., 1996, Improved Fourier terrain correction, Part II: Geophysics, **61**, no. 2, 365–372, https://doi.org/10.1190/1.1443965.

Parker, R. L., 1995, Improved Fourier terrain correction, Part I: Geophysics, **60**, no. 4, 1007–1017, https://doi.org/10.1190/1.1443829.

Parker, R. L., 1973, The rapid calculation of potential anomalies: Geophysical Journal of the Royal Astronomical Society, **31**, no. 4, 447–455, https://doi.org/10.1111/j.1365-246X.1973.tb06513.x.

Pašteka, R., F. P. Richter, R. Karcol, K. Brazda, and M. Hajach, 2009, Regularized derivatives of potential fields and their role in semi-automated interpretation methods: Geophysical Prospecting, **57**, no. 4, 507–516, https://doi.org/10.1111/j.1365-2478.2008.00780.x.

Pavlis, N. K., 2011a, Gravity, Global Models, *in* K. Gupta, ed., Encyclopedia of Solid Earth Geophysics: Springer, 533–547, https://doi.org/10.1007/978-90-481-8702-7_76.

Pavlis, N. K., 2011b, Spherical harmonic analysis applied to potential fields, *in* K. Gupta, ed., Encyclopedia of Solid Earth Geophysics: Springer, 1382–1392, https://doi.org/10.1007/978-90-481-8702-7_138.

Pawlowski, R. S., 2008, The use of gravity anomaly data for offshore continental margin demarcation: The Leading Edge, **27**, no. 6, 722–727, https://doi.org/10.1190/1.2944156.

Pawlowski, R. S., 1995, Preferential continuation for potential-field anomaly enhancement: Geophysics, **60**, no. 2, 390–398, https://doi.org/10.1190/1.1443775.

Pawlowski, R. S., 1992, Tutorial: Gravity anomalies for nonspecialists: The Leading Edge, **11**, no. 9, 41–43, https://doi.org/10.1190/1.1436903.

Pawlowski, R. S., and R. O. Hansen, 1990, Gravity anomaly separation by Wiener filtering: Geophysics, **55**, no. 5, 539–548, https://doi.org/10.1190/1.1442865.

Pearson, S. C. P., P. Franz, and J. Clearwater, 2015, Strategy of 4D microgravity survey for monitoring of fluid dynamics in the subsurface: Proceedings of the World Geothermal Congress 2015.

Pedersen, L. B., 1978, Wavenumber domain expressions for potential fields from arbitrary 2-, 2-1/2, 2-1/2: Geophysics, **43**, no. 3, 626–630, https://doi.org/10.1190/1.1440841.

Pedersen, L. B., and T. M. Rasmussen, 1990, The gradient tensor of potential field anomalies: Some implications on data collection and data processing of maps: Geophysics, **55**, no. 12, 1558–1566, https://doi.org/10.1190/1.1442807.

Pedersen, L. B., T. M. Rasmussen, and D. Dyrelius, 1990, Construction of component maps from aeromagnetic total field anomaly maps: Geophysical Prospecting, **38**, no. 7, 795–804, https://doi.org/10.1111/j.1365-2478.1990.tb01875.x.

Peirce, J. W., S. A. Goussev, R. A. Charters, H. J. Abercrombie, and G. R. DePaoli, 1998, Intra-sedimentary magnetization by vertical fluid flow and exotic geochemistry: The Leading Edge, **17**, no. 1, 89–92, https://doi.org/10.1190/1.1437840.

Peirce, J. W., and L. Lipkov, 1988, Structural interpretation of the Rukwa Rift, Tanzania: Geophysics, **53**, no. 6, 824–836, https://doi.org/10.1190/1.1442517.

Peters, L. J., 1949, The direct approach to magnetic interpretation and its practical application: Geophysics, **14**, no. 3, 290–320, https://doi.org/10.1190/1.1437537.

Phillips, J. D., 2005, Can we estimate total magnetization directions from aeromagnetic data using Helbig's integrals?: Earth, Planets, and Space, **57**, no. 8, 681–689, https://doi.org/10.1186/BF03351848.

Phillips, J. D., 1996, Potential field continuation: Past practice vs. modern methods: 66th Annual International Meeting, SEG, Expanded Abstracts, https://doi.org/10.1190/1.1826376.

Phillips, J. D., R. O. Hansen, and R. J. Blakely, 2007, The use of curvature in potential-field interpretation: Exploration Geophysics, **38**, no. 2, 111–119, https://doi.org/10.1071/EG07014.

Pierpaoli, C., and P. J. Basser, 1996, Toward a quantitative assessment of diffusion anisotropy: Magnetic Resonance in Medicine, **36**, no. 6, 893–906, https://doi.org/10.1002/mrm.1910360612.

Pilkington, M., 2014, Evaluating the utility of gravity gradient tensor components: Geophysics, **79**, no. 1, G1–G14, https://doi.org/10.1190/geo2013-0130.1.

Pilkington, M., M. E. Gregotski, and J. P. Todoeschuck, 1994, Using fractal crustal magnetization models in magnetic interpretation: Geophysical Prospecting, **42**, no. 6, 677–692, https://doi.org/10.1111/j.1365-2478.1994.tb00235.x.

Pilkington, M., and P. Keating, 2006, The relationship between local wavenumber and analytic signal in magnetic interpretation: Geophysics, **71**, no. 1, L1–L3, https://doi.org/10.1190/1.2163911.

Pilkington, M., and P. Shamsipour, 2014, Noise reduction procedures for gravity-gradiometer data: Geophysics, **79**, no. 5, G69–G78, https://doi.org/10.1190/geo2014-0084.1.

Pilkington, M., and J. B. Thurston, 2001, Draping corrections for aeromagnetic data: Line- versus grid-based approaches: Exploration Geophysics, **32**, no. 2, 95–101, https://doi.org/10.1071/EG01095.

Pilkington, M., and V. Tschirhart, 2017, Practical considerations in the use of edge detectors for geological mapping using magnetic data: Geophysics, **82**, no. 3, J1–J8, https://doi.org/10.1190/geo2016-0364.1.

Pool, D. R., 2008, The utility of gravity and water-level monitoring at alluvial aquifer wells in southern Arizona: Geophysics, **73**, no. 6, WA49–WA59, https://doi.org/10.1190/1.2980395.

Power, M., G. Belcourt, and E. Rockel, 2004, Geophysical methods for kimberlite exploration in northern Canada:

The Leading Edge, **23**, no. 11, 1124–1129, https://doi.org/10.1190/1.1825939.

Pratt, D. A., and Z. Shi, 2004, An improved pseudo-gravity transform technique for investigation of deep magnetic source rocks: ASEG Technical Program Extended Abstracts, https://doi.org/10.1071/ASEG2004ab116.

Prothero, D. R., 2011, Paleomagnetism, Magnetostratigraphy, *in* K. Gupta, ed., Encyclopedia of Solid Earth Geophysics: Springer, 925–933, https://doi.org/10.1007/978-90-481-8702-7_124.

Prouty, M. D., M. Tchernychev, and R. R. Mhaskar, 2015, Applications of micro-fabricated atomic magnetometers: SAGEEP Technical Program Expanded Abstracts, https://doi.org/10.4133/SAGEEP.28-067.

Purves, S., 2014, Phase and the Hilbert transform: The Leading Edge, **33**, no. 10, 1164–1166, https://doi.org/10.1190/tle33101164.1.

Rahmoune, R., M. Sulaimani, J. Stammeijer, S. Azri, R. van Gilst, A. Mahruqi, R. Aghbari, and A. Belghache, 2021, Multitemporal SAR interferometry for monitoring of ground deformations caused by hydrocarbon production in an arid environment: Case studies from the Sultanate of Oman: The Leading Edge, **40**, no. 1, 45–51, https://doi.org/10.1190/tle40010045.1.

Rao, V. B., and B. V. S. Murty, 1973, Note on Parasnis' method for surface rock densities: Pure and Applied Geophysics, **110**, no. 1, 1927–1931, https://doi.org/10.1007/BF00876555.

Ravat, D., 2011, Magnetic Methods, Satellite, *in* K. Gupta, ed., Encyclopedia of Solid Earth Geophysics: Springer, 771–774, https://doi.org/10.1007/978-90-481-8702-7_120.

Ravat, D., 1996, Analysis of the Euler method and its applicability in environmental magnetic investigations: Journal of Environmental & Engineering Geophysics, **1**, no. 3, 229–238, https://doi.org/10.4133/JEEG1.3.229.

Ray, A., S. Witte, and E. MacAllister, 2010, Sicily Channel – integrated interpretation of gravity and magnetic data: EGM Capri 2010 International Workshop Expanded Abstracts.

Reeves, C., 2005, Aeromagnetic Surveys: Principles, Practice and Interpretation: Geosoft Publication.

Reford, S., 2006, Gradient enhancement of the total magnetic field: The Leading Edge, **25**, no. 1, 59–66, https://doi.org/10.1190/1.2164757.

Reid, A. B., 2012, Forgotten truths, myths and sacred cows of Potential Fields Geophysics - II: 82nd Annual International Meeting, SEG, Expanded Abstract, https://doi.org/10.1190/segam2012-0178.1.

Reid, A. B., 2010, Forgotten truths, myths and sacred cows of Potential Fields Geophysics: 80th Annual International Meeting, SEG, Expanded Abstracts, 1198–1201, https://doi.org/10.1190/1.3513058.

Reid, A. B., 1980, Aeromagnetic survey design: Geophysics, **45**, no. 5, 973–976, https://doi.org/10.1190/1.1441102.

Reid, A. B., J. M. Allsop, H. Granser, A. J. Milett, and I. W. Somerton, 1990, Magnetic interpretation in three dimensions using Euler deconvolution: Geophysics, **55**, no. 1, 80–91, https://doi.org/10.1190/1.1442774.

Reid, A. B., J. Ebbing, and S. J. Webb, 2014, Avoidable Euler errors – the use and abuse of Euler deconvolution applied to potential fields: Geophysical Prospecting, **62**, no. 5, 1162–1168, https://doi.org/10.1111/1365-2478.12119.

Reid, A. B., D. Fitzgerald, and P. McInerny, 2003, Euler deconvolution of gravity data: 73rd Annual International Meeting, SEG, Expanded Abstracts, https://doi.org/10.1190/1.1817993.

Reid, A. B., and J. B. Thurston, 2014, The structural index in gravity and magnetic interpretation: Errors, uses, and abuses: Geophysics, **79**, no. 4, J61–J66, https://doi.org/10.1190/geo2013-0235.1.

Reitz, A., R. A. Krahenbuhl, Y. Li, and D. DiFrancesco, 2012, Feasibility analysis of time-lapse gravity gradiometry for reservoir monitoring: 82nd Annual International Meeting, SEG, Expanded Abstracts, https://doi.org/10.1190/segam2012-0775.1.

Reynolds, R. L., M. B. Goldhaber, and M. L. Tuttle, 1993, Sulfidization and magnetization above hydrocarbon reservoirs: Application of Paleomagnetism to Sedimentary Geology, SEPM Special Publication.

Rhodes, J., and J. Peirce, 1999, MaFIC – magnetic interpretation in 3D using a seismic workstation: 69th Annual International Meeting, SEG, Expanded Abstracts, https://doi.org/10.1190/1.1821015.

Ribeiro-Filho, N., R. Bijani, and C. Ponte-Neto, 2020, Improving crosscorrelation method to estimate the total magnetization direction vector of isolated sources: A space-domain approach for unstable inclination values: Geophysics, **85**, no. 4, J59–J70, https://doi.org/10.1190/geo2019-0008.1.

Ridsdill-Smith, T. A., 2000, The application of the wavelet transform to the processing of aeromagnetic data: Thesis for the degree of Doctor of Philosophy, University of Western Australia, Department of Geology and Geophysics.

Ridsdill-Smith, T. A., and M. C. Dentith, 1999, The wavelet transform in aeromagnetic processing: Geophysics, **64**, no. 4, 1003–1013, https://doi.org/10.1190/1.1444609.

Roberts, A., 2001, Curvature attributes and their application to 3D interpreted horizons: First Break, **19**, no. 2, 85–100, https://doi.org/10.1046/j.0263-5046.2001.00142.x.

Roest, W. R., and M. Pilkington, 1993, Identifying remanent magnetization effects in magnetic data: Geophysics, **58**, no. 5, 653–659, https://doi.org/10.1190/1.1443449.

Roest, W. R., J. Verhoef, and M. Pilkington, 1992, Magnetic interpretation using 3-D analytic signal: Geophysics, **57**, no. 1, 116–125, https://doi.org/10.1190/1.1443174.

Rosi, G., L. Cacciapuoti, F. Sorrentino, M. Menchetti, M. Prevedelli, and G. M. Tino, 2015, Measurement of the gravity-field curvature by atom interferometry: Physical Review Letters, **114**, no. 1, 013001, https://doi.org/10.1103/PhysRevLett.114.013001.

Ross, H. E., R. J. Blakely, and M. D. Zoback, 2006, Testing the use of aeromagnetic data for the determination of Curie depth in California: Geophysics, **71**, no. 5, L51–L59, https://doi.org/10.1190/1.2335572.

Rossi, L., 2017, Bayesian gravity inversion by Monte Carlo methods: Doctoral dissertation, 820437, Politecnico di Milano.

Rothleitner, C., T. M. Niebauer, and O. Francis, 2014, Measurement of the speed-of-light perturbation of free-fall absolute gravimeters: Metrologia, **51**, no. 3, L9–L13, https://doi.org/10.1088/0026-1394/51/3/L9.

Ruder, M., 1997, Introduction to magnetic theory: CSEG HRAM Forum Expanded Abstracts.

Saad, A. H., 2018a, Meter reader: Sedimentary magnetic anomalies: Part 2 – Examples, sources, and geologic origin of SEDMAG anomalies: The Leading Edge, **37**, no. 111, 830–837, https://doi.org/10.1190/tle37110830.1.

Saad, A. H., 2018b, Sedimentary magnetic anomalies: Part 1 – The validity of short-wavelength, low-amplitude SEDMAG anomalies: The Leading Edge, **37**, no. 10, 774–779, https://doi.org/10.1190/tle37100774.1.

Saad, A. H., 2006, Understanding gravity gradients – a tutorial: The Leading Edge, **25**, no. 8, 942–949, https://doi.org/10.1190/1.2335167.

Sabaka, T. J., N. Olsen, and M. F. Purucker, 2004, Extending comprehensive models for the Earth's magnetic field with Ǿrsted and CHAMP data: Geophysical Journal International, **159**, no. 2, 521–547, https://doi.org/10.1111/j.1365-246X.2004.02421.x.

Sagnotti, L., 2011, Magnetic anisotropy, *in* K. Gupta, ed., Encyclopedia of Solid Earth Geophysics: Springer, 717–729, https://doi.org/10.1007/978-90-481-8702-7_113.

Sailhac, P., D. Gibert, and H. Boukerbout, 2009, The theory of the continuous wavelet transform in the interpretation of potential fields: A review: Geophysical Prospecting, **57**, no. 4, 517–525, https://doi.org/10.1111/j.1365-2478.2009.00794.x.

Salem, A., C. Green, S. Cheyney, J. D. Fairhead, E. Aboud, and S. Campbell, 2014, Mapping the depth to magnetic basement using inversion of pseudogravity: Application to the Bishop model and the Stord Basin, northern North Sea: Interpretation, **2**, no. 2, T69–T78, https://doi.org/10.1190/INT-2013-0105.1.

Salem, A., C. Green, M. Stewart, and D. DeLerma, 2014, Inversion of gravity data with isostatic constraints: Geophysics, **79**, no. 6, A45–A50, https://doi.org/10.1190/geo2014-0220.1.

Salem, A., S. Masterton, S. Campbell, J. D. Fairhead, J. Dickinson, and C. Murphy, 2013, Interpretation of tensor gravity data using an adaptive tilt angle method: Geophysical Prospecting, **61**, no. 5, 1065–1076, https://doi.org/10.1111/1365-2478.12039.

Salem, A., and D. Ravat, 2003, A combined analytic signal and Euler method (AN-EUL) for automatic interpretation of magnetic data: Geophysics, **68**, no. 6, 1952–1961, https://doi.org/10.1190/1.1635049.

Salem, A., D. Ravat, R. Smith, and K. Ushijima, 2005, Interpretation of magnetic data using an enhanced local wavenumber (ELW) method: Geophysics, **70**, no. 2, L7–L12, https://doi.org/10.1190/1.1884828.

Salem, A., S. Williams, J. D. Fairhead, D. Ravat, and R. Smith, 2007, Tilt-depth method: A simple depth estimation method using first-order magnetic derivatives: The Leading Edge, **26**, no. 12, 1502–1505, https://doi.org/10.1190/1.2821934.

Salem, A., S. Williams, J. D. Fairhead, R. Smith, and D. Ravat, 2008, Interpretation of magnetic data using tilt-angle derivatives: Geophysics, **73**, no. 1, L1–L10, https://doi.org/10.1190/1.2799992.

Sampson, L. M., and N. C. Williams, 2021, The geophysical response of the Goldrush-Fourmile orebody and implications for camp-scale Carlin-type deposit exploration, Cortez District, Nevada: The Leading Edge, **40**, no. 2, 122–128, https://doi.org/10.1190/tle40020122.1.

Sander, S., M. Argyle, S. Elieff, S. Ferguson, V. Lavoie, and L. Sander, 2004, The AIRGrav airborne gravity system: ASEG-PESA Convention Technical Program Extended Abstracts.

Sander, L., and S. Elieff, 2008, Quality control metrics for airborne gravity data: KEGS Airborne Geophysics Symposium Expanded Abstracts.

Sandham, W. A., and D. J. Hamilton, 2011, Inverse Theory, Artificial Neural Networks, *in* K. Gupta, ed., Encyclopedia of Solid Earth Geophysics: Springer, 618–625, https://doi.org/10.1007/978-90-481-8702-7_35.

Sandwell, D. T., and W. H. F. Smith, 1997, Marine gravity anomaly from Geosat and ERS-1 satellite altimetry: Journal of Geophysical Research, **102**, B5, 10039–10054, https://doi.org/10.1029/96JB03223.

Santos, D. F., J. B. C. Silva, V. C. F. Barbosa, and L. F. S. Braga, 2012, Deep-pass: An aeromagnetic data filter to enhance deep features in marginal basins: Geophysics, **77**, no. 3, J15–J22, https://doi.org/10.1190/geo2011-0146.1.

Schaefer, L., and L. T. Elkins-Tanton, 2018, Magma oceans as a critical stage in the tectonic development of rocky planets: Philosophical Transactions Series A, Mathematical, physical and engineering sciences, **376**, no. 2132, 20180109, https://doi.org/10.1098/rsta.2018.0109.

Schmidt, P., and D. Clark, 2006, The magnetic gradient tensor: Its properties and uses in source characterization: The Leading Edge, **25**, no. 1, 75–78, https://doi.org/10.1190/1.2164759.

Schmidt, P., and D. Clark, 1998, The calculation of magnetic components and moments from TMI: A case study from the Tuckers igneous complex, Queensland: Exploration Geophysics, **29**, no. 3–4, 609–614, https://doi.org/10.1071/EG998609.

Schmidt, P., D. Clark, K. Leslie, M. Bick, D. Tilbrook, and C. Foley, 2004, GETMAG – a SQUID magnetic tensor gradiometer for mineral and oil exploration: Exploration Geophysics, **35**, no. 4, 297–305, https://doi.org/10.1071/EG04297.

Schouten, J. A., 1971, Fundamental analysis of magnetic anomalies over oceanic ridges: Marine Geophysical Researches, **1**, no. 2, 111–144, https://doi.org/10.1007/BF00305291.

Schumacher, D., and R. S. Foote, 2014, Seepage-induced magnetic anomalies associated with oil and gas fields: onshore and offshore examples: AAPG Convention Technical Program Expanded Abstracts.

Schwartz, M., 2011, Geomagnetic Excursions, *in* K. Gupta, ed., Encyclopedia of Solid Earth Geophysics: Springer, 371–373, https://doi.org/10.1007/978-90-481-8702-7_110.

Shamsipour, P., D. Marcotte, M. Chouteau, and P. Keating, 2010, 3D stochastic inversion of gravity data using cokriging and cosimulation: Geophysics, **75**, no. 1, I1–I10, https://doi.org/10.1190/1.3295745.

Sharma, P. V., 1968, Demagnetization effect of a rectangular prism: Geophysics, **33**, no. 1, 132–134, https://doi.org/10.1190/1.1439915.

Sheriff, R. E., 2002, Encyclopedic Dictionary of Applied Geophysics, 4th edition: Society of Exploration Geophysicists.

Sideris, M. G., 2011, Geoid Determination, Theory and Principles, *in* K. Gupta, ed., Encyclopedia of Solid

Earth Geophysics: Springer, 356–362, https://doi.org/10.1007/978-90-481-8702-7_154.

Siegert, A. J. F., 1942, Determination of the Bouguer correction constant: Geophysics, **7**, no. 1, 29–34, https://doi.org/10.1190/1.1444990.

Silva, J. C. B., 1986, Reduction to the pole as an inverse problem and its application to low-latitude anomalies: Geophysics, **51**, no. 2, 369–382, https://doi.org/10.1190/1.1442096.

Simpson, R. W., R. C. Jachens, and R. J. Blakely, 1983, AIRYROOT: A FORTRAN program for calculating the gravitational attraction of an Airy isostatic root out to 166.7 km: USGS Open-File Report.

Simpson, R. W., R. C. Jachens, R. J. Blakely, and R. W. Saltus, 1986, A new isostatic residual gravity map of the conterminous United States with a discussion on the significance of isostatic residual anomalies: Journal of Geophysical Research, **91**, B8, 8348–8372, https://doi.org/10.1029/JB091iB08p08348.

Singh, K. H., 2011, Magnetic, Global Anomaly Map, *in* K. Gupta, ed., Encyclopedia of Solid Earth Geophysics: Springer, 796–810, https://doi.org/10.1007/978-90-481-8702-7_136.

Skalbeck, J. D., R. E. Karlin, L. Shevenell, and M. C. Widmer, 2005, Gravity and aeromagnetic modeling of alluvial basins in the southern Truckee Meadows adjacent to the Steamboat Hills geothermal area, Washoe County, Nevada: Geophysics, **70**, no. 3, B1–B9, https://doi.org/10.1190/1.1925739.

Slotnick, M. M., 1932, Curvature of equipotential surfaces: AAPG Bulletin, **16**, 1250–1259.

Smith, R. S., and A. Salem, 2005, Imaging depth, structure, and susceptibility from magnetic data: The advanced source-parameter imaging method: Geophysics, **70**, no. 4, L31–L38, https://doi.org/10.1190/1.1990219.

Smith, R. S., J. B. Thurston, T.-F. Dai, and I. N. MacLeod, 1998, iSPI™ – the improved source parameter imaging method: Geophysical Prospecting, **46**, no. 2, 141–151, https://doi.org/10.1046/j.1365-2478.1998.00084.x.

Sokolov, A. V., 2011, High accuracy airborne gravity measurements. Methods and equipment: IFAC 18th World Congress Expanded Abstracts, 1889–1891.

Spielman, J. B., and D. A. Ponce, 1984, HANDTC, a Fortran program to calculate inner-zone terrain corrections: USGS Open-File Report.

Stasinowsky, W., 2010, The advantages of the full tensor over Gz: ASEG-PESA Airborne Gravity 2010 Workshop Abstracts, Geoscience Australia, Record, **2010**, no. 23.

Stavrev, P., 2006, Inversion of elongated magnetic anomalies using magnitude transforms: Geophysical Prospecting, **54**, no. 2, 153–166, https://doi.org/10.1111/j.1365-2478.2006.00528.x.

Stavrev, P., and A. Reid, 2010, Euler deconvolution of gravity anomalies from thick contact/fault structures with extended negative structural index: Geophysics, **75**, no. 6, I51–I58, https://doi.org/10.1190/1.3506559.

Stavrev, P., and A. Reid, 2007, Degrees of homogeneity of potential fields and structural indices of Euler deconvolution: Geophysics, **72**, no. 1, L1–L12, https://doi.org/10.1190/1.2400010.

Stone, V. C. A., J. D. Fairhead, and W. H. Oterdoom, 2004, Micromagnetic seep detection in the Sudan: The Leading Edge, **23**, no. 8, 734–737, https://doi.org/10.1190/1.1786893.

Stummer, C., T. Gruber, J. Bouman, and S. Rispens, 2008, GOCE gradiometry – A guide for users: IAG International Symposium Poster Presentation.

Sugihara, M., and T. Ishido, 2008, Geothermal reservoir monitoring with a combination of absolute and relative gravimetry: Geophysics, **73**, no. 6, WA37–WA47, https://doi.org/10.1190/1.2991105.

Swain, C. J., 1996, Horizontal acceleration corrections in airborne gravimetry: Geophysics, **61**, no. 1, 273–276, https://doi.org/10.1190/1.1443948.

Swain, C. J., and J. F. Kirby, 2011, Gravity Data, Advanced Processing, *in* K. Gupta, ed., Encyclopedia of Solid Earth Geophysics: Springer, 461–466, https://doi.org/10.1007/978-90-481-8702-7_89.

Talwani, M., 1998, Errors in the total Bouguer reduction: Geophysics, **63**, no. 4, 1125–1130, https://doi.org/10.1190/1.1444412.

Talwani, M., and M. Ewing, 1960, Rapid computation of gravitational attraction of three-dimensional bodies of arbitrary shape: Geophysics, **25**, no. 1, 203–225, https://doi.org/10.1190/1.1438687.

Talwani, M., J. L. Worzel, and M. Landisman, 1959, Rapid gravity computations for two-dimensional bodies with application to the Mendocino submarine fracture zone: Journal of Geophysical Research, **64**, no. 1, 49–59, https://doi.org/10.1029/JZ064i001p00049.

Tapley, B. D., S. Bettadpur, J. C. Ries, P. F. Thompson, and M. M. Watkins, 2004, GRACE measurements of mass variability in the Earth system: Science, **305**, no. 5683, 503–505, https://doi.org/10.1126/science.1099192.

Tarling, D. H., 2011, Archaeomagnetism, *in* K. Gupta, ed., Encyclopedia of Solid Earth Geophysics: Springer, 8–11, https://doi.org/10.1007/978-90-481-8702-7_108.

Telford, W. M., L. P. Geldart, and R. E. Sheriff, 1990, Applied Geophysics: Cambridge University Press, https://doi.org/10.1017/CBO9781139167932.

Tenzer, R., K. Hamayun, and P. Vajda, 2009, Global maps of the CRUST 2.0 crustal components stripped gravity disturbances: Journal of Geophysical Research, **114**, B5, B05408, https://doi.org/10.1029/2008JB006016.

Thébault, E., C. C. Finlay, C. D. Beggan, P. Alken, J. Aubert, O. Barrois, F. Bertrand, T. Bondar, A. Boness, L. Brocco, E. Canet, A. Chambodut, A. Chulliat, P.-D. Coisson, F. Civet, A. Du, A. Fournier, I. Fratter, N. Gillet, B. Hamilton, M. Hamoudi, G. Hulot, T. Jager, M. Korte, W. Kuang, X. Lalanne, B. Langlais, J.-M. Leger, V. Lesur, F. J. Lowes, S. MacMillan, M. Mandea, C. Manoj, S. Maus, N. Olsen, V. Petrov, V. Ridley, M. Rother, T. J. Sabaka, D. Saturnino, R. Schachtschneider, O. Sirol, A. Tangborn, A. Thomson, L. Toffner-Clausen, P. Vigneron, I. Wardinski, and T. Zvereva, 2015, International Geomagnetic Reference Field: The 12th generation: Earth, Planets and Space, **67**, no. 1, 79, 1–20.

Thébault, E., 2011, Magnetic Anomalies, Interpretation, *in* K. Gupta, ed., Encyclopedia of Solid Earth Geophysics: Springer, 729–736, https://doi.org/10.1007/978-90-481-8702-7_118.

Thompson, D. T., 1982, EULDPH: A new technique for making computer-assisted depth estimates from magnetic data: Geophysics, **47**, no. 1, 31–37, https://doi.org/10.1190/1.1441278.

Thurston, J. B., and R. S. Smith, 1997, Automatic conversion of magnetic data to depth, dip, susceptibility contrast using the SPI™ method: Geophysics, **62**, no. 3, 807–813, https://doi.org/10.1190/1.1444190.

Thurston, J. B., R. S. Smith, and J.-C. Guillon, 2002, A multimodel method for depth estimation from magnetic data: Geophysics, **67**, no. 2, 555–561, https://doi.org/10.1190/1.1468616.

Thybo, H., and I. M. Artemieva, 2013, Moho and magmatic underplating in continental lithosphere: Tectonophysics, **609**, 605–619, https://doi.org/10.1016/j.tecto.2013.05.032.

Tikoo, S. M., J. Gattacceca, N. L. Swanson-Hysell, B. P. Weiss, C. Suavet, and C. Cournede, 2015, Preservation and detectability of shock-induced magnetization: Journal of Geophysical Research, Planets, **120**, no. 9, 1461–1475, https://doi.org/10.1002/2015JE004840.

Tiwari, V. M., and J. Hinderer, 2011, Gravity Field, Time Variations from Surface Measurements, *in* K. Gupta, ed., Encyclopedia of Solid Earth Geophysics: Springer, 489–494, https://doi.org/10.1007/978-90-481-8702-7_236.

Torge, W., 1989, Gravimetry: Walter de Gruyter.

Toushmalani, R., and A. Rahmati, 2014, A new inversion method of estimation of simultaneous near surface bulk density variations and terrain correction across the Bandar Charak (Hormozgan-Iran): SpringerPlus, **3**, no. 1, 135, https://doi.org/10.1186/2193-1801-3-135.

Ugalde, H., and B. Morris, 2008, An assessment of topographic effects on airborne and ground magnetic data: The Leading Edge, **27**, no. 1, 76–79, https://doi.org/10.1190/1.2831683.

Ugalde, H., and W. A. Morris, 2010, Cluster analysis of Euler deconvolution solutions: New filtering techniques and geologic strike determination: Geophysics, **75**, no. 3, L61–L70, https://doi.org/10.1190/1.3429997.

Uieda, L., V. C. Oliveira Jr., and V. C. F. Barbosa, 2014, Geophysical tutorial: Euler deconvolution of potential-field data: The Leading Edge, **33**, no. 4, 448–450, https://doi.org/10.1190/tle33040448.1.

Vallée, M. A., L. Newitt, R. Dumont, and P. Keating, 2005, Correlation between aeromagnetic data rejection and geomagnetic indices: Geophysics, **70**, no. 5, J33–J38, https://doi.org/10.1190/1.2057982.

Vaníček, P., 2011, Geoid Undulation, Interpretation, *in* K. Gupta, ed., Encyclopedia of Solid Earth Geophysics: Springer, 362–366, https://doi.org/10.1007/978-90-481-8702-7_86.

Včelak, J., and P. Kašpar, 2006, Sensors for vector magnetometers: Journal of Electrical Engineering, **57**, 178–180.

Veloski, G. A., R. G. Hammack, J. I. Sams III, L. D. Wylie, and K. Heirendt, 2018, Evaluation of the microfabricated atomic magnetometer deployed from a small autonomous rotorcraft for locating legacy oil and gas wells: SAGEEP Technical Program Expanded Abstracts, https://doi.org/10.4133/sageep.31-002.

Verduzco, B., J. D. Fairhead, C. M. Green, and C. MacKenzie, 2004, New insights into magnetic derivatives for structural mapping: The Leading Edge, **23**, no. 2, 116–119, https://doi.org/10.1190/1.1651454.

von der Osten-Woldenburg, H., 2011, Magnetic gradiometry, *in* K. Gupta, ed., Encyclopedia of Solid Earth Geophysics: Springer, 758–761, https://doi.org/10.1007/978-90-481-8702-7_114.

von Frese, R. R. B., M. B. Jones, J. W. Kim, and J. H. Kim, 1997, Analysis of anomaly correlations: Geophysics, **62**, no. 1, 342–351, https://doi.org/10.1190/1.1444136.

Wahr, J. M., 2011, Earth Tides, *in* K. Gupta, ed., Encyclopedia of Solid Earth Geophysics: Springer, 129–133, https://doi.org/10.1007/978-90-481-8702-7_82.

Wahr, J., S. Swenson, V. Zlotnicki, and I. Velicogna, 2004, Time-variable gravity from GRACE: First results: Geophysical Research Letters, **31**, no. 11, https://doi.org/10.1029/2004GL019779.

Wan, L., M. Han, H. A. AlJanobi, and M. S. Zhdanov, 2018, Feasibility study of gravity gradiometry monitoring of CO_2 sequestration in deep reservoirs using surface and borehole data: SEG Technical Program Expanded Abstracts, 1450–1454, https://doi.org/10.1190/segam2018-2996382.1.

Watts, A. B., 2011, Isostasy, *in* K. Gupta, ed., Encyclopedia of Solid Earth Geophysics: Springer, 647–662, https://doi.org/10.1007/978-90-481-8702-7_81.

Watts, A. B., 2001, Isostasy and Flexure of the Lithosphere: Cambridge University Press.

Westin, C.-F., S. E. Maier, B. Khidir, P. Everett, F. A. Jolesz, and R. Kikinis, 1999, Image processing for diffusion tensor magnetic resonance imaging, *in* C. Taylor and A. Colchester, eds., Medical Image Computing and Computer-Assisted Intervention – MICCAI'99: Springer, **1678**, 441–453.

While, J., A. Jackson, D. Smit, and E. Biegert, 2006, Spectral analysis of gravity gradiometry profiles: Geophysics, **71**, no. 1, J11–J22, https://doi.org/10.1190/1.2169848.

Wijns, C., C. Perez, and P. Kowalczyk, 2005, Theta map: Edge detection in magnetic data: Geophysics, **70**, no. 4, L39–L43, https://doi.org/10.1190/1.1988184.

Wildan, D., A. M. Akbar, K. M. S. Novranza, R. Sobirin, A. N. Permadi, and Supriyanto, 2017, The importance of bulk density determination in gravity data processing for structure interpretation: AIP Conference Proceedings, **1862**, 030175, https://doi.org/10.1063/1.4991279.

Williams, S. E., J. D. Fairhead, and G. Flanagan, 2005, Comparison of grid Euler deconvolution with and without 2D constraints using a realistic 3D magnetic basement model: Geophysics, **70**, no. 3, L13–L21, https://doi.org/10.1190/1.1925745.

Williams, W. C., E. Meissl, J. Madrid, and B. C. de Machura, 1999, The Sun Jorge porphyry copper deposit, Mendoza, Argentina: A combination of orthomagmatic and hydrothermal mineralization: Ore Geology Reviews, **14**, 185–201.

Woollard, G. P., 1979, The new gravity system – changes in international gravity base values and anomaly values: Geophysics, **44**, no. 8, 1352–1366, https://doi.org/10.1190/1.1441012.

Won, I. J., and M. Bevis, 1987, Computing the gravitational and magnetic anomalies due to a polygon: Algorithms and Fortran subroutines: Geophysics, **52**, no. 2, 232–238, https://doi.org/10.1190/1.1442298.

Xie, R., S. Xiong, S. Duan, J. Wang, P. Wang, Y. Luo, H. Liu, and Z. Fan, 2020, Noise estimation in vector magnetic data derived from airborne vector magnetic system: Geophysics, **85**, no. 4, J71–J83, https://doi.org/10.1190/geo2019-0663.1.

Yale, M. M., D. T. Sandwell, and A. T. Herring, 1998, What are the limitations of satellite altimetry?: The Leading Edge, **17**, no. 1, 73–76, https://doi.org/10.1190/1.1437832.

Yu, N., J. M. Kohel, J. R. Kellogg, and L. Maleki, 2006, Development of an atom-interferometer gravity gradiometer for gravity measurements from space: Applied Physics, B, Lasers and Optics, **84**, no. 4, 647–652, https://doi.org/10.1007/s00340-006-2376-x.

Yule, D. E., M. K. Sharp, and D. K. Butler, 1998, Microgravity investigations of foundation conditions: Geophysics, **63**, no. 1, 95–103, https://doi.org/10.1190/1.1444331.

Zahorec, P., J. Papco, P. Vajda, F. Greco, M. Cantarero, and D. Carbone, 2018, Refined prediction of vertical gradient of gravity at Etna volcano gravity network (Italy): Contributions to Geophysics and Geodesy, **48**, no. 4, 299–317, https://doi.org/10.2478/congeo-2018-0014.

Zeng, H., and T. Wan, 2004, Clarification of the geophysical definition of a gravity field: Geophysics, **69**, no. 5, 1252–1254, https://doi.org/10.1190/1.1801941.

Zhang, L., T. Hao, and W. Jiang, 2009, Separation of potential field data using 3-D principal component analysis and textural analysis: Geophysical Journal International, **179**, no. 3, 1397–1413, https://doi.org/10.1111/j.1365-246X.2009.04357.x.

Zhang, C., M. F. Mushayandebvu, A. B. Reid, J. D. Fairhead, and M. E. Odegard, 2000, Euler deconvolution of gravity tensor gradient data: Geophysics, **65**, no. 2, 512–520, https://doi.org/10.1190/1.1444745.

Zhdanov, M. S., H. Cai, and G. A. Wilson, 2011, 3D inversion of full tensor magnetic gradiometry (FTMG) data: 81st Annual International Meeting, SEG, Expanded Abstracts, 791–795, https://doi.org/10.1190/1.3628195.

Zhong, D., T. M. Damiani, S. A. M. Preaux, and R. W. Kingdon, 2015, Comparison of airborne gravity processing results by GravPRO and Newton software packages: Geophysics, **80**, no. 4, G107–G118, https://doi.org/10.1190/geo2014-0519.1.

Zhou, W., X. Du, and J. Li, 2013, The limitation of curvature gravity gradient tensor for edge detection and method for overcoming it: Journal of Applied Geophysics, **98**, 237–242, https://doi.org/10.1016/j.jappgeo.2013.09.008.

Zuber, M. T., D. E. Smith, D. H. Lehman, T. L. Hoffman, S. W. Asmar, and M. M. Watkins, 2012, Gravity Recovery and Interior Laboratory (GRAIL): Mapping the lunar interior from crust to core, *in* T. Zuber and C. T. Russell, eds., GRAIL: Mapping the Moon's Interior: Springer, https://doi.org/10.1007/978-1-4614-9584-0_2.

Zumberge, M. A., 2011, Gravity Measurements, Absolute, *in* K. Gupta, ed., Encyclopedia of Solid Earth Geophysics: Springer, 494–497, https://doi.org/10.1007/978-90-481-8702-7_92.

Zumberge, M. A., J. R. Ridgway, and J. A. Hildebrand, 1997, A towed marine gravity meter for near-bottom surveys: Geophysics, **62**, no. 5, 1386–1393, https://doi.org/10.1190/1.1444243.

Index

A. Total Index

B. Application Index

B1. Acquisition

B2. Attributes

B3. Corrections

B4. Filters

B5. Geology

B6. Instruments

B7. Modeling

B8. Source Depth Estimation